Algorithms and Combinatorics

Volume 27

For further volumes:
http://www.springer.com/series/13

Stasys Jukna

Boolean Function Complexity

Advances and Frontiers

 Springer

Stasys Jukna
University of Frankfurt
Institute of Informatics
Robert-Mayer Str.11-15
60054 Frankfurt am Main
Germany

and

Vilnius University
Institute of Mathematics and Informatics
Akademijos 4
08663 Vilnius
Lithuania
jukna@thi.informatik.uni-frankfurt.de

ISSN 0937-5511
ISBN 978-3-642-43144-9   ISBN 978-3-642-24508-4 (eBook)
DOI 10.1007/978-3-642-24508-4
Springer Heidelberg Dordrecht London New York

Mathematics Subject Classification (2010): 68R05, 68Q17, 94C10

Printed on acid-free paper

Springer is part of Springer Science+Business Media (www.springer.com)

To Daiva and Indrė

Preface

> *Go to the roots of calculations! Group the operations. Classify*
> *them according to their complexities rather than their*
> *appearances! This, I believe, is the mission of future*
> *mathematicians.*
>
> – *Evariste Galois*

Computational complexity theory is the study of the inherent hardness or easiness of computational tasks. Research in this theory has two main strands.

One of these strands—*structural complexity*—deals with high-level complexity questions: is space a more powerful resource than time? Does randomness enhance the power of efficient computation? Is it easier to verify a proof than to construct one? So far we do not know the answers to any of these questions; thus most results in structural complexity are *conditional* results that rely on various unproven assumptions, like P $\neq$ NP.

The second strand—*concrete complexity* or *circuit complexity*—deals with establishing lower bounds on the computational complexity of specific problems, like multiplication of matrices or detecting large cliques in graphs. This is essentially a low-level study of computation; it typically centers around particular models of computation such as decision trees, branching programs, boolean formulas, various classes of boolean circuits, communication protocols, proof systems and the like. This line of research aims to establish *unconditional* lower bounds, which rely on no unproven assumptions.

This book is about the life on the second strand—circuit complexity—with a special focus on *lower bounds*. It gives self-contained proofs of a wide range of unconditional lower bounds for important models of computation, covering many of the gems of the field that have been discovered over the past several decades, right up to results from the last year or two. More than 20 years have passed since the well-known books on circuit complexity by Savage (1976), Nigmatullin (1983), Wegener (1987), and Dunne (1988) as well as a famous survey paper of Boppana and Sipser (1990) were written. I feel it is time to summarize the new developments in circuit complexity during these two decades.

The book is mainly devoted to mathematicians wishing to get an idea of what is actually going on in this one of the hardest, but also mathematically cleanest fields of computer science, to researchers in computer science wishing to refresh their knowledge about the state of art in circuit complexity, as well as to students wishing to try their luck in circuit complexity.

I have highlighted some of the most important proof arguments for circuit lower bounds, without trying to be encyclopedic. To keep the length of the book within reasonable limits, I was forced to focus on *classical* circuit models—results on their randomized or algebraic versions receive less attention here. Also, I often compromise the numerical tightness of results in favor of clarity of argument. My goal is to present the "big picture" of existing lower bound methods, in the hope that the reader will be motivated to find new ones. More than 40 open problems, marked as Research Problems, are mentioned along the way. Most of them are of a combinatorial or combinatorial-algebraic flavor and can be attacked by students with no background in computational complexity.

The book is meant to be approachable for graduate students in mathematics and computer science, and is self-contained. The text assumes a certain mathematical maturity but *no* special knowledge in the theory of computing. For non-mathematicians, all necessary mathematical background is collected in the appendix of the book. As in combinatorics or in number theory, the models and problems in circuit complexity are usually quite easy to state and explain, even for the layperson. Most often, their solution requires a clever insight, rather than fancy mathematical tools.

I am grateful to Miklos Ajtai, Marius Damarackas, Andrew Drucker, Anna Gál, Sergey Gashkov, Dmitry Gavinsky, Jonathan Katz, Michal Koucky, Matthias Krause, Andreas Krebs, Alexander Kulikov, Meena Mahajan, Igor Sergeev, Hans Ulrich Simon, György Turán, and Sundar Vishwanathan for comments and corrections on the draft versions of the book. Sergey Gashkov and Igor Sergeev also informed me about numerous results available only in Russian.

I am especially thankful to Andrew Drucker, William Gasarch, Jonathan Katz, Massimo Lauria, Troy Lee, Matthew Smedberg, Ross Snider, Marcos Villagra, and Ryan Williams for proofreading parts of the book and giving very useful suggestions concerning the contents. Their help was crucial when putting the finishing touches to the manuscript. The strong commitment of Andrew Drucker in organizing these final touches and proofreading more than a half of the book by himself cannot be acknowledged well enough. All remaining errors are entirely my fault. My sincere thanks to Georg Schnitger for his support during my stay in Frankfurt. Finally, I would like to acknowledge the German Research Foundation (Deutsche Forschungsgemeinschaft) for giving an opportunity to finish the book while working within the grant SCHN 503/5-1.

My deepest thanks to my wife, Daiva, and my daughter, Indrė, for their patience.

Frankfurt am Main/Vilnius   *Stasys Jukna*
August 2011

Contents

Part I
The Basics

Chapter 1
Our Adversary: The Circuit

Boolean (or switching) functions map each sequence of bits to a single bit 0 or 1. Bit 0 is usually interpreted as "false", and bit 1 as "true". The simplest of such functions are the product $x \cdot y$, sum $x \oplus y$ mod 2, non-exclusive Or $x \vee y$, negation $\neg x = 1 - x$. The central problem of boolean function complexity—the lower bounds problem—is:

> Given a boolean function, how many of these simplest operations do we need to compute the function on all input vectors?

The difficulty in proving that a given boolean function has high complexity lies in the nature of our adversary: the circuit. Small circuits may work in a counterintuitive fashion, using deep, devious, and fiendishly clever ideas. How can one prove that there is no clever way to quickly compute the function?

This is the main issue confronting complexity theorists. The problem lies on the border between mathematics and computer science: lower bounds are of great importance for computer science, but their proofs require techniques from combinatorics, algebra, analysis, and other branches of mathematics.

1.1 Boolean Functions

We first recall some basic concepts concerning boolean functions. The name "boolean function" comes from the boolean logic invented by George Boole (1815–1864), an English mathematician and philosopher. As this logic is now the basis of modern digital computers, Boole is regarded in hindsight as a forefather of the field of computer science.

Boolean values (or bits) are numbers 0 and 1. A *boolean function* $f(x) = f(x_1, \ldots, x_n)$ of n variables is a mapping $f : \{0, 1\}^n \to \{0, 1\}$. One says that f *accepts* a vector $a \in \{0, 1\}^n$ if $f(a) = 1$, and *rejects* it if $f(a) = 0$.

A boolean function $f(x_1, \ldots, x_n)$ need not to depend on all its variables. One says that f *depends* on its i-th variable x_i if there exist constants $a_1, \ldots, a_{i-1}, a_{i+1}, \ldots, a_n$ in $\{0, 1\}$ such that

S. Jukna, *Boolean Function Complexity*, Algorithms and Combinatorics 27,
DOI 10.1007/978-3-642-24508-4_1, © Springer-Verlag Berlin Heidelberg 2012

$$f(a_1, \ldots, a_{i-1}, 0, a_{i+1}, \ldots, a_n) \neq f(a_1, \ldots, a_{i-1}, 1, a_{i+1}, \ldots, a_n).$$

Since we have 2^n vectors in $\{0,1\}^n$, the total number of boolean functions $f : \{0,1\}^n \rightarrow \{0,1\}$ is doubly-exponential in n, is 2^{2^n}. A boolean function f is *symmetric* if it depends only on the number of ones in the input, and not on positions in which these ones actually reside. We thus have only 2^{n+1} such functions of n variables. Examples of symmetric boolean functions are:

- Threshold functions $\mathrm{Th}_k^n(x) = 1$ iff $x_1 + \cdots + x_n \geq k$.
- Majority function $\mathrm{Maj}_n(x) = 1$ iff $x_1 + \cdots + x_n \geq \lceil n/2 \rceil$.
- Parity function $\oplus_n(x) = 1$ iff $x_1 + \cdots + x_n \equiv 1 \bmod 2$.
- Modular functions $\mathrm{MOD}_k = 1$ iff $x_1 + \cdots + x_n \equiv 0 \bmod k$.

Besides these, there are many other interesting boolean functions. Actually, *any* property (which may or may not hold) can be encoded as a boolean function. For example, the property "to be a prime number" corresponds to a boolean function PRIME such that $\mathrm{PRIME}(x) = 1$ iff $\sum_{i=1}^n x_i 2^{i-1}$ is a prime number. It was a long-standing problem whether this function can be uniformly computed using a polynomial in n number of elementary boolean operations. This problem was finally solved affirmatively by Agrawal et al. (2004). The *existence* of small circuits for PRIME for every single n was known long ago.

To encode properties of graphs on the set of vertices $[n] = \{1, \ldots, n\}$, we may associate a boolean variable x_{ij} with each potential edge. Then any 0–1 vector x of length $\binom{n}{2}$ gives us a graph G_x, where two vertices i and j are adjacent iff $x_{ij} = 1$. We can then define $f(x) = 1$ iff G_x has a particular property. A prominent example of a "hard-to-compute" graph property is the *clique function* $\mathrm{CLIQUE}(n, k)$: it accepts an input vector x iff the graph G_x has a k-clique, that is, a complete subgraph on k vertices. The problem of whether this function can also be computed using a polynomial number of operations remains wide open. A negative answer would immediately imply that $\mathrm{P} \neq \mathrm{NP}$. Informally, the P vs. NP problem asks whether there exist mathematical theorems whose proofs are much harder to find than verify.

Roughly speaking, one of the goals of circuit complexity is, for example, to understand *why* the first of the following two problems is easy whereas the second is (apparently) very hard to solve:

1. Does a given graph contain at least $\binom{k}{2}$ edges?
2. Does a given graph contain a clique with $\binom{k}{2}$ edges?

The first problem is a threshold function, whereas the second is the clique function $\mathrm{CLIQUE}(n, k)$. We stress that the goal of circuit complexity is not just to give an "evidence" (via some indirect argument) that clique *is* much harder than majority, but to understand *why* this is so.

A *boolean matrix* or a 0–1 matrix is a matrix whose entries are 0s and 1s. If $f(x, y)$ is a boolean function of $2n$ variables, then it can be viewed as a boolean $2^n \times 2^n$ matrix A whose rows and columns are labeled by vector in $\{0,1\}^n$, and $A[x, y] = f(x, y)$.

$x\ y$	$x \wedge y$	$x \vee y$	$x \oplus y$	$x \to y$
0 0	0	0	0	1
0 1	0	1	1	1
1 0	0	1	1	0
1 1	1	1	0	1

x	$\neg x$
1	0
0	1

Fig. 1.1 Truth tables of basic boolean operations

We can obtain new boolean functions (or matrices) by applying boolean operations to the "simplest" ones. Basic boolean operations are:

- NOT (negation) $\neg x = 1 - x$; also denoted as $\overline{x}$.
- AND (conjunction) $x \wedge y = x \cdot y$.
- OR (disjunction) $x \vee y = 1 - (1 - x)(1 - y)$.
- XOR (parity) $x \oplus y = x(1 - y) + y(1 - x) = (x + y) \bmod 2$.
- Implication $x \to y = \neg x \vee y$ (Fig. 1.1).

If these operators are applied to boolean vectors or boolean matrices, then they are usually performed componentwise. Negation acts on ANDs and ORs via *DeMorgan rules*:

$$\neg(x \vee y) = \neg x \wedge \neg y \text{ and } \neg(x \wedge y) = \neg x \vee \neg y.$$

The operations AND and OR themselves enjoy the distributivity rules:

$$x \wedge (y \vee z) = (x \wedge y) \vee (x \wedge z) \text{ and } x \vee (y \wedge z) = (x \vee y) \wedge (x \vee z).$$

Binary cube The set $\{0, 1\}^n$ of all boolean (or binary) vectors is usually called the *binary n-cube*. A *subcube* of *dimension d* is a set of the form $A = A_1 \times A_2 \times \cdots \times A_n$, where each A_i is one of three sets $\{0\}$, $\{1\}$ and $\{0, 1\}$, and where $A_i = \{0, 1\}$ for exactly d of the i s. Note that each subcube of dimension d can be uniquely specified by a vector $a \in \{0, 1, *\}^n$ with d stars, by letting $*$ to attain any of two values 0 and 1. For example, a subcube $A = \{0\} \times \{0, 1\} \times \{1\} \times \{0, 1\}$ of the binary 4-cube of dimension $d = 2$ is specified by $a = (0, *, 1, *)$.

Usually, the binary n-cube is considered as a *graph Q_n* whose vertices are vectors in $\{0, 1\}^n$, and two vectors are adjacent iff they differ in exactly one position (see Fig. 1.2). This graph is sometimes called the *n-dimensional binary hypercube*. This is a regular graph of degree n with 2^n vertices and $n2^{n-1}$ edges. Moreover, the graph is bipartite: we can put all vectors with an odd number of ones on one side, and the rest on the other; no edge of Q_n can join two vectors on the same side.

Every boolean function $f : \{0, 1\}^n \to \{0, 1\}$ is just a coloring of vertices of Q_n in two colors. The bipartite subgraph G_f of Q_n, obtained by removing all edges joining the vertices in the same color class, accumulates useful information about the circuit complexity of f. If, for example, d_a denotes the average degree in G_f

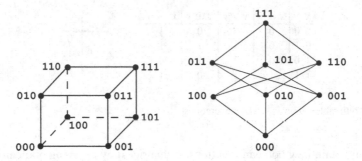

Fig. 1.2 The 3-cube and its Hasse-type representation (each level contains binary strings with the same number of 1s). There is an edge between two strings if and only if they differ in exactly one position

of vertices in the color-class $f^{-1}(a)$, $a = 0, 1$, then the product $d_0 \cdot d_1$ is a lower bound on the length of any formula expressing f using connectives $\wedge$, $\vee$ and $\neg$ (see Khrapchenko's theorem in Sect. 6.8).

CNFs and DNFs A trivial way to represent a boolean function $f(x_1, \ldots, x_n)$ is to give the entire *truth table*, that is, to list all 2^n pairs $(a, f(a))$ for $a \in \{0, 1\}^n$. More compact representations are obtained by giving a covering of $f^{-1}(0)$ or of $f^{-1}(1)$ by not necessarily disjoint subsets, each of which has some "simple" structure. This leads to the notions of CNFs and DNFs.

A *literal* is a boolean variable or its negation. For literals the following notation is often used: x_i^1 stands for x_i, and x_i^0 stands for $\neg x_i = 1 - x_i$. Thus, for every binary string $a = (a_1, \ldots, a_n)$ in $\{0, 1\}^n$,

$$x_i^1(a) = \begin{cases} 1 & \text{if } a_i = 1 \\ 0 & \text{if } a_i = 0 \end{cases} \quad \text{and} \quad x_i^0(a) = \begin{cases} 0 & \text{if } a_i = 1 \\ 1 & \text{if } a_i = 0. \end{cases}$$

A *monomial* is an AND of literals, and a *clause* is an OR of literals. A monomial (or clause) is *consistent* if it does not contain a contradicting pair of literals x_i and $\overline{x}_i$ of the same variable. We will often view monomials and clauses as *sets* of literals.

It is not difficult to see that the set of all vectors *accepted* by a monomial consisting of k (out of n) literals forms a binary n-cube of dimension $n - k$ (so many bits are not specified). For example, a monomial $\overline{x}_1 \wedge x_3$ defines the cube of dimension $n - 2$ specified by $a = (0, *, 1, *, \ldots, *)$. Similarly, the set of all vectors *rejected* by a clause consisting of k (out of n) literals also forms a binary n-cube of dimension $n-k$. For example, a clause $\overline{x}_1 \vee x_3$ rejects a vector a iff $a_1 = 1$ and $a_3 = 0$.

A *DNF* (disjunctive normal form) is an OR of monomials, and a *CNF* (conjunctive normal form) is an AND of clauses. Every boolean function $f(x)$ of n variables can be written both as a DNF $D(x)$ and as a CNF $C(x)$:

$$D(x) = \bigvee_{a: f(a)=1} \bigwedge_{i=1}^{n} x_i^{a_i} \qquad C(x) = \bigwedge_{b: f(b)=0} \bigvee_{i=1}^{n} x_i^{1-b_i} .$$

Indeed, $D(x)$ *accepts* a vector x iff x coincides with at least one vector a accepted by f, and $C(x)$ *rejects* a vector x iff x coincides with at least one vector b rejected by f.

A DNF is a *k-DNF* if each of its monomials has at most k literals; similarly, a CNF is a *k-CNF* if each of its clauses has at most k literals.

DNFs (and CNFs) are the simplest models for computing boolean functions. The *size* of a DNF is the total number of monomials in it. It is clear that every boolean function of n variables can be represented by a DNF of size at most $|f^{-1}(1)| \le 2^n$: just take one monomial for each accepted vector. This can also be seen via the following recurrence:

$$f(x_1, \ldots, x_{n+1}) = x_{n+1} \wedge f(x_1, \ldots, x_n, 1) \vee \neg x_{n+1} \wedge f(x_1, \ldots, x_n, 0). \quad (1.1)$$

It is not difficult to see that some functions require DNFs of exponential size. Take, for example, the parity function $f(x_1, \ldots, x_n) = x_1 \oplus x_2 \oplus \cdots \oplus x_n$. This function accepts an input vector iff the number of 1s in it is odd. Every monomial in a DNF for f must contain n literals, for otherwise the DNF would be forced to accept a vector in $f^{-1}(0)$. Since any such monomial can accept only one vector, $|f^{-1}(1)| = 2^{n-1}$ monomials are necessary. Thus the lower bounds problem for this model is trivial.

Boolean functions as set systems By identifying subsets S of $[n] = \{1, \ldots, n\}$ with their characteristic 0–1 vectors v_S, where $v_S(i) = 1$ iff $i \in S$, we can consider boolean functions as set-theoretic predicates $f : 2^{[n]} \to \{0, 1\}$. We will often go back and forth between these notations. One can identify a boolean function $f : 2^{[n]} \to \{0, 1\}$ with the family $\mathcal{F}_f = \{S : f(S) = 1\}$ of subsets of $[n]$. That is, there is a 1-to-1 correspondence between boolean functions and families of subsets of $[n]$:

boolean functions of n variables = families of subsets of $\{1, \ldots, n\}$.

Minterms and maxterms A 1-term (resp., 0-term) of a boolean function is a smallest subset of its variables such that the function can be made the constant 1 (resp., constant 0) function by fixing these variables to constants 0 and 1 in some way. Thus after the setting, the obtained function does not depend on the remaining variables. Minterms (maxterms) are 1-terms (0-terms) which are minimal under the set-theoretic inclusion.

Note that one and the same set of variables may be a 1-term and a 0-term at the same time. If, for example, $f(x_1, x_2, x_3) = 1$ iff $x_1 + x_2 + x_3 \ge 2$, then $\{x_1, x_2\}$ is a 1-term of f because $f(1, 1, x_3) \equiv 1$, and is a 0-term of f because $f(0, 0, x_3) \equiv 0$.

If all minterms of a boolean function f have length at most k then f can be written as a k-DNF: just take the OR of all these minterms. But the converse does not hold! Namely, there are boolean functions f such that f *can* be written as a k-DNF even though some of its minterms are much longer than k (see Exercise 1.7).

Duality The *dual* of a boolean function $f(x_1, \ldots, x_n)$ is the boolean function f^* defined by:

$$f^*(x_1, \ldots, x_n) := \neg f(\neg x_1, \ldots, \neg x_n).$$

For example, if $f = x \vee y$ then $f^* = \neg(\neg x \vee \neg y) = x \wedge y$. The dual of every
threshold function $\mathrm{Th}_k^n(x)$ is the threshold function $\mathrm{Th}_{n-k+1}^n(x)$. A function f is
self-dual if $f^*(x) = f(x)$ holds for all $x \in \{0, 1\}^n$. For example, the threshold-k
function $f(x) = \mathrm{Th}_k^{2k-1}(x)$ of $2k - 1$ variables is self-dual. Hence, if the number n
of variables is odd, then the majority function Maj_n is also self-dual.

In set-theoretic terms, if $\overline{S} = [n] \backslash S$ denotes the complement of S, then the values
of the dual of f are obtained by: $f^*(S) = 1 - f(\overline{S})$. Thus a boolean function f is
self-dual if and only if $f(\overline{S}) + f(S) = 1$ for all $S \subseteq [n]$.

Monotone functions For two vectors $x, y \in \{0, 1\}^n$ we write $x \leq y$ if $x_i \leq y_i$ for all
positions i. A boolean function $f(x)$ is *monotone*, if $x \leq y$ implies $f(x) \leq f(y)$.
If we view f as a set-theoretic predicate $f : 2^{[n]} \rightarrow \{0, 1\}$, then f is monotone iff
$f(S) = 1$ and $S \subseteq T$ implies $f(T) = 1$. Examples of monotone boolean functions
are AND, OR, threshold functions $\mathrm{Th}_k^n(x)$, clique functions $\mathrm{CLIQUE}(n, k)$, etc. On
the other hand, such functions as the parity function $\oplus_n(x)$ or counting functions
$\mathrm{Mod}_k^n(x)$ are not monotone.

Monotone functions have many nice properties not shared by other functions.
First of all, their minterms as well as maxterms are just subsets of variables (no
negated variable occurs in them). In set-theoretic terms, a subset $S \subseteq [n]$ is a
minterm of a monotone function f if

$$f(S) = 1 \text{ but } f(S \setminus \{i\}) = 0 \text{ for all } i \in S,$$

and is a maxterm of f if

$$f(\overline{S}) = 0 \text{ but } f(\overline{S \setminus \{i\}}) = 1 \text{ for all } i \in S.$$

Let $\mathrm{Min}(f)$ and $\mathrm{Max}(f)$ denote the set of all minterms and the set of all maxterms
of f. Then we have the following *cross-intersection property*:

$$S \cap T \neq \emptyset \text{ for all } S \in \mathrm{Min}(f) \text{ and all } T \in \mathrm{Max}(f).$$

Indeed, if S and T were disjoint, then for the vectors x with $x_i = 1$ for all $i \in S$,
and $x_i = 0$ for all $i \notin S$, we would have $f(x) = 1$ (because S is a minterm) and at
the same time $f(x) = 0$ (because $T \subseteq \overline{S}$ is a maxterm of f).

The next important property of monotone boolean functions is that every such
function f has a *unique* representation as a DNF as well as a CNF:

$$f(x) = \bigvee_{S \in \mathrm{Min}(f)} \bigwedge_{i \in S} x_i = \bigwedge_{T \in \mathrm{Max}(f)} \bigvee_{i \in T} x_i .$$

Moreover, for every monotone boolean function f we have the following three
equivalent conditions of their self-duality:

- $\mathrm{Min}(f) = \mathrm{Max}(f)$.
- Both families $\mathrm{Min}(f)$ and $\mathrm{Max}(f)$ are intersecting: $S \cap S' \neq \emptyset$ for all $S, S' \in$
  $\mathrm{Min}(f)$, and $T \cap T' \neq \emptyset$ for all $T, T' \in \mathrm{Max}(f)$.

- The family $\mathrm{Min}(f)$ is intersecting and, for every partition of $[n]$ into two parts, at least one minterm lies in one of these parts.

Equivalence of the first condition $\mathrm{Min}(f) = \mathrm{Max}(f)$ with the definition of self-duality ($f(\overline{S}) = 1 - f(S)$ for all $S \subseteq [n]$) is not difficult to see. To show that also the second and the third conditions are equivalent, needs a bit more work.

In the rest of this section we recall some facts that turn out to be very useful when analyzing circuits. We include them right here both because they have elegant proofs and because we will use them later several times.

Functions with many subfunctions A *subfunction* of a boolean function $f(x_1, \ldots, x_n)$ is obtained by fixing some of its variables to constants 0 and 1. Since each of the n variables has three possibilities (to be set to 0 or to 1 or remain unassigned), one function can have at most 3^n subfunctions.

If Y is some subset of variables, then a *subfunction of f on Y* is a boolean function of variables Y obtained from f by setting all the variables outside Y to constants 0 and 1, in some way. Some settings may lead to the same subfunction. So let $N_Y(f)$ denote the number *distinct* subfunctions of f on Y. It is not difficult to see that, if $|Y| = m$, then

$$N_Y(f) \leq \min\{2^{n-m}, 2^{2^m}\}.$$

Indeed, we have at most 2^{n-m} possibilities to assign constants to $n - |Y|$ variables, and there are at most 2^{2^m} distinct boolean functions on the same set Y of m variables. But some functions f may have fewer distinct subfunctions. For example, the parity function $\oplus_n(x) = x_1 \oplus x_2 \oplus \cdots \oplus x_n$ has only $N_Y(\oplus_n) = 2$ different subfunctions. On the other hand, we will show later (in Sect. 6.5) that functions with many subfunctions cannot be "too easy". So what functions have many subfunctions?

The simplest known example of a function with almost maximal possible number of distinct subfunctions is the *element distinctness function* $\mathrm{ED}_n(x)$ suggested by Beame and Cook (unpublished). This is a boolean function of [1] $n = 2m \log m$ variables divided into m consecutive blocks $Y_1, \ldots, Y_m$ with $2 \log m$ variables in each of them; m is assumed to be a power of 2. Each of these blocks encode a number in $[m^2] = \{1, 2, \ldots, m^2\}$. The function accepts an input $x \in \{0, 1\}^n$ if and only if all these numbers are distinct.

Lemma 1.1. *On each block,* ED_n *has at least* $2^{n/2}/n$ *subfunctions.*

Proof. It suffices to prove this for the first block Y_1. So let $N = N_{Y_1}(\mathrm{ED}_n)$, and consider the function f of m variables, each taking its value in $[m^2]$. The function accepts a string $(a_1, \ldots, a_m)$ of numbers in $[m^2]$ iff all these numbers are distinct. Thus $\mathrm{ED}_n(x)$ is just a boolean version of f.

---

[1]If not said otherwise, all logarithms in this book are to the basis of 2.

For a string $a = (a_2, \ldots, a_m)$ of numbers $[m^2]$, let $f_a : [m^2] \to \{0, 1\}$ be the function $f_a(x) := f(x, a_2, \ldots, a_m)$ obtained from f by fixing its last $m - 1$ variables. Note that N is exactly the number of distinct functions f_a.

The number of ways to choose a string $a = (a_2, \ldots, a_m)$ with all the a_i distinct is $\binom{m^2}{m-1}(m - 1)!$: each such string is obtained by taking an $(m - 1)$-element subset of $[m^2]$ and permuting its elements. If $b = (b_2, \ldots, b_m)$ is another such string, and if b is not a permutation of a, then there must be an a_i such that $a_i \notin \{b_2, \ldots, b_m\}$. But for such an a_i, we have that $f_a(a_i) = 0$ whereas $f_b(a_i) = 1$; hence, $f_a \neq f_b$. Since there are only $(m - 1)!$ permutations of a, we obtain that $N \geq \binom{m^2}{m-1} \geq m^{m-1} \geq 2^{n/2}/n$. $\square$

Matrix decomposition A matrix B is *primitive* if it is boolean (has only entries 0 and 1) and has rank 1 over the reals. Each such matrix consists of one all-1 submatrix and zeros elsewhere. The *weight*, $w(B)$, of such a matrix is $r + c$, where r is the number of nonzero rows, and c the number of nonzero columns in B. Here is a primitive 4×5 matrix of weight $2 + 3 = 5$:

$$\begin{pmatrix} 1 & 0 & 1 & 1 & 0 \\ 0 & 0 & 0 & 0 & 0 \\ 1 & 0 & 1 & 1 & 0 \\ 0 & 0 & 0 & 0 & 0 \end{pmatrix}$$

Primitive matrices are important objects—we will use them quite often.

A *decomposition* of a boolean $m \times n$ matrix A is a set $B_1, \ldots, B_r$ of primitive $m \times n$ matrices such that A can be written as the sum $A = B_1 + B_2 + \cdots + B_t$ of these matrices over the reals. That is, each 1-entry of A is a 1-entry in exactly one of the matrices B_i, and each 0-entry is a 0-entry in all matrices. The *weight* of such a decomposition is the sum $\sum_{i=1}^{t} w(B_i)$ of weights of the B_i. Let $\mathrm{Dec}(A)$ denote the minimum weight of a decomposition of a boolean matrix A, and let $|A|$ denote the number of 1-entries in A.

Note that $\mathrm{Dec}(A) \leq mn$: just decompose A into m primitive matrices corresponding to the rows of A. In fact, we have a better upper bound.

Lemma 1.2. (Lupanov 1956) *For every boolean $m \times n$ matrix,*

$$\mathrm{Dec}(A) \leq (1 + o(1)) \frac{mn}{\log m}.$$

Proof. We first prove that for every boolean $m \times n$ matrix A and for every integer $1 \leq k \leq m$,

$$\mathrm{Dec}(A) \leq \frac{mn}{k} + n2^{k-1}. \tag{1.2}$$

We first prove (1.2) for $k = n$, that is, we prove the upper bound

$$\mathrm{Dec}(A) \leq m + n2^{n-1}. \tag{1.3}$$

Split the rows of A into groups, where the rows in one group all have the same values. This gives us a decomposition of A into $t \leq 2^n$ primitive matrices. For the i-th of these matrices, let r_i be the number of its nonzero rows, and c_i the number of its nonzero columns. Hence, $r_i + c_i$ is the weight of the i-th primitive matrix. Since each nonzero row of A lies in exactly one of the these matrices, the total weight of the decomposition is

$$\sum_{i=1}^{t} r_i + \sum_{i=1}^{t} c_i \leq m + \sum_{j=0}^{n} \sum_{i:c_i=j} j \leq m + \sum_{j=0}^{n} \binom{n}{j} \cdot j = m + n2^{n-1},$$

where the last equality is easy to prove: just count in two ways the number of pairs (x, S) with $x \in S \subseteq \{1, \ldots, n\}$.

To prove (1.2) for arbitrary integer $1 \leq k \leq n$, split A into submatrices with k columns in each (one submatrix may have fewer columns). For each of these n/k submatrices, (1.3) gives a decomposition of weight at most $m + k2^{k-1}$. Thus, for every $1 \leq k \leq n$, every $m \times n$ matrix has a decomposition of weight at most $mn/k + n2^{k-1}$.

To finish the proof of the theorem, it is enough to apply (1.2) with k about $\log m - 2 \log \log m$. □

Using a counting argument, Lupanov (1956) also showed that the upper bound given in Lemma 1.2 is almost optimal: $m \times n$ matrices A requiring weight

$$\mathrm{Dec}(A) \geq (1 + o(1)) \frac{mn}{\log(mn)}$$

in any decomposition exist, even if the 1-entries in primitive matrices are allowed to overlap (cf. Theorem 13.18). Apparently, this paper of Lupanov remained unknown in the West, because this result was later proved by Tuza (1984) and Bublitz (1986).

Splitting a graph When trying to "balance" some computational models (decision trees, formulas, communication protocols, logical derivations) the following two structural facts are often useful.

Let G be a directed acyclic graph with one source node (the root) from which all leaves (nodes of outdegree 0) are reachable. Suppose that each non-leaf node has outdegree k. Suppose also that each vertex is assigned a non-negative weight which is *subadditive*: the weight of a node does not exceed the sum of the weights of its successors. Let r be the weight of the root, and suppose that each leaf has weight at most $l < r$.

Lemma 1.3. *For every real number ϵ between l/r and 1, there exists a node whose weight lies between $\epsilon r/k$ and ϵr. In particular, every binary tree with r leaves has a subtree whose number of leaves lies between $r/3$ and $2r/3$.*

Proof. Start at the root and traverse the graph until a node u of weight $> \epsilon r$ is found such that each of its successors has weight at most ϵr. Such a node u exists because each leaf has weight at most $l \leq \epsilon r$. Due to subadditivity of the weight function,

the (up to k) successors of u cannot *all* have weight $\leq \epsilon r/k$, since then the weight
of u would be $\leq \epsilon r$ as well. Hence, the weight of at least one successor of u must
lie between $\epsilon r/k$ and ϵr, as desired.

To prove the second claim, give each leaf of the tree weight 1, and define the
weight of an inner node as the number of leaves in the corresponding subtree. Then
apply the previous claim with $k = 2$ and $\epsilon = 2/3$.   □

The *length* of a path we will mean the number of nodes in it. The *depth* of a graph
is the length of a longest path in it. The following lemma generalizes and simplifies
an analogous result of Erdős et al. (1976). Let $d = 2^k$ and $1 \leq r \leq k$ be integers.

Lemma 1.4. (Valiant 1977) *In any directed graph with S edges and depth d it is
possible to remove rS/k edges so that the depth of the resulting graph does not
exceed $d/2^r$.*

Proof. A *labeling* of a graph is a mapping of the nodes into the integers. Such a
labeling is *legal* if for each edge (u, v) the label of v is strictly greater than the label
of u. A *canonical labeling* is to assign each node the length of a longest directed path
that terminates at that node. If the graph has depth d then this gives us a labeling
using only d labels $1, \ldots, d$. It is easy to verify that this is a legal labeling: if (u, v)
is an edge then any path terminating in u can be prolonged to a path terminating
in v. On the other hand, since in any legal labeling, all labels along a directed path
must be distinct, we have that the depth of a graph does not exceed the number of
labels used by any legal labeling.

After these preparations, consider now any directed graph with S edges and
depth d, and consider the canonical labeling using labels $1, \ldots, d$. For $i = 1, \ldots, k$
(where $k = \log d$), let E_i be the set of all edges, the binary representations of labels
of whose endpoints differ in the i-th position (from the left) for the *first time*.

If E_i is removed from the graph, then we can relabel the nodes using integers
$1, \ldots, d/2$ by simply deleting the i-th bit in the binary representations of labels.
It is not difficult to see that this is a legal labeling (of a new graph): if an edge
(u, v) survived, then the first difference between the binary representations of the
old labels of u and v were *not* in the i-th position; hence, the new label of u remains
strictly smaller than that of v. Consequently, if any $r \leq k$ of the *smallest* sets E_i are
removed, then at most rS/k edges are removed, and a graph of depth at most $d/2^r$
remains.   □

1.2 Circuits

In this section we recall the most fundamental models for computing boolean
functions.

General circuits Let Φ be a set of some boolean functions. A *circuit* (or a *straight
line program*) of n variables over the basis Φ is just a sequence $g_1, \ldots, g_t$ of
$t \geq n$ boolean functions such that the first n functions are input variables $g_1 =$

$x_1, \ldots, g_n = x_n$, and each subsequent g_i is an application $g_i = \varphi(g_{i_1}, \ldots, g_{i_d})$ of some basis function $\varphi \in \Phi$ (called the *gate* of g_i) to some previous functions.

That is, the value $g_i(a)$ of the i-th gate g_i on a given input $a \in \{0, 1\}^n$ is the value of the boolean function φ applied to the values $g_{i_1}(a), \ldots, g_{i_d}(a)$ computed at the previous gates. A circuit computes a boolean function (or a set of boolean functions) if it (or they) are among the g_i.

Each circuit can be viewed as a directed acyclic graph whose fanin-0 nodes (those of zero in-degree) correspond to variables, and each other node v corresponds to a function φ in Φ. One (or more) nodes are distinguished as outputs. The value at a node is computed by applying the corresponding function to the values of the preceding nodes (see Fig. 1.3).

> In the literature circuits are usually drawn in a "bottom-up" manner: the first (lowest) level consists of inputs, and the last (highest) level consists of output gates. We will, however, mostly draw circuits in a more natural "top-down" manner: inputs at the top, and outputs at the bottom. Only where there already are established terms "top gate" and "bottom level" we will use bottom-up drawings.

The *size* of the circuit is the total number $t - n$ of its gates (that is, we do not count the input variables), and its *depth* is the length of a longest path from an input to an output gate. More precisely, input variables have depth 0, and if $g_i = \varphi(g_{i_1}, \ldots, g_{i_d})$ then the depth of the gate g_i is 1 plus the maximum depth of the gates $g_{i_1}, \ldots, g_{i_d}$. We will assume that every circuit can use constants 0 and 1 as inputs for free.

Formulas A *formula* is a circuit all whose gates have fanout at most 1. Hence, the underlying graph of a formula is a tree. The *size* of a formula is also the number of gates, and the *leafsize* of a is the number of input gates, that is, the number of leaves in its tree, and the *depth* of a formula is the depth of its tree. Note that the only (but crucial) difference of formulas from circuits is that in the circuit model a result computed at some gate can be used many times with no need to recompute it again and again, as in the case of formulas.

DeMorgan circuits A *DeMorgan circuit* is a circuit over the basis $\{\wedge, \vee\}$ but the inputs are variables and their negations. That is, these are the circuits over the basis

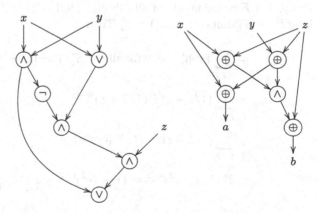

Fig. 1.3 On the left is a circuit with six gates over the basis $\{\wedge, \vee, \neg\}$ computing the majority function $\mathrm{Maj}_3(x, y, z) = 1$ iff $x + y + z \geq 2$. Its depth is five. On the right is a circuit with five gates over the basis $\{\oplus, \wedge\}$ computing the binary representation (a, b) of the (real) sum $x + y + z$ of three 0–1 bits

$\{\wedge, \vee, \neg\}$, where NOT gates are only applied to input variables; these gates do not contribute to the circuit size. Such circuits are also called *circuits with tight negations*. If there are no negated variables as inputs, then the circuit is *monotone*. By using DeMorgan rules $\neg(x \vee y) = \neg x \wedge \neg y$ and $\neg(x \wedge y) = \neg x \vee \neg y$, it can be easily shown that any circuit over $\{\wedge, \vee, \neg\}$ can be reduced to this form by at most doubling the total number of gates; the depth of the circuit remains the same. In the case of formulas, even the leafsize remains the same.

Probabilistic circuits   Such circuits have, besides standard input variables $x_1, \ldots, x_n$, some specially designed inputs $r_1, \ldots, r_m$ called random inputs. When these random inputs are chosen from a uniform distribution on $\{0, 1\}$, the output $C(x)$ of the circuit is a random 0–1 variable. A probabilistic circuit $C(x)$ computes a boolean function $f(x)$ if

$$\mathrm{Prob}[C(x) = f(x)] \geq 3/4 \quad \text{for each} \quad x \in \{0, 1\}^n.$$

There is nothing special about using the constant $3/4$ here—one can take any constant $> 1/2$ instead. The complexity would not change by more than a constant factor.

Can probabilistic circuits have much smaller size than usual (deterministic) circuits? We will answer this question *negatively* using the following simple (but often used) "majority trick". It implies that if a random circuit errs on a fixed input with probability $< 1/2$, then the majority of not too many independent copies of such a circuit will err on this input with *exponentially* small probability. A Bernoulli random variable with success probability p is a 0–1 random variable taking the value 1 with probability p.

Lemma 1.5. (Majority trick) *If $x_1, \ldots, x_m$ are independent Bernoulli random variables with success probability $1/2 + \epsilon$, then*

$$\mathrm{Prob}[\mathrm{Maj}(x_1, \ldots, x_m) = 0] \leq e^{-2\epsilon^2 m}.$$

Proof. Let $\mathcal{F}$ be the family of all subsets of $[m] = \{1, \ldots, m\}$ of size $> m/2$, and let $q := \mathrm{Prob}[\mathrm{Maj}(x_1, \ldots, x_m) = 0]$. Then

$$q = \sum_{S \in \mathcal{F}} \mathrm{Prob}[x_i = 0 \text{ for all } i \in S] \cdot \mathrm{Prob}[x_i = 1 \text{ for all } i \notin S]$$

$$= \sum_{S \in \mathcal{F}} (1/2 - \epsilon)^{|S|} (1/2 + \epsilon)^{m - |S|}$$

$$\leq \sum_{S \in \mathcal{F}} (1/2 - \epsilon)^{m/2} (1/2 + \epsilon)^{m/2}$$

$$\leq 2^m (1/4 - \epsilon^2)^{m/2} = (1 - 4\epsilon^2)^{m/2} \leq e^{-2\epsilon^2 m}.$$

The first inequality here follows by multiplying each term by

$$(1/2 - \epsilon)^{m/2-|S|}(1/2 + \epsilon)^{|S|-m/2} \geq 1 .$$ □

Theorem 1.6. (Adleman 1978) *If a boolean function f of n variables can be computed by a probabilistic circuit of size M, then f can be computed by a deterministic circuit of size at most $8nM$.*

Proof. Let C be a probabilistic circuit that computes f. Take m independent copies $C_1, \ldots, C_m$ of this circuit (each with its own random inputs), and consider the probabilistic circuit C' that computes the majority of the outputs of these m circuits. Fix a vector $a \in \{0, 1\}^n$, and let x_i be an indicator random variable for the event "$C_i(a) = f(a)$". For each of these random variables we have that $\text{Prob}[x_i = 1] \geq 1/2 + \epsilon$ with $\epsilon = 1/4$. By the majority trick, the circuit C' will err on a with probability at most $e^{-2\epsilon^2 m} = e^{-m/8}$. By the union bound, the probability that the new circuit C' makes an error on at least one of all 2^n possible inputs a is at most $2^n \cdot e^{-m/8}$. If we take $m = 8n$, then this probability is smaller than 1. Therefore, there must be a setting of the random inputs which gives the correct answer for all inputs. The obtained circuit is no longer probabilistic, and its size is at most $8n$ times larger than the size of the probabilistic circuit.

□

Average time of computations Let $C = (g_1, \ldots, g_s)$ be a circuit computing some boolean function $f(x)$ of n variables; hence, $g_s(x) = f(x)$. The number s of gates is the size of the circuit. One can also consider a notion of "computation time" on a given input $a \in \{0, 1\}^n$. For this, let us introduce one special boolean variable z, the *output* variable. Some of the gates may reset this variable, that is, set $z = g_i(a)$. In particular, gates of the form $z = 0$ and $z = 1$ are allowed. The last gate g_s always does this, that is, sets $z = g_s(a)$. Our goal however is to interrupt the computation sequence $g_1(a), \ldots, g_s(a)$ as soon as the output variable already has the correct value $z = f(a)$.

To realize this goal, we declare some gates as "stop-gates". Such a gate g stops the computation on an input a if $g(a) = 1$. Now, given an input $a \in \{0, 1\}^n$, a computation $g_1(a), g_2(a), \ldots, g_i(a)$ continues until the first gate g_i is found such that g_i is a stop-gate and $g_i(a) = 1$. The computation on a then stops, and the output $C(a)$ of the circuit is the actual value of the output variable z at this moment (see Fig. 1.4). The computation time $t_C(a)$ of the circuit C on a is the number i of gates evaluated until the value was computed. The *average time* of the circuit C is

$$t(C) = 2^{-n} \sum_{a \in \{0,1\}^n} t_C(a) .$$

If we have no stop-gates at all, then $t_C(a) = s$ for all inputs a, and hence, the average time $t(C)$ of the circuit C is just the size s of C.

This model of *stop-circuits* was introduced by Chashkin (1997, 2000, 2004); he calls this model "non-branching programs with conditional stop".

$$z = 1 \qquad\qquad z = x_1 \vee x_2 \text{ (stop)} \qquad\qquad g_1 = x_1 \vee x_2$$

$$g_1 = x_1 \text{ (stop)} \qquad\qquad z = x_3 \vee x_4 \qquad\qquad g_2 = x_3 \vee x_4$$

$$g_2 = x_2 \text{ (stop)} \qquad\qquad\qquad\qquad\qquad\qquad z = g_1 \vee g_2$$

$$g_3 = x_3 \text{ (stop)}$$

$$g_4 = x_4 \text{ (stop)}$$

$$z = 0$$

Fig. 1.4 Three circuits computing the OR $x_1 \vee \vee x_2 \vee x_3 \vee x_4$ of four variables. On input $a = (0, 1, 0, 0)$ the first circuit takes time $t_C(a) = 3$, the second takes time $t_C(a) = 1$, and the third (standard) circuit takes time $t_C(a) = 3$. The average time of the last circuit is $t(C) = 3$, whereas that of the middle circuit is $t(C) = \frac{1}{16}(12 \cdot 1 + 4 \cdot 2) = 5/4$

The average time, $\mathrm{t}(f)$, of a boolean function f is the minimum average time of a circuit computing f. We always have that $\mathrm{t}(f) \leq \mathrm{C}(f)$. Chashkin (1997) showed that boolean functions f of n variables requiring $\mathrm{t}(f) = \Omega(2^n/n)$ exist. But some functions have much smaller average time than $\mathrm{C}(f)$.

Example 1.7. Consider the threshold-2 function $\mathrm{Th}_2^n(x)$. Since every boolean function f, which depends on n variables, requires at least $n - 1$ gates, we have that $\mathrm{C}(\mathrm{Th}_2^n) \geq n - 1$. On the other hand, it is not difficult to show that $\mathrm{t}(\mathrm{Th}_2^n) = \mathcal{O}(1)$. To see this, let us first compute $z = \mathrm{Th}_2^3(x_1, x_2, x_3)$. This can be done using 6 gates (see Fig. 1.3), and hence, can be computed in time 6. After that we compute $z = \mathrm{Th}_2^3(x_4, x_5, x_6)$, and so on. Declare each gate re-setting the variable z as a stop-gate. This way the computations on $42^{n-3} = 2^{n-1}$ inputs will be stopped after 6 steps, the computations on $4^2 2^{n-6} = 2^{n-2}$ remaining inputs will be stopped after $6 \cdot 2 = 12$ steps and, in general, the computations on $4^t 2^{n-3t} = 2^{n-t}$ inputs will be stopped after $6t$ steps. Thus, the average computation times is at most $\sum_{t=1}^{n/3} 6t2^{-t} = \mathcal{O}(1)$.

An interesting aspect of stop-circuits is that one can compute non-monotone boolean functions using monotone operations! For example, the following circuit over $\{0, 1\}$ computes the negation $\neg x$ of a variable x:

$$z = 0; \ g_1 = x \text{ (stop)}; \ z = 1$$

and the following circuit over $\{\wedge, \vee, 0, 1\}$ computes the parity function $x \oplus y$:

$$z = 0; \ g_1 = x \wedge y \text{ (stop)}; \ z = 1; \ g_2 = x \vee y \text{ (stop)}; \ z = 0.$$

Let $\mathrm{t}_m(f)$ denote the minimum average time of a circuit over $\{\wedge, \vee, 0, 1\}$ computing f. Chashkin (2004) showed that there exist boolean functions f of n variables such that $\mathrm{t}(f) = \mathcal{O}(1)$ but $\mathrm{t}_m(f) = \Omega(\sqrt{2^n/n})$.

Arithmetic circuits Such circuits constitute the most natural and standard model for computing polynomials over a ring R. In this model the inputs are variables $x_1, \ldots, x_n$, and the computation is performed using the arithmetic operations $+, \times$ and may involve constants from R. The output of an arithmetic circuit is thus a polynomial (or a set of polynomials) in the input variables. Arithmetic circuits are a highly structured model of computation compared to boolean circuits. For example, when studying arithmetic circuits we are interested in syntactic computation of polynomials, whereas in the study of boolean circuits we are interested in the semantics of the computation. In other words, in the boolean case we are not interested in any specific polynomial representation of the function, but rather we just want to compute some representation of it, while in the arithmetic world we focus on a specific representation of the function. As such, one may hope that the P vs. NP question will be easier to solve in the arithmetical model. However, in spite of many efforts, we are still far from understanding this fundamental problem. In this book we will not discuss arithmetic circuits: a comprehensive treatment can be found in a recent survey by Shpilka and Yehudayoff (2010).

1.3 Branching Programs

Circuits and formulas are "parallel" models: given an input vector $x \in \{0, 1\}^n$, we process some pieces of x in parallel and join the results by AND or OR gates. The oldest "sequential" model for computing boolean functions, introduced already in pioneering work of Shannon (1949) and extensively studied in the Russian literature since about 1950, is that of switching networks; a modern name for these networks is "branching programs."

Nondeterministic branching programs The most general of "sequential" models is that of *nondeterministic branching programs* (n.b.p.). Such a program is a directed acyclic graph with two specified nodes[2] s (source) and t (target). Each wire is either unlabeled or is labeled by a literal (a variable x_i or its negation $\neg x_i$). A labeled wire is called a *contact*, and an unlabeled wire is a *rectifier*.

The graph may be a multigraph, that is, several wires may have the same endpoints. The *size* of a program is defined as the number of contacts (labeled wires).

Each input $a = (a_1, \ldots, a_n) \in \{0, 1\}^n$ switches the labeled wires On or Off by the following rule: the wire labeled by x_i is switched On if $a_i = 1$ and is switched Off if $a_i = 0$; the wire labeled by $\neg x_i$ is switched On if $a_i = 0$ and is switched Off if $a_i = 1$. The rectifiers are always considered On.

A nondeterministic branching program computes a boolean function in a natural way: it accepts the input a if and only if there exists a path from s to t which is

---

[2] We prefer to use the word "node" instead of "vertex" as well as "wire" instead of "edge" while talking about branching programs.

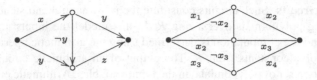

Fig. 1.5 A nondeterministic branching program computing the majority function $\mathrm{Maj}_3(x, y, z) = 1$ iff $x + y + z \geq 2$, and a non-monotone switching network computing the threshold function $\mathrm{Th}_2^4(x_1, x_2, x_3, x_4) = 1$ iff $x_1 + x_2 + x_3 + x_4 \geq 2$

consistent with a, that is, along which all wires are switched On by a. That is, each input switches the wires on or off, and we accept that input if and only if after that there is a nonzero conductivity between the nodes s and t (see Fig. 1.5). Note that we can have many paths consistent with one input vector a; this is why a program is nondeterministic.

An n.b.p. is *monotone* if it does not have negated contacts, that is, wires labeled by negated variables. It is clear that every such program can only compute a monotone boolean function. For a monotone boolean function f, let $\mathrm{NBP}_+(f)$ denote the minimum size of a monotone n.b.p. computing f, and let $\mathrm{NBP}(f)$ be the non-monotone counterpart of this measure. Let also $l(f)$ denote the minimum length of its minterm, and $w(f)$ the minimum length of its maxterm.

Theorem 1.8. (Markov 1962) *For every monotone boolean function f,*

$$\mathrm{NBP}_+(f) \geq l(f) \cdot w(f).$$

Proof. Given a monotone n.b.p. program, for each node u define $d(u)$ as the minimum number of variables that need to be set to 1 to establish a directed path from the source node s to u. In particular, $d(t) = l(f)$ for the target node t.

For $0 \leq i \leq l(f)$, let S_i be the set of nodes u such that $d(u) = i$. If u is connected to v by an unlabeled wire (i.e., not a contact) then $d(u) \geq d(v)$, hence there are no unlabeled wires from S_i to S_j for $i < j$. Thus for each $0 \leq i < l(f)$, the set E_i of contacts out of S_i forms a cut of the branching program. That is, setting these contacts to 0 disconnects the graph, and hence, forces the program output value 0 regardless on the values of the remaining variables. This implies that the set $X(E_i)$ of labels of contacts in E_i must contain a maxterm of f, hence $|X(E_i)| \geq w(f)$ distinct variables. $\square$

For the threshold function Th_k^n we have $l(\mathrm{Th}_k^n) = k$ and $w(\mathrm{Th}_k^n) = n - k + 1$, so every monotone n.b.p. has at least $k(n - k + 1)$ contacts. Actually, this bound is tight, as shown in Fig. 1.6. Thus we have the following surprisingly tight result.

Corollary 1.9. (Markov 1962) $\mathrm{NBP}_+(\mathrm{Th}_k^n) = k(n - k + 1)$.

In particular, $\mathrm{NBP}_+(\mathrm{Maj}_n) = \Theta(n^2)$.

It is also worth noting that the famous result of Szelepcsényi (1987) and Immerman (1988) translates to the following very interesting simulation: there

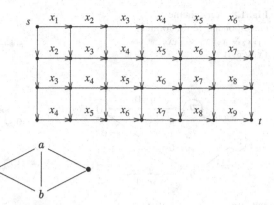

Fig. 1.6 The naive monotone n.b.p. for Th_k^n has $k(n-k+1)$ contacts; here $n=9, k=6$

Fig. 1.7 A graph which is *not* parallel-serial: it has a "bridge" $\{a, b\}$ which is traversed in different directions

exists a constant c such that for every sequence (f_n) of boolean functions,

$$\mathrm{NBP}(\neg f_n) \leq \mathrm{NBP}(f_n)^c.$$

This is a "NP = co-NP" type result for branching programs.

A *parity branching program* is a nondeterministic branching program with the "counting" mode of acceptance: an input vector a is accepted iff the number s-t paths consistent with a is odd.

Switching networks A *switching network* (also called a *contact scheme*) is defined in the same way as an n.b.p. with the only difference that now the underlying graph is *undirected*. Note that in this case unlabeled wires (rectifiers) are redundant since we can always contract them.

A switching network is a *parallel-serial network* (or π-*scheme*) if its underlying graph consists of parallel-serial components (see Fig. 1.8). Such networks can be equivalently defined as switching networks satisfying the following condition: it is possible to direct the wires in such a way that every s-t path will turn to a directed path from s to t; see Fig. 1.7 for an example of a switching network which is *not* parallel-serial.

It is important to note that switching networks include DeMorgan formulas as a special case!

Proposition 1.10. *Every DeMorgan formula can be simulated by a π-scheme of the same size, and vice versa.*

Proof. This can be shown by induction on the leafsize of a DeMorgan formula F. If F is a variable x_i or its negation $\neg x_i$, then F is equivalent to a π-scheme consisting of just one contact. If $F = F_1 \wedge F_2$ then, having π-schemes S_1 and S_2 for subformulas F_1 and F_2, we can obtain a π-scheme for F by just identifying the target node of S_1 with the source node of S_2 (see Fig. 1.8). If $F = F_1 \vee F_2$ then,

Fig. 1.8 A π-scheme
corresponding to the formula
$x_1(x_2 \vee x_3)(x_3 \vee \overline{x}_4 x_5$
$(x_1 \vee \overline{x}_2))$

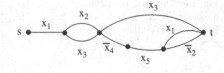

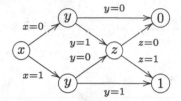

Fig. 1.9 A deterministic branching program computing the majority function $\text{Maj}_3(x, y, z) = 1$ iff $x + y + z \geq 2$, and such a program computing the parity function $\text{Parity}(x, y, z) = x + y + z \mod 2$; wires left without a label in the latter program make tests $y = 1$ and $z = 1$, respectively

having π-schemes S_1 and S_2 for subformulas F_1 and F_2, we can obtain a π-scheme for F by placing these two schemes in parallel and gluing their source nodes and their target nodes. $\square$

That the presence of unlabeled directed wires in a network makes a difference, can be seen on the example of the threshold function Th_2^n. Let $S(f)$ denote the minimum number of contacts in a switching network computing f, and let $S_+(f)$ denote the monotone counterpart of this measure. By Markov's theorem, $\text{NBP}_+(\text{Th}_2^n) = 2n - 3$, but it can shown that $S_+(\text{Th}_2^n) = \Omega(n \log_2 n)$ (see Exercise 1.12). In fact, if n is a power of 2, then we also have $S_+(\text{Th}_2^n) \leq n \log_2 n$, even in the class of π-schemes (see Exercise 1.11). It can also be easily shown that in the class of *non-monotone* switching networks we have that $S(\text{Th}_2^n) \leq 3n - 4$ (see Fig. 1.5 for a hint).

Deterministic branching programs In a nondeterministic branching program as well as in a switching network one input vector $a \in \{0, 1\}^n$ can be consistent with *many* s-t paths. The deterministic version forbids this: every input vector must be consistent with exactly one path.

Formally, a *deterministic branching program* for a given boolean function f of n variables $x_1, \ldots, x_n$ is a directed acyclic graph with one source node and two sinks, that is, nodes of out-degree 0. The sinks are labeled by 1 (accept) and by 0 (reject). Each non-sink node has out-degree 2, and the two outgoing wires are labeled by the tests $x_i = 0$ and $x_i = 1$ for some $i \in \{1, \ldots, n\}$; the node itself is labeled by the variable x_i (Fig. 1.9).

Such a program computes a boolean function $f : \{0, 1\}^n \to \{0, 1\}$ in a natural way: given an input vector $a \in \{0, 1\}^n$, we start in the source node and follow the unique path whose tests are consistent with the corresponding bits of a; this path is the *computation* on a. In this way we reach a sink, and the input a is accepted iff this is the 1-sink.

Thus, a deterministic branching program is a nondeterministic branching program with the restriction that each non-sink node has fanout 2, and the two outgoing wires from each such node are labeled by the tests $x_i = 0$ and $x_i = 1$ on the *same* variable x_i. The presence of the 0-sink is just to ensure that each input vector can reach a sink.

A *decision tree* is a deterministic branching program whose underlying graph is a binary tree. The *depth* of such a tree is the maximum number of wires in a path from the source node to a leaf.

> In the literature, branching programs are also called *binary decision diagrams* or shortly BDDs. This term is especially often used in circuit design theory as well as in other fields where branching programs are used to represent boolean functions. Be warned, however, that the term "BDD" in such papers is often used to denote a much weaker model, namely that of oblivious read-once branching programs (OBDD). These are deterministic branching programs of a very restricted structure: along every computation path all variables are tested in the same order, and no variable is tested more than once.

It is clear that $\mathrm{NBP}(f) \leq \mathrm{S}(f) \leq \mathrm{BP}(f)$, where $\mathrm{BP}(f)$ denotes the minimum size of a *deterministic* branching program computing f. An important result of Reingold (2008) translates to

$$\mathrm{BP}(f_n) \leq \mathrm{S}(f_n)^{\mathcal{O}(1)}.$$

This is a "P = NP" type result for branching programs.

1.4 Almost All Functions are Complex

We still cannot prove super-linear lower bounds for circuits with AND, OR and NOT gates. This is in sharp contrast with the fact, proved more than 60 years ago by Riordan and Shannon (1942) that most boolean functions require formulas of leafsize about $2^n/\log n$. Then Shannon (1949) showed a lower bound $2^n/n$ for circuits. Their arguments were the first applications of counting arguments in boolean function complexity: count how many *different* boolean functions of n variables can be computed using a given number of elementary operations, and compare this number with the total number 2^{2^n} of all boolean functions. After these works of Riordan and Shannon there were many results concerning the behavior of the so-called "Shannon function" in different circuit models.

Definition 1.11. (Shannon function) Given a circuit model with a particular their size-measure, the *Shannon function* for this model is $\mu(n) = \max \mu(f)$, where the maximum is taken over all boolean functions f of n variables, and $\mu(f)$ is the minimum size of a circuit computing f.

In other words, $\mu(n)$ is the smallest number t such that *every* boolean function of n variables can be computed by a circuit of size at most t.

Most bounds in circuit complexity ignore constant multiplicative factors. Moreover, boolean functions $f : \{0, 1\}^n \to \{0, 1\}$ are parameterized by their number of variables n. Hence, under a boolean function f we actually understand an infinite sequence $\{f_n : n = 1, 2, \ldots\}$ of boolean functions. So the claim "f requires $\Omega(\varphi(n))$ gates" means that there exists a constant $\epsilon > 0$ such that, for infinitely many values of n, the function f_n cannot be computed using fewer than $\epsilon \cdot \varphi(n)$ gates. We will also say that f requires a "super-polynomial" number of gates, if $\varphi(n) \geq n^\alpha$ for some $\alpha \to \infty$ as $n \to \infty$, and that f requires an "exponential" number of gates, if $\varphi(n) \geq 2^{n^\epsilon}$ for a constant $\epsilon > 0$.

Through this section, by a circuit (formula) we will understand a circuit (formula) over the basis $\{\wedge, \vee, \neg\}$; similar results, however, also hold when all 16 boolean functions of two variables are allowed as gates. By B_n we will denote the set of all 2^{2^n} boolean functions of n variables $x_1, \ldots, x_n$.

1.4.1 Circuits

Let $C(f)$ denote the minimum size of a fanin-two circuit over $\{\wedge, \vee, \neg\}$ computing f. Let also

$$\phi(n, t) := |\{f \in B_n : C(f) \leq t\}|$$

denote the number of distinct boolean functions $f \in B_n$ computable by circuits of size at most t. As before, we assume that the function computed by a circuit $g_1, g_2, \ldots, g_t$ is the function computed at its last gate g_t. So we now assume that every circuit computes only one boolean function. This implies that every class $F \subseteq B_n$ of $|F| > \phi(n, t)$ functions must contain a function requiring circuits of size $>t$. This was the main idea of Riordan–Shannon's argument.

Lemma 1.12. $\phi(n, t) \leq t^t e^{2t+4n}$. *In particular,* $\phi(n, t) \leq 2^{t^2}$ *for* $t \geq n \geq 16$.

Proof. Clearly, we may suppose $n, t \geq 2$. Let $g_1, \ldots, g_t$ be names of the gates in a circuit. To describe a concrete circuit, it is sufficient to attach to each gate one of the connectives $\wedge, \vee, \neg$ and an unordered pair of names of two other gates or literals. There are at most

$$\left(3 \binom{t-1+2n}{2}\right)^t \leq 2^t (t + 2n)^{2t}$$

such descriptions. Clearly, some of these descriptions do not represent a circuit satisfying all requirements, but every correct circuit may be described in this way. Note that the output does not have a special name. In a correct circuit, it is determined by the fact that it is the only gate not used in any other gate. It is easy to see that every function representable by a circuit of size at most t is also representable by a circuit of size exactly t satisfying the additional requirement that no two of its gates compute the same function. It is also easy to see that in a circuit satisfying the last mentioned property, each of the $t!$ permutations of the names of the gates leads to a different description of a circuit computing the same function. So using estimates $t! \geq (t/3)^t$ and $1 + x \leq e^x$, we can upper bound $\phi(n, t)$ by

$$\frac{2^t(t+2n)^{2t}}{t!} \leq \frac{2^t 3^t(t+2n)^{2t}}{t^t} = 6^t t^t \left(1 + \frac{2n}{t}\right)^{2t} \leq t^t 6^t e^{4n}. \qquad \square$$

Lemma 1.13. (Kannan 1981) *For every integer $k \geq 1$, there is a boolean function of n variables such that f can be written as a DNF with n^{2k+1} monomials, but $C(f) > n^k$.*

Proof. We view a circuit computing a boolean function f as accepting the set of vectors $f^{-1}(1) \subseteq \{0,1\}^n$, and rejecting the remaining vectors. Fix a subset $T \subseteq \{0,1\}^n$ of size $|T| = nt^2 = n^{2k+1}$. By Lemma 1.12, we know that at most $2^{n^{2k}} < 2^{|T|}$ distinct subsets of T can be accepted by circuits of size at most n^k. Thus, some subset $S \subseteq T$ cannot be accepted by a circuit of size n^k. But this subset S can be accepted by a trivial DNF with $|S| \leq |T| = n^{2k+1}$ monomials: just take one monomial for each vector in S. $\qquad \square$

Since we have 2^{2^n} distinct boolean functions of n variables, setting $t := 2^n/n$ in Lemma 1.12 immediately implies the following lower bound on the Shannon function $C(n)$ in the class of circuits.

Theorem 1.14. *For every sufficiently large n, $C(n) > 2^n/n$.*

On the other hand, it is easy to see that $C(n) = \mathcal{O}(n2^n)$: just take the DNFs. Muller (1956) proved that $C(n) = \Theta(2^n/n)$ for any finite complete basis. Lupanov (1958a) used an ingenious construction to prove an asymptotically tight bound.

Theorem 1.15. (Lupanov 1958a) *For every boolean function f of n variables,*

$$C(f) \leq (1+\alpha_n)\frac{2^n}{n} \text{ where } \alpha_n = \mathcal{O}\left(\frac{\log n}{n}\right). \qquad (1.4)$$

Proof. We assume that the number n of variables is large enough. For a boolean vector $a = (a_1, \ldots, a_n)$, let $\mathrm{bin}(a) := \sum_{i=1}^{n} a_i \cdot 2^{n-i}$ be the unique natural number between 0 and $2^n - 1$ associated with a; we call $\mathrm{bin}(a)$ the *code* of a.

Let $H_{n,m}(i)$ denote the set of all boolean functions $h(x)$ of n variables such that $h(a) = 0$ if $\mathrm{bin}(a) \leq m(i-1)$ or $\mathrm{bin}(a) > mi$. That is, we arrange the vectors of $\{0,1\}^n$ into a string of length 2^n according to their codes, split this string into consecutive intervals of length m, and let $H_{n,m}(i)$ to contain all boolean functions h that take value 0 outside the i-th interval:

$$\ldots, 0, 0, \quad \underbrace{*, \ldots, *}_{\text{values on } m \text{ vectors}}, 0, 0, \ldots .$$

Thus,[3] for each $i = 1, \ldots, 2^n/m$, each function in $H_{n,m}(i)$ can only accept a subset of a fixed set of m vectors, implying that

---

[3] An apology to purists: for simplicity of presentation, we will often ignore ceilings and floors.

$$|H_{n,m}(i)| \leq 2^{m+1}$$

for all i. Since every input vector a has its unique weight, every boolean function $f(x)$ of n variables can be represented as a disjunction

$$f(x) = \bigvee_{i=1}^{2^n/m} f_i(x), \tag{1.5}$$

where $f_i \in H_{n,m}(i)$ is the functions such that $f_i(a) = f(a)$ for every a such that $m(i-1) < \mathrm{bin}(a) \leq mi$. We can associate with every $a \in \{0,1\}^n$ the *elementary conjunction*

$$K_a = x_1^{a_1} x_2^{a_2} \cdots x_n^{a_n}.$$

Recall that $x_i^\sigma = 1$ if $a_i = \sigma$, and $x_i^\sigma = 0$ otherwise. Hence, $K_a(b) = 1$ if and only if $b = a$, and we have 2^n such elementary conjunctions of n variables.

Claim 1.16. All elementary conjunctions of n variables can be simultaneously computed by a circuit with at most $2^n + 2n2^{n/2}$ gates.

Proof. Assume for simplicity that n is even. We first compute all $2^{n/2}$ elementary conjunctions of the first $n/2$ variables using a trivial circuit with at most $(n/2)2^{n/2}$ gates, and do the same for the conjunctions of the remaining $n/2$ variables. We now can compute every elementary conjunction of n variables by taking an AND of the corresponding outputs of these two circuits. This requires $2^{n/2} \cdot 2^{n/2} = 2^n$ additional gates, and the entire circuit has size at most $2^n + n2^{n/2}$ To include the case when n is odd, we just multiply the last term by 2.   □

We now turn to the actual construction of an efficient circuit for a given boolean function $f(x)$ of n variables. Let $1 \leq k, m \leq n$ be integer parameters (to be specified latter). By (1.5), we can write $f(x)$ as a disjunction

$$f(x) = \bigvee_a K_a(x_1, \ldots, x_k) \wedge \bigvee_{i=1}^{2^n/m} f_{a,i}(x_{k+1}, \ldots, x_n),$$

where a ranges over $\{0,1\}^k$, and each $f_{a,i}$ belongs to $H_{n-k,m}(i)$. We will use this representation to design the desired circuit for f. The circuit consists of five subcircuits (see Fig. 1.10). The first subcircuit F_1 computes all elementary conjunctions of the first k variables. By Claim 1.16, this circuit has size

$$L(F_1) \leq 2^k + 2k2^{k/2}.$$

The second subcircuit F_2 also computes all elementary conjunctions of the remaining $n - k$ variables. By Claim 1.16, this circuit has size

$$L(F_2) \leq 2^{n-k} + 2(n-k)2^{(n-k)/2}.$$

Fig. 1.10 The structure of
Lupanov's circuit

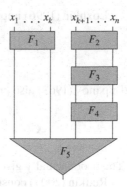

The third subcircuit F_3 computes all functions $f_{a,i}$ from the sets $H_{n-k,m}(i)$ using elementary conjunctions computed by F_2. Since every function in $H_{n-k,m}(i)$ is an OR of at most m elementary conjunctions, each of length $n-k$, and since we have at most $2^{m+1} \cdot 2^{n-k}/m$ such functions, the subcircuit F_3 has size

$$L(F_3) \leq m2^{n-k+m+1}/m = 2^{n-k+m+1}.$$

The fourth subcircuit F_4 computes all functions

$$f_a(x_{k+1}, \ldots, x_n) = \bigvee_{i=1}^{2^n/m} f_{a,i}(x_{k+1}, \ldots, x_n)$$

using the functions $f_{a,i}$ computed by F_3. Since we have at most 2^k such functions f_a, each of which is an OR of at most $2^{n-k}/m$ of the functions $f_{a,i}$, the subcircuit F_4 has size

$$L(F_4) \leq 2^k \cdot 2^{n-k}/m \leq \frac{2^n}{m} + 2^k.$$

The last subcircuit F_5 multiplies functions computed by F_3 by elementary conjunctions computed by F_1, and computes the disjunction of these products. This subcircuit has size

$$L(F_5) \leq 2 \cdot 2^k.$$

Thus, the entire circuit F computes $f(x)$ and has size

$$L(F) \leq \frac{2^n}{m} + 4 \cdot 2^k + 2^{n-k} + 2n2^{k/2} + 2n2^{n-k} + 2^{n-k+m+1}.$$

Now set $k = n - 2\log n$ and $m = n - 4\log n$. Then all but the first terms are at most $\mathcal{O}(2^n/n^2)$, and we obtain that $L(F) \leq 2^n/m + \mathcal{O}(2^n/n^2)$. After simple computations, this implies $L(F) \leq (1 + \alpha)2^n/n$ where $\alpha \leq c(\log n)/n$ for a constant c. $\qquad\square$

Lozhkin (1996) improved (1.4) to

$$\alpha_n = \frac{\log n + \log \log n + \mathcal{O}(1)}{n}.$$

Lupanov (1963) also proved a lower bound

$$C(n) \geq (1 + \beta_n)\frac{2^n}{n} \text{ where } \beta_n = (1 - o(1))\frac{\log n}{n}. \tag{1.6}$$

The proof actually gives that the $o(1)$ factor is equal to $\mathcal{O}(1/\log n)$.

Redkin (2004) considered the behavior of the Shannon function when restricted to boolean functions accepting a small number of input vectors. Let $C(n, K)$ denote the smallest number t such that every boolean function f of n variables such that $|f^{-1}(1)| = K$ can be computed by a circuit over $\{\wedge, \vee, \neg\}$ of size at most t. Redkin (2004) proved that, if $2 \leq K \leq \log_2 n - c \log_2 \log_2 n$ holds for some constant $c > 1$, then

$$C(n, K) \sim 2n.$$

For the Shannon function $M(n)$ restricted to the class of all *monotone* boolean functions of n variables, Ugol'nikov (1976) and Pippenger (1976b) independently proved that

$$M(n) \sim \frac{1}{n}\binom{n}{\lfloor n/2 \rfloor}.$$

This holds for circuits with AND, OR and NOT gates. An important improvement by Andreev (1988b) shows that the upper bound is actually achieved by *monotone* circuits with only AND and OR gates!

1.4.2 Approximation Complexity

In a standard setting, a circuit $F(x)$ must compute a given boolean function $f(x)$ correctly on *all* input vectors $x \in \{0, 1\}^n$. We can relax this and only require that F computes f correctly on some given subset $D \subseteq \{0, 1\}^n$ of vectors; on other input vectors the circuit may output arbitrary values, 0 or 1. That is, we are asking for the smallest size $C(f)$ of a circuit computing a *partial* boolean function $f : \{0, 1\}^n \to \{0, 1, *\}$ defined on

$$D = f^{-1}(0) \cup f^{-1}(1).$$

Let $N = |D|$ be the size of the domain, and $N_1 = |f^{-1}(1)|$. It is clear that $C(f) = \mathcal{O}(nN)$. Actually, we have a much better upper bound:

$$C(f) \le (1 + o(1)) \frac{N}{\log_2 N} + \mathcal{O}(n). \qquad (1.7)$$

For functions with $\log_2 N \sim n$ this was (implicitly) proved already by Nechiporuk (1963, 1965, 1969a) in a series of papers devoted to rectifier networks; Pippenger (1976) gave an independent proof. Then Sholomov (1969) proved this for all $N \ge n \log^{1+\Omega(1)} n$, and finally Andreev (1988) proved this for arbitrary N. It is also known that

$$C(f) \le (1 + o(1)) \frac{\log_2 \binom{N}{N_1}}{\log_2 \log_2 \binom{N}{N_1}} + \mathcal{O}(n).$$

For $\log_2 N_1 \sim n$ this was (implicitly) proved by Nechiporuk in the above mentioned papers, and by Pippenger (1976). Andreev et al. (1996) proved this in the case when $(1 + \epsilon) \log n < \log N_1 = \mathcal{O}(\log n)$ and $\log N = \Omega(n)$. Finally, Chashkin (2006) proved this for arbitrary N_1.

Counting arguments (similar to those above) show that these upper bounds are asymptotically tight. The proofs of the upper bounds are, however, non-trivial: it took more than 40 years to find them!

Let us call a partial boolean function $f : D \to \{0, 1\}$ of n variables *dense* if the size $N = |D|$ of its domain satisfies $\log_2 N \sim n$. The proof of (1.7) for dense functions uses arguments similar to that we used in the proof of Theorem 1.15. Moreover, for dense functions, (1.7) holds without the additive factor $\mathcal{O}(n)$. The proof of (1.7) for functions that are not necessarily dense used interesting ideas which we will sketch right now. We will follow a simplified argument due to Chashkin (2006).

Let $f(x)$ be a partial boolean function which is not dense, that is, for which $\log_2 N \ll n$ holds. If f takes value 1 on fewer than N/n^2 input vectors, then we can compute f by a DNF using at most $n(N/n^2) = N/n$ gates. Thus, the difficult case is when f is not dense but is "dense enough". The idea in this case is to express f as $f(x) = h(x) \oplus g(L(x))$, where h accepts only few vectors, $g : \{0, 1\}^m \to \{0, 1\}$ is a dense partial function, and $L : \{0, 1\}^n \to \{0, 1\}^m$ is an "almost" injective linear operator. Being *linear* means that $L(x) = Ax$ over GF(2) for some boolean $m \times n$ matrix A. Both h and L have small circuits, and for g we can use the upper bound for dense functions.

Say that an operator $L : \{0, 1\}^n \to \{0, 1\}^m$ is *almost injective* on a subset $D \subseteq \{0, 1\}^n$ if $L(x) = L(y)$ for at most $2^{-m} \binom{|D|}{2}$ pairs $x \ne y$ of distinct vectors in D.

Lemma 1.17. *Let $D \subseteq \{0, 1\}^n$ be a set of vectors, and m a positive integer. Then there exists a linear operator $L : \{0, 1\}^n \to \{0, 1\}^m$ which is almost injective on D.*

Proof. We will use a simple (but useful) fact about random vectors in $GF(2)^n$. A *random vector* a in $GF(2)^n$ is obtained by flipping n times a fair 0–1 coin. Hence, $\text{Prob}[a = x] = 2^{-n}$ for each vector $x \in GF(2)^n$. It is easy to show (see Appendix A) that $\text{Prob}[\langle a, x \rangle = \langle a, y \rangle] = 1/2$ holds for every two vectors $x \neq y$ in $\text{GF}(2)^n$, where $\langle a, x \rangle = \sum_{i=1}^{n} a_i x_i \bmod 2$ is the scalar product of a and x over $\text{GF}(2)$.

Now consider a random operator $L(x) = Ax$ where A is a random $m \times n$ matrix whose rows are random vectors in $\text{GF}(2)^n$. By the previous fact, every pair (x, y) of vectors $x \neq y$ in D is not separated by L with probability 2^{-m}. By the linearity of expectation, at most a fraction 2^{-m} of such pairs will not be separated by L. □

Now let f be a partial boolean function of n variables defined on some domain $D \subseteq \{0, 1\}^n$ of size $N = |D|$.

Lemma 1.18. *If* $\log N \geq n/3$ *then* $C(f) \leq (1 + o(1))N/\log N$.

Proof. Let $D_0 = \{x \in D : f(x) = 0\}$ and $D_1 = \{x \in D : f(x) = 1\}$; hence, $D = D_0 \cup D_1$ is the set on which our function f is defined, and $N = |D|$. Set also $m = \lceil \log N + 3 \log n \rceil$.

Lemma 1.17 gives us a linear operator $L : \{0, 1\}^n \to \{0, 1\}^m$ which is almost injective on D. Consider a partial boolean function $g : \{0, 1\}^m \to \{0, 1\}$ defined on $L(D)$ by: $g(z) = 0$ if $z \in L(D_0)$, and $g(z) = 1$ otherwise. If necessary, specify arbitrary values of g on some vectors outside $L(D)$ until the domain of g has exactly N vectors.

And now comes the trick. We can write our function $f(x)$ as

$$f(x) = h(x) \oplus g(L(x)),$$

where

$$h(x) := f(x) \oplus g(L(x))$$

is a partial function defined on D. Thus, we only need to show that all three functions h, g and L can be computed by small circuits.

The operator $L(x)$ is just a set of $m \leq n$ parity functions, and hence, can be computed by a trivial circuit of size $\mathcal{O}(n^2)$, which is $o(N/n)$ because $\log N = \Omega(n)$, by our assumption.

The function h can be computed by a small circuit just because it accepts at most N/n^3 vectors $x \in D$. Indeed, $h(x) = 0$ for all $x \in D_0$ because then $L(x) \in L(D_0)$. Hence, h can accept a vector $x \in D$ only if $x \in D_1$ and $g(L(x)) = 0$, that is, if $x \in D_1$ and $L(x) = L(y)$ for some $y \in D_0$. Since the operator L is almost injective, and since $2^m \geq Nn^3$, there are at most $2^{-m} \binom{N}{2} \leq N/n^3$ pairs $(y, x) \in D_0 \times D_1$ such that $L(x) = L(y)$. Thus, the function h can accept at most N/n^3 vectors. By taking a DNF, this implies that h can be computed by a circuit of size $n(N/n^3) = o(N/n)$.

It remains therefore to compute the function g. Recall that g is a partial function of m variables defined on N vectors. Since $\log N \sim m$, the function g is dense, implying that $C(g) \leq (1 + o(1))N/\log N$. □

We can now easily prove (1.7) for *any* partial function f. If $\log N \geq n/3$ then Lemma 1.18 gives the desired upper bound (without any additive term). Now suppose that $\log N \leq n/3$. In this case we take $m := \lceil 2 \log N \rceil$.

Lemma 1.17 gives us a linear operator $L : \{0,1\}^n \to \{0,1\}^m$ which is almost injective on D. But by our choice of m, the operator L is actually *injective* on D, because $2^{-m}\binom{N}{2} \leq 1/2 < 1$. Thus, in this case we do not need any "error correction" function h because now we have that $f(x) = g(L(x))$ for all $x \in D$, where g is defined as above using our new operator L. The function g has m variables and is defined on $|L(D)| = |D| = N$ vectors.

Since $m \leq \lceil 2 \log N \rceil \leq 3 \log N$, we can apply Lemma 1.18 to g and obtain $C(g) \leq (1 + o(1))N/\log N$. Since $C(L) = \mathcal{O}(n \log N)$, we obtain (1.7) with an additive factor $\mathcal{O}(n^2)$. One can reduce this factor to $\mathcal{O}(n)$ by using the existence of good linear codes computable by circuits of linear size; see Chashkin (2006) for details.

1.4.3 The Circuit Hierarchy Theorem

By using estimates of Shannon–Lupanov it is not difficult to show that one can *properly* increase the number of computed functions by "slightly" increasing the size of circuits. For a function $t : \mathbb{N} \to \mathbb{N}$, let Circuit[$t$] denote the set of all sequences f_n, $n = 1, 2, \ldots$ of boolean functions of n variables such that $C(f_n) \leq t(n)$.

Theorem 1.19. (Circuit Hierarchy Theorem) *If* $n \leq t(n) \leq 2^{n-2}/n$ *then*

$$\text{Circuit}[t] \subsetneq \text{Circuit}[4t].$$

Proof. Fix the maximal $m \in \{1, \ldots, n\}$ such that $t(n) \leq 2^m/m \leq 2 \cdot t(n)$. This is possible: if m is the largest number with $2^m/m \leq 2 \cdot t(n)$, then $2^{m+1}/(m+1) > 2 \cdot t(n)$, which implies $t(n) \leq 2^m/m$. Consider the set $B_{n,m}$ of all boolean functions of n variables that depend only on m bits of their inputs. By the Shannon–Lupanov lower bound, there exists $f_n \in B_{n,m}$ such that $C(f_n) > 2^m/m \geq t(n)$. On the other hand, Lupanov's upper bound yields $C(f_n) \leq 2 \cdot 2^m/m \leq 4 \cdot t(n)$.   □

Remark 1.20. Theorem 1.19 implies that $\phi(n, 4t) \geq \phi(n, t) + 1$; recall that $\phi(n, t)$ is the number of boolean functions of n variables computable by circuits of size at most t. Recently, Chow (2011) gave the following tighter lower bound: there exist constants c and $K > 1$ such that for all $t(n) \leq 2^{n-2}/n$ and all sufficiently large n,

$$\phi(n, t + cn) \geq K \cdot \phi(n, t). \tag{1.8}$$

That is, when allowing an additional cn gates, the number of computable functions is multiplied by at least some constant factor $K > 1$. In particular, if $t(n) \gg n \log n$, then for any fixed d, $\phi(n, t) \geq n^d \cdot \phi(n, t/2)$ for all sufficiently large n. To prove (1.8), Chow sets $N = 2^n$ and lets $A \subseteq \{0,1\}^N$ to be the set of all truth tables of boolean functions $f \in B_n$ computable circuits of size at most t.

A *truth table* is a 0–1 vector $a = (a_1, \ldots, a_N)$, and it describes the unique function $f_a \in B_n$ defined by $f_a(x) = a_{\text{bin}(x)}$ where $\text{bin}(x) = \sum_{i=1}^{n} x_i 2^{i-1}$ is the number whose binary code is vector $x \in \{0,1\}^n$. The *boundary* $\delta(A)$ of $A \subset \{0,1\}^N$ is the set of all vectors $b \notin A$ that differ from at least one $a \in A$ in exactly one position. The discrete isoperimetric inequality (see, for example, Bezrukov (1994)) states that,

$$\sum_{i=0}^{k} \binom{N}{i} \leq |A| < \sum_{i=0}^{k+1} \binom{N}{i} \quad \text{implies } |\delta(A)| \geq \binom{N}{k+1}.$$

Using this and some simple properties of binomial coefficients, Chow shows that the boundary $\delta(A)$ of the set A of truth tables contains at least $\epsilon |A|$ vectors, for a constant $\epsilon > 0$. Now, if $b \in \delta(A)$, then there exists a vector $a \in A$ such that f_b differs from f_a on only one input vector x_0. One can thus take a circuit for f_a, add additional cn gates to test the equality $x = x_0$, and obtain a circuit for f_b. Thus, using additional cn gates we can compute at least $K \cdot |A| = K \cdot \phi(n,t)$ boolean functions, where $K = (1 + \epsilon) > 1$.

Chow (2011) uses this result to show that the so-called "natural proofs barrier" in circuit lower bounds can be broken using properties of boolean functions of lower density; we shortly discuss the phenomenon of natural proofs in the Epilogue.

1.4.4 Switching Networks and Formulas

Let us now consider the Shannon function $S(n)$ in the class of switching networks. The worst-case complexity of switching networks is similar to that of circuits, and can be lower bounded using the following rough upper bound on the number of directed graphs with a given number of wires. Recall that multiple wires joining the same pair of nodes are here allowed.

Lemma 1.21. *There exist at most $(9t)^t$ graphs with t edges.*

Proof. Every set of t edges is incident with at most $2t$ nodes. Using these nodes, at most $r = (2t)^2$ their pairs (potential edges) can be built. Since $x_1 + \ldots + x_r = t$ has $\binom{r+t-1}{t}$ integer solutions $x_i \geq 0$, and since $t! \geq (t/3)^t$ (by Stirling's formula), the number of graphs with t edges is at most

$$\binom{r+t-1}{t} \leq \frac{(r+t-1)^t}{t!} \leq \frac{3^t (r+t-1)^t}{t^t} \leq \frac{3^{2t} t^{2t}}{t^t} = 3^{2t} t^t. \qquad \square$$

Theorem 1.22. *For every constant $\epsilon > 0$ and sufficiently large n,*

$$S(n) \geq (1 - \epsilon) \frac{2^n}{n}.$$

Proof. If is clear that if a boolean function can be computed by a network with at most t contacts then it can also be computed using exactly t contacts. By Lemma 1.21 we know that there are $(9t)^t$ graphs with t edges. Since we only have $2n$ literals, there are at most $(2n)^t$ ways to turn each such graph into a switching network by assigning literals to edges. Since every switching network computes only one boolean function, at most $(18nt)^t$ different boolean functions can be computed by switching networks with at most t contacts. Comparing this number when $t = (1 - \epsilon)2^n/n$ with the total number 2^{2^n} of all boolean functions, yields the result. □

Shannon (1949) proved that $(1 - \epsilon)2^n/n < S(n) < 2^{n+3}/n$ holds for an arbitrarily small constant $\epsilon > 0$. Lupanov (1958b) obtained much tighter bounds:

$$\left(1 + \frac{2\log n - \mathcal{O}(1)}{n}\right)\frac{2^n}{n} \leq S(n) \leq \left(1 + \frac{\mathcal{O}(1)}{\sqrt{n}}\right)\frac{2^n}{n}.$$

In the class of formulas over $\{\wedge, \vee, \neg\}$, that is, fanout-1 circuits constituting a subclass of switching networks (see Proposition 1.10), the behavior of Shannon's function is somewhat different: for some boolean functions, their formulas are at least $n/\log n$ times larger than circuits and switching networks.

When counting formulas, we have to count full binary tree, that is, binary trees where every vertex has either two children or no children. It is well known that the number of such trees with $n + 1$ leaves is exactly the n-th Catalan number:

$$C_n := \frac{1}{n+1}\binom{2n}{n} = \frac{(2n)!}{(n+1)!n!} \sim \frac{4^n}{n^{3/2}\sqrt{\pi}}.$$

Let $L(f)$ denote the smallest number of gates in a formula over $\{\wedge, \vee, \neg\}$ computing f, and let $L(n)$ be the corresponding Shannon function.

Theorem 1.23. *For every constant $\epsilon > 0$ and sufficiently large n,*

$$L(n) \geq (1 - \epsilon)\frac{2^n}{\log_2 n}.$$

Proof. We can assume that all negations are only applied to input gates (leaves). There are at most 4^t binary trees with t leaves, and for each such tree, there are at most $(2n + 2)^t$ possibilities to turn it into a DeMorgan formula: $2n$ input literals and two types of gates, AND and OR. Hence, the number of different formulas of leafsize at most t is at most $4^t(2n + 2)^t \leq (9n)^t$ for $n \geq 8$. Since, we have 2^{2^n} different boolean functions, the desired lower bound on t follows. □

Using more precise computations, tighter estimates can be proved. Riordan and Shannon (1942) proved that

$$L(n) > (1 - \delta_n)\frac{2^n}{\log n} \quad \text{where } \delta_n = \mathcal{O}\Big(\frac{1}{\log n}\Big).$$

On the other hand, Lupanov (1960) showed that

$$L(n) \leq (1 + \gamma_n)\frac{2^n}{\log n} \quad \text{where } \gamma_n = \frac{2\log\log n + \mathcal{O}(1)}{\log n}.$$

Lozhkin (1996) improved this to

$$\gamma_n = \mathcal{O}\Big(\frac{1}{\log n}\Big).$$

Interestingly, Lupanov (1962) showed (among other things) that $L(n)$ drops down from $2^n / \log n$ to

$$L(n) = \mathcal{O}(2^n/n),$$

if we allow just one of the basis functions AND, OR or NOT to have fanout 2. If we allow all three basis functions to have fanout 2, then even the asymptotic

$$L(n) \sim 2^n/n$$

holds. If only NOT gates are allowed to have fanout 2, then

$$L(n) \sim 2^{n+1}/n.$$

Savický and Woods (1998) gave tight estimates on the number of boolean functions computable by formulas of a given size. In particular, they proved that, for every constant k, almost all boolean functions of formula size n^k require circuits of size at least n^k/k.

Nechiporuk (1962) considered the behavior of the Shannon function in cases when some of the gates are given for free. He proved that the smallest number of gates that is enough to compute any boolean function of n variables is asymptotically equal to:

- $2^n/n$ for formulas over $\{\vee, \neg\}$ when $\vee$-gates are for free;
- $\sqrt{2^{n+1}}$ for circuits over $\{\vee, \neg\}$ when $\vee$-gates are for free;
- $2^n/2n$ for formulas over $\{\oplus, \wedge\}$ when $\oplus$-gates are for free;
- $\sqrt{2^n}$ for circuits over $\{\oplus, \wedge\}$ when $\oplus$-gates are for free,

Concerning the Shannon functions $\oplus BP(n)$ for parity branching programs and $NBP(n)$ for nondeterministic branching programs, Nechiporuk (1962) proved that

$$\oplus BP(n) \sim \sqrt{2^{n+1}}$$

Fig. 1.11 Construction of a
nondeterministic branching
program for an arbitrary
boolean function on n
variables. The program is
read-once (along every s-t
path, each variable is tested
only once), and is *oblivious*
(on each level, tests on the
same variable are made)

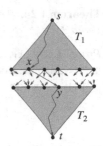

and

$$\sqrt{2^{n+1}} \leq \text{NBP}(n) \leq 2\sqrt{2^n}. \tag{1.9}$$

The upper bound $\text{NBP}(n) \leq 4\sqrt{2^n}$ for an even n is easy to prove. Take a boolean function $f(x_1, \ldots, x_n)$, and assume that $n = 2m$ is even. Let T_1 be a full decision tree on the first m variables, and T_2 a full decision tree on the remaining m variables. Turn T_2 "on its head", and reverse the orientation of its wires. Draw a switch (unlabeled wire) from the leaf of T_1 reached by a vector $x \in \{0, 1\}^m$ to the leaf of T_2 reached by a vector $y \in \{0, 1\}^m$ if and only if $f(x, y) = 1$ (see Fig. 1.11). We have $|f^{-1}(1)|$ switches, but they are for free. The number of contacts in the trees T_1 and T_2 is smaller than $2 \cdot 2^{m+1} = 4\sqrt{2^n}$. Note that the constructed program is "read-once": along each s-t path, each variable is tested only once. If the number of variables is odd, $n = 2m + 1$, then the above construction gives a program with at most $2(2^m + 2^{m+1}) = 3 \cdot 2^{m+1} = 3\sqrt{2^{n+1}}$ contacts. To obtain a better upper bound $2\sqrt{2^n}$, one can use more efficient contact schemes constructed by Lupanov (1958b).

The best known asymptotic bounds on the Shannon function restricted to *monotone* boolean functions can be found in a survey by Korshunov (2003).

1.4.5 Invariant Classes

Let B be the class of all boolean functions. A class $Q \subseteq B$ is *invariant* if together with every function $f(x_1, \ldots, x_n)$ in Q it contains

- all subfunctions of f, and
- all function $f(x_{\pi(1)}, \ldots, x_{\pi(n)})$ where $\pi : [n] \to [n]$ is a permutation.

For example, classes of all symmetric, all linear or all monotone functions are invariant. The class B itself is a trivial invariant class.

Let $Q(n)$ denote the set of all boolean functions $f \in Q$ of n variables; the functions need not depend on all their variables. Denote

$$\text{Lim}(Q) := \lim_{n \to \infty} |Q(n)|^{1/2^n}.$$

Theorem 1.24. *For every invariant class Q, $\mathrm{Lim}(Q)$ exists and lies between 1 and 2.*

Proof. Let $f(x_1, \ldots, x_{n+1})$ be an arbitrary boolean function in Q depending on $n+1$ variables. Recurrence (1.1) yields $|Q(n+1)| \leq |Q(n)|^2$. Hence, the sequence $|Q(n)|^{1/2^n}$ is non-increasing. If $Q \neq \emptyset$, then

$$1 = 1^{1/2^n} \leq |Q(n)|^{1/2^n} \leq (2^{2^n})^{1/2^n} = 2 \,.$$

Thus $\mathrm{Lim}(Q)$ exists and is a number in the interval $[1, 2]$. $\qquad\qquad\qquad\square$

By Theorem 1.24, every invariant class Q of boolean functions defines the unique real number $0 \leq \sigma \leq 1$ such that $\mathrm{Lim}(Q) = 2^\sigma$. This number is an important parameter of the invariant class characterizing its cardinality. It also characterizes the maximum circuit complexity of functions in Q. We will therefore denote this parameter by writing Q_σ if σ is the parameter of Q.

For example, if P is the class of all linear boolean functions (parity functions), then $|P(n)| \leq 2^{n+1}$, implying that $\mathrm{Lim}(P) = 1$, and hence, $\sigma = 0$. The same holds for the class S of all symmetric boolean functions. If M is the class of all monotone boolean functions, then

$$\binom{n}{n/2} \leq \log_2 |M(n)| \leq (1 + o(1)) \binom{n}{n/2} \,.$$

The lower bound here is trivial: consider monotone boolean functions whose minterms have length $n/2$. The upper bound was proved by Kleitman and Markowsky (1975) with the $o(1)$ factor being $\mathcal{O}(\log n / n)$. The number $|M(n)|$ is known as the *Dedekind number*, and was considered by many authors. Korshunov (1977, 1981) proved an asymptotically tight estimate

$$\log_2 |M(n)| \sim (1 + \alpha) \binom{n}{n/2} \quad \text{where} \quad \alpha = \Theta(n^2/2^n) \,.$$

Since $\binom{n}{n/2} = \Theta(2^n / \sqrt{n})$, we again have that $\mathrm{Lim}(M) = 1$, and $\sigma = 0$. On the other hand, $\mathrm{Lim}(B) = (2^{2^n})^{1/2^n} = 2$, and $\sigma = 1$.

Do there exist invariant classes Q with σ strictly between 0 and 1? Yablonskii (1959) showed that, for every real number $0 \leq \sigma \leq 1$ there exists an invariant class Q with $\mathrm{Lim}(Q) = 2^\sigma$.

Example 1.25. As an example let us construct an invariant class with $\sigma = \frac{1}{2}$. For this, let $Q(n)$ consist of all boolean functions of the form $f(x_1, \ldots, x_n) = l_S(x) \wedge g(x)$ where $l_S(x)$ is the parity function $\bigoplus_{i \in S} x_i$ or its negation, and g is an arbitrary boolean function depending only on variables x_i with $i \in S$. It is easy to see that Q is an invariant class. If we take $S = \{1, \ldots, n\}$, then $l_S(x) = 1$ for 2^{n-1} vectors x. Hence, $|Q(n)| \geq 2^{2^{n-1}}$. On the other hand, for a fixed $S \subseteq [n]$, there are at most $2^{2^{|S|-1}} \leq 2^{2^{n-1}}$ functions $f \in Q(n)$. Since we have only 2^{n+1} different linear functions on n variables, $|Q(n)| \leq 2^{n+1}2^{2^{n-1}}$. Thus $\mathrm{Lim}(Q) = \sqrt{2} \cdot \lim_{n \to \infty} 2^{n/2^n} = \sqrt{2}$.

Let $L_Q(n)$ denote the maximum, over all functions $f \in Q(n)$, of the minimum size of a DeMorgan circuit computing f. Yablonskii (1959) extended results of Shannon and Lupanov to all invariant classes.

Theorem 1.26. (Yablonskii 1959) *Let Q be an invariant class of boolean functions, and let $0 \leq \sigma \leq 1$ be its parameter. Then, for every constant $\epsilon > 0$,*

$$(1 - \epsilon)\sigma \frac{2^n}{n} \leq L_Q(n) \leq (1 + o(1))\sigma \frac{2^n}{n}.$$

The lower bound uses Shannon's counting argument and the fact that $Q(n)$ has about $2^{\sigma 2^n}$ boolean functions. The upper bound uses a construction similar to that used by Lupanov (1958a).

It is not difficult to verify that $\sigma < 1$ for every invariant class $Q \neq B$. Indeed, for some fixed m, there exists a boolean function $g(x_1, \ldots, x_m) \notin Q$. Since the sequence $|Q(n)|^{1/2^n}$ is non-increasing, we have that

$$\lim_{n \to \infty} |Q(n)|^{1/2^n} \leq |Q(m)|^{1/2^n} \leq (2^{2^m} - 1)^{1/2^m} < 2.$$

Now suppose we have an algorithm constructing a sequence $F = (f_n : n = 1, 2, \ldots)$ of boolean functions. Call such an algorithm *honest* if, together with the sequence F, it constructs some invariant class of boolean functions containing F. Specifying F as an element of an invariant class means that the sequence F is specified by its *properties*.

Theorem 1.27. (Yablonskii 1959) *Every honest algorithm constructing a sequence of most complex boolean functions must construct all boolean functions.*

Proof. Let us assume the opposite. That is, assume that some sequence $F = (f_n : n = 1, 2, \ldots)$ of most complex boolean functions is a member of some invariant class $Q_\sigma \neq B$. Then $\sigma < 1$, and Theorem 1.26 implies that *every* boolean function $g_n(x_1, \ldots, x_n) \in Q$ has a DeMorgan circuit of size at most $(1 - \lambda)2^n/n$ for some constant $\lambda > 0$. But the lower bound (1.6) implies that $C(f_n) > 2^n/n$. Comparing these bounds, we can conclude that the sequence F cannot be contained in any invariant class Q_σ with $\sigma < 1$. $\square$

This result serves as an indication that there (apparently) is no other way to construct a most-complex sequence of boolean function other than to do a "brute force search" (or "perebor" in Russian): just try all 2^{2^n} boolean functions.

1.5 So Where are the Complex Functions?

Unfortunately, the results above are not quite satisfactory: we know that almost
all boolean functions are complex, but no *specific* (or *explicit*) complex function
is known. This is a strange situation: we know that almost all boolean functions
are complex, but we cannot exhibit any single example of a complex function!
We also face a similar situation in other branches of mathematics. For example,
in combinatorics it is known that a random graph on n vertices is a Ramsey-graph,
that is, has no cliques or independent sets on more than $t = 2\log n$ vertices. But
where are these "mystical" graphs?

> The best known explicit construction of non-bipartite t-Ramsey graphs due to Frankl and
> Wilson only achieves a much larger value t about $\exp(\sqrt{\log n \log \log n})$. In the bipartite case,
> t-Ramsey graphs with $t = n^{1/2}$ can be obtained from Hadamard matrices: Lindsey's Lemma
> (see Appendix A) implies that such a matrix can have a monochromatic $a \times b$ submatrix only
> if $ab \leq n$. But even going below $t = n^{1/2}$ was only recently obtained by Pudlák and Rödl
> (2004), Barak et al. (2010), and Ben-Sasson and Zewi (2010). The paper of Barak et al. (2010)
> constructs bipartite t-Ramsey graphs with $t = n^\delta$ for an arbitrarily small constant $\delta > 0$.

The main goal of boolean complexity theory is to prove lower bounds on
the complexity of computing explicitly given boolean functions in interesting
computational models. By "explicitly given" researchers usually mean "belonging
to the class NP". This is a plausible interpretation since, on the one hand, this class
contains the overwhelming majority of interesting boolean functions, and on the
other hand, it is a sufficiently restricted class in which counting arguments seem
not to apply. The second point is illustrated by a result of Kannan (1981) showing
that already the class $\Sigma_2 \cap \Pi_2$, next after NP in the complexity hierarchy, contains
boolean functions whose circuit size is $\Omega(n^k)$ for any fixed $k > 0$. The proof of
this fact essentially uses counting arguments; we will present it in the Epilogue (see
Theorem 20.13).

1.5.1 On Explicitness

We are not going to introduce the classes of the complexity hierarchy. Instead, we
will use the following simple definitions of "explicitness". Say that a sequence of
boolean functions $g_{n,m}(x, y)$ of $n + m$ variables is "simple" if there exists a Turing
machine (or any other algorithm) which, given n, m and a vector (x, y), outputs
the value $g_{n,m}(x, y)$ in time polynomial in $n + m$. Then we can treat a sequence
of boolean functions $f_n(x)$ as "explicit" if there exists a sequence $g_{n,m}$ of simple
functions with $m = n^{\mathcal{O}(1)}$ such that

$$f_n(x) = 1 \text{ if and only if } g_{n,m}(x, y) = 1 \text{ for at least one } y \in \{0, 1\}^m.$$

In this case, simple functions correspond to the class P, and explicit functions form
the class NP. For example, the parity function $x_1 \oplus \cdots \oplus x_n$ is "very explicit": to

determine its value, it is enough just to sum up all bits and divide the result by 2. A classical example of an explicit function (a function in NP) which is not known to be in P is the clique function. It has $n = \binom{v}{2}$ variables $x_{u,v}$, each for one possible edge $\{u, v\}$ on a given set V of n vertices. Each 0–1 vector x of length $\binom{v}{2}$ defines a graph $G_x = (V, E_x)$ in a natural way: $\{u, v\} \in E_x$ iff $x_{u,v} = 1$. The function itself is defined by:

$$\text{CLIQUE}(x) = 1 \text{ iff the graph } G_x \text{ contains a clique on } \sqrt{n} \text{ vertices.}$$

In this case, $m = n$ and the graphs G_y encoded by vectors y are k-cliques for $k = \sqrt{n}$. Since one can test whether a given k-clique in present in G_x in time about $\binom{k}{2} \leq n$, the function is explicit (belongs to NP). Thus a proof that CLIQUE requires circuits of super-polynomial size would immediately imply that P $\neq$ NP.

Unfortunately, at the moment we are even not able to prove that CLIQUE requires, say, $10n$ AND, OR and NOT gates! The problem here is with NOT gates—we can already prove that the clique function requires $n^{\Omega(\sqrt{n})}$ gates, if no NOT gates are allowed; this is a celebrated result of Razborov (1985a) which we will present in Chap. 9.

1.5.2 Explicit Lower Bounds

The strongest known lower bounds for non-monotone circuits (with NOT gates) computing explicit boolean functions of n variables have the form:

- $4n - 4$ for circuits over $\{\wedge, \vee, \neg\}$, and $7n - 7$ for circuits over $\{\wedge, \neg\}$ and $\{\vee, \neg\}$ computing $\oplus_n(x) = x_1 \oplus x_2 \oplus \cdots \oplus x_n$; Redkin (1973). These bounds are tight.
- $5n - o(n)$ for circuits over the basis with all fanin-2 gates, except the parity and its negation; Iwama and Morizumi (2002).
- $3n - o(n)$ for general circuits over the basis with all fanin-2 gates; Blum and Micali (1984).
- $n^{3-o(1)}$ for formulas over $\{\wedge, \vee, \neg\}$; Håstad (1998).
- $\Omega(n^2/\log n)$ for general fanin-2 formulas, $\Omega(n^2/\log^2 n)$ for deterministic branching programs, and $\Omega(n^{3/2}/\log n)$ for nondeterministic branching programs; Nechiporuk (1966).

We have only listed the strongest bounds for unrestricted circuit models we currently have (some other known bounds are summarized in Tables 1.1–1.4 at the end of this chapter). The bounds for circuits and formulas were obtained by gradually increasing previous lower bounds.

A lower bound $2n$ for general circuits was first proved by Kloss and Malyshev (1965), and by Schnorr (1974). Then Paul (1977) proved a $2.5n$ lower bound, Stockmeyer (1977) gave the same $2.5n$ lower bound for a larger family of boolean functions including *symmetric* functions, Blum and Micali (1984) proved the lower bound $3n - o(n)$. A simpler proof of this lower bound, but for much more complicated functions, was recently found by Demenkov and Kulikov (2011). They prove such a bound for any boolean function which is not constant on any affine subspace of $GF(2)^n$ of dimension $o(n)$. A rather involved construction of such functions was given earlier by Ben-Sasson and Kopparty (2009).

For circuits over the basis with all fanin-2 gates, except the parity and its negation, a lower bound of $4n$ was obtained earlier by Zwick (1991b) (for a symmetric boolean function), then Lachish and Raz (2001) proved a $4.5n - o(n)$ lower bound, and finally Iwama and Morizumi (2002) extended this bound to $5n - o(n)$.

For formulas, the first nontrivial lower bound $n^{3/2}$ was proved by Subbotovskaya (1961), then a lower bound $\Omega(n^2)$ was proved by Khrapchenko (1971), and a lower bound of $\Omega(n^{2.5})$ by Andreev (1985). This was enlarged to $\Omega(n^{2.55})$ by Impagliazzo and Nisan (1993), and to $\Omega(n^{2.63})$ by Paterson and Zwick (1993), and finally to $n^{3-o(1)}$ by Håstad (1998).

The boolean functions for which these lower bounds are proved are quite "simple". For general circuits, a lower bound $3n - o(n)$ is achieved by particular symmetric functions, that is, functions whose value only depends on the number of ones in the input vector.

The lower bound $5n-o(n)$ holds for any k-mixed boolean function with $k = n - o(n)$; a function is k-mixed if for any two different restrictions fixing the same set of k variables must induce different functions on the remaining $n-k$ variables. We will construct an explicit k-mixed boolean function for $k = n - \mathcal{O}(\sqrt{n})$ in Sect. 16.1. Amano and Tarui (2008) showed that some highly mixed boolean functions *can* be computed by circuits of size $5n + o(1)$; hence, the property of being mixed alone is not enough to improve this lower bound.

Almost-quadratic lower bounds for general formulas and branching programs are achieved by the element distinctness function (see Sects. 6.5 and 15.1 for the proofs).

The strongest known lower bounds, up to $n^{3-o(1)}$, for DeMorgan formulas are achieved by the following somewhat artificial function $A_n(x, y)$ (see Sect. 6.4). The function has $n = 2^b + bm$ variables with $b = \log(n/2)$ and $m = n/(2b)$. The last bm variables are divided into b blocks $y = (y_1, \ldots, y_b)$ of length m, and the value of A_n is defined by $A_n(x, y) = f_x(\oplus_m(y_1), \ldots, \oplus_m(y_b))$.

1.6   A 3n Lower Bound for Circuits

Existing lower bounds for general circuits were proved using the so-called "gate-elimination" argument. The proofs themselves consist of a rather involved case analysis, and we will not present them here. Instead of that we will demonstrate the main idea by proving weaker lower bounds.

The *gate-elimination* argument does the following. Given a circuit for the function in question, we first argue that some variable 1 (or set of variables) must fan out to several gates. Setting this variable to a constant will eliminate several gates. By repeatedly applying this process, we conclude that the original circuit must have had many gates.

To illustrate the basic idea, we apply the gate-elimination argument to threshold functions

$$\text{Th}_k^n(x_1, \ldots, x_n) = 1 \text{ if and only if } x_1 + x_2 + \cdots + x_n \geq k \,.$$

Theorem 1.28. *Even if all boolean functions in at most two variables are allowed as gates, the function* Th_2^n *requires at least* $2n - 4$ *gates.*

Proof. The proof is by induction on n. For $n = 2$ and $n = 3$ the bound is trivial. For the induction step, take an optimal circuit for Th_2^n, and suppose that the bottom-most gate g acts on variables x_i and x_j with $i \neq j$. This gate has the form $g = \varphi(x_i, x_j)$ for some $\varphi : \{0, 1\}^2 \to \{0, 1\}$. Notice that under the four possible settings of these two variables, the function Th_2^n has *three* different subfunctions Th_0^{n-2}, Th_1^{n-2} and Th_2^{n-2}. It follows that either x_i or x_j fans out to another gate h, for otherwise our circuit would have only *two* inequivalent sub-circuits under the settings of x_i and x_j. Why? Just because the gate $g = \varphi(x_i, x_j)$ can only take *two* values, 0 and 1.

Now suppose that it is x_j that fans out to h. Setting x_j to 0 eliminates the need of both gates g and h. The resulting circuit computes Th_2^{n-1}, and by induction, has at least $2(n-1) - 4$ gates. Adding the two eliminated gates to this bound shows that the original circuit has at least $2n - 4$ gates, as desired. □

Theorem 1.28 holds for circuits whose gates are any boolean functions in at most two variables. For circuits over the basis $\{\wedge, \vee, \neg\}$ one can prove a slightly stronger lower bound. For this, we consider the parity function

$$\oplus_n(x) = x_1 \oplus x_2 \oplus \cdots \oplus x_n \,.$$

Theorem 1.29. (Schnorr 1974) *The minimal number of AND and OR gates in a circuit over* $\{\wedge, \vee, \neg\}$ *computing* $\oplus_n$ *is* $3(n - 1)$.

Proof. The upper bound follows since $x \oplus y$ is equal to $(x \wedge \neg y) \vee (\neg x \wedge y)$. For the lower bound we prove the existence of some x_i whose replacement by a suitable constant eliminates 3 gates. This implies the assertion for $n = 1$ directly and for $n \geq 3$ by induction.

Let g be the *first* gate of an optimal circuit for $\oplus_n(x)$. Its inputs are different variables x_i and x_j (see Fig. 1.12). If x_i had fanout 1, that is, if g were the only gate for which x_i is acting as input, then we could replace x_j by a constant so that gate g would be replaced by a constant. This would imply that the output became independent of the i-th variable x_i in contradiction to the definition of parity. Hence, x_i must have fanout at least 2. Let g' be the other gate to which x_i is an input.

We now replace x_i by such a constant that g becomes replaced by a constant. Since under this setting of x_i the parity is not replaced by a constant, the gate g cannot be an output gate. Let h be a successor of g. We only have two possibilities: either h coincides with g' (that is, g has no other successors besides g') or not.

Fig. 1.12 The two cases in the proof of Theorem 1.29

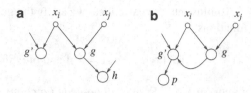

Case (a): $g' = h$. In this case g has fanout 1. We can set x_i to a constant so that g' will become set to a constant. This will eliminate the need for all three gates g, g' and p.

Case (b): $g' \neq h$. Then we can set x_i to a constant so that g will become set to a constant. This will eliminate the need for all three gates g, g' and h.

In either case we eliminate at least 3 gates.   □

Note that the same argument works if we allow as gates any boolean functions $\phi(x, y)$ with the following property: there exist constants $a, b \in \{0, 1\}$ such that both $\phi(a, y)$ and $\phi(x, b)$ are constants. The only two-variable functions that do not have this property is the parity function $x \oplus y$ and its negation $x \oplus y \oplus 1$.

1.7 Graph Complexity

As pointed out by Sipser (1992), one of the impediments in the lower bounds area is a shortage of problems of *intermediate* difficulty which lend insight into the harder problems. Most of known problems (boolean functions) are either "easy" (parity, majority, etc.) or are "very hard" (clique problem, satisfiability of CNFs, and all other NP-hard problems).

On the other hand, there are fields—like graph theory or matrix theory—with a much richer spectrum of known objects. It therefore makes sense to look more carefully at the graph structure of boolean functions: that is, to move from a "bit level" to a more global one and consider a given boolean function as a matrix or as a bipartite graph. The concept of graph complexity, as we will describe it below, was introduced by Pudlák et al. (1988), and was later considered by Razborov (1988, 1990), Chashkin (1994), Lokam (2003), Jukna (2006, 2010), Drucker (2011), and other authors.

A circuit for a given boolean function f generates this function starting from simplest "generators"—variables and their negations. It applies some boolean operations like AND and OR to these generators to produce new "more complicated" functions, then does the same with these functions until f is generated. Note however that there was nothing special to restrict ourselves to boolean functions—one can define, say, the complexity of graphs or matrices analogously.

A basic observation connecting graphs and boolean functions is that boolean functions can be treated as graphs. Namely, every boolean function $f(x_1, \ldots, x_m, y_1, \ldots, y_m)$ of $2m$ variables can be viewed as a bipartite $n \times n$

graph[4] $G_f \subseteq V_1 \times V_2$ with $n = 2^m$, whose vertex-sets $V_1 = V_2 = \{0, 1\}^m$ are binary vectors, and $(u, v) \in G_f$ iff $f(u, v) = 1$. In particular, literals x_i^a and y_j^a for $a \in \{0, 1\}$ then turn to *bicliques* (bipartite complete graphs):

1. If $f = x_i^a$ then $G_f = \{u \in V_1 : u_i = a\} \times V_2$.
2. If $f = y_j^a$ then $G_f = V_1 \times \{v \in V_2 : v_j = a\}$.

Boolean operations AND and OR turn to set-theoretic operations:

$$G_{f \wedge g} = G_f \cap G_g \text{ and } G_{f \vee g} = G_f \cup G_g.$$

Thus, every (non-monotone!) DeMorgan formula (or circuit) for the function f turns to a formula (circuit) which can use any of $4m$ bicliques defined above, and apply the union and intersection operations to produce the entire graph G_f.

We thus can take a "vacation" from boolean functions, and consider the computational complexity of graphs: how many $\cup$ and $\cap$ operations do we need to produce a given bipartite graph G starting from bicliques?

Remark 1.30. In the context of arbitrary bipartite graphs, restriction to these special bicliques (1) and (2) as generators looks somewhat artificial. And indeed, if we use only these $4m = 4 \log n$ generators, then the complexity of *isomorphic* graphs may be exponentially different. In particular, there would exist a perfect matching of formula size $\mathcal{O}(m) = \mathcal{O}(\log n)$, namely that corresponding to the equality function defined by $f(x, y) = 1$ iff $(x = y)$, as well as a perfect matching requiring $\Omega(n)$ formula size; the existence can be shown by comparing the number $m^{\mathcal{O}(t)}$ of formulas of size t with the total number $n!$ of perfect matchings.

1.7.1 Clique Complexity of Graphs

In view of the previous remark, let us allow all 2^{2n} bicliques $P \times V_2$ and $V_1 \times Q$ with $P \subseteq V_1$ and $Q \subseteq V_2$ as generators. The *bipartite formula complexity*, $L_{bip}(G)$, of a bipartite $n \times n$ graph $G \subseteq V_1 \times V_2$, is then the minimum number of leaves in a formula over $\{\cap, \cup\}$ which produces the graph G starting from these generators.

By what was said above, we have that every boolean function f of $2m = 2 \log n$ variables requires non-monotone DeMorgan formulas with at least $L_{bip}(G_f)$ leaves. Thus any explicit bipartite $n \times n$ graph G with $L_{bip}(G) = \Omega(\log^K n)$ would immediately give us a an explicit boolean function of $2m$ variables requiring non-monotone formulas of size $\Omega(m^K)$. Recall that the best known lower bound for formulas has the form $\Omega(m^3)$.

Note however that even if we have "only" to prove poly-logarithmic lower bounds for graphs, such bounds may be extremely hard to obtain. For example, we will prove later in Sect. 6.8 that, if f is the parity function of $2m$ variables, then

---

[4]Here and in what follows we will often consider graphs as *sets* of their edges.

any non-monotone DeMorgan formula computing f must have at least $\Omega(m^2) = \Omega(\log^2 n)$ leaves. But the graph G_f of f is just a union of two bicliques, implying that $\mathsf{L}_{\mathrm{bip}}(G) \leq 4$.

Another way to view the concept of bipartite complexity of graphs $G \subseteq V_1 \times V_2$ is to associate with subsets $P \subseteq V_1$ and $Q \subseteq V_2$ boolean variables (we call them *meta-variables*) $z_P, z_Q : V_1 \times V_2 \to \{0, 1\}$ interpreted as

$$z_P(u, v) = 1 \text{ iff } u \in P, \text{ and } z_Q(u, v) = 1 \text{ iff } v \in Q.$$

Then the set of edges accepted by z_P is exactly the biclique $P \times V_2$, and similarly for variables z_Q.

Remark 1.31. Note that in this case we do not need negated variables: for every $P \subseteq V_1$, the variable $z_{V_1 \setminus P}$ accepts exactly the same set of edges as the negated variable $\neg x_P$. Thus $\mathsf{L}_{\mathrm{bip}}(G)$ is exactly the minimum leafsize of a *monotone* DeMorgan formula of these meta-variables which accepts all edges and rejects all nonedges of G. Also, the depth of a decision tree for the graph G_f, querying the meta-variables, is *exactly* the communication complexity of the boolean function $f(x, y)$, a measure which we will introduce in Chap. 3.

1.7.2 Star Complexity of Graphs

Now we consider the complexity of graphs when only special bicliques—stars—are used as generators. A *star* is a bipartite graph formed by one vertex connected to all vertices on the other side of the bipartition. In this case the complexity of a given graph turns into a monotone complexity of monotone boolean functions "representing" this graph in the following sense.

Let $G = (V, E)$ be an n-vertex graph, and let $z = \{z_v : v \in V\}$ be a set of boolean variables, one for each vertex (not for each subset $P \subseteq V$, as before). Say that a boolean function (or a circuit) $g(z)$ *represents* the graph G if, for every input $a \in \{0, 1\}^n$ with exactly two 1s in, say, positions $u \neq v$, $g(a) = 1$ iff u and v are adjacent in G:

$$f(0, \ldots, 0, \overset{u}{1}, 0, \ldots, 0, \overset{v}{1}, 0, \ldots, 0) = 1 \quad \text{if and only if} \quad \{u, v\} \in E.$$

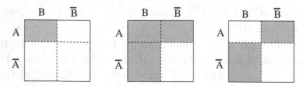

Fig. 1.13 The adjacency matrices of: (**a**) a complete bipartite graph $A \times B$ represented by $g = \left(\bigvee_{u \in A} z_u\right) \wedge \left(\bigvee_{v \in B} z_v\right)$, (**b**) a bipartite graph represented by an OR function $g = \bigvee_{v \in A \cup B} z_v$, and (**c**) a bipartite graph represented by a Parity function $g = \bigoplus_{v \in A \cup B} z_v$

If the graph is bipartite then we only require that this must hold for vertices u and v from different color classes. Note that in both cases (bipartite or not), on input vectors with fewer than two 1s as well as on vectors with more than two 1s the function g can take arbitrary values!

Another way to treat this concept is to view edges as 2-element sets of vertices, and boolean functions (or circuits) as accepting/rejecting subsets $S \subseteq V$ of vertices. Then a boolean function $f : 2^V \to \{0, 1\}$ represents a graph if it accepts all edges and rejects all non-edges. On subsets S with $|S| \neq 2$ the function can take arbitrary values.

Thus a single variable z_v represents a complete star around the vertex v, that is, the graph consisting of all edges connecting v with the remaining vertices. If we consider bipartite graphs with bipartition $V_1 \cup V_2$, then each single variable x_v with $v \in V_i$ represents the star consisting of all edges connecting v with vertices in V_{3-i}. If $A \subseteq V_1$ and $B \subseteq V_2$, then the boolean function

$$\left(\bigvee_{u \in A} z_u \right) \wedge \left(\bigvee_{v \in B} z_v \right)$$

represents the complete bipartite graph $A \times B$ (Fig. 1.13). Note also that every graph $G = (V, E)$ is represented by $\bigvee_{uv \in E} z_u \wedge z_v$. But this representation of n-vertex graphs is not quite compact: the number of gates in them may be as large as $\Theta(n^2)$. If we allow unbounded fanin OR gates then already $2n - 1$ gates are enough: we can use the representation

$$\bigvee_{u \in S} z_u \wedge \left(\bigvee_{v:uv \in E} z_v \right),$$

where $S \subseteq V$ is an arbitrary vertex-cover of G, that is, a set of vertices such that every edge of G has is endpoint in S.

We have already seen how non-monotone circuit complexity of boolean functions is related to biclique complexity of graphs. A similar relation is also in the case of star complexity.

As before, we consider a boolean function $f(x, y)$ of $2m$ variables as a bipartite $n \times n$ graph $G_f \subseteq U \times V$ with color classes $U = V = \{0, 1\}^m$ of size $n = 2^m$, in which two vertices (vectors) x and y are adjacent iff $f(x, y) = 1$. In the following lemma, by a "circuit" we mean an arbitrary boolean circuit with literals—variables and their negations—as inputs.

Lemma 1.32. (Magnification Lemma) *In every circuit computing $f(x, y)$ it is possible to replace its input literals by ORs of new variables so that the resulting monotone circuit represents the graph G_f.*

Proof. Any input literal x_i^a in a circuit for $f(x, y)$ corresponds to the biclique $U_i^a \times V$ with $U_i^a = \{u \in U : u_i = a\}$. Every such biclique is represented by an OR $\bigvee_{u \in U_i^a} z_u$ of $2^{m-1} = n/2$ new variables. $\square$

Instead of replacing input literals by ORs one can also replace them by any other boolean functions that compute 0 on the all-0 vector, and compute 1 on any input

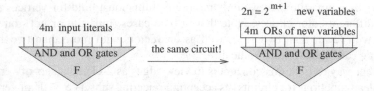

Fig. 1.14 Having a circuit F computing a boolean function f of $2m$ variables, we obtain a (monotone) circuit representing the graph G_f by replacing each input literal in F by an appropriate OR of new variables

vector with exactly one 1. In particular, parity functions also have this property, as well as any function $g(Z) = \varphi(\sum_{w \in S} z_w)$ with $\varphi : \mathbb{N} \to \{0, 1\}$, $\varphi(0) = 0$ and $\varphi(1) = 1$ does.

The Magnification Lemma is particularly appealing when dealing with circuits containing *unbounded* fanin OR (or unbounded fanin Parity gates) on the next to the input layer (Fig. 1.14). In this case the total number of gates in the circuit computing f is exactly the number of gates in the obtained circuit representing the graph G_f! Thus if we could prove that some explicit bipartite $n \times n$ graph with $n = 2^m$ cannot be represented by such a circuit of size n^ϵ, then this would immediately imply that the corresponding boolean function $f(x, y)$ in $2m$ variables cannot be computed by a (non-monotone!) circuit of size $n^\epsilon = 2^{\epsilon m}$, which is already exponential in the number of variables of f. We will use Lemma 1.32 in Sect. 11.6 to prove truly exponential lower bounds for unbounded-fanin depth-3 circuits with parity gates on the bottom layer.

It is important to note that moderate lower bounds for graphs even in very weak circuit models (where strong lower bounds for boolean functions are easy to show) would yield impressive lower bounds for boolean circuits in rather nontrivial models. To demonstrate this right now, let $\mathrm{cnf}(G)$ denote the smallest number of clauses in a monotone CNF (AND of ORs of variables) representing the graph G.

A bipartite graph is $K_{2,2}$-free if it does not have a cycle of length 4, that is, if its adjacency matrix does not have a 2×2 all-1 submatrix.

■ **Research Problem 1.33.** Does there exist a constant $\epsilon > 0$ such that $\mathrm{cnf}(G) \geq D^\epsilon$ for every bipartite $K_{2,2}$-free graph G of average degree D?

We will see later in Sect. 11.6 that a positive answer would give an explicit boolean function f of n variables such that any DeMorgan circuit of depth $\mathcal{O}(\log n)$ computing f requires $\omega(n)$ gates (cf. Research Problem 11.17). Thus graph complexity is a promising tool to prove lower bounds for boolean functions. Note, however, that even small lower bounds for graphs may be very difficult to prove. If, say, $n = 2^m$ and if $f(x, y)$ is the parity function of $2m$ variables, then any CNF for f must have at least $2^{2m-1} = n^2/2$ clauses. But the bipartite $n \times n$ graph G_f corresponding to this function consists of just two complete bipartite subgraphs; hence, G_f can be represented by a monotone CNF consisting of just four clauses.

1.8   A Constant Factor Away From P ≠ NP?

Having warned about the difficulties when dealing with the graph complexity, in this section we sketch a potential (albeit very hard to realize) approach to proving strong lower bounds on circuit complexity of boolean functions using the graph complexity.

Recall that a DeMorgan circuit consists of fanin-2 AND and OR gates, and has all variables as well as their negations as inputs. A monotone circuit is a DeMorgan circuit without negated variables as inputs.

Proposition 1.34. *Almost all bipartite $n \times n$ graphs require monotone circuits of size $\Omega(n^2/\log n)$ to represent them.*

Proof. Easy counting (as in the proof of Theorem 1.14) shows that there are at most $(nt)^{O(t)}$ monotone circuits with at most t gates. Since we have 2^{n^2} graphs, and different graphs require different circuits, the lower bound follows.   □

Thus the overwhelming majority of graphs require an almost-quadratic number of gates to represent. On the other hand, we are now going to show (Corollary 1.36 below) that any *explicit* bipartite $n \times n$ graph which cannot be represented by a monotone circuit with fewer than $7n$ gates would give us an explicit boolean function f in $2m$ variables which cannot be computed by a non-monotone(!) DeMorgan circuit with fewer than 2^m gates. That is, linear lower bounds on the monotone complexity of graphs imply exponential lower bounds on the non-monotone complexity of boolean functions.

When constructing the circuit for the graph G, as in the Magnification Lemma, we replace $4m$ input literals in a circuit for f_G by $4m = 4\log n$ disjunctions of $2n = 2^{m+1}$ (new) variables. If we compute these disjunctions separately then we need about $mn = n\log n$ fanin-2 OR gates. The disjunctions can, however, be computed much more efficiently using only about n OR gates, if we compute all these disjunctions simultaneously. This can be shown using the so-called "transposition principle".

Let $A = (a_{ij})$ be a boolean $p \times q$ matrix. Our goal is to compute the transformation $y = Ax$ over the boolean semiring. Such a transformation computes p boolean sums (disjunctions) of q variables $x_1, \ldots, x_q$:

$$y_i = \bigvee_{j=1}^{q} a_{ij} x_j = \bigvee_{j:a_{ij}=1} x_j \quad \text{for} \quad i = 1, \ldots, p.$$

Thus, our question reduces to estimating the *disjunctive complexity*, $\mathrm{OR}(A)$, of A defined as the minimum number of fanin-2 OR gates required to simultaneously compute all these p disjunctions.

By computing all p disjunctions separately, we see that $\mathrm{OR}(A) < pq$. However, in some situations (as in the graph complexity) we have that the number p of disjunctions (rows) is much smaller than the number q of variables (columns). In the context of graph complexity, we have $p = 4m$ and $q = 2^{m+1}$; hence, $p < 4\log_2 q$. In such situations, it would be desirable to somehow "replace" the roles of rows and columns. That is, it would be desirable to relate the disjunctive complexity of a matrix A with the disjunctive complexity of the transposed matrix A^T; recall that the transpose of a matrix $A = (a_{ij})$ is the matrix $A^T = (b_{ij})$ with $b_{ij} = a_{ji}$.

Transposition Principle. If A is a boolean matrix with p rows and q columns, then

$$\mathrm{OR}(A^T) = \mathrm{OR}(A) + p - q \,.$$

This principle was independently pointed out by Bordewijk (1956) and Lupanov (1956) in the context of rectifier networks. Mitiagin and Sadovskii (1965) proved the principle for boolean circuits, and Fiduccia (1973) proved it for bilinear circuits over any commutative semiring.

Proof. Let $A = (a_{ij})$ be a $p \times q$ boolean matrix, and take a circuit F with fanin-2 OR gates computing $y = Ax$. This circuit has q input nodes $x_1, \ldots, x_q$ and p output nodes $y_1, \ldots, y_p$. At y_i the disjunction $\vee_{j:a_{ij}=1} x_j$ is computed.

Let $\alpha(F)$ be the number of gates in F. Since each non-input node in F has fanin 2, we have that $\alpha(F) = e - v + q$, where e is the total number of wires and v is the total number of nodes (including the q input nodes). Since the circuit F computes $y = Ax$ and has only OR gates, we have that $a_{ij} = 1$ if and only if there exists a directed path from the j-th input x_j to the i-th output y_i.

We now transform F to a circuit F' for $x = A^T y$ such that the difference $e' - v'$ between the numbers of wires and nodes in F' does not exceed $e - v$. First, we transform F so that no output gate is used as an input to another gate; this can be achieved by adding nodes of fanin 1. After that we just reverse the orientation of wires in F, contract all resulting fanin-1 edges, and replace each node of fanin larger than 2 by a binary tree of OR gates (see Fig. 1.15). Finally, assign OR gates to all n input gates of F (now the output gates of F').

It is easy to see that the new circuit F' computes $A^T y$: there is a path from y_i to x_j in F' iff there is a path from x_j to y_i in F. Moreover, since $e' - v' \le e - v$, the new circuit F' has

$$\alpha(F') = e' - v' + p \le e - v + p = \alpha(F) + p - q$$

gates. This shows that $\mathrm{OR}(A^T) \le \mathrm{OR}(A) + p - q$, and by symmetry, that $\mathrm{OR}(A) \le \mathrm{OR}(A^T) + q - p$. □

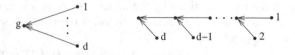

Fig. 1.15 We replace a node (an OR gate) g of fanin d by $d - 1$ nodes each of fanin 2. In the former circuit we have $e - v = d - 1$, and in the latter $e' - v' = 2(d-1) - (d-1) = d - 1 = e - v$

Corollary 1.35. *Let A be a boolean $p \times q$ matrix. Then, for every positive integer s dividing p,*

$$\text{OR}(A) \leq sq + s2^{p/s} - 2p - s.$$

Proof. The proof is similar to that of Lemma 1.2. We want to compute a set Ax of p disjunctions on q variables. Split the transposed $q \times p$ matrix A^T into s submatrices, each of dimension $q \times (p/s)$. By taking a circuit computing all possible disjunction of p/s variables, we can compute disjunctions in each of these submatrices using at most $2^{p/s} - p/s - 1$ OR gates. By adding $q(s - 1)$ gates to combine the results of ORs computed on the rows of the submatrices, we obtain that $\text{OR}(A^T) \leq s2^{p/s} - p - s + q(s - 1)$ and, by the Transposition Principle,

$$\text{OR}(A) \leq \text{OR}(A^T) + q - p = sq + s2^{p/s} - 2p - s. \qquad \square$$

In particular, taking $s = 1$, we obtain an upper bound $\text{OR}(A) \leq q + 2^p - 2p - 1$ which, as shown by Chashkin (1994) is optimal for $p \leq \log q$. Using a different argument (without applying the Transposition Principle), Pudlák et al. (1988) proved a slightly worse upper bound $\text{OR}(A) \leq q + 2^{p+1} - p - 2$.

Now we are able to give one consequence of the Transposition Principle for non-monotone circuits. Given a boolean function $f_{2m}(x, y)$ in $2m$ variables, its *graph* is a bipartite $n \times n$ graph G_f with $n = 2^m$ whose vertices are vectors in $\{0, 1\}^m$, and two vertices x and y from different parts are adjacent iff $f_{2m}(x, y) = 1$.

Corollary 1.36. *If a boolean function f_{2m} can be computed by a non-monotone DeMorgan circuit of size M, then its graph G_f can be represented by a monotone circuit of size $M + (6 + o(1))n$.*

Proof. Let $G_f = (V_1, V_2, E)$ be the graph of $f_{2m}(x, y)$. By Magnification Lemma, each of $2m = 2 \log n$ x-literals in a circuit computing f_{2m} is replaced by a disjunction on the set $\{z_u : u \in V_1\}$ of n variables. By Corollary 1.35 (with $p = 2 \log n$, $q = n$ and $s = 3$), all these disjunctions can be simultaneously computed using fewer than $3n + 3n^{2/3}$ fanin-2 OR gates. Since the same also holds for y-literals, we are done. $\qquad \square$

■ **Research Problem 1.37.** What is the smallest constant c for which the conclusion of Corollary 1.36 holds with $M + (6 + o(1))n$ replaced by $M + cn$?

By Corollary 1.36, any bipartite $n \times n$ graph requiring, say, at least $7n$ AND and OR gates to represent it gives a boolean function of $2m = 2 \log n$ variables requiring at least $\Omega(n) = \Omega(2^m)$ AND, OR and NOT gates to compute it. It is therefore not surprising that proving even linear lower bounds cn for explicit graphs may be a very difficult task. Exercise 1.10 shows that at least for $c = 2$ this task is still tractable.

■ **Research Problem 1.38.** Exhibit an explicit bipartite $n \times n$ graph requiring at least cn AND and OR gates to represent it, for $c > 2$.

Readers interested in this problem might want to consult the paper of Chashkin (1994) giving a somewhat tighter connection between lower bounds for graphs and the resulting lower bounds for boolean functions. In particular, he shows that the constant 6 in Corollary 1.36 can be replaced by 4, and even by 2 if the graph is unbalanced.

Exercises

1.1. Let, as before, $\mathrm{Dec}(A)$ denote the minimum weight of a decomposition of a boolean matrix A. Suppose that A does not contain an $a \times b$ all-1 submatrix with $a + b > k$. Show that $\mathrm{Dec}(A) \geq |A|/k$.

1.2. Let s_n be the smallest number s such that every boolean function of n variables can be computed by a DeMorgan formula of leafsize at most s. Show that $s_n \leq 4 \cdot 2^n - 2$. *Hint*: Use the recurrence (1.1) to show that $s_n \leq 4 \cdot 2^n - 2$, and apply induction on n.

1.3. Let $m = \lceil \log_2(n + 1) \rceil$, and consider the function $\mathrm{Sum}_n : \{0, 1\}^n \to \{0, 1\}^m$ which, given a vector $x \in \{0, 1\}^n$ outputs the binary code of the sum $x_1 + x_2 + \cdots + x_n$. Consider circuits where all boolean functions of two variables are allowed as gates, and let $C(f)$ denote the minimum number of gates in such a circuit computing f.

(a) Show that $C(\mathrm{Sum}_n) \leq 5n$. *Hint*: Fig. 1.3.
(b) Show that $C(f_n) \leq 5n + o(n)$ for every symmetric function f_n of n variables.
     Hint: Every boolean function g of m variables has $C(g) \leq 2^m/m$.

1.4. (Circuits as linear programs) Let $F(x)$ be a circuit over $\{\wedge, \vee, \neg\}$ with m gates. Show that there is a system $L(x, y)$ of $\mathcal{O}(m)$ linear constraints (linear inequalities with coefficients ± 1) with m y-variables such that, for every $x \in \{0, 1\}^n$, $F(x) = 1$ iff there is 0–1 vector y such that all constraints in $L(x, y)$ are satisfied.

Hint: Introduce a variable for each gate. For an $\wedge$-gate $g = u \wedge v$ use the constraints $0 \leq g \leq u \leq 1$, $0 \leq g \leq v \leq 1$, $g \geq u + v - 1$. What constraints to take for $\neg$-gates and for $\vee$-gates? For the output gate g add the constraint $g = 1$. Show that, if the x-variables have values 0 and 1, then all other variables are forced to have value 0 or 1 equal to the output value of the corresponding gate.

1.5. Write $g \leq h$ for boolean functions of n variables, if $g(x) \leq h(x)$ for all $x \in \{0, 1\}^n$. Call a boolean function h a *neighbor* of a boolean function g if either $g \oplus a \leq h \oplus a \oplus 1$ for some $a \in \{0, 1\}$, or $g \oplus x_i \leq g \oplus h$ for some $i \in \{1, \ldots, n\}$. Show that:

(a) Constants 0 and 1 are neighbors of all non-constant functions.
(b) Neighbors of the OR gate $\vee$ are all the two variable boolean functions, except the parity $\oplus$ and the function $\vee$ itself.

1.6. (Minimal circuits are very unstable) Let F be a circuit over some basis computing a boolean function f, and assume that F is *minimal*, that is, no circuit with a smaller number of gates can compute f. In particular, minimal circuits are "unstable" with respect to deletion of its gates: the resulting circuit must make an error. The goal of this exercise is to prove that, in fact, minimal circuits are unstable in a much stronger sense: we cannot even *replace* a gate by another one. That is, the size of the resulting circuit remains the same but, nevertheless, the function computed by a new circuit differs from that computed by the original one.

Let F be a minimal circuit, v a gate in it of fanin m, and h be a boolean function of m variables. Let $F_{v \to h}$ be the circuit obtained from F as follows: replace the boolean function g attached to the gate v by h and remove all the gates that become redundant in the resulting circuit. Prove that, if h is a neighbor of g, then $F_{v \to h} \neq F$.

Hint: Since F is minimal, we cannot replace the gate v by a constant a, that is, there must be at least one vector $x \in \{0, 1\}^n$ such that $F_{v \to a}(x) \neq F(x)$.

1.7. Let $n = 2^r$ and consider two sequences of variables $x = (x_1, \ldots, x_n)$ and $y = (y_1, \ldots, y_r)$. Each assignment $a \in \{0, 1\}^r$ to the y-variables gives us a unique natural number $\mathrm{bin}(a) = 2^{r-1}a_1 + \cdots + 2a_{r-1} + a_r + 1$ between 1 and n; we call $\mathrm{bin}(a)$ the *code* of a. The *storage access function* $f(x, y)$ is a boolean function of $n + r$ variables defined by: $f(x, y) := x_{\mathrm{bin}(y)}$.

Show that the monomial $K = x_1 x_2 \cdots x_n$ is a minterm of f, but still f can be written as an $(r + 1)$-DNF. *Hint*: For the second claim, observe that the value of $f(x, y)$ depends only on $r + 1$ bits $y_1, \ldots, y_r$ and $x_{\mathrm{bin}(y)}$.

1.8. Let $G = ([n], E)$ be an n-vertex graph, and d_i be the degree of vertex i in G. Then G can be represented by a monotone formula $F = F_1 \vee \cdots \vee F_n$, where

$$F_i = x_i \wedge \left(\bigvee_{j : \{i,j\} \in E} x_j \right).$$

A special property of this formula is that the i-th variable occurs at most $d_i + 1$ times. Prove that, if G has no complete stars, then *any* minimal monotone formula representing G must have this property.

Hint: Take a minimal formula F for G, and suppose that some variable x_i occurs $m_i > d_i + 1$ times in it. Consider the formula $F' = F_{x_i = 0} \vee F_i$, where $F_{x_i = 0}$ is the formula obtained from F by setting to 0 all m_i occurrences of the variable x_i. Show that F' represents G, and compute its leafsize to get a contradiction with the minimality of F.

1.9. Say that a graph is *saturated*, if its complement contains no triangles and no isolated vertices. Show that for every saturated graph $G = (V, E)$, its quadratic function $f_G(x) = \bigvee_{uv \in E} x_u x_v$ is the unique(!) monotone boolean function representing the graph G.

1.10. Let $G_n = K_{n-1} + E_1$ be a complete graph on $n - 1$ vertices $1, 2 \ldots, n - 1$ plus one isolated vertex n. Let $F(x_1, \ldots, x_n)$ be an arbitrary monotone circuit with fanin-2 AND and OR gates representing G_n. Show that G_n cannot be represented by a monotone circuit using fewer than $2n - 6$ gates.

Hint: Show that if $n \geq 3$ then every input gate x_i for $i = 1, \ldots, n - 1$ has fanout at least 2.

1.11. Let $n = 2^m$ be a power of 2. Show that Th_2^n can be computed by a monotone DeMorgan formula with at most $n \log_2 n$ leaves.

Hint: Associate with each index $i \in [n]$ its binary code in $\{0, 1\}^m$. For $k \in [m]$ and $a \in \{0, 1\}$, let $F_{k,a}$ be the OR of all variables x_i such that the binary code of i has a in the k-th position. Show that the monotone formula $F = \bigvee_{k=1}^{m} F_{k,0} \wedge F_{k,1}$ computes Th_2^n.

1.12. (Hansel 1964) The goal of this exercise is to show that

$$S_+(\text{Th}_2^n) \geq \tfrac{1}{2} n \log_2 n \,.$$

Let $F(x)$ be a monotone switching network computing Th_2^n with the start node s and the target node t. Say that F is *canonical* if it has the following property: if a node v is joined to s or to t by a contact x_i, then no other edge incident with v has x_i as its label.

(a) Suppose that $F(x) = 0$ for all input vectors x with at most one 1. Show that F can be made canonical without increasing the number of contacts.

   Hint: Assume that some node u is joined to the source node s and to some other node v by edges with the same label x_i. Then $v \neq t$ (why?). Remove the edge $\{u, v\}$ and add the edge $\{s, v\}$ labeled by x_i. Show that the obtained network computes the same function.

(b) Let F be a *minimal* canonical monotone network computing the threshold-2 function Th_2^n. Show that every node $u \notin \{s, t\}$ is adjacent with both nodes s and t.

   Hint: If we remove a label of any contact in a minimal network, then the new network must make an error.

(c) Let m be the number of contacts in a network F from (b). Show that $\text{Th}_2^n(x)$ can be written as an OR $F_1 \vee F_2 \vee \cdots \vee F_t$ of ANDs

$$F_k(x) = \left(\bigvee_{i \in A_k} x_i \right) \wedge \left(\bigvee_{i \in B_k} x_i \right)$$

such that $A_k \cap B_k = \emptyset$ and $w \leq 2m$, where $w := \sum_{k=1}^{t}(|A_k| + |B_k|)$ is the total number of occurrences of variables in the formula.

(d) Show that any expression of Th_2^n as in (c) must contain $w \geq n \log_2 n$ occurrences of variables.

   Hint: For a variable x_i, let m_i be the number of ANDs F_k containing this variable. Show that $w = \sum_{i=1}^{n} m_i$. To lower bound this sum, throw a fair 0–1 coin for each of the ANDs F_k and remove all occurrences of variables x_i with $i \in A_k$ from the entire formula if the outcome is 0; if the outcome is 1, then remove all occurrences of variables x_i with $i \in B_k$. Let $X = X_1 + \cdots + X_n$, where X_i is the indicator variable for the event "the variable x_i survives". Since at most one variable can survive at the end (why?), we have that $E[X] \leq 1$. On the other hand, each variable x_i will survive with probability 2^{-m_i} (why?). Now use the linearity of expectation together with the arithmetic-geometric mean inequality $(\sum_{i=1}^{n} a_i)/n \geq (\prod_{i=1}^{n} a_i)^{1/n}$ with $a_i = 2^{-m_i}$ to obtain the desired lower bound on $\sum_{i=1}^{n} m_i$.

Table 1.1 Upper bounds for *any* symmetric boolean function f_n of n variables

$\mathrm{BP}(f_n) \leq cn^2/\log_2 n$	where $c = 2 + o(1)$; Lupanov (1965b)
$\mathrm{NBP}(f_n) \preceq n^{3/2}$	Lupanov (1965b)
$\mathrm{L}(f_n) \preceq n^{4.93}$	Khrapchenko (1972)
$\mathrm{C}_*(f_n) \leq 4.5n + o(n)$	Demenkov et al. (2010); this improves a simple upper bound $\mathrm{C}_*(f_n) \leq 5n + o(n)$ which follows from a construction used by Lupanov (1965); see Exercise 1.3

Appendix: Known Bounds for Symmetric Functions

Here we summarize some (not all!) known results concerning bounds on the complexity of symmetric functions in various circuit models. Recall that a boolean function $f(x_1, \ldots, x_n)$ is *symmetric* if its value only depends on the sum $x_1 + \cdots + x_n$. Examples of symmetric functions are the parity function

$$\oplus_n(x) = 1 \text{ if and only if } x_1 + \cdots + x_n \text{ is odd},$$

all threshold functions

$$\mathrm{Th}_k^n(x) = 1 \text{ if and only if } x_1 + \cdots + x_n \geq k,$$

as well as the majority function

$$\mathrm{Maj}_n(x) = 1 \text{ if and only if } x_1 + \cdots + x_n \geq \lceil n/2 \rceil.$$

Let $\mathrm{C}(f)$ and $\mathrm{L}(f)$ denote, respectively, the minimum number of gates in a circuit and in a formula over $\{\wedge, \vee, \neg\}$ computing f. Let also $\mathrm{S}(f)$, $\mathrm{BP}(f)$ and $\mathrm{NBP}(f)$ denote, respectively, the minimum number of contacts (labeled edges) in a switching network, in a deterministic and in a nondeterministic branching program computing f. Subscript "+" denotes the *monotone* versions of these measures, and subscript "∗" means that all boolean functions of two variables can be used as gates.

Some relations between these basic measures are summarized in the following chain of inequalities (we will use $f \preceq g$ to denote $f = \mathcal{O}(g)$):

$$\mathrm{C}(f)^{1/3} \preceq \mathrm{NBP}(f) \preceq \mathrm{S}(f) \leq \mathrm{BP}(f) \preceq \mathrm{L}(f) \leq \mathrm{NBP}(f)^{\mathcal{O}(\log \mathrm{NBP}(f))}.$$

Proofs are easy and can be found, for example, in Pudlák (1987).

Table 1.2 Bounds for the parity function

$S(\oplus_n) = 4n - 4$	Cardot (1952); apparently, this was the first nontrivial lower bound at all!
$C(\oplus_n) = 4n - 4$	Redkin (1973)
$L(\oplus_n) \leq \frac{9}{8}n^2$	Yablonskii (1954); see Theorem 6.29 below
$L(\oplus_n) \geq n^{3/2}$	Subbotovskaya (1961); see Sect. 6.3 below
$L(\oplus_n) \geq n^2$	Khrapchenko (1971); n is power of 2; see Sect. 6.8 below
$L(\oplus_n) \geq n^2 + c$	Rychkov (1994); $c = 3$ for odd $n \geq 5$, and $c = 2$ for even $n \geq 6$ which are not powers of 2

Table 1.3 Bounds for threshold functions in *non-monotone* models

$L(\mathrm{Th}_2^n) \geq \frac{1}{4}n \log_2 n$	Krichevskii (1964)[a]
$L(\mathrm{Th}_2^n) \geq n \lfloor \log_2 n \rfloor$	Lozhkin (2005)
$L_+(\mathrm{Th}_2^n) \leq n \log_2 n$	if n is a power of 2; see Exercise 1.11
$L(\mathrm{Th}_k^n) \geq k(n - k + 1)$	Khrapchenko (1971); see Sect. 6.8
$L_*(\mathrm{Maj}_n) = \Omega(n \ln n)$	Fischer et al. (1982)
$L_*(\mathrm{Th}_2^n) = \Omega(n \ln \ln n)$	Pudlák (1984)
$L_*(\mathrm{Th}_k^n) \precsim n^{3.13}$	Paterson et al. (1992)
$L(\mathrm{Maj}_n) \precsim n^{4.57}$	Paterson and Zwick (1993b)
$BP(\mathrm{Th}_k^n) \precsim n^{3/2}$	Lupanov (1965b)
$S(\mathrm{Th}_k^n) \precsim \frac{1}{p}n \ln^4 n$	where $p = (\ln \ln n)^2$; Krasulina (1987, 1988)
$BP(\mathrm{Th}_k^n) \precsim \frac{1}{p}n \ln^3 n$	where $p = (\ln \ln n)(\ln \ln \ln n)$; Sinha and Thathachar (1997)
$BP(\mathrm{Maj}_n) = \Omega(np)$	where $p = \ln \ln n / \ln \ln \ln n$; Pudlák (1984)
$BP(\mathrm{Maj}_n) = \Omega(np)$	where $p = \ln n / \ln \ln n$; Babai et al. (1990)
$S(\mathrm{Maj}_n) = \omega(n)$	Grinchuk (1987, 1989)
$NBP(\mathrm{Maj}_n) = \omega(n)$	Razborov (1990b)

[a]Krichevskii (1964) actually proved an intriguing *structural* result: among minimal formulas computing Th_2^n there is a monotone formula of the form $F(x) = \vee_{k=1}^{t}(\vee_{i \in S_k} x_i) \wedge (\vee_{i \in T_k} x_i)$, where $S_k \cap T_k = \emptyset$ for all $k = 1, \ldots, t$; see also Sect. 6.12

Table 1.4 Bounds for threshold functions in *monotone* models

$NBP_+(Th_k^n) = k(n-k+1)$	Markov (1962); see Theorem 1.8 above
$NBP_+(Th_k^n) \preceq pk(n-k)$	where $p = \ln(n-k)$, if no unlabeled edges (rectifiers) are allowed; Halldórsson et al. (1993)
$NBP_+(Th_k^n) = \Omega(pkn)$	where $p = \ln\frac{n}{k}$, if no unlabeled edges (rectifiers) are allowed; Radhakrishnan (1997)
$S_+(Th_2^n) = np + 2(n-2^p)$	where $p := \lfloor \log_2 n \rfloor$; Krichevskii (1965), Hansel (1966)
$S(Th_2^n) \leq 3n-4$	easy exercise, see Fig. 1.5
$S_+(Maj_n) \preceq n^{4.99}$	Dubiner and Zwick (1992)
$L_+(Maj_n) \preceq n^{5.3}$	Valiant (1984). As observed by Lozhkin and Semenov (1988), the proof actually gives $\mathcal{O}(k^{4.3}n\log^2 n)$ for every k.
$L_+(Th_k^n) \preceq k^{6.3}n\log n$	Friedman (1986)
$L_+(Th_k^n) \preceq k^{4.27}n\log n$	Boppana (1986)
$C(Th_k^n) \leq kn+p$	where $p = \mathcal{O}(n^{1-1/k})$; Dunne (1984)
$C(Th_k^n) \preceq n\log k$	Kochol (1989); the proof is a simple application of a rather non-trivial result of Ajtai et al. (1983) stating that *all* threshold functions $Th_k^n, k=1,\ldots,n$, can be simultaneously computed by a monotone circuit of size $\mathcal{O}(n\log n)$ and depth $\mathcal{O}(\log n)$

Chapter 2
Analysis of Boolean Functions

We finish this introductory part with some *algebraic* properties of boolean functions: their expression and approximation by polynomials. We will use these properties later to prove lower bounds for several circuit models. It is, however, convenient to have all these properties collected at one place. The impatient reader who wishes to begin proving lower bounds immediately may safely skip this section, and return later when the properties in question are used.

2.1 Boolean Functions as Polynomials

Fix a field $\mathbb{F}$, and let $x_1, \ldots, x_n$ be variables taking values in this field. A (multilinear) *monomial* is a product $X_S = \prod_{i \in S} x_i$ of variables, where $S \subseteq [n] = \{1, \ldots, n\}$; we assume that $X_\emptyset = 1$. The degree of this monomial is the cardinality of S. A *multilinear polynomial* of n variables is a function $p : \mathbb{F}^n \to \mathbb{F}$ that can be written as $p(x) = \sum_{S \subseteq [n]} c_S X_S$ for some coefficients $c_S \in \mathbb{F}$. The *degree* of p is the degree of its largest monomial: $\deg(p) = \max\{|S| : c_S \neq 0\}$. Note that if we restrict attention to the boolean domain $\{0, 1\}^n$, then $x_i^k = x_i$ for all $k > 1$, so considering only multilinear monomials is no restriction when dealing with boolean functions.

It is not difficult to see that each boolean function f of n variables can be written as a polynomial over $\mathbb{F}$ of degree at most n. For this, associate with each vector $a = (a_1, \ldots, a_n)$ in $\{0, 1\}^n$ the following polynomial over $\mathbb{F}$ of degree n:

$$p_a(x) = \prod_{i : a_i = 1} x_i \prod_{j : a_j = 0} (1 - x_j).$$

Then $p_a(a) = 1$ while $p_a(x) = 0$ for all $x \in \{0, 1\}^n \setminus \{a\}$. Thus, we can write each boolean function f of n variables as

$$f(x) = \sum_{a \in \{0,1\}^n} f(a) \cdot p_a(x) = \sum_{a \in f^{-1}(1)} \prod_{i : a_i = 1} x_i \prod_{j : a_j = 0} (1 - x_j).$$

S. Jukna, *Boolean Function Complexity*, Algorithms and Combinatorics 27,
DOI 10.1007/978-3-642-24508-4_2, © Springer-Verlag Berlin Heidelberg 2012

By multiplying out via the distributive law, we obtain that

$$f(x) = \sum_{S \subseteq [n]} c_S \prod_{i \in S} x_i$$

for some coefficients $c_S \in \mathbb{F}$. Moreover, this representation is unique (see Exercise 2.23). For example, take $n = 3$ and $f(x, y, z) = x \vee y$. Then

$$\begin{aligned} f &= xyz + xy(1-z) + x(1-y)z + (1-x)yz \\ &\quad + z(1-y)(1-z) + (1-x)y(1-z) \\ &= x + y - xy. \end{aligned}$$

The coefficients c_S can be computed by the *Möbius inversion formula*:

$$c_S = \sum_{T \subseteq S} (-1)^{|S|-|T|} f(T), \tag{2.1}$$

where $f(T)$ is the value of f on the input where exactly the variables in T are 1. This can be shown using the fact that for any two sets $T \subseteq R$,

$$\sum_{T \subseteq S \subseteq R} (-1)^{|S|-|T|} = \begin{cases} 1 & \text{if } T = R, \\ 0 & \text{if } T \neq R \end{cases} \tag{2.2}$$

whose proof we leave as an exercise. Using (2.2), we can prove the correctness of (2.1) as follows:

$$\begin{aligned} f(R) &= \sum_{S \subseteq [n]} c_S \prod_{i \in S} x_i = \sum_{S \subseteq R} c_S \\ &= \sum_{S \subseteq R} \sum_{T \subseteq S} (-1)^{|S|-|T|} f(T) \qquad\qquad \text{by (2.1)} \\ &= \sum_{T \subseteq R} f(T) \sum_{S : T \subseteq S \subseteq R} (-1)^{|S|-|T|} \qquad \text{double counting} \\ &= f(R) \qquad\qquad\qquad\qquad\qquad\qquad \text{by (2.2).} \end{aligned}$$

2.2 Real Degree of Boolean Functions

We now consider representations of boolean functions by polynomials over the field of real numbers. We already know that every function $f : \{0, 1\}^n \to \mathbb{R}$ has the unique representation as a multilinear polynomial over $\mathbb{R}$.

The *real degree*, $\deg(f)$, of a boolean function f is the degree of the unique multilinear real polynomial p that represents f in the sense that $p(a) = f(a)$ for all $a \in \{0, 1\}^n$. The *approximate degree*, $\mathrm{adeg}(f)$, of a boolean function f is the minimum degree of a multilinear real polynomial p that approximates f in the sense that $|p(a) - f(a)| \leq 1/3$ for all $a \in \{0, 1\}^n$.

The AND of n variables $x_1, \ldots, x_n$ is represented by the polynomial consisting of just one monomial $\prod_{i=1}^{n} x_i$, and the OR is represented by the polynomial $1 - \prod_{i=1}^{n}(1 - x_i)$. Hence, both of these functions have degree n.

There is a handy criterion for determining if we have $\deg(f) = n$. Let $\mathrm{even}(f)$ (resp., $\mathrm{odd}(f)$) denote the number of vectors in $f^{-1}(1)$ with an even (resp., odd) number of ones. The following result is due to Yaoyun Shi and Andrew Yao (unpublished); see the survey of Buhrman and de Wolf (2002).

Lemma 2.1. $\deg(f) = n$ *if and only if* $\mathrm{even}(f) \neq \mathrm{odd}(f)$.

Proof. Applying the Möbius inversion formula (2.1) with $S = [n]$, we get

$$
c_{[n]} = \sum_{T \subseteq [n]} (-1)^{n-|T|} f(T)
$$

$$
= (-1)^n \sum_{x \in f^{-1}(1)} (-1)^{x_1 + \cdots + x_n}
$$

$$
= (-1)^n (\mathrm{even}(f) - \mathrm{odd}(f)).
$$

Since $\deg(f) = n$ iff the monomial $x_1 \cdots x_n$ has nonzero coefficient, the lemma follows. $\qquad\Box$

When restricted to the domain $\{0, 1\}^n$, multilinear polynomials may be replaced by univariate polynomials, that is, polynomials of just one variable.

Let S_n denote the set of all $n!$ permutations $\pi : [n] \to [n]$. Given a vector $x = (x_1, \ldots, x_n)$, let $\pi(x) = (x_{\pi(1)}, \ldots, x_{\pi(n)})$. If $p : \mathbb{R}^n \to \mathbb{R}$ is a multivariate polynomial, then the symmetrization of p is defined as follows:

$$
p^{sym}(x) = \frac{\sum_{\pi \in S_n} p(\pi(x))}{n!}.
$$

Note that p^{sym} is a polynomial of degree at most the degree of p. Symmetrization can actually lower the degree: if $p = x_1 - x_2$, then $p^{sym} = 0$.

An important point is that if we are only interested in inputs $x \in \{0, 1\}^n$, then p^{sym} turns out to depend only upon $x_1 + \cdots + x_n$. We can thus represent it as a univariate polynomial of $x_1 + \cdots + x_n$.

Theorem 2.2. (Minsky and Papert 1988) *If* $p : \mathbb{R}^n \to \mathbb{R}$ *is a multilinear polynomial, then there exists a univariate polynomial* $\tilde{p} : \mathbb{R} \to \mathbb{R}$ *of degree at most the degree of* p *such that*

$$
p^{sym}(x_1, \ldots, x_n) = \tilde{p}(x_1 + \cdots + x_n) \quad \text{for all } x \in \{0, 1\}^n.
$$

Proof. Let d be the degree of p^{sym}. Let P_k denote the sum of all $\binom{n}{k}$ products $\prod_{i \in S} x_i$ of $|S| = k$ different variables. Since the polynomial p^{sym} is symmetric, it can be shown by induction that it can be written as

$$p^{sym}(x) = c_0 + c_1 P_1(x) + c_2 P_2(x) + \cdots + c_d P_d(x)$$

with $c_i \in \mathbb{R}$. Note that on $x \in \{0,1\}^n$ with $z := x_1 + \cdots + x_n$ ones, P_k assumes value

$$P_k(x) = \binom{z}{k} = \frac{z(z-1)\cdots(z-k+1)}{k!},$$

which is a polynomial of degree k of z. Therefore the univariate polynomial $\tilde{p}(z)$ defined by

$$\tilde{p}(z) := c_0 + c_1 \binom{z}{1} + c_2 \binom{z}{2} + \cdots + c_d \binom{z}{d}$$

has the desired property. $\square$

The next result from Approximation Theory gives a lower bound on the degree of univariate polynomials. The result was shown by Ehlich and Zeller (1964) and by Rivlin and Cheney (1966).

Theorem 2.3. *Let $p : \mathbb{R} \to \mathbb{R}$ be a polynomial such that $b_1 \le p(k) \le b_2$ for every integer $0 \le k \le n$, and its derivative has $|p'(\xi)| \ge c$ for some real $0 \le \xi \le n$. Then*

$$\deg(p) \ge \sqrt{\frac{cn}{c + b_2 - b_1}}.$$

Using these two tools, Nisan and Szegedy (1994) gave a lower bound on the degree of boolean functions. Let $e_i \in \{0,1\}^n$ denote the boolean vector with exactly one 1 in the i-th position. Say that f is *fully sensitive at* **0** if $f(\mathbf{0}) = 0$ and $f(e_i) = 1$ for all $i = 1, \ldots, n$.

Lemma 2.4. (Nisan and Szegedy 1994) *Let f be a boolean function of n variables. If f is fully sensitive at* **0**, *then* $\deg(f) \ge \sqrt{n/2}$ *and* $\mathrm{adeg}(f) \ge \sqrt{n/6}$.

Proof. Let p be the polynomial representing f, and let $\tilde{p}$ be an univariate polynomial guaranteed by Theorem 2.2. Since p represents f, for every $x \in \{0,1\}^n$ we have that

$$p^{sym}(x) = \frac{1}{n!} \sum_{\pi \in S_n} p(\pi(x)) = \frac{1}{n!} \sum_{\pi \in S_n} f(\pi(x)).$$

Thus, for any boolean x, the value $p^{sym}(x)$ lies between 0 and 1, implying that $0 \le \tilde{p}(k) \le 1$ for every integer $0 \le k \le n$, and $\tilde{p}(0) = 0$. Moreover, since $\pi(e_i) = e_{\pi(i)}$, our assumption implies that $p^{sym}(e_i) = 1$ for all unit vectors e_i, and hence, $\tilde{p}(1) = 1$. Finally, since $\tilde{p}(0) = 0$ and $\tilde{p}(1) = 1$, there must exist a real number $\xi \in [0, 1]$ at which the derivative $\tilde{p}'(\xi) \geq 1$. Thus, applying Theorem 2.3 with $c = b_2 = 1$ and $b_1 = 0$ we obtain that $\deg(p) \geq \deg(p^{sym}) \geq \deg(\tilde{p}) \geq \sqrt{n/2}$.

The proof of $\mathrm{adeg}(f) \geq \sqrt{n/6}$ is the same. The only difference is that, in this case we have that $-1/3 \leq \tilde{p}(k) \leq 4/3$ instead of $0 \leq \tilde{p}(k) \leq 1$ $\square$

Nisan and Szegedy (1994) also give an example of a fully sensitive function with degree significantly smaller than n.

Lemma 2.5. *There exists an (explicitly given) boolean function f of n variables which is fully sensitive at $\mathbf{0}$ and $\deg(f) = n^\alpha$ for $\alpha = \log_3 2 = 0.631\ldots$. Furthermore, the polynomial of f has at most $2^{\mathcal{O}(n^\alpha)}$ monomials.*

Proof. Consider the boolean function $E_3(x, y, z) = 1$ iff $x + y + z \in \{1, 2\}$. This function is represented by the following degree-2 multilinear polynomial:

$$E_3(x, y, z) = x + y + z - xy - xz - yz.$$

Define a boolean function f_m of $n = 3^m$ variables obtained by building a complete ternary tree of depth m, where the 3^m leaves are distinct variables and each node is the E_3 function of its three children. It is easy to see that flipping any variable in the input $\mathbf{0} = (0, \ldots, 0)$ flips the function value from 0 to 1, hence, f_m is fully sensitive at $\mathbf{0}$. On the other hand, for $m > 1$, the representing polynomial for f_m is obtained by substituting independent copies of the f_{m-1}-polynomial in the above polynomial for $f_1 = E_3$. This shows that $\deg(f_m) \leq 2^m = n^{\log_3 2} = n^{0.631\ldots}$, and the total number of monomials does not exceed $6^m = 2^{\mathcal{O}(n^\alpha)}$. $\square$

2.3   The Fourier Transform

Consider the 2^n-dimensional vector space of all functions $f : \{0, 1\}^n \to \mathbb{R}$. We define an inner product in this space by

$$\langle f, g \rangle := \frac{1}{2^n} \sum_{x \in \{0,1\}^n} f(x)g(x) = E_x\left[f(x) \cdot g(x)\right],$$

where the latter expectation is taken uniformly over all $x \in \{0, 1\}^n$. This defines the L_2-norm

$$\|f\| := \sqrt{\langle f, f \rangle} = \sqrt{E[f^2]}.$$

For each $S \subseteq [n] = \{1, \ldots, n\}$, define a function $\chi_S : \{0, 1\}^n \to \{-1, +1\}$ by

$$\chi_S(x) := (-1)^{\sum_{i \in S} x_i} = \prod_{i \in S}(-1)^{x_i} = \prod_{i \in S}(1 - 2x_i);$$

for $S = \emptyset$ we set $\chi_\emptyset = 1$. Note that each χ_S is a multilinear polynomial of degree $|S|$, and is the ± 1 version of the parity function $\bigoplus_{i \in S} x_i$:

$$\chi_S(x) = \begin{cases} -1 & \text{if } \bigoplus_{i \in S} x_i = 1, \\ +1 & \text{if } \bigoplus_{i \in S} x_i = 0. \end{cases}$$

It is easy to see that

$$\langle \chi_S, \chi_T \rangle = \begin{cases} 1 & \text{if } S = T, \\ 0 & \text{if } S \neq T. \end{cases}$$

Indeed,

$$\langle \chi_S, \chi_T \rangle = \mathrm{E}_x \Big[\prod_{i \in S} x_i \cdot \prod_{j \in T} x_j \Big] = \mathrm{E}_x \Big[\prod_{i \in S \oplus T} x_i \Big],$$

because $x_i^2 = 1$; here $S \oplus T = (S \setminus T) \cup (T \setminus S)$ is the symmetric difference of sets S and T. Thus, if S and T are identical, then $\langle \chi_S, \chi_T \rangle = \mathrm{E}_x [\chi_\emptyset] = \mathrm{E}_x [1] = 1$. If, however, $S \neq T$ then $S \oplus T \neq \emptyset$, and we obtain:

$$\langle \chi_S, \chi_T \rangle = \prod_{i \in S \oplus T} \mathrm{E}[x_i] = \prod_{i \in S \oplus T} \Big[\frac{1}{2} \cdot (+1) + \frac{1}{2} \cdot (-1) \Big] = 0 \,.$$

Hence the set of all χ_S is an orthonormal basis (called the *Fourier basis*) for the space of all real-valued functions on $\{0, 1\}^n$. In particular, every function $f : \{0, 1\}^n \to \mathbb{R}$ can be written as a linear combination of these basis functions: there exist coefficients $c_S \in \mathbb{R}$ such that for every $x \in \{0, 1\}^n$,

$$f(x) = \sum_{S \subseteq [n]} c_S \chi_S(x) \,. \tag{2.3}$$

Of course, there are many different bases for the space of all functions $f : \{0, 1\}^n \to \mathbb{R}$. For example, the 2^n boolean functions $e_a : \{0, 1\}^n \to \{0, 1\}$ with $e_a(x) = 1$ iff $x = a$ is also a basis. What makes the Fourier basis particularly useful is that the basis functions themselves have a simple computational interpretation, namely as parity functions: $\chi_S(x) = -1$ if the number of S-variables having value 1 in the input x is odd, and $\chi_S(x) = +1$ if that number is even.

For any $f : \{0, 1\}^n \to \mathbb{R}$, viewed as a function $f : 2^{[n]} \to \mathbb{R}$, we can define another function $\widehat{f} : 2^{[n]} \to \mathbb{R}$ by

$$\widehat{f}(S) := \langle f, \chi_S \rangle = \mathrm{E}_x [f(x) \cdot \chi_S(x)] \,.$$

Due to the orthonormality of our basis $\{\chi_S \mid S \subseteq [n]\}$, we have that $\widehat{f}(S)$ is exactly the coefficient c_S in the representation (2.3):

$$\widehat{f}(S) = \langle f, \chi_S \rangle = \sum_{T \subseteq [n]} c_T \langle \chi_T, \chi_S \rangle = c_S \langle \chi_S, \chi_S \rangle = c_S \,.$$

Thus, for every function $f : \{0, 1\}^n \to \mathbb{R}$, we have that

$$f(x) = \sum_{S \subseteq [n]} \widehat{f}(S) \chi_S(x) \,.$$

The linear map $f \mapsto \widehat{f}$ is called the *Fourier transform*. The function $\widehat{f} : 2^{[n]} \to \mathbb{R}$ is the Fourier transform of f, and $\widehat{f}(S)$ is the *Fourier coefficient* of f at S. The order of a Fourier coefficient $\widehat{f}(S)$ is the cardinality $|S|$ of the subset S. The degree of a boolean function f is the maximum order of the nonzero Fourier coefficient, that is, $\deg(f) = \max\{|S| : \widehat{f}(S) \neq 0\}$. Note that this definition of degree coincides with our earlier definition in terms of the polynomial representation of boolean functions over the reals.

Because the χ_S form an orthonormal basis, we immediately get the following equality known as *Plancharel's Identity*:

$$\langle f, g \rangle = \sum_{S,T} \widehat{f}(S) \widehat{g}(T) \langle \chi_S, \chi_T \rangle = \sum_S \widehat{f}(S) \widehat{g}(S) = \langle \widehat{f}, \widehat{g} \rangle \,.$$

Taking $f = g$, we obtain an important identity, known as *Parseval's Identity*:

$$\|\widehat{f}\|^2 = \sum_S \widehat{f}(S)^2 = \langle f, f \rangle = \|f\|^2 = \mathrm{E}_x \left[f(x)^2 \right] \,. \qquad (2.4)$$

In particular, for $f : \{0, 1\}^n \to \{-1, 1\}$, we have that

$$\sum_{S \subseteq [n]} \widehat{f}(S)^2 = 1 \,.$$

The interpretation of Fourier coefficients of boolean functions $f : \{0, 1\}^n \to \{0, 1\}$ is that they measure the correlation under the uniform distribution between the function and parities of certain subsets of variables.

- The coefficient $\widehat{f}(\emptyset)$ is simply the probability that f takes the value 1.
- If $f_S(x) = \sum_{i \in S} x_i \bmod 2$ is a parity function with $S \neq \emptyset$, then $\widehat{f_S}(\emptyset) = 1/2$ because $\chi_\emptyset(x) = 1$ for all x, and $f_S(x) = 1$ for the half of all 2^n vectors x. Moreover, $\widehat{f_S}(S) = -1/2$ because $\chi_S(x) = -1$ as long as $f_S(x) \neq 0$. All other coefficients are 0.
- The first order coefficient $\widehat{f}(\{i\})$ measures the correlation of the function f with its i-th variable. Let A denote the event $f(x) = 1$, and B the event $x_i = 1$. Since $\chi_{\{i\}}(x) = -1$ if $x_i = 1$, and $\chi_{\{i\}}(x) = +1$ if $x_i = 0$, we can use the equality $\mathrm{Prob}[A \cap B] + \mathrm{Prob}[\overline{A} \cap B] = \mathrm{Prob}[B] = 1/2$, to obtain that

$$\widehat{f}(\{i\}) = \text{Prob}[A \cap \overline{B}] - \text{Prob}[A \cap B]$$

$$= \frac{1}{2} - \left(\text{Prob}[A \cap B] + \text{Prob}[\overline{A} \cap \overline{B}]\right)$$

$$= \frac{1}{2} - \text{Prob}[f(x) = x_i].$$

There is no correlation if $\widehat{f}(\{i\}) = 0$ and maximum correlation if $|\widehat{f}(\{i\})| = \frac{1}{2}$. The sign of the coefficient indicates if the correlation is actually with the variable x_i $(\widehat{f}(\{i\}) = -\frac{1}{2})$ or with its negation $\neg x_i$ $(\widehat{f}(\{i\}) = \frac{1}{2})$.

- The coefficients $\widehat{f}(S)$ of higher orders ($|S| \geq 2$) measure the correlation of the function f with the parity function $\bigoplus_{i \in S} x_i$:

$$\widehat{f}(S) = \frac{1}{2} - \text{Prob}[f(x) = \bigoplus_{i \in S} x_i].$$

Again, there is no correlation if $\widehat{f}(S) = 0$ and maximum correlation if $|\widehat{f}(S)| = \frac{1}{2}$. The sign of the coefficient indicates if the correlation is actually with the parity $(\widehat{f}(S) = -\frac{1}{2})$ or with its negation $(\widehat{f}(S) = \frac{1}{2})$.

The motivation to consider Fourier transforms in the context of proving lower bounds on the circuit size of f is that leading Fourier coefficients (those with $|S|$ large enough) contain useful information about the complexity of the function f itself (cf. Sects. 12.4 and 14.9).

2.4 Boolean 0/1 Versus Fourier ±1 Representation

It is often useful to represent the values *false* = 0 and *true* = 0 by *false* = −1 and *true* = +1. This representation of boolean values is called *Fourier representation*. Note that $\neg x = 1 - x$ in the standard representation, whereas $\neg x = -x$ in the Fourier representation. To convert from the standard 0/1 representation to the Fourier ±1 representation map $x \mapsto 1 - 2x = (-1)^x$. To convert from the Fourier ±1 representation to the standard 0/1 representation map $x \mapsto (1 - x)/2$; this is possible only in rings where 2 has an inverse, e.g., the reals, the rationals, and $\mathbb{Z}_m$ for odd m. Thus, every boolean function $f : \{0, 1\}^n \to \{0, 1\}$ can be transformed into an equivalent function $\tilde{f} : \{+1, -1\}^n \to \{+1, -1\}$ using the following conversion:

$$\tilde{f}(x_1, \ldots, x_n) = 1 - 2 \cdot f\left(\frac{1 - x_1}{2}, \ldots, \frac{1 - x_n}{2}\right).$$

Hence if the value of x_i in $\tilde{f}$ is set to +1, it is set to 0 in f and if x_i is set to −1 it is set to 1 in f.

The advantage of the ± 1 representation is that in this case the function χ_S is simply the product of the S-variables, and the Fourier representation is simply an n-variate multilinear polynomial over the reals, with $\widehat{f}(S)$ as the coefficient of the *monomial* $X_s = \prod_{i \in S} x_i$. Thus, in the ± 1 representation, for every function $f : \{0,1\}^n \to \mathbb{R}$, we have that

$$f(x) = \sum_{S \subseteq [n]} \widehat{f}(S) X_S$$

with

$$\widehat{f}(S) = \langle f, X_S \rangle = \frac{1}{2^n} \sum_{x \in \{-1,+1\}^n} f(x) X_S(x).$$

Similarly, depending on what is more convenient, we can treat the value of a boolean function as $0/1$-valued or as ± 1-valued. An advantage of the latter is that $\sum_S \widehat{f}(S)^2 = \mathrm{E}\left[f^2\right] = 1$ (by Parseval), which allows us to treat the squared Fourier coefficients as probabilities.

2.5 Approximating the Values 0 and 1

The AND of n variables $x_1, \ldots, x_n$ is represented by the polynomial consisting of just one monomial $\prod_{i=1}^{n} x_i$, and the OR is represented by the polynomial $1 - \prod_{i=1}^{n} (1 - x_i)$. These polynomials correctly compute AND and OR on all input vectors x, but the degree of these polynomials is n. Fortunately, the degree can be substantially reduced if we settle for an "imperfect" approximation.

For a boolean function f of n variables, let $\mathrm{Deg}_\epsilon(f)$ denote the minimum degree of a multivariate polynomial $p : \mathbb{R}^n \to \mathbb{R}$ such that $|f(x) - p(x)| \le \epsilon$ for all $x \in \{0,1\}^n$.

Although the OR function $\mathrm{OR}_n(x) = x_1 \vee x_2 \vee \cdots \vee x_n$ requires degree-n polynomial to represent it exactly, it can be approximated by a polynomial of much smaller degree.

Lemma 2.6. (Nisan and Szegedy 1994) *For every constant $k > 2$,*

$$\mathrm{Deg}_{1/k}(\mathrm{OR}_n) = \mathcal{O}(\sqrt{n} \ln k).$$

Proof. Nisan and Szegedy (1994) proved the lemma for $k = 3$. The extension to arbitrary k is due to Hayes and Kutin. We construct a univariate real polynomial $q(z)$ of degree at most $r := c\sqrt{n} \ln(2k)$ (for a constant c) such that $q(n) = 1$ and $|q(z)| \le 1/k$ for $0 \le z \le n - 1$. Then defining

$$p(x_1, \ldots, x_n) := 1 - q(n - x_1 - \ldots - x_n)$$

proves the theorem. The polynomial q is essentially a normalized Chebyshev polynomial. The degree-r Chebyshev polynomial T_r is given by

$$T_r(x) = \frac{1}{2}(x + \sqrt{x^2 - 1})^r + \frac{1}{2}(x - \sqrt{x^2 - 1})^r .$$

Define

$$q(z) := \frac{T_r(z/(n-1))}{T_r(n/(n-1))} .$$

Clearly, $q(n) = 1$. We want to select r such that

$$T_r\left(\frac{n}{n-1}\right) \geq k .$$

Then, since $|T_r(x)| \leq 1$ for $-1 \leq x \leq 1$, we will have $|q(z)| \leq 1/k$ for every $z \in \{0, \ldots, n-1\}$. For $x = n/(n-1)$ we have

$$x + \sqrt{x^2 - 1} = \frac{n}{n-1} + \sqrt{\frac{n^2}{(n-1)^2} - 1} \geq 1 + \sqrt{\frac{2}{n-1}} ,$$

and we get

$$T_r\left(\frac{n}{n-1}\right) \geq \frac{1}{2}\left(1 + \sqrt{\frac{2}{n-1}}\right)^r .$$

The right-hand side quantity is greater than or equal to k if

$$r \geq \frac{\ln(2k)}{\ln(1 + \sqrt{2/(n-1)})} .$$

This last inequality is satisfied if $r \geq c\sqrt{n}\ln(2k)$ for a suitable constant c. $\square$

2.6   Approximation by Low-Degree Polynomials

In the setting above, we were trying to approximate the *values* $f(x) \in \{0, 1\}$ of a given boolean function by the values $p(x)$ of a polynomial on *all* input vectors $x \in \{0, 1\}^n$. In some applications, however, it is desirable to have a low-degree polynomial $p(x)$ computing the function f correctly on most input vectors. For this purpose, define the *distance* between f and p as the number

$$\text{dist}(p, f) = |\{x \in \{0, 1\}^n : p(x) \neq f(x)\}|$$

of boolean inputs x on which the polynomial p outputs a wrong value. Let $\text{OR}_n(x) = x_1 \vee x_2 \vee \cdots \vee x_n$.

Lemma 2.7. (Razborov 1987) *For every integer* $r \geq 1$, *and every prime power* $q \geq 2$, *there exists a multivariate polynomial* $p(x)$ *of n variables and of degree at most* $r(q - 1)$ *over* $\mathrm{GF}(q)$ *such that* $\mathrm{dist}(p, \mathrm{OR}_n) \leq 2^{n-r}$.

Proof. Let $c = (c_1, \ldots, c_n)$ be a random vector in $\mathrm{GF}(q)^n$ with uniform distribution; hence, c takes each value in $\mathrm{GF}(q)^n$ with the same probability q^{-n}. Let $S \subseteq [n]$ and assume that $S \neq \emptyset$. We claim that

$$\mathrm{Prob}\Big[\sum_{i \in S} c_i = 0\Big] \leq \frac{1}{q}. \tag{2.5}$$

To show this, pick a position $k \in S$. Then $\sum_{i \in S} c_i = 0$ implies that the value c_k is uniquely determined by the values of the remaining coordinates: $c_k = -\sum_{i \in S \setminus \{k\}} c_i$. Thus, $\mathrm{Prob}[\sum_{i \in S} c_i = 0] \leq q^{n-1}/q^n = 1/q$, as claimed.

Now consider a random polynomial

$$g(x) = (c_1 x_1 + c_2 x_2 + \cdots + c_n x_n)^{q-1}$$

of degree $q - 1$ over $\mathrm{GF}(q)$. The only reason to rise the sum to the power of $q - 1$ is that we want this polynomial to only take values 0 and 1. This is guaranteed by Fermat's Little Theorem: $a^{q-1} = 1$ for all $a \in \mathrm{GF}(q)$, $a \neq 0$. Since $g(\mathbf{0}) = 0 = \mathrm{OR}_n(\mathbf{0})$, (2.5) implies that $\mathrm{Prob}[g(x) \neq \mathrm{OR}_n(x)] \leq 1/q$ for every $x \in \{0, 1\}^n$. To decrease this error probability to q^{-r}, we just take r independent copies $g_1(x), \ldots, g_r(x)$ of $g(x)$ and consider the polynomial

$$p(x) := 1 - \prod_{i=1}^{r} (1 - g_i(x)).$$

Note that $p(x) = 0$ if and only if $g_j(x) = 0$ for all $j = 1, \ldots, r$. Since each $g_j(x)$ has degree at most $q - 1$, the degree of $p(x)$ is at most $r(q - 1)$. For $x = \mathbf{0}$ we have that $p(x) = 1 - 1 = 0 = \mathrm{OR}_n(x)$. If $x \neq \mathbf{0}$, then

$$\mathrm{Prob}[p(x) \neq \mathrm{OR}_n(x)] = \mathrm{Prob}[p(x) = 0] = \prod_{i=1}^{r} \mathrm{Prob}[g_i(x) = 0] \leq q^{-r}.$$

Thus, for every fixed vector $x \in \{0, 1\}^n$ we have that $\mathrm{Prob}[p(x) \neq \mathrm{OR}_n(x)] \leq q^{-r}$, and the expected number of boolean inputs x on which $p(x) \neq \mathrm{OR}_n(x)$ is at most $2^n/q^r \leq 2^{n-r}$. Hence, there must be a choice of the coefficients c_i such that the resulting (non-random) polynomial $p(x)$ differs from $\mathrm{OR}_n(x)$ on at most a 2^{-r} fraction of all inputs $x \in \{0, 1\}^n$, as desired. $\square$

Over the field $\mathbb{R}$ of real numbers we have the following result.

Lemma 2.8. (Aspnes et al. 1994) *For every integer* $r \geq 1$ *there exists a real multivariate polynomial* $p(x)$ *of n variables and of degree* $\mathcal{O}(r \log n)$ *such that* dist$(p, \mathrm{OR}_n) \leq 2^{n-r}$. *In fact,* Prob$[p(x) \neq \mathrm{OR}_n(x)] \leq 2^{-r}$ *for any probability distribution on inputs* x.

Proof. Assume that $n = 2^m$ is a power of 2, and consider $m + 1$ random subsets $S_0, S_1, \ldots, S_m$ of $[n] = \{1, \ldots, n\}$ where each $j \in [n]$ is included in S_k with probability 2^{-k}; in particular, $S_0 = [n]$. For each $k = 0, 1, \ldots, m$ let $q_k(x) := \sum_{i \in S_k} x_i$, and consider the random polynomial

$$q(x) := 1 - \prod_{k=0}^{m}(1 - q_k(x))$$

of degree $m + 1 = 1 + \log n$. Fix a vector $x \in \{0, 1\}^n$. If $x = \mathbf{0}$, then clearly $q(x) = 0$. If $x \neq \mathbf{0}$ then Prob$[q(x) = 1] \geq 1/6$. To see this, let $w = x_1 + \cdots + x_n$ be the number of ones in x; hence, $1 \leq w \leq n$. Take a k for which $2^{k-1} < w \leq 2^k$. Then

$$\mathrm{Prob}[q_k(x) = 1] = \mathrm{Prob}\left[\sum_{i \in S_k} x_i = 1\right]$$

$$= \mathrm{Prob}[x_i = 1 \text{ for exactly one } i \in S_k]$$

$$= w \cdot 2^{-k}(1 - 2^{-k})^{w-1}$$

$$\geq \frac{1}{2}(1 - 2^{-k})^{2^k - 1}$$

$$\geq \frac{1}{2e} > \frac{1}{6}.$$

Hence, for each vector x, the random polynomial $q(x)$ agrees with $\mathrm{OR}_n(x)$ with probability at least $1/6$. To increase this probability to $1 - 2^{-r}$, we just apply the same trick: take $4r$ independent copies $p_1, \ldots, p_{4r}$ of the polynomial $q(x)$, and consider the polynomial

$$p(x) := 1 - \prod_{i=1}^{4r}(1 - p_i(x))$$

of degree $4r(m + 1) \leq 5r \log n$. If $x = \mathbf{0}$, then we again have that $p(x) = 0$ with probability 1. If $x \neq \mathbf{0}$ then we have

$$\mathrm{Prob}[p(x) = 1] \geq \mathrm{Prob}[p_i(x) = 1 \text{ for some } i]$$

$$= 1 - \mathrm{Prob}[p_i(x) \neq 1 \text{ for all } i]$$

$$= 1 - \mathrm{Prob}[q(x) \neq 1]^{4r} \geq 1 - \left(\frac{5}{6}\right)^{4r} \geq 1 - 2^{-r}.$$

We have therefore constructed a random polynomial $p(x)$ of degree $\mathcal{O}(r \log n)$ such that, for each vector x, $p(x) \neq \mathrm{OR}_n(x)$ with probability at most 2^{-r}. By averaging, there must be a realization of $p(x)$ which differs from $OR_n(x)$ on at most a 2^{-r} fraction of all inputs $x \in \{0, 1\}^n$, as desired. It also follows that $\text{Prob}[p(x) \neq OR_n(x)] \leq 2^{-r}$ for any probability distribution on inputs x.   □

2.7 Sign-Approximation

We now consider boolean functions as functions from $\{-1, +1\}^n$ to $\{-1, +1\}$, where a 0/1 bit b is represented as a ± 1 bit $(-1)^b$. In this representation, the parity function $\text{Parity}_n(x)$ on the set of variables $x = (x_1, \ldots, x_n)$ is simply the monomial $\text{Parity}_n(x) = \prod_{i=1}^n x_i$.

As before, the set of all real-valued functions on $\{-1, +1\}^n$ can be thought as a 2^n-dimensional vector space where $(f + g)(x) = f(x) + g(x)$ and $(af)(x) = af(x)$ for any functions f and g and scalar a. We already know that the set of all monomials $\chi_S(x) = \prod_{i \in S} x_i$ forms an orthonormal basis of this space. The set of all multilinear polynomials of degree $\leq k$ is a subspace of dimension $\sum_{i=0}^k \binom{n}{i}$.

We are interested in the degree of polynomials $p(x)$ that *signum-represents* a given boolean function f in the sense that $p(x) \cdot f(x) > 0$ for all $x \in \{-1, +1\}^n$. The minimum degree of such a polynomial p is called the *strong degree* of f, and is denoted by $d_s(f)$.

The *weak degree*, $d_w(f)$, is the minimum degree of a polynomial p such that $p \neq 0$ and $p(x) \cdot f(x) \geq 0$ for all $x \in \{-1, +1\}^n$. That is, in this case we only require that $\text{sgn}(p(x)) = \text{sgn}(f(x))$ for inputs x where $p(x) \neq 0$. Recall that the function $\text{sgn}(z)$, defined on the real numbers and called the *signum function*, is 1 for positive numbers $z > 0$, is -1 for negative numbers $z < 0$, and is 0 for $z = 0$.

The notion of weak degree is useful because for functions f whose weak degree is large it is possible to give a lower bound on the distance to *any* low-degree polynomial approximation. Let $\text{Error}(f, g)$ denote the set of all $x \in \{-1, +1\}^n$ for which $\text{sgn}(f(x)) \neq \text{sgn}(g(x))$.

Lemma 2.9. *Let p be a degree-k polynomial and f any boolean function of n variables. If $d_w(f) > k$ then*

$$|\text{Error}(p, f)| \geq \sum_{i=0}^{\Delta} \binom{n}{i} \quad \text{where} \quad \Delta := \lfloor (d_w - k - 1)/2 \rfloor.$$

Proof. We need the following auxiliary fact.

Claim 2.10. For every set $S \subset \{-1, +1\}^n$ of size $|S| < \sum_{i=0}^k \binom{n}{i}$ there exists a nonzero polynomial $q(x)$ of degree $\leq 2k$ that is 0 for all $x \in S$, and is non-negative elsewhere.

Proof. Any degree k polynomial has $\sum_{i=0}^k \binom{n}{i}$ coefficients (some of which may be zero), and its value on any particular input is a linear combination of those coefficients. Thus, if r stands for a degree k polynomial, the constraints $r(x) = 0$

for all $x \in S$ form a homogeneous system of $|S|$ linear equations in $\sum_{i=0}^{k} \binom{n}{i}$ variables, and have a nontrivial solution r since $|S| < \sum_{i=0}^{k} \binom{n}{i}$. Then $q = r^2$ is a desired nonzero polynomial.   □

Now let $S := \text{Error}(p, f)$, and suppose that $|S| < \sum_{i=0}^{\Delta} \binom{n}{i}$. Then, by Claim 2.10, there exists a nonzero polynomial q of degree $\leq d_w(f) - k - 1$ which is 0 on S and non-negative elsewhere. Consider the polynomial pq. Since $\text{sgn}(pq(x)) = \text{sgn}(p(x)) = \text{sgn}(f(x))$ for all $x \notin S$, and since $pq(x) = 0$ for all $x \in S$, the polynomial pq weakly represents f. But this polynomial has degree at most $d_w(f) - 1$, contradicting the definition of $d_w(f)$.   □

Lemma 2.11. $d_w(\text{Parity}_n) = n$.

Proof. Suppose p weakly represents $f(x) := \text{Parity}_n(x) = \prod_{i=1}^{n} x_i$. Consider the scalar product $\langle p, f \rangle = 2^{-n} \sum_x p(x) f(x)$. We have that $\langle p, f \rangle > 0$, since each term in $\sum_x p(x) f(x)$ is non-negative and at least one term is nonzero. But the parity function f is orthogonal to all other monomials. Thus, if p had degree $\leq n - 1$, we would have that $\langle p, f \rangle = 0$.   □

Corollary 2.12. *Let p be a polynomial of n variables and of degree $k \leq \delta \sqrt{n} + 1$ for some constant $0 < \delta < 1/2$. Then*

$$|\text{Error}(p, \text{Parity}_n)| \geq (1/2 - \delta)2^n.$$

Proof. Since $d_w(\text{Parity}_n) = n$ and $\binom{n}{n/2} \sim 2^n / \sqrt{\pi n/2}$, Lemma 2.9 implies that

$$|\text{Error}(f, \text{Parity}_n)| \geq \sum_{i=0}^{n/2 - \delta\sqrt{n}} \binom{n}{i} = \sum_{i=0}^{n/2} \binom{n}{i} - \sum_{i=n/2-\delta\sqrt{n}+1}^{n/2} \binom{n}{i}$$

$$\geq 2^{n-1} - \delta\sqrt{n} \binom{n}{n/2} \geq (1/2 - \delta)2^n.   □$$

Using a well known fact from linear algebra (Stiemke's Transposition Theorem), Aspnes et al. (1994) showed that the parity function is central to the connection between strong and weak degrees.

Lemma 2.13. (Aspnes et al. 1994) *For any boolean function f of n variables, $d_s(f) + d_w(f \cdot \text{Parity}_n) = n$.*

2.8 Sensitivity and Influences

A boolean function f of n variables *depends* on its i-th variable if there *exists* at least one vector $x \in \{0, 1\}^n$ such that $f(x \oplus e_i) \neq f(x)$, where $x \oplus e_i$ is x with the i-th bit flipped. The more such "witnessing" vectors x we have, the more "influential" this variable is. The *influence*, $\text{Inf}_i(f)$, of the i-th variable of f is defined as the fraction of vectors witnessing its importance. Thus, if we introduce the indicator variable

$$s_i(f, x) := \begin{cases} 1 & \text{if } f(x \oplus e_i) \neq f(x) \\ 0 & \text{otherwise}, \end{cases}$$

then influence of the i-th variable of f is

$$\text{Inf}_i(f) = \frac{1}{2^n} \sum_{x \in \{0,1\}^n} s_i(f, x) = \text{Prob}[f(x \oplus e_i) \neq f(x)] = E_x\left[s_i(f, x)\right],$$

where the probability is over uniform $x \in \{0, 1\}^n$; hence, $\text{Prob}[x] = 2^{-n}$. Thus, large $\text{Inf}_i(f)$ means that the value of f depends on the i-th variable on many input vectors. The notion of influence of a variable on a boolean function was introduced by Ben-Or and Linial (1990). It has since found many applications in discrete mathematics, theoretical computer science and social choice theory; we refer the reader to the survey by Kalai and Safra (2006).

For example, if $f(x) = x_1 \oplus (x_2 \vee x_3 \vee \cdots \vee x_n)$, then $\text{Inf}_1(f) = 1$ because, for every vector $x \in \{0, 1\}^n$, flipping its first bit flips the value of f. But the influence of its second variable (as well as of each remaining variable) is very small: $\text{Inf}_2(f) \leq 4/2^n$ because $f(x \oplus e_2) \neq f(x)$ can only hold if $x_3 = \ldots = x_n = 0$.

In the definition of the influence we fix a variable and look for how many inputs x the value $f(x)$ depends on this variable. Similarly, we can fix an input vector $x \in \{0, 1\}^n$ and look at how many variables the value $f(x)$ depends on. This leads to the concept of "sensitivity".

The *sensitivity of f on input x* is defined as the number

$$s(f, x) = \sum_{i=1}^{n} s_i(f, x) = |\{i : f(x \oplus e_i) \neq f(x)\}|$$

of Hamming distance-1 neighbors y of x such that $f(y) \neq f(x)$. For example, if $f(x) = x_1 \vee x_2 \vee \cdots \vee x_n$, then $s(f, \mathbf{0}) = n$ but $s(f, x) = 0$ for every vector x with at least two 1s. The *sensitivity* (or *maximum sensitivity*) of f is defined as

$$s(f) = \max_x s(f, x).$$

In the literature, $s(f)$ is also called the *critical complexity* of f. The *average sensitivity*, $\text{as}(f)$, of f is the expected sensitivity of f on a random assignment:

$$\text{as}(f) := 2^{-n} \sum_x s(f, x) = E_x\left[s(f, x)\right] = \sum_{i=1}^{n} E_x\left[s_i(f, x)\right] = \sum_{i=1}^{n} \text{Inf}_i(f). \quad (2.6)$$

For example, if $f(x) = \text{Parity}(x)$ is the parity of n variables, then $s(f, x) = n$ for every assignment x, implying that $as(f) = n$. But for some boolean functions the average sensitivity can be tiny, for example, $as(f) = n/2^{n-1}$ if f is an OR or an AND of n variables.

Functions with large average sensitivity require large unbounded-fanin circuits of constant depth (see Boppana's theorem in Sect. 12.4). By the theorem of Khrapchenko (Theorem 6.27 in Sect. 6.8), the square $as(f)^2$ of the average sensitivity is a lower bound on the minimum number of leaves in a DeMorgan formula computing f. High average sensitivity also implies that the function cannot be approximated by low-degree polynomials (see Lemma 2.19 below).

All these concepts (influence, sensitivity and average sensitivity) can be defined in graph-theoretic terms as follows. Every boolean function f of n variables defines a bipartite graph G_f with parts $f^{-1}(0)$ and $f^{-1}(1)$, where $(x, y) \in f^{-1}(0) \times f^{-1}(1)$ is an edge of G_f iff $y = x \oplus e_i$ for some $i \in [n]$. An edge (x, y) is an edge in the i-th direction if $y = x \oplus e_i$. It is easy to see that:

- $s(f, x) = $ degree of vertex x in G_f.
- $s(f) = $ maximum degree of a vertex in G_f.
- $as(f) = $ average degree of G_f.
- $\text{Inf}_i(f) = $ number of edges of G_f in the i-th direction divided by 2^n.

Hence, if $|G_f|$ denotes the number of edges in G_f, then $|G_f| \le s(f)2^{n-1}$, because one of the color classes must have at most 2^{n-1} vertices. On the other hand, we have the following somewhat counterintuitive lower bound: the smaller the degree $s(f)$ of the graph G_f is, the more edges we can force into it.

A boolean function f is *nondegenerate* if it depends on all its variables, that is, if the graph G_f has at least one edge in each direction.

Theorem 2.14. (Simon 1983) *For every nondegenerate boolean function of n variables, we have that $|G_f| \ge n2^{n-2s(f)+1}$.*

Proof. The graph G_f is a subgraph of the n-dimensional binary hypercube Q_n. Recall that Q_n is the graph whose vertices are vectors in $\{0, 1\}^n$, and two vectors are adjacent iff they differ in exactly one position. The i-th neighbor of a vertex $x \in \{0, 1\}^n$ in Q_n is the vertex $y = x \oplus e_i$. Thus, each vertex has exactly n neighbors in Q_n. A neighbor y of x is a *proper neighbor* if y is a neighbor of x in the graph $G_f \subseteq Q_n$, that is, if $f(y) \ne f(x)$. Call an edge $\{x, y\}$ of Q_n *colorful* if x and y are proper neighbors, that is, if $f(x) \ne f(y)$. Our goal is to show that there must be at least $n2^{n-2s(f)+1}$ such edges.

For this, fix an arbitrary position $i \in [n]$, and split the set of vertices of the hypercube Q_n into two subsets $X_0 = \{x \mid x_i = 0\}$ and $X_1 = \{x \mid x_i = 1\}$. We are going to show that at least 2^{n-2s+1} edges lying between these two set must be colorful, where $s := s(f)$. Consider the set

$$V_0 = \{y \in X_0 \mid f(y) = 0 \text{ and } f(y \oplus e_i) = 1\} \subseteq X_0 .$$

Fig. 2.1 The set Y of $|Y| \geq n - s$ neighbors y of β in X_0 such that $f(y) = 0$. Since $f(\beta \oplus e_i) = 1$, for at most $s - 1$ of them we can have $f(y \oplus e_i) = 0$

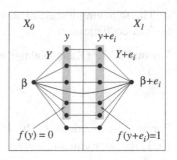

Note that V_0 is non-empty, because f depends on all variables, and hence, also on the i-th variable. The key of the whole argument is the following fact.

Claim 2.15. Every vertex in V_0 has at least $n - 2s + 1$ neighbors in V_0.

Proof. Fix an arbitrary $\beta \in V_0$, and let Y be the set of all neighbors y of β such that $y \in X_0$ and $f(y) = 0$. Since β belongs to $V_0 \subseteq X_0$, its i-th neighbor $\beta \oplus e_i$ belongs to X_1 and is a proper neighbor of β, that is, $f(\beta \oplus e_i) = 1 \neq 0 = f(\beta)$. All remaining $n - 1$ neighbors $\beta \oplus e_j$ of β lie in X_0 (because $\beta_i = 0$), and at most $s - 1$ of them can be proper (see Fig. 2.1). Hence, $|Y| \geq (n - 1) - (s - 1) = n - s$. Since vertices $y \in Y$ are neighbors of β, all vertices in

$$Y \oplus e_i := \{y \oplus e_i \mid y \in Y\} \subseteq X_1$$

are neighbors of $\beta \oplus e_i$. Since (again) at most $s - 1$ of them can be proper (one proper neighbor β of $\beta \oplus e_i$ does not belong to $Y \oplus e_i$), and since $f(\beta \oplus e_i) = 1$, there must be a subset $Y' \subseteq Y$ such that $|Y'| \geq |Y| - (s - 1) \geq n - 2s + 1$ and $f(y \oplus e_i) = 1$ for all $y \in Y'$. Thus, every $y \in Y'$ is a neighbor of β such that $f(y) = 0$ and $f(y \oplus e_i) = 1$. Since all these neighbors belong to V_0, we are done. □

Claim 2.15 implies that every vertex of the subgraph G of Q_n induced by V_0 has degree at least $d = n - 2s + 1$. Exercise 2.25 implies that $|V_0| \geq 2^d = 2^{n-2s+1}$. Thus, at least 2^{n-2s+1} of the edges in the i-th direction are colorful. Since this analysis is true for every fixed position $i \in [n]$ (as f depends on all its variables), the theorem follows. □

Together with a trivial upper bound $|G_f| \leq s(f)2^{n-1}$, Theorem 2.14 immediately yields the following general lower bound on the sensitivity. Wegener (1985a) showed that this lower bound is optimal up to an additive factor $\mathcal{O}(\log \log n)$.

Corollary 2.16. *If a boolean function f depends on all n variables, then*

$$s(f) \geq \frac{1}{2} \log_2 n - \frac{1}{2} \log_2 \log_2 n .$$

The following lemma relates the average sensitivity of a boolean function to its Fourier coefficients.

Lemma 2.17. (Kahn–Kalai–Linial 1988) *For every boolean function* $f : \{0, 1\}^n \to \{0, 1\}$,

$$\mathrm{as}(f) = 4 \sum_{S \subseteq [n]} |S| \widehat{f}(S)^2.$$

For boolean functions $f : \{0, 1\}^n \to \{-1, +1\}$ taking values in $\{-1, +1\}$ instead of $\{0, 1\}$ the same equality holds without the multiplicative factor 4 (see Exercise 2.36).

Proof. Fix a position $i \in [n]$, and consider the difference function $g_i(x) := f(x) - f(x \oplus e_i)$. This function takes its values in $\{-1, 0, +1\}$. The Fourier coefficients of g are closely related to those of f:

$$\widehat{g_i}(S) = \begin{cases} 2\widehat{f}(S) & \text{if } i \in S, \\ 0 & \text{otherwise.} \end{cases} \tag{2.7}$$

(Moreover, this holds for any function $f : \{0, 1\}^n \to \mathbb{R}$, not only for boolean functions.) To show (2.7), we use the linearity of expectation to obtain

$$\widehat{g_i}(S) = \mathrm{E}_x\left[g_i(x) \cdot \chi_S(x)\right] = \mathrm{E}_x\left[f(x) \cdot \chi_S(x)\right] - \mathrm{E}_x\left[f(x \oplus e_i) \cdot \chi_S(x)\right].$$

Since $x \oplus e_i$ has the same distribution as x, and since $\chi_S(x \oplus e_i) = -\chi_S(x)$ for $i \in S$, and $\chi_S(x \oplus e_i) = \chi_S(x)$ for $i \notin S$, (2.7) follows. Now, the Parseval Identity (2.4) gives

$$\|g_i\|^2 = \sum_{S \subseteq [n]} \widehat{g_i}(S)^2 = 4 \sum_{S : i \in S} \widehat{f}(S)^2.$$

On the other hand, $\mathrm{Inf}_i(f) = |X_i|/2^n$ where $X_i = \{x \mid f(x) \neq f(x \oplus e_i)\}$. Since $g_i(x) = \pm 1$ for $x \in X_i$, and $g_i(x) = 0$ otherwise, we obtain that

$$\mathrm{Inf}_i(f) = 2^{-n}|X_i| = 2^{-n} \sum_x g_i(x)^2 = \|g_i\|^2 = 4 \sum_{S : i \in S} \widehat{f}(S)^2.$$

Summing this over all $i = 1, \ldots, n$ we conclude that

$$\mathrm{as}(f) = \sum_{i=1}^n \mathrm{Inf}_i(f) = 4 \sum_{i=1}^n \sum_{S : i \in S} \widehat{f}(S)^2 = 4 \sum_{S \subseteq [n]} |S| \widehat{f}(S)^2. \qquad \square$$

Using Lemma 2.17, Kahn et al. (1988) derived the following general lower bounds on average sensitivity of boolean functions in terms of their density.

Theorem 2.18. *Let f be a boolean function of n variables, and p be a fraction of input vectors on which f takes value 1. Assume that $p \leq 1/2$. Then*

$$\sum_{i=1}^{n} \text{Inf}_i(f)^2 \geq \frac{0.2 p^2 \log^2 n}{n}$$

and

$$\sum_{i=1}^{n} \text{Inf}_i(f) \geq p \cdot \log \frac{1}{p}.$$

Consequently there always exists a variable whose influence is at least $0.2 p \log n / n$. Moreover, the second bound is tight, and the first is tight up to the constant factor 0.2.

Average sensitivity can also be used to lower-bound the degree of polynomials approximation the value of a given boolean function. For a boolean function f of n variables, let (as before) $\text{Deg}_\epsilon(f)$ denote the minimum degree of a multivariate polynomial $p : \mathbb{R}^n \to \mathbb{R}$ such that $|f(x) - p(x)| \leq \epsilon$ for all $x \in \{0, 1\}^n$.

Lemma 2.19. (Shi 2000) *For every boolean function f,*

$$\text{Deg}_\epsilon(f) \geq (1 - 2\epsilon)^2 \text{as}(f).$$

Proof. (Due to de Wolf 2008) The proof is similar to that of Lemma 2.17. Suppose a degree-d n-variate real polynomial $p : \{0, 1\}^n \to [0, 1]$ approximates $f : \{0, 1\}^n \to \{0, 1\}$, in the sense that there is an $\epsilon \in [0, 1/2)$ such that $|f(x) - p(x)| \leq \epsilon$ for every $x \in \{0, 1\}^n$. Let q be the degree-d polynomial $1 - 2p$. This has range $[-1, 1]$, hence, Parseval's Identity (2.4) yields

$$\sum_{S} \hat{q}(S)^2 = \langle q, q \rangle = 2^{-n} \sum_{x} g(x)^2 \leq 1.$$

Note that $q(x) \in [-1, -1 + 2\epsilon]$ if $f(x) = 1$, and $q(x) \in [1 - 2\epsilon, 1]$ if $f(x) = 0$. Fix a position $i \in [n]$, and consider the function $g(x) := q(x) - q(x \oplus e_i)$. Let $X_i := \{x \mid f(x \oplus e_i) \neq f(x)\}$. Then $\text{Inf}_i(f) = |X_i|/2^n$ and $|g(x)| \geq 2 - 4\epsilon > 0$ for all $x \in X_i$. So,

$$\text{E}_x\left[g(x)^2\right] = 2^{-n} \sum_{x \in \{0,1\}^n} g(x)^2 \geq 2^{-n} \sum_{x \in X_i} g(x)^2 \geq 2^{-n} \sum_{x \in X_i} (2 - 4\epsilon)^2$$

$$= (2 - 4\epsilon)^2 \cdot |X_i|/2^n = (2 - 4\epsilon)^2 \cdot \text{Inf}_i(f).$$

Using Parseval's Identity (2.4) and (2.7), we have

$$(2 - 4\epsilon)^2 \text{Inf}_i(f) \leq \text{E}_x\left[g(x)^2\right] = \sum_{S} \hat{g}(S)^2 = 4 \sum_{S : i \in S} \hat{q}(S)^2.$$

Dividing by 4 and summing over all i gives the desired lower bound on the degree:

$$(1 - 2\epsilon)^2 \mathrm{as}(f) = (1 - 2\epsilon)^2 \sum_{i=1}^{n} \mathrm{Inf}_i(f) \le \sum_{i=1}^{n} \sum_{S:i \in S} \widehat{q}(S)^2$$

$$= \sum_{S} |S| \widehat{q}(S)^2 \le d \sum_{S} \widehat{q}(S)^2 \le d . \qquad \square$$

A polynomial p represents f *exactly*, if $p(x) = f(x)$ for all $x \in \{0,1\}^n$. Let $d(f)$ denote the minimum degree of a multivariate polynomial representing f exactly. That is, $d(f) = \mathrm{Deg}_0(f)$. Together with (2.6), Lemma 2.19 (with $\epsilon = 0$) implies that

$$d(f) \ge \mathrm{as}(f) = \sum_{i=1}^{n} \mathrm{Inf}_i(f) . \tag{2.8}$$

Using this, we can show that every boolean function depending on n variables must have degree at least about $\log n$ (cf. Corollary 2.16).

Theorem 2.20. (Nisan and Szegedy 1994) *If a boolean function f depends on all its n variables, then $d(f) 2^{d(f)} \ge n$, and hence, $d(f) \ge \log_2 n - \mathcal{O}(\log \log n)$.*

Proof. We need the following simple fact about Reed-Muller codes; in computer science this fact is known as the Schwartz-Zippel Lemma.

Claim 2.21. Let $p(x)$ be a nonzero multilinear polynomial of degree at most d. If we choose $x_1, \ldots, x_n$ independently at random in $\{0, 1\}$, then

$$\mathrm{Prob}[p(x_1, \ldots, x_n) \ne 0] \ge 2^{-d} .$$

Proof. The proof is by induction on the number n of variables. For $n = 1$, we just have a linear function of one variable which can have only one zero. For the induction step, let $y := (x_1, \ldots, x_{n-1})$ and write $p(x)$ as

$$p(x) = x_n \cdot g(y) + h(y) ,$$

where g has degree at most $d - 1$. We consider three possible cases.

Case 1: $h \equiv 0$. In this case $p(x) = x_n \cdot g(y) \not\equiv 0$. Since $\mathrm{Prob}[x_n = 1] = 1/2$ and g has only $n - 1$ variables, the induction hypothesis yields

$$\mathrm{Prob}[p(x) \ne 0] = \frac{1}{2} \cdot \mathrm{Prob}[g(y) \ne 0] \ge \frac{1}{2} \cdot 2^{-(d-1)} = 2^{-d} .$$

Case 2: $h \not\equiv 0$ but $g + h \equiv 0$. In this case we have that $p(x) = p(y, 0) = h(y)$. Since h has only $n - 1$ variables, the induction hypothesis yields $\mathrm{Prob}[p(x) \ne 0] = \mathrm{Prob}[h(x) \ne 0] \ge 2^{-d}$.

Case 3: $h \not\equiv 0$ and $g + h \not\equiv 0$. Since both polynomials have only $n - 1$ variables and both are nonzero, induction hypothesis together with the rule of total probability yields

$$\text{Prob}[p(x) \neq 0] = \text{Prob}[x_n = 0] \cdot \text{Prob}[p(x) \neq 0 | x_n = 0]$$
$$+ \text{Prob}[x_n = 1] \cdot \text{Prob}[p(x) \neq 0 | x_n = 1]$$
$$= \text{Prob}[x_n = 0] \cdot \text{Prob}[h(x) \neq 0 | x_n = 0]$$
$$+ \text{Prob}[x_n = 1] \cdot \text{Prob}[g(y) + h(y) \neq 0 | x_n = 1]$$
$$= \frac{1}{2} \cdot \text{Prob}[h(y) \neq 0] + \frac{1}{2} \cdot \text{Prob}[g(y) + h(y) \neq 0]$$
$$\geq \frac{1}{2} \cdot 2^{-d} + \frac{1}{2} \cdot 2^{-d} = 2^{-d} . \qquad \square$$

We can now finish the proof of Theorem 2.20 as follows. Let $p(x)$ be a multilinear polynomial of degree $d = d(f)$ exactly representing f. For each $i = 1, \ldots, n$ consider the polynomial of $n - 1$ variables:

$$p_i(x_1, \ldots, x_{i-1}, x_{i+1}, \ldots, x_n) := p(x_1, \ldots, x_{i-1}, 1, x_{i+1}, \ldots, x_n)$$
$$- p(x_1, \ldots, x_{i-1}, 0, x_{i+1}, \ldots, x_n) .$$

Using this notation, it is clear that

$$\text{Inf}_i(f) = \text{Prob}[p_i(x_1, \ldots, x_{i-1}, x_{i+1}, \ldots, x_n) \neq 0] .$$

Since, by our assumption, f depends on all of its variables, we have that $p_i \not\equiv 0$ for every i. Thus, Claim 2.21 implies that $\text{Inf}_i(f) \geq 2^{-d}$ for all $i = 1, \ldots, n$. Together with (2.8), this implies that

$$\frac{n}{2^d} \leq \sum_{i=1}^{n} \text{Inf}_i(f) \leq d .$$

Thus $d 2^d \geq n$, and the theorem follows. $\qquad \square$

As Nisan and Szegedy observed, the *address function* shows that this bound is tight up to the $\mathcal{O}(\log \log n)$ term. This function $f(x, y)$ has $n = m + 2^m$ variables: m variables $x_1, \ldots, x_m$, and 2^m variables y_α indexed by binary strings $\alpha \in \{0, 1\}^m$. On input (x, y) with $x \in \{0, 1\}^m$ and $y \in \{0, 1\}^{2^m}$, the function outputs the bit y_x. The function depends on all n variables. It is represented by the polynomial

$$p(x, y) = \sum_{\alpha \in \{0,1\}^m} y_\alpha \prod_{j : \alpha_j = 1} x_j \prod_{j : \alpha_j = 0} (1 - x_j)$$

and hence has degree $m \leq \log n$.

Finally, let us mention an interesting result of Friedgut (1998) stating that boolean functions of small average sensitivity do not depend on many variables, and hence, can be approximated by low-degree polynomials.

Theorem 2.22. (Friedgut 1998) *For every boolean function f and every ϵ there exists a boolean function g depending on at most $2^{\mathcal{O}(\mathrm{as}(f)/\epsilon)}$ variables such that g differs from f on at most an ϵ fraction of input vectors.*

Exercises

2.23. Let $p, q : \mathbb{F}^n \to \mathbb{F}$ be multilinear polynomials of degree at most d. Show that, if $p(x) = q(x)$ for all $x \in \{0, 1\}^n$ with $x_1 + \cdots + x_n \leq d$, then $p = q$.

Hint: Suppose that the polynomial $r(x) = p(x) - q(x)$ is not identically zero. Let $c_S \prod_{i \in S} x_i$ be the minimal-degree monomial in r with $c_S \neq 0$. Show that $r(x) \neq 0$ for the characteristic vector $x \in \{0, 1\}^n$ of S.

2.24. Let $f : \{-1, +1\}^n \to \{-1, +1\}$ be a symmetric boolean function. Let k be the number of times f changes sign when expressed as a univariate function in $x_1 + \cdots + x_n$. Show that $d_s(f) = d_w(f) = k$. *Hint*: Lemma 2.13.

2.25. Let Q_n be the undirected graph whose vertices are vectors $\{0, 1\}^n$, and two vectors are adjacent iff they differ in exactly one position. For a set of vertices $X \subseteq \{0, 1\}^n$, let $d(X)$ denote the minimal number d such that every vertex $x \in X$ has at least d neighbors in X. That is, $d(X)$ is the minimum degree of the induced subgraph $Q_n[X]$ of Q_n. Show that for every non-empty X, $|X| \geq 2^{d(X)}$.

Hint: Induction on $m = |X|$. Base cases $m = 1, 2$ are trivial. For the induction step, choose a coordinate i such that both the two sub-cubes that correspond to $x_i = 0$ and $x_i = 1$ have nonempty subsets X_0 and X_1 of X. Use the arithmetic-geometric mean inequality $(a + b)/2 \geq \sqrt{ab}$.

2.26. (Due to Shi 2002) For a set $X \subseteq \{0, 1\}^n$ of vertices of the graph Q_n, let $e(X)$ denote the number of edges of Q_n that joint two vertices in X. Show that $e(X) \leq \frac{1}{2}|X| \log_2 |X|$.

Hint: Argue by induction as in Exercise 2.25, and use the fact that $e(X) \leq e(X_0) + e(X_1) + \min\{|X_0|, |X_1|\}$.

2.27. For a set $X \subseteq \{0, 1\}^n$ of vertices of the graph Q_n, let $\overline{d}(X)$ denote the *average* degree of the induced subgraph $Q_n[X]$ of Q_n, that is, $\overline{d}(X) = (\sum_{x \in X} d_x)/|X|$, where d_x is the number of neighbors of x lying in X. Improve Exercise 2.25 to $|X| \geq 2^{\overline{d}(X)}$.

Hint: Combine Exercise 2.26 with Euler's theorem stating that the sum of degrees of all vertices in a graph with m edges is equal to $2m$.

2.28. Show that $\mathrm{as}(f) = |G_f|/2^{n-1}$.

Hint: By Euler's theorem, the sum of degrees in every graph is equal to 2 times the number of its edges.

2.29. We have defined the average sensitivity $\mathrm{as}(f)$ of a boolean function f of n variables as 2^{-n} times the sum of $\mathrm{s}(f, x)$ over *all* vectors $x \in \{0, 1\}^n$. Show that

$$\mathrm{as}(f) = \frac{1}{2^{n-1}} \sum_{x \in f^{-1}(1)} \mathrm{s}(f, x) = \frac{1}{2^{n-1}} \sum_{x \in f^{-1}(0)} \mathrm{s}(f, x) .$$

Hint: Exercise 2.28.

2.30. Let f be a boolean function of n variables, and let $p = |f^{-1}(1)|/2^n$ be its density. Show that $\mathrm{as}(f) \geq 2p(1 - p)$.

2.31. Given a *monotone* boolean function $f : \{0, 1\}^n \to \{0, 1\}$ and a vector $x \in \{0, 1\}^n$, say that the i-th bit x_i of x is "correct" for f if $x_i = f(x)$. Let $c(f)$ denote the expected number of "correct" bits in a random string x. Show that $c(f) = (n + \mathrm{as}(f))/2$.

Hint: Observe that $c(f) = \sum_{i=1}^n \mathrm{Prob}[A_i]$ where A_i is the event "$x_i = f(x)$". When x is chosen randomly, there is an $\mathrm{Inf}_i(f)$ chance that x_i is influential for f. Use the monotonicity of f to show that in this case the expected number of correct bits is 1. With probability $1 - \mathrm{Inf}_i(f)$ the bit x_i is not influential for f; show that in this case the expected number of correct bits x_i is $1/2$. Conclude that $\mathrm{Prob}[A_i] = 1 \cdot \mathrm{Inf}_i(f) + \frac{1}{2} \cdot (1 - \mathrm{Inf}_i(f))$.

2.32. Show that $\mathrm{as}(\mathrm{Maj}_n) = \Theta(\sqrt{n})$.
Hint: $\binom{n}{n/2} \sim 2^n / \sqrt{\pi n/2}$.

2.33. Show that the Majority function has the highest average sensitivity among all *monotone* boolean functions: for every monotone boolean function $f : \{0, 1\}^n \to \{0, 1\}$, $\mathrm{as}(f) \leq \mathrm{as}(\mathrm{Maj}_n)$.
Hint: Take $a \in f^{-1}(0)$ and define $f_a(x)$ by: $f_a(a) = 0$ and $f_a(x) = f(x)$ for all $x \neq a$. Show that $\mathrm{as}(f_a) \geq \mathrm{as}(f)$ as long as a has fewer than $n/2$ ones.

2.34. Let f be a monotone boolean function in n variables. Show that $\mathrm{Inf}_i(f) = 2 \cdot \widehat{f}(\{i\})$ for all $i = 1, \ldots, n$. *Hint*: Proof of Lemma 2.17.

2.35. Let $f, g : \{0, 1\}^n \to \mathbb{R}$ be real-valued functions, $S \subseteq [n]$ and $a \in \{0, 1\}^n$. Show that:

(a) $\chi_S(x \oplus a) = \chi_S(x) \cdot \chi_S(a)$.
(b) $f(x \oplus a) = \sum_S \widehat{f}(S) \chi_S(x \oplus a)$.
(c) $\mathrm{E}_x[f(x)g(x \oplus a)] = \sum_S \widehat{f}(S)\widehat{g}(S)\chi_S(a)$. This fact is also known as the *cross correlation lemma*.

2.36. Use Exercise 2.35(c) to prove the following version of Lemma 2.17 for a boolean function $f : \{0, 1\}^n \to \{-1, +1\}$ taking ± 1 values (instead of 0/1 values): $\mathrm{as}(f) = \sum_{S \subseteq [n]} |S|\widehat{f}(S)^2$.

Hint: Note that $\mathrm{Inf}_i(f) = \frac{1}{2}(1 - \mathrm{E}[f(x) \cdot f(x \oplus e_i)])$. Apply Exercise 2.35(c) with $g = f$ and $a = e_i$ to get $\mathrm{E}[f(x) \cdot f(x \oplus e_i)] = 1 - 2\sum_{S:i \in S} \widehat{f}(S)^2$.

Part II
Communication Complexity

Chapter 3
Games on Relations

The communication complexity of boolean functions is an *information theoretic* measure of their complexity. Besides its own importance, this measure is closely related to the *computational complexity* of functions: it corresponds to the smallest *depth* of circuits computing them. Thus, this measure can be used to prove circuit lower bounds. Communication complexity is appealing not only for its elegance and relation to circuit complexity, but also because its study involves the application of diverse tools from algebra, combinatorics and other fields of mathematics.

Communication complexity has a comprehensive treatment in an excellent book by Kushilevitz and Nisan (1997). To avoid intersections, we will mainly concentrate on results not included in that book (including results obtained after that book was published).

3.1 Communication Protocols and Rectangles

The basic (deterministic) model of communication was introduced in the seminal paper by Yao (1979). We have two players, traditionally called Alice and Bob, who have to evaluate a given function $f(x, y)$ for every given input (x, y). The function f itself is known to both players. The complication that makes things interesting is that Alice holds the first part x of their shared input, while Bob holds another part y. They do have a two-sided communication channel, but it is something like a transatlantic phone line or a beam communicator with a spacecraft orbiting Mars. Communication is expensive, and Alice and Bob are trying to minimize the number of bits exchanged while computing $f(x, y)$.

A general scenario of a deterministic communication game is the following. We are given some function $f : X \times Y \rightarrow Z$ or, more generally, some relation $f \subseteq X \times Y \times Z$. There are two players, Alice and Bob, who wish to evaluate $f(x, y)$ for all inputs $(x, y) \in X \times Y$. In the case when f is an arbitrary relation, their goal is to find some $z \in Z$ such that $(x, y, z) \in f$, if there is one.

S. Jukna, *Boolean Function Complexity*, Algorithms and Combinatorics 27,
DOI 10.1007/978-3-642-24508-4_3, © Springer-Verlag Berlin Heidelberg 2012

Fig. 3.1 A communication
tree (protocol). Alice's
functions a_u, a_w do not
depend on y, and Bob's
functions b_v, b_p do not
depend on x. Besides this
(independence) there are no
other restriction on these
functions—they may be
arbitrary!

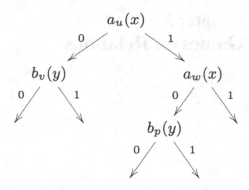

Both players know the entire function (relation) f. Also, the players are not
adversaries—they help and trust each other.[1] The difficulty, however, is that Alice
only knows x and Bob only knows y. Thus, to evaluate the function, they will need
to communicate with each other. The communication is carried according to some
fixed protocol (which depends only on f). Each player has unlimited computational
power. We are only concerned with the number of bits that have to be communicated
between them.

Before the game starts, the players agree on a protocol for exchanging messages.
After that, given an input pair (x, y), the protocol dictates to each player what
messages to send at each point, based on her/his input and the messages received
so far. It also dictates when to stop, and how to determine the answer from the
information received. There is no limit on the computational complexity of these
decisions, which are free of charge. The cost of the protocol is the number of bits
they have to exchange on the *worst case* choice of input pair (x, y). The goal is to
minimize this cost.

More formally, this measure can be defined as follows. A *protocol* (or a
communication tree) for a communication game is a binary tree T where each
internal node v is labeled either by a function $a_v : X \to \{0, 1\}$ or by a function
$b_v : Y \to \{0, 1\}$, and each leaf is labeled by an element $z \in Z$ (see Fig. 3.1). The
value of the protocol T on input (x, y) is the label of the leaf reached by starting
from the root, and walking on the tree. At each internal node v labeled by a_v we
walk left if $a_v(x) = 0$ and right if $a_v(x) = 1$. Similarly, at each internal node v
labeled by b_v we walk left if $b_v(y) = 0$ and right if $b_v(y) = 1$.

The *cost* of the protocol on input (x, y) is the length of the path taken on this
input. Then *cost* of the protocol T itself is its depth, that is, the length of its longest
path. The *communication complexity* of the relation f, denoted by $c(f)$, is the cost
of the best protocol for this game.

Intuitively, each internal node v labeled by a function a_v corresponds to a bit sent
by Alice (the bit being $a_v(x)$) and each internal node v labeled by b_v corresponds to a

---

[1]These are not "games" in the common sense where each of the players wants to win a game.

bit sent by Bob. Note that the value of each function a_v only depends on the part x of input (x, y) seen by Alice and on the results of the tests made (bits communicated) along the unique path to node v; the most important restriction is that no a_v depends on bits y seen by Bob. Similarly for the functions b_v.

Since the functions used by the players may be *arbitrary*, the model seems to be too powerful to be interesting. But we have one important restriction on how the players access a given input (x, y): Alice cannot see y and Bob cannot see x. This gives the most important structural restriction: for every node v in a communication tree, the set $S_v \subseteq X \times Y$ of all inputs reaching this node must be "rectangular".

Definition 3.1. (Rectangles) A *combinatorial rectangle* or just a *rectangle* is a subset $R \subseteq X \times Y$ of the form $R = R_0 \times R_1$ with $R_0 \subseteq X$ and $R_1 \subseteq Y$.

That is, a subset $R \subseteq X \times Y$ is a rectangle iff it fulfills the following "cut-and-paste" condition:

$$\text{if } (x, y), (x', y') \in R \text{ then } (x, y'), (x', y) \in R.$$

We stress that a rectangle need not to be contiguous! That is, even in the case where X, Y are ordered sets, the sets R_0, R_1 defining a rectangle need not be intervals.

Definition 3.2. (Monochromatic rectangles) Given a relation $f \subseteq X \times Y \times Z$, say that a subset $S \subseteq X \times Y$ is *monochromatic* (relative to f) if there exists a $z \in Z$ such that $(x, y, z) \in f$ for all $(x, y) \in S$.

Proposition 3.3. *If T is a communication tree for some relation, and v its node then*

$$S_v = \{(x, y) \in X \times Y \mid input (x, y) \text{ reaches node } v \text{ in } T\}$$

is a rectangle. Moreover, if v is a leaf, then the rectangle S_v is monochromatic.

Proof. We will prove by induction on the depth of v that S_v is a rectangle. If v is the root, then $S_v = X \times Y$, which is a rectangle. Otherwise, let w be the parent of v and assume, without loss of generality, that v is the left son of w and that in w Alice speaks, that is, w is labeled by a function $a_w : X \to \{0, 1\}$. Then

$$S_v = S_w \cap \{(x, y) \mid a_w(x) = 0\}.$$

By the induction hypothesis, $S_w = A \times B$ is a rectangle, and thus

$$S_v = (A \cap \{x \mid a_w(x) = 0\}) \times B$$

which is a rectangle. If v is a leaf, then no further communication is possible. That is, there must be an answer $z \in Z$ which suites all inputs $(x, y) \in S_v$, meaning that $(x, y, z) \in f$ for all $(x, y) \in S_v$. $\qquad\square$

By Proposition 3.3, a communication protocol for a relation $f \subseteq X \times Y \times Z$ is a binary tree whose inner nodes are rectangles $R \subseteq X \times Y$:

- The root is labeled by the whole rectangle $X \times Y$.
- If a node u is labeled by a rectangle R, then the sons of u are labeled by the corresponding subrectangles of R. Moreover, these subrectangles are obtained from R by splitting the rows of R (if u is Alice's node) or by splitting the columns of R (if u is Bob's node).
- Leafs are labeled by monochromatic rectangles.

Since at each node, the rows (or columns) of the corresponding submatrix are split into *disjoint* parts, the protocol is *deterministic*: each edge $(x, y) \in S$ will reach precisely one leaf. The depth of a tree is the maximum number of edges from the root to a leaf. The minimum depth of a communication tree is the communication complexity, $\mathfrak{c}(f)$, of the game for the relation f.

Simple as it is, Proposition 3.3 gives us a handy tool to show that some functions require many bits of communication.

Example 3.4. Consider a boolean function $f(x, y)$ in $2n$ variables defined by: $f(x, y) = 1$ iff $x = y$. The corresponding relation in this case is $F \subseteq X \times Y \times Z$ with $X = Y = \{0, 1\}^n$, $Z = \{0, 1\}$, and $(x, y, a) \in F$ iff $f(x, y) = a$. Now, every input in $f^{-1}(1) = \{(x, x) \mid x \in \{0, 1\}^n\}$ must reach a 1-leaf. On the other hand, if v is a 1-leaf, then Proposition 3.3 tells us that the set S_v of inputs (x, y) reaching this leaf must form a rectangle. Since S_v must lie in $f^{-1}(1)$, we obtain that $|S_v| = 1$: had the set S_v contain two inputs (x, x) and (y, y) with $x \neq y$, then it would be forced to contain (x, y), and the protocol would make an error because $f(x, y) = 0$. Thus the protocol must have at least $|f^{-1}(1)| = 2^n$ leaves, and hence, must have depth at least $\log(2^n) = n$.

3.2 Protocols and Tiling

Let $f \subseteq X \times Y \times Z$ be a relation. Recall that a rectangle $R \subseteq X \times Y$ is *monochromatic* relative to f (or f-*monochromatic*) if there exists a $z \in Z$ such that $(x, y, z) \in f$ for all $(x, y) \in R$. Communication protocols for f lead to the following complexity measures:

- Communication complexity $\mathfrak{c}(f)$ = minimum depth of a communication protocol tree for f.
- Number of messages $\mathrm{L}(f)$ = minimum number of leaves in a communication protocol tree for f.
- Tiling number $\chi(f)$ = smallest t such that $X \times Y$ can be decomposed into t disjoint f-monochromatic rectangles.

Proposition 3.3 yields the following lower bounds on the communication complexity:

$$\mathfrak{c}(f) \geq \log \mathrm{L}(f) \geq \log \chi(f).$$

How tight are these bounds? We first show that the lower bound $c(f) \geq \log L(f)$ is almost tight.

Lemma 3.5. (Balancing Protocols) $c(f) \leq 2\log_{2/3} L(f)$.

Proof. It suffices to show that given any deterministic protocol tree T for f with $|T| = L$ leaves, we are able to create a new protocol tree for f of depth at most $2\log_{2/3} L$. We argue by induction on L. The basis case $L = 1$ is trivial. For the induction step, apply Lemma 1.3 to obtain a node v in T such that the number $|T_v|$ of leaves in the sub-tree rooted at v satisfies

$$L/3 \leq |T_v| \leq 2L/3 .$$

Both players know T, and hence, they also know the sub-tree T_v. Let $S_v \subseteq X \times Y$ be a rectangle corresponding to the node v. Now, both players decide if $(x, y) \in S_v$, by sending 1 bit each of them. If yes, then they use the protocol sub-tree T_v. If no, then they use the protocol tree T', where T' is the same tree as T, except that the sub-tree T_v is replaced by a leaf labeled by an empty rectangle; hence, T' also has

$$|T'| \leq |T| - |T_v| \leq L - L/3 = 2L/3$$

leaves. The new protocol is correct, since in this last case $(x, y) \notin S_v$ implies that input (x, y) cannot reach node v in the original tree T.

To estimate the cost of the new protocol, let $c(L)$ be the number of bits that are communicated by the new protocol when the original tree has L leaves. By construction, we have that $c(L) \leq 2 + c(2L/3)$, where 2 is the number of bits that are communicated at the current step and $c(2L/3)$ the number of bits that will be communicated in the next (recursive) step in the worse case. Also note that $c(1) = 0$ (at a leaf no communication is necessary). Applying this inequality repeatedly we get (by setting $i = \log_{3/2} L$) that

$$c(L) \leq \underbrace{2 + 2 + \cdots + 2}_{i} + c((2/3)^i L) = 2\log_{3/2} L . \qquad \Box$$

Thus, we always have that $c(f) = \Theta(\log L(f))$. That is, when estimating the smallest depth of a communication protocol we are actually estimating the smallest number of leaves in such a protocol.

The situation with the tiling number $\chi(f)$ is worse: here we only know a quadratic upper bound $c(f) = \mathcal{O}(\log^2 \chi(f))$. We will obtain this upper bound as a direct consequence of the following more general result which we will apply several times later. Its proof idea is essentially due to Aho et al. (1983).

Suppose, we have a collection $\mathcal{R}$ of (not-necessarily disjoint) rectangles covering the entire rectangle $X \times Y$. Suppose also that each rectangle $R \in \mathcal{R}$ has its label. The only requirement is that the labeling must be *legal* in the following sense:

If two rectangles R and S have different labels, then $R \cap S = \emptyset$.

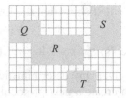

Fig. 3.2 The rectangle R intersects S in rows, intersects T in columns, and intersects Q in both rows and columns. We draw rectangles as consecutive blocks of entries only for simplicity: they need not be consecutive

That is, overlapping is only allowed for rectangles with the same label.

In the *find-a-rectangle game* for $\mathcal{R}$, Alice gets an $x \in X$, Bob gets an $y \in Y$, and their goal is to find a label of a rectangle containing (x, y). Note that the answer (label) is unique for each point (x, y): if the point belongs to more than one rectangle, then all these rectangles must have the same label.

Let $\mathfrak{c}(\mathcal{R})$ denote the deterministic communication complexity of such a game for $\mathcal{R}$.

Lemma 3.6. (Making Coverings Disjoint) $\mathfrak{c}(\mathcal{R}) \leq 2(\log |\mathcal{R}|)^2$.

Proof. Say that a rectangle $S = S_0 \times S_1$ intersects a rectangle $R = R_0 \times R_1$ in *rows*, if $S_0 \cap R_0 \neq \emptyset$, and intersects R in *columns*, if $S_1 \cap R_1 \neq \emptyset$ (see Fig. 3.2). Note that, $S \cap R \neq \emptyset$ if and only if S intersects R in rows *and* in columns. Given a rectangle R, we say that a rectangle R' is differently labeled if the label of R' is different from that of R.

Claim 3.7. No rectangle in $\mathcal{R}$ can intersect more than half of differently labeled rectangles in rows and more than half of these rectangles in columns.

Proof. Had this happen for some rectangle R, then R would intersect some other rectangle of different label, contradicting the legality of our labeling. $\quad\square$

Now set $r := \lceil \log |\mathcal{R}| \rceil$. The protocol consists of at most r rounds and in each round at most $1 + r$ bits are communicated. After each round the current set of rectangles is updated. Given an input (x, y), the players will be trying to decrease the number of rectangles in each round by at least one half. Say that a rectangle $R = R_0 \times R_1$ in $\mathcal{R}$ is Alice's rectangle if $x \in R_0$, and Bob's rectangle if $y \in R_1$. Thus, the goal of the players is to find a rectangle which is both Alice's and Bob's rectangle.

1. Alice checks whether all here rectangles have the same label. If yes, then the (unique) label of all these rectangles is a correct answer, and she announces it.
2. Otherwise, Alice tries to find among here rectangles a rectangle which intersects in *rows* at most half of the differently labeled rectangles. If such a rectangle R exists, then Alice sends its name (using r bits) to Bob and they both update $\mathcal{R}$ so that it only contains the rectangles that intersect with R in rows; each of

the remaining rectangles is not an Alice's rectangle, and hence, cannot contain (x, y).
3. If Alice is unable to find such a rectangle then she communicates this to Bob (using 1 bit).
4. Now is Bob's turn. Since Alice failed, Claim 3.7 ensures that among Bob's rectangles there must be a rectangle which intersects in *columns* at most half of the differently labeled rectangles (at least rectangles containing (x, y) must be such). Bob takes any such rectangle R and sends its name (using r bits) to Alice and they both update $\mathcal{R}$ so that it only contains the rectangles that intersect with R in columns (other rectangles cannot contain (x, y)). At this point the round is definitely over since they successfully eliminated at least half of the rectangles in $\mathcal{R}$ labeled differently than R, and we can proceed by induction.

After at most r rounds the players will agree on a rectangle containing (x, y), and the label of this rectangle is the correct answer. □

Lemma 3.8. *For every relation* $f \subseteq X \times Y \times Z$, $\mathfrak{c}(f) \leq 2(\log \chi(f))^2$.

Proof. Let $\mathcal{R}$ be an optimal covering of $X \times Y$ by $|\mathcal{R}| = \chi(f)$ disjoint f-monochromatic rectangles. Since each $R \in \mathcal{R}$ is monochromatic, there must exist a $z \in Z$ such that $(x, y, z) \in f$ for all $(x, y) \in R$. Fix one such z for each rectangle $R \in \mathcal{R}$, and let z be the label of R. Since all rectangles in $\mathcal{R}$ are disjoint, this is a legal labeling in the sense of Lemma 3.6. Hence, we can apply this lemma and obtain that, for every input (x, y), the players can find out the (unique) rectangle containing (x, y) by communicating at most $2(\log |\mathcal{R}|)^2 = 2(\log \chi(f))^2$ bits. □

Lemma 3.8 has the following purely combinatorial consequence about covering matrices by their submatrices. A submatrix of a matrix is *monochromatic* if all its entries have the same value. Suppose we can cover all entries of a given matrix by t its (possibly overlapping) monochromatic submatrices. How many *disjoint* monochromatic submatrices do we need then to cover all entries? Using communication complexity arguments one can show that $t^{\mathcal{O}(\log t)}$ disjoint submatrices are always enough!

Lemma 3.9. *If a matrix A can be covered by t (not necessarily disjoint) monochromatic submatrices, then A can be decomposed into at most $t^{2\log t}$ pairwise disjoint monochromatic submatrices.*

Proof. Let $\mathcal{R}$ be a set of t monochromatic submatrices covering A. Label each such submatrix by its unique element. This labeling is clearly legal. By Lemma 3.6, there is a communication protocol of depth at most $c = 2(\log t)^2$ which for every row x and every column y outputs the entry $A[x, y]$. Each leaf of the protocol corresponds to a monochromatic submatrix of A, and we obtain a decomposition of A into at most $2^c = t^{2\log t}$ pairwise disjoint monochromatic submatrices. □

3.3 Games and Circuit Depth

In this section we will see a surprisingly tight connection between communication and circuits that has proved very useful in boolean function complexity.

Let $S = X \times Y$ be a rectangle with $X, Y \subseteq \{0, 1\}^n$ and $X \cap Y = \emptyset$. For example, S could be an "ambient" rectangle $S = f^{-1}(1) \times f^{-1}(0)$ of a boolean function f. The *find-a-difference game* for a rectangle $S = X \times Y$ is the following game:

- Alice gets a vector $x \in X$.
- Bob gets a vector $y \in Y$.
- The goal is to find a position i such that $x_i \neq y_i$.

Let $\mathfrak{c}(S)$ denote the communication complexity of this game. Note that the find-a-difference game for $S = X \times Y$ is just a game for the relation $F \subseteq X \times Y \times [n]$ given by: $(x, y, i) \in F$ iff $x_i \neq y_i$. In particular, for this relation, a rectangle $R \subseteq X \times Y$ is *monochromatic* if there exists a position $i \in [n]$ such that $x_i \neq y_i$ for all $(x, y) \in R$. Thus, every protocol for the find-a-difference game on a rectangle $S = X \times Y$ gives a *partition* (or tiling) of S into monochromatic rectangles. This motivates the following purely combinatorial measure of rectangles.

Definition 3.10. (Tiling number of rectangles) The *tiling number* $\chi(S)$ of a rectangle S is the smallest number t such that S can be decomposed into t disjoint monochromatic rectangles.

Lemma 3.8 immediately yields the following connection of this measure with the communication complexity.

Lemma 3.11. *For every rectangle S,*

$$\log \chi(S) \le \mathfrak{c}(S) \le 2(\log \chi(S))^2.$$

This relation between the communication complexity and the tiling number is a useful combinatorial tool to prove lower bounds on the communication complexity. But it is also a handy tool to obtain efficient tiling of rectangles. To give an example, let us show how one can obtain a decomposition of the rectangle of the parity function $\oplus_n = x_1 \oplus x_2 \oplus \cdots \oplus x_n$ into a small number of monochromatic rectangles.

Proposition 3.12. $\chi(\oplus_n) \le 4n^2$ *and* $\mathfrak{c}(\oplus_n) \le 2 \log n + 2$.

Proof. We will only show that $\chi(\oplus_n) \le n^2$ if n is a power of 2. The general case then follows by adding redundant zeros to the strings so that their length is a power of 2. The resulting strings will have length at most $2n$, and the upper bound $\chi(\oplus_n) \le 4n^2$ follows.

Consider the communication game for $\oplus_n$. That is, given a pair (x, y) of binary strings of length n such that x has a odd and y and even number of 1s, the goal of Alice and Bob is to find an i with $x_i \neq y_i$.

The basic idea is binary search. Bob begins by saying the parity of the left half of y. Alice then says the parity of the left half of x. If these parities differ, then they continue playing on the left half, otherwise they continue playing on the right half. With each round they halve the size of the playing field, and use 2 bits of communication. Thus after $\log n$ rounds and $2 \log n$ bits of communication they determine an i on which x and y differ. This gives a decomposition into $2^{2 \log n} = n^2$ disjoint monochromatic rectangles.   □

Now we are going to prove one particularly important fact connecting communication with computation.

For a boolean function f, let $D(f)$ be the minimum depth of a DeMorgan circuit computing f. Let also $c(f)$ denote the communication complexity of the find-a-difference game on the rectangle $f^{-1}(1) \times f^{-1}(0)$.

Theorem 3.13. (Karchmer and Wigderson 1990) *For every boolean function f,*

$$D(f) = c(f).$$

We prove lower and upper bounds on $D(f)$ separately. The first claim is just a reformulation of Rychkov's lemma in terms of games.

Claim 3.14. (Circuit to protocol) $c(f) \le D(f)$.

Proof. We may assume that Alice and Bob have agreed on a circuit of smallest depth computing f. Now suppose Alice gets an input x such that $f(x) = 1$, and Bob gets an input y such that $f(y) = 0$. In order to find an i such that $x_i \ne y_i$, the players use the information provided by the underlying circuit. At OR gates speaks Alice, and at AND gates speaks Bob.

Suppose the output gate is an AND gate, that is, we can write $f = f_0 \wedge f_1$. Then Bob sends a bit i corresponding to a function f_i such that $f_i(y) = 0$; if both $f_0(y)$ and $f_1(y)$ output 0, then Bob sends 0. We know that we must have $f_i(x) = 1$. We can then repeat this step at the gate corresponding to the output gate of f_i, where Bob sends a bit if the gate is an AND gate and Alice sends a bit if the gate is an OR gate (she sends a bit corresponding to a function which outputs 1). Alice and Bob repeat this process until they reach a leaf of the circuit. This leaf is labeled by some variable z_i or its negation $\neg z_i$. Hence, $x_i \ne y_i$ implying that i is a correct answer.□

The other direction is more interesting: having a protocol we can build a circuit.

Claim 3.15. (Protocol to circuit) $D(f) \le c(f)$.

Proof. We will prove a more general claim: For every rectangle $S = A \times B$ there is a boolean function f such that $A \subseteq f^{-1}(1)$, $B \subseteq f^{-1}(0)$ and $D(f) \le c(S)$. It then remains to take $S = f^{-1}(1) \times f^{-1}(0)$.

We prove the claim by induction on $c = c(S)$. Suppose $c = 0$. Then we must have, for some index i, that $x_i \ne y_i$ for all pairs $(x, y) \in S$. Thus we may choose either $f = x_i$ or $f = \neg x_i$ according to which function satisfies $f(A) = 1$ and $f(B) = 0$.

Next, we prove the claim is true for c assuming it is true for $c - 1$. Consider a protocol for the communication game on S that uses at most c bits. Let us assume

Alice sends the first bit. Then there is a partition $A = A_0 \cup A_1$, $A_0 \cap A_1 = \emptyset$, such that for $x \in A_0$, Alice sends the bit 0 and for $x \in A_1$, Alice sends the bit 1. After that we are left with two disjoint rectangles $A_0 \times B$ and $A_1 \times B$ whose communication complexity is at most $c - 1$. Applying our induction hypothesis, we find there exists a function f_0 such that

$$f_0(A_0) = 1, \quad f_0(B) = 0 \quad \text{and} \quad \mathrm{D}(f_0) \leq c - 1,$$

and there exists a function f_1 such that

$$f_1(A_1) = 1, \quad f_1(B) = 0 \quad \text{and} \quad \mathrm{D}(f_1) \leq c - 1.$$

We define $f = f_0 \vee f_1$. Then $f(A) = 1$, $f(B) = 0$ and

$$\mathrm{D}(f) \leq 1 + \max\{\mathrm{D}(f_0), \mathrm{D}(f_1)\} \leq c$$

as desired. Note that, if Bob had sent the first bit, we would have partitioned B and defined $f = f_0 \wedge f_1$. This finishes the proof of the claim, and thus the proof of the theorem.   □

Remark 3.16. In fact, $c(f)$ is a lower bound for the parameter "depth times logarithm of the maximal fanin" of any circuit with unbounded-fanin AND and OR gates. If, say, f is computed by a circuit of depth d and fanin of every gate is at most S, then $c(f) = \mathcal{O}(d \log S)$. This holds because, at each step, one of the players can use $\log S$ bits to tell what of at most S gates feeding into the current gate to choose.

3.3.1 Monotone Depth

For monotone circuits we can give a modified version of Theorem 3.13 that captures, in a nice way, the restrictions of monotone computations. Recall that a *minterm* (resp., *maxterm*) of a monotone boolean function f is a minimal set of variables such that, if we set all these variables to 1 (resp., to 0), f will be 1 (resp., 0) regardless of the other variables. We will view minterms and maxterms as subsets of $[n] = \{1, \ldots, n\}$. Let $\mathrm{Min}(f)$ denote the set of all minterms, and $\mathrm{Max}(f)$ the set of all maxterms of f. It is easy to see that every minterm intersects every maxterm. This suggests the following communication game. Let $P, Q \subseteq 2^{[n]}$ be two families of sets such that $p \cap q \neq \emptyset$ for all $p \in P$ and $q \in Q$.

- Alice gets a set $p \in P$.
- Bob gets a set $q \in Q$.
- The goal is to find an element $i \in p \cap q = p \setminus \overline{q}$.

Note that in a non-monotone game (corresponding to non-monotone circuits) the goal of the players is to find an element i in $a \cap b$ or in $\overline{a} \cap \overline{b} = \overline{a \cup b}$. This is that "or" which makes the analysis of such protocols very difficult.

Let $c_+(P, Q)$ denote the communication complexity of this game. For a monotone boolean function f, let $c_+(f) := c_+(P, Q)$ with $P = \text{Min}(f)$ and $Q = \text{Max}(f)$. Finally, let $\text{Depth}_+(f)$ be the minimum depth of a monotone DeMorgan formula computing f. The same argument as in the proof of Theorem 3.13 gives

Theorem 3.17. *For every monotone boolean function f,*

$$\text{Depth}_+(f) = c_+(f).$$

Proof. Note that in the base case of Claim 3.14 (when players reach a leaf of the circuit), the monotonicity of the circuit (no negated variables) they find an $i \in [n]$ such that $x_i = 1$ and $y_1 = 0$. On the other hand, if a protocol always finds an i with this property, Claim 3.15 gives a monotone circuit.

Let $x \in f^{-1}(1)$ be the characteristic vector of a subset $p \subseteq [n]$. Similarly, let $y \in f^{-1}(0)$ be the characteristic vector of the *complement* of a subset $q \subseteq [n]$, that is, $i \in q$ iff $y_i = 0$. By the above argument, it is clear that the answer of the protocol will be an element of $p \cap q$. The theorem follows by noticing that it is enough to design a protocol for $\text{Min}(f), \text{Max}(f)$ because the players can always behave as if they got $p' \subseteq p$ and $q' \subseteq q$ where $p' \in \text{Min}(f)$ and $q' \in \text{Max}(f)$. $\square$

Exercises

3.18. Let f be a k-CNF formula with m clauses. Show that there exists a one-round Karchmer-Wigderson protocol for f where Bob sends $\log m$ bits and Alice responds with $\log k$ bits.

3.19. Show that for any boolean function f there exists a Karchmer-Wigderson protocol where at each round Bob sends 2^a bits while Alice responds with a bits such that the number r of rounds satisfies $r \leq D(f)/a$.

Hint: Take the best circuit for f, divide it into stages of depth a each and look at the subcircuits of each stage. Each one computes a function which depends on at most 2^a wires, and thus can be represented as a 2^a-CNF formula with at most $m = 2^{2^a}$ clauses. Use the previous exercise.

3.20. (Razborov 1990) The game FORMULA is a game of two players Up (upper) and Lo (lower), Up will try to prove an upper bound for the formula size of a boolean function; Lo will try to interfere him. A *position* in this game is a triplet (U, V, t) where $U, V \subseteq \{0, 1\}^n$, $U \cap V = \emptyset$ and $t \geq 1$ is an integer. Up begins the game. He obtains a position (U, V, t), chooses one of the two sets U, V (say, U), chooses some representations of U, t of the form

$$U = U' \cup U'' \quad t = t' + t'' \quad (t', t'' \geq 1)$$

and hands to Lo the two positions (U', V, t') and (U'', V, t''). If Up chooses the set V, the description of his actions is given in the analogous way.

Lo chooses one of the two positions offered to him and returns it to Up (the remaining position is thrown out). Then Up moves as above (in the new position) and so on. The game is over when Up receives a position of the form $(U^*, V^*, 1)$. Up wins if $U^* \times V^*$ forms a monochromatic rectangle, that is, if there is an $i \in [n]$ such that $x_i \neq y_i$ for all $x \in U^*$ and $y \in V^*$.

Prove that Up has a winning strategy in a position (U, V, t) iff there exists a boolean function $f : \{0, 1\}^n \to \{0, 1\}$ such that: $f(U) = 0$, $f(V) = 1$ and f has a DeMorgan formula of leafsize $\leq t$. *Hint*: Argue by induction on t as in the proof of Theorem 3.13.

3.21. (Brodal and Husfeld 1996) Prove that $D(f) = \mathcal{O}(\log n)$ for every symmetric boolean function f of n variables. For this, consider the following communication game: given a pair A, B of subsets of $[n]$ such that $|A| \neq |B|$, find an element in the symmetric difference $(A \setminus B) \cup (B \setminus A)$. Design a communication protocol with $\mathcal{O}(\log n)$ bits of communication for this game.

Hint: We already know (Proposition 3.12) how to design such a protocol if we the parities of $|A|$ and $|B|$ are different. In the general case, use that fact that the cardinality $|A|$ of a set A can be communicated using only $\log |A|$ bits. Look at restrictions $A^{l,s} = A \cap \{l, l+1, \ldots, s-l+1\}$ with $A^{1,n} = A$ and do an appropriate binary search, as in Proposition 3.12.

Chapter 4
Games on 0-1 Matrices

We have already seen that communication complexity of relations captures the depth of circuits. Protocols in this case are trying to solve *search* problems. In this chapter we consider the communication complexity of *decision* problems. That is, Alice gets a vector x, Bob gets a vector y, and their goal is to compute the value $f(x, y)$ of a given boolean function f.

In the *fixed-partition* communication game, the players are given a function f as well as some partition (x, y) of its variables into two disjoint blocks of equal size. The concept of the fixed-partition communication complexity was invented by Yao (1979, 1981). In the *best-partition* model of communication the players are allowed to choose a most suitable for this function balanced partition (x, y) of its input variables. The best-partition communication complexity was introduced by Lipton and Sedgewick (1981). Even trickier is the communication model where we have more than two players, each seeing all but a small piece of the input vector. We will consider this model later in Chap. 5.

4.1 Deterministic Communication

It will be convenient to consider a boolean function $f(x, y)$ of $2m$ variables as a boolean $n \times n$ matrix A with $n = 2^m$ such that $A[x, y] = f(x, y)$. Such a matrix A is usually referred to as the *communication matrix* of f. Recall that a *primitive matrix* is a 0-1 matrix of rank 1, that is, a boolean matrix consisting of one all-1 submatrix and zeros elsewhere. These are exactly the matrices of the form uv^T for two boolean vectors u and v. The *complement* of a boolean matrix A is the boolean matrix $\overline{A} = J - A$, where J is the all-1 matrix. A submatrix is *monochromatic* if all its entries are the same. Two submatrices of the same matrix are *disjoint* if they do not share a common entry.

The communication game for a given matrix A is as follows: Alice is given as input a row index x, Bob is given a column index y, and Bob must determine the value $A[x, y]$.

S. Jukna, *Boolean Function Complexity*, Algorithms and Combinatorics 27,
DOI 10.1007/978-3-642-24508-4_4, © Springer-Verlag Berlin Heidelberg 2012

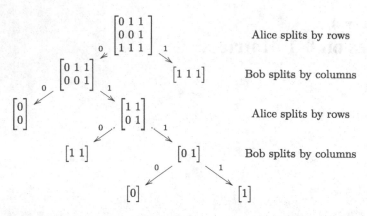

Fig. 4.1 An example of a communication tree for a boolean function $f : X \times Y \rightarrow \{0, 1\}$ represented as a matrix. The communication complexity of this protocol is 4

By sending bits 0 and 1, the players actually split the rows (if this bit is sent by Alice) or columns (if this bit is sent by Bob) into two disjoint parts. A *communication protocol* (or a *communication tree*) of a game is a binary tree, each inner node of which corresponds to a decision made by one of the players at this node. Each node of the tree is labeled by a submatrix of A so that the following holds (see Fig. 4.1).

- The root is labeled by the whole matrix A.
- If a node u is labeled by a matrix M, then the sons of u are labeled by the corresponding submatrices M_0 and M_1 of M. Moreover, these submatrices are obtained from M by splitting the rows of M (if u is Alice's node) or by splitting the columns of M (if u is Bob's node).
- If w is a leaf and R is its label, then R is *monochromatic*, that is, is either all-0 matrix or all-1 matrix.

Since at each node, the rows (or columns) of the corresponding submatrix are split into *disjoint* parts, the protocol is *deterministic*: each pair (x, y) will reach precisely one leaf. The depth of a tree is the maximum number of edges from the root to a leaf. The *deterministic communication complexity* $\mathfrak{c}(A)$ of a matrix A is defined as:

$$\mathfrak{c}(A) = \text{the minimum depth of a communication tree for } A.$$

It is clear that for any boolean $n \times n$ matrix A (n being a power of 2) we have that $\mathfrak{c}(A) \leq 1 + \log n$ since using $\log n$ bits Alice can just tell Bob the binary code of her row x, and Bob can announce the answer $A[x, y]$. Lower bounds on $\mathfrak{c}(A)$ can be shown using the rank as well as the "tiling number" of A.

Definition 4.1. (Tiling number of matrices) The *tiling number* $\chi(A)$ of a boolean matrix A is the smallest number of pairwise disjoint monochromatic submatrices of A covering all entries of A.

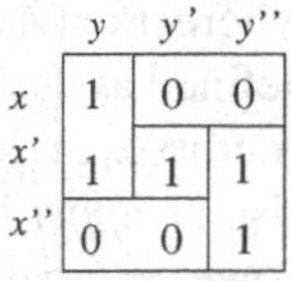

Fig. 4.2 A decomposition of
a 3×3 matrix that does not
correspond to any protocol

	y	y'	y''
x	1	0	0
x'	1	1	1
x''	0	0	1

If we are only interested in decomposing s-entries of A for some $s \in \{0,1\}$ into disjoint all-s submatrices, then the corresponding measure is denoted by $\chi_s(A)$. Note that $\chi(A) = \chi_0(A) + \chi_1(A)$ and $\chi_0(A) = \chi_1(\overline{A})$.

Since the submatrices occurring on the leaves of any communication tree for A must be disjoint, we immediately have a basic inequality

$$c(A) \geq \log \chi(A).$$

Note however, that unlike arbitrary decompositions of a given matrix A into mono-chromatic submatrices, decompositions arising from communication protocols have a special form: they are produced inductively by splitting the resulting submatrices only row-wise or column-wise. And indeed, there are decompositions that cannot be produced by any communication protocol, like one depicted in Fig. 4.2. Thus, $c(A)$ may be larger than $\log \chi(A)$. Kushilevitz et al. (1999) exhibited matrices A with $c(A) \geq (2 - o(1)) \log \chi(A)$.

On the other hand, $c(A)$ is the communication complexity of a relation $F \subseteq X \times Y \times Z$ where X is the set of rows of A, Y is the set of columns of A, and $Z = \{0,1\}$. The relation itself is given by: $(x,y,z) \in F$ iff $A[x,y] = z$. Thus, Lemma 3.8 gives the following upper bound:

$$c(A) \leq 2(\log \chi(A))^2. \tag{4.1}$$

Observe that $\chi_1(A)$ is just the least number t such that A can be written as a sum $A = \sum_{i=1}^{t} R_i$ of t primitive matrices $R_i = u_i v_i^T$ with $u_i, v_i \in \{0,1\}^n$. This is reminiscent of another matrix measure—their rank. Recall that a (real) rank of a matrix A is the least number r such that $A = \sum_{i=1}^{r} \alpha_i R_i$ for some $\alpha_i \in \mathbb{R}$ and primitive 0-1 matrices R_i. Thus, $\chi_1(A)$ corresponds to the case when all $\alpha_i = 1$. This immediately gives us another basic estimate:

$$c(A) \geq \log \chi_1(A) \geq \log \mathrm{rk}(A). \tag{4.2}$$

This bound was first observed by Mehlhorn and Schmidt (1982). Yet another characterization of rank is in terms of scalar products. Namely, $\mathrm{rk}(A)$ is the least r such that A can be written as a sum $A = \sum_{i=1}^{r} x_i y_i^T$ for some vectors $x_i, y_i \in \mathbb{R}^n$. Since the coefficients 0 and 1 of characteristic vectors are clearly non-negative, we thus actually have slightly stronger lower bound

$$c(A) \geq \log \chi(A) \geq \log \mathrm{rk}^+(A), \tag{4.3}$$

where $\mathrm{rk}^+(A)$ is the *non-negative rank* introduced by Yannakakis (1991): it is defined as the least r such that A can be written as a sum $A = \sum_{i=1}^r x_i y_i^T$ for some vectors $x_i, y_i \in \mathbb{R}^r$ with $x_i, y_i \geq \mathbf{0}$. It can be shown (Exercise 4.2) that $\log \mathrm{rk}^+(A)$ is also an *upper* bound on the nondeterministic communication complexity. It remains, however, open whether $c(A)$ is upper bounded by a polynomial in $\log \mathrm{rk}(A)$; see Sect. 4.6 for a discussion.

The rank lower bound (4.2) implies that implies that even some of the "simplest" matrices, like the identity matrix I_n, have maximal communication complexity $\log n$. The goal however is to understand what properties of a given matrix A actually force its communication complexity to be large. Having 1s on the diagonal and 0s elsewhere is just one of these properties.

The identity matrix I_n is very sparse: it has only n ones. Using the rank, one can show that any symmetric matrix with at most $n^{2-\epsilon}$ ones requires an almost maximal number $\Omega(\log n)$ bits of communication, as well.

For a matrix A, let $|A|$ denote the number of its nonzero entries. The following is a "folklore result" rediscovered by many authors.

Lemma 4.2. *If A is a symmetric $n \times n$ boolean matrix with 1s on the diagonal, then*

$$\chi(A) \geq \frac{n^2}{|A|}.$$

Proof. Let $\lambda_1, \ldots, \lambda_n$ be the eigenvalues of A. Then their sum $t = \sum_{i=1}^n \lambda_i$ is the trace of A (sum of diagonal entries of A), and at most $r = \mathrm{rk}(A)$ of them are nonzero. Thus, the Cauchy–Schwarz inequality yields

$$\mathrm{tr}(A^2) = \sum_{i=1}^n \lambda_i^2 \geq r(t/r)^2 = t^2/r.$$

Since A is a 0-1 matrix, we also have that $\mathrm{tr}(A^2) = |A|$: the i-th diagonal entry of A^2 is the number of 1s in the i-th row of A. This implies $\mathrm{rk}(A) = r \geq \mathrm{tr}(A)^2/|A|$, where $\mathrm{tr}(A) = n$ since A has 1s on the diagonal. $\square$

Lower bounds on communication complexity obtained using rank and other algebraic measures are discussed in a survey by Lee and Shraibman (2009).

4.2 Nondeterministic Communication

For a two-party communication problem specified by a matrix A, one way to view a *nondeterministic* communication protocol is as a scheme by which a third party, Carole, who knows Alice's input x and Bob's input y, can convince Alice and Bob of the value of the matrix entry $A[x, y]$. Of course, if Carole were trustworthy, she could simply tell the two players this value.

A more interesting case is when Carole is untrusted and has the goal of convincing the players that this value equals 1, whether or not this is true. For this purpose, she announces to both players some binary string, a *witness* for (or a *proof* of) the fact that "$A[x, y] = 1$". Having this witness, Alice and Bob verify it *independently* and respond with either Yes or No. Alice and Bob agree that $A[x, y] = 1$ (and accept the input (x, y)) if and only if they both replied with Yes. If $A[x, y] = 0$ then Alice and Bob must be able to detect that the witness is wrong no matter what Carole says. The protocol is correct if, for every input (x, y), Alice and Bob accept it if and only if $A[x, y] = 1$. The communication complexity of this game is the length of the witness in the worst case. That is, $nc(A) \leq t$ iff for every input (x, y) we have that:

1. If $A[x, y] = 1$ then there exists a witness $w \in \{0, 1\}^t$ on which both Alice and Bob answer with Yes.
2. If $A[x, y] = 0$ then, for every witness $w \in \{0, 1\}^t$, at least one player responds with No.

In other words, if X is the set of rows and Y the set of columns of A, then a nondeterministic protocol of cost t is a pair of functions $a : X \times \{0, 1\}^t \to \{0, 1\}$ and $b : \{0, 1\}^t \times Y \to \{0, 1\}$ such that, for all $(x, y) \in X \times Y$,

$$A[x, y] = 1 \text{ iff } a(x, w) \wedge b(w, y) = 1 \text{ for some } w \in \{0, 1\}^t.$$

The *nondeterministic communication complexity*, $nc(A)$, of a matrix A is the smallest number t for which such functions a and b exist. This measure has a very simple combinatorial description.

The *cover number*, $Cov(A)$, of a 0-1 matrix A is the smallest number of all-1 submatrices of A covering all its 1s; this time the matrices in a cover need not be disjoint. That is, $Cov(A)$ is the least positive integer t such that A can be written as componentwise OR $A = \bigvee_{i=1}^{t} R_i$ of t primitive matrices $R_1, \ldots, R_t$, that is, boolean matrices of rank 1.

Proposition 4.3. *For every boolean matrix A, we have that*

$$\log Cov(A) \leq nc(A) \leq \lceil \log Cov(A) \rceil.$$

Proof. To show $nc(A) \leq \lceil \log Cov(A) \rceil$, suppose that all 1s of A can be covered by $C = Cov(A)$ all-1 submatrices $R_1, \ldots, R_C$. Let $t = \lceil \log C \rceil$, and let R_w be the submatrix the binary code of whose index is $w \in \{0, 1\}^t$. Then define the functions $a(x, w)$ and $b(w, y)$ by: $a(x, w) = 1$ iff x is a row of R_w, and $b(w, y) = 1$ iff y is a column of R_w.

To show the converse direction $\log Cov(A) \leq nc(A)$, suppose that desired functions $a(x, w)$ and $b(w, y)$ with witnesses of length $t = nc(A)$ are given. Then, for every $w \in \{0, 1\}^t$, the set $R_w = \{(x, y) : a(x, w) \wedge b(w, y) = 1\}$ must be an all-1 submatrix of A, and their union must cover all 1s of A. Since there are only 2^t such rectangles, we can conclude that $\log Cov(A) \leq t = nc(A)$. $\square$

We don't want to worry about ceilings and floorings, so we adopt the following
definition of the nondeterministic communication complexity:

$$\mathrm{nc}(A) := \log_2 \mathrm{Cov}(A). \tag{4.4}$$

Example 4.4. Carole can easily convince Alice and Bob that two binary strings x
and y of length n are *not* equal: using only $\lceil \log n \rceil + 1$ bits she announces (the
binary code of) a position i with $x_i \neq y_i$ and the bit x_i; Alice checks whether the
bit she received is the i-th bit of the string she can see, and Bob checks whether
$y_i \neq x_i$. If however Carole wants to convince that $x = y$, then she is forced to send
n bits, just because $\mathrm{Cov}(I_n) = 2^n$ for a $2^n \times 2^n$ identity matrix I_n.

One can show that only sparse matrices can have large nondeterministic commu-
nication complexity. For a boolean matrix, let $|A|$ denote the number of its 1-entries.
The following lemma is a modification of a probabilistic argument used by Alon
(1986).

Lemma 4.5. *Let A be a boolean matrix. If every column or every row of A contains
at most d zeros, then $\mathrm{Cov}(A) = \mathcal{O}(d \ln |A|)$.*

Proof. We only consider the column case, the row case is the same. To cover the
ones of A we construct an all-1 submatrix B with row set I and column set J via
the following probabilistic procedure: pick every row of A with probability $p =
1/(d + 1)$ to get a random subset I of rows, and let J be the set of all columns of
A that have no zeros in the rows of B.

A 1-entry (x, y) of A is covered by B if x was chosen in I and none of (at most
d) rows with a 0 in the y-th column was chosen in I. Hence,

$$\mathrm{Prob}[(x, y) \text{ is covered by } B] \geq p(1 - p)^d \geq pe^{-pd} \geq p/e.$$

If we apply this procedure t times to get t all-1 submatrices, then the probability that
(x, y) is covered by *none* of these submatrices does not exceed $(1 - p/e)^t \leq e^{-tp/e}$.
Hence, the probability that some 1-entry of A remains uncovered is at most

$$|A| \cdot e^{-tp/e} = \exp(\ln |A| - t/(e(d + 1))),$$

which is < 1 for $t > e(d + 1) \ln |A|$.   □

4.2.1   Greedy Bounds

We now turn to *lower* bounds on the covering number $\mathrm{Cov}(A)$, and hence, on
the nondeterministic communication complexity $\mathrm{nc}(A) = \log \mathrm{Cov}(A)$. Recall that
the covering number $\mathrm{Cov}(A)$ of A is the smallest number of its all-1 submatrices
covering all 1-entries of A. If A has $|A|$ ones, then a trivial lower bound is
$\mathrm{Cov}(A) \geq |A|/r$, where r is the largest number of entries in an all-1 submatrix

of A. This bound, however, may be very far from the truth. Let, for example, $A = (a_{ij})$ be an upper triangular $n \times n$ matrix, that is $a_{ij} = 1$ iff $i \leq j$. Then $|A| = n + (n-1) + \cdots + 1 = n(n+1)/2$, but also $r \geq n^2/4$, and the resulting (trivial) lower bound on $\mathrm{Cov}(A)$ is even smaller than 2.

A much better way to show that $\mathrm{Cov}(A)$ must be large is to choose a particular subset S of "hard to cover" 1-entries in A, and to show that no all-1 submatrix of A can cover many entries in S. To formalize this, let us define $r_A(S)$ as the largest number of the selected "hard" 1-entries (those in S) that can be covered by an all-1 submatrix of A. This immediately implies that we need $\mathrm{Cov}(A) \geq |S|/r_A(S)$ all-1 submatrices of A to cover all entries in S. In the case of the triangular matrix Δ, we can take S to be, say, the set all its n diagonal entries. Then $r_\Delta(S) = 1$, and a much more respectful lower bound $\mathrm{Cov}(\Delta) \geq n$ follows. This motivates the following measure:

$$\mu(A) = \max_{S} \frac{|S|}{r_A(S)},$$

where the maximum is over all subsets S of 1-entries of A. By what was said above, we have that $\mathrm{Cov}(A) \geq \mu(A)$. Actually, this lower bound is already not very far from the truth.

Lemma 4.6. (Lovász 1975) $\mathrm{Cov}(A) \leq \mu(A) \cdot \ln |A| + 1$.

Proof. Let I be the set of all 1-entries of A. Consider a greedy covering of I by all-1 submatrices $R_1, \ldots, R_t$ of A. That is, in the i-th step we choose an all-1 submatrix R_i covering the largest number of all yet uncovered entries in I. Let $B_i = I \setminus (R_1 \cup \cdots \cup R_{i-1})$ be the set of entries in I that are left uncovered after the i-th step. Hence, $B_0 = I$ and $B_t = \emptyset$. Let $b_i = |B_i|$ and $r_i = r_A(B_i)$. That is, r_i is the maximum number of entries in B_i contained in an all-1 submatrix of A. Since, by the definition of $\mu := \mu(A)$, none of the fractions b_i/r_i can exceed μ, we have that $b_{i+1} = b_i - r_i \leq b_i - b_i/\mu$. This yields

$$b_i \leq b_0(1 - 1/\mu)^i \leq |A| \cdot e^{-i/\mu}.$$

For $i = t - 1$, we obtain $1 \leq b_{t-1} \leq |A| \cdot e^{-(t-1)/\mu}$, and the desired upper bound $\mathrm{Cov}(A) \leq t \leq \mu \ln |A| + 1$ follows. $\square$

4.2.2 Fooling-Set Bounds

Since any two 1-entries lying on one line (row or column) *can* be covered by one all-1 submatrix, a natural choice for a "difficult-to-cover" subset S of 1-entries of A is to take "independent" entries.

Namely, say that two 1-entries in a matrix are *independent* if they do not lie in one row or in one column. The *term-rank*, $\mathrm{trk}(A)$, of A is the largest number of its pairwise independent 1-entries. The *clique number*, $\omega(A)$, of A is the largest number r such that A contains an $r \times r$ all-1 submatrix. Finally, the *line weight*, $\lambda(A)$, of A is the largest number of 1s in a line (row or column). If G is the bipartite graph whose adjacency matrix [1] is A then:

- $\text{trk}(A)$ = maximum number of edges in a matching in G.
- $\omega(A)$ = maximum r such that G contains a complete $r \times r$ graph $K_{r,r}$ as a subgraph.
- $\lambda(A)$ = maximum degree of a vertex in G.

Using these parameters we can lower bound the cover number as follows:

$$\text{Cov}(A) \geq \frac{\text{trk}(A)}{\omega(A)} \geq \frac{|A|}{\lambda(A) \cdot \omega(A)}. \tag{4.5}$$

The first inequality follows since any $r \times r$ all-1 submatrix of A can have at most r independent 1s. The second inequality is a direct consequence of a classical result of König-Egervary saying that the term-rank $\text{trk}(A)$ of A is exactly the minimum number of lines (rows and columns) covering all 1s in A; hence, $\text{trk}(A) \geq |A|/\lambda(A)$.

Although simple, the first lower bound in (4.5)—known as the *fooling set bound*—is one of the main tools for proving lower bounds on the nondeterministic communication complexity of boolean functions.

For sparse matrices, we have a somewhat better bound. Lemma 4.2 implies that, for any symmetric matrix A, the fraction $\text{trk}(A)^2/|A|$ is a lower bound on the tiling number $\chi(A)$ of A. It can be shown that this fraction is also a lower bound on the covering number $\text{Cov}(A)$.

Lemma 4.7. (Jukna and Kulikov 2009) *For every nonzero boolean matrix A,*

$$\text{Cov}(A) \geq \frac{\text{trk}(A)^2}{|A|}.$$

Proof. Take a largest set I of $|I| = \text{trk}(A)$ independent 1-entries in A, and let $R_1, \ldots, R_t$ be a covering of the 1-entries in A by $t = \text{Cov}(A)$ all-1 submatrices. Define a mapping $f : I \rightarrow \{1, \ldots, t\}$ by $f(x, y) = \min\{i : R_i[x, y] = 1\}$, and let $I_i = \{(x, y) \in I : f(x, y) = i\}$. That is, I_i consists of those independent 1-entries in I that are covered by the i-th all-1 submatrix R_i for the *first time*. Note that some of the I_i's may be empty, so let $I_1, \ldots, I_k$ be the nonempty ones. Say that an entry (x, y) is *spanned* by I_i if $(x, y') \in I_i$ for some column y' and $(x', y) \in I_i$ for some row x'.

Let S_i be the submatrix of R_i spanned by I_i. Hence, $S_1, \ldots, S_k$ are disjoint all-1 submatrices of A covering all 1-entries in I. Moreover, each S_i is an $r_i \times r_i$ matrix

---

[1]A boolean $m \times n$ matrix $A = (a_{ij})$ is the *adjacency matrix* of a bipartite $m \times n$ graph $G = (L \cup R, E)$ with parts $L = \{u_1, \ldots, u_m\}$ and $R = \{v_1, \ldots, v_n\}$, if $a_{ij} = 1$ if and only if u_i and v_j are adjacent in G.

with $r_i = |I_i|$. Since the S_i's are disjoint, we have that

$$r_1 + \cdots + r_k = |I| = \text{trk}(A) \quad \text{and} \quad r_1^2 + \cdots + r_k^2 \leq |A|.$$

By the Cauchy–Schwarz inequality,

$$\text{trk}(A)^2 = (r_1 + \cdots + r_k)^2 \leq k \cdot (r_1^2 + \cdots + r_k^2) \leq k \cdot |A|,$$

and the desired lower bound $t \geq k \geq \text{trk}(A)^2/|A|$ follows. $\qquad\qquad\square$

Remark 4.8. For all boolean matrices A with $|A| < \text{trk}(A) \cdot \omega(A)$ ones, Lemma 4.7 yields somewhat better lower bounds than those given by the fooling set bound (4.5). If, for example, an $n \times n$ matrix A contains an identity matrix and some constant number c of $r \times r$ all-1 matrices with $r = \sqrt{n}$, then Lemma 4.7 yields $\text{Cov}(A) \geq n^2/(cr^2 + n) = \Omega(n)$, whereas the fooling set bound (4.5) only yields $\text{Cov}(A) \geq n/r = \sqrt{n}$.

4.3 P = NP ∩ co-NP for Fixed-Partition Games

Having two modes (deterministic and nondeterministic) and having the (far-fetched) analogy with the P versus NP question, it is natural to consider the relations between the corresponding complexity classes. Here for convenience (and added thrill) we use the common names for the analogs of the complexity classes:

Let P (resp., NP) consist of all boolean functions in $2m$ variables whose deterministic (resp., nondeterministic) communication complexity is polynomial in $\log m$.

The *complement* of a boolean matrix A is the matrix $\overline{A} = A - J$, where J is the all-1 matrix (of the same dimension). Note that in the case of *deterministic* protocols, there is no difference what of the two matrices A or $\overline{A}$ we consider: we always have that $\mathfrak{c}(A) = \mathfrak{c}(\overline{A})$, because each deterministic protocol must cover all 0s as well as all 1s of A. In the case of *nondeterministic* communication, the situation is different in two respects:

1. we only need to cover the 1s of A, and
2. the submatrices need not be disjoint.

This is where an asymmetry between nondeterministic protocols for A and $\overline{A}$ comes from. And indeed, we have already seen that the nondeterministic communication complexities of the identity matrix and its complement are exponentially different.

But what if *both* A and $\overline{A}$ have small nondeterministic communication complexity—what can be then said about the *deterministic* communication complexity of A? This is a version of the famous P versus NP ∩ co-NP question in communication complexity. To answer questions of this type (in the communication complexity frame), we now give a general upper bound on the deterministic communication complexity of a matrix A in terms of the nondeterministic communication complexity of A and $\overline{A}$. This shows that P = NP ∩ co-NP holds for the fixed-partition communication complexity.

Theorem 4.9. (Aho et al. 1983) *For every boolean matrix A,*

$$c(A) \le 2 \max\{nc(A), nc(\overline{A})\}^2.$$

Proof. Let $\mathcal{R} = \mathcal{R}_0 \cup \mathcal{R}_1$ where $\mathcal{R}_0$ is a set of $|\mathcal{R}_0| \le 2^{nc(\overline{A})}$ all-0 submatrices covering all zeros of A, and $\mathcal{R}_1$ is a set of $|\mathcal{R}_1| \le 2^{nc(A)}$ all-1 submatrices covering all ones of A. Assign label "0" to all rectangles in $\mathcal{R}_0$, and label "1" to all rectangles in $\mathcal{R}_1$. It is clear that this is a legal labeling in the sense of Lemma 3.6, since every rectangle in $\mathcal{R}_0$ must be disjoint from every rectangle in $\mathcal{R}_1$. Hence, on a given input (x, y), the players have only to find out the label of a rectangle containing (x, y). By Lemma 3.6, this can be done using at most $2(\log |\mathcal{R}|)^2 \le 2 \max\{nc(A), nc(\overline{A})\}^2$ bits of communication.   □

Theorem 4.9 itself cannot be substantially improved. To show this, consider the k-disjointness matrix $D_{n,k}$. This is a 0-1 matrix whose rows as well as columns are labeled by all $\sum_{i=0}^{k} \binom{n}{i}$ subsets a of $[n]$ of size at most k; the entry in the a-th row and b-th column is defined by:

$$D_{n,k}[a, b] = \begin{cases} 0 & \text{if } a \cap b \ne \emptyset, \\ 1 & \text{if } a \cap b = \emptyset. \end{cases}$$

Theorem 4.10. (Razborov 1990) *If* $A = D_{n,k}$ *with* $k = \log n$ *then both* $nc(A)$ *and* $nc(\overline{A})$ *are* $\mathcal{O}(\log n)$, *but* $c(A) = \Omega(\log^2 n)$.

The theorem is a direct consequence of the following two lemmas. In the first lemma, rk(A) stands for the rank of A over GF(2).

Lemma 4.11. *The k-disjointness matrix has full rank,* rk($D_{n,k}$) = $\sum_{i=0}^{k} \binom{n}{i}$.

Proof. The matrix $D = D_{n,k}$ has $N = \sum_{i=0}^{k} \binom{n}{i}$ rows and as many columns. We must show that the rows of D are linearly independent over GF(2), i.e., that for any nonzero vector $\lambda = (\lambda_{I_1}, \lambda_{I_2}, \ldots, \lambda_{I_N})$ in GF(2)^N, indexed by subsets of $[n]$ of size at most k, we have $\lambda^T D \ne 0$. For this, consider the following polynomial:

$$f(x_1, \ldots, x_n) := \sum_{|I| \le k} \lambda_I \prod_{i \in I} x_i.$$

Since $\lambda \ne 0$, at least one of the coefficients λ_I is nonzero, and we can find some I_0 such that $\lambda_{I_0} \ne 0$ and I_0 is *maximal* in that $\lambda_I = 0$ for all $I \supset I_0$. Assume w.l.o.g. that $I_0 = \{1, \ldots, t\}$, and make in the polynomial f the substitution $x_i = 1$ for all $i \notin I_0$. After this substitution has been made, a *nonzero* polynomial over the first

t variables $x_1, \ldots, x_t$ remains such that the term $x_1 x_2 \cdots x_t$ is left untouched (here we use the maximality of I_0). Hence, after the substitution we obtain a polynomial which is 1 for some assignment $(a_1, \ldots, a_t)$ to its variables. But this means that the polynomial f itself takes the value 1 on the assignment $b = (a_1, \ldots, a_t, 1, \ldots, 1)$. Hence,

$$1 = f(b) = \sum_{|I| \leq k} \lambda_I \prod_{i \in I} b_i.$$

Let $J_0 := \{i : a_i = 0\}$. Then $|J_0| \leq k$ and, moreover, $\prod_{i \in I} b_i = 1$ if and only if $I \cap J_0 = \emptyset$, which is equivalent to $D_{I,J_0} = 1$. Thus,

$$\sum_{|I| \leq k} \lambda_I D_{I,J_0} = 1,$$

meaning that the J_0-th coordinate of the vector $\lambda^T \cdot D$ is nonzero. □

The lemma implies that the deterministic communication complexity of disjointness matrices is large:

$$c(D_{n,k}) \geq \log \mathrm{rk}(D_{n,k}) = \Omega(k \log(n/k)).$$

We are now going to show that both the matrix $D_{n,k}$ and its complement have small nondeterministic communication complexity.

Lemma 4.12. *For all $1 \leq k \leq n$ we have that*

$$\mathrm{nc}(\overline{D_{n,k}}) \leq \log n \quad and \quad \mathrm{nc}(D_{n,k}) = \mathcal{O}(k + \log \log n).$$

Proof. The first upper bound is trivial: just guess a point in the intersection of a and b. To prove the second one, we use the probabilistic argument.

The rows as well as columns of $D_{n,k}$ are labeled by subsets of $[n]$ of size at most k. Say that a subset $Y \subseteq [n]$ *separates* pair (a, b) of two disjoint sets a and b if $a \subseteq Y$ and $b \cap Y = \emptyset$. Let Y be a random subset of $[n]$ chosen uniformly with probability 2^{-n}. Then for a fixed pair (a, b),

$$\mathrm{Prob}[\text{Y does not separate } (a,b)] = 1 - \mathrm{Prob}[a \subseteq \text{Y and } b \cap \text{Y} = \emptyset]$$

$$= 1 - \frac{2^{n-|a|-|b|}}{2^n} = 1 - 2^{-|a|-|b|}.$$

Let $l := 2k4^k \ln n$, and take l independent copies $Y_1, \ldots, Y_l$ of Y. Then the probability that none of them separates a given pair (a, b) is at most

$$\left(1 - 2^{-|a|-|b|}\right)^l \leq \left(1 - 2^{-2k}\right)^l < e^{-l \cdot 2^{-2k}}.$$

Since there are no more than n^{2k} pairs (a,b), the probability that at least one of the pairs (a,b) is left unseparated by all the sets $Y_1,\ldots,Y_l$, is smaller than

$$n^{2k}\cdot e^{-l\cdot 2^{-2k}} = n^{2k}\cdot e^{-2k\ln n} = 1.$$

So, there must exists a sequence $Y_1,\ldots,Y_l$ of subsets of $[n]$ such that $D_{n,k}[a,b]=1$ iff (a,b) is separated by at least one of these sets. Since the set $\{(a,b)\colon a\subseteq Y_i, b\cap Y_i=\emptyset\}$ of all pairs separated by the i-th set Y_i corresponds to an all-1 submatrix of $D_{n,k}$, this implies $\mathrm{Cov}(D_{n,k})\le l$, and the desired upper bound $\mathrm{nc}(D_{n,k})\le\log l = \mathcal{O}(k+\log\log n)$ follows.   □

4.4 Clique vs. Independent Set Game

Recall that the tiling number of a boolean matrix A is $\chi(A)=\chi_0(A)+\chi_1(A)$, where $\chi_\sigma(A)$ for $\sigma\in\{0,1\}$ is the minimum number of pairwise disjoint all-σ submatrices of A covering all its σ-entries.

By (4.1) we know that, if *all* entries of a boolean matrix A can be decomposed into m disjoint monochromatic submatrices, then $\mathfrak{c}(A)\le 2(\log m)^2$. This implies that $\mathfrak{c}(A)\le 2(\log\chi(A))^2$. But what if we only know that all 1-entries of A can be decomposed in a small number of disjoint all-1 submatrices—does then $\mathfrak{c}(A)$ is small? It turns out that this is indeed the case: we can replace $\chi(A)$ in the above upper bound on $\mathfrak{c}(A)$ by either $\chi_1(A)$ or $\chi_0(A)$.

Theorem 4.13. *Let $\sigma\in\{0,1\}$. If the σ-entries of a boolean matrix can be decomposed into m pairwise disjoint all-σ submatrices, then $\mathfrak{c}(A)=\mathcal{O}(\log^2 m)$.*

We will derive the theorem from a more general result about the communication complexity of the following *clique versus independent set* game CIS_G on a given graph G:

- Alice gets a clique $C\subseteq V$ of G.
- Bob gets an independent set $I\subseteq V$ of G.
- Answer "1" iff $C\cap I=\emptyset$.

Note that we always have that $|C\cap I|\le 1$.

Theorem 4.14. (Yannakakis 1991) *The clique versus independent set problem on every n-vertex graph can be deterministically solved by communicating at most $\mathcal{O}(\log^2 n)$ bits.*

Proof. Given an n-vertex graph $G=(V,E)$ we describe an appropriate communication protocol for the game CIS_G. The protocol works in $\log n$ rounds, and in each round at most $\mathcal{O}(\log n)$ bits are communicated. The idea is to do binary search for an intersection.

First Alice checks whether her set C contains a vertex of degree $<n/2$. If it does then she sends the name of such a vertex v to Bob. Now they both know that Alice's

set is contained among v and vertices adjacent to v, and so the problem is reduced to one for a graph on at most $n/2$ vertices.

If Alice does not find such a node, then Bob checks whether his set I contains a vertex of degree $\geq n/2$. If it does then he sends the name of such a vertex w to Alice. Then they both know that Bob's set is contained among the vertices nonadjacent to w (including w), and the problem is again reduced to one for a graph on at most $n/2$ vertices.

If no one was successful, they know that *every* node of Alice's set has degree at least $n/2$ while *every* vertex of Bob's set has degree smaller than $n/2$, and hence they know that the two sets are disjoint. $\quad\square$

Proof of Theorem 4.13. Let $R_1, \ldots, R_m$ be a decomposition of 1-entries of A into m pairwise disjoint all-1 submatrices. Consider the graph G_A on m vertices $1, \ldots, m$ in which

$$i \text{ and } j \text{ are adjacent iff } R_i \text{ and } R_j \text{ intersect in rows.}$$

Now, given an input (x, y), Alice and Bob transform them to sets

$$C_x = \{i : x \text{ labels a row of } R_i\} \quad \text{and} \quad I_y = \{i : y \text{ labels a column of } R_i\}.$$

Note that C_x is a clique in G_A. Moreover, since the submatrices $R_1, \ldots, R_m$ are disjoint, I_y is an independent set in G_A. Further, $C_x \cap I_y \neq \emptyset$ iff (x, y) is in a 1-rectangle. Thus, the players can use the protocol for the clique versus independent game on G_A. $\quad\square$

The clique vs. independent sets game CIS_G is important to understand the power of linear programming for NP-hard problems. Namely, Yannakakis (1991) showed that any n-vertex graph G, for which this game requires $\omega(\log n)$ bits of *nondeterministic* communication, gives a super-polynomial lower bound for the size of linear programs expressing Vertex Packing and Traveling Salesman Problem polytopes.

Note that $\text{nc}(\neg \text{CIS}_G) \leq \log n$ for any n-vertex graph: just guess a vertex in the intersection. But for the problem CIS_G itself only graphs G with $\text{nc}(\text{CIS}_G) = \Omega(\log n)$ are known. Apparently, the highest so far remains the lower bound $(2 - o(1)) \log_2 n$ proved by Kushilevitz et al. (1999). The lower bound $\log_2 n$ is a trivial one (see Exercise 4.5).

■ **Research Problem 4.15.** Exhibit an n-vertex graph G such that $\text{nc}(\text{CIS}_G) = \omega(\log n)$.

Some partial result towards this problem were recently obtained by Kushilevitz and Weinreb (2009a,b).

Remark 4.16. The measure $\text{nc}(\text{CIS}_G)$ has an equivalent graph-theoretic formulation. For a graph G, let $q(G)$ be the smallest number t with the following property: There is a sequence $S_1, \ldots, S_t$ of subsets of V such that, for every clique $C \subseteq V$ and every independent set $I \subseteq V$ of G such that $C \cap I = \emptyset$, there is an i such that

$C \subseteq S_i$ and $I \cap S_i = \emptyset$. It can be shown (see Exercise 4.4) that

$$\mathrm{nc}(\mathrm{CIS}_G) = \log q(G). \tag{4.6}$$

Remark 4.17. The clique versus independent set game is related to the so-called Alon-Saks-Seymour Conjecture. The *chromatic number* of G, which we denote[2] by $\mathrm{chr}(G)$, is the smallest number of colors that are enough to color the vertices of G in such a way that no adjacent vertices receive the same color. The conjecture states that, if a graph G can be written as an union of m edge-disjoint complete bipartite graphs, then $\mathrm{chr}(G) \leq m + 1$. In this strong form, the conjecture was recently disproved by Huang and Sudakov (2011) who showed that $\mathrm{chr}(G) = \Omega(m^{1.2})$ for an infinite sequence of graphs. This yields a lower bound $\mathrm{nc}(\mathrm{CIS}_G) \geq 1.2 \log n - \mathcal{O}(1)$.

4.5   Communication and Rank

There is yet another upper bound, similar in its form to that of Theorem 4.9. Instead of $\mathrm{Cov}(A)$ it uses the following matrix parameter. Say that a boolean $t \times t$ matrix is *triangular* if—after a suitable rearrangement of the rows and columns—there are all 1s in the main diagonal and all 0s everywhere above the main diagonal; entries below the main diagonal may be arbitrary. For a boolean matrix A, define

$$\Delta(A) = \max\{t: A \text{ contains a } t \times t \text{ triangular submatrix}\}.$$

Theorem 4.18. (Lovász and Saks 1993) *For every boolean matrix A,*

$$\mathrm{c}(A) \leq (2 + \mathrm{nc}(\overline{A})) \cdot \log \Delta(A) \leq (2 + \mathrm{nc}(\overline{A})) \cdot \min\{\log \mathrm{rk}(A), \mathrm{nc}(A)\}. \tag{4.7}$$

Proof. Let $r = \mathrm{nc}(\overline{A})$, and let $A_1, \ldots, A_{2^r}$ be the all-0 submatrices of A covering all 0s of A. For every matrix A_i, consider the matrix R_i formed by the rows of A intersecting A_i, and C_i be the matrix formed by the columns of A intersecting A_i. Since A_i consists only of 0s, we have that (see Fig. 4.3 for a proof):

$$\Delta(R_i) + \Delta(C_i) \leq \Delta(A). \tag{4.8}$$

The protocol consists of $\log \Delta(A)$ rounds, in each of which at most $2 + r = 2 + \mathrm{nc}(\overline{A})$ bits are communicated.

In each round, the players do the following. First, Alice checks whether there is an index i such that her row intersects A_i and $\Delta(R_i) \leq \frac{1}{2}\Delta(A)$. If yes, then (using $1 + r$ bits) she sends "1" and the index i of this submatrix to Bob. If not, then

---

[2]In the graph-theoretic literature, the chromatic number of a graph is usually denoted by $\chi(G)$, but we already use this symbol for the tiling number.

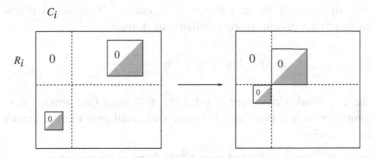

Fig. 4.3 Proof of (4.8): Since $A_i = R_i \times C_i$ is an all-0 submatrix, no triangular submatrix of R_i can share a row or a column with a triangular submatrix of C_i. Permute rows and columns of A to "glue" these triangular submatrices into a triangular submatrix of A

she sends "0". Now Bob checks whether there is an index i such that his column intersects A_i and $\Delta(C_i) \leq \frac{1}{2}\Delta(A)$. If yes, then (using $1 + r$ bits) he sends "1" and the index i to Alice. If not, then he sends "0".

If either Alice or Bob find a suitable index i in this round then, by communicating at most $2 + r$ bits, they have restricted the problem to a matrix A' ($= R_i$ or C_i) for which $\Delta(A') \leq \frac{1}{2}\Delta(A)$. Hence, in this case, the theorem follows by induction.

If both players have sent "0" in this round, then they can finish the protocol: the answer is "$A[x, y] = 1$". Indeed, if there were a 0 in the intersection of Alice's row and Bob's column, then this 0 would belong to some submatrix A_i. However, for this submatrix we have on the one hand $\Delta(R_i) > \frac{1}{2}\Delta(A)$ (since i did not suit Alice), on the other hand $\Delta(C_i) > \frac{1}{2}\Delta(A)$ since i did not suit Bob. But this contradicts (4.8).

Thus, we have shown that $c(M) \leq (2 + r) \cdot \log \Delta(A)$, as desired. To show the second inequality in (4.7), it suffices to observe that $\Delta(A) \leq \min\{\text{rk}(A), \text{Cov}(A)\}$ and $\text{nc}(A) = \log \text{Cov}(A)$. $\qquad\square$

Note that the proof of this theorem works not only for the matrix measure $\Delta(A)$ but also for any matrix measure $\mu(A)$ satisfying (4.8).

4.6 The Log-Rank Conjecture

We already know that $c(A) \geq \log \text{rk}(A)$ holds for any matrix A. But how tight is this lower bound? Lovász and Saks (1988) made the following conjecture.

Conjecture 4.19. (Log-Rank Conjecture) There is a constant $c > 0$ such that for every 0-1 matrix A,

$$c(A) \leq (\log \text{rk}(A))^c.$$

Here $\text{rk}(A)$ is the rank of A over the reals! If we consider rank over *finite* fields (instead of the reals), the conjecture does not hold (see Exercise 4.17).

The *area* of a matrix is the number of its entries. If mono(A) denotes the maximum area of a monochromatic submatrix of A, then

$$c(A) \geq \log \chi(A) \geq \log \frac{n^2}{\text{mono}(A)}.$$

Hence, the Log-Rank Conjecture implies the following (seemingly "easier" to tackle) conjecture stating that every 0-1 matrix of small rank must contain a large monochromatic submatrix.

Conjecture 4.20. (Nisan and Wigderson 1995) There is a constant $c > 0$ such that for every boolean $n \times n$ matrix A of rank r,

$$\text{mono}(A) \geq \frac{n^2}{2^{(\log r)^c}}.$$

In fact, Nisan and Wigderson (1995) showed that this last conjecture is also equivalent to the Log-Rank Conjecture (see Exercise 4.11). Moreover, they gave a support for Conjecture 4.20: every matrix of small rank must contain a submatrix of large "discrepancy" (see Theorem 4.25 below).

Since $\Delta(A) \leq \text{rk}(A)$ and $\text{nc}(A) \leq \log \text{rk}^+(A)$ (see Exercise 4.2), Theorem 4.18 implies that

$$c(A) \leq 2 \log \text{rk}^+(A) \cdot \log \text{rk}(A).$$

Thus the Log-Rank Conjecture is equivalent to the following purely mathematical question about the relation between the rank of boolean matrices and their non-negative rank.

Conjecture 4.21. (Positive Rank) There is a constant $c > 0$ such that for every 0-1 matrix A,

$$\log \text{rk}^+(A) \leq (\log \text{rk}(A))^c.$$

It is also known that the Log-Rank Conjecture is equivalent to the a conjecture stating that $\text{chr}(G) \leq 2^{(\ln r)^{O(1)}}$ for any graph G, where $\text{chr}(G)$ is the chromatic number of G and $r = \text{rk}(G)$ is the real rank of the adjacency matrix of G (see Theorem 4.28 below). Yet another algebraic analogue of the Log-Rank Conjecture was found by Valiant (2004).

The existence of so many seemingly unrelated but in fact equivalent formulations supports the importance of the Log-Rank Conjecture.

4.6.1 Known Gaps

For some time it was thought that $\text{chr}(G) \leq \text{rk}(G)$. This was conjectured by van Nuffelen (1976). The first counterexample to van Nuffelen's conjecture was obtained by Alon and Seymour (1989). They constructed a graph with chromatic number 32 and with an adjacency matrix of rank 29. Razborov (1992c) then showed that the gap between the chromatic number and the rank of the adjacency matrix can be super-linear, and Raz and Spieker (1995) showed that the gap can even be super-polynomial. The best result known so far is due to Nisan and Wigderson (1995). It gives an infinite family of graphs with rank r and with chromatic number $\mathrm{chr}(G) = 2^{\Omega(\log r)^{1+\epsilon}}$ for a constant $\epsilon > 0$.

Theorem 4.22. (Nisan and Wigderson 1995) *There exist explicitly given* 0-1 *matrices A of size $2^n \times 2^n$ such that $c(A) = \Omega(n)$, and $\log \mathrm{rk}(A) = \mathcal{O}(n^\alpha)$, where* $\alpha = \log_3 2 = 0.63\ldots$.

The same $\Omega(n)$ lower bound applies also to the randomized and to the non-deterministic communication complexities. The construction is based on boolean functions with high sensitivity and low degree. Such a function was constructed by Nisan and Wigderson (1994); see Lemma 2.5. The lower bound for the communication complexity relies on the known lower bounds for randomized communication complexity of disjointness matrices.

Proof. With every boolean function $f : \{0,1\}^n \to \{0,1\}$ we associate $2^n \times 2^n$ matrix A_f as follows:

$$A_f[x, y] := f(x_1 \cdot y_1, x_2 \cdot y_2, \ldots, x_n \cdot y_n).$$

We will need the following two properties if these matrices. Recall that a boolean function f is fully sensitive at $\mathbf{0}$ if $f(\mathbf{0}) = 0$ and $f(x) = 1$ for every vector with exactly one 1.   □

Claim 4.23. If f is fully sensitive at $\mathbf{0}$ then $c(A_f) = \Omega(n)$.

The same lower bound holds for the randomized and for the nondeterministic complexity of A_f.

Proof. This is a corollary of a deep result, due to Kalyanasundaram and Schnitger (1992) and Razborov (1992a), that the randomized (as well as deterministic and nondeterministic) communication complexity *unique disjointness matrix* UDISJ_n is $\Omega(n)$; see Theorem 4.37 and Exercise 4.20. The matrix UDISJ_n is a partial $2^n \times 2^n$ matrix whose rows and columns are labeled by distinct binary vectors of length n, and

$$\mathrm{UDISJ}_n[x, y] = \begin{cases} 1 & \text{if } \sum_{i=1}^n x_i \cdot y_i = 0 \\ 0 & \text{if } \sum_{i=1}^n x_i \cdot y_i = 1 \\ * & \text{otherwise.} \end{cases}$$

Now notice that if f is fully sensitive at $\mathbf{0}$ then any protocol for A_f is also a protocol for UDISJ_n.   □

Claim 4.24. If the polynomial of f has m monomials then $\mathrm{rk}(A_f) \le m$.

Proof. Let $f(z) = \sum_S \lambda_S \prod_{i \in S} z_i$ be the representation of f as a real polynomial. By the definition of A_f it follows that $A_f = \sum_S \lambda_S B_S$ where the matrix B_S is defined by $B_S[x, y] := \prod_{i \in S} x_i \cdot y_i$. Since each B_S has rank 1, the rank of A_f is at most the number m of all monomials $\lambda_S \prod_{i \in S} z_i$ with $\lambda_S \neq 0$. $\square$

To finish the proof of Theorem 4.22, take a boolean function $f : \{0,1\}^n \to \{0,1\}$ constructed in Lemma 2.5. This function is fully sensitive at $\mathbf{0}$, and contains $m \leq 2^{\mathcal{O}(n^\alpha)}$ monomials, where $\alpha = \log_3 2 \approx 0.631$. By Claim 4.23, we have that A_f has maximal communication complexity, $c(A_f) = \Omega(n)$. On the other hand, Claim 4.24 implies that $\log \text{rk}(A_f) = \mathcal{O}(n^\alpha)$. $\square$

4.6.2   Small Rank Implies Large Discrepancy

We now give a result supporting the Log-Rank Conjecture: every matrix of small rank must contain a submatrix of large "discrepancy".

The *discrepancy* of a ± 1 matrix is just the absolute value of the sum of its entries. Hence, small discrepancy means that the matrix is balanced: it has almost the same number of positive and negative entries. The *maximum discrepancy*, disc(A), of a matrix A is the maximum discrepancy of its submatrices. That is,

$$\text{disc}(A) = \max \left| x^T A y \right| = \max \left| \sum_{i,j=1}^n a_{ij} x_i y_j \right|, \tag{4.9}$$

where the maximum is over all 0-1 vectors x and y: each pair of such vectors corresponds to a submatrix of A determined by the 1-position of x and y. Note that $0 \leq \text{disc}(A) \leq n^2$ for every $n \times n$ matrix A.

The discrepancy of an $a \times b$ matrix does not exceed its area ab, and is equal to ab if the matrix is monochromatic. Hence, we always have that $\text{disc}(A) \geq \text{mono}(A)$. Interestingly, if we replace "maximal area of a monochromatic submatrix" by "maximal discrepancy of a submatrix", then the modified Conjecture 4.20 holds in a very strong sense!

Theorem 4.25. (Nisan and Wigderson 1995) *For every* $n \times n$ ± 1 *matrix* A *of rank* r,

$$\text{disc}(A) \geq \frac{n^2}{16r}.$$

Proof. We are given a ± 1 matrix $A = (a_{ij})$ of low rank $r = \text{rk}(A)$ and wish to find in it a submatrix of high discrepancy. By the definition (4.9) of the discrepancy, we only need to find 0-1 vectors x and y for which $x^T A y$ is large. As an intermediate step we shall consider the set

$$U = \{u \in \mathbb{R}^n : |u_i| \leq 1 \text{ for all } i\}$$

of real vectors of small maximum norm and show that $\text{disc}(A)$ can be lower bounded by the maximum of $u^T A v$ over the vectors $u, v \in U$.   □

Claim 4.26. For any $u, v \in U$ we have that

$$\text{disc}(A) \geq \frac{u^T A v}{4}.$$

Proof. Letting $z = Av$, we have $u^T A v = \sum_{i=1}^n u_i z_i$. Hence, $\sum_{i \in K} u_i z_i \geq u^T A v / 2$, where K is either the set of coordinates i where both u_i and z_i are positive or the set of coordinates in which both are negative. Assume the first case (the second case is similar by using vector $-v$ instead of v). Then letting $x \in \{0, 1\}^n$ to be the characteristic vector of K and using the fact that $|u_i| \leq 1$ for all i, we have

$$x^T A v \geq \sum_{i=1}^n x_i z_i = \sum_{i \in K} u_i z_i \geq u^T A v / 2.$$

Repeating this argument with $z = x^T A$, we can replace v with a 0-1 vector y obtaining that $x^T A y \geq u^T A v / 4$. Hence, $\text{disc}(A) \geq x^T A y \geq u^T A v / 4$, as claimed.   □

To finish the proof of the theorem, it is enough by Claim 4.26, to find two vectors $u, v \in U$ for which $u^T A v \geq n^2 / 4r$. For this, we will use a relation between spectral norm, Euclidean norm and the rank of a matrix.

The *spectral norm* $\|A\|$ of a matrix A is the maximum, over all unit vectors x, of the Euclidean norm $\|Ax\|$ of the matrix-vector product Ax. It is well known that $\|A\| = \max |y^T A x|$ over all real vectors $x, y \in \mathbb{R}^n$ whose Euclidean norm $\|x\| = \|y\| = 1$. The *Frobenius norm* $\|A\|_F$ of $A = (a_{ij})$ is just the Euclidean norm $(\sum_{i,j} a_{ij}^2)^{1/2}$ of A viewed as a vector in $\mathbb{R}^{n^2}$. For every real matrix A, we have the following relation of this norm with the spectral norm (see Lemma A.3 in Appendix A for the proof):

$$\frac{\|A\|_F}{\sqrt{\text{rk}(A)}} \leq \|A\| \leq \|A\|_F. \tag{4.10}$$

We will now construct the desired vectors $u, v \in U$ with $u^T A v \geq n^2 / 4r$. We start with two vectors $x, y \in \mathbb{R}^n$ of Euclidean norm $\|x\| = \|y\| = 1$ for which $x^T A y = \|A\|$. Let $p \geq 1$ be a parameter (to be specified later), and consider the sets of indices

$$I = \{i : |x_i| > 1/\sqrt{p}\} \quad \text{and} \quad J = \{j : |y_j| > 1/\sqrt{p}\}.$$

Since $1 = \|x\|^2 = \sum_{i=1}^n x_i^2 \geq |I|/p$, we have that $|I| \leq p$, and similarly, $|J| \leq p$. Consider the vectors a and b defined by:

$$a_i = \begin{cases} 0 & \text{if } i \in I, \\ x_i & \text{otherwise} \end{cases} \quad \text{and} \quad b_j = \begin{cases} 0 & \text{if } j \in J, \\ y_j & \text{otherwise}. \end{cases}$$

We claim that

$$a^T A b \geq \frac{n}{\sqrt{r}} - p. \tag{4.11}$$

To show this, consider the matrix B which agrees with A on all entries (i, j) with $i \in I$ and $j \in J$, and has 0s elsewhere. Then

$$a^T A b = x^T A y - x^T B y.$$

Since $\|A\|_F = n$, Eq. 4.10 yields $x^T A y \geq n/\sqrt{r}$. The same claim also yields $x^T B y \leq \|B\| \leq W(B) \leq p$, where the last inequality follows since B has at most p nonzero rows and p nonzero columns. So,

$$a^T A b = x^T A y - x^T B y \geq \frac{n}{\sqrt{r}} - p.$$

Set now $p := n/(2\sqrt{r})$, and consider the vectors $u := \sqrt{p} \cdot a$ and $v := \sqrt{p} \cdot b$. Since $|a_i|, |b_i| \leq 1/\sqrt{p}$ for all i, both vectors u, v belong to U, and we have

$$u^T A v = p \cdot a^T A b \geq p \cdot \frac{n}{\sqrt{r}} - p^2 = \frac{n^2}{4r}. \qquad \square$$

4.6.3   Rank and Chromatic Number

As mentioned above, the Log-Rank Conjecture in communication complexity is equivalent to the following conjecture for graphs. For a graph G, let $\mathrm{rk}(G)$ denote the rank of the adjacency matrix A_G of G. Recall that the *chromatic number* $\mathrm{chr}(G)$ of G is the smallest number of colors that are enough to color the vertices of G in such a way that no adjacent vertices receive the same color.

Conjecture 4.27. (Chromatic Number Conjecture) There exists a constant $c > 0$ such that $\log \mathrm{chr}(G) \leq (\log \mathrm{rk}(G))^c$ holds for every graph G.

Theorem 4.28. (Lovász and Saks 1988) *The log-rank conjecture and the chromatic number conjecture are equivalent.*

This result was only announced in Lovász and Saks (1988). The proof presented below was communicated to us by Michael Saks.

Proof. Let, as before, $\mathrm{rk}(G)$ denote the real rank of the adjacency matrix A_G of G. The two conjectures we are interested in state, respectively, that for every boolean matrix A and every graph G:

(a) $c(A)$ is at most poly-logarithmic in $\mathrm{rk}(A)$;
(b) $\log \mathrm{chr}(G)$ is at most poly-logarithmic in $\mathrm{rk}(G)$.

To show that (a) $\Rightarrow$ (b), recall that $\log \chi(A_G) \leq \mathfrak{c}(A_G)$. So, if $\mathfrak{c}(A_G)$ is poly-logarithmic in $\mathrm{rk}(A_G)$, then $\log \chi(A_G)$ is also poly-logarithmic in $\mathrm{rk}(A_G)$. It is therefore enough to show that $\mathrm{chr}(G) \leq \chi_0(A_G) \leq \chi(A_G)$. For this, let $R_1, \ldots, R_t$ be a decomposition of all 0-entries of the adjacency matrix A_G of $G = (V, E)$ into $t = \chi_0(A_G)$ all-0 submatrices. Define the mapping $c : V \to \{1, 2, \ldots, t\}$ by letting $c(v)$ be the unique i such that the diagonal entry (v, v) of A_G is covered by R_i. It is easy to see that each color-class $c^{-1}(i)$ is an independent set in G, implying that c is a legal coloring of G with $t = \chi_0(A_G)$ colors.

The proof of the other direction (b) $\Rightarrow$ (a) is less trivial. But it is a direct consequence of the following claim relating the chromatic number of graphs with the tiling number of boolean matrices. $\square$

Claim 4.29. For every boolean $m \times n$ matrix A there is a graph $G_A = (V, E)$ on $|V| = mn$ vertices such that

$$\mathrm{chr}(G_A) = \chi(A) \text{ and } \mathrm{rk}(G_A) = \mathcal{O}(\mathrm{rk}(A)^2).$$

Proof. For a graph G, let $\mathrm{cl}(G)$ denote its clique decomposition number, that is, the minimum number of vertex-disjoint cliques covering all vertices of G. Hence, $\mathrm{chr}(G) = \mathrm{cl}(\overline{G})$, where $\overline{G}$ is the complement of G, that is, two vertices are adjacent in $\overline{G}$ iff they are not adjacent in G.

Given a boolean $m \times n$ matrix A, consider the graph $H = (V, E)$ whose $|V| = mn$ vertices are entries $(i, j) \in [m] \times [n]$ of A. Two entries (i, j) and (k, l) are joined by an edge if and only if the $\{i, k\} \times \{j, l\}$ submatrix of A is monochromatic:

$$((i, j), (s, t)) \in E \text{ iff } A[i, j] = A[s, t] = A[i, t] = A[s, j]. \tag{4.12}$$

Note that $S \subseteq V$ is a clique in H iff the submatrix of A spanned by S (that is, the smallest submatrix covering all entries in S) is monochromatic. Hence, if we take the complement $G_A := \overline{H}$ of H, then $\mathrm{chr}(G_A) = \mathrm{cl}(H) = \chi(A)$.

To prove $\mathrm{rk}(G_A) = \mathcal{O}(\mathrm{rk}(A)^2)$, observe that $\mathrm{rk}(G_A) \leq \mathrm{rk}(H) + 1$. Hence, it is enough to upper-bound $\mathrm{rk}(H)$ in terms of $\mathrm{rk}(A)$. Recall that edges of H correspond to monochromatic $s \times t$ submatrices of A with $1 \leq s, t \leq 2$. Let $N = mn$, and consider the adjacency $N \times N$ matrix M of H with all diagonal entries set to 1. We will write M as the Hadamard (entry-wise) product of few $N \times N$ matrices. For $a \in \{0, 1\}$, let

$$C_a[(i, j), (s, t)] = 1 \text{ iff } A[i, j] = a$$

and

$$D_a[(i, j), (s, t)] = 1 \text{ iff } A[s, t] = a.$$

Observe that each row of C_a is either an all-1 (if $A[i, j] = a$) or an all-0 row (if $A[i, j] = 1 - a$). Similarly for columns of D_a. Hence, C_a and D_a have rank at most 1. Further, define

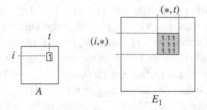

Fig. 4.4 Original $n \times n$ matrix A and the $n^2 \times n^2$ matrix E_1 with $E_1[(i, j), (s, t)] = 1$ iff $A[i, t] = 1$. Up to a permutation of rows and columns, the all-1 blocks of E_1 have the same distribution in E_1 as 1s in the original matrix A. Hence, $\mathrm{rk}(E_1) = \mathrm{rk}(A)$

$$E_a[(i, j), (s, t)] = 1 \text{ iff } A[i, t] = a$$

and

$$F_a[(i, j), (s, t)] = 1 \text{ iff } A[s, j] = a.$$

The matrices E_a and F_a consist of blocks of all-1 matrices, and the block structures are given by the matrix A (if $a = 1$) or by its complement $\overline{A}$ (if $a = 0$); see Fig. 4.4. So, their rank is equal to $\mathrm{rk}(A)$ (if $a = 1$) or to $\mathrm{rk}(\overline{A}) \leq 1 + \mathrm{rk}(A)$.

Recall that a componentwise product (or Hadamard product) of two matrices $A = (a_{ij})$ and $B = (b_{ij})$ is the matrix $A \circ B = (a_{ij} \cdot b_{ij})$. For such a product we have that $\mathrm{rk}(A \circ B) \leq \mathrm{rk}(A) \cdot \mathrm{rk}(B)$ (see Lemma A.1 in Appendix A for the proof). Now, by (4.12), we can write the matrix M as

$$M = C_0 \circ D_0 \circ E_0 \circ F_0 + C_1 \circ D_1 \circ E_1 \circ F_1,$$

implying that $\mathrm{rk}(M) \leq (1 + \mathrm{rk}(A))^2 + \mathrm{rk}(A)^2$.   □

To finish the proof of the implication (b) $\Rightarrow$ (a), assume (b) and take an arbitrary boolean matrix A. Take a graph G_A guaranteed by Claim 4.29. By (b), we have that $\log \chi(A) = \log \mathrm{chr}(G_A)$ is polylogarithmic in $\mathrm{rk}(A)$. Inequality (4.1) implies that $\mathfrak{c}(A)$ must also be polylogarithmic in $\mathrm{rk}(A)$.   □

4.7   Communication with Restricted Advice

Recall that $\mathrm{Cov}(A) \leq t$ iff all 1-entries of A can be covered by at most t all-1 submatrices. When doing this, one 1-entry of A may be covered many times. Let us now consider a version of this measure where the cover frequency is restricted. This corresponds to nondeterministic communication in which Carole cannot use one and the same witness for many inputs; this situation is usually referred to as a nondeterministic communication with a restricted number of advice bits.

Let $\mathrm{Cov}_k(A)$ be the smallest number of all-1 submatrices of A covering all its 1-entries in such a way that no 1-entry of A is covered by more than k of these submatrices. Let, as before, $\mathrm{rk}(A)$ denote the rank of A over the real numbers.

The following lemma is due to Grolmusz and Tardos (2003); a slightly weaker bound was proved earlier by Karchmer et al. (1994).

Lemma 4.30. *For every boolean matrix A and any integer positive integer k, we have $\mathrm{Cov}_k(A) \geq (k/3) \cdot \mathrm{rk}(A)^{1/k}$.*

Proof. Let $R_1, \ldots, R_t$ be $t = \mathrm{Cov}_k(A)$ boolean matrices of rank 1 such that $A = \bigvee_{i=1}^{t} R_i$ and $\sum_{i=1}^{t} R_i \leq kJ$, where J is the all-1 matrix. For a subset $I \subseteq \{1, \ldots, t\}$, let R_I be a boolean matrix with $R_I[x, y] = 1$ iff $R_i[x, y] = 1$ for *all* $i \in I$. By the inclusion-exclusion formula, we can write the matrix A as a linear ± 1 combination

$$A = \bigvee_{i=1}^{t} R_i = \sum_{I \neq \emptyset} (-1)^{|I|+1} R_I. \tag{4.13}$$

The condition $\sum_{i=1}^{t} R_i \leq kJ$ implies that $R_I = 0$ for all I of size $|I| > k$. Hence, the right hand of (4.13) has at most $\sum_{i=1}^{k} \binom{t}{i}$ nonzero terms. The subadditivity of rank yields

$$\mathrm{rk}(A) \leq \sum_{i=1}^{k} \binom{t}{i} \leq \left(\frac{et}{k} \right)^k,$$

from which the desired lower bound on $t = \mathrm{Cov}_k(A)$ follows. $\square$

Example 4.31. We show that the lower bound in Lemma 4.30 is almost tight. Let I be an identity $n \times n$ matrix with $n = 2^m$ for some m divisible by k, and let $\overline{I} = J - I$ be its complement. Then $\mathrm{rk}(\overline{I}) = n$, but we have that $\mathrm{Cov}_k(\overline{I}) \leq kn^{1/k}$. To see this, encode the rows and the columns by vectors $x \in \{0, 1\}^m$; hence, $\overline{I}[x, y] = 1$ iff $x \neq y$. Split the set $[m]$ into k disjoint subsets $S_1, \ldots, S_k$, each of size m/k. For every $j \in [k]$ and $a \in \{0, 1\}^{m/k}$, define the rectangle $R_{j,a}$ consisting of all pairs (x, y) such that

the projection of x onto S_j coincides with a and that of y doesn't.

These $k2^{m/k} = kn^{1/k}$ rectangles cover all 1s of $\overline{I}$, and each pair (x, y) with $x \neq y$ appears in at most k of them (since we take only k projections).

Together with Lemmas 4.30 and 4.18 implies that nondeterministic communication complexity with a small number k of witnesses cannot be much smaller than the deterministic communication complexity. Define

$$\mathrm{nc}_k(A) := \log \mathrm{Cov}_k(A).$$

Corollary 4.32. *For any boolean matrix A, $\mathfrak{nc}_k(A) = \Omega(\sqrt{\mathfrak{c}(A)/k})$.*

Proof. Since $\mathfrak{c}(A) = \mathfrak{c}(\overline{A})$ and $\mathrm{rk}(\overline{A}) \geq \mathrm{rk}(A) - 1$, Lemma 4.18 implies that $\mathfrak{c}(A)$ is at most about $\mathfrak{nc}(A) \cdot \log \Delta(\overline{A}) \leq \mathfrak{nc}_k(A) \cdot \log \Delta(\overline{A})$, and hence, at most about $\mathfrak{nc}_k(A) \cdot \log \mathrm{rk}(A)$. On the other hand, by Lemma 4.30, we have that $\mathfrak{nc}_k(A)$ must be at least about $(\log \mathrm{rk}(A))/k$. This implies $\log \mathrm{rk}(A) = \mathcal{O}(\mathfrak{nc}_k(A)/k)$, and hence, the desired lower bound on $\mathfrak{nc}_k(A)$ follows. $\square$

4.8   P $\neq$ NP $\cap$ co-NP for Best-Partition Games

If $f : \{0,1\}^{2n} \to \{0,1\}$ is a boolean function, then any balanced partition (x, y) of its variables into two blocks of equal size gives us a communication matrix M_f of f: this is a boolean $2^n \times 2^n$ matrix with $M_f[x, y] = f(x, y)$. The communication complexity of this matrix is then referred to as the communication complexity of f under this (particular) partition. Note however, that different partitions may result in different communication matrices of the same boolean function f.

The *deterministic* best-partition communication complexity $\mathfrak{c}^*(f)$ of f is the minimum, over all balanced partitions (x, y), of the deterministic communication complexity of M_f under partition (x, y). Let also $\mathfrak{nc}^*(f)$ denote the *nondeterministic* best-partition communication complexity of f. The best-partition communication complexity was introduced by Lipton and Sedgewick (1981).

Although historically the best-partition model of communication has received less attention than the fixed-partition model, the former one has larger applicability. This model naturally arises when dealing with time-space tradeoffs of VLSI chips; see, for example, Lengauer (1990). It (also naturally) arises in the context of branching programs. In fact, most of lower bounds for various restricted models of branching programs were obtained by proving (more or less explicitly) the corresponding lower bounds on the communication complexity of different types of best-partition protocols; see Wegener (2000) for a comprehensive description of such applications. Recently, Raz and Yehudayoff (2011) applied best-partition complexity to prove lower bounds for arithmetic circuits.

> For many functions, the possibility to choose a suitable partition can drastically reduce the number of communicated bits. For example, the equality function, defined by $f(x, y) = 1$ iff $x_i = y_i$ for all i, has maximal possible communication complexity equal to n (even nondeterministic), if the players are forced to use this "bad" partition (x, y). If, however, Alice receives the first half of x and y, and Bob receives the remaining variables, then they can locally test whether their pieces are equal and tell this the other player. Thus, under this "good" partition, just 2 bits of communication are enough!

Theorem 4.9 implies that P $=$ NP $\cap$ co-NP in the case of fixed partition games: if both the function f and its negation $\neg f$ have nondeterministic communication complexity at most t, then the deterministic communication complexity of f does not exceed $\mathcal{O}(t^2)$.

But what about best-partition complexity? The question is important because it exposes something about the power of lower bound arguments. We can prove a lower bound on the deterministic communication complexity of a function f by arguing about either f or $\neg f$. But if both the function and its negation have low nondeterministic complexity under *some* partitions of variables, other arguments are needed to show that the deterministic communication complexity must be large for *any* partition.

It turns out that *no* analogue of Theorem 4.9 holds in the best-partition case. Recall that in the best-partition case the players can choose different (most suitable) partitions for a function f and its negation $\neg f$. The following simple function, separating P from NP ∩ co-NP in the best-partition model of communication was used in (Jukna 2005).

To visualize the effect of the choice of a partition of input variables, we define our function f as a boolean function of n^2 variables, arranged into an $n \times n$ matrix $X = (x_{ij})$. Hence, inputs for f are 0/1 matrices $A : X \to \{0, 1\}$. We define $f(A)$ in such a way that a partition of the variables according to columns is suitable for computing f, and that according to rows is suitable for $\neg f$. Say that a row/column of a 0-1 matrix is *good* if it contains exactly two 1s, and *bad* otherwise. Define the function f by:

$f(A) = 1$ iff at least one row of A is good and all columns of A are bad.

Theorem 4.33. *Both* $\mathrm{nc}^*(f)$ *and* $\mathrm{nc}^*(\neg f)$ *are* $\mathcal{O}(\log n)$, *but* $\mathrm{c}^*(f) = \Omega(n)$.

Thus, for the best-partition games, we have P ≠ NP ∩ co-NP.

Proof. We first show that both $\mathrm{nc}^*(f)$ and $\mathrm{nc}^*(\neg f)$ are $\mathcal{O}(\log n)$. In the protocol for f Alice takes the first half of *columns* whereas in the protocol for $\neg f$ she takes the first half of *rows*. To compute $f(A)$ for a given matrix $A : X \to \{0, 1\}$, the protocol first guesses a row r (a candidate for a good row). Then, using 3 bits, Alice tells Bob whether all her columns are bad, and whether the first half of the row r contains none, one, two or more 1s. After that Bob has the whole information about the value $f(A)$ and can announce the answer. The negation $\neg f(A)$ can be computed in the same manner by replacing the roles of rows and columns.

Now we show that $\mathrm{c}^*(f) = \Omega(n)$ by a reduction to the disjointness function DISJ(x, y). Recall that this is a boolean function of $2n$ variables which outputs 1 iff $\sum_{i=1}^{n} x_i y_i = 0$. Since the general disjointness matrix has full rank (see Exercise 7.1), the lower bound (4.2) implies that the deterministic communication complexity of DISJ, as well as of ¬DISJ, under this partition is $\Omega(n)$. (In fact, even nondeterministic and randomized communication complexity of this function is $\Omega(n)$, but we will not use this fact.)

Take an arbitrary deterministic protocol for f. The protocol uses some balanced partition of the set X of variables into two halves where the first half is seen by Alice and the second by Bob. Recall that X is arranged into an $n \times n$ matrix.

Say that a column is seen by Alice (resp., by Bob) if Alice (resp., Bob) can see all its entries. A column is *mixed* if it is seen by none of the two players, that is, if each player can see at least one of its entries. Let m be the number of mixed columns. We consider two cases depending on how large this number m is. In both cases we

describe a "hard" subset of inputs, that is, a subset of input matrices on which the
players need to communicate many bits.

Case 1: $m < n/2$. In this case each player can see at least one column: if, say, Alice
had seen all $n - m$ non-mixed columns, then she would see than more than half of all
entries. Take one column x seen by Alice and another column y seen by Bob, and
let Y be the $(n - 3) \times 2$ submatrix of X formed by these two columns without the
last three rows. We restrict the protocol to input matrices $A : X \to \{0, 1\}$ defined as
follows. We first set all entries in the last three rows to 1. In this way we ensure that
all columns of A are already bad. Then we set all remaining entries of X outside Y
to 0. The columns x and y of Y may take arbitrary values. Such a matrix looks like:

$$\begin{pmatrix} x_1 & y_1 & \\ \vdots & \vdots & 0 \\ x_{n-4} & y_{n-4} & \\ 1 & \cdots & 1 \end{pmatrix}.$$

In each such matrix all columns are bad and, for $n \geq 3$, the last three all-1 rows
are also bad. Thus, given such a matrix, the players must determine whether any
of the remaining rows is good. Since all these rows have 0s outside the columns x
and y, this means that the players must determine whether $x_i = y_i = 1$ for some
$1 \leq i < n - 3$. That is, they must compute $\neg\text{DISJ}(x, y)$ which requires $\Omega(n)$ bits
of communication.

Case 2: $m \geq n/2$. Let Y be the $n \times m$ submatrix of Y formed by the mixed columns.
Select from the i-th column of Y one entry x_i seen by Alice and one entry y_i seen
by Bob. Since $m \leq n$ and we select only $2m$ entries, there must be a row r with
$t \leq 2$ selected entries. Let Y be the $n \times (m - t)$ submatrix consisting of the mixed
columns with no selected entries in the row r. We may assume that $m - t$ is odd and
that $m - t \leq n - 2$ (if not, then just include fewer columns in Y).

Now restrict the protocol to input matrices $A : X \to \{0, 1\}$ defined as follows.
First we set to 1 some two entries of the row r lying outside Y, and set to 0 all the
remaining entries of r. This ensures that the obtained matrices will already contain
a good row. After that we set all the remaining non-selected entries of X to 0. A
typical matrix looks like:

$$\begin{pmatrix} & & & & & 1 & 1 \\ x_1 & y_2 & & & & & \\ & & & & x_{n-t} & & \\ & x_2 & & & & & \\ y_1 & & & & & & \\ & & y_3 & & & & \\ & & x_3 & y_{n-t} & & & \end{pmatrix}$$

where r is the first row and all remaining entries are zeros.

Since each obtained matrix A contains a good row (such is the row r) and all columns outside the submatrix Y are bad (each of them can have a 1 only in the row r), the players must determine whether all columns of A in Y are also bad. Since all non-selected entries of Y are set to 0, the players must determine whether $x_i + y_i \leq 1$ for all $i = 1, \ldots, m - t$. Hence, the players must decide whether $\sum_{i=1}^{m-t} x_i y_i = 0$, that is, to compute the set-disjointness function $\mathrm{DISJ}(x, y)$, which again requires $\Omega(m - t) = \Omega(n)$ bits of communication.   □

4.9 Randomized Communication

In a *randomized* communication protocol, Alice and Bob are allowed to flip a coin. The coin can be public (seen by both players) or private. Alice and Bob are allowed to get a wrong result with probability smaller than some (fixed in advance) constant $\epsilon < 1/2$. That is, a randomized communication protocol P using a string r of random 0-1 bits is an ϵ-error protocol for a boolean matrix A if, for all entries (x, y), the probability that $P(x, y, r) \neq A[x, y]$ does not exceed ϵ. We will assume that the random string r is public (seen by both players); we will later show that this is not a restrictive assumption. We will also assume that the coin is *fair*, that is, each time 0 and 1 come with the same probability $1/2$. We assume this only for simplicity of presentation—most of the results also hold for any probability distribution.

For a boolean matrix A, let $c_\epsilon(A)$ denote the cost of the best randomized protocol for A that uses a public random string and errs with probability smaller than ϵ.

Example 4.34. A standard example of a matrix where randomization is much more powerful than nondeterminism is the $n \times n$ identity matrix I_n with $n = 2^m$. That is, $I_n[x, y] = 1$ iff $x = y$. Since the 1-entries of this matrix cannot be covered by fewer than 2^m all-1 submatrices, the nondeterministic communication complexity of I_n is m. On the other hand, the randomized communication complexity of I_n is constant!

Indeed, the players can pick a random string $r = (r_1, \ldots, r_m)$ in $\{0, 1\}^m$. Alice sends the scalar product $\langle r, x \rangle$, Bob checks whether $\langle r, y \rangle = \langle r, x \rangle$ and sends the answer. Since every nonzero 0-1 vector $v \neq \mathbf{0}$ is orthogonal over GF(2) to exactly half of all vectors, the error probability is $\epsilon = 1/2$: just take $v = x \oplus y$. To reduce the error to $\epsilon < 1/3$, just repeat the protocol several times and output the most frequent answer.

4.9.1 Distributional Complexity

Let us now look at how to prove that some matrices are hard for randomized protocols. Let A be a boolean matrix (a 0-1 matrix) with rows X and columns Y. The result of a randomized communication protocol of each input $(x, y) \in X \times Y$ is a *random variable*. To lower bound $c_\epsilon(A)$ from below, it is often easier to give a lower bound on a "dual" measure. Instead of requiring that, on each input (x, y), the randomized protocol can err with probability at most ϵ, we now consider *deterministic* protocols and require that they output correct value everywhere except an ϵ-fraction of inputs (x, y).

Namely, define the *ϵ-error distributional complexity*, $\text{distr}_\epsilon(A)$ of a matrix A as the smallest communication complexity of a deterministic protocol $P(x, y)$ computing $A[x, y]$ correctly on all but at most an ϵ-fraction of all inputs (x, y).

Proposition 4.35. $c_\epsilon(A) \geq \text{distr}_\epsilon(A)$.

Proof. Let P be a randomized protocol for A of cost $t = c_\epsilon(A)$, and let l be the number of random bits it uses. For every input (x, y), the protocol P must be correct for at least a $(1 - \epsilon)$ fraction of all 2^l choices of these random bits. Hence, there must be a choice $r \in \{0, 1\}^l$ after which the (deterministic) protocol P_r must be correct on at least a $(1 - \epsilon)$ fraction of all inputs.   □

As before, a rectangle is a set of the form $R = F \times G$ with $F \subseteq X$ and $G \subseteq Y$. Its *area* $|R|$ is the total number of entries in it. Let $A : X \times Y \to \{0, 1\}$ be a boolean matrix, and let $\mu > 0$ be its density, that is, the fraction of 1-entries in A.

We already know that, if $\text{nc}(A) \leq t$, then at least one 1-monochromatic submatrix occupies at least a fraction $\mu/2^t$ of the whole area of A. A similar result also holds for the randomized communication complexity. We only have to allow submatrices that are "nearly" 1-monochromatic in that almost all their entries are ones (see Exercise 4.19 for a weaker statement).

Fix an arbitrary constant $0 < \epsilon \leq \mu/4$, where μ is the density of A. Say that a rectangle R is *nearly 1-monochromatic* if $A[x, y] = 0$ for at most a fraction $4\epsilon/\mu$ of the entries (x, y) in R.

Lemma 4.36. (Yao 1983) *Let A be a boolean $m \times n$ matrix of density μ. If $c_\epsilon(A) \leq t$ then at least one nearly 1-monochromatic rectangle occupies at least a fraction $\mu/2^{t+2}$ of the whole area of A.*

Proof. Since $c_\epsilon(A) \leq t$, Proposition 4.35 implies that there exists a deterministic protocol P of cost at most t which is correct on all but a ϵ fraction of inputs. The protocol P decomposes our matrix A into at most $p = 2^t$ rectangles. On all entries of each of these rectangles the protocol outputs the same answer "0" or "1". We concentrate on only those rectangles on which the protocol gives answer "1". Let $Q = X \times Y$ be the set of all entries of A, and $T \subseteq Q$ the subset of these entries covered by the rectangles on which the protocol gives answer "1".

Since A has at least $\mu|Q|$ ones, and since $\epsilon < \mu/2$, the set T must cover at least half of ones of A. Indeed, otherwise more than half of the ones of A would be covered by rectangles giving wrong answer "0", which would result in more than $\frac{1}{2}\mu|Q| \geq \epsilon|Q|$ errors. Thus, $|T| \geq \frac{1}{2}\mu|Q|$.

Now let $T' \subseteq T$ be the set of entries covered by nearly 1-monochromatic rectangles. If the protocol gives answer "1" on a rectangle R, and if this rectangle is *not* nearly 1-monochromatic, then R wrongly covers more than $\frac{4\epsilon}{\mu}|R|$ zeros of A. Together with $|T| \geq \frac{1}{2}\mu|Q|$ and the disjointness of the rectangles, this already implies that $|T'| \geq \frac{1}{2}|T|$; otherwise, T would wrongly cover more than $\frac{1}{2} \cdot \frac{4\epsilon}{\mu}|T| \geq \epsilon|Q|$ zeros of A. Thus $|T'| \geq \frac{1}{2}|T| \geq \frac{\mu}{4}|Q|$ entries of A are covered by nearly 1-monochromatic rectangles. Since we have at most 2^t such rectangles, at least one of them must occupy at least a fraction of $\frac{\mu}{4}/2^t = \mu/2^{t+2}$ of the whole area $|Q|$, as claimed.   □

4.10   Lower Bound for the Disjointness Function

To give an explicit lower bound on the randomized communication complexity, let us consider the *disjointness matrix* D_n. This is a boolean $2^n \times 2^n$ matrix whose rows and columns are labeled by subsets $x \subseteq [n]$, and

$$D_n[x, y] = 1 \quad \text{iff} \quad x \cap y = \emptyset.$$

Note that the matrix D_n contains huge identity submatrices. If we take all rows labeled by subsets x of size $|x| = n/2$, and columns labeled by complements of such sets, then we obtain an $N \times N$ identity submatrix I_N with $N = \binom{n}{n/2}$. Thus,

$$\operatorname{nc}(D_n) \geq \log N = n - \mathcal{O}(\log n).$$

But this says nothing about the *randomized* communication complexity of D_n: we have already seen (Example 4.34) that the randomized communication complexity of the identity matrix is constant. This is why the following theorem is interesting.

Theorem 4.37. *For every sufficiently small constant $\epsilon > 0$, the ϵ-error randomized communication complexity of the disjointness matrix D_n is $\Omega(n)$.*

This result was first proved by Kalyanasundaram and Schnitger (1992); a simpler proof was then found by Razborov (1992a). In fact, Razborov's argument works for the *unique disjointness matrix* whose (x, y)-entry is 1 if $|x \cap y| = 0$, is 0 if $|x \cap y| = 1$, and is arbitrary otherwise (see Exercise 4.20 for a rough sketch of the proof). Razborov's proof is presented in the book by Kushilevitz and Nisan (1997). Using some ideas from Chakrabarti et al. (2001), Bar-Yossef et al. (2004) gave an information-theoretic proof.

We will present a simpler and intuitive proof of a weaker bound (but strong enough for most applications) $\Omega(\sqrt{n})$ obtained earlier by Babai et al. (1986).

Proof. We concentrate on the submatrix A of D_n with row-set X as well as column-set Y consist of all subsets of $[n]$ of size $s = \sqrt{n}$. The probability that two random s-element subsets x and y of $[n]$ are disjoint is

$$\binom{n}{s}\binom{n-s}{s}\binom{n}{s}^{-2} = \binom{n-s}{s}\binom{n}{s}^{-1} \geq \left(1 - \frac{s}{n}\right)^s.$$

Since the sets in X and in Y have size $s = \sqrt{n}$, a random pair (x, y) in $Q = X \times Y$ has probability about $1/e > 1/3$ to be disjoint. Thus, matrix A has density $\mu > 1/3$. Recall that $A[x, y] = 1$ iff $x \cap y = \emptyset$. Hence, if we set $\gamma := 4\epsilon/\mu$, then a rectangle $F \times G$ is nearly 1-monochromatic if

$$|\{(x, y) \in F \times G : x \cap y \neq \emptyset\}| \leq \gamma |G \times F|. \tag{4.14}$$

We will prove the lower bound by showing that there are no large nearly 1-rectangles. Specifically we will prove that for any rectangle $F \times G$, where at most a γ fraction of the pairs in $F \times G$ are intersecting, either $|F|$ or $|G|$ is small. By Lemma 4.36, it is enough to show that there is a constant $c > 0$ such that

$$|F \times G| \leq |X \times Y| \cdot 2^{-c\sqrt{n}} \tag{4.15}$$

holds for every nearly 1-rectangle. The argument is roughly the following. If F is small, then we are done. Otherwise, if $|F|$ is large, then there must be a large subset of F, where the union of these sets spans nearly all of $[n]$. But if $F \times G$ is nearly 1-monochromatic, this means that any subset $y \in G$ must avoid nearly all of $[n]$, and hence $|G|$ must be small. Now we proceed to the details. It suffices to show that

$$\text{if } |F| \geq |X| \cdot 2^{-c\sqrt{n}+1} \text{ then } |G| < |Y| \cdot 2^{-c\sqrt{n}}.$$

We focus on the set $F_1 \subseteq F$ of all rows $x \in F$ that intersect with at most a 2γ fraction of the y in G. Clearly $|F_1| \geq |F|/2$, for otherwise (4.14) would not hold. Since $|F_1|$ is still large, we claim that there exists $k = \sqrt{n}/3$ sets $x_1, \ldots, x_k$ in F_1 such that each x_p contains at least $r = \sqrt{n}/2$ *new* points relative to the union $z := x_1 \cup \cdots \cup x_{p-1}$ of the previous $p - 1$ sets. This can be proven by induction. Since $|x| = \sqrt{n}$ for each $x \in X$, we have that $|z| < p\sqrt{n} \leq n/3$. The number of $x \in X$ satisfying $|x \cap z| > r = \sqrt{n}/2$ is smaller than

$$n \binom{n/3}{r} \binom{2n/3}{r} \leq n \left(\frac{1}{3}\right)^r \left(\frac{2}{3}\right)^r \binom{n}{r}^2 \qquad \text{since } \binom{\delta n}{k} \leq \delta^k \binom{n}{k}$$

$$\leq n \left(\frac{2}{9}\right)^r 2^{2r} \binom{n}{2r}$$

$$= n \left(\frac{2\sqrt{2}}{3}\right)^{\sqrt{n}} \binom{n}{\sqrt{n}} \qquad \text{since } r = \sqrt{n}/2$$

$$< \binom{n}{\sqrt{n}} 2^{-c\sqrt{n}} \qquad \text{provided } 2^{-c} > \frac{2\sqrt{2}}{3}$$

$$= |X| 2^{-c\sqrt{n}}.$$

Since F_1 is large, this implies that $|x_p \cap z| < \sqrt{n}/2$ for some $x_p \in F_1$, as desired.

Now we have $k := \sqrt{n}/3$ sets $x_1, \dots, x_k$ in F_1 whose union is of size at least $k\sqrt{n}/2 = n/6$, and such that each of them intersects with only a few sets in G, namely each x_i intersects with at most a 2γ fraction of the y in G. Let $G_1 \subseteq G$ be the set of all columns $y \in G$ that intersect at most $l := 4\gamma k$ of the x_i. It is easy to verify that $|G| \leq 2|G_1|$. Indeed, if more than a half of the $y \in G$ were to intersect more than $4\gamma k$ of the x_i, then some x_i would intersect more than $2\gamma|G|$ of the $y \in G$, contradicting $x_i \in F_1$.

But this means that G_1 (and hence, also G) must be small. There are $\binom{k}{l}$ ways to select the l of the x_i which a set $y \in G_1$ is allowed to intersect. Then the union of the remaining x_i's has size at least

$$n/6 - l\sqrt{n} = n/6 - 4\gamma n/3 \geq n/9,$$

as long as $\gamma \leq 1/24$. Since this union must be avoided, we get

$$|G| \leq 2|G_1| \leq 2\binom{k}{l}\binom{n - n/9}{\sqrt{n}} = 2\binom{k}{4\gamma k}\binom{8n/9}{\sqrt{n}}$$

where, using $\binom{n}{k} \leq (en/k)^k$ and $\binom{n-r}{k}\binom{n}{k}^{-1} \leq e^{-kr/n}$,

$$\binom{k}{4\gamma k} \leq \left(\frac{ck}{4\gamma k}\right)^{4\gamma k} = \left(\frac{c}{4\gamma}\right)^{4\gamma\sqrt{n}/3} \leq e^{\sqrt{n}/18}$$

provided $\gamma \ln(e/4\gamma) \leq 1/24$, and

$$\binom{8n/9}{\sqrt{n}} = \binom{n - n/9}{\sqrt{n}} \leq e^{-\sqrt{n}/9}\binom{n}{\sqrt{n}}.$$

Hence, $|G|$ is at most $|Y| = \binom{n}{\sqrt{n}}$ times $e^{-\sqrt{n}/18+1} \leq 2^{-c\sqrt{n}}$, as desired. □

Remark 4.38. Beame and Lawry (1992) exhibited a boolean function f such that both f and $\neg f$ have small nondeterministic communication complexity whereas the randomized communication complexity of f is large:

$$c_\epsilon(f) = \Omega\left(\max\{nc(f), nc(\neg f)\}^2\right).$$

Remark 4.39. In Theorem 4.37 we required the error probability ϵ to be a sufficiently small positive constant. Actually, the same lower bound $\Omega(n)$ also holds for, say, $\epsilon = 1/3$. This can be shown using a general technique called amplification. Namely, assume that Alice and Bob have at their disposal a protocol of cost t that achieves $\epsilon = 1/3$. They can repeat it independently 1000 times and output at the end the most frequent answer. Then, by Chernoff bounds, the error probability of this repeated protocol of cost only $1000 \cdot t$ will not exceed 10^{-10}.

Remark 4.40. Håstad and Wigderson (2007) observed that the lower bound of Theorem 4.37 extends to submatrices of D_n as well: for every $k < n/2$, the randomized communication complexity of the submatrix $D_{n,k}$ of D_n, corresponding to k-element subsets of $[n]$, is $\Omega(k)$. They also proved that this lower bound is tight.

4.11 Unbounded Error Communication and Sign-Rank

So far we have considered randomized protocols where the error probability is bounded by some *fixed* constant $\epsilon < 1/2$. In an *unbounded-error* model of communication the error probability is not bounded by some constant given in advance. Instead of that it is only required that, for every input (x, y), the protocol outputs a correct value with probability *strictly larger* than $1/2$, for example, with probability at least $1/2 + 2^{-n}$. Let $R(A)$ denote the smallest cost of such a protocol for a matrix A.

An important restriction in unbounded-error communication model is that the random sources for both players must be *private*. This requirement is important because using *public* randomness we would have that $R(A) \leq 1$ for every(!) matrix A. Here is a communication protocol.

- The two players agree on an n-bit segment r of the public random coins.
- Alice compares her input x with r. If $x = r$, she sends Bob the bit 1, otherwise she sends him the bit 0.
- If Bob receives 1, he outputs $A[r, y]$. Otherwise, he outputs the result of a random unbiased coin flip.

Bob outputs the correct output with probability at least

$$2^{-n} \cdot 1 + (1 - 2^{-n}) \cdot \frac{1}{2} = \frac{1}{2} + 2^{-(n+1)} > \frac{1}{2}.$$

Therefore it is essential that the randomness source must be private.

Paturi and Simon (1986) established a surprisingly tight relation between $R(A)$ and the "sign rank" of A. In what follows, let $A = (a_{ij})$ denote an $m \times n$ ± 1 matrix. The function $\mathrm{sgn}(x)$, defined on the real numbers and called the *signum function* or sign function, is 1 for positive numbers $x > 0$, is -1 for negative numbers $x < 0$, and is 0 for $x = 0$. The *signum rank*, $\mathrm{signrk}(A)$, of A is the smallest possible rank over the reals of a matrix $B = (b_{ij})$ such that $\mathrm{sgn}(b_{ij}) = a_{ij}$ for all i, j. Thus, signum rank of A measures the robustness of the rank of A under sign-preserving changes; note that *every* entry is allowed to be changed!

Theorem 4.41. (Paturi and Simon 1986) *If* $\mathrm{signrk}(A) = r > 1$ *then*

$$\log_2 r \leq R(A) \leq \log_2 r + 1.$$

Due to this surprisingly tight connection, there were many attempts to find explicit matrices of high signum rank. In general, the signum rank of a matrix can be vastly smaller than its rank (see Example 4.21). Thus, bounding the signum rank from below is a considerable challenge.

That most of $n \times n$ ± 1 matrices have signum rank $\Theta(n)$ was first shown by Alon et al. (1985). Since then, finding an *explicit* matrix with signum rank more than logarithmic remained a challenge. Then, Forster (2002) achieved a breakthrough by proving that any $n \times n$ Hadamard matrix has signum rank at least $\Omega(\sqrt{n})$.

In this section we sketch this result. For this, we first recall that the (real) rank of a given real matrix A is the smallest number r such that B can be written as a matrix of scalar products of vectors in $\mathbb{R}^r$. More precisely, an *r-dimensional realization* of a matrix $A = (a_{ij})$ is a pair of sets $X = \{x_1, \ldots, x_m\}$ and $Y = \{y_1, \ldots, y_n\}$ of vectors in $\mathbb{R}^r$ such that

$$a_{ij} = \langle x_i, y_j \rangle = x_i^T y_j \text{ (scalar product)}.$$

If we relax this condition to

$$a_{ij} = \mathrm{sgn}(\langle x_i, y_j \rangle)$$

then we arrive to the concept of *r-dimensional arrangement* of A. From Linear Algebra we know that $\mathrm{rk}(A) \leq r$ iff A has an r-dimensional realization. Using this, one can easily show that $\mathrm{signrk}(A) \leq r$ iff A has an r-dimensional arrangement.

Recall that the *spectral norm* $\|A\|$ of a matrix A is the maximum, over all unit vectors x, of the Euclidean norm $\|Ax\|$ of the vector Ax. A vector x is a *unit vector* if its Euclidean norm $\|x\| = \sqrt{x_1^2 + \cdots + x_n^2}$ equals 1.

Theorem 4.42. (Forster 2002) *For every $m \times n$ ± 1 matrix A we have*

$$\mathrm{signrk}(A) \geq \frac{\sqrt{mn}}{\|A\|}.$$

Proof. Let $r = \mathrm{signrk}(A)$, and let $x_i, y_j \in \mathbb{R}^r$ be the corresponding vectors in an r-dimensional arrangement of A. We can assume that these vectors are unit vectors, and both the x_i and the y_i are in general position, that is, any r of them are linearly independent. The technical crux of Forster's argument is the fact (which we will not prove here) that the x_i can be assumed to be "nicely balanced" in the sense that they satisfy

$$\sum_{i=1}^{m} x_i x_i^T = \frac{m}{r} I_r \tag{4.16}$$

where I_r is the $r \times r$ identity matrix. (Recall the difference between the vector-products $x^T \cdot y$ and $x \cdot y^T$: the first is a number whereas the second is a matrix M with entries $M[i, j] = x_i y_j$.)

We will derive the theorem by giving upper and lower bounds on the quantity

$$\Delta := \sum_{j=1}^{n} \left(\sum_{i=1}^{m} |\langle x_i, y_j \rangle| \right)^2.$$

For a fixed column j we have:

$$\sum_{i=1}^{m} |\langle x_i, y_j \rangle| \geq \sum_{i=1}^{m} \langle x_i, y_j \rangle^2 \qquad \text{since } x_i, y_j \text{ are unit vectors}$$

$$= \sum_{i=1}^{m} y_j^T x_i x_i^T y_j \qquad \text{since } \langle x_i, y_j \rangle = x_i^T y_j = y_j^T x_i$$

$$= y_j^T \left(\sum_{i=1}^{m} x_i x_i^T \right) y_j$$

$$= \frac{m}{r} y_j^T I_r y_j = \frac{m}{r} \qquad \text{by (4.16)}$$

It follows that $\Delta \geq n(m/r)^2$. We will next show that $\Delta \leq m\|A\|$. Combining these two bounds, we obtain that $r \geq mn/\|A\|$ and the theorem is proved.

Since the vectors x_i and y_j form arrangement of $A = (a_{ij})$, we have that $|\langle x_i, y_j \rangle| = a_{ij} \cdot \langle x_i, y_j \rangle$. Hence, for any fixed column j,

$$\sum_{i=1}^{m} |\langle x_i, y_j \rangle| = \sum_{i=1}^{m} a_{ij} \cdot \langle x_i, y_j \rangle = \left\langle y_j, \sum_{i=1}^{m} a_{ij} x_i \right\rangle \leq \left\| \sum_{i=1}^{m} a_{ij} x_i \right\|,$$

by the Cauchy-Schwartz inequality $|\langle x, y \rangle| \leq \|x\| \cdot \|y\|$, since y_j is a unit vector. Thus,

$$\Delta \leq \sum_{j=1}^{n} \left\| \sum_{i=1}^{m} a_{ij} x_i \right\|^2 = \sum_{j=1}^{n} \left(\sum_{k=1}^{m} a_{kj} x_k^T \right) \left(\sum_{l=1}^{m} a_{lj} x_l \right)$$

$$= \sum_{k,l=1}^{m} (x_k^T x_l) \sum_{j=1}^{n} a_{kj} a_{lj} = \sum_{k,l=1}^{m} \langle x_k, x_l \rangle \cdot AA^T[k,l].$$

A symmetric $m \times m$ matrix P is *positive semi-definite matrix* if all its eigenvalues are non-negative. Equivalent definitions are: (i) $z^T P z \geq 0$ for all $z \in \mathbb{R}^m$, (ii) P is a *Gramian matrix* of some set of vectors $v_1, \ldots, v_n$, that is, $P[i,j] = \langle v_i, v_j \rangle$, and (iii) $P = AA^T$ for some matrix A. Of interest for us will be the following property of positive semi-definite matrices, known as Fejer's theorem:

A matrix P is positive semi-definite if and only if $\langle P, Q \rangle \geq 0$ for all positive semi-definite matrices Q.

Here $\langle P, Q \rangle$ is the scalar product of matrices P and Q when looked as vectors of length m^2. Now, the matrices $P[k,l] := \langle x_k, x_l \rangle$ and $Q = \|A\|^2 I_m - AA^T$ are positive semi-definite. By Fejer's Theorem, we have that

$$\sum_{k,l=1}^{m} P[k,l] \cdot Q[k,l] = \sum_{k,l=1}^{m} \langle x_k, x_l \rangle \cdot \left(\|A\|^2 I_m[k,l] - AA^T[k,l] \right) \geq 0.$$

Using this, the desired upper bound on Δ follows:

$$\Delta \leq \sum_{k,l=1}^{m} \langle x_k, x_l \rangle \|A\|^2 I_m[k,l] = \|A\|^2 \sum_{k=1}^{m} \langle x_k, x_k \rangle = \|A\|^2 m. \qquad \square$$

A *Hadamard matrix* is an $n \times n$ ± 1 matrix H such that $H^T H = n I_n$, where I_n is the $n \times n$ identity matrix (with ones on the diagonal and zeros elsewhere).

Corollary 4.43. *For every $n \times n$ Hadamard matrix H, signrk$(H) \geq \sqrt{n}$.*

Proof. By theorem 4.42, it is enough to verify that H has spectral norm $\|H\| \leq \sqrt{n}$. Recall that $\|H\|$ is the maximum, over all unit vectors x, of the Euclidean norm $\|Hx\|$ of the matrix-vector product Hx. On the other hand, for every unit vector x we have that

$$\|Hx\|^2 = \langle Hx, Hx \rangle = \langle x, (H^T H)x \rangle = \langle x, (nI_n)x \rangle = n\langle x, x \rangle = n. \qquad \square$$

The *inner product function* is a boolean function of $2m$ variables defined by $IP_m(x,y) = \sum_{i=1}^{m} x_i y_i$ mod 2. Since the $2^m \times 2^m$ ± 1 matrix M of this functions with entries $M[x,y] = (-1)^{IP_m(x,y)}$ is a Hadamard matrix, Corollary 4.43 implies that M has signum rank at least $2^{m/2}$. Together with Theorem 4.41, this implies that the randomized unbounded error communication complexity of the inner product function is $\Omega(m)$.

Recently, Razborov and Sherstov (2010) proved an important extension of Forster's result. Namely, they have exhibited a boolean function $f_m(x,y)$ of $n = 2m^3$ variables which *can* be computed by depth-3 circuits of size $\mathcal{O}(n)$ and such that its ± 1 matrix M with $M[x,y] = (-1)^{f_m(x,y)}$ has signum rank $2^{\Omega(n^{1/3})}$. The sets of variables x and y of this function are looked at as arranged into $m \times m^2$ matrices, and the function itself is defined by:

$$f_m(x,y) = \bigwedge_{i=1}^{m} \bigvee_{j=1}^{m^2} x_{ij} \wedge y_{ij}.$$

One ingredient in their proof is the following generalization of Theorem 4.42. They consider a more general notion of signum rank. For an $m \times n$ real matrix $A = (a_{ij})$, define its signum rank, signrk(A), as the minimum rank of a matrix $B = (b_{ij})$ such that $a_{ij} b_{ij} > 0$ for all i, j with $a_{ij} \neq 0$. That is, this time the matrix A may contain zero entries, and these entries may be arbitrarily manipulated.

Theorem 4.44. (Razborov and Sherstov 2010) *Let A be a real $m \times n$ matrix such that all but h of its entries have absolute value at least γ. Then*

$$\mathrm{signrk}(A) \geq \frac{\gamma m n}{\|A\|\sqrt{mn} + \gamma h}.$$

Note that, if A is a ± 1 matrix, then $\gamma = 1$ and $h = 0$.

Finally, note that if one takes an AND as a top gate, then the function $f_m(x, y)$ can be computed in depth three with $\mathcal{O}(n)$ gates. In Sect. 11.5 we will show that the situation changes drastically if we require the top gate be an OR gate: then an exponential number of gates is necessary.

4.12 Private vs. Public Randomness

In randomized protocols the players are allowed to use an additional information, namely the result r of random coin flips. A subtle question arises: how do players access this information? There are two possibilities:

1. *Public randomness*: the coins are flipped by a third player, and the result is seen by both players (with no additional communication).
2. *Private randomness*: the players must flip their coins privately; hence, some additional communication about the results of these flips may be necessary.

Example 4.34 shows that using public randomness the communication complexity of the identity matrix I_n can be reduced to a constant. If the random strings r are *private* (a much more realistic situation), the protocol is less trivial. Still, also in this case it is enough to communicate $\mathcal{O}(\log \log n)$ bits.

Alice picks a random prime number p between 1 and m^2, and sends $4 \log m$ bits encoding p as well as $x \bmod p$ to Bob. He checks whether $y \bmod p = x \bmod p$, and sends the answer to Alice. If $x = y$ the result is always correct. If $x \neq y$ the protocol may err. The protocol errs when Alice picks a prime number p such that p divides $|x - y|$. Since $|x - y| < 2^m$, there are at most $\log 2^m = m$ such "bad" primes numbers. On the other hand, the number of prime numbers in the interval $1, \ldots, k$ is at least $k / \ln k$. Hence, Alice is choosing her number p with equal probability from a collection of at least $\Omega(m^2 / \ln m^2)$ numbers. Therefore the error probability, that is, the probability to pick one of at most m "bad" primes is $\epsilon \leq (\ln m^2)/m \to 0$.

> We have completely ignored the subtle issue on *how* to choose a random prime number. But in the communication complexity the players are considered to be "superior beings", capable of performing any computation on their own data—only communication between the players is costly.

We have just seen that randomized protocols with private random bits have harder to do. Still, Newman (1991) proved that any randomized communication protocol with public random bits can be simulated by a protocol with private random bits at the cost of relatively small increase of the number of communicated bits. Let $c_\epsilon^{\mathrm{priv}}(A)$ denote the complexity of the best randomized protocol for A that uses private random strings and errs with probability smaller than ϵ.

Theorem 4.45. (Newman 1991) *For every boolean $n \times n$ matrix A and for every constant $\epsilon < 1/2$,*

$$c_{2\epsilon}^{\mathrm{priv}}(A) \leq c_\epsilon(A) + \mathcal{O}(\log \log n).$$

A similar argument to reduce the number of random bits was subsequently used by several authors, including Canetti and Goldreich (1993), Fleischer et al. (1995), and Sauerhoff (1999). Although the main trick is quite simple, it is usually hidden behind the technical details of a particular model of computation. Since the trick may be of independent interest, it makes sense to formulate it as a purely combinatorial lemma about the average density of 0-1 matrices. By a *row-density* of a boolean matrix H we will mean the maximum fraction of ones in each of its rows.

Lemma 4.46. *Let $\log m = o(\sqrt{n})$, $m > 4$, and let $0 \leq p < 1$ and $c > 0$ be constants. Let H be a boolean $m \times n$ matrix of row-density at most p. Then there is an $m \times l$ submatrix H' of H with $l = \mathcal{O}(\log m/c^2)$ columns and row-density at most $p + c$.*

Proof. Select $l = \lceil \log m/c^2 \rceil$ columns uniformly at random. Since two of the selected columns may coincide with probability at most $1/n$, and since we have only $\binom{l}{2} = o(n)$ pairs of selected columns, with probability $1 - o(1)$ all the selected columns are distinct. Next, fix a row x of H, and consider the 0-1 random variables $X_1, \ldots, X_l$ where X_j is the value of the bit of x in the j-th selected column; hence, $\mathrm{Prob}[X_i = 1] \leq p$ for all $j = 1, \ldots, l$. By Chernoff's inequality, the average density $(\sum X_i)/l$ of ones in the selected columns can exceed $p + \delta$ with probability at most $2e^{-2c^2 t} \leq 2/m^2$. Since we have only m rows, with probability at least $1 - 2/m > 1/2$, all the rows of the selected submatrix will have density at most $p + c$. $\qquad\square$

Proof of Theorem 4.45. Let P be an ϵ-error communication protocol for A using t public random bits. Let H be a boolean matrix whose $m = n^2$ rows correspond to inputs (x, y), 2^t columns corresponding to values $r \in \{0, 1\}^t$ of the random string, and $H[(x, y), r] = 1$ iff $P(x, y, r) \neq A[x, y]$. The matrix has row-density $p \leq \epsilon$. Taking $c = \epsilon$, Lemma 4.46 gives us $l = \mathcal{O}(\log n/c^2)$ strings $r_1, \ldots, r_l$ in $\{0, 1\}^t$ such that, for every input (x, y), the protocol P errs, that is, outputs value $P(x, y, r_i) \neq A[x, y]$ for at most an $\epsilon + c = 2\epsilon$ fraction for the r_i. Hence, by choosing an index $i \in \{1, \ldots, l\}$ uniformly at random, we obtain a 2ϵ-error communication protocol P' for A using only $\log l = \mathcal{O}(\log \log n)$ public random bits. Now, in the private randomness model of communication Alice can just flip that many random coins by herself, send the result of these flips to Bob, and then the two payers can proceed as in P'. $\qquad\square$

Exercises

4.1. There is a tree T with n nodes, known to both players. Alice has subtree T_A and Bob has subtree T_B. They wan to decide whether the subtrees have a common point. Show that $\mathcal{O}(\log n)$ bits of communication are enough.

Hint: Alice chooses a vertex x in her subtree and sends it to Bob. If x is not in Bob's subtree, he chooses the point y of T_B closest to x and sends it to Alice.

4.2. (Non-negative rank) A real matrix is *non-negative* if all its entries are non-negative. Recall that the *non-negative rank* of a non-negative $n \times n$ matrix A is the smallest number r such that A can be written as a product $A = B \cdot C$ of a non-negative $n \times r$ matrix B and a non-negative $r \times n$ matrix C. Show that $\operatorname{nc}(A) \leq \log \operatorname{rk}^+(A)$ holds for every boolean matrix A.

Hint: Note that $\operatorname{Cov}(A) \leq t$ holds if and only if there exist boolean matrices $B_1, \ldots, B_t$ of rank 1 such that the (x, y)-entry of the matrix $B = \sum_{i=1}^t B_i$ is 0 iff $A[x, y] = 0$.

4.3. (Threshold matrices) Let A be a boolean $n \times n$ matrix whose rows and columns are subsets of $[r] = \{1, \ldots, r\}$, and whose entries are defined by: $A[x, y] = 1$ iff $|x \cap y| \geq k$. Show that either (i) A contains an all-1 submatrix with at least $n^2/4\binom{r}{k}^2$ entries, or (ii) A contains an all-0 submatrix with at least $n^2/4$ entries.

Hint: Let $\alpha = 1/2\binom{r}{k}$ and call a subset $S \subseteq [r]$ row-popular (resp., column-popular) if S is contained in at least αn subsets corresponding to rows (resp., to columns) of A. Look at what happens if at least one k-element subset of $[r]$ is both row-popular *and* column-popular, at what happens when this is not the case.

4.4. (Clique vs. independent set game) Prove the equality (4.6).

Hint: To prove $\log q(G) \leq \operatorname{nc}(\operatorname{CIS}_G)$, take a matrix A whose rows correspond to cliques and columns to independent sets of G. Take a covering $R_1, \ldots, R_t$ of ones of A by all-1 submatrices of A. Let S_i be the union of all vertices appearing in at least one clique corresponding to the rows of R_i, and let T_i be the union of all vertices appearing in at least one independent set corresponding to the columns of R_i. Show that $S_i \cap T_i = \emptyset$.

4.5. Prove that $\operatorname{nc}(\operatorname{CIS}_G) \geq \log n$ for every graph G on n vertices.

Hint: Consider each single vertex as a clique as well as an independent set.

4.6. (Intersection dimension) For a boolean matrix A, define its *intersection dimension*, $\operatorname{Int}(A)$, as the smallest number d with the following property: the rows and columns x of A can be labeled by subsets $f(x) \subseteq \{1, \ldots, d\}$ such that $A[x, y] = 1$ iff $f(x) \cap f(y) \neq \emptyset$. Show that $\operatorname{Cov}(A) = \operatorname{Int}(A)$.

Hint: Let $A = \bigvee_{i=1}^d B_i$ be a covering of all 1-entries of A by boolean matrices of rank 1. Each B_i consists of an $I_i \times J_i$ submatrix of 1s, and 0s outside this submatrix. Assign to each row x of A the set $f(x) = \{i : x \in I_i\}$ and to each column y the set $f(y) = \{i : y \in J_i\}$.

4.7. (Generalized covering number) For a boolean matrix A, let $\operatorname{Cov}_\&(A)$ be the smallest number t such that A can be written as a componentwise AND $A = \bigwedge_{i=1}^t A_i$ of t boolean matrices such that $\operatorname{Cov}(A_i) \leq t$ for all i. Show that boolean $n \times n$ matrices A with $\operatorname{Cov}_\&(A) = \Omega(\sqrt{n})$ exist.

Hint: Exercise 4.6.

4.8. ■ **Research Problem.** Exhibit an explicit boolean $n \times n$ matrix A with $\text{Cov}_\&(A) = n^{\Omega(1)}$.

Comment: This would resolve at least two old problems in circuit complexity: give an explicit boolean function f_{2m} in $2m = 2\log n$ variables requiring: (i) depth-3 circuits of size $2^{\Omega(m)}$, and (ii) a superlinear number of fanin-2 gates in any log-depth circuit. Why this is so will be sketched later in Sect. 11.6.

4.9. ■ **Research Problem.** Say that a boolean matrix is *square-free* if it does not contain any 2×2 zero submatrix. If B is a boolean square-free $n \times n$ matrix with at least dn zeros, does then $\text{Cov}(B) = d^{\Omega(1)}$?

Comment: A positive answer would resolve Problem 4.8 because explicit square-free matrices A with $d = \Omega(\sqrt{n})$ zeros in each row and each column are known: such are, for example, complements of adjacency matrices of dense graphs without four-cycles (see Examples 7.4 and 7.5). Since adding new 1s cannot destroy the square-freeness, this would imply that $\text{Cov}_\&(A) = d^{\Omega(1)}$.

4.10. Let $L : \mathbb{N}^2 \to \mathbb{N}$ be a function such that $L(x,0) = L(0,y) = 1$ and $L(x,y) \le L(x,y/2) + L(x - x/K, y)$. Show that $L(x,y) \le K^{O(\log y)}$.

Hint: Use induction or show that $L(x,y)$ is at most the number of binary strings of length $\le K + \log x$ with $\le K$ ones and $\le \log x$ zeros.

4.11. (Conjecture 4.20 implies Log-Rank Conjecture) Suppose that Conjecture 4.20 is true. That is, assume that there exists a constant $c > 0$ such that every boolean matrix of area m and rank r contains a monochromatic submatrix of area at least $m/2^{(\log r)^c}$. Use this to show that $c(M) = O((\log r)^{c+1})$ for every boolean matrix M, where $r = \text{rk}(M)$.

Hint: Let A be a largest monochromatic submatrix of M. Up to permutation of rows and columns the matrix M has the form $M = \begin{bmatrix} A & B \\ C & D \end{bmatrix}$. Assume that $\text{rk}(B) \le \text{rk}(C)$. Use $\text{rk}(B) + \text{rk}(C) \le M + 1$ to show that $\text{rk}(A|B) \le 2 + \text{rk}(M)/2$. In the communication protocol let Alice send a bit saying if her input belongs to the rows of A or not. Then continue recursively with a protocol for the submatrix $[A|B]$, or for the submatrix $[C|D]$, according to the bit communicated. If $L(m,r)$ denotes the number of leaves of this protocol, starting from a matrix of area at most m and rank at most r, then $L(m,r) \le L(m,2+r/2) + L(m - \delta m, r)$ where $\delta = 2^{-(\log r)^c}$. Use Exercise 4.10 to show that $L(m,r)$ is at most exponential in $O((\log r)^{c+1})$.

4.12. (Quadratic forms) Let B be an $m \times m$ matrix over GF(2) of rank r, and let A be a boolean $n \times n$ matrix with $n = 2^m$ whose rows and columns correspond to vectors in GF(2)^m. The entries of A are defined by $A[x,y] = y^T B x$ over GF(2). Show that, if $s + t > 2m - r$, then the matrix A has no monochromatic $2^s \times 2^t$ submatrix.

Hint: Let X be a set of $|X| \ge 2^s$ rows and Y a set of $|Y| \ge 2^t$ columns of A, and suppose that A is monochromatic on $X \times Y$. We can assume that $A[x,y] = 0$ for all $(x,y) \in X \times Y$ (why?). Let H be the subspace generated by X, and G the subspace generated by Y. Argue that the subspaces BH and G are orthogonal, hence, $\dim(BH) + \dim(G) \le m$. Combine this with $\dim(BH) \ge \dim(H) - (m - r)$ to deduce that $\dim(H) + \dim(G) \le 2m - r$. Use $s + t > 2m - r$ to get a contradiction.

4.13. (Lower bounds via discrepancy) The maximum discrepancy, disc(A), of a boolean matrix A is the maximum discrepancy of its ± 1 version A' defined by $A'[x, y] = (-1)^{A[x,y]}$. That is, disc(A) is the maximum, over all submatrices B of A, of the absolute value of the difference between the number of 1s and the number of 0s in B. Prove that matrices of small discrepancy have large distributional, and hence, also randomized communication complexity: for every boolean $n \times n$ matrix A and for every constant $0 \le \epsilon < 1/2$,

$$\mathrm{distr}_\epsilon(A) \ge \log \frac{(1 - 2\epsilon)n^2}{\mathrm{disc}(A)}.$$

Hint: Fix a deterministic protocol $P(x, y)$ of cost $c = \mathrm{distr}_\epsilon(A)$ which correctly computes A on all but an ϵ-fraction of inputs. Let B be a ± 1 matrix of errors: $B[x, y] = +1$ if $P(x, y) = A[x, y]$, and $B[x, y] = -1$ otherwise. Show that the discrepancy of this matrix (the absolute value of the sum of all its entries) is at least $(1 - \epsilon)n^2 - \epsilon n^2 = (1 - 2\epsilon)n^2$. The protocol P decomposes A (and hence, also B) into $t = 2^c$ submatrices $R_1, \ldots, R_t$ such that the discrepancy of B is at most the sum of the discrepancies of these matrices.

4.14. (Discrepancy and spectral norm) Let A be an $n \times n$ matrix with real entries, and $\|A\|$ its spectral norm, that is, $\|A\| = \max |u^T A v|$ over all real vectors $u, v \in \mathbb{R}^n$ with $\|u\| = \|v\| = 1$. Show that disc(A) $\le \|A\| \cdot n$.

Hint: Consider an $S \times T$ submatrix of A of maximal discrepancy. Show that $|\chi_S^T A \chi_T| \le \|A\|\sqrt{|S||T|}$, where χ_S is the characteristic vector of the set $S \subseteq [n]$.

4.15. (Discrepancy and eigenvalues) Let A be an $n \times n$ ± 1-matrix. Suppose that A is symmetric ($A^T = A$), and let λ be the largest eigenvalue of A. Show that disc(B) $\le \lambda\sqrt{ab}$ for every $a \times b$ submatrix B of A.

Hint: If x and y are characteristic 0-1 vectors of the rows and columns of B, then disc(B) $= |x^T A y|$. Take an orthonormal basis $v_1, \ldots, v_n$ for $\mathbb{R}^n$ corresponding to the eigenvalues $\lambda_1, \ldots, \lambda_n$ of A, write x and y as linear combinations $x = \sum_i \alpha_i v_i$ and $y = \sum_i \beta_i v_i$ of these basis vectors, and show that $|x^T A y| = |\sum_i \alpha_i \beta_i \lambda_i| \le \lambda |\sum_i \alpha_i \beta_i|$. Observe that $\sum_i \alpha_i^2 = \|x\|^2 = a$ and $\sum_i \beta_i^2 = \|y\|^2 = b$, and apply the Cauchy–Schwarz inequality to derive the desired upper bound on disc(B).

4.16. (Inner product function) Let S_m be a $2^m \times 2^m$ Sylvester matrix. Its rows and columns are labeled by vectors in GF(2)^m, and the entries of S_m are the scalar products of these vectors over GF(2). Hence, the function corresponding to this matrix is the inner product function over GF(2). Show that disc(S_m) $\le 2^{m/2}$, and hence, that $c_\epsilon(S_m) = \Omega(m)$ for any constant $\epsilon > 0$.

Hint: Use Lindsey's Lemma.

4.17. (Log-Rank Conjecture) Use the Sylvester matrix S_m to conclude that if we consider rank over finite fields (instead of the reals) the gap between $c(S_m)$ and $\log \mathrm{rk}(S_m)$ may be exponential.

4.18. (The greater-than function) Consider the following $GT_n(x, y)$: Alice gets a non-negative m-bit integer x, Bob gets a non-negative m-bit integer y, and their goal is to decide whether $x \ge y$. Show that $c_{1/m}(GT_n) = \mathcal{O}(\log^2 m)$.

Hint: Let the players recursively examine segments of their strings until they find the lexicographi-
cally first bit in which they differ—this bit determines whether $x \geq y$. Alice can randomly select a
prime number $p \leq m^3$, compute x' mod p where x' is the first half of x, and send p and x' mod p
to Bob; this can be done using $\mathcal{O}(\log m)$ bits. If x' mod $p \neq y'$ mod p, then x' is different from
y', and the players can continue on the first half of their strings. Otherwise the players assume that
$x' = y'$, and they continue on the second half of their strings. The players err in this later case
when $x' \neq y'$ but x' mod $p = y'$ mod p. Estimate the probability of this error, keeping in mind
that there are $\Theta(n/\ln n)$ primes $p \leq n$.

4.19. (Almost monochromatic rectangles) Let $A : X \times Y \to \{0, 1\}$ be a boolean
matrix of area $m = |X| \cdot |Y|$. Suppose that $c_\epsilon(A) \leq t$. Show that there exists a
submatrix B of A of area at least $m/2^{t+1}$ and bit $a \in \{0, 1\}$ such that at least a
$1 - 2\epsilon$ fraction of the entries in B are equal to a.

Hint: Call a submatrix *large* if its area is at least $m/2^{t+1}$, and *small* otherwise. Take a deterministic
protocol P of cost $t = c_\epsilon(A)$ which is correct on all but a ϵ fraction of inputs. The protocol P
decomposes A into at most 2^t rectangles. Argue that more than half of the entries of A must
belong to large rectangles. Recall that at most an ϵ fraction of the entries (x, y) of A do not satisfy
$P(x, y) = A[x, y]$. Even if these entries are in large rectangles, their fraction in the set of all
entries of large rectangles is at most 2ϵ (since more than half of the entries of A belong to large
submatrices).

4.20. (Disjointness matrix) Let D_n be the $2^n \times 2^n$ disjointness matrix. Its rows and
columns are labeled by subsets $x \subseteq [n]$, and $D_n[x, y] = 1$ iff $x \cap y = \emptyset$. Let T_0
(resp., T_1) be the set of all pairs (x, y) of subsets of $[n]$ such that $|x \cap y| = 1$ (resp.,
$|x \cap y| = 0$). Note that $D_n[x, y] = 0$ for all $(x, y) \in T_0$, and $D_n[x, y] = 1$ for all
$(x, y) \in T_1$. For $a \in \{0, 1\}$ and a rectangle R, let $\mu_a(R) = |R \cap T_a|/|T_a|$ be the
fraction of elements of T_a in R. Razborov (1992a) proved that $\mu_1(R)$ is not much
smaller than $\mu_0(R)$, unless $\mu_0(R)$ is negligible:

For every rectangle R in D_n, if $\mu_0(R) > 2^{-n/100}$ then $\mu_1(R) > \mu_0(R)/100$.

This implies that $\mu_1(R) > \mu_0(R)/100 - 2^{-n/100}$. Use this fact to show that $c_\epsilon(D_n) =
\Omega(n)$ for every constant $\epsilon < 1/200$.

Hint: Let $t = c_\epsilon(D_n)$, and take a deterministic protocol P of cost $t = c_\epsilon(D_n)$ which is correct on
all but an ϵ fraction of inputs. Argue as in Exercise 4.19 to show that P outputs the right answer
for all but a 2ϵ fraction of inputs in T_0 and all but a 2ϵ fraction of inputs in T_1. The protocol P
decomposes A into at most 2^t rectangles. Let $R_1, \ldots, R_m$ be the rectangles where $P(x, y) = 0$.
Then $\sum_{i=1}^{m} \mu_0(R_i) \geq 1 - 2\epsilon$ (why?). Use the fact above to show that $\sum_{i=1}^{m} \mu_1(R_i) \geq (1 -
2\epsilon)/100 - 2^{t-n/100}$. What happens if (say) $t < n/200$?

4.21. (Rank versus sign-rank) Let A be an $n \times n$ ± 1 matrix of the greater-than
function, that is, $A[i, j] = 1$ if $i \leq j$, and $A[i, j] = -1$ if $i > j$. Show that
$\mathrm{rk}(A) \geq n - 1$ but $\mathrm{signrk}(A) \leq 2$.
Hint: Consider the matrix B with $B[i, j] = 2(j - i) + 1$.

4.22. (Randomization and seminorms) Let Φ be a measure assigning every real
matrix a real number. Call such a measure *seminorm* if for any two non-negative
matrices A, B of the same dimension and any real number $c \geq 0$ we have that
$\Phi(A + B) \leq \Phi(A) + \Phi(B)$ and $\Phi(cA) \leq c\Phi(A)$. A seminorm Φ is *normalized* if,
for every boolean matrix A, $\Phi(A)$ does not exceed the tiling number $\chi(A)$ of A. An

ϵ-*approximator* of A is a matrix B such that $|A[x, y] - B[x, y]| \leq \epsilon$ for all entries (x, y). Define

$$\Phi_\epsilon(A) := \min\{\Phi(B) \colon B \text{ is an } \epsilon\text{-approximator of } A\}.$$

Show that, for every normalized seminorm Φ, $c_\epsilon(A) \geq \log \Phi_\epsilon(A)$.

Hint: Let P be a randomized ϵ-error communication protocol for A of cost $c_\epsilon(A)$. If we fix the random string r used by P to some r, then what we get is a deterministic protocol; let A_r be a boolean matrix of answers of this protocol, and let $p_r = \text{Prob}[r = r]$. Consider the convex combination $A_P := \sum_r p_r A_r$. What are the entries of this matrix? Show that $\Phi_\epsilon(A) \leq \Phi(A_P) \leq \sum_r p_r \Phi(A_r) \leq \sum_r p_r \chi(A_r) \leq \sum_r p_r 2^{c(A_r)} \leq 2^{c_\epsilon(A)}$.

4.23. Let $A : X \times Y \rightarrow \{0, 1\}$ be a boolean $n \times n$ matrix, and suppose that all entries of A can be covered by m not-necessarily disjoint monochromatic rectangles $R_1, \ldots, R_m$. Thus, both $\text{nc}(A)$ and $\text{nc}(\overline{A})$ are at most $\log m$. For every row $x \in X$ define the vector $a_x \in \{0, 1\}^m$ by $a_x(i) = 1$ iff $x \in R_i$. Consider the monotone boolean function $f : \{0, 1\}^m \rightarrow \{0, 1\}$ by: $f(x) = 1$ iff $z \geq a_x$ for some $x \in X$. Show that $\text{Depth}_+(f) \geq c(A)$.

Hint: Take a communication protocol for the Karchmer-Wigderson game on the function f. Given $x' \in f^{-1}(1)$ and $y' \in f^{-1}(0)$, the goal is to find a position i such that $x'_i \neq y'_i$. Use this protocol to design a deterministic communication protocol for the matrix A, where Alice gets a row $x \in X$, and Bob a column $y \in Y$. Let Alice to construct $x' \in \{0, 1\}^m$ by assigning $x'_i = 1$ iff the row x belongs to R_i, and let Bob to construct $y' \in \{0, 1\}^m$ by assigning $y'_i = 0$ iff the column y belongs to R_i.

Chapter 5
Multi-Party Games

The rich mathematical theory of two-party communication naturally invites us to consider scenarios involving $k > 2$ players. In the simplest case, we have some function $f(x)$ whose input x is decomposed into k equally-sized parts $x = (x_1, \ldots, x_k)$. There are k players who wish to collaboratively evaluate a given function f on every input x. Each player has unlimited computational power and full knowledge of the function. As in the case of two players, the players are not adversaries—they help and trust each other. Depending on what parts of the input x each player can see, there are two main models of communication:

- *Number-in-hand* model: the i-th player can only see x_i.
- *Number-on-forehead* model: the i-th player can see all the x_j *except* x_i.

Players can communicate by writing bits 0 and 1 on a blackboard. The blackboard is seen by all players. The game starts with the empty blackboard. For each string on the blackboard, the protocol either gives the value of the output (in that case the protocol is over), or specifies which player writes the next bit and what that bit should be as a function of the inputs this player knows (and the string on the board). During the computation on one input the blackboard is never erased, players simply append their messages. The objective is to compute the function with as small amount of communication as possible.

The *communication complexity* of a k-party game for f is the minimal number c such that on every input x the players can decide whether $f(x) = 1$ or not, by writing at most c bits on the blackboard. Put differently, the communication complexity is the minimal number of bits written on the blackboard on the worst-case input. For simplicity, we will only consider deterministic protocols.

Note that for $k = 2$ (two players) there is no difference between these two models. The difference comes when we have $k \geq 3$ players. In this case the second model seems to be (and actually is) more difficult to analyze because players share some common information. For example, the first two players *both* can then see all inputs $x_3, \ldots, x_k$. Moreover, if the number k of players increases, the communication complexity in the "number-in-hand" model can only increase (the pieces of input each player can see are smaller and smaller), whereas it can only decrease in the "number-on-forehead" model (the pieces of seen input are larger and larger). This is why the first model attracted much less attention. Still, the model becomes interesting if instead of computing a given function f exactly, the players are only required to *approximate* its values. In particular, this model has found applications in so-called combinatorial auctions; see Nisan (2002).

5.1   The "Number-in-Hand" Model

The *disjointness problem* is, given a sequence $a = (a_1, \ldots, a_k)$ of subsets $a_i \subseteq [n]$, to decide whether the a_i are pairwise disjoint. In the *approximate disjointness* problem $Disj_n$ the k players are only required to distinguish between the following two extreme cases:

1. Answer "input is positive" if $a_i \cap a_j = \emptyset$ for all $i \neq j$.
2. Answer "input is negative" if $a_1 \cap \cdots \cap a_k \neq \emptyset$.
3. If neither of these two events happens, then any answer is legal.

Lemma 5.1. *In the "number-in-hand" model, the approximate disjointness problem $Disj_n$ requires $\Omega(n/k)$ bits of communication.*

Proof. (Due to Jaikumar Radhakrishnan and Venkatesh Srinivasan) Any c-bit communication protocol for the approximate disjointness problem partitions the space of inputs into at most 2^c "boxes", where a box is a Cartesian product $S_1 \times S_2 \times \cdots \times S_k$ with $S_i \subseteq 2^{[n]}$ for each i. Each box must be labeled with an answer, and thus the boxes must be "monochromatic" in the following sense: no box can contain both a positive instance and a negative instance. (There are no restrictions on instances that are neither negative nor positive.)

We will show that there are exactly $(k + 1)^n$ positive instances, but any box that does not contain a negative instance can contain at most k^n positive instances. It then follows that there must be at least

$$(k + 1)^n / k^n = (1 + 1/k)^n \approx e^{n/k}$$

boxes to cover all positive instances and thus the number of communicated bits must be at least the logarithm $\Omega(n/k)$ of this number, giving the desired lower bound.

To count the number of positive instances, note that any partition of the n items in $[n]$ between k players, leaving some items "unlocated", corresponds to a mapping $g : [n] \to [k+1]$, implying that the number of positive instances is exactly $(k+1)^n$.

Now consider a box $S = S_1 \times S_2 \times \cdots \times S_k$ that does not contain any negative instance. Note that for each item $x \in [n]$ there must be a player $i = i_x$ such that $x \notin a$ for all $a \in S_i$. This holds because otherwise there would be, in each S_i, a set $a_i \in S_i$ containing x, and we would have that $a_1 \cap \cdots \cap a_k \supseteq \{x\} \neq \emptyset$, a negative instance in the box S.

We can now obtain an upper bound on the number of positive instances in S by noting that any such instance corresponds to a partition of the n items among

k players and "unlocated", but now with an additional restriction that each item $x \in [n]$ can not be in the block given to the i_x-th player. Thus each item has only k (instead of $k + 1$) possible locations for it, and the number of such partitions is at most n^k. □

The same lower bound for randomized protocols was obtained by Chakrabarti et al. (2003), and Gronemeier (2009).

5.2 The Approximate Set Packing Problem

The *set packing* problem is, given a collection A of subsets of $[n] = \{1, \ldots, n\}$, to find the largest packing—that is, the largest collection of pairwise disjoint sets in A. The *packing number* of A, is the largest number of sets of A in a packing of $[n]$.

The *set packing* communication problem is as follows: we have k players each holding a collection A_i of subsets of $[n]$, and the players are looking for the largest packing in the union $A = A_1 \cup \cdots \cup A_k$ of their collections. The goal of players is to approximate the packing number of A to within a given multiplicative factor λ.

Proposition 5.2. *In the "number-in-hand" model, there is a k-player protocol approximating the packing number within a factor of $\lambda = \min\{k, \sqrt{n}\}$ and using $\mathcal{O}(kn^2)$ bits of communication.*

Proof. Getting an approximation factor k is easy by just picking the single player with the largest packing in her collection. If $k > \sqrt{n}$, we can do better by using the following simple greedy protocol: at each stage each player announces the smallest set $a_i \in A_i$ which is disjoint from all previously chosen sets; this requires n bits of communication from each of k players. The smallest such set is chosen to be in the packing. This is repeated until no more disjoint sets exist; hence, the protocol ends after at most n stages. It remains to verify that this packing is by at most a factor of $\sqrt{n}$ smaller than the number of sets in an optimal packing.

Let $a_1, \ldots, a_t$ be the sets in A chosen by our protocol. The collections

$$B_i = \{a \in A \mid a \cap a_i \neq \emptyset \text{ and } a \cap a_j = \emptyset \text{ for all } i = 1, \ldots, j - 1\}$$

form a partition of the whole collection A. Since all sets in B_i contain an element of a_i, the maximum number of disjoint sets in B_i is at most the cardinality $|a_i|$ of a_i. On the other hand, since a_i is the *smallest* member of A which is disjoint from all $a_1, \ldots, a_{i-1}$, every member of B_i is of size at least $|a_i|$, so the maximum number of disjoint sets in B_i is also at most $\lfloor n/|a_i| \rfloor$. Thus, the optimal solution can contain at most

$$\min\{|a_i|, \lfloor n/|a_i| \rfloor\} \leq \max_{x \in \mathbb{N}} \min\{x, \lfloor n/x \rfloor\} = \lfloor \sqrt{n} \rfloor$$

sets from each B_i. □

On the other hand we have the following lower bound.

Theorem 5.3. (Nisan 2002) *Any k-player protocol for approximating the packing number to within a factor smaller than k requires $2^{\Omega(n/k^2)}$ bits of communication.*

In particular, as long as $k \leq n^{1/2-\epsilon}$ for $\epsilon > 0$, the communication complexity is exponential in n.

Proof. We have k players, each holding a collection A_i of subsets of $[n]$. It is enough to prove a lower bound on the communication complexity needed in order to distinguish between the case where the packing number is 1 and the case where it is k. That is, to distinguish the case where there exist k disjoint sets $a_i \in A_i$, and the case where any two sets $a_i \in A_i$ and $a_j \in A_j$ intersect (packing number is 1).

Now suppose that ℓ bits of communication are enough to distinguish these two cases. We will show that then the approximate disjointness problem $Disj_N$ for $N = e^{\Omega(n/k^2)}$ can also be solved using at most ℓ bits of communication. Together with Lemma 5.1 this will immediately yield the desired lower bound $\ell = \Omega(N/k)$.

The reduction uses a collection of partitions $\mathcal{A} = \{a^s \mid s = 1, \ldots, N\}$, where each a^s is a partition $a^s = (a_1^s, \ldots, a_k^s)$ of $[n]$ into k disjoint blocks. Say that such a collection $\mathcal{A}$ of partitions is *cross-intersecting* if

$$a_i^s \cap a_j^r \neq \emptyset \quad \text{for all} \quad 1 \leq i \neq j \leq k \quad \text{and} \quad 1 \leq s \neq r \leq N,$$

that is, if different blocks from different partitions have non-empty intersection.

Claim 5.4. A cross-intersecting collection of $N = e^{n/(2k^2)}/k$ partitions exists.

Proof. Consider a random function $f : [n] \rightarrow [k]$ where $\text{Prob}[f(x) = i] = 1/k$ for every $x \in [n]$ and $i \in [k]$. Let $f_1, \ldots, f_N$ be independent copies of f. Each function f_s gives us a partition $a^s = (a_1^s, \ldots, a_k^s)$ of $[n]$ with

$$a_i^s = \{x \mid f_s(x) = i\}.$$

Now fix $1 \leq i \neq j \leq k$ and two indices of partitions $1 \leq s \neq r \leq N$. For every fixed $x \in [n]$, the probability that $f_s(x) \neq i$ or $f_r(x) \neq j$ is $1 - 1/k^2$. Since $a_i^s \cap a_j^r = \emptyset$ holds if and only if this happens for all n elements x, we obtain that

$$\text{Prob}[a_i^s \cap a_j^r = \emptyset] = (1 - 1/k^2)^n < e^{-n/k^2}.$$

Since there are at most $k^2 N^2$ such choices of indices, we get that the desired set of partitions exist, as long as $k^2 N^2 \leq e^{n/k^2}$. $\qquad\square$

We now describe the reduction of the approximate disjointness problem $Disj_N$ to the problem of distinguishing whether the packing number is 1 of k. Fix a collection $\mathcal{A}$ of partitions guaranteed by Claim 5.4. Player i, who gets as input a set $b_i \subseteq [N]$ in the problem $Disj_N$, constructs the collection $A_i = \{a_i^s \mid s \in b_i\}$ of subsets of $[n]$.

That is, the i-th player takes the i-th block a_i^s from each partition $a^s \in \mathcal{A}$ with $s \in b_i$.

Now, if there exists $s \in \bigcap_{i=1}^k b_i$, then a k-packing exists: $a_1^s \in A_1, \ldots a_k^s \in A_k$. On the other hand, if $b_i \cap b_j = \emptyset$ for all $i \neq j$, then for any two sets $a_i^s \in A_i$ and $a_j^r \in A_j$, we have that $s \neq r$, and thus $a_i^s \cap a_j^r \neq \emptyset$, meaning that the packing number is 1. $\qquad\square$

5.3 Application: Streaming Algorithms

Let $x = (x_1, \ldots, x_m) \in [n]^m$ be a string of length m with elements in the range 1 to n. Suppose we want to compute $f(x)$ for some function f, but m is huge, so that it is impractical to try to store all of x. In a streaming algorithm we assume that we see the input x one symbol at a time with no knowledge of what future symbols will be. The question then is how many bits must be kept in memory in order to successfully compute $f(x)$. An ultimate goal is to compute $f(x)$ using only $\mathcal{O}(\log n + \log m)$ bits of memory. The field of streaming algorithms was first formalized and popularized in a paper by Alon et al. (1999).

Example 5.5. Let $x \in [n]^{n-1}$ be a string of length $n-1$, and assume that x contains every element in $[n]$ except for the number $p \in [n]$. Let $f(x) = p$, that is, f outputs the unique missing element p. A streaming algorithm to compute $f(x)$ can maintain a sum $T_k = \sum_{i=1}^k x_i$. At the end of the stream, it outputs $f(x) = \frac{n(n+1)}{2} - T_{n-1}$. This algorithms uses $\mathcal{O}(\log n)$ memory.

The *replication number* of an element $i \in [n]$ in a string x is the number $r_i = |\{j \in [m] \mid x_j = i\}|$ of occurrences of symbol i in the string. The d-th *frequency moment* of the string S is defined as

$$f_d(x) = r_1^d + r_2^d + \cdots + r_n^d.$$

In particular, $f_0(x)$ is the number of distinct symbols in the string, and $f_1(x)$ is the length m of the string. For $d \geq 2$, $f_d(x)$ gives useful statistical information about the string.

Theorem 5.6. *For $d \geq 3$, any deterministic streaming algorithm computing f_d requires memory with $\Omega(n^{1-2/d})$ bits.*

Proof. Fix $d \geq 3$ and let $k = n^{1/d}$. Suppose there exists a streaming algorithm $\mathcal{A}$ to compute f_d using C bits of memory. Our goal is to show that there exists a k-party "number-in-hand" communication protocol which solves the approximate disjointness problem $Disj_n$ using at most $Ck = Cn^{1/d}$ bits of communication. Together with Lemma 5.1, this will imply that $Cn^{1/d} = \Omega(n/k)$, and hence, $C = \Omega(n^{1-2/d})$.

Let $x = (a_1, \ldots, a_k)$ be an input to $Disj_n$ with $a_i \subseteq [n]$. We can assume that $|\cup_i a_i| \geq 2$, for otherwise $\mathcal{O}(\log n)$ bits of communication are enough. The players look at x as a stream and run $\mathcal{A}$ on it. Whenever player i finishes running $\mathcal{A}$ on the portion of S corresponding to a_i, he writes the state of the algorithm on the blackboard using C bits. At the end of the algorithm the players have computed $f_d(x)$ using $Ck = Cn^{1/k}$ bits of communication. This is sufficient to approximate $Disj_n$: if $a_i \cap a_j = \emptyset$ for all $i \neq j$, then $r_i \leq 1$ for all i, and hence, $f_d(x) \leq n$. On the other hand, if there is an element $p \in [n]$ contained in all sets a_i, then $r_x^d \geq k^d = n$, and we have that $f_d(x) \geq n + 1$ (because $|\cup_i a_i| \geq 2$). $\qquad\square$

5.4   The "Number-on-Forehead" Model

The number-on-forehead model of multi-party communication games was introduced by Chandra et al. (1983). The model is related to many other important problems in circuit complexity, and is much more difficult to deal with than the previous one. Recall that in this model the information seen by players on a given input $x = (x_1, \ldots, x_k)$ can overlap: the i-th player has access to all the x_j *except* x_i. Recall also that each x_i is an element from some (fixed in advance) n-element set X_i. Thus, we have two parameters: the size $n = |X_i|$ of a domain for each players, and the number k of players.

We can imagine the situation as k players sitting around the table, where each one is holding a number to his/her forehead for the others to see. Thus, all players know the function f, but their access to the input string $(x_1, x_2, \ldots, x_k)$ is restricted: the first player sees the string $(*, x_2, \ldots, x_k)$, the second sees $(x_1, *, x_3, \ldots, x_k)$, ..., the k-th player sees $(x_1, \ldots, x_{k-1}, *)$ (Fig. 5.1).

Let $c_k(f)$ denote the minimum communication complexity of f in this "number-on-forehead" model. That is, $c_k(f)$ is the minimal number t such that on every input

$$x \in X := X_1 \times \cdots \times X_k$$

the players can determine the value $f(x)$ by writing at most t bits on the blackboard. It is clear that

Fig. 5.1 A communication tree (protocol) for a function $f(x_1, x_2, x_3)$. Each function p_i attached to a node may depend on the sequence of bits until this node, but cannot depend on x_i. Different nodes can be labeled by different functions of the same i-th player. Their independence on the i-th position is the only restriction

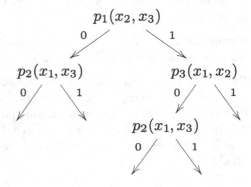

$$c_k(f) \leq \log n + 1 \text{ for any } f.$$

Namely, the first player can write the binary code of x_2, and the second player can announce the result. But what about the lower bounds? The twist is that (for $k \geq 3$) the players share some inputs, and (at least potentially) can use this overlap to encode the information in some wicked and nontrivial way (see Exercises 5.2 and 5.6).

Still, we know that the access of each player is restricted: the i-th player cannot distinguish inputs differing only in the i-th coordinate. This leads to the following concept.

A *combinatorial star*, or just a *star* in X around a vector $x = (x_1, \dots, x_k)$ is a set S of k vectors of the form:

$$x^1 = (x_1', x_2, \dots, x_k), \quad x^2 = (x_1, x_2', \dots, x_k), \quad \dots, \quad x^k = (x_1, x_2, \dots, x_k'),$$

where for each i, $x_i' \neq x_i$ and $x_i, x_i' \in X_i$. The vector x is a *center* of this star. Hence, each star contains exactly k vectors, and there are exactly $(n-1)^k$ stars around each vector x.

Say that a subset $T \subseteq X$ of $X = X_1 \times \cdots \times X_k$ is *closed* if, for every star $S_x \subseteq X$ around a vector $x \in X$, $S_x \subseteq T$ implies $x \in T$.

We have the following analogue of Proposition 3.3.

Proposition 5.7. *Every set $T \subseteq X$ of vectors reaching a leaf of a k-party communication protocol is closed.*

Proof. Take a k-party protocol of the communication game for f. Color each vector $x \in X$ by the string, which is written on the blackboard at the end of communication between the players on the input x. It is enough to show that each color class T is a closed set.

To show this, let $S = \{x^1, \dots, x^k\}$ be a star around some vector x, and assume that $S \subseteq T$. We have to show that then $x \in T$ as well. At each step, the player that needs to send the next message, say the i-th player, cannot distinguish between the input $x = (x_1, \dots, x_i, \dots, x_k)$ and $x^i = (x_1, \dots, x_i', \dots, x_k)$, because he does not see the i-th part of the input. Thus, the player will send the same message in both cases. Hence the whole communication on the center x is the same as on all elements of the star, as desired. $\square$

Unlike the definition of a rectangle, the definition of a closed set $T \subseteq X$ is only implicit: if T contains a star then it must also contain its center. It is clear that all sets $A \subseteq X$ of the form $A = A_1 \times \cdots \times A_k$ with $A_i \subseteq X_i$ are closed. Actually, for $k = 2$ all closed sets have this form. (Show this!) But for $k \geq 3$ not every closed set has this form. A simple counterexample is a set T consisting of one star together with its center. Still, we can give a more explicit definition of closed sets in terms of so-called "cylinders".

A subset $T_i \subseteq X$ is a *cylinder* in the i-th dimension if membership in T_i does not depend on the i-th coordinate. That is,

$(x_1, \ldots, x_i, \ldots, x_k) \in T_i$ implies $(x_1, \ldots, x_i', \ldots, x_k) \in T_i$ for all $x_i' \in X_i$.

A subset $T \subseteq X$ is a *cylinder intersection* if it is an intersection

$$T = T_1 \cap T_2 \cap \cdots \cap T_k,$$

where T_i is a cylinder in the i-th dimension.

Note that in the case of two players ($k = 2$), cylinder intersections are exactly rectangles, that is, subsets $T \subseteq X_1 \times X_2$ of the form $T = A_1 \times A_2$ with $A_i \subseteq X_i$. In this case, a cylinder in dimensions $i = 1, 2$ are sets of the form $T_1 = A_1 \times X_2$ and $T_2 = X_1 \times A_2$; hence, $T_1 \cap T_2 = A_1 \times A_2$. But this nice structure is lost when we go to games with $k \geq 3$ players. In this case the structure of cylinder intersections is more complicated; still, we have the following useful result:

Proposition 5.8. *A set $T \subseteq X$ is closed iff it is a cylinder intersection.*

Proof. The "only if" direction ($\Rightarrow$) is simple. Let $T = \cap_{i=1}^{k} T_i$ where T_i is a cylinder in the i-th dimension. If $S = \{x^1, \ldots, x^k\}$ is a star around some vector $x \in X$, and if $S \subseteq T$, then $x^i \in T \subseteq T_i$ and hence $x \in T_i$ for all $i = 1, \ldots, k$, implying that $x \in T$, as desired.

For the "if" direction ($\Leftarrow$), take an arbitrary subset $T \subseteq X$ and assume that T contains the center of every star it contains. For every $i = 1, \ldots, k$, let T_i be the set of all strings $x \in X$ such that x coincides with at least one string $x^i \in T$ in all but perhaps the i-th coordinate. By its definition, the set T_i is a cylinder in the i-th dimension. Hence, the set $T' = \cap_{i=1}^{k} T_i$ is a cylinder intersection. If a vector x belongs to T, then it also belongs to all the T_i, by their definition. This shows $T \subseteq T'$. To show that $T' \subseteq T$, take a vector $x \in T'$. Then $x \in T_i$ for all $i = 1, \ldots, k$. But $x \in T_i$ implies that there must be a vector $x^i \in T$ from which x differs in exactly the i-th coordinate. The vectors $x^1, \ldots, x^k$ form a star around x and are contained in T. Hence, vector x must belong to T as well. $\square$

Define the *k-tiling number* $\chi_k(f)$ of $f : X \to \mathbb{R}$ as the smallest number t such that the set X can be decomposed into t f-monochromatic cylinder intersections. Propositions 5.7 and 5.8 immediately yield the following general lower bound.

Lemma 5.9. $c_k(f) \geq \log \chi_k(f)$.

Since in the "number-on-forehead" model players share common information (for $k \geq 3$ players), proving lower bounds in this model is a difficult task. Actually, it was even not immediately clear whether the number $c_k(f)$ of communicated bits must grow with growing size $n = |X_i|$ of the domain for each player. That it must grow was first shown by Chandra et al. (1983). They considered the following *hyperplane problem*. The players get a sequence $(x_1, \ldots, x_k)$ of numbers in $[n] = \{1, \ldots, n\}$ and must decide whether $x_1 + \cdots + x_n = n$. Using some results from Ramsey theory, they proved that, for any fixed number $k \geq 2$ of players, the number of communicated bits in the hyperplane problem must go to infinity as n goes to infinity. Using one of the basic results of Ramsey theory—the Hales–Jewett theorem—Tesson (2003) obtained a similar result for the *partition problem*: players obtain a sequence $(x_1, \ldots, x_k)$ of subsets $x_i \subseteq [n]$, and the goal is to decide whether the x_i form a partition of $[n]$. Much stronger, almost optimal lower bounds of the form $\Omega(\log n)$ for any fixed number k of players were, however, obtained using discrepancy arguments.

5.5 The Discrepancy Bound

The (relative) *discrepancy* of a function $f : X \to \{-1, 1\}$ on a set $T \subseteq X$ is the absolute value of the sum of the values of f on points in T, divided by the total number $|X|$ of points:

$$\mathrm{disc}_T(f) = \frac{1}{|X|} \left| \sum_{x \in T} f(x) \right|.$$

Thus, large discrepancy means that one value is taken significantly more often than the other one. We have that $0 \le \mathrm{disc}_T(f) \le 1$ with $\mathrm{disc}_T(f) = |T|/|X|$ iff f is constant on T. The *discrepancy* of the function f itself is the maximum $\mathrm{disc}(f) = \max_T \mathrm{disc}_T(f)$ over all cylinder intersections $T \subseteq X$.

Intuitively, a function f has small discrepancy if it is "balanced enough" on all cylinder intersections. It turns out that such functions must have large multi-party communication complexity.

Proposition 5.10. *For every $f : X \to \{-1, 1\}$,*

$$c_k(f) \ge \log \frac{1}{\mathrm{disc}(f)}.$$

Proof. By Lemma 5.9, it is enough to show that $\chi_k(f) \ge 1/\mathrm{disc}(f)$. To do this, let $T \subseteq X$ be a cylinder intersection. Then T is f-monochromatic iff $\mathrm{disc}_T(f) = |T|/|X|$, implying that $|T| \le |X| \cdot \mathrm{disc}(f)$. Thus, at least $|X|/(|X| \cdot \mathrm{disc}(f)) = 1/\mathrm{disc}(f)$ f-monochromatic cylinder intersections T are necessary even to cover (not only to decompose) the whole set X. □

In fact, the logarithm of $\epsilon/\mathrm{disc}(f)$ is also a lower bound on the *randomized k-party communication complexity* with error-probability ϵ, if the random string is public.

However, this fact alone does not give immediate lower bounds for the multi-party communication complexity, because $\mathrm{disc}(f)$ is very difficult to estimate. Fortunately, the discrepancy can be bounded from above using the following more tractable measure.

A k-dimensional *cube* is defined to be a multi-set $D = \{a_1, b_1\} \times \cdots \times \{a_k, b_k\}$, where $a_i, b_i \in X_i$ (not necessarily distinct) for all i. Being a multi-set means that one element can occur several times. Thus, for example, the cube $D = \{a_1, a_1\} \times \cdots \times \{a_k, a_k\}$ has 2^k elements.

Given a function $f : X \to \{-1, 1\}$ and a cube $D \subseteq X$, define the *sign* of f on D to be the value

$$f(D) = \prod_{x \in D} f(x).$$

Hence, $f(D) = 1$ if and only if $f(x) = -1$ for an even number of vectors $x \in D$. We choose a cube D at random according to the uniform distribution. This can be done by choosing $a_i, b_i \in X_i$ for each i according to the uniform distribution. Let

$$\mathcal{E}(f) := \mathrm{E}\left[f(D)\right] = \mathrm{E}\left[\prod_{x \in D} f(x)\right]$$

be the expected value of the sign of a random cube D. To stress the fact that the expectation is taken over a particular random object (this time, over D) we will also write $\mathrm{E}_D\left[f(D)\right]$ instead of $\mathrm{E}\left[f(D)\right]$.

Example 5.11. The difference between the measures $\mathrm{disc}(f)$ and $\mathcal{E}(f)$ can best be seen in the case when $k = 2$. In this case $X = X_1 \times X_2$ is just a grid, and each function $f : X \to \{-1, 1\}$ is just a ± 1 matrix M_f. Cylinder intersections $T \subseteq X$ in this case correspond to submatrices of M_f, and $\mathrm{disc}_T(f)$ is the sum of all entries in T divided by $|X|$. Thus, to determine $\mathrm{disc}(f)$ we must consider *all* submatrices of M_f. In contrast, to determine $\mathcal{E}(f)$ it is enough to only consider all $s \times t$ submatrices with $1 \le s, t \le 2$.

The following result was first proved in Chung (1990) and generalizes a similar result from Babai et al. (1992). An elegant and relatively simple proof presented below was found by Raz (2000).

Theorem 5.12. *For every* $f : X \to \{-1, 1\}$, $\mathcal{E}(f) \ge \mathrm{disc}(f)^{2^k}$. *Hence,*

$$c_k(f) \ge \frac{1}{2^k} \log \frac{1}{\mathcal{E}(f)}.$$

The theorem is very useful because $\mathcal{E}(f)$ is a much simpler object than $\mathrm{disc}(f)$. For many functions f, it is relatively easy to compute $\mathcal{E}(f)$ exactly; we will demonstrate this in the next sections.

Proof. (Due to Raz 2000) We will only prove the theorem for $k = 2$; the general case is similar. So let $X = X_1 \times X_2$ and $f : X \to \{-1, 1\}$ be a given function. Our goal is to show that $\mathcal{E}(f) \ge \mathrm{disc}(f)^4$. To do this, pick at random (uniformly and independently) an element $x \in X$.

Claim 5.13. *For all functions* $h : X \to \{-1, 1\}$, $\mathcal{E}(h) \ge (\mathrm{E}_x\left[h(x)\right])^4$.

Proof. We will use two well-known facts about the mean value of random variables:

$$E\left[\xi^{2}\right] \geq E\left[\xi\right]^{2} \quad \text{for any random variable } \xi \tag{5.1}$$

and

$$E\left[\xi \cdot \xi'\right] = E\left[\xi\right] \cdot E\left[\xi'\right] \quad \text{if } \xi \text{ and } \xi' \text{ are independent.} \tag{5.2}$$

The first one is a consequence of the Cauchy–Schwarz inequality, and the second is a basic property of expectation.

Now take a random two-dimensional cube $D = \{a, a'\} \times \{b, b'\}$. Then

$$
\begin{aligned}
\mathcal{E}(h) = E_D\left[h(D)\right] = E_D &\left[\prod_{x \in D} h(x)\right] \\
= E_{a,a'} E_{b,b'} &\left[h(a,b) \cdot h(a,b') \cdot h(a',b) \cdot h(a',b')\right] \\
= E_{a,a'} &\left[\left(E_b\left[h(a,b) \cdot h(a',b)\right]\right)^2\right] && \text{by (5.2)} \\
\geq \left(E_{a,a'} E_b\right. &\left.\left[h(a,b) \cdot h(a',b)\right]\right)^2 && \text{by (5.1)} \\
= \left(E_a E_b\right. &\left.\left[h(a,b)^2\right]\right)^2 && \text{Prob}[a'] = \text{Prob}[a] \\
= \left(E_a\left(E_b\left[h(a,b)\right]\right)^2\right)^2 && \text{by (5.2)} \\
\geq \left(E_{a,b}\left[h(a,b)\right]\right)^4 && \text{by (5.1).} \qquad \square
\end{aligned}
$$

Claim 5.14. There exists h such that $\left|E_x\left[h(x)\right]\right| \geq \text{disc}(f)$ and $\mathcal{E}(h) = \mathcal{E}(f)$.

Proof. Let $T = A \times B$ be a cylinder intersection (a submatrix of X, since $k = 2$) for which $\text{disc}(f)$ is attained. We prove the existence of h by the probabilistic method. The idea is to define a *random* function $g : X_1 \times X_2 \to \{-1, 1\}$ such that the expected value $E\left[g(x)\right] = E_g\left[g(x)\right]$ is the characteristic function of T. For this, define g to be the product $g(x) = g_1(x) \cdot g_2(x)$ of two random functions, whose values are defined on the points $x = (a, b) \in X_1 \times X_2$ by:

$$
g_1(a,b) = \begin{cases} 1 & \text{if } a \in A; \\ \text{set randomly to } \pm 1 & \text{otherwise} \end{cases}
$$

and

$$
g_2(a,b) = \begin{cases} 1 & \text{if } b \in B; \\ \text{set randomly to } \pm 1 & \text{otherwise.} \end{cases}
$$

These function have the property that g_1 depends only on the rows and g_2 only on the columns of the grid $X_1 \times X_2$. That is, $g_1(a,b) = g_1(a,b')$ and $g_2(a,b) = g_2(a',b)$

for all $a, a' \in X_1$ and $b, b' \in X_2$. Hence, for $x \in T$, $g(x) = 1$ with probability 1, while for $x \notin T$, $g(x) = 1$ with probability $1/2$ and $g(x) = -1$ with probability $1/2$; this holds because the functions g_1, g_2 are independent of each other, and $x \notin T$ iff $x \notin A \times X_2$ or $x \notin X_1 \times B$. Thus, the expectation $\mathrm{E}\,[g(x)]$ takes the value 1 on all $x \in T$, and takes the value $\frac{1}{2} + \left(-\frac{1}{2}\right) = 0$ on all $x \notin T$, that is, $\mathrm{E}\,[g(x)]$ is the characteristic function of the set T:

$$\mathrm{E}\,[g(x)] = \begin{cases} 1 & \text{if } x \in T; \\ 0 & \text{if } x \notin T. \end{cases}$$

Now let x be a random vector uniformly distributed in $X = X_1 \times X_2$. Then

$$\mathrm{disc}_T(f) = \left| \mathrm{E}_x\left[f(x) \cdot \mathrm{E}_g\left[g(x)\right]\right]\right| = \left|\mathrm{E}_x\mathrm{E}_g\left[f(x) \cdot g(x)\right]\right|$$

$$= \left|\mathrm{E}_g\mathrm{E}_x\left[f(x) \cdot g(x)\right]\right|.$$

So there exists some choice of $g = g_1 \cdot g_2$ such that

$$\left|\mathrm{E}_x\left[f(x) \cdot g(x)\right]\right| \geq \mathrm{disc}_T(f)$$

and we can take $h(x) := f(x) \cdot g(x)$. Then $\left|\mathrm{E}_x\left[h(x)\right]\right| \geq \mathrm{disc}(f)$. Moreover, $\mathcal{E}(h) = \mathcal{E}(f)$ because g_1 is constant on the rows and g_2 is constant on the columns so the product $g(D) = \prod_{x \in D} g(x)$ cancels to 1. $\qquad \square$

Claims 5.13 and 5.13 imply that $\mathcal{E}(f) = \mathcal{E}(h) \geq (\mathrm{E}_x\,[h(x)])^4 \geq \mathrm{disc}(f)^4$. This completes the proof of Theorem 5.12 in case $k = 2$. To extend it for arbitrary k, just repeat the argument k times. $\qquad \square$

5.6 Generalized Inner Product

Say that a 0-1 matrix A is *odd* if the number of its all-1 rows is odd. Note that, if the matrix has only two columns, then it is odd iff the scalar (or inner) product of these columns over GF(2) is 1. For this reason, the boolean function which decides whether a given input matrix is odd is called the "generalized inner product" function. We will assume that input matrices have n rows and k columns.

That is, the *generalized inner product function* GIP(x) is a boolean function of kn variables, arranged in an $n \times k$ matrix $x = (x_{ij})$, and is defined by:

$$\mathrm{GIP}_{n,k}(x) = \bigoplus_{i=1}^{n} \bigwedge_{j=1}^{k} x_{ij}.$$

We consider k-party communication games for GIP(x), where the j-th player can see all but the j-th column of the input matrix x. The following lower bound was first proved by Babai et al. (1992). Similar lower bounds for other explicit functions were obtained by Chung (1990). Note that in this case the size of the domain for each player is $|X_i| = 2^n$, not n. Hence, $n + 1$ is a trivial upper bound on the number of communicated bits for any number k of players.

Theorem 5.15. *The k-party communication complexity of* $\mathrm{GIP}_{n,k}$ *is* $\Omega(n4^{-k})$.

It can be shown (see Exercise 5.6) that this lower bound is almost optimal: $c_k(\mathrm{GIP}) = \mathcal{O}(kn/2^k)$.

Proof. Since we want our function to have range $\{-1, 1\}$, we will consider the function

$$f(x) = (-1)^{\mathrm{GIP}(x)} = \prod_{i=1}^{n}(-1)^{x_{i1}x_{i2}\cdots x_{ik}}. \tag{5.3}$$

By Theorem 5.12, it is enough to prove that $\mathcal{E}(f) \leq 2^{-\Omega(n2^{-k})}$. In fact we will prove that

$$\mathcal{E}(f) = \left(1 - \frac{1}{2^k}\right)^n. \tag{5.4}$$

In our case, the function f is a mapping $f : X_1 \times X_2 \times \cdots X_k \to \{-1, 1\}$, where the elements of each set X_j are column vectors of length n. Hence, a cube D in our case is specified by two boolean $n \times k$ matrices $A = (a_{ij})$ and $B = (b_{ij})$. The cube D consists of all 2^k n-by-k matrices, the j-th column in each of which is either the j-th column of A or the j-th column of B. By (5.3), we have (with $x_{ij} \in \{a_{ij}, b_{ij}\}$) that

$$f(D) = \prod_{x \in D} f(x) = \prod_{x \in D}\prod_{i=1}^{n}(-1)^{x_{i1}x_{i2}\cdots x_{ik}} = \prod_{i=1}^{n}\prod_{x \in D}(-1)^{x_{i1}x_{i2}\cdots x_{ik}}$$

$$= \prod_{i=1}^{n}(-1)^{(a_{i1}+b_{i1})(a_{i2}+b_{i2})\cdots(a_{ik}+b_{ik})}.$$

Note that the exponent $(a_{i1} + b_{i1})(a_{i2} + b_{i2}) \cdots (a_{ik} + b_{ik})$ is even if $a_{ij} = b_{ij}$ for at least one $1 \leq j \leq k$, and is equal to 1 in the unique case when $a_{ij} \neq b_{ij}$ for all $j = 1, \ldots, k$, that is, when the i-th row of B is complementary to the i-th row of A. Thus,

$f(D) = -1$ iff the number of complementary rows in A and B is odd.

Now, $\mathcal{E}(f)$ is the average of the above quantity over all choices of matrices A and B. We fix the matrix A and show that the expectation over all matrices B is precisely the right-hand side of (5.4). Let $\mathbf{a}_1, \ldots, \mathbf{a}_n$ be the rows of A and $\mathbf{b}_1, \ldots, \mathbf{b}_n$ be the rows of B. Then $f(D) = \prod_{i=1}^{n} g(\mathbf{b}_i)$, where

$$g(\mathbf{b}_i) := (-1)^{(a_{i1}+b_{i1})(a_{i2}+b_{i2})\cdots(a_{ik}+b_{ik})} = \begin{cases} +1 & \text{if } \mathbf{b}_i \neq \mathbf{a}_i \oplus \mathbf{1}, \\ -1 & \text{if } \mathbf{b}_i = \mathbf{a}_i \oplus \mathbf{1}. \end{cases}$$

Thus, for every fixed matrix A, we obtain

$$E_B\left[\prod_{i=1}^{n} g(\mathbf{b}_i)\right] = \prod_{i=1}^{n} E_{\mathbf{b}_i}[g(\mathbf{b}_i)] \qquad\qquad \text{by (5.2)}$$

$$= \prod_{i=1}^{n} \frac{1}{2^k} \sum_{\mathbf{b}_i} g(\mathbf{b}_i) = \prod_{i=1}^{n} \frac{1}{2^k}\left(2^k - 1\right) = \left(1 - \frac{1}{2^k}\right)^n. \qquad \square$$

Remark 5.16. Similar in form to $\text{GIP}_{n,k}$ is the (generalized) *disjointness function* $\text{DISJ}(x)$. This function also has kn variables arranged into an $n \times k$ matrix $x = (x_{ij})$, and is defined by:

$$\text{DISJ}(x) = \bigvee_{i=1}^{n} \bigwedge_{j=1}^{k} x_{ij}.$$

That is, given a boolean $n \times k$ matrix x, $\text{DISJ}(x) = 1$ iff it has an all-1 row. If we interpret columns as characteristic vectors of subsets of $[n]$, then the function accepts a sequence of k subsets if and only if their intersection is non-empty. The discrepancy argument fails for this function because DISJ is constant on huge cylinder intersections. Using different arguments, Tesson (2003) and Beame et al. (2006) were able to prove a lower bound $\Omega((\log n)/k)$ for $\text{DISJ}(x)$, even for randomized protocols. Lee and Shraibman (2009) and Chattopadhyay and Ada (2008) proved that $\text{DISJ}(x)$ requires $\Omega(n^{1/(k+1)}/2^{2^k})$ bits of communication in randomized k-party protocols. More applications of algebraic methods in communication complexity can be found in a survey by Lee and Shraibman (2007).

5.7   Matrix Multiplication

Let $X = X_1 \times \cdots X_k$, where each X_i is the set of all $n \times n$ matrices over the field GF(2); hence, the domain X_i for each player has size $|X_i| = 2^{n^2}$. For $x_1 \in X_1, \ldots, x_k \in X_k$, denote by $x_1 \cdots x_k$ the product of $x_1, \ldots, x_k$ as matrices over GF(2). Let $F(x_1, \ldots, x_k)$ be a boolean function whose value is the element in the first row and the first column of the product $x_1 \cdots x_k$.

Theorem 5.17. (Raz 2000) $c_k(F) = \Omega(n/2^k)$.

The theorem is a direct consequence of Theorem 5.12 and the following lemma. Define the function $f : X \to \{-1, 1\}$ by

$$f(x_1, \ldots, x_k) = (-1)^{F(x_1, \ldots, x_k)} = 1 - 2F(x_1, \ldots, x_k).$$

Lemma 5.18. $\mathcal{E}(f) \le (k-1)2^{-n}$.

Proof. For every cube $D = \{a_1, b_1\} \times \cdots \times \{a_k, b_k\}$,

$$f(D) = \prod_{x \in D} f(x) = \prod_{x \in D} (-1)^{F(x)} = (-1)^{\oplus_{x \in D} F(x)}.$$

Since F is linear in each variable,

$$f(D) = (-1)^{F(a_1 \oplus b_1, \ldots, a_k \oplus b_k)} = 1 - 2F(a_1 \oplus b_1, \ldots, a_k \oplus b_k),$$

where $a_i \oplus b_i$ denotes the sum of matrices a_i and b_i over GF(2). If we choose D at random according to the uniform distribution, then $(a_1 \oplus b_1, \ldots, a_k \oplus b_k)$ is a random vector $x = (x_1, \ldots, x_k)$ uniformly distributed over X. Therefore,

$$\mathcal{E}(f) = \mathrm{E}_D \left[f(D) \right] = \mathrm{E} \left[1 - 2F(a_1 \oplus b_1, \ldots, a_k \oplus b_k) \right]$$
$$= \mathrm{E}_x \left[1 - 2F(x) \right] = \mathrm{E}_x \left[f(x) \right].$$

To estimate the expectation $\mathrm{E}_x \left[f(x) \right]$, where $x = (x_1, \ldots, x_k)$ is uniformly distributed over X sequence of $n \times n$ matrices, let E_d denote the event that the first row of the matrix $x_1 \cdots x_d$ contains only 0s. Define $p_d = \mathrm{Prob}[E_d]$. Since p_1 is determined by x_1 and since x_1 is uniformly distributed, we have $p_1 = \mathrm{Prob}[E_1] = 2^{-n}$. Clearly we also have $\mathrm{Prob}[E_{d+1}|E_d] = 1$. On the other hand, since x_{d+1} is uniformly distributed, $\mathrm{Prob}[E_{d+1}|\neg E_d] = 2^{-n}$. Therefore, for all $1 \le d < k$,

$$p_{d+1} = \mathrm{Prob}[E_{d+1}|E_d] \cdot \mathrm{Prob}[E_d] + \mathrm{Prob}[E_{d+1}|\neg E_d] \cdot \mathrm{Prob}[\neg E_d]$$
$$= p_d + (1 - p_d) \cdot 2^{-n} \le p_d + 2^{-n},$$

implying that $p_d \le d \cdot 2^{-n}$ for all $d = 1, \ldots, k$.

If E_{k-1} occurs then $F(x_1, \ldots, x_k)$ is always 0, and hence, $f(x_1, \ldots, x_k)$ is always 1. If E_{k-1} does not occur then, since the first column of x_k is uniformly distributed, the value $F(x_1, \ldots, x_k)$ is uniformly distributed over $\{0, 1\}$, and hence, $f(x_1, \ldots, x_k)$ is uniformly distributed over $\{-1, 1\}$. Therefore,

$$\mathcal{E}(f) = \mathrm{E}_x \left[f(x) \right] = \mathrm{Prob}[E_{k-1}] = p_{k-1} \le (k-1) \cdot 2^{-n}. \qquad \square$$

Remark 5.19. We have seen that some "simple" functions (like GIP or DISJ) have large multi-party communication complexity. On the other hand, Chattopadhyay et al. (2007) showed that there exist boolean functions of arbitrarily large circuit complexity which can be computed with constant(!) communication by k players for $k \ge 3$.

Remark 5.20. Note that lower bounds on the multi-party communication complexity given above are only non-trivial if the number k of players is much smaller than $\log n$. To prove good lower bounds for $k \geq \log n$ players is a long-standing problems whose solution would have great consequences in circuit complexity; see Sect. 12.6 for some of these consequences.

5.8 Best-Partition k-Party Communication

Let $f : \{0, 1\}^n \to \{0, 1\}$ be a boolean function on $n = km$ variables. The "number-on-forehead" communication protocols work with a *fixed* partition $x = (x_1, \ldots, x_k)$ of the input vector $x \in \{0, 1\}^n$ into k blocks $x_i \in \{0, 1\}^m$.

We now consider the situation where, given a function f, the players are allowed to choose the balanced partition of input variables that is best-suited for computing f. (We say that a partition of a finite set into k disjoint blocks is balanced if the sizes of blocks differ by at most one.) Let $c_k^*(f)$ denote the smallest possible k-party communication complexity of f over all balanced partitions of its input vector.

Recall that the generalized inner product function $\text{GIP}_{m,k}$ is a boolean function of $n = km$ variables which takes a boolean $m \times k$ matrix x as its input, and outputs 1 iff the number of all-1 rows in it is odd:

$$\text{GIP}_{m,k}(x) = \bigoplus_{i=1}^{n} \bigwedge_{j=1}^{k} x_{ij}.$$

We have already shown in Sect. 5.6 that, if we split the input matrix in such a way that the i-th player can see all its columns but the i-th one, then

$$c_k(\text{GIP}_{m,k}) = \Omega(n/k4^k). \tag{5.5}$$

On the other hand, the best-partition communication complexity of this function is very small: for every $k \geq 2$ we have that $c_k^*(\text{GIP}_{m,k}) \leq 2$. To see this, split the rows of the input $m \times k$ matrix x into m/k blocks and give to the i-th player all but the i-th block of these rows. Then the first player can write the parity of the number of all-1 rows she can see, and the second player can announce the answer.

So, what boolean functions have large k-party communication complexity under the best-partition of their inputs? To answer this question we use a graph-theoretic approach.

Let H be a hypergraph on an n-vertex set $V = V(H)$, that is, a family of subsets $e \subseteq V$; the members of H are called *hyperedges*. Thus, graphs are special hypergraphs with each hyperedge containing just two vertices. Associate with each vertex $v \in V$ a boolean variable x_v and consider the following boolean function of these variables:

Fig. 5.2 A four-matching on
$n = 16$ vertices. The
matching is an induced
matching if no other
hyperedge lies in the set of
these 16 vertices

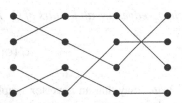

$$\mathrm{GIP}_H(x) = \bigoplus_{e \in H} \bigwedge_{v \in e} x_v.$$

A k-*matching* is a hypergraph consisting of pairwise disjoint hyperedges, each of
size k (Fig. 5.2). Note that, if M is a k-*matching*, then (up to renaming of variables),

$$\mathrm{GIP}_M(x) = \mathrm{GIP}_{m,k}(x), \qquad\qquad (5.6)$$

We have however just shown that for such hypergraphs, $c_k^*(\mathrm{GIP}_M) \leq 2$. But what
if for some hypergraph H we could show that H contains an *induced* k-matching
on sufficiently many vertices, for any balanced partition of vertices into k parts? It
turns out that this is enough to force $c_k^*(\mathrm{GIP}_H)$ to be large.

Let us formalize our terms. First, a k-*partition* of a set V of n vertices is its
partition $V_1 \cup \cdots \cup V_k$ into k disjoint blocks V_i, each of size n/k. Given such a
partition, say that a k-matching M is *consistent* with it if every edge e of M has
exactly one vertex in each of the blocks, that is, if $|e \cap V_i| = 1$ for all $i = 1, \ldots, k$.
Let $V(M) = \cup_{e \in M} e$ be the set of vertices incident to at least one hyperedge of M.

We say that M is an *induced* matching in a hypergraph H on V if for every
hyperedge $e \in H$, $e \subseteq V(H)$ implies $e \in M$. That is, no hyperedge $e \in H \setminus M$
can be covered by hyperedges of M. It is easy to see that then the function GIP_M
is a subfunction of GIP_H, and hence, the communication complexity of GIP_H is
lower bounded by the communication complexity of GIP_M: just set $x_v = 0$ for all
$v \notin V(M)$. We fix this observation as

Proposition 5.21. *Let H be a hypergraph. Suppose that for every k-partition of its
vertex-set, there exists an induced k-matching with m hyperedges which is consistent
with that partition. Then*

$$c_k^*(\mathrm{GIP}_H) = \Omega\left(\frac{m}{k\,4^k}\right).$$

We will construct the desired hypergraphs H starting from (ordinary) "well-
mixed" graphs $G = (V, E)$. Namely, call a graph s-*mixed* if, for any pair of disjoint
s-element subsets of vertices, there is at least one edge between these sets. A k-*star*
of a graph G is a set of its k vertices such that at least one of them is adjacent to all
of the remaining $k - 1$ of these vertices.

Theorem 5.22. (Hayes 2001) *Let G be an s-mixed regular graph of degree $d \geq$
2 on n vertices. Let $2 \leq k \leq \min\{d, n/s\}$ and let H be the hypergraph whose*

hyperedges are all k-stars of G. Then

$$c_k^*(\text{GIP}_H) = \Omega\left(\frac{n-sk}{dk^2 4^k}\right).$$

Proof. Say that an n-vertex graph $G = (V, E)$ is *s-starry* if for any $2 \leq k \leq n/s$ and for any pairwise disjoint sets $S_1, \ldots, S_k$, each of size $|S_i| \geq s$, there exist vertices $v_1 \in S_1, \ldots, v_k \in S_k$ such that $\{v_1, \ldots, v_k\}$ forms a k-star of G. Note that every s-starry graph is also s-mixed, since we can let $k = 2$. Interestingly, the converse is also true:

Claim 5.23. Every s-mixed graph is s-starry.

Proof. Let G be an s-mixed graph, and let $S_1, \ldots, S_k$ be pairwise disjoint subsets of its vertices each of size $|S_i| \geq s$. For $i \in \{1, \ldots, k\}$, let T_i be the set of all vertices $v \in V \setminus S_i$ that are *not* adjacent to *any* vertex in S_i. Since $|S_i| \geq s$ and since G is s-mixed, we have that $|T_i| \leq s - 1$. Hence, the set $T = \bigcup_{i=1}^{k} T_i$ can have at most

$$|T| \leq (s-1)k < sk \leq \sum_{i=1}^{k} |S_i| = \left|\bigcup_{i=1}^{k} S_i\right|$$

vertices. Thus, there must exist a vertex $v \in (\bigcup_{i=1}^{k} S_i) \setminus T$. That is, v belongs to some S_i and does not belong to T. By the definition of T, $v \in S_i$ and $v \notin T$ means that v must be connected by an edge with at least one vertex in each of the sets S_j, $j \neq i$. But then v is a center of the desired star. □

Now let G be a graph satisfying the conditions of Theorem 5.22, and let H be the hypergraph of its k-stars. To finish the proof of Theorem 5.22, it is enough to prove that this hypergraph satisfies the conditions of Proposition 5.21 with $m \geq (n - sk)/dk$.

To prove this, let $V = V_1 \cup \cdots \cup V_k$ be an arbitrary balanced partition of the set V into k blocks. Hence, $|V_j| \geq n/k \geq s$ for all j. We construct the desired k-matching $M \subseteq H$ recursively. Initially, let M be empty. In each stage, apply Claim 5.23 to find a k-star $e = \{v_1, \ldots, v_k\}$ with vertices in each set of the partition. Add this set to M. Delete the k vertices $v_1, \ldots, v_k$ and all their neighbors from G. Repeat the procedure restricting the given partition to the remaining vertices of G.

After i stages, at most id vertices have been removed from each block V_j, which means that each block in the partition (of the remaining vertices) has size at least $n/k - id$. Since G is s-mixed, Claim 5.23 will apply as long as $n/k - id \geq s$. Thus, the algorithm will run for at least $i \geq (n - sk)/(dk)$ stages. Let $M = \{e_1, \ldots, e_m\}$ be the constructed k-matching. Since $m \geq (n - sk)/(dk)$, it is enough to show that M is an *induced* matching of H.

To show this, observe that, by the construction, no two vertices $u \in e_i$ and $v \in e_j$ for $i < j$ can be adjacent in G: when a hyperedge is added to M, all its neighbors in G are removed from the further consideration. Assume now that M is not an induced matching. Then there exists a k-star e of G such that $e \subseteq V(M) = \bigcup_{i=1}^{m} e_i$, but $e \notin M$. Let $u \in e$ be its vertex adjacent (in G) to all the remaining $k - 1$ vertices of e. Since $e \subseteq V(M)$, the vertex u must lie in some hyperedge e_i of our matching. Moreover, $e \notin M$ implies that some other vertex $v \in e$ must lie in some other hyperedge e_j of the matching. Since u and v *are* adjacent in G, we obtain a contradiction with the observation we just made. $\square$

Thus, what we need are explicit graphs satisfying the following two conditions:

1. The graph must have small degree, but
2. Any two moderately large subsets of vertices must be joined by an edge.

Graphs with these properties are known as *expander graphs*. By the Expander Mixing Lemma (see Appendix A for its proof), if G is a d-regular graph on n vertices and if $s > \lambda n/d$, where λ is the second largest eigenvalue of the adjacency matrix of G, then G is s-mixed. Important examples of such graphs are the *Ramanujan graphs* $RG(n, q)$ (see Appendix A). These are $(q + 1)$-regular graphs with the property that $\lambda \leq 2\sqrt{q}$. Since $2n/\sqrt{q} > 2\sqrt{q}n/(q + 1)$, we have that the Ramanujan graphs are well-mixed.

Corollary 5.24. *Ramanujan graphs $RG(n, q)$ are s-mixed for $s = 2n/\sqrt{q}$.*

Explicit constructions of Ramanujan graphs on n vertices for every prime $q \equiv 1 \bmod 4$ and infinitely many values of n were given in Margulis (1973), Lubotzky et al. (1988); these were later extended to the case where q is an arbitrary prime power in Morgenstern (1994) and Jordan and Livné (1997).

Let q be a prime number lying between $16k^2$ and $32k^2$. Then the Ramanujan graph $RG(n, q)$ has degree $d = q + 1$ and (by Corollary 5.24) is s-mixed for $s = 2n/\sqrt{q} \leq n/2k$. Using such graphs, Theorem 5.22 yields the following

Corollary 5.25. *If H is the hypergraph of k-stars in the Ramanujan graph $RG(n, q)$, then $c_k^*(\mathrm{GIP}_H) = \Omega(n/k^4 4^k)$.*

It can be shown that, this bound is tight with respect to the number k of players: for any balanced partition of n vertices into $k + 1$ parts, we have that $c_k(\mathrm{GIP}_H) \leq k + 1$ (Exercise 5.8). Thus, for every constant $k \geq 2$ there is an explicit boolean function $f = \mathrm{GIP}_H$ such that $c_k^*(f) = \Omega(n)$ but $c_{k+1}^*(f) = O(1)$.

Exercises

5.1. The *set cover* communication problem is as follows: we have k players each holding a collection A_i of subsets of $[n] = \{1, \ldots, n\}$, and the players are looking for the smallest covering of $[n]$ using the sets in their collections. That is, the goal is to find the smallest number r of subsets $a_1, \ldots, a_r$ of $[n]$ such that each a_j belongs to at least one A_i, and $a_1 \cup \cdots \cup a_r = [n]$. Show that $O(kn^2)$ bits of communication are enough to construct a covering using at most $\ln n + 1$ times larger number of sets than an optimal covering algorithm would do.

Hint: Use a greedy protocol, as in the proof of Lemma 4.6.

5.2. Three players want to compute the following boolean function $f(x, y, z)$ in $3m$ variables. Inputs x, y, z are vectors in $\{0, 1\}^m$, and the function is defined by: $f(x, y, z) = \bigoplus_{i=1}^m \mathrm{Maj}(x_i, y_i, z_i)$. Prove that $c_3(f) \leq 3$.

Hint: Show that the following protocol is correct. Each player counts the number of i's such that she can determine that $\mathrm{Maj}(x_i, y_i, z_i) = 1$ by examining the bits available to her. She writes the parity of this number on the blackboard, and the final answer is the parity of the three written bits.

5.3. ■ Research Problem. Three players want to compute the following boolean function $f(x, y, z)$ in $3m$ variables. Inputs x, y, z are vectors in $\{0, 1\}^m$, and the function is defined by: $f(x, y, z) = 1$ iff there exists an index $i \in [m]$ such that $x_i = y_i = z_i = 1$. Does $c_3(f) = \omega(\log^3 m)$?

Comment: As shown by Beame et al. (2007), this would imply super-polynomial lower bounds for some cutting plane proof systems.

5.4. Given an operator $f : \{0, 1\}^{2n} \to \{0, 1\}^n$, consider the boolean function $g_f(x, y, j)$ whose value is the value $f_j(x, y)$ of the j-th component of f. Consider the three-party communication game for g_f, where Alice gets (x, j), Bob gets (y, j) and Charlie gets (x, y). Prove the following: if f can be computed by a circuit of depth $\mathcal{O}(\ln n)$ using $\mathcal{O}(n)$ fanin-2 gates, then $\mathcal{O}(n / \log \log n)$ bits of communication are enough.

Hint: Use Valiant's lemma (Lemma 1.4). Take an input (x, y, j) for g_f. Seeing the entire input (x, y) of f, Charlie can compute the values of removed wires. Alice and Bob both know j. The value $f_j(x, y)$ can be computed from the values of the removed wires and from the values of at most n^ϵ input variables. Alice and Bob can just write down these n^ϵ values.

5.5. Consider the following k-party communication game. Input is a boolean $n \times k$ matrix A, and the i-th player can see all A except its i-th column. Suppose that the players a priori know that some string $v = (0, \ldots, 0, 1, \ldots, 1)$ with the first 1 in position $t + 1$, does not appear among the rows of A. Show that the players can decide if the number of all-1 rows is even or odd by communicating only t bits.

Hint: Let y_i denote the number of rows of A of the form $(0, \ldots, 0, 1, \ldots, 1)$, where the first 1 occurs in position i. For every $i = 1, \ldots, t$, the i-th player announces the parity of the number of rows of the form $(0, \ldots, 0, *, 1, \ldots, 1)$, where the $*$ is at place i. Observe that this number is $y_i + y_{i+1}$. Subsequently, each player privately computes the mod 2 sum of all numbers announced. The result is $(y_1 + y_{t+1}) \bmod 2$, where $y_{t+1} = 0$.

5.6. (Grolmusz 1994) Prove that $c_k(\mathrm{GIP}) = \mathcal{O}(kn/2^k)$.

Hint: Use the previous protocol to show that (without any assumption) k-players can decide if the number of all-1 rows in a given boolean $n \times k$ matrix is even or odd by communicating only $\mathcal{O}(kn/2^k)$ bits. To do this, divide the matrix A into blocks with at most $2^{k-1} - 1$ rows in each. For each block there will be a string v' of length $k - 1$ such that neither $(0, v')$ nor $(1, v')$ occurs among the rows in that block. Using k bits the first player can make the string $(0, v')$ known to all players, and we are in the situation of the previous exercise.

Comment: Using similar arguments, Grolmusz (1999) proved the following general upper bound. The L_1-norm of a function $f : \{-1, 1\}^n \to \{-1, 1\}$ is the sum

$$L_1(f) := \sum_{S \subseteq [n]} \left| \widehat{f}(S) \right| = 2^{-n} \sum_{S \subseteq [n]} \left| \sum_x f(x) \prod_{i \in S} x_i \right|$$

of the absolute values of its Fourier coefficients $\widehat{f}(S)$. If $L_1(f) = M$ then $c_k(f)$ is at most about $k^2 2^{-k} n M \log(n M)$. In particular, for k about $\log M$, $c_k(f)$ is at most about $(\log(n M))^3$.

5.7. Consider the following multiparty game with a *referee*. As before, we have an $n \times k$ 0-1 matrix A, and the i-th player can see all A except its i-th column. The restriction is that now the players do not communicate with each other but simultaneously write their messages on the blackboard. Using only this information (and without seeing the matrix A), an additional player (the referee) must compute the string $P(A) = (x_1, \ldots, x_n)$, where x_i is the sum modulo 2 of the number of 1's in the i-th row of A. Let N be the maximal number of bits which any player is allowed to write on any input matrix. Prove that $N \geq n/k$.

Hint: For a matrix A, let $f(A)$ be the string $(p_1, \ldots, p_k)$, where $p_i \in \{0, 1\}^N$ is the string written by the i-th player on input A. For each possible answer $x = (x_1, \ldots, x_n)$ of the referee, fix a matrix A_x for which $P(A_x) = x$. Argue that $f(A_x) \neq f(A_y)$ for all $x \neq y$.

5.8. Let H be a hypergraph on n vertices, and $2 \leq k \leq n$ be a divisor of n Suppose that $|e| \leq k - 1$ for all $e \in H$. Show that for any balanced partition of the input into k parts, there is a k-party communication protocol evaluating GIP_H using at most k bits of communication.

Hint: Given a partition of n vertices into k blocks, each $e \in H$ must lie outside at least one of these blocks.

5.9. Let us consider simultaneous messages n-party protocols for the parity function $f(x) = x_1 \oplus x_2 \oplus \cdots \oplus x_n$ in which an additional player (the referee) just outputs the majority of the answers of players. Consider the following strategy for players:

> Each player looks around at everybody else. If a player sees as many 0's as 1's, she sends a value 0. Otherwise, she assumes that the bit on her forehead is the same as the majority of the bits she sees; she then she sends a value consistent with this assumption.

Show that this strategy has a success probability $1 - \frac{1}{\theta(\sqrt{n})}$, that is, will correctly compute the parity for all 2^n but a fraction $1/\theta(\sqrt{n})$ of input vectors.

5.10. (The card-flipping game, due to J. Edmonds and R. Impagliazzo)   Suppose that we have two 0-1 vectors $u = (u_1, \ldots, u_n)$ and $v = (v_1, \ldots, v_n)$ of length n. We want to decide whether $u = v$, but our access to the bits is very limited: at any moment we can see at most one bit of each pair of the bits u_i and v_i. We can imagine the corresponding bits to be written on two sides of a card, so that we can see all the cards, but only one side of each card. A *probe* consists in flipping of one or more of the cards. After every probe we can write down some information but the memory is not reusable: after the next probe we have to use new memory (that is, we cannot wipe it out). Moreover, this is the only memory for us: seeing the information written here (but not the cards themselves), we ask to flip some of the cards; seeing the actual values of the cards and using the current information from the memory, we either give an answer or we write some additional bits of information in the memory; after that the cards are closed for us, and we make the next probe. Suppose we are charged for every bit of memory that we use as well as for the number of probes. The goal is to decide if both sides of all cards are the same using as little of memory and as few probes as possible.

Of course, n bits of memory are always enough: simply write u in the memory, and flip all the cards to see v:

(a) Let $n = r^2$ for some $r \geq 1$. Show that it is possible to test the equality of two vectors in $\{0, 1\}^n$ using only $r + 1$ probes and writing down only r bits in the memory.

   Hint: Split the given vectors u and v into r pieces of length r: $u = (u^1, \ldots, u^r)$ and $v = (v^1, \ldots, v^r)$. During the i-th probe flip the cards of the i-th piece; compute the vector $w_i := u^1 \oplus \cdots \oplus u^{i-1} \oplus v^i \oplus u^{i+1} \oplus \cdots \oplus u^r$ and just test if the obtained vector w_i coincides with the vector $w_0 := u^1 \oplus \cdots \oplus u^r$ (written in the memory).

(b) Improve this to $\mathcal{O}(\log n)$ probes at each of which only $\mathcal{O}(\log n)$ bits written in the memory.

   Hint: (Due to Pudlák and Sgall 1997) Think of u and v as 0-1 vectors in real vector space $\mathbb{R}^n$. Compute (a square of) the Euclidean distance $\|u - v\|^2 = \langle u, u \rangle + \langle v, v \rangle - 2\langle u, v \rangle$ of u and v, and check if it is 0.

Part III
Circuit Complexity

Chapter 6
Formulas

Although for general non-monotone *circuits* no super-linear lower bounds are
known, the situation for *formulas* is somewhat better: here we are able to prove
quadratic and even super-quadratic lower bounds. These bounds are achieved using
remarkably diverse arguments: counting, random restrictions, set covering and
graph entropy.

6.1   Size Versus Depth

If not stated otherwise, by a formula we will mean a DeMorgan formula, that is,
a formula with fanin-2 AND and OR gates whose inputs are variables and their
negations. By $L(f)$ we denote the minimal leafsize and by $D(f)$ the minimal depth
of a DeMorgan formula computing a given boolean function f.

Since the underlying graph of a DeMorgan formula is a binary tree, any formula
of depth d can have at most 2^d leaves. This implies that, for every boolean
function f,

$$D(f) \geq \log_2 L(f).$$

A natural question is: can formulas be balanced? More precisely: does there exist
a constant c such that every formula of leafsize L can be transformed into an
equivalent formula of depth at most $c \log_2 L$?

The question was first affirmatively answered by Khrapchenko: boolean formulas
over any finite complete basis can be balanced; see Yablonskii and Kozyrev (1968,
p. 5). For the formulas over the basis $\{\wedge, \vee, \neg\}$, this result was independently proved
by Spira (1971) with $c < 3.42$, and by Brent et al. (1973) with $c < 2.47$. The
constant c was then improved[1] to $c < 2.16$ by Preparata and Muller (1975), to

---

[1] The results of Brent et al. (1973) and Preparata and Muller (1975) actually hold for formulas over
any commutative ring with the multiplicative identity, not only for boolean formulas.

S. Jukna, *Boolean Function Complexity*, Algorithms and Combinatorics 27,
DOI 10.1007/978-3-642-24508-4_6, © Springer-Verlag Berlin Heidelberg 2012

$c \leq 2$ by Barak and Shamir (1976), then to $c \leq 1.82$ by Preparata and Muller (1976), and finally to $c \leq 1.73$ by Khrapchenko (1978).

All these improvements of the constant c were obtained by going deeper and deeper in the actual structure of a given formula. But the main idea is the same: choose a particular subformula Y, balance it, then balance the rest of the formula, and finally combine these balanced formulas to obtain a balanced version of the original formula. We will demonstrate this idea with a proof for $c = 1.82$.

Lemma 6.1. (Formula Balancing Lemma) *For every boolean function f,*

$$D(f) \leq 1.82 \log_2 L(f).$$

Proof. First, observe that it is enough to consider *monotone* formulas, that is, formulas over the basis $\{\wedge, \vee\}$. Indeed, if we replace all leaves labeled by a negated variable $\neg x_i$ ($i = 1, \dots, n$) in a DeMorgan formula $F(x)$ by a new variable y_i, then we obtain a monotone formula $F'(x, y)$ with the property that $F(x) = F'(x, \neg x)$. We then can replace the y-variables in a balanced equivalent of F' by the corresponding negations of x-variables to obtain a balanced formula for $F(x)$.   □

So we only need to show that every monotone formula with m leaves can be transformed into an equivalent monotone formula of depth at most $c \log_2 m$. To warm-up, we first prove this with a worse constant $c = 3$; this argument is due to Brent et al. (1973).

We argue by induction on m. The claim is trivially true for $m = 2$. Now assume that the claim holds for all formulas with fewer than m leaves, and prove it for formulas with m leaves. Take an arbitrary monotone formula F with m leaves. By walking from the output-gate of F we can find a subformula Y such that Y has $\geq m/2$ leaves but its left as right subformulas each have $< m/2$ leaves. Now replace the subformula Y of F by constants 0 and 1, and let F_0 and F_1 be the resulting formulas. The key observation is that, due to the monotonicity, $F_1(x) = 0$ implies $F_0(x) = 0$. Thus the following formula

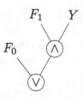

$$(6.1)$$

is equivalent to F. The formulas F_0 and F_1 as well as the left and right subformulas of Y each have at most $m/2$ leaves. By the induction hypothesis, F_0 and F_1 can be replaced by formulas of depth at most $3 \log_2(m/2)$, and the formula Y can be replaced by a formula of depth at most $1 + 3 \log_2(m/2)$. Thus, the resulting entire formula is equivalent to F and has depth at most $2 + 1 + 3 \log_2(m/2) = 3 \log_2 m$, as desired.

To improve the constant c to 1.82, we must choose the subformulas F_0 and F_1 more carefully. The point is that we can take the *larger* of these subformulas to be the last subformula in the constructed formula. The formula $F(x_1, \ldots, x_n)$ has the form $F = H(Y, x)$ where $x = (x_1, \ldots, x_n)$ and $H(y, x)$ is some monotone formula of $n + 1$ variables. Now look at the path from Y to the root of F:

Then F can be written as

$$H(Y, x) = H_t *_t (H_{t-1} *_{t-1} (\ldots (H_2 *_2 (H_1 *_1 Y))))$$

where each $*_i$ is either $\wedge$ or $\vee$. Let $H^\wedge$ be the AND of all H_i feeding into $\wedge$-gates, and $H^\vee$ be the OR of all H_i feeding into $\vee$-gates:

$$H^\wedge := \bigwedge_{i: *_i = \wedge} H_i \quad \text{and} \quad H^\vee := \bigvee_{i: *_i = \vee} H_i.$$

We set $H^\wedge = 1$ if there are no $\wedge$-gates, and set $H^\vee = 0$ if there are no $\vee$-gates along the path. It can be verified that

$$H^\wedge \vee H(0, x) = H(1, x), \tag{6.2}$$

$$H^\vee \wedge H(1, x) = H(0, x). \tag{6.3}$$

For this, take a vector $x \in \{0, 1\}^n$. If $H(1, x) = 0$ then $H(0, y) = 0$ by the monotonicity of H, and $H^\wedge(x) = 0$ because otherwise H would output 1 on input $(1, x)$. On the other hand, if $H(1, x) = 1$ but $H(0, x) = 0$, then H depends on its first input Y. Since neither $H^\wedge$ not $H^\vee$ depend on this input, $H^\wedge(x) = 1$ must hold. This proves (6.2); (6.3) follows by a dual argument.

The equalities (6.2) and (6.3) imply that

$$H(Y, x) = (Y \wedge H^\wedge(x)) \vee H(0, x), \tag{6.4}$$

$$H(Y, x) = (Y \vee H^\vee(x)) \wedge H(1, x). \tag{6.5}$$

After these preparations, we construct the desired balanced formula by induction on the leafsize $m = L(F)$ of a given formula F. Let $a > 1$ be a parameter satisfying the following two inequalities:

$$1 - \frac{1}{a^3} \le \frac{1}{a} \quad \text{and} \quad \frac{1}{2}\left(1 - \frac{1}{a^3}\right) \le \frac{1}{a^2}. \tag{6.6}$$

Our goal is to show that F can be transformed into an equivalent formula of depth at most $\log_a m$. Assume that $L(H^\wedge) \le L(H^\vee)$. (The case when $L(H^\wedge) > L(H^\vee)$ is treated similarly by using (6.5) instead of (6.4).) As before, we can find a subformula Y of F such that Y has $> m/a^3$ leaves but its left as well as right subformula has $\le m/a^3$ leaves. By (6.4), the formula

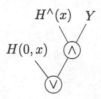

is equivalent to F. By (6.6),

$$L(H(0, x)) \le L(H) \le m\left(1 - \frac{1}{a^3}\right) \le \frac{m}{a}, \tag{6.7}$$

$$L(H^\wedge)) \le \frac{1}{2}L(H) \le \frac{m}{2}\left(1 - \frac{1}{a^3}\right) \le \frac{m}{a^2}. \tag{6.8}$$

By the induction hypothesis,

- $H(0, x)$ has an equivalent formula of depth $\le \log_a(m/a) = \log_a m - 1$;
- $H^\wedge$ has an equivalent formula of depth $\le \log_a(m/a^2) = \log_a m - 2$;
- Y has an equivalent formula of depth $\le \log_a(m/a^3) + 1 = \log_a m - 2$.

Thus, the entire formula has depth at most $\log_a m$, as desired. It remains to choose the parameter $a > 1$ satisfying (6.6) (the larger a is, the better upper bound we obtain). In particular, we can take $a = 1.465$. This results in the upper bound on the depth of the form $c \log_2 m$ where $c = 1/\log_2 a = 1.82$. □

When reducing the depth of a formula with m leaves, the leafsize m' of the new formula increases. The reduction above (for the case $c = 3$) leads to $m' = \mathcal{O}(m^\alpha)$ with α about 2. Bshouty et al. (1991) showed that m' can be made smaller at the cost of a somewhat larger depth: for every $k \ge 2$, every DeMorgan formula with m leaves can be transformed to an equivalent formula of depth at most $(3k \ln 2) \log_2 m \approx 2.08k \log_2 m$ and of leafsize at most m^α, where $\alpha = 1 + (1 + \log_2(k - 1))^{-1}$. A simpler proof was later found by Bonet and Buss (1994).

Lemma 6.1 states that formulas can be balanced: $D(f) = \Theta(\log L(f))$. Nothing similar is known for *circuits*. Let $C(f)$ denote the minimum number of gates in a circuit over the basis $\{\wedge, \vee, \neg\}$ computing f. It can be easily shown that $D(f) = \Omega(\log C(f))$. But the best *upper* bound on depth in terms of circuit size is

$$D(f) = \mathcal{O}(C(f)/\log C(f)). \qquad (6.9)$$

This was first proved by Paterson and Valiant (1976), and later proved by Dymond and Tompa (1985) using another method. For functions whose circuits are of exponential size, this is essentially a tight bound. However, for functions that can be computed by subexponential-size circuits, there is still a large gap between $\log C(f)$ and $C(f)/\log C(f)$. In the class of *layered* circuits a better bound $D(f) = \mathcal{O}(\sqrt{C(f)\log C(f)})$ was recently proved by Gál and Jang (2011). A circuit is layered if its set of gates can be partitioned into subsets called layers, such that every wire in the circuit is between adjacent layers.

Let $D(n) = \max D(f)$ where the maximum is over all boolean functions of n variables. By considering DNFs or CNFs (depending on whether $|f^{-1}(1)| \le 2^{n-1}$ or not), we see that $D(n) \le n + \log_2 n$. McColl and Paterson (1977) improved this to $D(n) \le n + 1$, then Gashkov (1978) improved this to $D(n) \le n - \log_2\log_2 n + 2 + o(1)$, and finally, Lozhkin (1983) improved this to

$$D(n) \le n - \lfloor \log_2\log_2 n - o(1) \rfloor,$$

which is already optimal up to the $o(1)$ factor. The optimality follows from $D(f) \ge \log_2 L(f)$ together with the fact (Theorem 1.23) that most boolean functions require formulas of leafsize at least $(1 - \epsilon)2^n / \log_2 n$.

Lozhkin (1981) has also proved the following interesting result: if a boolean function f can be computed by a depth-d DeMorgan formula using *unbounded* fanin AND and OR gates and having S leaves, then

$$D(f) \le d - 1 + \lceil \log_2 S \rceil.$$

Note that a trivial upper bound, obtained by simulating each gate by a tree, is only $D(f) = \mathcal{O}(d \log S)$. A similar result was independently proved by Hoover et al. (1984). They consider a dual question of reducing the fanout of gates, not their fanin. What they proved is the following. Take an arbitrary fanin-2 circuit with n inputs and m outputs. Let S be the size and D the depth of that circuit. Then there is an equivalent fanin-2 circuit of fanout-2 whose size is at most $3S - 2n$ and depth is at most $2D + \log m$.

The original proofs of these results were somewhat involved. Gashkov (2007) observed that the proof idea in both papers is actually based on the following simple fact which may be of independent interest. Let $q \ge 2$ be an integer. A q-ary tree T is a rooted tree, each node of which has at most q children.

Theorem 6.2. *Every tree with S leaves and depth d can be transformed into a q-ary tree with S leaves and depth D satisfying $q^D \le q^d S$.*

Proof. A well-known (and easy to prove) result of Kraft (1949) and McMillan (1956) states that a q-ary tree with m leaves at depths $\ell_1, \ldots, \ell_m$ exist if and only if $\sum_{i=1}^{m} q^{-\ell_i} \le 1$. We will need the following "weighted" version of this fact. Let T be a q-ary tree with t leaves weighted by natural numbers $w_1, \ldots, w_t$. Let l_i be the depth of the i-th leaf. Define the weighted depth of T as $\max\{\ell_1 + w_1, \ldots, \ell_t + w_t\}$. □

Claim 6.3. It is possible to transform T into a weighted q-ary tree with the same number of leaves and weighted depth D satisfying $q^D < q \sum_{i=1}^{t} q^{w_i}$.

Proof. Set $D := \lceil \log_q \sum_{i=1}^{t} q^{w_i} \rceil$; hence, $\sum_{i=1}^{t} q^{w_i} \leq q^D < q \sum_{i=1}^{t} q^{w_i}$. Consider the full q-ary tree T' of depth D. Let $w_1 \geq \ldots \geq w_t$. First declare any node v_1 of T' at depth $D - w_1 = \ell_1$ a leaf, remove all of its descendants, and give v_1 weight w_1. This removes $q^{D-\ell_1} = q^{w_1}$ leaves of T' from being considered for the rest of the remaining leaves. In the next iteration choose any node v_2 from the remaining tree at depth $D - w_2 = \ell_2$, not on the path to v_1, remove all of its descendants, and give it weight w_2. This removes a $q^{D-\ell_2} = q^{w_2}$ leaves of T' from being considered for the rest of the remaining leaves. Since $\sum_{i=1}^{t} q^{w_i} \leq q^D$, we can continue this for t iterations, thus a desired tree with t leaves can be constructed.

We now prove the theorem by induction on d. The case $d = 1$ follows from Claim 6.3 with all $w_i = 0$. Now let $d \geq 2$, and suppose that the theorem holds for all trees of depth at most $d - 1$. Let $v_1, \ldots, v_m$ be the children of the root, and $T_1, \ldots, T_m$ subtrees of T rooted in these children. Let S_i be the number of leaves in T_i, and d_i the depth of T_i. By the induction hypothesis, each T_i can be transformed into a q-ary tree of depth w_i such that $q^{w_i} \leq q^{d-1} S_i$. By Claim 6.3. the entire tree T can be transformed into a q-ary tree with S leaves and depth D satisfying $q^D \leq q \sum_{i=1}^{m} q^{w_i} \leq q^d \sum_{i=1}^{m} S_i = q^d S$. □

Earlier, Lozhkin (1976) considered the behavior of the Shannon function $D(n)$ of the depth of circuits over arbitrary bases (including infinite ones) in the case where some of the basis functions are assumed to have zero delay, that is, do not contribute to the depth of a circuit. He proved that, depending on what basis we take and what gates have zero delay, only one of the following three situations can occur: either $D(n) = \alpha$, or $D(n) \sim \beta \log_2 n$, or $D(n) \sim \gamma n$. Here α, β and γ are non-negative constants depending only on the used basis and on the delays assigned to its functions; the dependence is explicitly given.

6.2 A Quadratic Lower Bound for Universal Functions

In order to prove super-linear lower bounds for formulas, one can use the idea of "universal functions", going back to Nechiporuk (1966).

Let $n = 2^r$ and consider two sequences of variables $z = (z_1, \ldots, z_r)$ and $y = (y_1, \ldots, y_n)$. Each assignment $a \in \{0, 1\}^r$ to the z-variables gives us a unique natural number $\text{bin}(a) = 2^{r-1}a_1 + \cdots + 2a_{r-1} + a_r + 1$ between 1 and n; we call $\text{bin}(a)$ the *code* of a. Consider the boolean function on $r + n$ variables defined by:

$$U_n(z, y) := y_{\text{bin}(z)}.$$

That is, given two vectors $z \in \{0, 1\}^r$ and $y \in \{0, 1\}^n$, we just compute the code $i = \mathrm{bin}(z)$ of z, and output the i-th bit y_i of y. This function is also known as *multiplexer* or a *storage access function*, and is "universal" in the following sense.

Proposition 6.4. *For every boolean function $h(z_1, \ldots, z_r)$ there is an assignment $b \in \{0, 1\}^n$ such that $U_n(z, b) = h(z)$.*

Proof. For each $a \in \{0, 1\}^r$ replace the variable $y_{\mathrm{bin}(a)}$ by the constant $h(a)$.   □

By Theorem 1.23, we know that boolean functions $h(z_1, \ldots, z_r)$ requiring DeMorgan formulas with $\mathrm{L}(h) \geq 2^{r-1}/\log r = n/2 \log \log n$ leaves exist. Thus, the function $U_n(z, y)$ also requires so many z-leaves. Of course, this lower bound is trivial because U_n depends on all its y-variables, and hence, requires at least n leaves. One can, however, boost the complexity of $U_n(z, y)$ by considering boolean functions of $2n$ variables of the form $f(x, y) = U_n(g(x), y)$, where $g : \{0, 1\}^n \to \{0, 1\}^k$ is some easily computable boolean operator. For this, take a set x of n variables, and arrange them into a $r \times m$ matrix

$$x = \begin{pmatrix} x_{11} & x_{12} & \cdots & x_{1m} \\ x_{21} & x_{22} & \cdots & x_{2m} \\ & & \cdots & \\ x_{r1} & x_{r2} & \cdots & x_{rm} \end{pmatrix}$$

where $m = n/r$; we assume that n is divisible by $r = \log_2 n$. If we take a boolean function $\varphi(u_1, \ldots, u_m)$ of m variables, and apply it to the rows of x, then we obtain r bits $z_i = \varphi(x_{i1}, x_{i2}, \ldots, x_{im})$, $i = 1, \ldots, r$. The *universal function induced by φ* is the following boolean function of $2n$ variables:

$$U_n^{\varphi}(x, y) := U_n(z_1, \ldots, z_r, y) \quad \text{where} \quad z_i = \varphi(x_{i1}, x_{i2}, \ldots, x_{im}).$$

We call φ the *generating function* of $U_n^{\varphi}(x, y)$.

   Our goal is now to exhibit a boolean function requiring about n^2 leaves in any *binary* formula, that is, in a formula where all $2^4 = 16$ boolean functions in two variables are allowed as gates. For this purpose, we take as generating function φ the OR of variables:

$$\mathrm{OR}(u_1, \ldots, u_m) = u_1 \vee u_2 \vee \cdots \vee u_m.$$

Note that the resulting boolean function $U_n^{\mathrm{OR}}(x, y)$ of $2n$ variables is explicitly given. The function first computes the ORs $z_i = x_{i1} \vee \cdots \vee x_{im}$ of variables along the rows of x, then computes the code $i = \mathrm{bin}(z)$ of the resulting vector $z = (z_1, \ldots, z_r)$, and finally outputs the i-th bit y_i of y.

Theorem 6.5. *Every binary formula computing U_n^{OR} must have at least $n^{2-o(1)}$ leaves.*

Proof. Let $F(x, y)$ be a binary formula computing $U_n^{OR}(x, y)$. Fix a boolean function $h(z_1, \ldots, z_r)$ of r variables requiring the largest binary formula. By Theorem 1.23, we know that any such formula must have at least

$$\frac{c2^r}{\log r} = \frac{cn}{\log \log n} \tag{6.10}$$

leaves, for a constant $c > 0$. By Proposition 6.4, there is an assignment $b \in \{0, 1\}^n$ to the y-variables of $U_n(z, y)$ such that $U_n(z, b) = h(z)$. Thus, the boolean function

$$f(x) := U_n^{OR}(x, b) = h\left(\bigvee_{j=1}^m x_{1j}, \bigvee_{j=1}^m x_{2j}, \ldots, \bigvee_{j=1}^m x_{rj}\right)$$

of n variables is a subfunction of $U_n^{OR}(x, y)$; recall that $m = n/r = n/\log n$.

In each row of the matrix x of variables, mark a variable which appears as a leaf of F the *smallest* number of times (when compared to the remaining variables of that row). Set all non-marked variables in x to 0. After this setting we obtain a formula F' on r marked variables such that F' computes h. Since we have fixed only the "most popular" variables of each row of x, the number of leaves in F must be at least m times larger than the number $L(F')$ of leaves in the obtained formula F'. Together with (6.10), this implies that the number of leaves in the formula $F(x, b)$, and hence also in the original formula $F(x, y)$, must be at least

$$m \cdot L(F') \geq m \cdot \frac{cn}{\log \log n} = \frac{cn^2}{\log n \log \log n}. \qquad \square$$

Remark 6.6. Note that the same argument works for the universal function $U_n^\varphi(x, y)$ induced by any(!) boolean function φ depending on all its m variables. Recall that a boolean function $\varphi(u_1, \ldots, u_m)$ depends on its i-th variable if there are constants $a_1, \ldots, a_{i-1}, a_{i+1}, \ldots, a_m$ in $\{0, 1\}$ such that the subfunction $\varphi(a_1, \ldots, a_{i-1}, u_i, a_{i+1}, \ldots, a_m)$ is either u_i or $\neg u_i$. Having a boolean function φ which depends on all its m variables, we can replace the unmarked variables by the corresponding constants. What we obtain may (apparently) not be the function $h(x_1, \ldots, x_r)$ itself but some function of the form $h'(x_1, \ldots, x_r) = h(x_1 \oplus \sigma_1, \ldots, x_r \oplus \sigma_r)$ where $\sigma_i \in \{0, 1\}$. This, however, does not matter because h' still requires $c2^r/\log r$ leaves.

Remark 6.7. In the definition of the universal function $U_n(z, y)$ of $r + n$ variables with $n = 2^r$ we used binary representation of positions i of vector y. One can also use other encodings of boolean functions. Recall that an *elementary conjunction* of r variables $z = (z_1, \ldots, z_r)$ is a boolean function of the form $K(z) = z_1^{a_1} z_2^{a_2} \cdots z_n^{a_n}$, where $a_1, \ldots, a_r \in \{0, 1\}$, $z_i^\sigma = 1$ if $a_i = \sigma$, and $z_i^\sigma = 0$ otherwise. Since each such conjunction accepts exactly one vector $a = (a_1, \ldots, a_n)$, every boolean function $f(z)$ of r variables can be represented as an OR of at most 2^r elementary conjunctions. Thus, if $K_1, \ldots, K_n$ are all $n = 2^r$ elementary conjunctions of variables z, then the boolean function

$$V_n(z, y) = \bigvee_{i=1}^{n} y_i \wedge K_i(z)$$

is also universal in the sense of Proposition 6.4. In particular, if we take this function instead of $U_n(z, y)$ and use the OR as the generating function φ, then the same argument as in the prof of Theorem 6.5 yields that resulting function $V_n^{OR}(x, y)$ of $2n$ variables also requires $n^{2-o(1)}$ leaves in any binary formula. Note that the function V_n^{OR} can be computed by a DeMorgan formula with $\mathcal{O}(n)$ unbounded fanin AND and OR gates.

6.3 The Effect of Random Restrictions

The quadratic lower bound, given in Theorem 6.5, is not quite satisfying because it only holds for some specially designed boolean functions. In fact, this lower bound does not give much more information about the power of formulas than the counting lower bounds we derived in Sect. 1.4.4. In particular, such a counting-based argument says nothing about the complexity of other boolean functions.

As early as 1961, Bella Abramovna Subbotovskaya, a student of Oleg Borisovitch Lupanov, found a more subtle lower bounds argument for DeMorgan formulas. Given a formula F computing some function f, her idea was to set randomly some of the variables to constants and show that this restriction reduces the size of F considerably whereas the resulting subfunction of f is not much easier. Subbotovskaya was actually the inventor of the "method of random restrictions", one of the most powerful tools for proving lower bounds.

Lemma 6.8. (Subbotovskaya 1961) *For every boolean function f of n variables, it is possible to fix one of its variables such that the resulting boolean function f' of $n - 1$ variables satisfies*

$$L(f') \leq \left(1 - \frac{1}{n}\right)^{3/2} \cdot L(f).$$

Proof. Let F be a minimal DeMorgan formula which computes $f(x_1, \ldots, x_n)$ and has $s = L(f)$ leaves (input gates). Since f has only n variables, at least one variable x_i must appear at least s/n times as a leaf, that is, at least s/n leaves are labeled by x_i or $\neg x_i$. Thus if we set x_i to a constant $c \in \{0, 1\}$, what we obtain is a formula of $n - 1$ variables with at most $s - s/n = (1 - 1/n)s$ leaves. But this is not the whole truth: when setting a variable to a constant we can expect to "kill off" not only the variable itself but also some other leaves labeled by other variables.

To show this, say that a subformula G of F is a *neighbor* of a leaf z, if $z \wedge G$ or $z \vee G$ is a subformula of F. □

Claim 6.9. *If* $z \in \{x_i, \neg x_i\}$ *is a leaf of* F, *then the neighbor of* z *does not contain the variable* x_i.

Proof. Let F be a minimal non-constant formula, $z \in \{x_i, \neg x_i\}$ one of its leaves, and G the neighbor of z in F. Hence, $H = z \wedge G$ (or $H = z \vee G$) is a subformula of F. For the sake of contradiction, assume that G contains a leaf $z' \in \{x_i, \neg x_i\}$. Replace this leaf by that constant $c \in \{0, 1\}$ for which the literal z get value 1. That is, replace the leaf z' by 1 if $z = x_i$, and by 0 if $z = \neg x_i$. (If $H = z \vee G$, then we set z' so that the literal z gets value 0.)

After this setting, the resulting subformula G' has one leaf fewer than G, but the resulting subformula $H' = x_i \wedge G'$ computes the same boolean function as $x_i \wedge G$. To see this, take an input vector $a \in \{0, 1\}^n$. If $a_i = c$ then $z(a) = 1$ and $G'(a) = G(a)$, implying that $H'(a) = H(a)$. If $a_i = c \oplus 1$ then $z(a) = 0$, and we again have that $H'(a) = 0 = H(a)$. Thus, we obtained a formula F' which computes the same boolean function as the original formula F but has one leaf fewer. This contradicts the minimality of F. □

Take now a variable x_i which appears $t \geq s/n$ times as a leaf of F. Let $z_1, \dots, z_t$ be the leaves labeled by x_i or $\neg x_i$. By Claim 6.9, for every $i = 1, \dots, t$ there is a constant $c_i \in \{0, 1\}$ such that, after setting $x_i = c_i$, the neighbor G_i of z_i will disappear from F, thereby erasing at least one more leaf which is *not* among the leaves $z_1, \dots, z_t$. Let $c \in \{0, 1\}$ be the constant which appears most often in the sequence $c_1, \dots, c_t$. If we set $x_i = c$, then all the leaves $z_1, \dots, z_t$ will disappear from the formula, and at least $t/2$ additional leaves will disappear because of these secondary effects. In total, we thus eliminate at least $t + t/2 \geq 3s/2n$ leaves, and the resulting formula has at most

$$s - \frac{3s}{2n} = s \cdot \left(1 - \frac{3}{2n}\right) \leq s \cdot \left(1 - \frac{1}{n}\right)^{3/2}$$

leaves, as claimed. □

Theorem 6.10. (Subbotovskaya 1961) *For every boolean function* f *of* n *variables and for every integer* $1 \leq k \leq n$, *it is possible to fix* $n - k$ *variables such that the resulting boolean function* f' *of* k *variables satisfies*

$$L(f') \leq \left(\frac{k}{n}\right)^{3/2} L(f).$$

Proof. Let $s = L(f)$. By applying Lemma 6.8 $n - k$ times, we obtain a formula of k variables with at most

$$s \cdot \left(1 - \frac{1}{n}\right)^{3/2} \left(1 - \frac{1}{n-1}\right)^{3/2} \cdots \left(1 - \frac{1}{k+1}\right)^{3/2} = s \cdot \left(\frac{k}{n}\right)^{3/2}$$

leaves. □

Example 6.11. (Parity function) Let $f = x_1 \oplus x_2 \oplus \cdots \oplus x_n$. If we fix all but one variables of f, we obtain a boolean function f' (a variable or its negation) requiring formula of leafsize 1. Thus, we can apply Theorem 6.10 with $k = 1$ and obtain that

$$1 \le L(f') \le \left(\frac{1}{n}\right)^{3/2} L(f),$$

which gives the lower bound $L(f) \ge n^{3/2}$.

In order to prove larger lower bounds, it is useful to restate Subbotovskaya's argument in probabilistic terms. Let f be a boolean function, and $X = \{x_1, \ldots, x_n\}$ the set of its variables. A *partial assignment* (or *restriction*) is a function ρ : $X \to \{0, 1, *\}$, where we understand $*$ to mean that the corresponding variable is unassigned. Each such partial assignment ρ yields a *restriction* (or a *subfunction*) f_ρ of f in a natural way: $f_\rho = f(\rho(x_1), \ldots, \rho(x_n))$. Note that f_ρ is a function of the variables x_i for which $\rho(x_i) = *$. For example, if

$$f = (x_1 \vee x_2 \vee x_3) \wedge (\neg x_1 \vee x_2) \wedge (x_1 \vee \neg x_3)$$

and $\rho(x_1) = 1$, $\rho(x_2) = \rho(x_3) = *$, then $f_\rho = x_2$.

Let $\mathcal{R}_k$ be the set of all partial assignments which leave exactly k variables unassigned. What we will be interested in is the *random* restrictions f_ρ that results from choosing a *random* partial assignment from $\mathcal{R}_k$. The probability distribution of restrictions in $\mathcal{R}_k$ we will use is the following: randomly assign k variables to be $*$, and assign all other variables to be 0 or 1 randomly and independently.

Theorem 6.10 states that, for every boolean function f, there *exists* a restriction $\rho \in \mathcal{R}_k$ such that $L(f_\rho) \le (k/n)^{3/2}L(f)$. We now show that, in fact, this happens for at least a 3/4 fraction of all restrictions in $\mathcal{R}_k$.

Lemma 6.12. *Let f be a boolean function of n variables, and let ρ be a random restriction from $\mathcal{R}_k$. Then, with probability at least 3/4,*

$$L(f_\rho) \le 4 \left(\frac{k}{n}\right)^{3/2} L(f).$$

Proof. The argument is actually the same as in the proof of Theorem 6.10. Let F be an optimal DeMorgan formula for the function f of size $s = L(f)$. Construct the restriction ρ in $n - k$ stages as follows: At any stage, choose a variable randomly from the remaining ones, and assign it 0 or 1 randomly. We analyze the effect of this restriction to the formula F, stage-by-stage.

Suppose the first stage chooses the variable x_i. When this variable is set to a constant, then all the leaves labeled by the literals x_i and $\neg x_i$ will disappear from the formula F. By averaging, the expected number of such literals is s/n. Since x_i is assigned 0 or 1 randomly with equal probability 1/2, we can expect (by Claim 6.9) at least $s/2n$ *additional* leaves to disappear. In total, we thus expect at least $s/n + s/2n = 3s/2n$ leaves to disappear in the first stage, yielding a new formula with expected size at most $s - 3s/2n \leq s(1 - 1/n)^{3/2}$. Subsequent stages of the restriction can be analyzed in the same way. After each stage the number of variables decrements by one. Hence, after $n - k$ stages, the expected leafsize of the final formula is at most $s(k/n)^{3/2}$. By Markov's inequality, the probability that the random variable $L(f_\rho)$ is more than four times its expected value is smaller than $1/4$, which completes the proof. $\Box$

Subbotovskaya's result can be stated for more general probability distributions. Suppose that p is a real number between 0 and 1. A *p-random restriction* independently assigns each variable x_i the value 0 or 1 with equal probabilities $(1 - p)/2$, and with the remaining probability p keeps x_i unfixed. Thus, the distributions we have considered above correspond to $p = k/n$. What will be the expected formula size of the induced function when we apply a random restriction? The obvious answer is that this size will be at most pL.

What Subbotovskaya actually shows is that formulas shrink more. Namely, she establishes an upper bound $\mathcal{O}(p^{3/2}L)$ on the expected formula size of the induced function. Her work and the subsequent result of Andreev (Theorem 6.14 below) motivated consideration of the *shrinkage exponent* Γ of DeMorgan formulas. This number Γ is defined as a largest number such that, if a boolean function f has a DeMorgan formula of size L, then the expected formula size of the induced function is $\mathcal{O}(p^\Gamma L)$.

Impagliazzo and Nisan (1993) showed that $\Gamma \geq 1.55$, then Paterson and Zwick (1993) showed $\Gamma \geq 1.63$, and finally Håstad (1998) proved that $\Gamma = 2$.

Theorem 6.13. (Håstad 1998) *If we apply a p-random restriction to a DeMorgan of leafsize L, then the expected remaining leafsize is at most $\mathcal{O}(p^2 L)$.*

6.4 A Cubic Lower Bound

Andreev (1987a) was the first to prove a super-quadratic lower bound for DeMorgan formulas. His idea was to combine the method of Subbotovskaya with Nechiporuk's method of universal functions (discussed in Sect. 6.2) For this purpose, Andreev considered the universal function generated by the parity function:

$$\oplus(u_1, \ldots, u_m) = u_1 \oplus u_2 \oplus \cdots \oplus u_m.$$

The resulting function is then a boolean function $U_n^\oplus(x, y)$ of $2n$ variables, where $n = 2^r$ is a power of 2. The first variable x are arranged into an $r \times m$ matrix

$$x = \begin{pmatrix} x_{11} & x_{12} & \cdots & x_{1m} \\ x_{21} & x_{22} & \cdots & x_{2m} \\ & & \cdots & \\ x_{r1} & x_{r2} & \cdots & x_{rm} \end{pmatrix}$$

where $m = n/r$; we assume that n is divisible by $r = \log_2 n$. The function first computes the parities $z_i = x_{i1} \oplus \cdots \oplus x_{im}$ of variables along the rows of x, then computes the code $i = \text{bin}(z)$ of the resulting vector $z = (z_1, \ldots, z_r)$, and finally outputs the i-th bit y_i of y.

Theorem 6.14. (Andreev 1987a) $L(U_n^{\oplus}) \geq n^{5/2 - o(1)}$.

Proof. Fix a boolean function $h(z_1, \ldots, z_r)$ of r variables requiring the largest DeMorgan formula. By Theorem 1.23, we know that any such formula must have at least

$$\frac{2^{r-1}}{\log r} = \frac{n}{2 \log \log n} \tag{6.11}$$

leaves. Note that, if we replace some variable z_i in h by their negations, the resulting boolean function will require this number of leaves. By Proposition 6.4, there is an assignment $b \in \{0, 1\}^n$ to the y-variables of $U_n(z, y)$ such that $U_n(z, b) = h(z)$. Thus, the boolean function

$$f(x) := U_n^{\oplus}(x, b) = h\left(\bigoplus_{j=1}^{m} x_{1j}, \bigoplus_{j=1}^{m} x_{2j}, \ldots, \bigoplus_{j=1}^{m} x_{rj}\right)$$

of n variables is a subfunction of $U_n^{\oplus}(x, y)$. Recall that $x = (x_{ij})$ is an $r \times m$ matrix of boolean variables for $r = \log_2 n$ and $m = n/r$. Let ρ be a random restriction from $\mathcal{R}_k$ on the x-variables where $k = \lceil r \ln(4r) \rceil$. Our first goal is to show that, with a large probability, at least one variable in each row of x will remain unfixed by ρ:

$$\text{Prob}[\rho \text{ assigns an } * \text{ to each row of } x] \geq \frac{3}{4}. \tag{6.12}$$

To show this, observe that the restriction ρ assigns an $*$ to each single variable with probability $\binom{n-1}{k-1} / \binom{n}{k} = \frac{k}{n}$. By the union bound, the probability that some of r rows will get no $*$ is at most

$$r \cdot \left(1 - \frac{k}{n}\right)^m \leq r \cdot e^{-\frac{km}{n}} \leq r \cdot e^{-\ln(4r)} = 1/4.$$

On the other hand, Lemma 6.12 implies that

$$\text{Prob}[\, L(f_\rho) \leq 4 \left(\tfrac{k}{n}\right)^{3/2} L(f)\,] \geq \frac{3}{4}. \tag{6.13}$$

Some restriction ρ will thus satisfy both conditions (6.12) and (6.13). Fix such a restriction ρ. By (6.12), the function h is a subfunction of f_ρ, whereas by (6.13), $L(f_\rho)$ is at least $4(k/n)^{3/2}$ times smaller than $L(f)$. Recalling that $k = \lceil r \ln(4r) \rceil = \mathcal{O}(\log n \log \log n)$ and using (6.11), for such a restriction ρ we obtain

$$L(U_n^{\oplus}) \geq L(f) \geq \frac{1}{4} \left(\frac{n}{k}\right)^{3/2} L(f_\rho)$$

$$\geq \frac{1}{4} \left(\frac{n}{k}\right)^{3/2} L(h)$$

$$\geq \frac{1}{4} \left(\frac{n}{k}\right)^{3/2} \frac{n}{2 \log \log n}$$

$$\geq n^{5/2 - o(1)}.$$ $\qquad\qquad$ □

The proof of Theorem 6.14 actually gives the lower bound

$$L(U_n^{\oplus}) = \Omega \left(n^{\Gamma + 1 - o(1)}\right),$$

where Γ is the shrinking exponent of DeMorgan formulas. Andreev used Subbotovskaya's lower bound $\Gamma \geq 3/2$. Using Håstad's improvement $\Gamma \geq 2$ (see Theorem 6.13), we immediately obtain a larger lower bound for Andreev's function.

Theorem 6.15. (Håstad 1998) $L(U_n^{\oplus}) \geq n^{3 - o(1)}$.

6.5 Nechiporuk's Theorem

Nechiporuk (1966) found another argument which works for *binary* formulas where all $2^4 = 16$ boolean functions in two variables are allowed as gates. Let $L_B(f)$ denote the minimum leafsize of a binary formula computing f.

A *subfunction* of a boolean function $f(X)$ on $Y \subseteq X$ is a function obtained from f by setting all the variables of $X \setminus Y$ to constants.

Theorem 6.16. (Nechiporuk 1966) *Let f be a boolean function on a variable set X, let $Y_1, Y_2, \ldots, Y_m$ be disjoint subsets of X, and let s_i be the number of distinct subfunctions of f on Y_i. Then*

$$L_B(f) \geq \frac{1}{4} \sum_{i=1}^{m} \log s_i.$$

Proof. Let F be an optimal binary formula for f and let l_i be the number of leaves labeled by variables in Y_i. It is sufficient to prove that $l_i \geq (1/4) \log s_i$. Let L_i be the set of all leaves in F labeled by variables from Y_i. Consider the subtree T_i consisting of all these leaves and all paths from these leaves to the output of F. Each node of T_i has in-degree 0, 1 or 2. Let W_i be the set of nodes of in-degree 2. Since $|W_i| \leq l_i - 1$, it is enough to lower-bound the number $|W_i|$ of such nodes in terms of the number s_i of different Y_i-subfunctions of f.

Let P_i be the set of all paths in T_i starting from a leaf in L_i or a node in W_i and ending in a node in W_i or at the root of T_i and containing no node in W_i as inner node. There are at most $|W_i| + 1$ different end-points of paths in P_i. Moreover, at most two of these path can end in the same node v. These paths can be found by starting in v and going backwards until a node in W_i or a leaf is reached. Hence

$$|P_i| \le 2(|W_i| + 1). \tag{6.14}$$

Assignments to the variables outside Y_i must lead to s_i different subformulas. Fix such an assignment α. If we remove from F all gates that are evaluated to a constant 0 or 1 by α, then what we obtain is precisely the tree T_i. That is, the subfunction $F(Y_i, \alpha)$ is computed by the gates of T_i whose fanin-1 gates correspond to fanin-2 gates of F with one of the input gates replaced by a constant. So, if p is a path in P_i, and h is a function computed at the first gate of p (after an assignment α) then the function computed at the last edge of p is 0, 1, h or $\neg h$. Thus, under different assignments α at most $4^{|P_i|}$ subfunctions can be computed, implying that $s_i \le 4^{|P_i|}$. Together with (6.14), this implies that $|W_i| \ge (1/4) \log s_i - 1$. Since $|W_i| \le l_i - 1$, this gives the desired lower bound $l_i \ge (1/4) \log s_i$ on the number of leaves labeled by variables in Y_i. □

Recall that the *element distinctness function* ED_n is a boolean function of $n = 2m \log m$ variables divided into m consecutive blocks with $2 \log m$ variables in each of them. Each of these blocks encode a number in $[m^2]$. The function accepts an input $x \in \{0, 1\}^n$ iff all these numbers are distinct.

By considering $\binom{m}{2}$ subformulas, each testing the distinctness of one pair of blocks, we see that ED_n can be computed by a DeMorgan formula with about $\binom{m}{2} \log m = \mathcal{O}(n^2 / \log n)$ leaves. On the other hand, we have already shown that ED_n has $2^{\Omega(n)}$ subfunctions on each of these $m = \Omega(n / \log n)$ blocks (see Lemma 1.1). Thus, Nechiporuk's theorem immediately yields.

Theorem 6.17. *A minimal binary formula computing* ED_n *has* $\Theta(n^2 / \log n)$ *leaves.*

Thus, Nechiporuk's theorem can be used to prove almost quadratic lower bounds.

Remark 6.18. Unfortunately, the theorem is inherently unable to give a lower bound larger than $\Theta(n^2 / \log n)$. To see this, take an arbitrary partition $Y_1, \dots, Y_m$ of the set X of n variables of a boolean function. Let $t_i := |Y_i|$ and let s_i be the number of subfunctions of f on Y_i. Then $\log s_i \le \min\{n - t_i, 2^{t_i}\}$. Assume w.l.o.g. that only the first k of the block-lengths $t_1, \dots, t_m$ are larger than $\log n$. Since the blocks are disjoint, this implies that $k < n / \log n$, and the contribution of the first k blocks is

$$\sum_{i=1}^{k} \log s_i \le \sum_{i=1}^{k} (n - t_i) \le n^2 / \log n.$$

Each of the remaining $m - k$ blocks can contribute at most $\log s_i \le 2^{t_i} \le n$. Since the function $x \mapsto 2^x$ is convex, the sum along these blocks is maximized when as many of the t_i as possible are near to their upper bound $\log n$, that is, when

$m - k = O(n/\log n)$. Thus, the total contribution of the remaining blocks is also at most $mn = O(n^2/\log n)$.

Remark 6.19. Nechiporuk's theorem has no analogue for *circuits*. Namely, Uhlig (1991) showed that there exist boolean functions f of n variables such that f has about 3^n subfunctions, but f can be computed by a circuit of depth $\log n$ using at most $2n$ fanin-2 boolean functions as gates (cf. Exercise 6.1). Interestingly, earlier Uhlig (1974) showed that, for every $\gamma \in [0, 1]$, the class of all boolean functions with at least $\gamma 3^n$ subfunctions *contains* functions requiring circuits of size $c_\gamma 2^n/n$.

6.6 Lower Bounds for Symmetric Functions

Recall that a boolean function $f(x_1, \ldots, x_n)$ is *symmetric* if its value only depends on the number $|x| = x_1 + x_2 + \cdots + x_n$ of 1s in the input vector. Thus, every such function is specified by boolean vector $v = (v_0, v_1, \ldots, v_n)$ in $\{0, 1\}^{n+1}$ such that $f(x) = 1$ iff $v_{|x|} = 1$; the vector v is the *characteristic vector* of f. If f is a symmetric boolean function, then f can have at most $n - |Y| + 1$ distinct subfunctions on any set Y of variables. Thus, Nechiporuk's method cannot yield superlinear lower bounds for symmetric functions. Khrapchenko's method, which we will present in the next section, can yield even quadratic lower bounds for symmetric functions, but it only works for DeMorgan formulas.

Superlinear lower bounds $\mathrm{L}_\Omega(f) = \Omega(n\alpha(n))$ for formulas over an *arbitrary* finite complete basis Ω computing symmetric functions f were proved by Hodes and Specker (1968). Here $\alpha(n)$ is a very slowly growing function, slower than a k-fold logarithm of n, for every k. Their method was applied by Khrapchenko (1976) and Paterson and Valiant (1976), who proved superlinear lower bounds for all symmetric boolean functions of n variables, except for 16 functions which can be computed with linear-size formulas. The 16 exceptional functions have the form $f(x) = \alpha_0 \oplus \alpha_1 P(x) \oplus \alpha_2 K(x) \oplus \alpha_3 \overline{K}(x)$, where the coefficients α_i take values 0 and 1, $P(x) = x_1 \oplus \cdots \oplus x_n$, $K(x) = x_1 \wedge \cdots \wedge x_n$, and $\overline{K}(x) = \overline{x}_1 \wedge \cdots \wedge \overline{x}_n$.

Pudlák (1984b) substantially enhanced their method to prove that $\mathrm{L}_\Omega(f) \geq c_\Omega n \log \log n$ holds for all symmetric boolean functions of n variables, with an exception of only 16 functions described above; here $c_\Omega > 0$ is a constant depending only on the basis Ω.

Pudlák's argument is based on the following structural property of boolean functions computable by small formulas. Say that a symmetric boolean function f is *homogeneous* if its characteristic vector v has the same values in all odd positions, and has the same values in all even position, except possibly in position 0. That is, when ignoring the all-0 vector, such a function is either a constant function, or the parity function, or the negation of the parity function. In other words, every homogeneous function has the forms $f(x) = \alpha_0 \oplus \alpha_1 P(x) \oplus \alpha_2 \overline{K}(x)$.

A boolean function $f(X)$ of n variables is *r-homogeneous* if there exists a subset $Y \subseteq X$ of $|Y| = n - r$ variables such that the subfunction of f obtained by setting

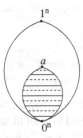

Fig. 6.1 After setting to 0 all variables in Y, the function $f(X)$ turns to a homogeneous function $f'(Y)$. The function $f'(Y)$ has the same values in all odd levels, and has the same values in all even levels, except perhaps the zero-level. Here a is the boolean vector with ones in position in Y, and zeros elsewhere

to 0 all variables in Y is a homogeneous symmetric boolean function of r variables (see Fig. 6.1).

Theorem 6.20. (Pudlák 1984b) *Let $f(X)$ be a boolean function of n variables. If f is not r-homogeneous for some integer $r \geq 3$, then*

$$L_\Omega(f) \geq c_\Omega n(\log \log n - \log r).$$

The proof uses Ramsey-type arguments and is sketched, for example, in the book by Dunne (1988). To see how this theorem works, let us consider the threshold-2 function Th_2^n and take $r = 3$. If we set to 0 any $n - 3$ variables, then the obtained symmetric subfunction Th_2^3 is not homogeneous. Theorem 6.20 implies that $L_\Omega(\mathrm{Th}_2^n) \geq c_\Omega n \log \log n$.

Yet another lower bounds argument for the formula size of symmetric boolean functions was proposed by Fischer et al. (1982). Their argument works only for binary formulas (over the basis containing all boolean function of two variables) but the resulting lower bounds are stronger, up to $\Omega(n \log n)$.

To state their result, call a subfunction of a boolean function f *balanced* if $n_1 - n_0 \in \{0, 1\}$, where n_1 is the number of variables replaced by ones, and n_0 is the number of variables replaced by zeros . Let $a(f)$ denote the maximum number d such that the set of all balanced subfunctions of f contains a parity function of d variables or its negation.

Theorem 6.21. (Fischer et al. 1982) *There is a constant $c > 0$ such that for every boolean function f of n variables,*

$$L_B(f) \geq cn \log \frac{n}{a(f)}.$$

The proof can also be found in the book by Dunne (1988). We will only prove an important consequence of this theorem, giving us a simple lower-bounds criterion. Say that a boolean function f is *m-separated* if there exists a constant $a \in \{0, 1\}$

such that $f(x) = a$ for all input vectors x with exactly m ones, and $f(x) = a \oplus 1$ for all input vectors x with exactly $m + 2$ ones.

Theorem 6.22. *There is a constant $c > 0$ such that for every $0 \leq m \leq n - 2$ and every m-separated boolean function f of n variables, $L_B(f)$ is at least the minimum of $cn \log m$ and $cn \log(n - m)$.*

Proof. We can assume w.l.o.g. that $m \leq \lfloor n/2 \rfloor$. Since f is m-separated, we know that f takes some fixed value $a \in \{0, 1\}$ on all inputs with exactly m ones, and takes value $a \oplus 1$ on all inputs with exactly $m + 2$ ones.

We first prove the assertion for $m = \lfloor n/2 \rfloor$. It is sufficient to prove that $\mathrm{a}(f) \leq 2$. Let us consider an arbitrary balanced subfunction f' of f on three variables. That is, f' is obtained from f by setting some $n-3$ variables to constants 0 and 1. Moreover, $\lceil (n - 3)/2 \rceil$ variables have been replaced by ones, and $\lfloor (n - 3)/2 \rfloor$ variables have been replaced by zeros . Since $m - \lceil (n - 3)/2 \rceil = 1$, we have that $f'(1, 0, 0) = a$ but $f'(1, 1, 1) = a \oplus 1$. Thus, f' can be neither the parity of three variables nor its negation, and Theorem 6.21 implies that $L_B(f) \geq cn \log(n/2) \geq cn \log m$.

We now consider the case when $m < \lfloor n/2 \rfloor$. Let F be an optimal formula for f, and let r_i be "replication" of the i-variable in F, that is, the number of leaves of F labeled by x_i. Hence, F has $L(F) = \sum_{i=1}^{n} r_i$ leaves. Set to zero $n - 2m$ variables of the largest replication, and let F' be the resulting subformula of $2m$ variables. Since the function f was m-separated, its subfunction computed by F' is m-separated as well. By what we just proved above, F' must have at least $c(2m) \log m$ leaves. Thus, there must be a variable x_i such that at least $c \log m$ leaves of F' (and hence, also of F) are labeled by x_i. But then we must also have that $r_j \geq c \log m$ for all $n - 2m$ variables x_j which we replaced by zeros. Thus,

$$L_B(f) = L(F) \geq (n - 2m)c \log m + c(2m) \log m = cn \log m. \qquad \square$$

Theorem 6.22 yields superlinear lower bounds $L_B(f) = \Omega(n \log n)$ for *every* symmetric boolean function f of n variables whose characteristic vector $v = (v_0, v_1, \ldots, v_n)$ has a position m such that $v_m \neq v_{m+2}$ and both m and $n - m$ are at least $n^{\Omega(1)}$. In particular, we have such a lower bound for the majority function Maj_n, as well as for every counting function $\mathrm{Mod}_k^n(x) = 1$ iff $x_1 + \cdots + x_n \equiv 0 \bmod k$, as long as $k > 2$ and k is not too large.

6.7 Formulas and Rectangles

For DeMorgan formulas, that is, for formulas over the basis $\{\wedge, \vee, \neg\}$, we have yet another lower bounds argument, due to Khrapchenko (1971). He used this argument to prove an n^2 lower bound for the parity function. Later, Rychkov (1985) observed that the essence of Khrapchenko's argument is more general: it reduces the lower bounds problem for DeMorgan formulas to a combinatorial problem about the covering of rectangles by monochromatic subrectangles.

We have already arrived to the concept of "rectangles" and "monochromatic rectangles" in Chap. 3 when dealing with the communication complexity of relations, see Definitions 3.1 and 3.2. Since we will use these concepts extensively, let us recall them.

An n-dimensional *combinatorial rectangle*, or just a *rectangle*, is a Cartesian product $R = A \times B$ of two disjoint subsets A and B of vectors in $\{0, 1\}^n$. A *subrectangle* of R is a subset $S \subseteq R$ which itself forms a rectangle.

A boolean function $f : \{0, 1\}^n \to \{0, 1\}$ *separates* a rectangle $A \times B$ if $f(A) = 1$ and $f(B) = 0$, that is, if $f(a) = 1$ for all $a \in A$ and $f(b) = 0$ for all $b \in B$. Rectangles separated by a boolean variable x_i or by its negation $\neg x_i$ are called *monochromatic*. That is, a rectangle R is monochromatic, if there exists an index i such that

$$a_i \neq b_i \text{ for all edges } (a, b) \in R.$$

Rectangles separated by non-negated variables x_i are called *positively monochromatic*. For such rectangles we additionally have that

$$a_i = 1 \text{ and } b_i = 0 \text{ for all edges } (a, b) \in R.$$

For example, if $A = \{000\}$ and $B = \{100, 010, 001\}$ then the rectangle $A \times B$ is *not* monochromatic, whereas if $A = \{001, 111\}$ and $B = \{000, 110\}$ then the rectangle $A \times B$ is monochromatic, and even positively monochromatic: it is separated by x_3.

The *tiling number* $\chi(R)$ of a rectangle R is the smallest number t such that R can be decomposed into t disjoint monochromatic rectangles. The tiling number $\chi(f)$ of a boolean function is the tiling number of the rectangle $f^{-1}(1) \times f^{-1}(0)$. The monotone tiling numbers, where only positively monochromatic rectangles are allowed in decompositions, are denoted by $\chi_+(R)$ and $\chi_+(f)$. In general, these two numbers may not be defined: the simplest example is the function such that $f(0) = 1$ and $f(1) = 0$. But if f is a monotone boolean function f, then $f(a) = 1$ and $f(b) = 0$ implies that there must be an i for which $a_i = 1$ and $b_i = 0$. Hence, $\chi_+(f)$ is well-defined for all monotone boolean functions.

Recall that *any* n-dimensional rectangle R *can* be covered by at most $2n$ monochromatic rectangles, if we do not insist on their *disjointness*. For this, it is enough to take the $2n$ rectangles

$$M_{i,a} = \{(x, y) \in R : x_i = a, y_i = 1 - a\} \qquad (a = 0, 1 \text{ and } i = 1, \ldots, n)$$

called *canonical monochromatic rectangles*. Thus, it is the *disjointness* constraint which makes the tiling number $\chi(R)$ nontrivial.

The following lemma reduces the (computational) problem of proving a lower bound on the formula size to a (combinatorial) problem about decomposition of rectangles.

Let $L(f)$ be the smallest leafsize of a DeMorgan formula computing f. A formula is monotone if it has no negated variables as input literals. If f is a monotone boolean function, then $L_+(f)$ denotes the smallest leafsize of a monotone DeMorgan formula computing f.

Lemma 6.23. (Rychkov 1985) *For every boolean function f and for every monotone boolean function g, $L(f) \geq \chi(f)$ and $L_+(g) \geq \chi_+(g)$.*

Proof. We prove the first inequality by induction on $L(f)$. If $L(f) = 1$ then f is just a single variable x_i or its negation. In this case R itself is a monochromatic rectangle.

For the induction step, let $t = L(f)$ and assume that the theorem holds for all boolean functions g with $L(g) \leq t - 1$. Take a minimal formula for f, and assume that its last gate is an And gate (the case of an Or gate is similar). Then $f = f_0 \wedge f_1$ for some boolean functions f_0 and f_1 such that $L(f_0) + L(f_1) = L(f)$.

Suppose that f separates a rectangle $R = A \times B$, that is, $f(A) = 1$ and $f(B) = 0$. Consider the set $B_0 = \{b \in B : f_0(b) = 0\}$. Then f_0 separates the rectangle $R_0 = A \times B_0$, and f_1 separates the rectangle $R_1 = A \times (B_1 \setminus B_0)$. By the induction hypothesis, $\chi(R_i) \leq L(f_i)$ for both $i = 0, 1$. Hence,

$$\chi(R) \leq \chi(R_0) + \chi(R_1) \leq L(f_0) + L(f_1) = L(f).$$

The proof of the lemma in the monotone case is the same with the basis case replaced by $L_+(g) = 1$. In this case R itself is a positively monochromatic rectangle. The induction step is the same.   □

It is not known whether some polynomial inverse of Rychkov's lemma holds.

■ **Research Problem 6.24.** Does $L(f) \leq \chi(f)^{O(1)}$?

What we know is only a "quasi-polynomial" inverse

$$L(f) \leq \chi(f)^{2\log \chi(f)}$$

which is a direct consequence of Lemma 3.8. Still, since boolean functions with $L(f) = \Omega(2^n / \log n)$ exist (see Theorem 1.23), the latter inequality implies that boolean functions f of n variables such that $\chi(f) \geq 2^{(1-o(1))\sqrt{n}}$ exist. Hence, in principle, the tiling number can also achieve super-polynomial lower bounds on the formula size.

Remark 6.25. Khrapchenko (1971) himself considered not formulas but an equivalent model of π-schemes (see Proposition 1.10). These are switching networks of a special parallel-serial structure. A *path* in a π-scheme is a set P of contacts in a simple path from the source to the target node. A *cut* is a set C of contacts such that the removal of C cuts off all paths, and no proper subset of C does this. Khrapchenko observed that π-schemes (unlike general switching networks) have the following special property: $|P \cap C| = 1$ for every path P and every cut C (we leave the proof of this observation as an exercise). Now, with every vector $x \in f^{-1}(1)$ we can associate one path, all contacts of which *accept* x, and with every vector $y \in f^{-1}(0)$ we can associate one cut, all contacts of which *reject* y.

By the observation above, every pair $(x, y) \in f^{-1}(1) \times f^{-1}(0)$ will be associated with *exactly one contact* in the scheme, and the set of pairs associated with a single contact form a monochromatic rectangle. In this way we obtain a partition of $f^{-1}(1) \times f^{-1}(0)$ into monochromatic rectangles whose number is exactly the number of contacts in the scheme.

6.8 Khrapchenko's Theorem

As early as 1971, Khrapchenko suggested one way to prove lower bounds on the tiling number. Define the set

$$A \otimes B = \{(a, b) : a \in A \text{ and } b \in B \text{ and } a \sim b\},$$

where $a \sim b$ means that inputs a and b differ on exactly 1 bit. The main property of the set $A \otimes B$ is accumulated in the following

Lemma 6.26. *No monochromatic $s \times t$ rectangle can cover more than $\sqrt{st}$ elements of $A \otimes B$.*

Proof. Let $S \times T$ be a monochromatic $s \times t$ subrectangle of $A \times B$. Since the rectangle is monochromatic, each element of S differs from each element in T in one particular position j, whereas (a, b) is in $A \otimes B$ only if a and b differ in exactly one position. Hence, for any given $a \in S$, the only possible $b \in T$ for which $a \sim b$ is one which differs from a exactly in position j. As a result, we have that $S \times T$ can cover at most $\min\{|S|, |T|\} = \min\{s, t\} \leq \sqrt{st}$ entries of $A \otimes B$. $\quad\square$

Intuitively, if $A \otimes B$ is large, then every formula separating A and B should be large, since the formula must distinguish many pairs of adjacent inputs. The following theorem says just how large.

Theorem 6.27. (Khrapchenko 1971) *If a boolean function f separates a rectangle $A \times B$, then*

$$L(f) \geq \frac{|A \otimes B|^2}{|A| \cdot |B|}.$$

Viewing $A \otimes B$ as the set of edges of a bipartite graph with parts A and B, the theorem states that the leafsize of any formula separating A and B must be at least the product of the average degrees of these two parts.

Proof. Suppose we have a decomposition of $A \times B$ into r monochromatic rectangles of dimensions $s_i \times t_i$, $i = 1, \ldots, r$. Let c_i be the number of elements of $A \otimes B$ in the i-th of these rectangles. By Lemma 6.26, we know that $c_i^2 \leq s_i t_i$. Since the rectangles are disjoint and cover the whole rectangle $A \times B$, we also have that $|A \otimes B| = \sum_{i=1}^{r} c_i$ and $|A \times B| = \sum_{i=1}^{r} a_i b_i$. Applying the Cauchy–Schwarz inequality $(\sum x_i y_i)^2 \le (\sum x_i^2) \cdot (\sum y_i^2)$ with $x_i = c_i$ and $y_i = 1$, we obtain

$$|A \otimes B|^2 = \left(\sum_{i=1}^{r} c_i\right)^2 \le r \sum_{i=1}^{r} c_i^2 \le r \cdot \sum_{i=1}^{r} a_i b_i = r \cdot |A \times B|. \qquad \square$$

Khrapchenko's theorem can be used to show that some explicit boolean functions require formulas of quadratic size. Consider, for example, the parity function $\oplus_n(x) = x_1 \oplus \cdots \oplus x_n$ and threshold-k functions $\text{Th}_k^n(x) = 1$ iff $x_1 + \cdots + x_n \ge k$.

Theorem 6.28. $L(\oplus_n) \ge n^2$.

Proof. Let A be the set of all vectors with an odd number of ones, and B the set of all vectors with an even number of ones. Then $|A| = |B| = 2^{n-1}$, whereas $|A \otimes B| = n2^{n-1}$. Hence,

$$L(\oplus_n) \ge \frac{n^2 2^{2(n-1)}}{2^{2(n-1)}} = n^2. \qquad \square$$

This lower bound is almost optimal. Let p_n denote the largest power of 2 not exceeding n, that is, $p_n = 2^{\lfloor \log_2 n \rfloor}$.

Theorem 6.29. (Yablonskii 1954) *For every $n \ge 2$,*

$$L(\oplus_n) \le 3np_n - 2p_n^2 \le \frac{9}{8}n^2.$$

In particular, $L(\oplus_n) = n^2$ if $n = 2^k$ is a power of 2.

Proof. Let $\lambda(n) = L(\oplus_n)$ denote the smallest leafsize of a DeMorgan formula computing $\oplus_n(x) = x_1 \oplus x_2 \oplus \cdots \oplus x_n$. Write n as $n = 2^m + k$ where $0 \le k < 2^m$. Then $p_n = 2^m$ and

$$3np_n - 2p_n^2 = p_n^2 + 3kp_n = 2^{2m} + 3k2^m.$$

So, it is enough to show that $\lambda(n) \le 2^{2m} + 3k2^m$. We do this by induction on $n = 2^m + k$. Basis cases $n = 1, 2$ are trivial. For the induction step, we use that fact that a parity $f = g \oplus h$ of two functions can be computed as $f = (g \wedge \neg h) \vee (\neg g \wedge h)$. Thus, $L(f) \le 2L(g) + 2L(g)$. Since $\lfloor n/2 \rfloor = 2^{m-1} + \lfloor k/2 \rfloor$ and $\lceil n/2 \rceil = 2^{m-1} + \lceil k/2 \rceil$, we obtain

$$\lambda(n) \le 2\left(\lambda(\lfloor n/2 \rfloor) + \lambda(\lceil n/2 \rceil)\right)$$
$$= 2\left(2^{2(m-1)} + 3\lfloor k/2 \rfloor 2^{m-1} + 2^{2(m-1)} + 3\lceil k/2 \rceil 2^{m-1}\right)$$
$$= 2^{2m} + 3k2^m = 3np_n - 2p_n^2.$$

It remains to show that $3np_n - 2p_n^2 \leq (9/8)n^2$:

$$\tfrac{9}{8}n^2 - \left(3np_n - 2p_n^2\right) = \frac{9n^2 - 24np_n + 16p_n^2}{8} = \frac{(3n - 4p_n)^2}{8} \geq 0. \qquad \square$$

Remark 6.30. It is conjectured that the upper bound given in Theorem 6.29 is optimal. Zdobnov (1987) confirmed this conjecture in the class of so-called *null-path-free* formulas. When viewed as a parallel-serial network (see Proposition 1.10), the restriction is that no *s-t* path contains edges labeled by a variable and by its negation. In other words, if $F \wedge G$ is a subformula, and if F contains a leaf labeled by x_i, then G cannot contain a leaf labeled by $\neg x_i$. Note that the formula constructed in the proof of Theorem 6.29 is null-path-free. Zdobnov shows that this formula is also an *optimal* one in this class of formulas. For unrestricted formulas, Rychkov (1994) showed a lower bound $L(\oplus_n) \geq n^2 + 3$ for odd $n \geq 5$, and $L(\oplus_n) \geq n^2 + 2$ for even $n \geq 6$ which are not powers of 2.

Theorem 6.31. $L(Th_k^n) \geq k(n - k + 1)$.

Proof. Let A be the set of all vectors $a \in \{0, 1\}^n$ with exactly $k - 1$ ones, and B the set of all vectors $b \in \{0, 1\}^n$ with exactly k ones. Then every element of A is at Hamming distance 1 from exactly $n - k + 1$ elements of B. Similarly every element of B is at Hamming distance 1 from exactly k elements of A. It follows that $|A \otimes B| = (n - k + 1)|A| = k|B|$, and we obtain

$$L(Th_k^n) \geq \frac{(n - k + 1)|A| \cdot k|B|}{|A||B|} = k(n - k + 1). \qquad \square$$

A nice application of communication complexity is the following depth analogue of Khrapchenko's theorem (Theorem 6.27). Since we always have that $D(f) \geq \log L(f)$, the theorem itself is a direct consequence of Khrapchenko's theorem. We give here an alternative information-theoretic proof which is interesting in its own right.

Theorem 6.32. (Karchmer and Wigderson 1990) *Let f be a boolean function, $A \subseteq f^{-1}(0)$ and $B \subseteq f^{-1}(1)$. Then*

$$D(f) \geq \log \frac{|A \otimes B|^2}{|A||B|}.$$

Proof. By Theorem 3.13, it is enough to prove the corresponding lower bound on the number of communicated bits in the "find a difference" game on the rectangle $A \times B$: Alice gets $a \in A$, Bob gets $b \in B$, and their goal is to find a position $i \in [n]$ such that $a_i \neq b_i$.

We again consider $Y := A \otimes B$ as a bipartite graph with parts A and B. For a node a, let $N(a)$ denote the set of its neighbors in this graph. Hence $d(a) = |N(a)|$ is the degree of a in Y. Then $|Y|/|A|$ and $|Y|/|B|$ are average degrees of nodes in A and B, respectively. We claim that the number of bits Alice sends is at least the logarithm of the average degree of nodes in B (similarly with Bob). Intuitively, this is so because

Even if Alice knows $b \in B$, she needs $\log d(b)$ bits to tell Bob which $a \in N(b)$ she has.

That is, Alice needs to tell Bob in which bit, out of $d(b)$ possible bits, her vector a differs from b. Thus, if $A(a,b)$ and $B(a,b)$ are the numbers of bits sent by Alice and Bob on input $(a,b) \in A \times B$, then

$$A(a,b) \geq \log d(b) \quad \text{and} \quad B(a,b) \geq \log d(a). \tag{6.15}$$

Now take (a, b) uniformly at random from Y, and let $C(\mathrm{a}, \mathrm{b})$ be the number of bits communicated on this input. The expectation of this random variable (the average number of bits communicated on this input) is

$$
\begin{aligned}
\mathrm{E}\left[C(\mathrm{a}, \mathrm{b})\right] &= \frac{1}{|Y|} \sum_{(a,b) \in Y} (A(a,b) + B(a,b)) \\
&= \frac{1}{|Y|} \sum_{b \in B} \sum_{a \in N(b)} A(a,b) + \frac{1}{|Y|} \sum_{a \in A} \sum_{b \in N(a)} B(a,b) \\
&\geq \frac{1}{|Y|} \sum_{b \in B} d(b) \log d(b) + \frac{1}{|Y|} \sum_{a \in A} d(a) \log d(a) \qquad \text{by (6.15).}
\end{aligned}
$$

Since $\sum_{b \in B} d(b) = |Y|$ and since $f(x) = x \log x$ is a convex function, Jensen's inequality $\sum \lambda_b f(x_b) \geq f\left(\sum \lambda_b x_b\right)$ with $\lambda_b = 1/|B|$ and $x_b = d(b)$ yields

$$\frac{1}{|Y|} \sum_{b \in B} d(b) \log d(b) \geq \log \frac{|Y|}{|B|},$$

and the desired lower bound follows:

$$\mathrm{E}\left[C(\mathrm{a}, \mathrm{b})\right] \geq \log \frac{|Y|}{|B|} + \log \frac{|Y|}{|A|} = \log \frac{|Y|^2}{|A||B|}. \qquad \square$$

6.9 Complexity is not Convex

Khrapchenko's measure (see Theorem 6.27) is of the form

$$\mu(R) := \frac{|Y|^2}{|R|} = |R| \cdot \varphi\left(\frac{|Y|}{|R|}\right) \tag{6.16}$$

where $\varphi(x) = x^2$, and Y is the set of all pairs of vectors $(x, y) \in R$ differing in exactly 1 bit. Exercise 6.2 shows that this measure cannot yield larger than $\Omega(n^2)$

lower bounds. All subsequent attempts to modify his measure with the goal of braking the "n^2 barrier" have failed so far. So, what is bad about this measure? Perhaps larger lower bounds can be obtained by taking other subsets Y of special entries and/or using some other functions $\varphi(x)$ instead of x^2?

The answer, given by Hrubes et al. (2010), is somewhat disappointing. Namely, it turns out that the reason for the failure of Khrapchenko-type measures is much deeper than expected: for *every* choice of the set Y of special entries and for *every* convex function $\varphi(x)$, the resulting measure is *convex*, and convex measures cannot yield super-quadratic lower bounds! To show this, we first define what is meant under a "convex" rectangle measure.

Recall that a rectangle of dimension n is a set S of the form $S = A \times B$ with $A, B \subseteq \{0, 1\}^n$ and $A \cap B = \emptyset$. A subset $R \subseteq S$ is a *subrectangle* of S if R itself forms a rectangle. A *rectangle function* on S is a mapping μ that assigns to each subrectangle R of S a real number $\mu(R)$. Such a function is a *rectangle measure* if it is:

- *Subadditive*: $\mu(R) \le \mu(R_1) + \mu(R_2)$, for every rectangle $R \in \mathcal{R}$ and for each of its partitions into disjoint union of rectangles $R_1, R_2 \in \mathcal{R}$.
- *Normalized*: $\mu(M) \le 1$ for every monochromatic rectangle M.

Rychkov's lemma (Lemma 6.23) can then be restated as:

Lemma 6.33. *Let f be a boolean function and $R = A \times B$ be a rectangle with $A \subseteq f^{-1}(1)$ and $B \subseteq f^{-1}(0)$. Then for every rectangle measure μ we have that $L(f) \ge \mu(R)$.*

Thus, every rectangle measure gives a lower bound on formula size. The most restricted class of rectangle measures is that of *additive* ones. Such measures must satisfy $\mu(R) = \sum_{e \in R} \mu(e)$. By letting $\mu(e) := |Y|^2/|R|^2$ for all $e \in R$, we see that Khrapchenko's measure (6.16) is additive, and it cannot break the "n^2 barrier", that is, cannot yield super-quadratic lower bounds on formula size. We now will identify a much larger class of "bad" measures.

Let S be a rectangle, $R_1, \ldots, R_t$ its subrectangles, and $r_1, \ldots, r_m$ weights from $[0, 1]$. We say that the rectangles $R_1, \ldots, R_m$ with the weights $r_1, \ldots, r_m$ form a *fractional partition* of the rectangle S if $\sum_{i : e \in R_i} r_i = 1$ for all $e \in S$. We will shorten this condition as $S = \sum_{i=1}^t r_i \cdot R_i$. Notice that if all $r_i \in \{0, 1\}$ then a fractional partition is a partition.

A rectangle function is *convex* if, for every rectangle S and every fractional partition $S = \sum_i r_i R_i$,

$$\mu(S) \le \sum_{i=1}^t r_i \cdot \mu(R_i). \tag{6.17}$$

A *fractional partition number*, $\pi(S)$, of a rectangle S is the minimum

$$\pi(S) = \min \sum_{i=1}^t r_i$$

over all fractional partitions $R_1, \ldots, R_t, r_1, \ldots, r_t$ of S where each rectangle R_i is monochromatic. It is easy to see that $\pi(S)$ is a rectangle measure. It is not difficult to show that the fractional partition number π is the largest convex measure.

Proposition 6.34. *The measure $\pi(S)$ is convex, and for every convex measure μ, $\mu(S) \leq \pi(S)$ for all rectangles S.*

Proof. First we will show that $\pi(S)$ is convex. Let $S = \sum_{j \in J} r_j R_j$ be a fractional partition of S and, for every j, let $R_j = \sum_{i \in I_j} s_{ij} M_{ij}$ be a fractional partition of R_j such that M_{ij} are monochromatic and $\pi(R_j) = \sum_i s_{ij}$. Then, clearly, $S = \sum_{ij} r_j s_{ij} M_{ij}$ is a fractional partition of R into monochromatic rectangles. Hence

$$\pi(S) \leq \sum_{ij} r_j s_{ij} = \sum_j r_j \pi(R_j).$$

Now we will show the second part. Let μ be a convex measure. Let $S = \sum_i r_i M_i$ be a fractional partition of S into monochromatic rectangles such that $\pi(S) = \sum_i r_i$. Using convexity and normality of μ we get

$$\mu(S) \leq \sum_i r_i \mu(M_i) \leq \sum_i r_i = \pi(S). \qquad \square$$

The following is an analogue of Khrapchenko's theorem using the fractional partition number.

Theorem 6.35. (Karchmer et al. 1995) *Let R be a rectangle, and Y the set of all pairs of vectors $(x, y) \in R$ differing in exactly 1 bit. Then $\pi(R) \geq |Y|^2/4|R|$.*

Proof. Applying the duality theorem for linear programs, one can write the fractional partition number as $\pi(R) = \max_w \sum_{e \in S} w(e)$, where the maximum is over all functions $w : R \to \mathbb{R}$ satisfying the constraints $\sum_{e \in M} w(e) \leq 1$ for all monochromatic rectangles $M \subseteq R$. Hence, in order to prove a lower bound $\pi(R) \geq t$ it is enough to find at least one weight function $w : R \to \mathbb{R}$ such that $\sum_{e \in R} w(e) \geq t$ and the weight of each monochromatic rectangle does not exceed 1.

We define the weight $w(e)$ of each edge $e \in R$ by:

$$w(e) = \begin{cases} p^{-1} & \text{if } e \in Y, \\ -p^{-2} & \text{otherwise,} \end{cases}$$

where $p > 0$ is a parameter to be specified soon. Since only entries of Y have positive weights, the heaviest monochromatic rectangles M are the square ones with exactly one entry from Y in each row and column. If M is such a $k \times k$ square rectangle, then

$$\sum_{e \in M} w(e) = \frac{k}{p} - \frac{k(k-1)}{p^2} \leq \frac{k}{p}\left(1 - \frac{k-1}{p}\right) \leq 1.$$

Indeed, if $k \geq p+1$ then the expression in the parenthesis is at most 0, and if $k \leq p$ then both terms are at most 1. Hence, w is a legal weight function, and we obtain

$$\pi(R) \geq \sum_{e \in R} w(e) = \frac{|Y|}{p} - \frac{|R| - |Y|}{p^2} = \frac{|Y|}{p}\left(1 - \frac{|R| - |Y|}{p|Y|}\right).$$

For $p = 2|R|/|Y|$, the expression in the parenthesis is at least $1/2$, and we obtain $\pi(R) \geq |Y|^2/4|R|$.   □

Hence, one can obtain quadratic lower bounds using the fractional partition number, as well. We now show that this is actually the best we can get using *any* convex rectangle measure.

Karchmer et al. (1995) proved that $\pi(S) \leq 4n^2$ holds for every n-dimensional rectangle S. Together with Proposition 6.34, this implies that $\mu(S) \leq 4n^2$ holds for every convex rectangle measure μ. Using similar arguments, this upper bound can be improved to an almost optimal one.

Theorem 6.36. (Hrubes et al. 2010) *If a rectangle measure μ is convex then, for every n-dimensional rectangle S, $\mu(S) \leq \frac{9}{8}(n^2 + n)$*

Proof. Following Karchmer et al. (1995), associate with each subset $I \subseteq [n] = \{1, \ldots, n\}$ the following two I-*parity rectangles*:

$$S_I = \{x : \oplus_{i \in I} x_i = 0\} \times \{y : \oplus_{i \in I} y_i = 1\},$$

$$T_I = \{x : \oplus_{i \in I} x_i = 1\} \times \{y : \oplus_{i \in I} y_i = 0\}.$$

Note that monochromatic rectangles correspond to the case when $|I| = 1$. There are exactly 2^{n+1} parity rectangles.   □

Claim 6.37. *Every edge $(x, y) \in \{0, 1\}^n \times \{0, 1\}^n$ such that $x \neq y$ belongs to 2^{n-1} parity rectangles.*

Proof. For $I \subseteq [n]$, let $v_I \in \{0, 1\}^n$ be its incidence vector. If $x \neq y$, then $x \oplus y$ is not a zero vector. Since each nonzero vector is orthogonal over GF(2) to exactly half of the vectors in $\{0, 1\}^n$, this implies that precisely 2^{n-1} of the vectors v_I are non-orthogonal to $x \oplus y$. This means that (x, y) belongs to precisely 2^{n-1} of the sets $S_I \cup T_I$. Since $S_I \cap T_I = \emptyset$, we are done.   □

Claim 6.38. *Let μ be a rectangle measure. Then for every $I \subseteq [n]$, both $\mu(S_I)$ and $\mu(T_I)$ are at most $\frac{9}{8}|I|^2$.*

Proof. A parity rectangle S_I can be viewed as a rectangle corresponding to the parity function, and T_I as a rectangle corresponding to the negation parity function in $|I|$ variables. We already know (see Theorem 6.29) that the parity of m variables can be computed by a DeMorgan formula of size $\frac{9}{8}m$. Since μ is a lower bound to the formula size (see Lemma 6.33), the desired upper bound on $\mu(S_I)$ and $\mu(T_I)$ follows.   □

Now let S be an n-dimensional rectangle. Let $\mathcal{R}$ be the set of all parity rectangles $S_I \cap S$ and $T_I \cap S$ restricted to S. Let also $\mathcal{R}_i \subseteq \mathcal{R}$ be the set of all parity rectangles corresponding to subsets $I \subseteq [n]$ of size $|I| = i$. For counting reasons, we shall understand $\mathcal{R}$ as a multi-set, elements of $\mathcal{R}$ corresponding to different parity rectangles are considered as different. Under this provision, $\mathcal{R}$ has size 2^{n+1} and, by Claim 6.37, every edge in S is contained in exactly 2^{n-1} elements of $\mathcal{R}$. Hence $\mathcal{R}$ forms a fractional partition of S with each rectangle $R \in \mathcal{R}$ of weight $r_R = 2^{-(n-1)}$. By Claim 6.38, we know that $\mu(R) \le ci^2$ for every $R \in \mathcal{R}_i$, where $c = 9/8$. The convexity of μ implies that

$$\mu(S) \le \sum_{R \in \mathcal{R}} r_R \cdot \mu(R) = 2^{-(n-1)} \sum_{R \in \mathcal{R}} \mu(R) = 2^{-(n-1)} \sum_{i=0}^{n} \sum_{R \in \mathcal{R}_i} \mu(R)$$

$$\le 2^{-(n-1)} \sum_{i=1}^{n} 2 \binom{n}{i} ci^2 = 2^{-(n-1)} 2c \sum_{i=1}^{n} \binom{n}{i} i^2$$

$$= 2^{-(n-2)} c \sum_{i=1}^{n} \binom{n}{i} i^2.$$

The identity $\binom{n}{k} \cdot k = n \cdot \binom{n-1}{k-1}$ gives

$$\sum_{i=1}^{n} \binom{n}{i} i^2 = n \cdot \sum_{i=1}^{n} \binom{n-1}{i-1} i = n \cdot \sum_{i=1}^{n} \binom{n-1}{i-1} + n \cdot \sum_{i=1}^{n} \binom{n-1}{i-1}(i-1)$$

$$= n \cdot \sum_{i=1}^{n} \binom{n-1}{i-1} + n \cdot \sum_{i=2}^{n} \binom{n-1}{i-1}(i-1)$$

$$= n \cdot \sum_{i=1}^{n} \binom{n-1}{i-1} + n(n-1) \cdot \sum_{i=2}^{n} \binom{n-2}{i-2}$$

$$= n2^{n-1} + n(n-1)2^{n-2} = (n^2 + n)2^{n-2}.$$

Thus, $\mu(S) \le 2^{-(n-2)} c(n^2 + n)2^{n-2} = c(n^2 + n) = \frac{9}{8}(n^2 + n)$. $\qquad\qquad \square$

We now show that many Khrapchenko-like measures (6.16) are convex, and hence, cannot yield super-quadratic lower bounds. This observation was made in Hrubes et al. (2010).

Theorem 6.39. *Let $s(R)$ be an additive rectangle function such that $s(R) > 0$ for every non-empty rectangle R. Let $\varphi : \mathbb{R} \to \mathbb{R}$ be a convex function. Then the rectangle function*

$$\mu(R) := s(R) \cdot \varphi\left(\frac{w(R)}{s(R)}\right),$$

is convex if either $w(R)$ is additive, or $w(R)$ is convex and φ is nondecreasing.

Proof. To prove the first claim, assume that both $w(R)$ and $s(R)$ are additive, and let $R_1, \ldots, R_m, r_1 \ldots, r_m$ be a fractional partition of R. Set $s_i = s(R_i)$ and $w_i = w(R_i)$. By Exercise 6.8, we have that $w(R) = \sum_i r_i \cdot w_i$ and $s(R) = \sum_i r_i \cdot s_i$.

For a real convex function φ, numbers x_i in its domain, and positive weights a_i, Jensen's inequality states that

$$\varphi\left(\frac{\sum a_i x_i}{\sum a_i}\right) \leq \frac{\sum a_i \varphi(x_i)}{\sum a_i}. \tag{6.18}$$

Applying this we obtain (where the sums are over all i with $r_i > 0$):

$$
\begin{aligned}
\mu(R) \quad &= s(R) \cdot \varphi\left(\frac{w(R)}{s(R)}\right) \\
&= \left(\sum_i r_i s_i\right) \cdot \varphi\left(\frac{\sum_i r_i w_i}{\sum_i r_i s_i}\right) \quad \text{Exercise 6.8} \\
&\leq \sum_i r_i s_i \cdot \varphi\left(\frac{w_i}{s_i}\right) \quad \text{(6.18) with } a_i = r_i s_i \text{ and } x_i = w_i/s_i \\
&= \sum_i r_i \mu(R_i).
\end{aligned}
$$

If $w(R)$ is convex and φ is nondecreasing, then we can replace the second equality by inequality, and the desired inequality $\mu(R) \leq \sum_i r_i \cdot \mu(R_i)$ still holds. $\square$

6.10 Complexity is not Submodular

In order to prove that some boolean function f requires large formulas, one tries to find some clever "combinatorial" measure μ on the set of all boolean functions satisfying two conditions: $\mu(f)$ is a lower bound on the size of any formula computing f, and $\mu(f)$ can be nontrivially bounded from below at some *explicit* boolean functions f. One class of such measures, proposed by Mike Paterson, is the following.

Let B_n be the set of all boolean function of n variables. A *formal complexity measure* of boolean functions is a mapping $\mu : B_n \to \mathbb{R}$ which assigns positive values to each boolean function. The requirements are that μ is *normalized*, that is, assigns each literal a value ≤ 1, and is *subadditive*, that is, for all $f, g \in B_n$:

$$\mu(f \vee g) \leq \mu(f) + \mu(g); \tag{6.19}$$
$$\mu(f \wedge g) \leq \mu(f) + \mu(g). \tag{6.20}$$

Note that the minimal formula size $\mathrm{L}(f)$ itself is a formal measure. Moreover, by induction on formula size it is easy to show that $\mathrm{L}(f) \geq \mu(f)$ for *every* formal complexity measure μ. It can also be shown (see Exercise 6.6) that Khrapchenko's measure is a formal complexity measure.

In order to understand what measures are "good" (can lead to large lower bounds) it is important to understand what measures are "bad". We have already seen that convex measures are bad. There is another class of bad measures—submodular ones.

A formal complexity measure $\mu : B_n \to \mathbb{R}$ is *submodular* if it is normalized and for all $f, g \in B_n$,

$$\mu(f \wedge g) + \mu(f \vee g) \leq \mu(f) + \mu(g). \tag{6.21}$$

Note that this condition is stronger than both (6.19) and (6.20). The following result shows that submodular measures cannot even yield super-linear lower bounds.

Theorem 6.40. (Razborov 1992b) *If μ is a submodular measure on B_n, then* $\mu(f) \leq 4(n + 1)$ *for each* $f \in B_n$.

Proof. Let g_n be a random boolean function of n variables $x_1, \ldots, x_n$. That is, we choose g_n randomly and uniformly from B_n. We are going to prove by induction on n that

$$E\left[\mu(g_n)\right] \leq n + 1. \tag{6.22}$$

Given a variable x_i, set $x_i^1 := x_i$ and $x_i^0 := \neg x_i$.

Base: $n = 1$. Here we have $\mu(g(x_1)) \leq 2$ for any $g(x_1)$. This follows from the normalization condition if g is a variable x_1 or its negation $\neg x_1$. By the subadditivity we also have

$$\mu(0) + \mu(1) = \mu(x_1 \wedge \neg x_1) + \mu(x_1 \vee \neg x_1) \leq \mu(x_1) + \mu(\neg x_1) \leq 2$$

which proves $\mu(g(x_1)) \leq 2$ in the remaining case when g is a constant.

Inductive step: Assume that (6.22) is already proved for n. Let the symbol $\approx$ mean that two random functions have the same distribution. Note that

$$g_{n+1} \approx \left(g_n^0 \wedge x_{n+1}^0\right) \vee \left(g_n^1 \wedge x_{n+1}^1\right), \tag{6.23}$$

where g_n^0 and g_n^1 are two independent copies of g_n. By duality,

$$g_{n+1} \approx \left(g_n^0 \vee x_{n+1}^0\right) \wedge \left(g_n^1 \vee x_{n+1}^1\right). \tag{6.24}$$

By the linearity of expectation, we obtain from (6.23) and (6.19) (remember that the latter is a consequence of the submodularity condition) that

$$E\left[\mu(g_{n+1})\right] \leq E\left[\mu\left(g_n^0 \wedge x_{n+1}^0\right)\right] + E\left[\mu\left(g_n^1 \wedge x_{n+1}^1\right)\right] \tag{6.25}$$

and similarly from (6.24) and (6.20),

$$E\left[\mu(g_{n+1})\right] \le E\left[\mu\left(g_n^0 \vee x_{n+1}^0\right)\right] + E\left[\mu\left(g_n^1 \vee x_{n+1}^1\right)\right]. \qquad (6.26)$$

Summing (6.25), (6.26) and applying consecutively (6.21), normalization of μ, and the inductive assumption, we obtain

$$
\begin{aligned}
2 \cdot E\left[\mu(g_{n+1})\right] &\le E\left[\mu\left(g_n^0 \wedge x_{n+1}^0\right)\right] + E\left[\mu\left(g_n^0 \vee x_{n+1}^0\right)\right] \\
&\quad + E\left[\mu\left(g_n^1 \wedge x_{n+1}^1\right)\right] + E\left[\mu\left(g_n^1 \vee x_{n+1}^1\right)\right] \\
&\le E\left[\mu(g_n^0)\right] + \mu(x_{n+1}^0) + E\left[\mu(g_n^1)\right] + \mu(x_{n+1}^1) \\
&\le 2 \cdot E\left[\mu(g_n)\right] + 2 \le 2d + 4.
\end{aligned}
$$

This completes the proof of (6.22). But this inequality only says that the expected value of $\mu(g_n)$ does not exceed $n+1$ for a *random* function g_n, whereas our goal is to give an upper bound on $\mu(f_n)$ for *each* function f_n. So, we must somehow "derandomize" this result. To achieve this goal, observe that every function $f_n \in F_n$ can be expressed in the form

$$f_n = (g_n \wedge (g_n \oplus f_n \oplus 1)) \vee ((g_n \oplus 1) \wedge (g_n \oplus f_n)). \qquad (6.27)$$

But $g_n \approx g_n \oplus f_n \oplus 1 \approx g_n \oplus 1 \approx g_n \oplus f_n$. So, applying to (6.27) the inequalities (6.19) and (6.20), averaging the result over g_n and applying (6.22), we obtain $\mu(f_n) = E\left[\mu(f_n)\right] \le 4 \cdot E\left[\mu(g_n)\right] \le 4n + 4$, as desired. $\qquad \square$

6.11 The Drag-Along Principle

Suppose we want to prove that a boolean function f has high complexity, say, requires large DeMorgan formulas over $\wedge, \vee, \neg$. If the function is indeed hard, then it should have some *specific* properties forcing its formulas be large, that is, forcing every small formula to make an error.

It turns out that formal complexity measures cannot capture any specific properties of boolean functions. When using such measures, every lower bound for a given function f must also prove that many other unrelated functions have large complexity. Thus, we cannot use any special properties of our function! Namely, Razborov and Rudich (1997) proved the following fact; the name "drag-along principle" was suggested by Lipton (2010).

Theorem 6.41. (The Drag-Along Principle) *Suppose μ is a formal complexity measure and there exists a function $f \in B_n$ such that $\mu(f) > s$. Then, for at least $1/4$ of all g in B_n, $\mu(g) > s/4$.*

Proof. Let g be any function in B_n. Define $f = h \oplus g$ where $h = f \oplus g$. Then,

$$\mu(f) \leq \mu(g) + \mu(\neg g) + \mu(h) + \mu(\neg h). \tag{6.28}$$

This follows from (6.19) and (6.20) and the definition of parity,

$$f = (f \oplus g) \oplus g = h \oplus g = (h \wedge g) \vee (\neg h \wedge \neg g).$$

By way of contradiction assume that the set $\mathcal{G} = \{g \in B_n : \mu(g) < s/4\}$ contains *more* than $3/4$ of all function in B_n. If we pick the above function g randomly in B_n with probability $|B_n|^{-1}$, then $\neg g, h, \neg h$ are also random elements of B_n (though not independent) each with the same probability. Using the trivial union bound we have

$$\text{Prob[one or more of } h, \neg h, g, \neg g \text{ is not in } \mathcal{G}] < 4 \cdot \frac{1}{4} = 1.$$

Thus, there must be at least one choice for g such that all four functions $h, \neg h, g, \neg g$ belong to $\mathcal{G}$, that is, have measure $< s/4$. By (6.28), this implies that $\mu(f) < s$, which is a contradiction.   □

Theorem 6.41 above shows that for any lower bound proof for formulas based on some formal complexity measure $\mu(f)$, essentially the same lower bound automatically applies to almost all boolean functions. The important "natural proofs" concept, discussed in the Epilogue, reveals a possible barrier facing attempts to prove lower bounds of this type.

6.12 Bounds Based on Graph Measures

In view of graph complexity—the concept we have introduced in Sect. 1.7—it is important to have lower-bound arguments for monotone boolean functions f that only explore the structure of the set of length-2 minterms of f. In this section we present one approach in this direction. It was suggested by Newman and Wigderson (1995). In its current form, the method cannot yield lower bounds larger than $n \log n$. (The reason is the "monotonicity" condition of the graph measures used.) Still, the approach is applicable to functions for which other arguments fail, and has the potential to be extended to other graph measures.

Let $K_n = \binom{[n]}{2}$ be the set of all unordered pairs of members of $[n] = \{1, \dots, n\}$, that is, the set of all edges of a complete graph on $[n]$. Each subset $E \subseteq K_n$ gives us a graph on these vertices.

Let μ be a measure which assigns to each such graph E a non-negative real number $\mu(E)$. Say that such a measure μ is a *good graph measure* if

- $\mu(\emptyset) = 0$;
- μ is subadditive: $\mu(E \cup F) \leq \mu(E) + \mu(F)$;
- μ is monotone: $E \subseteq F$ implies $\mu(E) \leq \mu(F)$;

- μ respects cliques: if E forms a complete bipartite graph on m (out of n) vertices, then $\mu(E) \leq m/n$.

With every monotone boolean function $f(x_1, \ldots, x_n)$ such that $f(0) = 0$ we associate the graph $E_f \subseteq K_n$ where vertices i and j are adjacent [2] iff $f(e_i + e_j) = 1$ and $f(e_i) = f(e_j) = 0$, as well as set $V_f \subseteq [n]$ of vertices $i \in [n]$ such that $f(e_i) = 1$. Note that the edges of E_f correspond to monotone length-2 minterms of f: vertices i and j are adjacent in E_f iff $x_i x_j$ is a minterm of f. Similarly, V_f corresponds to monotone length-1 minterms: vertex i belongs to V_f iff x_i is a minterm of f. For example, if $f(x) = \text{Th}_2^n(x)$ is the threshold-2 function of n variables, then $E_f = K_n$ and $V_f = \emptyset$.

Lemma 6.42. (Newman and Wigderson 1995) *For every monotone boolean function f of n variables, and every good graph measure μ, we have that* $\text{L}_+(f) \geq n \cdot \mu(E_f)$.

In fact, we will prove a slightly stronger bound $\text{L}_+(f) \geq n \cdot \mu(E_f) + |V_f|$. But, for simplicity, we will ignore the last term $|V_f|$ since it never exceeds n, and we are interested in lower bounds that are super-linear in n.

Proof. Define the following cost function $c(f)$ of boolean functions f of n variables:

$$c(f) := \mu(E_f) + \frac{|V_f|}{n}.$$

If $f = x_i$ is a variable (a leaf of a formula), then $E_f = \emptyset$ and we get $c(x_i) = 1/n$. Moreover, the monotonicity of μ implies that the cost function is monotone with respect to inclusion: if $V_g \subseteq V_h$ and $E_g \subseteq E_h$, then $c(g) \leq c(f)$.   □

Claim 6.43. $c(g \vee h) \leq c(g) + c(h)$ and $c(g \wedge h) \leq c(g) + c(h)$.

Note that this claim already implies the theorem since the cost of every leaf in a formula is $1/n$ and, by Claim 6.43, the cost of the output function does not exceed the sum of the costs of all the leaves. Thus $c(f) \leq \text{L}_+(f)/n$, implying that $\text{L}_+(f) \geq n \cdot c(f) \geq n \cdot \mu(E_f) + |V_f|$. So, it remains to prove the claim.

Case 1: $f = g \vee h$. Then $V_f = V_g \cup V_h$ and $E_f = E_g \cup E_h$. The subadditivity of μ yields

$$c(f) = \mu(E_g \cup E_h) + \frac{|V_g \cup V_h|}{n}$$

$$\leq \mu(E_g) + \mu(E_h) + \frac{|V_g|}{n} + \frac{|V_h|}{n} = c(g) + c(h).$$

Case 2: $f = g \wedge h$. Denote $A = V_g$ and $B = V_h$. Since $V_f = A \cap B$ and

---

[2] Here and throughout, e_i is a 0-1 vector with exactly one 1 in the i-th position.

$$E_f = (E_g \cap E_h) \cup K_{A,B} \subseteq E_g \cup E_h \cup K_{A,B},$$

where $K_{A,B} := (A \setminus B) \times (B \setminus A)$, we get:

$$c(f) \le \mu(E_g \cup E_h \cup K_{A,B}) + \frac{|A \cap B|}{n} \qquad \text{(monotonicity of } \mu)$$

$$\le \mu(E_f) + \mu(E_f) + \mu(K_{A,B}) + \frac{|A \cap B|}{n} \qquad \text{(subadditivity of } \mu)$$

$$\le \mu(E_g) + \mu(E_h) + \frac{|A \setminus B| + |B \setminus A|}{n} + \frac{|A \cap B|}{n} \quad (\mu \text{ respects cliques})$$

$$= \mu(E_g) + \mu(E_h) + \frac{|A|}{n} + \frac{|B|}{n} = c(g) + c(h).$$

This completes the proof of the claim, and thus the proof of the lemma. □

To extend this lemma to *non-monotone* formulas, we need the following result. Say that a boolean function f *rejects singletons* if f rejects all inputs with at most one 1. Note that in any DNF formula of such a function each monomial must have at least two non-negated variables.

Lemma 6.44. (Krichevskii 1964) *Let f be a boolean function which rejects singletons. Then there exists a monotone boolean function φ_f on the same variables such that:*

(a) φ_f *rejects singletons;*
(b) $\varphi_f(e_i + e_j) \ge f(e_i + e_j)$ *for all $i \ne j$, and*
(c) $L(f) \ge L_+(\varphi_f)$.

Proof. The proof is by induction on the leafsize $L(f)$. For $L(f) = 2$ the claim is true since in this case f must be an AND of two variables. Now let F be an optimal formula for f. If $F = G \vee H$, where G and H are optimal formulas for g and h, then by induction there are φ_g and φ_h satisfying all three conditions (a)–(c). It is easy to see that then the function $\varphi_f = \varphi_g \vee \varphi_h$ also satisfies these three conditions.

The case when $F = G \wedge H$ is less trivial. The problem is that both formulas G and H *may* accept singletons, even if their AND rejects them: if $G(e_i) = H(e_j) = 1$ and $G(e_j) = H(e_i) = 0$ then $F(e_i) = F(e_j) = 0$. To overcome this obstacle, define

$$I_g = \{i \in [n] : g(e_i) = 1 \text{ and } x_i \text{ appears in } G\},$$

and define I_h similarly. Since F rejects singletons, these two sets must be disjoint. Let G_0 be the subformula of G obtained by setting to 0 all variables x_i with $i \in I_g$, and let H_0 be defined similarly. Consider the formula $F' = G' \wedge H'$, where

$$G' = G_0 \vee \bigvee_{i \in I_g} x_i \qquad \text{and} \qquad H' = H_0 \vee \bigvee_{i \in I_h} x_i.$$

Note that the leafsize of F' is at most that of F, since all leaves corresponding to variables x_i with $i \in I_g$ appear in G, and are set to 0 in G_0; similarly with H and H_0.

Since $I_g \cap I_h = \emptyset$, the formula F' must reject singletons. Further, we claim that $F'(e_i + e_j) = 1$ as long as $F(e_i + e_j) = 1$. To show this, suppose that F accepts the input $e_i + e_j$. Then both G and H must accept it. If i or j belongs to I_g, then G' accepts this vector as well. If $\{i, j\} \cap I_g = \emptyset$, then $G_0(e_i + e_j) = G(e_i + e_j) = 1$, since $e_i + e_j$ has only zeros in all bits in I_g. So, G' must accept this vector in both cases. Since the same argument also holds for H', we are done.

Now, G_0 and H_0 are formulas of some boolean functions g_0 and h_0 that meet the requirement of the lemma. So, by induction, there are monotone boolean functions φ_{g_0} and φ_{h_0} with monotone formulas G_+ and H_+ as required. By plugging these monotone formulas into F' we get a monotone formula F_+ for φ_f that satisfies all three conditions (a)–(c). □

Now we can extend Lemma 6.42 to *non-monotone* formulas.

Theorem 6.45. *For every boolean function f of n variables which rejects singletons, and every good graph measure μ, we have $L(f) \geq n \cdot \mu(E_f)$.*

Proof. By Krichevskii's lemma there is a monotone boolean function $g = \varphi_f$ for which $L(f) \geq L_+(g)$. From Lemma 6.42 we get $L_+(g) \geq n\mu(E_g) + |V_g|$. Since $g(a) \geq f(a)$ for all inputs a with at most two 1s, the monotonicity of μ implies the result. ⊔

6.13 Lower Bounds via Graph Entropy

In order to use Theorem 6.45 we have to define some good measure of graphs. For this purpose, Newman and Wigderson (1995) used the measure of graph entropy introduced by Körner (1973).

Let E be a graph on $|V| = n$ vertices. The *graph entropy* $H(E)$ of E is the minimum

$$H(E) = \frac{1}{n} \cdot \min_Y \sum_{v \in V} \log \frac{1}{\text{Prob}[v \in Y]} = -\frac{1}{n} \cdot \min_Y \sum_{v \in V} \log \text{Prob}[v \in Y]$$

taken over all (arbitrarily distributed) random variables Y over independent sets in E. If $E = \emptyset$, then we set $H(E) = 0$.

Lemma 6.46. *Graph entropy is a good graph measure.*

We have to show that the graph entropy is monotone, subadditive and respects cliques.

Claim 6.47. (Monotonicity) If $F \subseteq E$ are graphs on the same set of vertices, then $H(F) \leq H(E)$.

Proof. Let Y be the random variable taking values in independent sets of E, which attains the minimum in the definition of the entropy $H(E)$. Since an independent set in E is also an independent set in F, we have

$$H(F) \leq -\frac{1}{n} \sum_{v \in V} \log \text{Prob}[v \in Y] = H(E). \qquad \square$$

Claim 6.48. (Subadditivity) If E and F are graphs on the same set of vertices, then $H(E \cup F) \leq H(E) + H(F)$.

Proof. Let Y_1, Y_2 be random variables taking values in independent sets of E and F, respectively, which attain the minimum in the definition of entropy. We can assume that Y_1, Y_2 are independent. Also note that $Y_1 \cap Y_2$ is a random variable taking values in independent sets of $E \cup F$. We therefore have

$$H(E) + H(F) = -\frac{1}{n} \sum_{v \in V} \log \text{Prob}[v \in Y_1] - \frac{1}{n} \sum_{v \in V} \log \text{Prob}[v \in Y_2]$$

$$= -\frac{1}{n} \sum_{v \in V} \log(\text{Prob}[v \in Y_1] \cdot \text{Prob}[v \in Y_2])$$

$$= -\frac{1}{n} \sum_{v \in V} \log \text{Prob}[v \in Y_1 \cap Y_2]$$

$$\geq H(E \cup F). \qquad \square$$

Claim 6.49. (Respecting cliques) If E is a bipartite graph with m (out of n) vertices, then $H(E) \leq m/n$.

Proof. Let $A, B \subseteq V$ be the parts of E; hence, $|A \cup B| = m$ and $|V| = n$. By the monotonicity, we can assume that E is a complete bipartite graph, $E = A \times B$. Define a random independent set Y by letting $\text{Prob}[Y = A] = \text{Prob}[Y = B] = 1/2$ and $\text{Prob}[Y = C] = 0$ for all remaining independent sets. Then

$$H(E) \leq -\frac{1}{n} \sum_{v \in V} \log \text{Prob}[v \in Y]$$

$$= -\frac{1}{n} \sum_{v \in A \cup B} \log \text{Prob}[v \in Y]$$

$$= -\frac{1}{n} \sum_{v \in A \cup B} -1$$

$$= \frac{|A \cup B|}{n} = \frac{m}{n}.$$

This completes the proof of Claim 6.49, and thus of Lemma 6.46. $\qquad \square$

Together with Theorem 6.45 we obtain the following general lower bound on the formula size in terms of graph entropy.

Corollary 6.50. *For every boolean function f of n variables which rejects single-tons, we have that $L(f) \geq n \cdot \log H(E_f)$.*

Graph entropy $H(E)$ can be lower-bounded in terms of the independence number $\alpha(E)$ of a graph E, that is, the maximum number of vertices in E no two of which are adjacent.

Proposition 6.51. *For every graph E on n vertices, we have that*

$$H(E) \geq \log \frac{n}{\alpha(E)}.$$

Proof. Let Y be a random independent set in E which attains the minimum in the definition of the entropy $H(E)$. For a vertex v, let $p_v := \text{Prob}[v \in Y]$. Then $\sum_{v=1}^{n} p_v$ is the expected value of $|Y|$, and hence, cannot exceed $\alpha(E)$. On the other hand, since $\log x$ is a concave function, we can apply Jensen's inequality and obtain

$$H(E) = -\sum_{v=1}^{n} \frac{1}{n} \log p_v \geq -\log \left(\sum_{v=1}^{n} \frac{1}{n} p_v \right)$$

$$\geq -\log \frac{\alpha(E)}{n} = \log \frac{n}{\alpha(E)}. \qquad \qquad \square$$

Corollary 6.52. *For every boolean function f of n variables which rejects single-tons, we have that*

$$L(f) \geq n \cdot \log \frac{n}{\alpha(E_f)}.$$

Let $f = \text{Th}_2^n$ be the threshold-2 function of n variables. Then $E_f = K_n$ and $V_f = \emptyset$. If n is even, then we can cover all edges in K_n by $t \leq \lceil \log n \rceil$ bipartite complete graphs $A_i \times B_i$ with $A_i \cap B_i = \emptyset$ and $|A_i| = |B_i| = n/2$. So, Th_2^n can be computed by a *monotone* DeMorgan formula

$$\bigvee_{i=1}^{t} \left(\bigvee_{j \in A_i} x_j \right) \wedge \left(\bigvee_{k \in B_i} x_k \right)$$

of leafsize at most $n \lceil \log n \rceil$.

Theorem 6.53. (Krichevskii 1964; Hansel 1964) *Every DeMorgan formula computing Th_2^n must have at least $n \log n$ leaves.*

Proof. For $f = \text{Th}_2^n$ we have $E_f = K_n$ and $V_f = \emptyset$. The only independent sets in K_n are sets consisting of just one vertex; hence, $\alpha(E_f) = 1$, By Corollary 6.52, we obtain that $L(f) \geq n \log n$, as desired. $\qquad \qquad \square$

Remark 6.54. Lupanov (1965) proved that for any k, $2 \leq k \leq n$, the threshold-k function Th_k^n can be computed by a DeMorgan circuit with $\mathcal{O}(n)$ gates. This shows a logarithmic gap between circuit and formula complexity of some functions.

As another example, consider the following boolean function f_n in $n = \binom{m}{2}$ variables corresponding to the edges of K_m. Each assignment $a \in \{0,1\}^n$ to these variables specifies a graph $G_a \subseteq K_m$. The values of the function f_n are defined by: $f_n(a) = 1$ iff the graph G_a has a vertex of degree at least 2.

Theorem 6.55. *Every DeMorgan formula computing f_n must have at least $\Omega(n \ln n)$ leaves.*

Proof. Let $f = f_n$. Note that the graph E_f in this case is a line graph $L = \mathrm{L}(K_m)$ of K_m, that is, the graph whose vertices are the edges of K_m and two vertices are connected iff they have a common endpoint in K_m. Since independent sets in L are matchings in K_m, we have that the maximum size $\alpha(L)$ of an independent set in L is $\alpha(L) \leq m/2 = O(\sqrt{n})$. By Corollary 6.52, we obtain that $\mathrm{pdim}_{\mathbb{F}}(n) = \Omega(n \log n)$, as desired. $\square$

The bad news is that good graph measures μ cannot yield lower bounds on the formula size larger than $\Omega(n \log n)$. Indeed, the upper bound $\mathrm{L}_+(\mathrm{Th}_2^n) \leq 2n \log n$ together with the lower bound $\mathrm{L}(\mathrm{Th}_2^n) \geq n \cdot \mu(K_n)$, given by Theorem 6.45, implies that $\mu(K_n) \leq 2 \log n$ must hold for *any* good graph measure μ. The main reason for this failure is the *monotonicity* condition: one of the "simplest" graphs—the complete graph K_n—has the largest measure.

6.14   Formula Size, Rank and Affine Dimension

The approach of Khrapchenko and Rychkov can be used in the graph-theoretic setting as well. Recall that the *bipartite complexity*, $\mathrm{L}_{\mathrm{bip}}(G)$, of a bipartite $n \times n$ graph $G \subseteq V_1 \times V_2$ is the minimum number of leaves in a formula over $\{\cap, \cup\}$ which produces the graph G using any of the complete bipartite graphs $P \times V_2$ and $V_1 \times Q$ with $P \subseteq V_1$ and $Q \subseteq V_2$ as inputs.

We already know (see Sect. 1.7) that if we encode the vertices of G by binary vectors of length $m = \log n$, and define the adjacency boolean function f_G of $2m$ variables by $f_G(u, v) = 1$ iff $(u, v) \in G$, then $\mathrm{L}(f) \geq \mathrm{L}_{\mathrm{bip}}(G)$. Thus, any lower bound $\mathrm{L}_{\mathrm{bip}}(G) = \Omega(\log^K n)$ for $K > 3$ would improve the best known lower bound $\Omega(m^3)$ for non-monotone formula size.

Given a bipartite graph $G \subseteq V_1 \times V_2$, let X be the set of its edges, and Y the set of its nonedges (non-adjacent pairs of vertices from different parts). Consider the rectangle $X \times Y$; its elements are (edge, nonedge) pairs. Consider the collection $\mathcal{C} = \mathcal{C}_1 \cup \mathcal{C}_2$ of *canonical rectangles* in $X \times Y$, where

$$\mathcal{C}_1 := \{(P \times V_2) \times (\overline{P} \times V_2) : P \subseteq V_1\},$$
$$\mathcal{C}_2 := \{(V_1 \times Q) \times (V_1 \times \overline{Q}) : Q \subseteq V_2\}. \qquad (6.29)$$

Note that every entry $(x, y) \in X \times Y$ lies in at least one of these rectangles just because x and y cannot share a common endpoint in *both* parts V_1 and V_2. Say that a rectangle $R \subseteq X \times Y$ is *monochromatic* if there is a canonical rectangle containing all elements of R.

Define the *edge-nonedge tiling number*, $\chi(G)$, of the graph G as the minimum number of pairwise disjoint monochromatic rectangles covering all entries of $X \times Y$.

Lemma 6.56. $L_{\mathrm{bip}}(G) \geq \chi(G)$.

Proof. Another way to look at the concept of bipartite complexity of graphs $G \subseteq V_1 \times V_2$ is to associate boolean variables $z_P, z_Q : V_1 \times V_2 \to \{0, 1\}$ with subsets $P \subseteq V_1$ and $Q \subseteq V_2$ interpreted as $z_P(u, v) = 1$ iff $u \in P$, and $z_Q(u, v) = 1$ iff $v \in Q$. Then the set of edges accepted by z_P is exactly the biclique $P \times V_2$, and similarly for variables z_Q. Thus, $L_{\mathrm{bip}}(G)$ is exactly the minimum leafsize of a monotone DeMorgan formula of these variables which accepts all edges and rejects all nonedges of G.

Arguing as in the proof of Rychkov's lemma, we obtain that $L_{\mathrm{bip}}(G)$ is at least the minimum number $\mu(G)$ of pairwise disjoint rectangles R such that: (i) their union covers all entries of the rectangle $X \times Y$, and (ii) each of the rectangles R is separated by one of the variables z_P (or z_Q), that is, $z_P(x) = 1$ and $z_P(y) = 0$ for all $(x, y) \in R$. To show that $\mu(G) \geq \chi(G)$, it remains to observe that each of the rectangles R separated by a variable z_P (or z_Q) is monochromatic, that is, is contained in at least one rectangle from $\mathcal{C} = \mathcal{C}_1 \cup \mathcal{C}_2$.

Indeed, if $x = (u, v)$ is an edge, $y = (u', v')$ a nonedge of G, then $z_P(x) = 1$ implies that $x = (u, v)$ belongs to $P \times V_2$, and $z_P(y) = 0$ implies that $y = (u', v')$ belongs to $\overline{P} \times V_2$. Therefore, every entry (x, y) of the rectangle $X \times Y$ separated by a variable z_P belongs to the rectangle $(P \times V_2) \times (\overline{P} \times V_2)$ from $\mathcal{C}_1$; similarly for entries separated by variables z_Q. $\square$

■ **Research Problem 6.57.** Exhibit an explicit bipartite $n \times n$ graph G with $\chi(G) = \Omega(\log^K n)$.

By Lemma 6.56, this would give a lower bound $\Omega(m^K)$ for the non-monotone formula size of an explicit boolean function in $2m$ variables. Unfortunately, no explicit graphs even with $\chi(G) = \Omega(\log^{1+\epsilon} n)$ for $\epsilon > 0$ are known.

The edge-nonedge tiling number $\chi(G)$ of a bipartite graph G can be lower bounded by the minimum rank of a special partial matrix associated with G, as well as by so-called "affine dimension" of G.

6.14.1 Affine Dimension and Formulas

Let W be a vector space of dimension d over some field $\mathbb{F}$. An *affine representation* of a graph G associates an affine space $S_v \subseteq W$ with every vertex v in such a way that two vertices u and v are adjacent in G iff $S_u \cap S_v \neq \emptyset$. The *affine dimension*, $\text{adim}_\mathbb{F}(G)$, of G is the minimum d such that G has a d-dimensional affine representation.

A *partial matrix* over $\mathbb{F}$ is a usual matrix with the exception that some entries can be left empty (marked by $*$) without placing into them any elements of the underlying field $\mathbb{F}$. An *extension* of such a matrix is a fully defined matrix obtained by filling the unspecified entries by some elements of $\mathbb{F}$. The *rank* of a partial matrix is the minimum rank of its extension.

Given a bipartite graph $G \subseteq V_1 \times V_2$, we can associate with it the following partial *edge-nonedge matrix* A_G whose rows correspond to edges x and columns to nonedges y of G. Fix any two elements $a_1 \neq a_2$ of $\mathbb{F}$, and define the entries of A_G by:

$$A_G[x,y] = \begin{cases} a_1 & \text{if } x \text{ and } y \text{ share a vertex in } V_1; \\ a_2 & \text{if } x \text{ and } y \text{ share a vertex in } V_2; \\ * & \text{if } x \cap y = \emptyset. \end{cases}$$

Theorem 6.58. (Razborov 1990) *For every bipartite graph G,*

$$L_{\text{bip}}(G) \geq \chi(G) \geq \text{rk}(A_G) \geq \text{adim}_\mathbb{F}(G).$$

Proof. The first inequality is Lemma 6.56. To prove the second inequality $\chi(G) \geq \text{rk}(A_G)$, let X be the set of all edges, and Y the set of all nonedges of G. Take a partition $\mathcal{R}$ of $X \times Y$ into $|\mathcal{R}| = \chi(G)$ monochromatic rectangles. Thus, each rectangle $R \in \mathcal{R}$ lies entirely in some rectangle of the collection $\mathcal{C} = \mathcal{C}_1 \cup \mathcal{C}_2$ of canonical rectangles defined by (6.29). Split $\mathcal{R}$ into two collections $\mathcal{R} = \mathcal{R}_1 \cup \mathcal{R}_2$, where $\mathcal{R}_1 = \{R \in \mathcal{R} : R \subseteq C \text{ for some } C \in \mathcal{C}_1\}$ and $\mathcal{R}_2 = \mathcal{R} \setminus \mathcal{R}_1$.

We want to fill in the $*$-entries of A_G so that the resulting full matrix B satisfies $\text{rk}(B) \leq |\mathcal{R}|$. Rectangles in $\mathcal{R}$ partition the entire set of entries. Take a non-$*$ entry (x,y); hence, x and y have a common vertex v. Let $R \in \mathcal{R}$ be the unique rectangle containing (x,y). If $R \in \mathcal{R}_1$, then $R \subseteq C$ for some $C \in \mathcal{C}_1$, meaning that $v \in V_2$ and hence $A_G[x,y] = a_2$. Similarly, if $R \in \mathcal{R}_2$ then $A_G[x,y] = a_1$. Thus, for every rectangle $R \in \mathcal{R}$, all non-$*$ entries of A_G lying in R have the same value; this common value is a_1 if $R \in \mathcal{R}_2$, and is a_2 if $R \in \mathcal{R}_1$. We can therefore fill all $*$-entries of every $R \in \mathcal{R}_i$ by the corresponding to R value. In other words, if J_R denotes a boolean matrix with $J_R[x,y] = 1$ for $(x,y) \in R$, and $J_R[x,y] = 0$ for $(x,y) \notin R$, then the matrix

$$B := a_1 \sum_{R \in \mathcal{R}_2} J_R + a_2 \sum_{R \in \mathcal{R}_1} J_R$$

is an extension of A_G. By subadditivity of rank, we have that $\text{rk}(A_G) \leq \text{rk}(B) \leq |\mathcal{R}| = \chi(G)$, as desired. This completes the proof of $\chi(G) \geq \text{rk}(A_G)$.

Now we prove the last inequality $\text{rk}(A_G) \geq \text{adim}_\mathbb{F}(G)$. Let A be an extension of the partial edge-nonedge matrix A_G such that $\text{rk}(A) = \text{rk}(A_G)$. Let a_x be the row of A corresponding to edge x of G. Assign to each vertex v of G an affine space S_v spanned by all rows a_x with $v \in x$, that is, S_v is the set of all affine combinations of these rows. If two vertices u and v are adjacent, then the spaces S_u and S_v contain the vector a_{uv}, and hence $S_u \cap S_v \neq \emptyset$.

Now suppose that u and v are not adjacent, and consider the y-th column of A, where $y = uv$. Since $v \in V_2$, all rows a_x with $v \in x$ must have a_2 in the y-th position (in the partial matrix A_G, and hence also in its extension A), implying that their affine combination (with coefficients summing up to 1) must also have a_2 in that position. Thus, all vectors in S_v have a_2 in the y-th position. But $u \in V_1$ implies that all vectors in S_u must have a_1 in the y-th position. Thus, $S_u \cap S_v = \emptyset$. We have therefore constructed an affine representation of G of dimension rk(A).   □

6.14.2 Projective Dimension and Branching Programs

Another measure of a graph's dimensionality, called the "projective dimension", was introduced by Pudlák and Rodl (1992). Let W be a vector space of dimension d over some field $\mathbb{F}$. A *projective representation* of a graph G associates a vector space $S_v \subseteq W$ with every vertex v in such a way that two vertices u and v are adjacent in G iff $S_u \cap S_v \neq \{\mathbf{0}\}$. (Note that $\mathbf{0}$ belongs to every vector space.) The *projective dimension*, $\mathrm{pdim}_{\mathbb{F}}(G)$, of G is the minimum d such that G has a d-dimensional projective representation.

This definition could be restated using representation by subspaces of the *projective space* of dimension $d - 1$; adjacent vertices would then correspond to non-disjoint subspaces of $PG(\mathbb{F}, d - 1)$.

Pudlák and Rodl (1992) showed that $\mathrm{pdim}_{\mathbb{F}}(G)$ is a lower bound on the size of *branching programs* computing the graph G in the following sense. At each node the program may test any of variables x_P and x_Q, as defined in the proof of Lemma 6.56. Thus, given a pair of vertices $(u, v) \in V_1 \times V_2$, we have that $x_P(u, v) = 1$ iff $u \in P$, and $x_Q(u, v) = 1$ iff $v \in Q$. In this way, every pair (u, v) defines the unique path in the program. The program *computes* a given bipartite $n \times n$ graph $G \subseteq V_1 \times V_2$ if, for every two vertices $u \in V_1$ and $v \in V_2$, u and v are adjacent in G iff the unique path followed by the pair (u, v) in the program ends in a one-leaf. The size of a program is the total number of nodes in it.

Let BP(G) denote the minimum size of a deterministic branching program computing the graph G. For a boolean function f, let also BP(f) denote the minimum size of a standard deterministic branching program which computes f by testing its variables. Just like in the case of formulas, we have that BP$(f_{2m}) \geq$ BP(G), where $f_{2m} : \{0, 1\}^{2m} \rightarrow \{0, 1\}$ is the adjacency function of G with $m = \log n$. Thus, any lower bound BP$(G) = \Omega(\log^K n)$ would give us a lower bound $\Omega(m^K)$ for the branching program size of an explicit boolean function f_G in $2m$ variables. Recall that the best known lower bound is BP$(f_{2m}) = \Omega(m^2 / \log^2 m)$ proved by Nechiporuk (1966) by counting subfunctions.

On the other hand, Pudlák and Rodl (1992) showed that $\mathrm{BP}(G) \geq \mathrm{pdim}_{\mathbb{F}}(G)$. Unfortunately, no lower bounds larger than $\mathrm{pdim}_{\mathbb{F}}(G) = \Omega(\log n)$ for explicit graphs G are known. The situation with affine dimension is even worse: here even logarithmic lower bounds are not known. The good news, however, is that graphs with high projective (and affine) dimensions *exist*. Let $\mathrm{pdim}_{\mathbb{F}}(n)$ denote the maximum of $\mathrm{pdim}_{\mathbb{F}}(G)$ over all bipartite $n \times n$ graphs G.

If the underlying field $\mathbb{F}$ has a finite number q of elements, then there are at most $\sum_{i=0}^{d} \binom{q^d}{i} \leq q^{d^2}$ possibilities to assign a vector space $S_v \subseteq \mathbb{F}^d$ of dimension $\leq d$ to each of the $2n$ vertices. Thus, there are at most $q^{2d^2 n}$ different projective realizations of a graph. On the other hand, we have 2^{n^2} graphs in total. By comparing these bounds, we obtain that $\mathrm{pdim}_{\mathbb{F}}(n) = \Omega(\sqrt{n})$.

As shown by Pudlák and Rodl (1992), a comparable lower bound $\mathrm{adim}_{\mathbb{F}}(n) = \Omega(\sqrt{n/\log n})$ also holds for the infinite field $\mathbb{F} = \mathbb{R}$ of real numbers. They have also shown that $\mathrm{adim}_{\mathbb{F}}(G) \leq \mathrm{pdim}_{\mathbb{F}}(G)^2$ for every field, and $\mathrm{adim}_{\mathbb{F}}(G) \leq \mathrm{pdim}_{\mathbb{F}}(G) - 1$ if the field is infinite. A partial inverse $\mathrm{pdim}_{\mathbb{F}}(G) \leq \mathrm{adim}_{\mathbb{F}}(G)^{\mathcal{O}(\mathrm{adim}_{\mathbb{F}}(G))}$ for finite fields was shown by Razborov (1990). This does not hold for infinite fields because, say, a complement M of a perfect matching has $\mathrm{adim}_{\mathbb{F}}(M) = 2$ and $\mathrm{pdim}_{\mathbb{F}}(M) = \Omega(\log n)$ over the field $\mathbb{R}$. The lower bound $\mathrm{pdim}_{\mathbb{F}}(M) = \Omega(\log n)$ over any field $\mathbb{F}$ is a direct consequence of the following result due to Lovász (1977): if $U_1, \ldots, U_n$ are r-dimensional and $V_1, \ldots, V_n$ s-dimensional subspaces of a linear space over $\mathbb{F}$ such that $U_i \cap V_i = \{\mathbf{0}\}$ and $U_i \cap V_j \neq \{\mathbf{0}\}$ whenever $i < j$, then $n \leq \binom{r+s}{r}$.

Exercises

6.1. (Due to Augustinovich 1980) Let $K_1, \ldots, K_p$ be all $p = 2^k$ possible monomials in variables $x_1, \ldots, x_k$. Let $z_1, \ldots, z_p$ be new variables, and consider the boolean function $\varphi(x_1, \ldots, x_k, z_1, \ldots, z_p) := \bigvee_{j=1}^{p} K_j \wedge z_j$. Note that any boolean function of variables $x_1, \ldots, x_k$ is a subfunction of φ, that is, can be obtained from φ by setting its z-variables to constants 0 and 1. Now replace each variable x_i in φ by an AND $x_i^1 \wedge x_i^2 \wedge \cdots \wedge x_i^m$ of $m = 2^k/k$ new variables. Let f be the resulting boolean function of $n = km + p = 2^{k+1}$ variables. Let $c_j(f)$ be the number of distinct subfunctions of f of the variables in $Y_i = \{x_1^i, x_2^i, \ldots, x_k^i\}$ obtained by fixing the remaining variables to constants in all possible ways. Show that:

(a) $\log c_i(f) \geq 2^k = n/2$ for all $i = 1, \ldots, m$.

   Hint: Fix an i, $1 \leq i \leq m$, and set to 1 all variables x_i^j different from $x_1^i, x_2^i, \ldots, x_k^i$. Show that *every* boolean function with variables $x_1^i, x_2^i, \ldots, x_k^i$ is a subfunction of the obtained function f_n^i.

(b) f can be computed by a circuit of size $\mathcal{O}(n)$.

   Hint: Construct a circuit computing all ANDs of new variables, and take its outputs as inputs to a circuit computing all monomials K_j.

6.2. Show that Khrapchenko's theorem cannot yield larger than quadratic lower bounds.

Hint: Each vector in $\{0, 1\}^n$ has only n neighbors, that is, vectors y with $\text{dist}(x, y) = 1$.

6.3. Suppose that f can be represented by an s-CNF and by a t-DNF. Show that then Khrapchenko's theorem cannot yield a lower bound larger than st.

6.4. Let $R = A \times B$ be a rectangle with $A, B \subseteq \{0, 1\}^n$, $A \cap B = \emptyset$, and $|A| \geq |B|$. Let $Q = (q_{a,b})$ be the boolean "distance-1" matrix of this rectangle with $q_{a,b} = 1$ iff a and b differ in exactly 1 bit. Let s_b denote the number of ones in the b-th column of Q; hence, $s_b \leq n$. In these terms, the Khrapchenko measure is

$$\mu(R) = \frac{1}{|A| \cdot |B|} \left(\sum_{b \in B} s_b \right)^2.$$

Koutsoupias (1993) proposed the following measure

$$v(R) := \frac{1}{|A|} \sum_{b \in B} s_b^2.$$

Show that $v(R) \geq \mu(R)$ with an equality iff all the s_b's are equal, and that $v(R)$ is a convex measure.

Hint: Cauchy–Schwarz inequality and Theorem 6.39.

6.5. The *spectral norm* of A is can be defined as

$$\|A\| = \max_{x,y \neq 0} \frac{|x^T A y|}{\|x\| \|y\|},$$

where $\|x\|_2 = (\sum_i x_i^2)^{1/2}$ is the Euclidean norm of x. Associate with every matrix A the following rectangle measure proposed by Laplante et al. (2006):

$$\mu_A(R) = \frac{\|A_R\|^2}{\max_M \|A_M\|^2},$$

where A_R denotes the restriction of A to the rectangle R obtained by setting to 0 all entries outside R, and the maximum is over all monochromatic subrectangles M of R. Prove that this rectangle measure is convex.

Hint: Show that the rectangle function $s(R) = \|x_R\|^2 \cdot \|y_R\|^2$ is additive, and use Theorem 6.39.

6.6. (Due to Mike Paterson) Let $A = f^{-1}(0)$ and $B = f^{-1}(1)$. Prove that the Khrapchenko measure

$$\mu(f) = \frac{|A \otimes B|^2}{|A| \cdot |B|},$$

is a formal complexity measure (see Sect. 6.10).

Hint: Argue by induction as in the proof of Rychkov's lemma (Lemma 6.23). In the induction step use the inequality

$$\frac{c_1^2}{a_1 \cdot b} + \frac{c_2^2}{a_2 \cdot b} \geq \frac{(c_1 + c_2)^2}{(a_1 + a_2) \cdot b}$$

which can be checked by a cross-multiplication.

6.7. (Zwick 1991) The unweighted size of a formula is the number of occurrences of variables in it. If the variables $x_1, \ldots, x_n$ are assigned non-negative costs $c_1, \ldots, c_n \in \mathbb{R}$ then the weighted size of the formula is the sum of the costs of all occurrences of variables in the formula. Let $L_c(f)$ denote the smallest weighted size of a DeMorgan formula computing f. For vectors $a, b \in \{0, 1\}^n$, let $c(a, b) = \sqrt{c_i}$ if a and b differ in exactly the i-th bit, and $c(a, b) = 0$ otherwise. For $A \subseteq f^{-1}(0)$ and $B \subseteq f^{-1}(1)$, define $c(A, B) = \sum_{a \in A} \sum_{b \in B} c(a, b)$. Show that, for every weighting $c_1, \ldots, c_n \in \mathbb{R}$,

$$L_c(f) \geq \frac{c(A, B)^2}{|A| \cdot |B|}.$$

Hint: Induction. If the formula is a variable x_i or its negation $\neg x_i$, then $c(A, B)$ is at most $|A|\sqrt{c_i}$ as well as at most $|B|\sqrt{c_i}$. In the induction step use the hint to Exercise 6.6.

6.8. Show that, if μ is an additive rectangle function then, for every fractional partition $R = \sum_i r_i \cdot R_i$, we have that $\mu(R) = \sum_{i=1}^t r_i \cdot \mu(R_i)$.

6.9. Show that any linear combination of convex rectangle functions is a convex rectangle function.

6.10. Let $a(R)$ and $b(R)$ be arbitrary additive non-negative rectangle functions, and consider the rectangle function $\mu(R) = f(a(R))/g(b(R))$, where $f, g : \mathbb{R} \to \mathbb{R}$ are non-decreasing, and f is *sub-multiplicative* in that $f(x \cdot y) \leq f(x) \cdot f(y)$. Show that, if μ is normalized then, for every n-dimensional rectangle R, we have that $\mu(R) \leq f(2n)$.

Hint: Consider a covering of R by $2n$ (overlapping) monochromatic rectangles.

6.11. Consider rectangle measures of the form $\mu(R) = w(R)^k / |R|^{k-1}$, where $w(R)$ is an arbitrary subadditive rectangle function: if $R = R_1 \cup \cdots \cup R_t$ is a partition of R, then $w(R) \leq w(R_1) + \cdots + w(R_t)$. Recall that Khrapchenko's measure has this form with $k = 2$ and $w(R)$ being the number of pairs $(x, y) \in R$ with $\mathrm{dist}(x, y) = 1$. The goal of this exercise is to show that, for $k > 2$, such measures fail badly: they cannot yield even non-constant lower bounds! Namely, let S_n be the rectangle of the parity function of n variables. Show that, for every constant $k > 2$ there is a constant $c = c_k$ (depending on k, but not on n) such that $\mu(S_n) \leq c$.

Hint: Consider the following decomposition of S_n. For $1 \leq i < n$, $\sigma \in \{0, 1\}$ and a string $u \in \{0, 1\}^i$, let $R_{u,\sigma}^i$ be the rectangle consisting of all pairs (x, y) such that $x_{i+1} = \sigma$, $y_{i+1} = 1 - \sigma$

and $x_j = y_j = u_j$ for all $j = 1, \ldots, i$. Use the normalization condition $\mu(R_{u,\sigma}^i) \leq 1$ and geometric series to show that the sum of μ-measures of these rectangles is constant.

6.12. (Rank-measures are not convex) Given an $n \times n$ matrix A (over some field), associate with it the following measure for n-dimensional rectangles:

$$\mu_A(R) = \frac{\text{rk}(A_R)}{\max_M \text{rk}(A_M)}, \tag{6.30}$$

where A_R is the restriction of A to the rectangle R (obtained by setting to 0 all entries outside R), and the maximum is over all monochromatic sub-rectangles of R. If $\text{rk}(A_R) = 0$ then we set $\mu_A(R) = 0$. Subadditivity of rank implies that these measures are subadditive.

Let n be even. Take a rectangle $R = X \times Y$ with $X = \{x_1, \ldots, x_n\}$ and $Y = \{y_1, \ldots, y_n\}$ where $x_i = e_i$, $y_i = e_i + e_{i+1}$ and $e_i \in \{0, 1\}^{n+1}$ is the i-th unit vector. Let A be the complement of the $n \times n$ unit matrix. We define the fractional partition of the rectangle R as follows. For every $i \in [n]$ we take the size-1 rectangle $R_i = \{(x_i, y_i)\}$ and give it weight $r_i = 1$. To cover the rest of the rectangle R, we use rectangles $R_I = \{(x_i, y_j): i \in I, j \notin I\}$ for all $I \subseteq [n]$ of size $|I| = n/2$, and give them weight $r_I = (4 - 4/n)\binom{n}{n/2}^{-1}$. Show that:

(a) This is indeed a fractional partition of R.
(b) The right-hand of the convexity inequality (6.17) is ≤ 4, but the right-hand is $\mu_A(R) \geq (n-1)/2$.

Chapter 7
Monotone Formulas

We have seen that proving lower bound for general circuits is a very difficult task. Thus it is natural to try to obtain large lower bounds for a more restricted class of circuits, the class of monotone circuits. Monotone circuits consist only of AND and OR gates and have no NOT gates. Of course, such circuits cannot compute all boolean functions. What they compute are *monotone* functions, that is, functions f such that $f(x) = 1$ implies that $f(y) = 1$ for all vectors y obtained from x by flipping some of the 0s of x to 1s.

In this chapter we present two general arguments—a rank argument and a game theoretic argument—that allow us to prove super-polynomial lower bounds on monotone formula size. Recall that a formula is a circuit whose underlying graph is a tree; the *leafsize* of a formula is the number of leaves in its underlying tree.

7.1 The Rank Argument

For a monotone boolean function f, let $L_+(f)$ denote the minimum leafsize of a monotone formula for f consisting of fanin-2 AND and OR gates. By results of Khrapchenko and Rychkov, we already know that $L_+(f) \geq \chi_+(f)$, where $\chi_+(f)$ is the monotone tiling number of f defined as the minimum number of pairwise disjoint positively monochromatic rectangles covering all edges of the rectangle $S_f = f^{-1}(1) \times f^{-1}(0)$.

Recall that a rectangle $A \times B$ is *monochromatic* if there is a literal z (a variable x_i or its negation $\neg x_i$) such that $z(a) = 1$ for all $a \in A$, and $z(b) = 0$ for all $b \in B$. A rectangle $A \times B$ is *positively monochromatic* if it is separated by a variable x_i, that is, if there exists an index i such that $a_i = 1$ and $b_i = 0$ for all $a \in A$ and $b \in B$.

One approach to lower-bound the tiling number of f is to choose appropriate subsets of vectors $A \subseteq f^{-1}(1)$ and $B \subseteq f^{-1}(0)$, to choose an appropriate matrix $M : A \times B \to \mathbb{F}$ of large rank over some field $\mathbb{F}$, and to show that the submatrix

S. Jukna, *Boolean Function Complexity*, Algorithms and Combinatorics 27, DOI 10.1007/978-3-642-24508-4_7, © Springer-Verlag Berlin Heidelberg 2012

M_R of M, corresponding to any positively monochromatic subrectangle R, has rank at most some given number r. Since rectangles in each decomposition must be *pairwise disjoint*, the subadditivity of rank implies that

$$L_+(f) \geq \chi_+(f) \geq \frac{\text{rk}(M)}{r}. \tag{7.1}$$

The proof of this last conclusion is simple. Given a decomposition $A \times B = R_1 \cup \cdots \cup R_t$ into t pairwise disjoint monochromatic rectangles, let M_i be the matrix M with all entries outside R_i set to 0. Since the R_i are disjoint, we have that $M = M_1 + \cdots + M_t$. By our assumption, we also have that $\text{rk}(M_i) \leq r$ for all i. The subadditivity of rank gives $\text{rk}(M) \leq \sum_{i=1}^{t} \text{rk}(M_i) \leq t \cdot r$, from which $t \geq \text{rk}(M)/r$ follows.

Of course, we have a similar lower bound for *non-monotone* formulas as well: if every submatrix corresponding to a monochromatic (not necessarily positively monochromatic) subrectangle R has rank at most r, then $L(f) \geq \text{rk}(M)/r$. Unfortunately, in this case the measure $\text{rk}(M)/r$ is submodular, and we already know (see Theorem 6.40) that submodular measures cannot yield even super-linear lower bounds. Still, we will now show that in the monotone case the rank argument *can* yield strong lower bounds.

7.2 Lower Bounds for Quadratic Functions

The quadratic function of an n-vertex graph $G = ([n], E)$ is a monotone boolean function

$$f_G(x_1, \ldots, x_n) = \bigvee_{\{i,j\} \in E} x_i x_j. \tag{7.2}$$

It is often more convenient to consider boolean functions $f(x_1, \ldots, x_n)$ as set-theoretic predicates $f : 2^{[n]} \to \{0, 1\}$. In this case we say that f accepts a set $a \subseteq \{1, \ldots, n\}$ if and only if f accepts its characteristic vector $v_a \in \{0, 1\}^n$ with $v_a(i) = 1$ if and only if $i \in a$. Hence, the quadratic function of a graph G is the unique monotone boolean function f_G such that, for every set of vertices I, we have that

$$f_G(I) = 0 \text{ if and only if } I \text{ is an independent set in } G.$$

Representation (7.2) shows that $L_+(f_G) \leq 2|E|$ for any graph $G = (V, E)$, but for some graphs this trivial upper bound may be very far from the truth.

Example 7.1. Let $G = ([n], E)$ be a complete bipartite graph with $E = S \times T$, $S \cap T = \emptyset$ and $|S| = |T| = n/2$. Then $|E| = n^2/4$, but f_G can be computed by a monotone formula

$$F(x_1, \ldots, x_n) = \left(\bigvee_{i \in S} x_i \right) \wedge \left(\bigvee_{j \in T} x_j \right)$$

of leafsize $|S| + |T| = n$.

So, a natural question is: what quadratic functions require monotone formulas of super-linear size? It turns out that such are dense graphs without triangles and without four-cycles, that is, dense graphs that do not contain cycles with three or four vertices; this was shown in Jukna (2008).

Theorem 7.2. *If* $G = (V, E)$ *is a triangle-free graph without four-cycles, then* $L_+(f_G) \geq |E|$.

Proof. We consider vertices as one-element and edges as two-element sets. For a vertex $y \in V$, let I_y be the set of its neighbors. For an edge $y \in E$, let I_y be the set of all its *proper* neighbors; that is, $v \in I_y$ precisely when $v \not\in y$ and v is adjacent with an endpoint of y. Let $\mathcal{I} = \{ I_y \mid y \in V \cup E \}$. Since G has no triangles and no four-cycles, the sets in $\mathcal{I}$ are independent sets, and must be rejected by f. We will concentrate on only these independent sets.

Let M be a boolean matrix of the rectangle $E \times \mathcal{I}$ defined as follows. The rows are labeled by edges and columns by edges and vertices of G; a column labeled by y corresponds to the independent set I_y. The entries are defined by:

$$M[x, y] = \begin{cases} 1 & \text{if } x \cap y \neq \emptyset, \\ 0 & \text{if } x \cap y = \emptyset. \end{cases}$$

Claim 7.3. *If* R *is a positively monochromatic rectangle, then* $\mathrm{rk}(M_R) = 1$.

Proof. Let $R = S \times T$. Since R is positively monochromatic, there must be a vertex $v \in V$ such that all edges $x \in S$ and all edges or vertices $y \in T$,

$$v \in x \text{ and } v \not\in I_y \text{ for all } x \in S \text{ and } y \in T. \tag{7.3}$$

Thus, for each $y \in T$, we have two possible cases: either v is in y or not.

Case 1: $v \in y$. Since $v \in x$ for all $x \in S$, in this case we have that $x \cap y \supseteq \{v\} \neq \emptyset$, implying that $M_R[x, y] = 1$ for all $x \in S$. That is, in this case the y-th column of M_R is the all-1 column.

Case 2: $v \not\in y$. We claim that in this case the y-th column of M_R must be the all-0 column. To show this, assume that $M_R[x, y] = 1$ for some edge $x \in S$. Then $x \cap y \neq \emptyset$, implying that x and y must share a common vertex $u \in x \cap y$ (see Fig. 7.1). Moreover, $u \neq v$ since $v \not\in y$. Together with $v \in x$, this implies that $y = \{u, v\}$. But then $v \in I_y$, a contradiction with (7.3). $\square$

Fig. 7.1 The cases when
$y \in V$ (*left*) and when $y \in E$
(*right*)

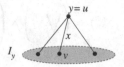 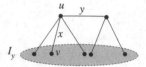

By (7.1), it remains to show that the entire matrix M has full row-rank $|E|$ over
GF(2). For this, take an arbitrary subset $\emptyset \neq F \subseteq E$ of edges. We have to show
that the columns of the submatrix M' of M corresponding to the rows labeled by
edges in F cannot sum up to the all-0 column over GF(2).

If F is not an even factor, that is, if the number of edges in F containing some
vertex v is odd, then the column of v in M' has an odd number of 1s, and we are
done.

So, we may assume that F *is* an even factor. Take an arbitrary edge $y = uv \in F$,
and let $H \subseteq F$ be the set of edges in F incident to at least one endpoint of y.
Since both vertices u and v have even degree (in F), the edge y has a nonempty
intersection with an *odd* number of edges in F: one intersection with itself and an
even number of intersections with the edges in $H \setminus \{y\}$. Thus, the y-th column of
M' contains an odd number of 1s, as desired.   □

Explicit constructions of dense triangle-free graphs without four-cycles are
known.

Example 7.4. (Point-line incidence graph) For a prime power q, a projective plane
$PG(2, q)$ has $n = q^2 + q + 1$ points and n subsets of points (called lines). Every
point lies in $q + 1$ lines, every line has $q + 1$ points, any two points lie on a unique
line, and any two lines meet is the unique point. Here is a $PG(2, 2)$, known as the
Fano plane (with 7 lines and 3 points on a line):

Now, if we put points on the left side and lines on the right, and joint a point x with
a line L by an edge if and only if $x \in L$, then the resulting bipartite $n \times n$ graph will
have $(q + 1)n = \Theta(n^{3/2})$ edges and contain no four-cycles. The graph clearly has
no triangles, since it is bipartite.

Example 7.5. (Sum-product graph) Let p be a prime number and take a bipartite
$n \times n$ graph with vertices in both its parts being pairs (a, b) of elements of a finite
field $\mathbb{Z}_p$; hence, $n = p^2$. We define a graph G on these vertices, where (a, b) and
(c, d) are joined by an edge if and only if $ac = b + d$ (all operations modulo p).
For each vertex (a, b), its neighbors are all pairs $(x, ax - b)$ with $x \in \mathbb{Z}_p$. Thus,
the graph is p-regular, and has $n = np = p^3 = n^{3/2}$ edges. Finally, the graph
cannot have four-cycles, because every system of two equations $ax = b + y$ and
$cx = d + y$ has at most one solution.

If G is any of these two constructed graphs, then Theorem 7.2 implies that any monotone formula computing its quadratic function must have an almost maximal number $\Omega(n^{3/2})$ of leaves.

The quadratic function $f_G(x)$ of a graph G represents this graph: it accepts all its edges, and rejects all its non-edges (see Sect. 1.7). If we could show that *any* monotone boolean function f representing a bipartite $n \times n$ graph G requires $L_+(f) = \Omega(n \log^k n)$, then the Magnification Lemma would give us an explicit boolean function of $2m = 2 \log n$ variables requiring non-monotone(!) formulas of leafsize $\Omega(m^k)$. Unfortunately, Unfortunately, this argument does not work here: it was important that the function f_G rejects all sufficiently large independent sets, not just non-edges. Actually, for a large class of graphs G (so-called *saturated* graphs), we have that f_G is the only(!) monotone boolean function representing G (see Exercise 1.9). Unfortunately, the argument above does not work for saturated graphs either.

7.3 A Super-Polynomial Size Lower Bound

Let $f : 2^{[n]} \to \{0, 1\}$ be a monotone boolean function, $A \subseteq 2^{[n]}$ some subset of its 1-terms and $B \subseteq 2^{[n]}$ some subset of its 0-terms. We thus have $f(a) = 1$ for all $a \in A$, and $f(\overline{b}) = 0$ for all $b \in B$; here $\overline{b} = [n] \setminus b$ is the complement of b. In particular, the pair of families A, B is *cross-intersecting*: $a \cap b \neq \emptyset$ for all $a \in A$ and $b \in B$. Indeed, if $a \cap b = \emptyset$ then $a \subseteq \overline{b}$, the monotonicity of f together with $f(a) = 1$ would imply that $f(\overline{b}) = 1$, a contradiction.

In order to apply the rank argument to lower-bound the monotone formula size of f, we have to associate the rectangle $A \times B$ with an appropriate matrix M of large rank so that each its submatrix, corresponding to a positively monochromatic subrectangle R, has small rank. Recall that R is positively monochromatic if there is an $i \in [n]$ such that $i \in a \cap b$ for all $(a, b) \in R$. In general, the choice of a matrix M is not an easy task. But things are easier if the families A and B have the following "local intersection" property.

Definition 7.6. (Local intersection) A pair A, B of families is *locally intersecting* if every set $b \in B$ can be divided into two nonempty parts $b = b_0 \cup b_1$ such that every $a \in A$ has a nonempty intersection with *exactly one* of these parts.

The *disjointness matrix* of such a pair A, B is an $|A|$ by $|B|$ matrix $D_{A,B}$, with its rows indexed by sets $a \in A$ and its columns indexed by sets $b \in B$, such that the entries of $D = D_{A,B}$ are defined by

$$D[a, b] = \begin{cases} 0 & \text{if } a \cap b_0 \neq \emptyset, \\ 1 & \text{if } a \cap b_1 \neq \emptyset. \end{cases}$$

For a 0-1 matrix M, let $\text{rk}(M)$ denote its rank over GF(2).

Lemma 7.7. (Gál and Pudlák 2003) *Let f be a monotone boolean function, A some set of its 1-terms and B some set of its 0-terms. If the pair A, B is locally intersecting, then $L_+(f) \geq \mathrm{rk}(D_{A,B})$.*

Proof. We are going to apply the rank lower bound (7.1). For this purpose, we take the matrix $M = D_{A,B}$. By (7.1), it is enough to show that $\mathrm{rk}(M_R) \leq 1$ for every positively monochromatic subrectangle $R = A' \times B'$ of $A \times B$. We know that $f(a) = 1$ for all $a \in A'$ and $f(\overline{b}) = 0$ for all $b \in B'$. Since the rectangle R is positively monochromatic, there must exist an $i \in [n]$ such that

$$i \in a \setminus \overline{b} = a \cap b \text{ for all } a \in A' \text{ and } b \in B'.$$

Since the pair A, B is locally intersecting, we know that each $b \in B'$ is divided into two nonempty parts $b = b_0 \cup b_1$ such that each $a \in A'$ intersects exactly one of these parts. Depending on which of these two parts our element i lies in, we divide the set B' into two sets $B_0' := \{b \in B' \mid i \in b_0\}$ and $B_1' := \{b \in B' \mid i \in b_1\}$. Then the submatrix of M_R, corresponding to the rectangle $A' \times B_0'$ is an all-0 matrix, and that corresponding to the rectangle $A' \times B_1'$ is an all-1 matrix. Thus, the submatrix M_R has rank at most 1, as desired.   □

7.3.1 Rank of Disjointness Matrices

In order to get a large lower bound on the formula size, it is therefore enough to find an explicit locally intersecting pair A, B of families and to show that its intersection matrix has large rank. The starting point when doing this is the fact that disjointness matrices of some *single* families have large rank.

Let A be a family of subsets of $[n]$, and let $T : A \to 2^{[n]}$ be a mapping which associates with each member $a \in A$ a subset $T(a) \subseteq a$ such that $T(a) \not\subseteq a'$ for all $a' \in A$, $a' \neq a$. If such a mapping T exists, then we call it a *contractor* of A. In particular, if A is an *antichain*, that is, if no member of A is a subset of another member of A, then the trivial mapping $T(a) = a$ is a contractor of A, but there may also be other contractors as well.

Definition 7.8. Given an antichain A and its contractor T, the *disjointness matrix* of A is a boolean matrix D_A whose rows are labeled by members of A and columns are labeled by all subsets $b \subseteq T(a)$ for all $a \in A$. That is, for every $a \in A$ and for every $b \subseteq T(a)$ there is a column labeled by b. The entry in the a-th row and b-th column is defined by: $D_A[a, b] = 1$ if and only if $a \cap b = \emptyset$.

Formally the matrix D_A also depends on the contractor T: different contractors T may yield different matrices D_A. However, we suppress this for notational convenience: for us it will only be important that all resulting matrices D_A have full row-rank.

Lemma 7.9. *For every antichain A, and for each of its disjointness matrices D_A, we have that $\mathrm{rk}(D_A) = |A|$.*

Proof. The *general disjointness matrix* D_m is a boolean $2^m \times 2^m$ matrix whose rows and columns are labeled by the subsets a of a fixed m-element set, and the (a, b)-th entry is 1 if and only if $a \cap b = \emptyset$. It is not difficult to show that this matrix has full rank over any field, $\text{rk}(D_m) = 2^m$ (see Exercise 7.1).

Take now a disjointness matrix D_A of A corresponding to some contractor T of A. Our goal is to show that all its rows are linearly independent. Fix an $a \in A$ and let M_a be the submatrix of D_A consisting of the columns indexed by subsets $b \subseteq T(a)$, and each label a' of M replaced by $a' \cap T(a)$. Since, by definition of D_A, *every* subset of $T(a)$ appears as a column of D_A, the rows of M_a are rows of the general disjointness matrix D_m with $m = |T(a)|$, some of them repeated. We already know that D_m has full rank. Since $T(a) \subseteq a$, the row with index a is 1 in the column indexed by the empty set, and 0 in every other column of M_a. But the row indexed by a occurs in M_a only once, because $T(a) \not\subseteq a'$ for all $a' \in A$, $a' \neq a$. This implies that this row cannot be a linear combination of other rows in M_a: since the matrix M_a has full rank, all its *distinct* rows must be linearly independent. Thus, the a-th row of the entire matrix D_A cannot be a linear combination of others, either. □

7.3.2 A Lower Bound for Paley Functions

We will consider boolean functions defined by bipartite graphs. Say that a bipartite $n \times n$ graph $G = (U \cup V, E)$ is *k-separated* if for every two disjoint subsets X, Y of U of size at most k there exists a vertex $v \in V$ such that every vertex $u \in X$ is connected with v and no vertex $u \in Y$ is connected with v.

For a bipartite graph satisfying this condition we define A to be the family of sets $a \subseteq U \cup V$ such that $|a \cap U| = k$ and $a \cap V$ is the set of all vertices that are joined to every vertex of $a \cap U$, that is, maximal complete bipartite graphs with the part in U of size k. Associate with each vertex $i \in U \cup V$ a boolean variable x_i, and consider the monotone boolean function

$$f_{G,k}(x) = \bigvee_{a \in A} \bigwedge_{i \in a} x_i .$$

Let, as before, $L_+(f)$ denote the smallest leafsize of a monotone formula computing f. Note that, for every graph G on $2n$ vertices, and for every $1 \leq k \leq n$, we have that

$$L_+(f_{G,k}) \leq 2n \binom{n}{k}.$$

Theorem 7.10. *If the graph G is k-separated, then* $L_+(f_{G,k}) \geq \binom{n}{k}$.

Proof. Let $f = f_{G,k}$. Define B to be the family of sets $b = b_0 \cup b_1$ such that $b_0 \subseteq U$, $|b_0| \leq k$ and b_1 consists of all vertices of V that have no neighbors in b_0 (Fig. 7.2). Since each $a \in A$ induces a complete bipartite graph and $b = b_0 \cup b_1$ an empty graph, a cannot intersect both b_0 and b_1. Moreover, the condition that the underlying graph is k-separated guarantees that $a \cap b_0 = \emptyset$ if and only if $a \cap b_1 \neq \emptyset$. That is, the pair of families A, B is locally intersecting (and hence, also cross-intersecting). Moreover, this pair must be separated by f: if we had $f(\overline{b}) = 1$ for some $b \in B$ then, by the definition of f, some set a would lie in the complement of b, implying that $a \cap b = \emptyset$, and contradicting the cross-intersection property of A, B. Thus, Lemma 7.7 implies that every monotone formula separating the pair A, B, and hence, any such formula computing f_G must have size at least $\mathrm{rk}(D_{A,B})$. To lower bound the rank of $D_{A,B}$, relabel each row $a \in A$ of $D_{A,B}$ by $a' := a \cap U$ (a k-element subset of U), and column $b \in B$ of $D_{A,B}$ by $b' := b_0$ (an at most k-element subset of U), and let M be the resulting matrix; this matrix differs from $D_{A,B}$ only in labelings of rows and columns—the entries remain the same. Since b_0 ranges over *all* at most k-element subsets of U, and since we have:

$$D_{A,B}[a, b] = 1 \text{ iff } a \cap b_1 \neq \emptyset \text{ iff } a \cap b_0 = \emptyset \text{ iff } a' \cap b' = \emptyset,$$

the matrix M is the disjointness matrix $D_{A'}$ of the family A' of all sets $a' = a \cap U$ with $a \in A$. (The contractor in this case is a trivial one $T(a') = a'$.) By Lemma 7.9, $\mathrm{rk}(D_{A,B}) = \mathrm{rk}(D_{A'}) = |A'| = \binom{n}{k}$ and we are done.   □

Paley graphs give an example of explicit bipartite k-separated graphs for $k = \Omega(\log n)$. Let n be an odd prime congruent to 1 modulo 4. A *Paley graph* is a bipartite graph $G = (U \cup V, E)$ with parts $U = V = \mathrm{GF}(n)$ where two nodes, $i \in U$ and $j \in V$, are joined by an edge if and only if $i - j$ is a nonzero square in $\mathrm{GF}(n)$, that is, if $i - j = z^2 \bmod n$ for some $z \in \mathrm{GF}(n)$, $z \neq 0$. The condition $n \equiv 1 \bmod 4$ is to ensure that -1 is a square in the field, making the resulting graph undirected.

Let G be a bipartite $n \times n$ Paley graph. Define the *Paley function* of $2n$ variables by:

$$\mathrm{Paley}_n(x) := f_{G,k}(x) \quad \text{where} \quad k := \lfloor (\log n)/3 \rfloor. \tag{7.4}$$

Theorem 7.11. $\mathrm{L}_+(\mathrm{Paley}_n) = n^{\Theta(\log n)}$.

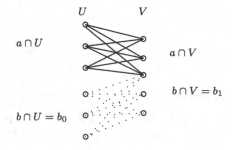

Fig. 7.2 Each a shares a common vertex with each $b = b_0 \cup b_1$, but only in one of the parts U and V

Proof. Let $G = (U \cup V, E)$ be a bipartite $n \times n$ Paley graph with n sufficiently large. Given two disjoint sets of nodes $A, B \subseteq U$ of size $|A| = |B| = k$, let $v(A, B)$ denote the number of nodes in V joined to each node of A and to no node of B. Based on a deep theorem of Weil (1948) regarding sums of quadratic characters $\chi(x) = x^{(n-1)/2}$ over GF(n), Graham and Spencer (1971), and Bollobás and Thomason (1981) proved that

$$|v(A, B) - 2^{-k}n| \leq k\sqrt{n}.$$

This implies that $v(A, B) > 0$ as long as $k2^k < \sqrt{n}$. Thus, G is k-separated as long as $k2^k < \sqrt{n}$, and in particular, is k-separated for $k := \lfloor (\log n)/3 \rfloor$. The desired lower bound for Paley$_n$ now follows from Theorem 7.10. □

Finally, we show that Lemma 7.7 cannot yield larger than $n^{\Omega(\log n)}$ lower bounds.

Lemma 7.12. *If $A, B \subseteq 2^{[n]}$ is a locally intersecting pair, then the rank of the disjointness matrix $D_{A,B}$ does not exceed $n^{O(\log n)}$.*

Proof. We will use the following fact (Lemma 3.9 proved in the first chapter): if a boolean matrix can be covered by t (not necessarily disjoint) monochromatic submatrices, then the matrix can be decomposed into at most $t^{2\log t}$ monochromatic submatrices, and hence, has rank at most $t^{2\log t}$.

Let $D = D_{A,B}$ be the intersection matrix of the pair (A, B). Since the pair is locally intersecting, every set $b \in B$ can be divided into two nonempty parts $b = b_0 \cup b_1$ so that every $a \in A$ has a nonempty intersection with exactly one of these parts. Moreover, $D[a, b] = \alpha$ if and only if $a \cap b_\alpha \neq \emptyset$, $\alpha \in \{0, 1\}$. Thus, each matrix $M_{i,\alpha}$ consisting of all pairs (a, b) with $i \in a \cap b_\alpha$ is monochromatic (is an all-α submatrix), and their union covers all entries of D. Since we only have $2n$ such submatrices, Lemma 3.9 yields the desired upper bound on the rank of D. □

7.4 A log^2 n Depth Lower Bound for Connectivity

We already know (Proposition 1.10) that switching networks are not weaker than DeMorgan formulas. In this section we will show that *monotone* switching networks can be even exponentially more powerful than monotone formulas. To show this, we consider directed graphs on $n + 2$ vertices with two special vertices s (the source) and t (the target). There is one boolean variable x_e for each potential edge e. Each assignment α of values 0 and 1 to these variables defines a directed graph G_α. The *st-connectivity problem* is a monotone boolean function defined by:

$$\text{STCON}_n(\alpha) = 1 \text{ iff } G_\alpha \text{ contains a path from } s \text{ to } t.$$

Note that this function is indeed monotone: if we add edges we cannot disconnect an existing path from s to t.

It can be shown (Exercises 7.2 and 7.3) that STCON_n can be computed by a monotone nondeterministic branching program of size $\mathcal{O}(n^2)$ as well as by a monotone DeMorgan circuit of depth $\mathcal{O}(\log^2 n)$.

We will use the communication complexity approach to show that any monotone circuit solving this problem must have depth $\Omega(\log^2 n)$, and hence, any monotone DeMorgan formula must have super-polynomial leafsize $n^{\Omega(\log n)}$. This lower bound was first proved by Karchmer and Wigderson (1990); a simpler proof was then found by Grigni and Sipser (1995).

The corresponding communication game for STCON_n is the following one: Alice gets a graph G with s-t path, Bob gets a graph H with no s-t paths, and their goal is to find an edge which is present in G but is absent in H.

Note that this is a *monotone* game: an edge which is present in H but absent in G is *not* a correct answer. Since we are interested in proving *lower* bounds on the communication complexity of this game, we can restrict our attention to special inputs. Thus, the game STCON_n corresponding to the st-connectivity problem is the following one:

- Alice gets a directed path p from s to t.
- Bob gets a coloring c of vertices by the colors 0 and 1 such that $c(s) = 0$ and $c(t) = 1$.
- Find an edge $(u, v) \in p$ such that $c(u) = 0$ and $c(v) = 1$.

Note that the path p must have at least one such edge (u, v) because it starts in the node s colored 0 and ends in the node t colored 1.

Let $\mathfrak{c}(\text{STCON}_n)$ denote the communication complexity of this last game. Note that every protocol for the original game can be used to solve this (restricted) game: given a coloring c, Bob converts it into a graph H in which u and v are adjacent if and only if $c(u) = c(v)$.

So as it is, the game STCON_n is no longer "symmetric" since the players receive objects of different types: Alice receives paths and Bob colorings. Still, it is possible to reduce this game to a symmetric one.

7.4.1 Reduction to the Fork Game

Let $[m]^k$ be a grid consisting of all strings $a = (a_1, \ldots, a_k)$ with $a_i \in [m] = \{1, \ldots, m\}$. Given two paths (strings) a and b in $[m]^k$, say that $i \in [k]$ is a *fork position* of a, b if either $i = 1$ and $a_1 \neq b_1$, or $i > 0$ and $a_{i-1} = b_{i-1}$ but $a_i \neq b_i$. Note that any two *distinct* strings must have at least one fork position: either they differ in the first coordinate, or there must be a coordinate where they differ "for the first time". We will be interested in the following symmetric games $\text{FORK}(S)$ on subsets $S \subseteq [m]^k$.

- Alice gets a string $a \in S$ and Bob gets a string $b \in S$.
- Find a fork position i of a and b, if $a_k \neq b_k$.

- If $a_k = b_k$ then $i = k + 1$ is also a legal answer.

If, for example, $a = (1, 2, 4, 3, 4)$ and $b = (3, 2, 2, 5, 4)$ then $i = 1, i = 3$ and $i = 6$ are legal answers, and players can output any of them. In particular, if $a = b$ then $i = k + 1$ is the only legal answer.

Let $c(\text{FORK}_{m,k})$ denote the communication complexity of the fork game on the whole set $S = [m]^k$.

We can relate this game to the previous (s-t connectivity) game. When doing this, we restrict our attention to graphs on $n = mk$ vertices, where only edges connecting nodes from adjacent levels are allowed.

Lemma 7.13. $c(\text{FORK}_{m,k}) \leq c(\text{STCON}_n)$.

Proof. Suppose we have a protocol Π for STCON_n. We will show that this protocol can be used for the game $\text{FORK}_{m,k}$. To use the protocol Π, the players must convert their inputs $a = (a_1, \ldots, a_k)$ and $b = (b_1, \ldots, b_k)$ for the fork game to inputs for the s-t connectivity game.

Alice converts her input $(a_1, \ldots, a_k)$ into a path $p = (u_0, u_1, \ldots, u_k, u_{k+1})$ where $u_0 = s$, $u_{k+1} = t$, and $u_i = a_i$ for $1 \leq i \leq k$. Bob converts his input $(b_1, \ldots, b_k)$ into a coloring c by assigning color 0 to all vertices $s, b_1, \ldots, b_k$, and assigning color 1 to the remaining vertices; hence, $c(s) = 0$ and $c(t) = 1$ (see Fig. 7.3). The players now can use the protocol Π for STCON_n to find an edge (u_{i-1}, u_i) in p such that $c(u_{i-1}) = 0$ and $c(u_i) = 1$. This means that u_{i-1} is in the path $(s, b_1, \ldots, b_k)$ and u_i is not. We claim that i is a valid answer for the fork game on the pair a, b.

If $i = 1$ then $u_{i-1} = u_0 = s$ and $u_1 = a_1$. Therefore, $c(s) = 0$ and $c(a_1) = 1 \neq 0 = c(b_1)$, implying that $a_1 \neq b_1$ (no vertex can receive two colors).

Now let $1 < i \leq k$. Recall that, for each $j = 1, \ldots, k$, the coloring c assigns color 0 to exactly one vertex in the j-th layer, namely to the vertex b_j. Hence, the fact that $c(a_{i-1}) = c(u_{i-1}) = 0$ means that $a_{i-1} = b_{i-1}$, and the fact that $c(a_i) = c(u_i) = 1 \neq 0 = c(b_i)$ means that $a_i \neq b_i$.

Finally, let $i = k + 1$. Then $u_{i-1} = a_k$ and $u_i = t$. Since $c(a_k) = c(u_{i-1}) = 0$ and since only the vertex b_k on the k-th layer can receive color 0, this implies $a_k = b_k$. Since, in this case, $i = k + 1$ is a legal answer for the game, we are done. $\square$

Fig. 7.3 Alice extends her string $a = (a_1, \ldots, a_k)$ to an s-t path $(s, a_1, \ldots, a_k, t)$, and Bob turns his string $b = (b_1, \ldots, b_k)$ into a 2-coloring by assigning color "0" to vertices $s, b_1, \ldots, b_k$ and color "1" to all remaining vertices

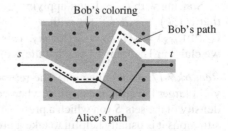

7.4.2   Lower Bound for the Fork Game

Let $c(\text{FORK}_n)$ denote the communication complexity of the game $\text{FORK}(S)$ on $S = [m]^m$ for $m = \sqrt{n}$. By Lemma 7.13 and Exercise 7.3 we know that $c(\text{FORK}_n) = \mathcal{O}(\log^2 n)$. We will show that this upper bound is almost optimal.

Theorem 7.14. (Grigni and Sipser 1995) $c(\text{FORK}_n) = \Omega(\log^2 n)$.

Proof. Call a two-player protocol an (α, k)-*protocol* if it is a protocol for the game $\text{FORK}(S)$ on some subset $S \subseteq [m]^k$ such that $|S| \geq \alpha m^k$. Denote by $c(\alpha, k)$ the minimum communication complexity of an (α, k)-protocol. That is, if $c(\alpha, k) \leq c$ then there exists a subset $S \subseteq [m]^k$ of $|S| \geq \alpha m^k$ strings and a protocol Π of communication complexity c such that Π works correctly on S. In particular, $c(1, k) = c(\text{FORK}_{m,k})$.

We start with two simple claims.

Claim 7.15. For any $k \geq 1$ and any $\alpha > 1/m$, $c(\alpha, k) > 0$.

Proof. Suppose that $c(\alpha, k) = 0$. Thus, there exists a subset of strings $S \subseteq [m]^k$ such that $|S| \geq \alpha m^k > m^{k-1}$ and the players must know the unique answer $i \in \{1, \ldots, k, k+1\}$ for all input pairs $a, b \in S$ without any communication. Since $|S|$ is strictly larger than m^{k-1}, there must be two strings $a, b \in S$ with $a_k \neq b_k$. Hence, on this input pair (a, b) the answer must be some $i \leq k$. But on input pair (a, a) the only legal answer is $i = k + 1$, a contradiction. $\quad\Box$

Claim 7.16. If $k \geq 1$ and $c(\alpha, k) > 0$ then $c(\alpha, k) \geq 1 + c(\alpha/2, k)$.

Proof. Let $c = c(\alpha, k)$. Thus, there exists a subset $S \subseteq [m]^k$ such that $|S| \geq \alpha m^k$ and there is a protocol Π such that for all $a, b \in S$, the protocol correctly solves the game on these inputs with c bits of communication. Assume w.l.o.g. that Alice speaks first (the case when Bob speaks first is similar). Hence Alice sends either 0 or 1. After this (first) bit is communicated, the set S is split into two parts S_0 and S_1. Assume w.l.o.g. that $|S_0| \geq |S_1|$. Let Π_0 be the rest of the protocol Π, after assuming that the first bit send by Alice was 0. That is, Π_0 works exactly like Π, but without sending the first bit, and continuing as if the value of the first bit was 0. The communication complexity of Π_0 is at most $c - 1$. Obviously, Π_0 must work correctly on S_0, because Π does this on the whole set $S = S_0 \cup S_1$. Hence, Π_0 is an $(\alpha/2, k)$-protocol, implying that $c(\alpha/2, k) \leq c - 1 = c(\alpha, k) - 1$. $\quad\Box$

Starting with $\alpha = 1$ and applying Claim 7.16 $t = (\log m)/2$ times, we obtain that $c(1, k) \geq c(\alpha, k) + t$ with $\alpha = 1/\sqrt{m}$. Since $\alpha > 1/m$, Claim 7.15 yields $c(\text{FORK}_{m,k}) = c(1, k) = \Omega(\log m)$. This lower bound is, however, too weak. What we claim in Theorem 7.14 is that the actual lower bound is about $\log k$ times larger.

Remark 7.17. (Amplification) The reason why Claims 7.15 and 7.16 alone cannot yield larger lower bounds is that, when compared to the entire "universe" $[m]^k$, the density of the sets S (on which a protocol is still correct) drops very quickly. In such situations it is usually helpful to take a projection S_I of S onto some subset $I \subseteq [k]$ and to work in the smaller universe $[m]^I$. The hope is that the relative density of S_I

within $[m]^l$ will be much larger than that of S within the whole universe $[m]^k$. This trick is usually called the "amplification" of density.

The point here is the following. Split the set $[k]$ of positions into two equal-sized parts $[k] = I \cup J$. Since $|S| \leq |S_I| \cdot |S_J|$, at least one set of projections, say, S_I must have $|S_I| \geq |S|^{1/2}$ strings. Hence, if $\mu(S) = |S|/m^k$ is the relative density of the original set S, and $\mu(S_I) = |S_I|/m^{k/2}$ the relative density of its projection, then $\mu(S_I) \geq |S|^{1/2}/m^{k/2} = \sqrt{\mu(S)}$, which is at least twice(!) larger than $\mu(S)$, if $\mu(S) \leq 1/4$.

We now turn to the amplification step: given an (α, k)-protocol with $k \geq 2$ and α not too small, we convert it into a $(\sqrt{\alpha}/2, k/2)$-protocol. Thus α may be amplified greatly while k is only cut in half. By amplifying α, after every about $\log k$ steps, we may keep $\alpha > 1/m$ until k reaches 1, showing the protocol must have a path of length at least $\log m$ times $\log k$.

We will need the following combinatorial fact about dense matrices. Say that a boolean vector or a boolean matrix is α-*dense* if at least an α-fraction of all its entries are 1s.

Lemma 7.18. *If A is 2α-dense then*

(a) *there exists a row which is $\sqrt{\alpha}$-dense, or*
(b) *at least a fraction $\sqrt{\alpha}$ of the rows are α-dense.*

Proof. Let A be a boolean $M \times N$ matrix, and suppose that neither case holds. We calculate the density of the entire matrix. Since (b) does not hold, fewer than $\sqrt{\alpha}M$ of the rows are α-dense. Since (a) does not hold, each of these rows has fewer than $\sqrt{\alpha}N$ 1s; hence, the fraction of 1s in α-dense rows is strictly smaller than $(\sqrt{\alpha})(\sqrt{\alpha}) = \alpha$. We have at most M rows which are not α-dense, and each of them has fewer than αN ones. Hence, the fraction of 1s in these rows is also smaller than α. Thus, the total fraction of 1s in the matrix is less than 2α, contradicting the 2α-density of A. $\square$

Lemma 7.19. (Amplification) *For every $k \geq 2$ and $\alpha \geq 16/m$,*

$$c(\alpha, k) \geq c(\sqrt{\alpha}/2, k/2).$$

Proof. We are given an (α, k)-protocol working correctly on some set $S \subseteq [m]^k$ of $|S| \geq \alpha m^k$ strings (paths). Consider a bipartite graph $G = (U \cup V, S)$ with parts U and V, where U consists of all $m^{k/2}$ possible strings on the first $k/2$ levels, and V consists of all $m^{k/2}$ possible strings on the last $k/2$ levels. We connect $u \in U$ and $v \in V$ if their concatenation $u \circ v$ is a string in S; in this case we say that v is an *extension* of u. Hence, G is a bipartite graph with parts of size $m^{k/2}$ and $|S| \geq \alpha m^k$ edges. When applied to the adjacency matrix of G, Lemma 7.18 implies that at least one of the following two must hold:

(a) Some string $u_0 \in U$ has at least $\sqrt{\frac{\alpha}{2}}m^{k/2}$ extensions.
(b) There is an $U' \subseteq U$ such that $|U'| \geq \sqrt{\frac{\alpha}{2}}m^{k/2}$ and each $u \in U'$ has at least $\frac{\alpha}{2}m^{k/2}$ extensions.

Fig. 7.4 Two cases for constructing a protocol for strings of length $k/2$

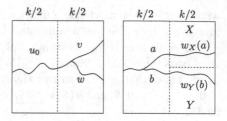

In both cases (a) and (b), we show how to construct a $(\sqrt{\alpha}/2, k/2)$-protocol. □

Case (a): In this case, we have one string u_0 on the left that has many extensions v on the right such that $u_0 \circ v \in S$ (see Fig. 7.4). Thus we can recover a $(\sqrt{\alpha}/2, k/2)$-protocol as follows: let $V' \subseteq V$ be the set of all extensions of u_0. Given two strings $v, w \in V'$, the players can play the S'-game on these inputs by following the S-protocol for the pair of strings $u_0 \circ v$ and $u_0 \circ w$. Since these strings are identical on the first $k/2$ coordinates, the answer i must be either $k+1$ or a point on the last $k/2$ coordinates where the paths v and w diverge.

Case (b): In this case, we take a random partition of the $km/2$ nodes in the right $k/2$ levels. More precisely, take $m/2$ nodes at random from each of the right $k/2$ levels, and call their union X; call the set of remaining $km/4$ right nodes Y. Say that a string $u \in U$ is *good* if it has an extension $v_X(u)$ lying entirely in X and another extension $v_Y(u)$ lying entirely in Y.

Claim 7.20. *The expected number of good strings in U' is at least $0.9|U'|$.*

Proof. We can construct a subset of X as follows. Take $m/2$ uniformly distributed paths $p_1, \ldots, p_{m/2}$ of the right $k/2$ layers, color their vertices red and let X be the union of these vertices. The paths are not necessarily vertex disjoint and some layers may have fewer than $m/2$ vertices. To correct the situation, we randomly color additional vertices red in each layer to ensure that all layers have exactly $m/2$ red vertices. Finally, we color all remaining vertices blue.

Now take a path $u \in U'$. By (b) we know that each red path p_i is an extension of u with probability at least $\alpha/2$. That is, p_i is *not* and extension of u with probability at most $1 - \alpha/2$. Since $\alpha \geq 12/m$, the union bound implies that the probability that none of $m/2$ red paths is an extension of u does not exceed

$$(1 - \alpha/2)^{m/2} \leq (1 - 6/m)^{m/2} \leq e^{-3} < 0.05.$$

Since the red and blue vertices are identically distributed, the same also holds for blue paths. Therefore, each $u \in U'$ is good with probability at least $1 - 2 \cdot 0.05 = 0.9$, implying that the expected fraction of good strings in U' is at least 0.9. □

This yields a $(\sqrt{\alpha}/2, k/2)$-protocol as follows. Let $U'' \subseteq U'$ be the set of all good strings in U'. By Claim 7.20 and since $0.9/\sqrt{2} > 0.5$, the density of the set U'' within $[m]^{k/2}$ is at least $0.9\sqrt{\alpha/2} \geq \sqrt{\alpha}/2$, as desired. Given strings $a, b \in U''$, the players follow the S-protocol on the inputs $a \circ v_X(a)$ and $b \circ v_Y(b)$. Since the

S-protocol is correct on these strings, and since they share no vertices in the right $k/2$ levels, the protocol must return an answer i in the first $k/2$ levels, hence the answer is in fact valid for a and b.

This completes the proof of Lemma 7.19.   □

Now we can finish the proof of Theorem 7.14 as follows.

Let $m = k = \sqrt{n}$. By $r := \lfloor \log(\sqrt{m}/8) \rfloor = \Theta(\log n)$ applications of Claim 7.16 and one application of Lemma 7.19, we obtain that

$$c(2/\sqrt{m}, k) \geq c(16/m, k) + r \geq c(2/\sqrt{m}, k/2) + r.$$

Applying the last inequality $s := \lfloor \log k \rfloor = \Omega(\log n)$ times, we obtain

$$c(2/\sqrt{m}, k) \geq c(2/\sqrt{m}, 1) + r \cdot s \geq r \cdot s.$$

Hence, $c(\text{FORK}_{m,k}) = c(1, k) \geq c(2/\sqrt{m}, k) \geq r \cdot s = \Omega(\log^2 n)$.   □

7.5 An $n^{1/6}$ Depth Lower Bound for Clique Function

The *clique function* $f_n = \text{CLIQUE}(n, k)$ has $\binom{n}{2}$ variables x_{ij}, one for each potential edge in a graph on a fixed set $V = \{1, 2, \ldots, n\}$ of n vertices; the function outputs 1 if and only if the associated graph contains a clique (complete subgraph) on some k vertices. The clique function is monotone because setting more edges to 1 can only increase the size of the largest clique. The corresponding to this function monotone Karchmer-Wigderson game is the following *clique-coloring game*:

- Alice gets a clique $q \subseteq V$ on k vertices.
- Bob gets a coloring $c : V \rightarrow \{1, \ldots, k-1\}$.
- The goal is to find an edge $\{u, v\} \subseteq q$ such that $c(u) = c(v)$.

Since no clique on k vertices can be properly (that is, in a 1-1 manner) colored by $k - 1$ colors, at least one edge of Alice's clique q must be left monochromatic under Bob's coloring; the goal is to find such an edge. Thus, if a leaf of a communication protocol for this game is reached by some set $Q \times C$ of clique/coloring pairs, then there must be an edge $\{u, v\}$ such that

$$\{u, v\} \subseteq q \text{ for all } q \in Q, \text{ and } c(u) = c(v) \text{ for all } c \in C. \tag{7.5}$$

That is, the edge must belong to all cliques in Q and be monochromatic under all colorings in C. In order to force a long path in a communication tree, our strategy is to achieve that the following invariant holds:

(*) If an edge is monochromatic in *all* colorings $c \in C$, then at least one of its two endpoints lies in *none* of the cliques $q \in Q$.

It is clear that, if $Q \times C \neq \emptyset$ and the invariant (*) holds, then (7.5) cannot hold, and hence, the node cannot be a leaf.

In a round the players are allowed to send 1 bit each rather than only one of them sending a bit. Since we are interested in the number of rounds rather than the number of bits transferred, this can only make life easier for them. Each bit that Alice or Bob sends decreases the set of possible cliques or colorings.

Remark 7.21. As in the case of the fork game above, it will be convenient to work with densities of sets in a given universe, rather than with their actual cardinalities (cf. Remark 7.17). Namely, if A is a subset of some given universe B, then its density in B is the fraction $\mu(A) = |A|/|B|$. This definition is useful when we want to describe how much is known about some element $x \in B$. Suppose that we know that $x \in A \subseteq B$. Suppose further that we know the structure of B but that the structure of A is unknown or very complicated. Then the amount of information we (this time, the players) have about x is given by the structure of B and $\mu(A)$. The smaller $\mu(A)$ is the more we know about x.

Example 7.22. To illustrate the usefulness of dealing with densities instead of cardinalities, let B be the family of all $|B| = \binom{n}{k}$ k-element subsets of $[n]$. Each element $x \in [n]$ appears in a $\binom{n-1}{k-1}\binom{n}{k}^{-1} = k/n$ fraction of sets in B. Take a subfamily $A \subseteq B$, and say that an element x is "popular" for A if it appears at least *twice* as often as the average element, that is, if the family $A_x \subseteq A$ of all sets in A containing x has density

$$\frac{|A_x|}{|B|} = \mu(A_x) \geq \frac{2k}{n} \cdot \mu(A) = \frac{2k}{n} \cdot \frac{|A|}{|B|}$$

relative to the underlying set B. Now remove the element x from all sets, and consider the resulting families A' and B' of subsets of the smaller set $[n] \setminus \{x\}$. We only know that $|A'| = |A_x| \geq \frac{2k}{n}|A|$. But A' is a subset of a *smaller* underlying family B' and, relative to this (new) underlying family, has density at least *twice* larger than that of the original family A:

$$\mu(A') = \frac{|A'|}{|B'|} = \frac{|A_x|}{\binom{n-1}{k-1}} = \frac{n}{k} \cdot \frac{|A_x|}{\binom{n}{k}} \geq \frac{n}{k} \cdot \frac{2k}{n} \cdot \frac{|A|}{|B|} = 2 \cdot \mu(A).$$

Theorem 7.23. (Goldmann and Håstad 1992) *Let* $3 \leq k \leq (n/2)^{2/3}$ *and* $t \leq \sqrt{k}/4$. *Then every monotone DeMorgan circuit computing the clique function* CLIQUE(n, k) *must have depth at least* t.

Note that, in terms of the total number $N = \binom{n}{2}$ of *variables*, the depth is $\Omega(N^{1/6})$.

The proof uses an adversary argument. Given a communication protocol tree for the clique-coloring game, our goal is to show that it must have a long path. To analyze the behavior of the protocol, Q_t will denote the set of cliques that remain after round t, C_t the set of colorings that remain after round t, M_t the set of vertices that belong to *all* cliques in Q_t ("fixed" vertices), L_t the set of vertices that belong to *none* of the cliques in Q_t ("forbidden" vertices), $m_t = |M_t|$ and $l_t = |L_t|$.

All these sets are known to both players. Recalling Remark 7.21 we consider Q_t and C_t as subsets of *smaller* universes so that their relative densities may (and will) increase:

- Q_t is a subset of all cliques $q \setminus M_t$ such that $M_t \subseteq q \subseteq V \setminus L_t$.
- C_t is a subset of all colorings of $V \setminus L_t$.

This is because both players already know that Alice's clique must contain M_t, and a monochromatic edge they are looking for cannot have an endpoint in L_t. Thus, m_t and $\mu(Q_t)$ tells us how much Bob knows about Alice's clique after round t. Similarly, l_t and $\mu(C_t)$ measure what Alice knows about the Bob's coloring.

At the beginning ($t = 0$) we have $M_t = L_t = \emptyset$. The t-th round proceeds in two sub-rounds: first speaks Alice, then Bob.

Sub-round 1: Alice sends 1 bit, that is, she splits the current set of cliques Q_{t-1} into two parts, and let Q be the larger of these parts. We now fix some additional vertices (add them to M). On average, a vertex appears in a k/n fraction of k-cliques on n vertices. Say that a vertex $v \in V \setminus M$ is "popular" if it appears at least *twice* as often as the average vertex, that is, if

$$\mu(\{q \in Q \mid v \in q\}) \geq \frac{2k}{n} \cdot \mu(Q).$$

We fix such a vertex (add it to M), remove all cliques that do not contain v, and look for a new popular vertex in the shrunken set of cliques. Each time when we find a popular vertex, the density of the new set of cliques increases by at least factor two since this set is now a subset of a smaller set (cf. Example 7.22). We proceed in this way until no popular vertices exist. If M_t is the new (extended) set of fixed vertices, we set

$$Q_t' := \{q \in Q \mid q \supseteq M_t\} \quad \text{and} \quad C_t' := \{c \in C_{t-1} \mid c \text{ is 1-1 on } M_t\}.$$

Thus, the desired edge cannot lie in M_t.

Sub-round 2: Bob sends 1 bit, that is, he splits the current set of colorings into two parts. In particular, he splits the set C_t' into two parts, and let C be the larger of these parts. We now remove some additional vertices, that is, add them to L. On average, one pair $u \neq v$ of vertices is left monochromatic by a fraction $1/(k-1)$ of all colorings. Say that a pair $u \neq v$ is "popular" if it is left monochromatic by the colorings in C at least *twice* as often as the average pair of vertices, that is, if

$$\mu(\{c \in C \mid c(u) = c(v)\}) \geq \frac{2}{k-1} \cdot \mu(C).$$

Since c is 1-1 on M_t, at least one of u or v must lie outside M_t. We add this endpoint to the current set L of "forbidden" vertices, and restrict C to colorings with $c(u)$

$= c(v)$. Again, the density $\mu(C)$ increases by at least the factor two since the domain $V \setminus L$ of colorings is now smaller. We repeat this step until no desired edges can be found, and let

$$C_t := \{c \in C \mid c(u) = c(v) \text{ for all popular pairs } u \neq v\}.$$

Let L_t be the resulting set of "forbidden" vertices, and set

$$Q_t := \{q \in Q_t' \mid q \cap L_t = \emptyset\}.$$

Thus, the desired edge cannot have an endpoint in L_t.

When the t-th round is over, the resulting set of inputs is the set $Q_t \times C_t$, where all cliques $q \in Q_t$ satisfy $M_t \subseteq q \subseteq V \setminus L_t$, and all colorings $c \in C_t$ are 1-1 on M_t and leave all popular edges (that is, edges touched by L_t) monochromatic. Moreover, if an edge is monochromatic in *all* colorings $c \in C_t$, then at least one of its two endpoints must lie in L_t, and hence, is in *none* of the cliques $q \in Q_t$. Thus, the set of inputs $Q_t \times C_t$ satisfies our invariant (*). It remains therefore to show that, if the number t of rounds is small enough, then this set is still non-empty.

Recall that the set $Q_t \times C_t$ is constructed from $Q_{t-1} \times C_{t-1}$ in two sub-rounds:

$$Q_{t-1} \times C_{t-1} \mapsto Q_t' \times C_t' \mapsto Q_t \times C_t.$$

Lemma 7.24. *Suppose that* $3 \leq k \leq (n/2)^{2/3}$ *and* $t \leq \sqrt{k}/4$. *Then*

$$\mu(C_t') \geq \frac{1}{2}\mu(C_{t-1}) \qquad\qquad \text{if } m_t \leq 2t; \qquad\qquad (7.6)$$

$$\mu(Q_t) \geq \frac{1}{2}\mu(Q_t') \qquad\qquad \text{if } l_t \leq 2t. \qquad\qquad (7.7)$$

Proof. To prove the first inequality (7.6), recall that C_t' consists of all colorings in C_{t-1} that are 1-1 on M_t. By the definition of L_{t-1}, for every edge $e \subseteq V \setminus L_{t-1}$ the relative density of all colorings in C_{t-1} leaving e monochromatic is $< 2\mu(C_{t-1})/(k-1)$. Since $M_t \subseteq V \setminus L_{t-1}$, the same must hold for all edges $e \subseteq M_t$ as well. Since we only have $\binom{m_t}{2}$ edges $e \subseteq M_t$, the density of the set of colorings in C_{t-1} leaving at least one edge in M_t monochromatic is at most

$$\binom{m_t}{2} \frac{2\mu(C_{t-1})}{k-1} \leq \frac{4t^2}{k-1}\mu(C_{t-1}) \leq \frac{1}{2}\mu(C_{t-1}).$$

To prove the second inequality (7.7), recall that, by the definition of Q_t', the density of cliques in Q_t' containing any fixed vertex $v \notin M_t$ must be smaller than $2k/n$ times the density $\mu(Q_t')$ of the set Q_t' itself. Since $L_t \cap M_t = \emptyset$, we have the same bound for all $v \in L_t$. The set Q_t was obtained from Q_t' by removing all cliques containing a vertex in L_t. Hence, the density of removed cliques is at most

$(2k/n)\mu(Q_t')$ times $|L_t| = l_t$. By our assumptions $k \leq (n/2)^{2/3}$ and $l_t = |L_t| \leq 2t \leq \sqrt{k}/2$, we have that the density of removed cliques does not exceed

$$\frac{2kl_t}{n}\mu(Q_t') \leq \frac{k\sqrt{k}}{n}\mu(Q_t') = \frac{k^{3/2}}{n}\mu(Q_t') \leq \frac{1}{2}\mu(Q_t'). \qquad \square$$

We can now show that, if the number t of rounds is not too large, then both sets Q_t and C_t have a nontrivial density.

Lemma 7.25. *Let $k \leq (n/2)^{2/3}$. Then for every $t \leq \sqrt{k}/4$ we have that*

$$\mu(Q_t) \geq 2^{m_t - 2t} \quad and \quad \mu(C_t) \geq 2^{l_t - 2t}.$$

Proof. Since we have $\mu(Q_0) = \mu(C_0) = 1$ and $m_0 = l_0 = 0$, the lemma holds for $t = 0$ (at the beginning of the game). Now assume that the lemma holds for the first $t - 1$ rounds. First we give *explicit* lower bounds on $\mu(Q_t')$ and $\mu(C_t')$. The number of new vertices fixed in round t is $m_t - m_{t-1}$. Since after fixing each of these vertices the density $\mu(Q_t')$ increases by a factor at least two, the induction hypothesis yields

$$\mu(Q_t') \geq \frac{1}{2}2^{m_t - m_{t-1}}\mu(Q_{t-1}) = 2^{m_t - 2t + 1}. \qquad (7.8)$$

Since the density $\mu(Q_t')$ cannot exceed 1, we obtain that $m_t \leq 2t - 1 < 2t$. With this upper bound on m_t, (7.6) yields $\mu(C_t') \geq \mu(C_{t-1})/2$. Together with the induction hypothesis, we obtain that

$$\mu(C_t') \geq 2^{l_{t-1} - 2t + 1}. \qquad (7.9)$$

The sets Q_t and C_t we finally determined in the second part of round t, where Bob (the "color player") sends 1 bit. The bounds obtained for $\mu(Q_t')$ and $\mu(C_t')$ allow us to finish the proof. Since after the removal of all colorings c with $c(u) \neq c(v)$ for a popular pair $u \neq v$ the density $\mu(C_t')$ increases by a factor at least two, the lower bound (7.9) implies

$$\mu(C_t) \geq \frac{1}{2}2^{l_t - l_{t-1}}\mu(C_t') \geq \frac{1}{2}2^{l_t - l_{t-1}}2^{l_{t-1} - 2t + 1} = 2^{l_t - 2t}.$$

Since $\mu(C_t) \leq 1$ we have that $l_t \leq 2t$. Thus, we can apply (7.7) and obtain that $\mu(Q_t) \geq \mu(Q_t')/2$. Together with (7.8) this gives the second desired lower bound

$$\mu(Q_t) \geq \frac{1}{2}\mu(Q_t') \geq 2^{m_t - 2t}. \qquad \square$$

After all these preparations we are now ready to finish the proof of Theorem 7.23 itself.

Proof of Theorem 7.23. Take an arbitrary communication protocol for the monotone Karchmer-Wigderson game corresponding to CLIQUE(n, k). Run this protocol for $t = \sqrt{k}/4$ rounds. The adversary's strategy gives us sets Q_t, C_t, M_t and L_t. Since $t \leq \sqrt{k}/4$, Lemma 7.25 ensures that the sets Q_t and C_t are both non-empty (their densities are not zero). So take some pair $(q, c) \in Q_t \times C_t$. We have that $q \cap L_t = \emptyset$.

If t rounds were sufficient, Alice would know an edge $\{u, v\} \subseteq q$ which is monochromatic under *all* colorings in C_t, including the coloring c. But then, by the definition of the set L_t of "forbidden vertices", at least one of the vertices u and v must lie in L_t, contradicting $q \cap L_t = \emptyset$. $\square$

7.6 An $n^{1/2}$ Depth Lower Bound for Matching

Let $\text{MATCH}_n(x)$ be a monotone boolean function of $\binom{n}{2}$ variables encoding the edges of a graph G_x on $n = 3m$ vertices. The function computes 1 if and only if the graph G_x contains an m-matching, that is, a set of m vertex disjoint edges.

Theorem 7.26. (Raz and Wigderson 1992) *Every monotone circuit computing* $\text{MATCH}_n(x)$ *must have depth at least* $\Omega(n)$.

Note that, in terms of the total number $N = \binom{n}{2}$ of *variables*, the depth is $\Omega(N^{1/2})$.

Proof. Every m-matching p is clearly a 1-term (in fact, a minterm) of MATCH_n. What are 0-terms? For a subset q of $m - 1$ vertices, let c_q be the complete graph on the remaining $2m + 1$ vertices. It is not difficult to see that the complement of c_q can contain no m-matching. Hence, the graphs c_q are 0-terms of MATCH_n.

In a monotone version of Karchmer-Wigderson game for MATCH_n, Alice (holding a minterm p) and Bob (holding a 0-term c_q) must find an edge $e \in p \cap c_p$. It will be convenient to give Bob not graphs c_q but rather the sets q themselves; then $e \in p \cap c_q$ if and only if $e \in p$ and $e \cap q = \emptyset$. Hence, the monotone Karchmer-Wigderson game for MATCH_n must solve the following "find edge" problem:

FE_m: Alice gets an m-matching p and Bob gets an $(m - 1)$-element set q of
           vertices. The goal is to find an edge $e \in p$ such that $e \cap q = \emptyset$.

Let $\mathfrak{c}_+(FE_m)$ be the deterministic communication complexity of this game. By Theorem 3.17, it is enough to show that $\mathfrak{c}_+(FE_m) = \Omega(n)$.

The game FE_m corresponds to a *search* problem: find a desired edge. Our proof strategy is first to reduce this problem to a *decision* problem: given two subsets of $[m]$ decide whether they are disjoint. Since this last problem has randomized communication complexity at least $\Omega(m)$, we will be done. To construct the desired reduction, we consider several intermediate decision problems.

M_m: Alice gets an m-matching p and Bob gets an m-element set q' of vertices.
           Is there an edge e such that $e \in p$ and $e \cap q' = \emptyset$?

DIST_m: Alice gets $x \in \{0, 1, 2\}^m$ and Bob gets $y \in \{0, 1, 2\}^m$.
Is $x_i \neq y_i$ for all $i = 1, \ldots, n$?

DISJ_m: Alice gets $x \in \{0, 1\}^m$ and Bob gets $y \in \{0, 1\}^m$.
Is $x_i \wedge y_i = 0$ for all $i = 1, \ldots, m$?

Through a series of reductions we will show (by ignoring multiplicative constants) that

$$\Omega(m) \overset{(a)}{=} c_{1/3}(\text{DISJ}_m) \overset{(b)}{\leq} c_{1/3}(\text{DIST}_m) \overset{(c)}{\leq} c_{1/3}(M_m) \overset{(d)}{\leq} c_+(FE_m).$$

The lower bound (a) is due to Kalyanasundaram and Schnitger (1992). A proof of a weaker lower bound of $\Omega(\sqrt{n})$ is given in Sect. 4.10. We use randomized protocols because we will need randomness in the last reduction (d).

Proof of (b): $c_{1/3}(\text{DISJ}_m) \leq c_{1/3}(\text{DIST}_m)$. Transform an input $(x, y) \in \{0, 1\}^{2m}$ for DISJ_m into an input $(x, y') \in \{0, 1, 2\}^{2m}$ for DIST_m by setting $y_i' = 1$ if $y_i = 1$, and $y_i' = 2$ if $y_i = 0$. Then $\exists i\ x_i = y_i = 1$ if and only if $\exists i\ x_i = y_i'$.

Proof of (c): $c_{1/3}(\text{DIST}_m) \leq c_{1/3}(M_m)$. Since each randomized protocol for a function f is also a randomized protocol for its negation $\neg f$, it is enough to reduce DIST_m to $\neg M_m$. For this, we need to encode inputs for DIST_m as inputs for M_m. To do this, split all $n = 3m$ vertices into m vertex-disjoint triples, and number the three vertices in each triple by $0, 1, 2$. Given a vector $x \in \{0, 1, 2\}^m$, Alice chooses from the i-th triple the edge $e = \{0, 1, 2\} \setminus \{x_i\}$. Similarly, given a vector $y \in \{0, 1, 2\}^m$, Bob chooses from the i-th triple the vertex y_i. Since the triples are vertex-disjoint, Alice obtains an m-matching p_x, and Bob obtains an m-element set q_y' of vertices. It remains to observe that an edge e with $e \in p_x$ and $e \cap q_y' = \emptyset$ exists if and only if $x_i = y_i$ for some $i \in [m]$ (see Fig. 7.5).

Proof of (d): $c_{1/3}(M_m) \leq c_+(FE_m)$. This is the only nontrivial reduction. In both games M_m and FE_m Alice gets an m-matching p. In the game M_m Bob gets an m-element set q' of vertices, and the goal is to decide whether some edge of p does not touch the set q'. If we pick a vertex $v \in q'$ and remove it from q' to obtain the $(m-1)$-set $q = q' \setminus \{v\}$, then each protocol for FE_m will definitely find an edge $e \in p$ such that $e \cap q = \emptyset$. If we are lucky and the removed vertex v is not an endpoint of the found edge e, then we know the answer: the edge e *is* disjoint from q'. But if $v \in e$, then the answer e of the protocol $FE_m(p, q)$ is useless. The idea

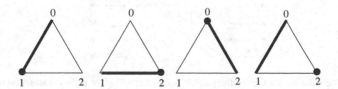

Fig. 7.5 Alice's vector $x = (2, 0, 1, 2)$ and Bob's vector $y = (1, 2, 0, 2)$. From the i-th triple, Alice chooses the edge $e = \{0, 1, 2\} \setminus \{x_i\}$, and Bob chooses the vertex y_i.

therefore is to introduce randomness in the protocol for FE_m to make the probability of this "bad" event small.

Having a protocol P for FE_m, we construct a randomized protocol for M_m as follows.

- Given an m-matching p (for Alice) and an m-element set q' of vertices (for Bob), Bob chooses a *random* vertex $v \in q'$ and defines $q := q' \setminus \{v\}$.
- Alice and Bob flip coins publicly and choose a *random* permutation $\pi : [n] \to [n]$ on the set of vertices of the graph. Then they execute the protocol P on $\pi(p)$ and $\pi(q)$. If $e_1, \dots, e_k \in p$ were the edges in p which do not intersect q, then P returns each edge from $\{e_1, \dots, e_k\}$ with equal probability. Note that $k \geq 1$ since $|q| \leq m - 1$.
- The players eventually agree on an edge e such that $e \in p$ and $e \cap q = \emptyset$. Bob checks whether $v \in e$ and reports this to Alice.
- If $v \notin e$ then $e \cap q' = \emptyset$, and the players know that the answer is "1". Otherwise they gamble on "0".

It remains to show that the gamble can only be wrong with probability at most $1/2$. Let E be the set of all edges in p that contain no endpoint in q'. The gamble is wrong if $E \neq \emptyset$ and $v \in e$. But the protocol outputs each edge in $E \cup \{e\}$ with the same probability $1/|E \cup \{e\}| \leq 1/2$. In particular, it will pick the edge e (and not some edge e' in E) with such a probability (see Fig. 7.6). Since vertex v cannot be an endpoint of more than one edge of the matching p, the probability of error is at most $1/2$. To decrease the error probability, just repeat the protocol twice.

This completes the reductions, and thus the proof of Theorem 7.26.   □

Theorem 7.26, together with the Formula Balancing Lemma (Lemma 6.1), gives an exponential lower bound on the monotone size of DeMorgan formulas. Recall that MATCH_n is a monotone boolean function of $\binom{n}{2} = \Theta(n^2)$ variables.

Corollary 7.27. *Every monotone formula for* MATCH_n *must have size* $2^{\Omega(n)}$.

Borodin et al. (1982) observed that a randomized algorithm for matching, proposed by Lovász (1979b), can be implemented by non-monotone circuits of depth $\mathcal{O}(\log^2 n)$. Together with the lower bound $\Omega(n)$ on the monotone depth, this

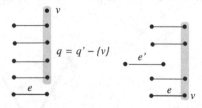

Fig. 7.6 In the situation on the *left* the gamble is correct: $e \cap q' = \emptyset$. In the situation on the *right* the gamble is wrong since $e' \cap q' = \emptyset$, although $e \cap q' \neq \emptyset$. But the error probability is then $\leq 1/2$ since, in this case, the protocol will *not* choose e with probability at least $1/|\{e, e'\}| = 1/2$

gives an exponential gap between the *depth* of monotone and non-monotone circuits, just like Theorem 9.28 gave such a gap for the *size* of circuits.

Yet another consequence of Theorem 7.26 is for switching networks. Such a network is *monotone* if it has no negated variables as contacts.

Corollary 7.28. *Every monotone switching network for* MATCH$_n$ *must have* $2^{\Omega(\sqrt{n})}$ *contacts.*

Proof. Every switching network with s contacts can be simulated by a DeMorgan circuit of depth $\mathcal{O}(\log^2 s)$. We leave this as an exercise. (Hint: binary search.) □

Finally, the proof of Theorem 7.26 also gives a stronger than Theorem 7.23 depth lower bound for the clique function.

Corollary 7.29. *For every* $k \leq n/2$, *every monotone DeMorgan circuit computing the clique function* CLIQUE(n, k) *must have depth at least* $\Omega(k)$.

Proof. Consider graphs on a fixed set V of $|V| = n = 3m$ vertices. Observe that every set $c_q \subseteq V$ of $|c_q| = 2m + 1$ vertices is a minterm, and every m-matching is a maxterm of CLIQUE$(n, 2m + 1)$. Hence, a protocol for the "find edge" game FE_m is in fact a protocol for CLIQUE$(n, 2m + 1)$, with the names of players switched. Theorem 7.26 gives an $\Omega(n)$ depth lower bound for CLIQUE$(n, 2m+1)$. By restricting the inputs to graphs containing a fixed clique of an appropriate size, this also gives lower bounds $\Omega(k)$ for detecting k-cliques. □

Exercises

7.1. The *general disjointness matrix* D_n is a $2^n \times 2^n$ 0-1 matrix whose rows and columns are labeled by the subsets of an n-element set, and the (a, b)-th entry is 1 if and only if $a \cap b = \emptyset$. Prove that this matrix has full rank over any field, i.e., that $\text{rk}(D_n) = 2^n$.

Hint: Use the induction on n together with the following recursive construction of D_n:

$$D_1 = \begin{pmatrix} 1 & 1 \\ 1 & 0 \end{pmatrix}, \qquad D_n = \begin{pmatrix} D_{n-1} & D_{n-1} \\ D_{n-1} & 0 \end{pmatrix}$$

7.2. Show that STCON$_n$ can be computed by a monotone nondeterministic branching program of size $\mathcal{O}(n^2)$. *Hint:* Take one contact for each potential edge.

7.3. Prove that $c(\text{STCON}_m) = O(\log^2 n)$. *Hint:* Use binary search; in fact one of the players may do most of the talking, with the other player communicating only $\mathcal{O}(\log n)$ bits overall.

7.4. Say that a string $x \in [m]^k$ is a *limit* for a subset $S \subseteq [m]^k$ of strings if $x \in S$ and for every position $i = 1, \ldots, k$ there is a string $y \in S$ such that $x \neq y$ and $x_i = y_i$. Prove: if $S \subseteq [m]^k$ and $|S| > km$ then S has a limit for itself.
Hint: What does it mean that S *does not* have a limit for itself?

7.5. The definition of the fork game is somewhat artificial in that the players need not necessarily output a fork position, even when $a \neq b$ (note that then at least one fork position must exist). Instead, they are also allowed to answer "$k + 1$", if $a_k = b_k$. It makes therefore sense to look at what happens if we consider the following modified fork game:

- Alice gets a string $a \in S$ and Bob gets a string $b \in S$.
- Find a fork position i of a and b, if there is one.

That is, the only difference from the original fork game is the following: if $a_k = b_k$ but $a \neq b$, then "$i = k+1$" is no longer a legal answer. In this case the players must output some other position $i \leq k$ (such a fork position exists since $a \neq b$). Prove that the modified fork game on $[m]^k$ has communication complexity $\Omega(k \cdot \log m)$.

Hint: Assume that d bits of communication are enough, where $2^d < m^k/(km)$. Use the previous exercise to get a contradiction.

7.6. (Johan Håstad) Let $A_1, \ldots, A_k$ be finite sets (alphabets), and H be some set of strings $(a_1, \ldots, a_k)$ with $a_i \in A_i$. On average, each letter of A_i appears in the i-th position of $d_i = |H|/|A_i|$ such strings. Let $B_i \subseteq A_i$ be the set of all letters $a \in A_i$ that appear in the i-th position of at least $d_i/2k$ of the strings in H. Let $B = B_1 \times \cdots \times B_k$. Prove that $|B| \geq |H|/2$.

Hint: It might be simpler to prove that $|H \setminus B| < |H|/2$ using the fact that no letter $a \in A_i \setminus B_i$ can appear in more than $d_i/2k$ of the strings. Thus, the number of strings not containing a letter from B_i in the i-th position cannot exceed $|H|/2k$.

7.7. Prove the following analogue of Lemma 7.18: In any 2α-dense matrix either a $\sqrt{\alpha}$-fraction of its rows or a $\sqrt{\alpha}$-fraction of its columns (or both) are $(\alpha/2)$-dense.

Hint: Argue directly or solve Exercise 7.6 and apply it for $k = 2$.

Chapter 8
Span Programs

In 1993 Karchmer and Wigderson introduced an interesting linear algebraic model for computing boolean functions—the *span program*. A span program is just a matrix over some field with rows labeled by literals. (In this chapter we will only work over the field GF(2), but the results hold for any field.) The span program accepts an input assignment if and only if the all-1 vector can be obtained as a linear combination of the rows whose labels are satisfied by the input. The size of the span program is the number of rows in the matrix. A span program is *monotone* if only positive literals are used as labels of the rows, that is, negated variables are not allowed.

The model turns out to be quite strong: classical models for computing boolean functions—like switching networks or DeMorgan formulas—can be simulated by span programs without any increase in size. Moreover, the size of span programs lower bounds the size of parity branching programs—a model where no larger than $n^{3/2}/\log n$ lower bounds are known even in the simplest, read-once case (along each s-t path, each variable can be tested at most once). It is therefore not surprising that proving lower bounds on the size of span programs is a hard task, even in the monotone case.

In this chapter we will show how this task can be solved using linear algebra arguments.

8.1 The Model

First we describe the model more precisely. Let $\mathbb{F}$ be a field. A *span program* over $\mathbb{F}$ is a linear algebraic model that computes a boolean function $f(x_1, \ldots, x_n)$ as follows. Literals are variables $x_i^1 = x_i$ and their negations $x_i^0 = \neg x_i$. Fix a vector space W over $\mathbb{F}$ with a nonzero vector $w \in W$, and associate with each of $2n$ literals x_i^σ a subspace X_i^σ of W. Any truth assignment $a \in \{0, 1\}^n$ to the variables makes exactly n literals "true". We demand that the n associated subspaces span the fixed vector w if and only if $f(a) = 1$. The size measure for this model is the sum

S. Jukna, *Boolean Function Complexity*, Algorithms and Combinatorics 27, 229
DOI 10.1007/978-3-642-24508-4_8, © Springer-Verlag Berlin Heidelberg 2012

of dimensions of the $2n$ subspaces. In a *monotone* span program we do not allow
negated variables.

It is convenient to consider a span program as a matrix M over $\mathbb{F}$ with its rows
labeled by $2n$ literals; one literal may label several rows. If only positive literals
$x_1, \ldots, x_n$ are used, then the program is called *monotone*. The *size* of a span program
M is the number of rows in it. For an input $a = (a_1, \ldots, a_n) \in \{0, 1\}^n$, let
M_a denote the submatrix of M obtained by keeping those rows whose labels are
satisfied by a. That is, M_a contains rows labeled by those x_i for which $a_i = 1$ and
by those $\neg x_i$ for which $a_i = 0$.

The program M *accepts* the input a if the all-1 vector $\mathbf{1}$ (or any other, fixed
in advance *target vector*) belongs to the span of the rows of M_a. A span program
computes a boolean function f if it accepts exactly those inputs a where $f(a) = 1$.
That is,

$$f(a) = 1 \quad \text{if and only if} \quad \mathbf{1} \in \mathrm{Span}(M_a). \tag{8.1}$$

In what follows we will work over the field $\mathbb{F} = GF(2)$. In this case there is the
following equivalent definition of the acceptance condition (8.1). Say that a vector
v is *odd* if the number of 1s in it is odd. Then

$$f(a) = 0 \quad \text{iff there exists an odd vector } v \text{ such that } M_a \cdot v = \mathbf{0}. \tag{8.2}$$

That is, a vector a is rejected if and only if some odd vector v (vector with an
odd number of 1s) is orthogonal to all *rows* of M_a. This follows from the simple
observation that $\mathbf{1} \in \mathrm{Span}(M_a)$ if and only if all vectors in $\mathrm{Span}(M_a)^\perp$ are even;
here, as customary, $V^\perp$ is the orthogonal complement of V, and is defined as the
set of vectors orthogonal to every vector in V; $\mathrm{Span}(V)$ is the set of all linear
combinations of vectors in V.

Remark 8.1. Note also that the number of columns is not counted as a part of the
size. It is always possible to restrict the matrix of a span program to a set of linearly
independent columns without changing the function computed by the program,
therefore it is not necessary to use more columns than rows. However, it is usually
easier to design a span program with a large number of columns, many of which
may be linearly dependent.

Let M be a span program computing f over GF(2). Such a program is *canonical*
if the columns of M are in one-to-one correspondence with the vectors in $f^{-1}(0)$,
and for every $b \in f^{-1}(0)$, the column corresponding to b in M_b is an all-0 column.

Lemma 8.2. *Every span program can be converted to a canonical span program
of the same size and computing the same function.*

Proof. Take a vector $b \in f^{-1}(0)$. By (8.2), there is an odd vector $r = r_b$ for which
$M_b \cdot r_b = 0$. Define the column corresponding to b in a new span program N to be
$M \cdot r_b$. Doing this for all $b \in f^{-1}(0)$ defines the program N and guarantees that it
rejects every $b \in f^{-1}(0)$. To see that M' accepts all ones of f, fix an $a \in f^{-1}(1)$,

and let v_a be such that $v_a^T M_a = \mathbf{1}$. But since r_b is odd for every column $b \in f^{-1}(0)$, we have that $v_a^T N_a = v_a^T M_a r_b = \langle \mathbf{1}, r_b \rangle = 1$. $\square$

8.2 The Power of Span Programs

Together with Proposition 1.10, the following fact shows that span programs are not weaker than DeMorgan formulas.

Proposition 8.3. *If a boolean function can be computed by a switching network of size S then it can also be computed by a span program of size at most S. The same holds for their monotone versions.*

Proof. Let $G = ([n], E)$ be a switching network for a function f, with $s, t \in [n]$ its special vertices. Take the standard basis $\{e_i \mid i \in [n]\}$ of the n-dimensional space over GF(2), that is, e_i is a binary vector of length n with exactly one 1 in the i-th coordinate.

The span program M is constructed as follows. For every edge $\{i, j\}$ in E add the row $e_i \oplus e_j = e_i + e_j$ to M and label this row by the label of this edge (see Fig. 8.1). It is easy to see that there is an s-t path in G, all whose labeled edges are switched on by an input vector a, if and only if the rows of M_a span the target vector $v := e_s \oplus e_t$. Therefore, M computes f, and its size is $|E|$. $\square$

Proposition 8.3 shows that span programs are not weaker than switching networks, and hence, than DeMorgan formulas and deterministic branching programs. What span programs capture is the size of parity branching programs. These are switching networks with the "parity-mode": an input a is accepted if and only if the number of s-t paths consistent with a is odd (see Sect. 1.2). Namely, if SP(f) denotes the complexity of a boolean function in the class of span programs, and $\oplus$BP(f) in the class of parity branching programs, then SP$(f) \leq 2 \cdot \oplus$BP(f) and $\oplus$BP$(f) \leq$ SP$(f)^{\mathcal{O}(1)}$; see Karchmer and Wigderson (1993a) for details.

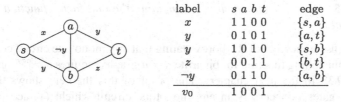

label	s a b t	edge
x	1 1 0 0	$\{s, a\}$
y	0 1 0 1	$\{a, t\}$
y	1 0 1 0	$\{s, b\}$
z	0 0 1 1	$\{b, t\}$
$\neg y$	0 1 1 0	$\{a, b\}$
v_0	1 0 0 1	

Fig. 8.1 A switching network for the threshold-2 function $\mathrm{Th}_2^3(x, y, z)$ in three variables (which outputs 1 if and only if $x + y + z \geq 2$) and the corresponding span program

8.3 Power of Monotone Span Programs

We will exhibit a monotone boolean function f of n variables such that f can be computed by a monotone span program of linear size, but any monotone circuit for f requires $n^{\Omega(\log n)}$ gates.

A *spanning subgraph* of a graph $G = (V, E)$ is a graph $G' = (V, F)$ where $F \subseteq E$; the set of vertices remains the same. A (connected) *component* of a graph is a maximal set of its vertices such that there is a path between any two of them. A graph is *connected* if it consists of just one component. The *degree* $d_F(i)$ of a vertex i is the number of edges of F which are incident to i. An *odd factor* in a graph is a spanning subgraph with all degrees odd.

Lemma 8.4. *If a graph is connected then it has an odd factor if and only if the number of its vertices is even.*

Proof. Suppose that G has an odd factor $G' = (V, F)$. Hence, all degrees $d_F(i)$ are odd. By Euler's theorem, the sum $\sum_{i \in V} d_F(i)$ equals $2|F|$ and is even. Thus, the number $|V|$ of summands must be even, as claimed.

For the other direction, suppose that the graph $G = (V, E)$ is connected and has an even number of vertices, say $V = \{x_1, \ldots, x_{2m}\}$. For every $i = 1, \ldots, m$, fix any one path $P_i = (V_i, E_i)$ connecting x_i to x_{i+m}. Let F be the set of those edges from E which appear in an odd number of the sets $E_1, \ldots, E_m$.

We claim that the subgraph (V, F) is the desired odd factor. Indeed, observe that if a vertex x appears in a path P_i then either $d_{E_i}(x)$ is even or $d_{E_i}(x) = 1$, and this last event happens if and only if x is a leaf of this path, that is, if $x = x_i$ or $x = x_{i+m}$. Since each vertex $x \in V$ is a leaf of exactly one of the paths $P_1, \ldots, P_m$, we have that the sum of degrees $D(x) := \sum_{i=1}^{m} d_{E_i}(x)$ is odd. It remains to observe that, by the definition of F, this sum $D(x)$ is congruent modulo 2 to the degree $d_F(x)$ of x in the graph (V, F). $\square$

The *odd factor function* has $n = m^2$ variables encoding a boolean $m \times m$ matrix representing a bipartite graph with m vertices in each part; the graph is accepted if it has an odd factor.

Lemma 8.5. *Every monotone circuit computing the odd factor function requires* $n^{\Omega(\log n)}$ *gates.*

Proof. We use a theorem of Razborov stating that any monotone circuit computing the perfect matching function for bipartite $n \times n$ graphs requires $n^{\Omega(\log n)}$ gates (see Theorem 9.38 in the next chapter). The proof of this theorem shows that such number of gates is necessary in any monotone circuit which: (1) accepts every perfect matching, and (2) rejects a constant fraction of all unbalanced two-colorings of vertices; each two-coloring is identified with the graph of all monochromatic edges.

Every perfect matching is an odd factor, and should be accepted. On the other hand, an odd two-coloring (in which each color occupies an odd number of vertices) has two odd components, and thus must be rejected: by Lemma 8.4, none of them can have an odd factor. As odd two-colorings constitute half of all two-colorings we are done.   □

It is therefore somewhat surprising that the odd factor function can be computed by a very small monotone span program.

Theorem 8.6. (Babai et al. 1999) *The odd factor function can be computed by a monotone span program of linear size.*

Proof. We construct the desired span program for the odd factor function on $n = m^2$ variables as follows. Let $V = V_1 \cup V_2$ be the vertex set with $|V_1| = |V_2| = m$, and let $X = \{x_{i,j}\}$ with $i \in V_1$ and $j \in V_2$ be the corresponding set of boolean variables (one for each potential edge). Take the standard basis $\{e_i \mid i \in V\}$ of the $2m$-dimensional space over GF(2), that is, e_i is a binary vector of length $2m$ with exactly one 1 in the i-th coordinate. Let M be the m^2 by $2m$ matrix whose rows are vectors $e_i + e_j$ labeled by the corresponding variables $x_{i,j}$. We claim that this span program computes the odd factor function. To verify this we have to show that the all-1 vector $\mathbf{1} = (1, \ldots, 1)$ is a sum over GF(2) of vectors of the form $e_i + e_j$ precisely when the corresponding edges $\{i, j\}$ form an odd factor.

Take an arbitrary graph $E \subseteq V_1 \times V_2$ (viewed as a set of edges). Then the program M accepts E if and only if there exists a subset $F \subseteq E$ such that $\sum_{\{i,j\} \in F}(e_i + e_j) = \mathbf{1} \bmod 2$. Since for each vector $i \in V$, the vector e_i occurs exactly $d_F(i)$ times in this sum, this happens if and only if $\sum_{i \in V} d_F(i) e_i = \mathbf{1} \bmod 2$. That is, the program M accepts a graph E if and only if there exists a subgraph $F \subseteq E$ in which all vertices have odd degrees. Thus, M accepts exactly graphs containing odd factors as claimed.   □

We have seen that, for some monotone boolean functions, their monotone span program size is exponentially smaller than their monotone circuits size. The converse direction remains open.

■ **Research Problem 8.7.** Do there exist functions admitting polynomial-size monotone circuits which require super-polynomial size monotone span programs?

8.4 Threshold Functions

A curious thing about the model of monotone span programs is that it uses non-monotone operations (linear algebra over a field) to compute monotone functions. Karchmer and Wigderson (1993a) detected yet another curious property unique to this model: except the trivial AND and OR, all threshold functions have almost the same complexity! Recall that a *k-threshold function* $\mathrm{Th}_k^n(x_1, \ldots, x_n)$ outputs 1 if and only if $x_1 + \cdots + x_n \geq k$.

Lemma 8.8. *Any monotone span program computing* Th_2^n *over the field* GF(2) *has size at least $n \log n$.*

Proof. Let M be a monotone span program for Th_2^n. Let t be the number of columns in M, and R the set of all odd vectors in $\mathrm{GF}(2)^t$. Clearly $|R| = 2^{t-1}$. Let X_i be the span of the rows of M labeled by the i-th variable, and let $d_i = \dim(X)_i$. Recall that M rejects an input a if and only if at least one vector from R is orthogonal to all the vectors from those sets X_i for which $a_i = 1$. For every $i = 1, \ldots, n$, let $R_i := R \cap (X_i)^\perp$. Since M rejects vectors of weight one, $\mathbf{1} \notin X_i$, and so for every i, $|R_i| = 2^{t-d_i-1}$ (see Exercise 8.1).

We now claim that $R_i \cap R_j = \emptyset$. To see this, observe that for every pair $i \neq j$ we must have vector $\mathbf{1}$ in the span of $X_i \cup X_j$ (as M accepts the vector with 1s in positions i and j). Therefore, for some vectors $u_i \in X_i$ and $u_j \in X_j$, $u_i \oplus u_j = \mathbf{1}$. If there is a vector $r \in R_i \cap R_j$, then $\langle r, \mathbf{1} \rangle = 1$ while $\langle r, u_i \rangle = \langle r, u_j \rangle = 0$, a contradiction.

The previous two paragraphs imply that

$$\sum_{i=1}^n 2^{t-d_i-1} = \sum_{i=1}^n |R_i| = \left| \bigcup_{i=1}^n R_i \right| \leq |R| = 2^{t-1},$$

and hence, $\sum_{i=1}^n 2^{-d_i} \leq 1$. Jensen's inequality states that, if $0 \leq \lambda_i \leq 1$, $\sum_{i=1}^n \lambda_i = 1$ and f is convex, then $f(\sum_{i=1}^n \lambda_i x_i) \leq \sum_{i=1}^n \lambda_i f(x_i)$. Applying this inequality with $\lambda_i = 1/n$ and $f(d_i) = 2^{-d_i}$ we obtain

$$\frac{1}{n} \geq \frac{1}{n} \sum_i 2^{-d_i} = \sum_i \lambda_i f(d_i) \geq f\left(\sum_i \lambda_i d_i \right) = 2^{-(\sum_i d_i)/n},$$

implying that the matrix M must have at least $\sum_i d_i \geq n \log n$ rows. $\qquad\square$

Lemma 8.9. *Let $\mathbb{F}$ be a field with more than $n+1$ elements. Then, for any $1 \leq k \leq n$, the function Th_k^n can be computed by a monotone span program over $\mathbb{F}$ of size n.*

Proof. Since the field $\mathbb{F}$ has more than n elements, we can find a set $\{v_0, v_1, \ldots, v_n\} \subset \mathbb{F}^k$ of $n + 1$ vectors in *general position*, that is, any k of these vectors are linearly independent (see Exercise 8.3). Moreover, we may assume w.l.o.g. that $v_0 = \mathbf{1}$ (the all-1 vector). This suggests the following span program M over $\mathbb{F}$: M is an $n \times k$ matrix whose i-th row ($1 \leq i \leq n$) is v_i and is labeled by x_i. It is now straightforward to check that for any input $a \in \{0, 1\}^n$, the vector $\mathbf{1}$ is spanned by the rows of M_a if and only if $|a| \geq k$. Indeed, if $|a| \geq k$ then the vectors $\{v_i \mid a_i = 1\}$ contain a basis for $\mathbb{F}^k$ (because some k of them are linearly independent), thus vector $\mathbf{1}$ is a linear combination of them. If $|a| \leq k - 1$ then all the vectors $\{v_0\} \cup \{v_i \mid a_i = 1\}$ are linearly independent, and hence, $v_0 = \mathbf{1}$ cannot be a linear combination of $\{v_i \mid a_i = 1\}$. $\qquad\square$

Taking $\mathbb{F} = \mathrm{GF}(2^l)$ with l the smallest integer $> \log n$ (which corresponds to a binary encoding of the field elements), it is possible to reduce the constructed span program over $\mathbb{F}$ to a program over the field $\mathrm{GF}(2)$ of size $O(n \log n)$; see Karchmer and Wigderson (1993a) for details.

8.5 The Weakness of Monotone Span Programs

Lower bounds of order $n^{\Omega(\log n/\log\log n)}$ on the size of monotone span programs were first obtained by Babai et al. (1996, 1999) for explicit boolean functions defined by bipartite graphs with certain properties. Then, using a different property of the underlying bipartite graphs, Gál (2001) simplified and improved these bounds to $n^{\Omega(\log n)}$. All these proofs were based on a general combinatorial lower bound for such programs, found earlier by Beimel et al. (1996); we will present this lower bound in Sect. 8.6.

In this section we will give a different and simpler argument due to Gál and Pudlák (2003). We saw their rank argument in Sect. 7.3 for monotone formulas. Recall that a pair (A, B) of families of sets is *locally intersecting* if every set $b \in B$ can be divided in two nonempty parts $b = b_0 \cup b_1$ so that every $a \in A$ has a nonempty intersection with *exactly one* of these parts. The *disjointness matrix* of such a pair (A, B) is an $|A|$ by $|B|$ matrix $D_{A,B}$, with rows indexed by sets $a \in A$ and columns indexed by sets $b \in B$. The entries of $D = D_{A,B}$ are defined by

$$D[a,b] = \begin{cases} 0 & \text{if } a \cap b_0 \neq \emptyset, \\ 1 & \text{if } a \cap b_1 \neq \emptyset. \end{cases}$$

The following lemma extends a general lower bound for monotone formulas (given in Lemma 7.7) to monotone span programs.

Lemma 8.10. (Gál and Pudlák 2003) *Let f be a monotone boolean function, and let A, B be sets of 1-terms and 0-terms of f respectively. If the pair A, B is locally intersecting, then any monotone span program over* GF(2) *separating this pair must have size at least* $\mathrm{rk}(D_{A,B})$.

Proof. Let M be a monotone span program separating (A, B). Let r be the number of rows and c the number of columns in M. The idea is to show that the disjointness matrix $D = D_{A,B}$ of A, B is a matrix of scalar products of vectors of dimension at most r. This will yield $\mathrm{rk}(D) \leq r$, as desired.

For every $a \in A$, let $v_a \in \mathrm{GF}(2)^r$ be a vector witnessing the fact that a must be accepted, which means that

$$v_a^T \cdot M = \mathbf{1},$$

where $\mathbf{1}$ is the all-1 vector and v_a is a vector which is nonzero only in coordinates corresponding to elements of a, that is, $v_a(i) \neq 0$ implies $i \in a$.

Let $b = b_0 \cup b_1 \in B$. Since the complement $\bar{b}$ of b cannot be accepted, no linear combination of the rows of $M_{\bar{b}}$ can give $\mathbf{1}$; recall that $M_{\bar{b}}$ contains rows of M labeled by those variables x_i for which $i \in \bar{b}$. Hence, by the dual acceptance condition (8.2), for each $b \in B$ there is a vector u_b in $\mathrm{GF}(2)^c$ such that

$$\langle \mathbf{1}, u_b \rangle = 1 \text{ and } M_{\bar{b}} \cdot u_b = \mathbf{0}.$$

Let w_b be the vector in $\mathrm{GF}(2)^r$ obtained from the vector Mu_b by replacing to 0 all its elements, corresponding to the rows labeled by elements of b_0:

$$Mu_b = \overbrace{0\ldots0}^{\overline{b}}\;\overbrace{*\ldots*}^{b_0}\;\overbrace{*\ldots*}^{b_1}$$
$$w_b = 0\ldots0\;\;0\ldots0\;\;*\ldots*$$

Note that $w_b(i) \neq 0$ only if $i \in b_1$. We claim that $D[a,b] = \langle v_a, w_b \rangle$. Indeed, if $a \cap b_0 \neq \emptyset$ then $a \cap b_1 = \emptyset$, and hence, the vectors v_a and w_b have no element on which they both are nonzero, thus $\langle v_a, w_b \rangle = 0$. If $a \cap b_1 \neq \emptyset$ then $a \cap b_0 = \emptyset$, and hence, $\langle v_a, w_b \rangle = \langle v_a, Mu_b \rangle$, implying that

$$\langle v_a, w_b \rangle = \langle v_a, Mu_b \rangle = \langle v_a^T M, u_b \rangle = \langle \mathbf{1}, u_b \rangle = 1\,.$$

This shows that D is a matrix of scalar products of vectors of dimension r, implying that $\mathrm{rk}(D) \leq r$. $\square$

Thus, the Paley function Paley_n defined by Eq. 7.4 in Sect. 7.3.2 requires monotone span programs of super-polynomial size.

Corollary 8.11. *Every monotone span program computing* Paley_n *must have size* $n^{\Omega(\log n)}$.

In the next chapter we will show that some explicit monotone boolean functions (the perfect matching function and some clique-like functions) require *monotone* circuits of super-polynomial size whereas their *non-monotone* circuit size is polynomial. The existence of such a gap between monotone and non-monotone span programs remains open.

■ **Research Problem 8.12.** Do there exist monotone functions that have span programs of polynomial size but require monotone span programs of super-polynomial size?

8.6   Self-Avoiding Families

Let A be a family of subsets of $[n]$. Given a subset $y \subseteq [n]$, define its *spread* as the union

$$S(y) := \bigcup_{a \in A,\, a \cap y \neq \emptyset} a$$

of all members of A that "touch" the set y. Call the family A *self-avoiding* if it is possible to associate a subset $T(a) \subseteq a$ with each set $a \in A$ such that:

- T is a *contractor* of A, that is, no other set in the family A contains $T(a)$ as a subset.

- For every $a \in A$ and every subset $y \subseteq T(a)$, the set $S(y) \setminus y$ contains no member of A.

In particular, the first condition implies that no member of A can be a subset of an another member.

Theorem 8.13. (Beimel et al. 1996) *Every monotone span program computing a monotone boolean function with the minterms forming a self-avoiding family has size at least the number of minterms.*

Using this theorem, they were able to prove the first nontrivial lower bound for monotone span programs. Namely, they proved that any such program detecting the presence of a clique on six vertices in a given n-vertex graph must have size $\Omega(n^5)$. Applying Theorem 8.13 for explicit boolean functions defined by bipartite graphs with certain properties, Babai et al. (1996, 1999) obtained a super-polynomial lower bound of $n^{\Omega(\log n / \log\log n)}$. Then, using a different property of the underlying bipartite graphs, Gál (2001) simplified and improved these bounds to $n^{\Omega(\log n)}$.

Their proofs actually give a lower bound for every monotone span program that accepts all sets $a \in A$ and rejects all sets of the form $[n] \setminus S(y)$ for $y \subseteq T(a)$, $a \in A$. We shall now show that this can also be derived using the rank criterion, thus proving that this criterion is at least as general as the method of self-avoiding families

Lemma 8.14. *For every self-avoiding family A there is a family B such that the pair (A, B) is locally intersecting and $\mathrm{rk}(D_{A,B}) = |A|$.*

Proof. Given a self-avoiding family A, we construct a locally intersecting pair (A, B) as follows: for every $a \in A$, include in B all pairs (b_1, b_2) (more precisely, all sets $b = b_1 \cup b_2$) such that $b_1 \subseteq T(a)$ and $b_2 = [n] \setminus S(b_1)$. Observe that $D_{A,B}$ is just the disjointness matrix of A corresponding to the contractor T (see Definition 7.8). Thus, Lemma 7.9 implies that $\mathrm{rk}(D_{A,B}) = \mathrm{rk}(D_A) = |A|$. So, it remains to show that the pair (A, B) is locally intersecting.

To show this, take an arbitrary set a' in A and $b = b_1 \cup b_2$ in B. Our goal is to show that a' has a nonempty intersection with *exactly one* of the parts b_1 and b_2. If $a' \cap b_1 \neq \emptyset$, then $a' \subseteq S(b_1)$, hence $a' \cap b_2 = \emptyset$. Thus a' intersects at most one of the sets. If $a' \cap b_1 = \emptyset$, then $a' \cap b_2 \neq \emptyset$, because otherwise we would have that $a' \subseteq S(b_1) \setminus b_1$ contradicting the definition of A. Thus a' intersects at least one of the sets. □

Babai et al. (1999) asked whether self-avoiding families of *exponential* size exist. Lemmas 7.12 and 8.14 give a *negative* answer: $|A| \leq n^{O(\log n)}$ for every self-avoiding family of subsets of $[n]$.

8.7   Characterization of Span Program Size

Let f be a boolean function of n variables, and consider the rectangle $S_f = f^{-1}(1) \times f^{-1}(0)$. Recall that a subrectangle $R \subseteq S_f$ is *monochromatic* if there exists a position $i \in [n]$ such that $a_i \neq b_i$ for all $(a, b) \in R$, that is, there exists a value $\sigma \in \{0, 1\}$ such that $a_i = \sigma$ and $b_i = 1 - \sigma$ for all $(a, b) \in R$. Note that every monochromatic rectangle is a subrectangle of at least one of $2n$ *canonical rectangles*:

$$R_{\sigma,i} := \{(a, b) \in S_f \mid a(i) = \sigma, b(i) = 1 - \sigma\}.$$

Fix a field $\mathbb{F}$, and consider matrices $A : S_f \to \mathbb{F}$ over $\mathbb{F}$ whose rows are labeled by vectors in $f^{-1}(1)$, and columns by vectors in $f^{-1}(0)$. Say that such a matrix A is *monochromatic* if there is a canonical rectangle $R_{\sigma,i}$ containing all nonzero entries of A.

Define the *algebraic tiling number* $\chi_{\mathbb{F}}(f)$ of f as the smallest number t such that there exist t monochromatic matrices of rank-1 summing up (over $\mathbb{F}$) to the all-1 matrix $\mathbf{1}$.

Recall that the tiling number $\chi(f)$ is the smallest number of pairwise disjoint monochromatic rectangles covering the entire ambient rectangle S_f. With each rectangle in such a decomposition we can associate a 0-1 matrix of rank 1. Since the rectangles are disjoint and cover all entries of S_f, the sum of these matrices over any field $\mathbb{F}$ is the all-1 matrix. This shows that $\chi_{\mathbb{F}}(f) \leq \chi(f)$ holds in any field $\mathbb{F}$.

We already know (see Lemma 6.23) that $\chi(f)$ is a lower bound for the formula size $L(f)$ of f, but no converse better than $L(f) \leq \chi(f)^{2 \log \chi(f)}$ is known so far. This is why the following tight(!) characterization of the span program size is particularly interesting.

Theorem 8.15. (Gál 2001) *For every boolean function f and every field $\mathbb{F}$,*

$$\mathrm{SP}_{\mathbb{F}}(f) = \chi_{\mathbb{F}}(f).$$

The same also holds for *monotone* span programs and monotone algebraic tiling number, where only n canonical rectangles $R_{1,i}$, $i = 1, \ldots, n$ are used.

We will prove the upper and lower bounds on $\chi_{\mathbb{F}}(f)$ separately. For definiteness, we restrict our attention to the field $\mathbb{F} = \mathrm{GF}(2)$, but the argument works for every field. Let M be a span program with s rows $m_1, \ldots, m_s$ computing f over $\mathrm{GF}(2)$. The rows are labeled by literals $z_1, \ldots, z_s$, with each z_j being a variable x_i or its negation $\neg x_i$, so that for every input vector $a \in \{0, 1\}^n$, $f(a) = 1$ if and only if the set of rows $\{m_j \mid z_j(a) = 1\}$ spans the all-1 vector $\mathbf{1}$ (we use $\mathbf{1}$ as the target vector.)

By Lemma 8.2, we can assume that our program M is *canonical*: the columns of M are in one-to-one correspondence with the vectors in $f^{-1}(0)$, and for every $b \in f^{-1}(0)$, the column corresponding to b in M_b is an all-0 column. In other words, for every $b \in f^{-1}(0)$ there is a column $v_b \in \mathbb{F}^s$ of M with nonzero entries only at rows whose label take the value 0 on b. On the other hand, since M computes

f, for every $a \in f^{-1}(1)$ there is a vector $u_a \in \mathbb{F}^s$ with nonzero entries only at rows whose label take the value 1 on a, such that $u_a^T M = \mathbf{1}$. Thus, a canonical span program of size s gives us sets of vectors $U = \{u_a \in \mathbb{F}^s \mid a \in f^{-1}(1)\}$ and $V = \{v_b \in \mathbb{F}^s \mid b \in f^{-1}(0)\}$ such that

$$\langle u_a, v_b \rangle = 1 \text{ for every } (a,b) \in f^{-1}(1) \times f^{-1}(0), \tag{8.3}$$

and for every position $j \in [s]$,

$$u_a(j) \neq 0 \Rightarrow z_j(a) = 1 \quad \text{and} \quad v_b(j) \neq 0 \Rightarrow z_j(b) = 0. \tag{8.4}$$

Claim 8.16. $\chi_\mathbb{F}(f) \leq \mathrm{SP}_\mathbb{F}(f)$.

Proof. For every $\sigma \in \{0,1\}$ and $1 \leq i \leq n$, let $J_{\sigma,i} := \{j \in [s] : z_j = x_i^\sigma\}$ be the set of positions corresponding to rows of M labeled by the literal x_i^σ. Thus,

$$\sum_{\sigma,i} |J_{\sigma,i}| = s \ \ (= \mathrm{SP}_\mathbb{F}(f)). \tag{8.5}$$

Let $Q_{\sigma,i}$ be the matrix over the ambient rectangle $f^{-1}(1) \times f^{-1}(0)$ whose (a,b)-entry is the scalar product of vectors u_a and v_b restricted to the positions in $J_{\sigma,i}$, that is,

$$Q_{\sigma,i}[a,b] = \sum_{j \in J_{\sigma,i}} u_a(j) \cdot v_b(j).$$

By (8.4), all nonzero entries of $Q_{\sigma,i}$ lie in the canonical rectangle $R_{\sigma,i}$. Clearly, the rank of each matrix $Q_{\sigma,i}$ is at most $|J_{\sigma,i}|$. Hence, we can write each matrix $Q_{\sigma,i}$ as a sum of at most $|J_{\sigma,i}|$ rank-1 matrices, all nonzero entries of each of which lie in $R_{\sigma,i}$. Thus, each of these rank-1 matrices is monochromatic. By (8.5), the matrix $Q := \sum_{\sigma,i} Q_{\sigma,i}$ is a sum of at most $s = \mathrm{SP}_\mathbb{F}(f)$ monochromatic matrices of rank-1. It remains therefore to verify that $Q = \mathbf{1}$. But this follows directly from (8.3), because the sets $J_{\sigma,i}$ form a partition of the entire set $[s]$ of coordinates of the vectors u_a and v_b, and the scalar product of these vectors is equal to 1. $\qquad\square$

Claim 8.17. $\mathrm{SP}_\mathbb{F}(f) \leq \chi_\mathbb{F}(f)$.

Proof. Let $A_1, \ldots, A_t$ be $t = \chi_\mathbb{F}(f)$ monochromatic $|f^{-1}(1)| \times |f^{-1}(0)|$ matrices of rank-1 that sum up to the all-1 matrix $\mathbf{1}$. We will construct a canonical span program of size t computing f. In fact, we will construct sets of vectors $\{u_a \in \mathbb{F}^t \mid a \in f^{-1}(1)\}$ and $\{v_b \in \mathbb{F}^t \mid b \in f^{-1}(0)\}$ satisfying (8.3) and (8.4). Then we can take the vectors v_b as columns of our canonical span program for f.

We know that each of the matrices $A_1, \ldots, A_t$ has rank 1, and all nonzero entries of each A_j are contained in one of $2n$ canonical monochromatic rectangles $R_{\sigma,i}$. Now, collect together the rank-1 matrices contained in $R_{\sigma,i}$ (resolve ambiguities arbitrarily, but uniquely) and add them up to form a matrix $Q_{\sigma,i}$. Clearly, $\mathrm{rk}(Q_{\sigma,i})$ is at most the number of these rank-1 matrices. Write $Q_{\sigma,i}$ as a product

$Q_{\sigma,i} = S_{\sigma,i} \cdot T_{\sigma,i}$ where $S_{\sigma,i}$ (resp., $T_{\sigma,i}$) has $\mathrm{rk}(Q_{\sigma,i})$ columns (resp., rows). For $a \in f^{-1}(1)$, let u_a be the concatenation of the a-th rows of $S_{\sigma,i}$ for all σ, i. Similarly, for $b \in f^{-1}(0)$, let v_b be the concatenation of the b-th columns of $T_{\sigma,i}$ for all σ, i. Note that u_a and v_b are vectors in $\mathbb{F}^s$ with $s \leq \sum_{\sigma,i} \mathrm{rk}(A_{\sigma,i}) \leq \sum_{i=1}^{t} \mathrm{rk}(A_i) = t$. So, it remains to verify that these vectors satisfy the conditions (8.3) and (8.4). The first condition (8.3) follows from the fact that $\sum_{\sigma,i} Q_{\sigma,i} = \sum_{i=1}^{t} A_i = \mathbf{1}$. The second condition (8.4) follows from the fact that all nonzero entries of every matrix $Q_{\sigma,i}$ lie in the canonical monochromatic rectangle $R_{\sigma,i}$.   □

8.8 Monotone Span Programs and Secret Sharing

Monotone span programs capture in an elegant way secret sharing schemes in the information theoretic model. Informally, a secret sharing scheme for a monotone function f prescribes a way for a "sender" having a secret $s \in \mathbb{F}$ to assign n strings ("pieces of secret") $s_i \in \mathbb{F}^{d_i}$ satisfying the following. Let $a \subseteq [n]$ be a subset of (indexes of) the pieces, and denote by $f(a)$ the function f evaluated on the characteristic vector of a. Then if $f(a) = 1$ the pieces $\{s_i \mid i \in a\}$ determine the secret s, while if $f(a) = 0$ these pieces give no information whatsoever about s. The size of such scheme is $\sum_{i=1}^{n} d_i$. Let $\mathrm{mSP}(f)$ denote the minimum size of a monotone span program computing f.

Theorem 8.18. (Karchmer and Wigderson 1993a) *For every prime p, every monotone function has a secret sharing scheme over* $\mathrm{GF}(p)$ *of size* $\mathrm{mSP}(f)$.

Thus, $\mathrm{mSP}(f)$ is an upper bound on the size of secret sharing schemes. Beimel and Chor (1994) showed that $\mathrm{mSP}(f)$ is also a *lower* bound for so-called "linear" secret sharing schemes.

Proof. Fix a prime p, set $\mathbb{F} = \mathrm{GF}(p)$ and let M be a monotone span program for a monotone function f. Let d_i be the number of rows in M labeled x_i, and M_i the submatrix of M consisting of these rows. Let t be the number of columns in M.

Let $s \in \mathbb{F}$ be the secret, and let $W = \{w \in \mathbb{F}^t \mid \langle w, \mathbf{1} \rangle = s\}$. Let $w \in W$ be chosen uniformly at random, and define the "random pieces" $q_i \in \mathbb{F}^{d_i}$ for every $i \in [n]$ by $q_i := M_i w$. Further, for any subset $a \subseteq [n]$ let $q_a := M_a w$, where M_a is the matrix associated with the characteristic vector of a. Note that q_a is just the concatenation of the vectors $\{q_i \mid i \in a\}$. The theorem follows from the following two claims:

(a) If $f(a) = 1$ then s can be efficiently determined from q_a.
(b) If $f(a) = 0$ then for every $r \in \mathbb{F}$, $\mathrm{Prob}[s = r \mid q_a] = 1/p$.

To prove (a), assume that $f(a) = 1$. Then, by definition, there is a vector v such that $v^T M_a = \mathbf{1}$. Then $s = \langle w, \mathbf{1} \rangle = v^T M_a w = \langle v, q_a \rangle$. To prove (b), assume $f(a) = 0$. By (8.2), there exists a vector $z \in \mathbb{F}^t$ such that $M_a z = \mathbf{0}$ but $\langle z, \mathbf{1} \rangle \neq 0$. Then for any q, we can associate with any w such that $M_a w = q$, p vectors $w_j := w + jz$, $j \in \mathbb{Z}_p$.

Note that $M_a w_j = q$ as well, but the values $\langle w_j, 1 \rangle$ are all distinct and exhaust $GF(p)$. This breaks up the probability space $\{w \mid M_a w = q\}$ into p equiprobable classes, each giving s a different value, which concludes the proof of the second claim (b), and thus the proof of the theorem. $\qquad \square$

Exercises

8.1. Let $V \subseteq GF(2)^n$ be a linear space and $y \in GF(2)^n$ be a vector. Assume that $y \notin V^{\perp}$. Show that $v \cdot y = 0$ for precisely one-half of the vectors v in V.

Hint: Split V into V_0 and V_1 according to whether $v \cdot y = 0$ or $v \cdot y = 1$. Take $x \in V$ such that $x \cdot y = 1$; hence $x \in V_1$. Then, show that $x + V_0 \subseteq V_1$, $x + V_1 \subseteq V_0$, $|x + V_0| = |V_0|$ and $|x + V_1| = |V_1|$.

8.2. The *Vandermonde matrix* is the $n \times n$ matrix X_n whose i-th row is $(1, x_i, x_i^2, \ldots, x_i^{n-1})$. Prove that $\det(X_n) = \prod_{1 \le i < j \le n}(x_j - x_i)$.

Hint: Argue by induction on n. Multiply each column by x_1 and subtract it from the next column on the right starting from the right-hand side. This yields $\det(X_n) = (x_n - x_1) \cdots (x_2 - x_1) \det(X_{n-1})$.

8.3. Let $\mathbb{F}$ be a field with more than $n + 1$ elements. Let $a_0, a_1, \ldots, a_n$ be distinct nonzero elements of $\mathbb{F}$. For each $0 \le i \le n$ define the vector $v_i := (a_i^0, a_i^1, \ldots, a_i^{k-1})$. Show that any k, the v_i are linearly independent over $\mathbb{F}$.

Hint: $\det(A) \ne 0$ implies $\mathrm{rk}(A) = n$.

8.4. (Subfunctions) Let $f(x_1, \ldots, x_n)$ be a boolean function, and g its subfunction obtained by setting the variable x_1 to 1. Let M be a canonical span program computing f. Let $U_1 \subseteq f^{-1}(0)$ be the subset of vectors whose first coordinate (corresponding to x_1) is 1. Remove from M all rows labeled by x_1 and $\neg x_1$, and all columns corresponding to vectors in $f^{-1}(0) \setminus U_1$. Show that the resulting span program computes g.

8.5. (Due to Serge Fehr) Let $f : 2^{[n]} \to \{0, 1\}$ be a non-constant monotone boolean function. Recall that the *dual* of f is a monotone boolean function $f^* : 2^{[n]} \to \{0, 1\}$ defined by $f^*(a) = 1 - f(\bar{a})$. Let M be a monotone $s \times t$ ($s \ge t$) span program over $GF(2)$ which computes f using the target vector $e_1 = (1, 0, \ldots, 0)$. Let v_0 be a solution of the system of linear equations $x^T M = e_1$ and $w_1, \ldots, w_{s-t}$ a basis of the linear space $\{w \mid w^T M = 0\}$. Let M^* be the $s \times (s - t + 1)$ matrix with columns $v_0, w_1, \ldots, w_{s-t}$, and $e_1^* = (1, 0, \ldots, 0) \in GF(2)^{s-t+1}$. Show that:

(a) $f(a) = 0$ if and only if there exists a vector u such that $M_a \cdot u = 0$ and $\langle u, e_1 \rangle = 1$.

(b) Every solution x of $x^T M = e_1$ is a linear combination of the columns of M^* in which the first column, v_0, occurs exactly once.

(c) M^* computes f^*.

Hint: If $f(a) = 1$ then there exists a vector v such that $v^T M = e_1$ and $v(i) = 0$ for all $i \in \bar{a}$. Use (b) to conclude that $v = M^* u$ for some vector u with $u(1) = 1$. Show that $M_{\bar{a}}^* u = 0$ and

$\langle u, e_1^* \rangle = 1$, and use (a) to show that $\overline{a}$ is not accepted by M^*. For the other direction, assume that the complement $\overline{a}$ of some set a is not accepted by M^*. Take a vector u guaranteed by (a), and set $b := M^* u$. Show that $b(i) = 0$ for all $i \in \overline{a}$ and $b^T M = M^T M^* u = Eu = e_1$.

8.6. (Beimel et al. 1996) Let $L_1, \ldots, L_n$ be subsets of $[n]$ such that $|L_i \cap L_j| \le 1$ for all $i \ne j$. Let A be the family of all two-element subsets $a = \{x_i, y_j\}$ of the $(2n)$-element set $X = \{x_1, \ldots, x_n, y_1, \ldots, y_n\}$ such that $j \in L_i$. By Example 7.4, explicit families of this form with $\Omega(n^{3/2})$ members exist. Prove that the family A is self-avoiding.

Hint: Define $T(a) := a$.

8.7. ■ **Research Problem.** Let k be the minimal number for which the following holds: there exist n colorings $c_1, \ldots, c_n$ of the n-cube $\{0, 1\}^n$ in k colors such that for every triple of vectors x, y, z there exists a coordinate i on which not all three vectors agree, and the three colors $c_i(x), c_i(y), c_i(z)$ are distinct. Bound the smallest number k of colors for which such a good collection of colorings $c_1, \ldots, c_n$ exists.

Comment: This problem is connected with proving lower bounds on the size of *non-monotone* span programs, see Wigderson (1993).

8.8. (Wigderson 1993) Consider the version of the problem above where we additionally require that the colorings c_i are *monotone*, that is, $x < y$ implies $c_i(x) \le c_i(y)$. Prove that in this case $k = \Omega(n)$.

The goal of the next exercises is to show that we cannot replace the acceptance condition "accept vector a if and only if the rows of M_a span vector **1**" of span programs by "accept vector a if and only if the rows of M_a are linearly dependent". This is because in that case very simple boolean functions require programs of exponential size. A *monotone dependency program* over a field $\mathbb{F}$ is given by a matrix M over $\mathbb{F}$ with its rows labeled by variables $x_1, \ldots, x_n$. For an input $a = (a_1, \ldots, a_n) \in \{0, 1\}^n$, let (as before) M_a denote the submatrix of M obtained by keeping those rows whose labels are satisfied by a. The program M *accepts* the input a if and only if the rows of M_a are linearly dependent (over $\mathbb{F}$). A program computes a boolean function f if it accepts exactly those inputs a where $f(a) = 1$. The *size* of a dependency program is again the number of rows in it.

Comment: Beimel and Gál (1999) showed that the minimum size of a dependency program for f is polynomially related to the minimum size of an *arithmetic branching program* computing $\neg f$. Such branching programs are just extension of parity branching programs from GF(2) to any field $\mathbb{F}$.

8.9. Suppose that a boolean function $f \not\equiv 1$ is computed by a monotone dependency program M of size smaller than the number of minterms of f. Prove that then there exists a set of minterms A, $|A| \ge 2$, such that for any nontrivial partition $A = A_0 \cup A_1$, the set $S(A_0, A_1) := \left(\bigcup_{a \in A_0} a \right) \cap \left(\bigcup_{b \in A_1} b \right)$ contains at least one minterm of f.

Hint: For every minterm a of f choose some linear dependence v_a of the rows of M, that is, v_a is a vector such that $v_a \cdot M = 0$, and v_a has nonzero coordinates only at rows labeled by variables in a. The vectors v_a are linearly dependent (why?). Let A be a *minimal* set of minterms such that $\{v_a \mid a \in A\}$ are linearly dependent. Thus, $\sum_{a \in A} \alpha_a v_a = 0$ for some coefficients $\alpha_a \neq 0$. Observe that for any nontrivial partition $A = A_0 \cup A_1$,

$$v := \sum_{a \in A_0} \alpha_a v_a = - \sum_{a \in A_1} \alpha_a v_a \neq 0.$$

Let b be the set of variables labeling the rows of M corresponding to nonzero coordinates of v. This set lies in $S(A_0, A_1)$ and contains at least one minterm of f.

8.10. (Pudlák and Sgall 1998) Use Exercise 8.9 to show that the function $f = (x_1 \vee x_2) \wedge (x_3 \vee x_4) \wedge \cdots \wedge (x_{2n-1} \vee x_{2n})$ cannot be computed by a monotone dependency program of size smaller than 2^n. Show that this function has a small monotone span program.

Hint: Each minterm a of f has precisely one variable from each of the sets $\{x_{2i-1}, x_{2i}\}$, $i = 1, \ldots, n$. Hence, there are 2^n minterms. Suppose that f has a program of size smaller than 2^n, and let A be the set of minterms guaranteed by Exercise 8.9. Pick i such that both sets of minterms $A_0 = \{a \in A \mid x_{2i-1} \notin a\}$ and $A_1 = \{a \in A \mid x_{2i} \notin a\}$ are non-empty (why is this possible?). By Exercise 8.9, the set $S(A_0, A_1)$ must contain at least one minterm b of f. But, by the definition of A_0 and A_1, this minterm can contain neither x_{2i-1} nor x_{2i}, a contradiction.

Chapter 9
Monotone Circuits

We now consider monotone circuits, that is, circuits with fanin-2 AND and OR gates. As monotone formulas, such circuits can only compute monotone boolean functions. Recall that a boolean function f is *monotone* if $f(x) \leq f(y)$ as long as $x_i \leq y_i$ for all i. The difference from formulas is that now the fan-outs of gates may be arbitrary, not just 1. That is, a result computed at some gate can be used many times with no need to recompute it again and again. This additional feature makes the lower bounds problem more difficult.

Until 1985, the largest known lower bound on the size of such circuits for an explicit boolean function of n variables was only $4n$ (Tiekenheinrich 1984). A breakthrough was achieved in 1985 when two mathematicians from Lomonosov University in Moscow—Andreev (1985) and Razborov (1985a)—almost simultaneously proved super-polynomial lower bounds for monotone circuits.

In this chapter we present Razborov's method of approximations as well as another, simpler argument yielding exponential lower bounds even for circuits with monotone *real-valued* functions as gates.

As in the entire book, here our focus is on proving lower bounds. A comprehensive exposition of known upper bounds for monotone circuits and monotone switching networks can be found in a survey by Korshunov (2003).

9.1 Large Cliques are Hard to Detect

We will first demonstrate Razborov's method of approximations for the case of monotone circuits computing the clique function. Later, in Sect. 9.10, we describe his method in its full generality, and apply it to the perfect matching function.

The *clique function* $f_n = \text{CLIQUE}(n, k)$ has $\binom{n}{2}$ variables x_{ij}, one for each potential edge in a graph on n vertices $[n] = \{1, \ldots, n\}$; the function outputs 1 iff the associated graph contains a clique (complete subgraph) on some k vertices. The clique function is monotone because setting more edges to 1 can only increase the size of the largest clique.

S. Jukna, *Boolean Function Complexity*, Algorithms and Combinatorics 27,
DOI 10.1007/978-3-642-24508-4_9, © Springer-Verlag Berlin Heidelberg 2012

Theorem 9.1. (Razborov 1985a; Alon and Boppana 1987) *For* $3 \leq k \leq n^{1/4}$, *the monotone circuit complexity of* CLIQUE(n, k) *is* $n^{\Omega(\sqrt{k})}$.

We will analyze the behavior of circuits for f_n on two types of input graphs:

* *Positive graphs* are k-cliques, that is, graphs consisting of a clique on some k vertices and $n - k$ isolated vertices; we have $\binom{n}{k}$ such graphs and they all must be accepted by f_n.
* *Negative graphs* are $(k - 1)$-cocliques formed by assigning each vertex a color from the set $\{1, 2, \ldots, k - 1\}$, and putting edges between those pairs of vertices with different colors; we have $(k - 1)^n$ such graphs and they must be rejected by f_n. (Different colorings can lead to the same graph, but we will consider them as different for counting purposes.)

The main goal of Razborov's method is to show that, if a circuit is "too small", then it must make a lot of errors, that is, must either reject most of positive graphs or accept most of negative graphs. Circuits can be amorphous, so analyzing their behavior directly is difficult. Instead, every monotone circuit will be *approximated* by another monotone circuit of a very special type—namely, a short DNF that is tailor-made to represent collections of cliques.

Now we define these DNFs, our so-called "approximators". For a subset X of vertices, the *clique indicator* of X is the monotone boolean function $\lceil X \rceil$ of $\binom{n}{2}$ variables such that $\lceil X \rceil(E) = 1$ if and only if the graph E contains a clique on the vertices X. Note that $\lceil X \rceil$ is just a monomial

$$\lceil X \rceil = \bigwedge_{i,j \in X; i < j} x_{ij}$$

depending on only $\binom{|X|}{2}$ variables.

An (m, l)-*approximator* is an OR of at most m clique indicators, whose underlying vertex-sets each have cardinality at most l:

$$A = \bigvee_{t=1}^{r} \lceil X_t \rceil = \bigvee_{t=1}^{r} \bigwedge_{i \neq j \in X_t} x_{ij} \qquad (r \leq m, \ |X_t| \leq l).$$

Here $l \geq 2$ and $m \geq 2$ are parameters depending only on values of k and n; the values of these parameters will be fixed later.

The main combinatorial tool used in the proof of Theorem 9.1 is the well-known *Sunflower Lemma* discovered by Erdős and Rado (1960). A *sunflower* with p petals and a core T is a collection of sets $S_1, \ldots, S_p$ such that $S_i \cap S_j = E$ for all $i \neq j$. In other words, each element belongs either to none, or to exactly one, or to *all* of the S_i (Fig. 9.1). Note that a family of pairwise disjoint sets is a sunflower (with an empty core).

Fig. 9.1 A sunflower with
eight petals

Sunflower Lemma. Let $\mathcal{F}$ be family of non-empty sets each of size at most l. If $|\mathcal{F}| > l!(p-1)^l$ then $\mathcal{F}$ contains a sunflower with p petals.

In particular, every graph with at least $2(p-1)^2 + 1$ edges must have p vertex-disjoint edges of a star with p edges.

Proof. We proceed by induction on l. For $l = 1$, we have more than $p - 1$ points (disjoint 1-element sets), so any p of them form a sunflower with p petals (and an empty core). Now let $l \geq 2$, and take a maximal family $\mathcal{S} = \{S_1, \ldots, S_t\}$ of pairwise disjoint members of $\mathcal{F}$.

If $t \geq p$, these sets form a sunflower with $t \geq p$ petals (and empty core), and we are done.

Assume that $t \leq p - 1$, and let $S = S_1 \cup \cdots \cup S_t$. Then $|S| \leq l(p-1)$. By the maximality of $\mathcal{S}$, the set S intersects every member of $\mathcal{F}$. By the pigeonhole principle, some point $x \in S$ must be contained in at least

$$\frac{|\mathcal{F}|}{|S|} > \frac{l!(p-1)^l}{l(p-1)} = (l-1)!(p-1)^{l-1}$$

members of $\mathcal{F}$. Let us delete x from these sets and consider the family

$$\mathcal{F}_x := \{F \setminus \{x\} \ : \ F \in \mathcal{F}, x \in F\}.$$

By the induction hypothesis, this family contains a sunflower with p petals. Adding x to the members of this sunflower, we get the desired sunflower in the original family $\mathcal{F}$. $\qquad\square$

9.1.1 Construction of the Approximated Circuit

Given a monotone circuit F for the clique function f_n, we will construct the approximator for F in a "bottom-up" manner, starting from the input variables. An input variable is of the form x_{ij}, where i and j are different vertices; it is equivalent to the clique indicator $\lceil \{i, j\} \rceil = x_{ij}$.

Suppose at some internal node of the circuit, say at an OR gate, the two subcircuits feeding into this gate already have their (m, l)-approximators $A = \bigvee_{i=1}^{r} \lceil X_i \rceil$ and $B = \bigvee_{i=1}^{s} \lceil Y_i \rceil$, where r and s are at most m. We could approximate this OR gate by just $A \vee B$, but this could potentially give us a $(2m, l)$-approximator, while we want to stay at (m, l).

At this place the Sunflower Lemma comes to our rescue. To apply the Sunflower Lemma to the present situation, consider the family

$$\mathcal{F} = \{X_1, \dots, X_r, Y_1, \dots, Y_s\}$$

and set

$$m := l!(p-1)^l .$$

If $r + s > m$ then some p of the sets in $\mathcal{F}$ form a sunflower. We then replace these p sets by their core; this operation is called a *plucking*. Repeatedly perform such pluckings until no more are possible. The entire procedure is called the *plucking procedure*. Since the number of vertex sets decreases with each plucking, after at most $|\mathcal{F}| = r + s \le 2m$ pluckings we will obtain an (m, l)-approximator for our OR gate, which we denote by $A \sqcup B$.

If the gate was an AND gate (not an OR gate) then forming the AND of the two approximators $A = \bigvee_{i=1}^{r} \lceil X_i \rceil$ and $B = \bigvee_{i=1}^{s} \lceil Y_i \rceil$ yields the expression $\bigvee_{i=1}^{r} \bigvee_{j=1}^{s} (\lceil X_i \rceil \wedge \lceil Y_i \rceil)$. Two reasons why this expression itself might not be an (m, l)-approximator are that the terms $\lceil X_i \rceil \wedge \lceil Y_i \rceil$ might not be clique indicators and that there can be as many as m^2 terms.

To overcome these difficulties, apply the following three steps:

1. Replace the term $\lceil X_i \rceil \wedge \lceil Y_i \rceil$ by the clique indicator $\lceil X_i \cup Y_i \rceil$;
2. Erase those clique indicators $\lceil X_i \cup Y_i \rceil$ for which $|X_i \cup Y_j| \ge l + 1$;
3. Apply the plucking procedure (described above for OR gates) to the remaining clique indicators; there will be at most m^2 pluckings.

These three steps guarantee that an (m, l)-approximator is formed; we denote it by $A \sqcap B$. (Note an "asymmetry" in the argument: AND gates need more work to approximate than OR gates.)

9.1.2   Bounding Errors of Approximation

Now fix a monotone circuit F computing $f_n = \text{CLIQUE}(n, k)$, and let F' be the approximated circuit, that is, an (m, l)-approximator of the last gate of F. We will show that

1. Every approximator (including F') must make a lot of errors, that is, disagree with f_n on many negative and positive graphs.
2. If size(F) is small, then F' cannot make too many errors.

This will already imply that size(F) must be large.

Lemma 9.2. *Every approximator either rejects all graphs or wrongly accepts at least a fraction $1 - l^2/(k-1)$ of all $(k-1)^n$ negative graphs.*

Proof. Let $A = \bigvee_{i=1}^{r}\lceil X_i\rceil$ an (m,l)-approximator, and assume that A accepts at least one graph. Then $A \geq \lceil X_1\rceil$. A negative graph is rejected by the clique indicator $\lceil X_1\rceil$ iff its associated coloring assigns some two vertices of X_1 the same color. We have $\binom{|X_1|}{2}$ pairs of vertices in X_1, and for each such pair at most $(k-1)^{n-1}$ colorings assign the same color. Thus, at most $\binom{|X_1|}{2}(k-1)^{n-1} \leq \binom{l}{2}(k-1)^{n-1}$ negative graphs can be rejected by $\lceil X_1\rceil$, and hence, by the approximator A. $\square$

Thus, every approximator (including F') must make a lot of errors. We are now going to show that, if size(F) is small, then the number of errors cannot be large, implying that size(F) must be large.

Lemma 9.3. *The number of positive graphs wrongly rejected by F' is at most* size$(F) \cdot m^2\binom{n-l-1}{k-l-1}$.

Proof. We shall consider the errors introduced by the approximator of a single gate, and then apply the union bound to get the claimed upper bound on the total number of errors.

If g is an OR gate and A, B are the approximators of subcircuits feeding into this gate, then our construction of the approximator $A \sqcup B$ for g involves taking an OR $A \vee B$ (which does not introduce any errors) and then repeatedly plucking until we get down our number of clique indicators. Each plucking replaces a clique indicator $\lceil X_i\rceil$ by some $\lceil X\rceil$ with $X \subseteq X_i$ which can accept only more graphs. Hence, on positive graphs, $A \sqcup B$ produces no errors at all.

Now suppose that g is an AND gate. The first step in the transformation from $A \wedge B$ to $A \sqcap B$ is to replace $\lceil X_i\rceil \wedge \lceil Y_j\rceil$ by $\lceil X_i \cup Y_j\rceil$. These two functions behave identically on positive graphs (cliques). The second step is to erase those clique indicators $\lceil X_i \cup Y_j\rceil$ for which $|X_i \cup Y_j| \geq l+1$. For each such clique indicator, at most $N := \binom{n-l-1}{k-l-1}$ of the positive graphs are lost. Since there are at most m^2 such clique indicators, at most $m^2 N$ positive graphs are lost in the second step. The third and final step, applying the plucking procedure, only enlarges the class of accepted graphs, as noted in the previous paragraph. Summing up the three steps, at most $m^2 N$ positive graphs can be lost by approximating one AND gate. Since we have at most size(F) such gates, the lemma is proved. $\square$

Lemma 9.4. *The number of negative graphs wrongly accepted by F' is at most* size$(F) \cdot m^2 l^{2p}(k-1)^{n-p}$.

Proof. Again, we shall analyze the errors introduced at each gate.

If g is an OR gate and A, B are the approximators of subcircuits feeding into this gate, then our construction of the approximator $A \sqcup B$ for g involves taking an OR $A \vee B$ (which does not introduce any errors) and then performing at most $2m$ pluckings until we get down our number of clique indicators.

Each plucking will be shown to accept only a few additional negative graphs. Color the vertices randomly, with all $(k-1)^n$ possible colorings equally likely, and let G be the associated negative graph. Let $Z_1,\ldots,Z_p$ be the petals of a sunflower with core Z. What is the probability that $\lceil Z\rceil$ accepts G, but none of the functions

$\lceil Z_1 \rceil, \ldots, \lceil Z_p \rceil$ accept G? This event occurs iff the vertices of Z are assigned distinct colors (called a proper coloring, or PC), but every petal Z_i has two vertices colored the same. We have

$$\text{Prob}[Z \text{ is PC and } Z_1, \ldots, Z_p \text{ are not PC}]$$

$$\leq \text{Prob}[Z_1, \ldots, Z_p \text{ are not PC}|Z \text{ is PC}]$$

$$= \prod_{i=1}^{p} \text{Prob}[Z_i \text{ is not PC}|Z \text{ is PC}]$$

$$\leq \prod_{i=1}^{p} \text{Prob}[Z_i \text{ is not PC}]$$

$$\leq \binom{l}{2}^{p} \cdot (k-1)^{-p} \leq l^{2p}(k-1)^{-p}.$$

The first inequality holds by the definition of the conditional probability. The second line holds because the sets $Z_i \setminus Z$ are disjoint and hence the events are independent. The third line holds because the event "Z_i is not a clique" is less likely to happen given the fact that $Z \subseteq Z_i$ is a clique. The fourth line holds because Z_i is not properly colored iff two vertices of Z_i get the same color.

Thus to the class of wrongly accepted negative graphs each plucking adds at most $l^{2p}(k-1)^{n-p}$ new graphs. There are at most $2m$ pluckings, so the total number of negative graphs wrongly accepted when approximating the gate OR g is at most $2m l^{2p}(k-1)^{n-p}$.

Next consider the case when g is an AND gate. In the transformation from $A \wedge B$ to $A \sqcap B$, the first step introduces no new violations, since $\lceil X_i \rceil \wedge \lceil Y_j \rceil \geq \lceil X_i \cup Y_j \rceil$. Only the third step, the plucking procedure, introduces new violations. This step was analyzed above; the only difference is that there can be m^2 pluckings instead of just $2m$. This settles the case of AND gates, thus completing the proof. □

Proof of Theorem 9.1. Set $l = \lfloor \sqrt{k-1}/2 \rfloor$ and $p = \Theta(\sqrt{k} \log n)$; recall that $m = l!(p-1)^l \leq (pl)^l$. Let F be a monotone circuit that computes CLIQUE(n,k). By Lemma 9.2, the approximator F' of F is either identically 0 or outputs 1 on at least a $(1 - l^2/(k-1)) \geq \frac{1}{2}$ fraction of all $(k-1)^n$ negative graphs. If the former case holds, then apply Lemma 9.3 to obtain

$$\text{size}(F) \cdot m^2 \cdot \binom{n-l-1}{k-l-1} \geq \binom{n}{k}.$$

Since $\binom{n}{k}/\binom{n-x}{k-x} \geq (n/k)^x$, simple calculation show that in this case size(F) is $n^{\Omega(\sqrt{k})}$. If the later case holds then apply Lemma 9.4 to obtain

$$\text{size}(F) \cdot m^2 \cdot 2^{-p} \cdot (k-1)^n \geq \frac{1}{2}(k-1)^n .$$

Since $2^p = n^{\Omega(\sqrt{k})}$, in this case we again have that $\text{size}(F)$ is $n^{\Omega(\sqrt{k})}$. □

Remark 9.5. Recently, Rossman (2010) gave lower bounds for the Clique function that apply to finding small cliques in *random graphs*. Let $G(n, p)$ denote a random graph on n vertices in which each edge appears at random and independently with probability p. Let k be a fixed natural number. It is well known that $p := n^{-2/(k-1)}$ is a threshold for appearance of k-cliques. Rossman showed that, for every constant k, no monotone circuit of size smaller than $\mathcal{O}(n^{k/4})$ can correctly compute (with high probability) the Clique function on $G(n, p)$ and on $G(n, 2p)$ *simultaneously*.

9.2 Very Large Cliques are Easy to Detect

By Theorem 9.1, we known that there exists a constant $c > 0$ such that every monotone circuit computing the clique function CLIQUE(n, k) requires at least $n^{c\sqrt{k}}$ gates. Moreover, it can be shown (see Theorem 9.19 below) that already for $k = 3$ at least $\Omega(n^3/\log^4 n)$ gates are necessary. In fact, Alon and Boppana (1987) showed that Razborov's lower bound can be improved to $\Omega((n/\log^2 n)^k)$ for any constant $k \geq 3$, and for growing k we need at least $2^{\Omega(\sqrt{k})}$ gates, as long as $k \leq (n/\log n)^{2/3}/4$. Thus, small cliques are hard to detect.

By a simple padding argument, this implies that even detecting cliques of size $n - k$ requires a super-polynomial number of gates, as long as $k \leq n/2$ grows faster than $\log^3 n$.

Proposition 9.6. *For $k \leq n/2$, every monotone circuit for* CLIQUE$(n, n - k)$ *requires $2^{\Omega(k^{1/3})}$ gates.*

Proof. Fix the integer m with $m - s = k$ where $s = \lfloor (m/\log m)^{2/3}/4 \rfloor$; hence $s = \Omega(k^{2/3})$. Then CLIQUE(m, s) is a sub-function of (that is, can be obtained by setting to 1 some variables in) CLIQUE$(n, n - k)$: just consider only the n-vertex graphs containing a fixed clique on $n - m$ vertices connected to all the remaining vertices (the rest may be arbitrary). On the other hand, according to the lower bound of Alon and Boppana (mentioned above) the function CLIQUE(m, s), and hence, also the function CLIQUE$(n, n - k)$ requires monotone circuits of size exponential in $\Omega(\sqrt{s}) = \Omega(k^{1/3})$. □

But what is the complexity of CLIQUE$(n, n - k)$ when k is very small, say, constant—can this function then be computed by a monotone circuit using substantially fewer than n^k gates? Somewhat surprisingly, for every(!) constant k, the CLIQUE$(n, n - k)$ function can be computed by a monotone circuit of size $\mathcal{O}(n^2 \log n)$. Moreover, the number of gates is polynomial, as long as $k = O(\sqrt{\log n})$. Recall that CLIQUE(n, k) requires $\Omega(n^k/\log^{2k} n)$ for every constant

k, and that already for $k = \omega(\log^3 n)$, any monotone circuit for CLIQUE$(n, n - k)$ requires a super-polynomial number of gates.

Theorem 9.7. (Andreev and Jukna 2008) *For every constant k, the function* CLIQUE$(n, n - k)$ *can be computed by a monotone DeMorgan formula containing at most* $\mathcal{O}(n^2 \log n)$ *gates. The number of gates remains polynomial in n as long as* $k = \mathcal{O}(\sqrt{\log n})$.

In this section we will prove Theorem 9.7. To do this, we need some preparations. First, instead of constructing a small formula for the Clique function, it will be convenient to construct a small formula for the dual function. Recall that the *dual* of a boolean function $f(x_1, \ldots, x_n)$ is the boolean function $f^*(x_1, \ldots, x_n) = \neg f(\neg x_1, \ldots, \neg x_n)$. If f is monotone, then its dual f^* is also monotone. For example,

$$(x \vee y)^* = \neg(\neg x \vee \neg y) = x \wedge y;$$
$$(x \wedge y)^* = \neg(\neg x \wedge \neg y) = x \vee y.$$

In particular, the dual of CLIQUE$(n, n-k)$ accepts a given graph G on n vertices iff G has no independent set with $n-k$ vertices, which is equivalent to $\tau(G) \geq k+1$, where $\tau(G)$ ist the vertex-cover number of G. Recall that a *vertex cover* in a graph G is a set of its vertices containing at least one endpoint of each edge; $\tau(G)$ is the minimum size of such a set. Hence, the dual of CLIQUE$(n, n - k)$ is a monotone boolean function VC(n, k) of $\binom{n}{2}$ boolean variables representing the edges of an undirected graph G on n vertices, whose value is 1 iff G does not have a vertex-cover of cardinality k.

We will construct a monotone formula for VC(n, k). Replacing OR gates by AND gates (and vice versa) in this formula yields a monotone formula for CLIQUE$(n, n - k)$, thus proving Theorem 9.7.

9.2.1 Properties of τ-Critical Graphs

A graph is τ-*critical* if removing any of its edges reduces the vertex-cover number. We will need some properties of such graphs.

Theorem 9.8. (Hajnal 1965) *In a τ-critical graph without isolated vertices every independent set S has at least $|S|$ neighbors.*

Proof. Let $G = (V, E)$ be a τ-critical graph without isolated vertices. Then G is also α-critical in that removing of any its edge increases its independence number $\alpha(G)$, that is, the maximum size of an independent set in G. An independent set T is *maximal* if $|T| = \alpha(G)$.

Let us first show that every vertex belongs to at least one maximal independent set but not to all such sets. For this, take a vertex x and an edge $e = \{x, y\}$. Remove

e from G. Since G is α-critical, the resulting graph has an independent set T of size $\alpha(G) + 1$. Since T was not independent in G, $x, y \in T$. Then $T \setminus \{x\}$ is an independent set in G of size $|T \setminus \{x\}| = \alpha(G)$, that is, is a maximal independent set avoiding the vertex x, and $T \setminus \{y\}$ is a maximal independent set containing x.

Hence, if X is an arbitrary independent set in G, then the intersection of X with *all* maximal independent sets in G is empty. It remains therefore to show that, if Y is an arbitrary independent set, and S is an intersection of Y with an arbitrary number of maximal independent sets, then

$$|N(Y)| - |N(S)| \geq |Y| - |S|,$$

where $N(Y)$ is the set of all neighbors of Y, that is, the set of all vertices adjacent to at least one vertex in Y. Since an intersection of independent sets is an independent set, it is enough to prove the claim for the case when T is a maximal independent set and $S = Y \cap T$. Since clearly $N(S) \subseteq N(Y) - T$, we have

$$
\begin{aligned}
|N(Y)| - |N(S)| &\geq |N(Y) \cap T| \\
&= |T| - |S| - |T \setminus (Y \setminus N(Y))| \\
&= \alpha(G) - |S| + |Y| - |(T \cup Y) \setminus N(Y)| \\
&\geq |Y| - |S|,
\end{aligned}
$$

where the last inequality holds because the set $(T \cup Y) - N(Y)$ is independent. □

In our construction of a small circuit for the Vertex Cover function, the following consequence of this theorem will be important.

Corollary 9.9. *Every τ-critical graph G has at most $2\tau(G)$ non-isolated vertices.*

Proof. Let $G = (V, E)$ be an arbitrary τ-critical graph, and let $U \subseteq V$ be the set of non-isolated vertices of G. The induced subgraph $G' = (U, E)$ has no isolated vertices and is still τ-critical with $\tau(G') = \tau(G)$. Let $S \subseteq U$ be an arbitrary vertex-cover of G' with $|S| = \tau(G)$. The complement $T = U - S$ is an independent set. By Hajnal's theorem, the set T must have at least $|T|$ neighbors. Since all these neighbors must lie in S, the desired upper bound $|U| = |S| + |T| \leq 2|S| \leq 2\tau(G)$ on the total number of non-isolated vertices of G follows. □

Finally, we will need a fact stating that τ-critical graphs cannot have too many edges. We will derive this fact from the following more general result.

Theorem 9.10. (Bollobás 1965) *Let $A_1, \ldots, A_m$ and $B_1, \ldots, B_m$ be two sequences of sets such that $A_i \cap B_j = \emptyset$ if and only if $i = j$. Then*

$$\sum_{i=1}^{m} \binom{|A_i| + |B_i|}{|A_i|}^{-1} \leq 1. \tag{9.1}$$

Proof. Let X be the union of all sets $A_i \cup B_i$. If A and B are disjoint subsets of X then we say that a permutation $(x_1, x_2, \ldots, x_n)$ of X *separates* the pair (A, B) if no element of B precedes an element of A, that is, if $x_k \in A$ and $x_l \in B$ imply $k < l$.

Each of the $n!$ permutations can separate at most one of the pairs (A_i, B_i), $i = 1, \ldots, m$. Indeed, suppose that $(x_1, x_2, \ldots, x_n)$ separates two pairs (A_i, B_i) and (A_j, B_j) with $i \neq j$, and assume that $\max\{k \mid x_k \in A_i\} \leq \max\{k \mid x_k \in A_j\}$. Since the permutation separates the pair (A_j, B_j),

$$\min\{l \mid x_l \in B_j\} > \max\{k \mid x_k \in A_j\} \geq \max\{k \mid x_k \in A_i\}$$

which implies that $A_i \cap B_j = \emptyset$, contradicting the assumption.

We now estimate the number of permutations separating one fixed pair. If $|A|=a$ and $|B| = b$ and A and B are disjoint then the pair (A, B) is separated by exactly

$$\binom{n}{a+b} a! b! (n - a - b)! = n! \binom{a+b}{a}^{-1}$$

permutations. Here $\binom{n}{a+b}$ counts the number of choices for the positions of $A \cup B$ in the permutation; having chosen these positions, A has to occupy the first a places, giving $a!$ choices for the order of A, and $b!$ choices for the order of B; the remaining elements can be chosen in $(n - a - b)!$ ways.

Since no permutation can separate two different pairs (A_i, B_i), summing up over all m pairs we get all permutations at most once

$$\sum_{i=1}^{m} n! \binom{a_i + b_i}{a_i}^{-1} \leq n!$$

and the desired bound (9.1) follows.   □

Theorem 9.11. (Erdős–Hajnal–Moon 1964) *Every τ-critical graph H has at most* $\binom{\tau(H)+1}{2}$ *edges.*

Proof. Let H be a τ-critical graph with $\tau(H) = t$, and let $E = \{e_1, \ldots, e_m\}$ be the edges of H. Since H is critical, $E \setminus \{e_i\}$ has a $(t-1)$-element vertex-cover S_i. Then $e_i \cap S_i = \emptyset$ while $e_j \cap S_i \neq \emptyset$, if $j \neq i$. We can therefore apply Theorem 9.10 and obtain that $m \leq \binom{2+(t-1)}{2} = \binom{t+1}{2}$, as desired.   □

Proof of Theorem 9.7

We consider graphs on vertex-set $[n] = \{1, \ldots, n\}$. We have a set X of $\binom{n}{2}$ boolean variables x_e corresponding to edges. Each graph $G = ([n], E)$ is specified by setting the values 0 and 1 to these variables: $E = \{e \mid x_e = 1\}$. The function $\mathrm{VC}(n, k)$ accepts G iff $\tau(G) \geq k + 1$.

Let $\mathrm{Crit}(n,k)$ denote the set of all τ-critical graphs on $[n] = \{1,\ldots,n\}$ with $\tau(H) = k + 1$. Observe that graphs in $\mathrm{Crit}(n,k)$ are exactly the minterms of $\mathrm{VC}(n,k)$, that is, the smallest with respect to the number of edges graphs accepted by $\mathrm{VC}(n,k)$.

Given a family F of functions $f : [n] \to [r]$, let $\Phi_F(X)$ be the OR over all graphs $H \in \mathrm{Crit}(r,k)$ and all functions $f \in F$ of the following monotone formulas

$$K_{f,H}(X) = \bigwedge_{\{a,b\}\in E(H)} \bigvee_{e\in f^{-1}(a)\times f^{-1}(b)} x_e.$$

The formula Φ_F accepts a given graph $G = ([n], E)$ iff there exists a graph $H \in \mathrm{Crit}(r,k)$ and a function $f \in F$ such that for each edge $\{a,b\}$ of H there is at least one edge in G between $f^{-1}(a)$ and $f^{-1}(b)$.

A family F of functions $f : [n] \to [r]$ is s-*perfect* if for every subset $S \subseteq [n]$ of size $|S| = s$ there is an $f \in F$ such that $|f(S)| = |S|$. That is, for every s-element subset of $[n]$ at least one function in F is one-to-one when restricted to this subset. Such families are also known in the literature as (n,r,s)-*perfect hash families*.

Lemma 9.12. *If F is an (n,r,s)-perfect hash family with $s = 2(k+1)$ and $r \geq s$, then the formula Φ_F computes $\mathrm{VC}(n,k)$.*

Proof. Since the formula is monotone, it is enough to show that:

(a) $\tau(G) \geq k + 1$ for every graph G accepted by Φ_F, and
(b) Φ_F accepts all graphs from $\mathrm{Crit}(n,k)$.

To show (a), suppose that Φ_F accepts some graph G. Then this graph must be accepted by some sub-formula $K_{f,H}$ with $f \in F$ and $H \in \mathrm{Crit}(r,k)$. That is, for every edge $\{a,b\}$ in H there must be an edge in G joining some vertex $i \in f^{-1}(a)$ with some vertex $j \in f^{-1}(b)$. Hence, if a set S covers the edge $\{i,j\}$, that is, if $S \cap \{i,j\} \neq \emptyset$, then the set $f(S)$ must cover the edge $\{a,b\}$. This means that, for any vertex-cover S in G, the set $f(S)$ is a vertex-cover in H. Taking a minimal vertex-cover S in G we obtain $\tau(G) = |S| \geq |f(S)| \geq \tau(H) = k + 1$.

To show (b), take an arbitrary graph $G = ([n], E)$ in $\mathrm{Crit}(n,k)$, and let U be the set of its non-isolated vertices. By Corollary 9.9, $|U| \leq 2\tau(G) = 2(k+1) \leq s$. By the definition of F, some function $f : [n] \to [r]$ must be one-to-one on U. For $i,j \in U$ join $a = f(i)$ and $b = f(j)$ by an edge iff $\{i,j\} \in E$. Since $G \in \mathrm{Crit}(n,k)$ and f is one-to-one on all non-isolated vertices of G, the resulting graph H belongs to $\mathrm{Crit}(r,k)$. Moreover, for every edge $\{a,b\}$ of H, the pair $e = \{i,j\}$ with $f(i) = a$ and $f(j) = b$ is an edge of G, implying that $x_e = 1$. This means that the sub-formula $K_{f,H}$ of Φ_F, and hence, the formula Φ_F itself must accept G. $\square$

Let us now estimate the number of gates in the formula Φ_F. Using a simple counting argument, Mehlhorn and Schmidt (1982) shows that (n,r,s)-perfect hash families F of size $|F| \leq se^{s^2/r} \log n$ exist for all $2 \leq s \leq r \leq n$. In our case we can take $r = s = 2(k+1)$. Hence, $|F| = \mathcal{O}(\log n)$ for every constant k.

If we allow unbounded fanin, then each sub-formula $K_{f,H}$ contributes just one AND gate. Hence, Φ_F has at most $|\mathrm{Crit}(r,k)| + |F|$ unbounded-fanin AND gates. The fanin of each AND gate is actually bounded by the number of edges in the corresponding graph $H \in \mathrm{Crit}(r,k)$ which, by Theorem 9.11, does not exceed $l :=$ $\binom{k+2}{2} = \mathcal{O}(1)$. Hence, $|\mathrm{Crit}(r,k)|$ does not exceed $\binom{r^2}{l} = \mathcal{O}(1)$. Thus, for every constant k, we have only $\mathcal{O}(|F|) = \mathcal{O}(\log n)$ fanin-2 AND gates in Φ_F. Each of these gates takes at most $\mathcal{O}(n^2)$ fanin-2 OR gates as inputs. Thus, the total size of our formula Φ_F is $\mathcal{O}(n^2 \log n)$, as desired. For growing k, the upper bound has the form $\mathcal{O}(kC^{k^2}n^2 \log n)$ for a constant C, which is polynomial as long as $k = \mathcal{O}(\sqrt{\log n})$.

We thus constructed a monotone formula Φ_F for the vertex cover function $\mathrm{VC}(n,k)$. Since this function is the dual function of the clique function $\mathrm{CLIQUE}(n, n - k)$, we can just replace OR gates by AND gates (and vice versa) in this formula to obtain a monotone formula for $\mathrm{CLIQUE}(n, n - k)$. This completes the proof of Theorem 9.7. $\qquad\square$

Remark 9.13. Observe that the formula Φ_F for $\mathrm{VC}(n,k)$ is multilinear, that is, inputs to each its AND gate are computed from disjoint sets of variables. On the other hand, Krieger (2007) shows that *every* monotone multilinear circuit for the dual function $\mathrm{CLIQUE}(n, n - k)$ requires at least $\binom{n}{k}$ gates. This gives an example of a boolean function, whose dual requires much larger multilinear circuits than the function itself.

Remark 9.14. Using *explicit* perfect hash families we can obtain explicit circuits. For fixed values of r and s, infinite classes of (n, r, s)-perfect hash families F of size $|F| = O(\log n)$ were constructed by Wang and Xang (2001) using algebraic curves over finite fields. With this construction Theorem 9.7 gives explicit monotone formulas.

The construction in Wang and Xang (2001) is almost optimal: the family has only a *logarithmic* in n number of functions. The construction is somewhat involved. On the other hand, perfect hash families of *poly-logarithmic* size can be constructed very easily.

Let $s \geq 1$ be a fixed integer and $r = 2^s$. Let $M = \{m_{a,i}\}$ be an $n \times b$ matrix with $b = \lceil \log n \rceil$ columns whose rows are distinct 0-1 vectors of length b. Let $h_1, \ldots, h_b$ be the family of functions $h_i : [n] \to \{0, 1\}$ determined by the columns of M; hence, $h_i(a) = m_{a,i}$. Let also $g : \{0, 1\}^s \to [r]$ be defined by $g(x) = \sum_{i=1}^s x_i 2^{i-1}$.

By Bondy's theorem (Bondy 1972), the projections of any set of $s + 1$ distinct binary vectors on some set of s coordinates must all be distinct. Hence, for any set $a_1, \ldots, a_{s+1}$ of $s+1$ rows there exist s columns $h_{i_1}, \ldots, h_{i_s}$ such that all $s+1$ vectors $(h_{i_1}(a_j), \ldots, h_{i_s}(a_j))$, $j = 1, \ldots, s+1$ are distinct. Therefore, the function $f(x) = g(h_{i_1}(x), \ldots, h_{i_s}(x))$ takes different values on all $s + 1$ points $a_1, \ldots, a_{s+1}$. Thus, taking the superposition of g with $\binom{b}{s} \leq \log^s n$ s-tuples of functions $h_1, \ldots, h_b$, we obtain a family F of $|F| \leq \log^s n$ functions $f : [n] \to [r]$ which is $(s + 1)$-perfect.

9.3 The Monotone Switching Lemma

In Razborov's method of approximations one only uses DNFs to approximate gates. In this way, OR gates can be easily approximated: an OR of DNFs is a DNF, and we only need to keep its small enough. The case of AND gates is, however, more complicated. So, a natural idea to try to approximate by *both* DNFs and CNFs. When appropriately realized, this idea leads to a general, and relatively simple lower-bounds criterion for monotone circuits. Due to the symmetry between DNFs and CNFs, this criterion is often much easier to apply and yields exponential lower bounds for many functions, including the clique function.

Still, there are functions—like the perfect matching function—for which the criterion seems to fail. This is why we will discuss Razborov's method later in Sect. 9.10 in full detail: unlike the general criterion, which we are going to present now, Razborov's method is much more subtle, tailor made for the *specific* function one deals with and can be applied in situations where the general criterion fails to produce strong lower bounds. Yet another reason to include Razborov's proof for the perfect matching function is that this function belongs to P, and the proof was never treated in a book.

Our goal is to show that, if a monotone boolean function can be computed by a small monotone circuit, then it can be approximated by small monotone CNFs and DNFs. Thus, in order to prove that a function requires large circuits it is enough to show that it does not have a small CNF/DNF approximation. The proof of this will be based on the "monotone switching lemma" allowing us to switch between CNFs and DNFs, and vice versa.

By a monotone k-CNF (conjunctive normal form) we will mean an And of an arbitrary number of monotone clauses, each being an Or of at most k variables. Dually, a monotone k-DNF is an Or of an arbitrary number of monomials, each being an And of at most k variables. In an *exact* k-CNF all clauses must have *exactly* k distinct variables; *exact* k-DNFs are defined similarly. For two boolean functions f and g of the same set of variables, we write $f \leq g$ if $f(x) \leq g(x)$ for all input vectors x. For a CNF/DNF C we will denote by $|C|$ the number of clauses/monomials in it.

The following lemma was first proved in Jukna (1999) in terms of so-called "finite limits", a notion suggested by Sipser (1985); we will also use this notion later (in Sect. 11.3) to prove lower bounds for depth-3 circuits. In terms of DNFs and CNFs the lemma was then proved by Berg and Ulfberg (1999). Later, a similar lemma was used by Harnik and Raz (2000) to improve the numerically strongest known lower bound $2^{\Omega(n^{1/3}/\log n)}$ of Andreev (1987b) to $2^{\Omega((n/\log n)^{1/3})}$. The idea of the lemma itself was also implicit in the work of Haken (1995).

Lemma 9.15. (Monotone Switching Lemma) *For every* $(s-1)$-*CNF* f_{cnf} *there is an* $(r-1)$-*DNF* f_{dnf} *and an exact* r-*DNF* D *such that*

$$f_{\mathrm{dnf}} \leq f_{\mathrm{cnf}} \leq f_{\mathrm{dnf}} \vee D \quad and \quad |D| \leq (s-1)^r. \qquad (9.2)$$

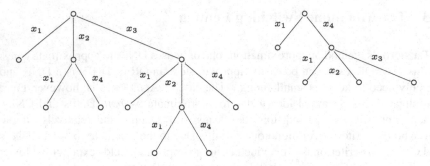

Fig. 9.2 Two DNF-trees of the same 3-CNF $f_{\text{cnf}} = (x_1 \vee x_2 \vee x_3) \wedge (x_1 \vee x_2 \vee x_4) \wedge (x_1 \vee x_4)$. The second tree is obtained by parsing the clauses of f_{cnf} in the reverse order

Dually, for every $(r-1)$-DNF f_{dnf} there is an $(s-1)$-CNF f_{cnf} and an exact s-CNF C such that

$$f_{\text{cnf}} \wedge C \leq f_{\text{dnf}} \leq f_{\text{cnf}} \quad \text{and} \quad |C| \leq (r-1)^s . \tag{9.3}$$

Proof. We prove the first claim (the second is dual). Let $f_{\text{cnf}} = q_1 \wedge \cdots \wedge q_l$ be an $(s-1)$-CNF; hence, each clause q_i has $|q_i| \leq s-1$ variables. It will be convenient to identify clauses and monomials with the *sets* of indices of their variables. We say that a monomial p *pierces* a clause q_i if $p \cap q_i \neq \emptyset$.

We associate with f_{cnf} the following "transversal" tree T of fan-out at most $s-1$ (see Fig. 9.2).

The first node of T corresponds to the first clause q_1, and the outgoing $|q_1|$ edges are labeled by the variables from q_1. Suppose we have reached a node v, and let p be the monomial consisting of the labels of edges from the root to v. If p pierces all the clauses of f_{cnf}, then v is a leaf. Otherwise, let q_i be the *first* clause such that $p \cap q_i = \emptyset$. Then the node v has $|q_i|$ outgoing edges labeled by the variables in q_i.

Note that the resulting tree T depends on what ordering of clauses of f_{cnf} we fix, that is, in which order we parse the clauses (see Fig. 9.2). Still, for any such tree we have that, for every assignment $x \in \{0, 1\}^n$, $f_{\text{cnf}}(x) = 1$ if and only if x is consistent with at least one path from the root to a leaf of T. This holds because $f_{\text{cnf}}(x) = 1$ iff the set $S_x = \{i \mid x_i = 1\}$ intersects all clauses $q_1, \ldots, q_l$.

Some paths in T may be longer than $r-1$. So, we now cut off these long paths. Namely, let f_{dnf} be the OR of all paths of length at most $r-1$ ending in leafs, and D be the OR of all paths of length exactly r. Observe that for every assignment $x \in \{0, 1\}^n$:

- $f_{\text{dnf}}(x) = 1$ implies $f_{\text{cnf}}(x) = 1$, and
- $f_{\text{cnf}}(x) = 1$ implies $f_{\text{dnf}}(x) = 1$ or $D(x) = 1$.

Thus, $f_{\text{dnf}} \leq f_{\text{cnf}} \leq f_{\text{dnf}} \vee D$. Finally, we also have that $|D| \leq (s-1)^r$, because every node of T has fan-out at most $s-1$. $\square$

Most important in the Switching Lemma is that the exact DNFs and CNFs correcting possible errors contain only $(s-1)^r$ monomials instead of all $\binom{n}{r}$ possible monomials, and only $(r-1)^s$ clauses instead of all $\binom{n}{s}$ possible clauses.

9.4 The Lower-Bounds Criterion

We now give a general lower-bounds criterion for monotone circuits.

Definition 9.16. Let f be a monotone boolean function of n variables. We say that f is t-*simple* if for every pair of integers $2 \le r, s \le n$ there exists an exact s-CNF C, an exact r-DNF D, and a subset $I \subseteq [n]$ of size $|I| \le s - 1$ such that

(a) $|C| \le t \cdot (r - 1)^s$ and $|D| \le t \cdot (s - 1)^r$, and
(b) Either $C \le f$ or $f \le D \vee \bigvee_{i \in I} x_i$ (or both) hold.

Theorem 9.17. *If a monotone boolean function can be computed by a monotone circuit of size t, then f is t-simple.*

Proof. Let $F(x_1, \ldots, x_n)$ be a monotone boolean function, and suppose that F can be computed by a monotone circuit of size t. Our goal is to show that the function F is t-simple. To do this, fix an arbitrary pair of integer parameters $2 \le s, r \le n$.

Let $f = g * h$ be a gate in our circuit. That is, f is a function computed at some node of the circuit, and g and h are functions computed at its inputs. By an *approximator* of this gate we will mean a pair $(f_{\text{cnf}}, f_{\text{dnf}})$, where f_{cnf} is an $(s - 1)$-CNF (a *left* approximator of f) and f_{dnf} is an $(r - 1)$-DNF (a *right* approximator of f) such that $f_{\text{dnf}} \le f_{\text{cnf}}$.

We say that such an approximator $f_{\text{cnf}}, f_{\text{dnf}}$ of f introduces a new error on input $x \in \{0, 1\}^n$ if the approximators of g and of h did not make an error on x, but the approximator of f does. That is, $g_{\text{cnf}}(x) = g_{\text{dnf}}(x) = g(x)$ and $h_{\text{cnf}}(x) = h_{\text{dnf}}(x) = h(x)$, but either $f_{\text{cnf}}(x) \ne f(x)$ or $f_{\text{dnf}}(x) \ne f(x)$.

We define approximators inductively as follows.

Case 1: f is an input variable, say, $f = x_i$. In this case we take $f_{\text{cnf}} = f_{\text{dnf}} := x_i$. It is clear that this approximator introduces no errors.

Case 2: f is an And gate, $f = g \wedge h$. In this case we take $f_{\text{cnf}} := g_{\text{cnf}} \wedge h_{\text{cnf}}$ as the left approximator of f; hence, f_{cnf} introduces no new errors. To define the right approximator of f we use Lemma 9.15 to convert f_{cnf} into an $(r - 1)$-DNF f_{dnf}; hence, $f_{\text{dnf}} \le f_{\text{cnf}}$. Let E_f be the set of inputs on which f_{dnf} introduces a new error, that is,

$$E_f := \{x \mid f(x) = f_{\text{cnf}}(x) = 1 \text{ but } f_{\text{dnf}}(x) = 0\}.$$

By Lemma 9.15, all these errors can be "corrected" by adding a relatively small exact r-DNF: there is an exact r-DNF D such that $|D| \le (s - 1)^r$ and $D(x) = 1$ for all $x \in E_f$.

Case 3: f is an Or gate, $f = g \vee h$. This case is dual to Case 2. We take $f_{\text{dnf}} := g_{\text{dnf}} \vee h_{\text{dnf}}$ as the right approximator of f; hence, f_{dnf} introduces no new errors. To define the left approximator of f we use Lemma 9.15 to convert f_{dnf} into an $(s - 1)$-CNF f_{cnf}; hence, $f_{\text{dnf}} \le f_{\text{cnf}}$. Let E_f be the set of inputs on which f_{cnf} introduces a new error, that is,

$$E_f := \{x \mid f(x) = f_{\text{dnf}}(x) = 0 \text{ but } f_{\text{cnf}}(x) = 1\}.$$

By Lemma 9.15, all these errors can be "corrected" by adding a relatively small exact s-CNF: there is an exact s-CNF C such that $|C| \leq (r-1)^s$ and $C(x) = 0$ for all $x \in E_f$.

Proceeding in this way we will reach the last gate of our circuit computing the given function F. Let $(F_{\text{cnf}}, F_{\text{dnf}})$ be its approximator, and let E be the set of all inputs $x \in \{0, 1\}^n$ on which F differs from at least one of the functions F_{cnf} or F_{dnf}. Since at input gates (= variables) no error was made, for every such input $x \in E$, the corresponding error must be introduced at some intermediate gate. That is, for every $x \in E$ there is a gate f such that $x \in E_f$ (approximator of f introduces an error on x for the first time). But we have shown that, for each gate, all these errors can be corrected by adding an exact s-CNF of size at most $(r-1)^s$ or an exact r-DNF of size at most $(s-1)^r$. Since we have only t gates, all such errors $x \in E$ can be corrected by adding an exact s-CNF C of size at most $t \cdot (r-1)^s$ and an exact r-DNF D of size at most $t \cdot (s-1)^r$, that is, for all inputs $x \in \{0, 1\}^n$, we have

$$C(x) \wedge F_{\text{cnf}}(x) \leq F(x) \leq F_{\text{dnf}}(x) \vee D(x),$$

where $F_{\text{dnf}} \leq F_{\text{cnf}}$. This already implies that the function F is t-simple. Indeed, if the CNF F_{cnf} is empty (that is, if $F_{\text{cnf}} \equiv 1$) then $C \leq F$, and we are done. Otherwise, F_{cnf} must contain some clause q of length at most $s - 1$, say, $q = \bigvee_{i \in I} x_i$ for some $I \subseteq [n]$ of size $|I| \leq s - 1$. Since clearly $F_{\text{cnf}} \leq q$, the condition $F_{\text{dnf}} \leq F_{\text{cnf}}$ implies $F \leq F_{\text{dnf}} \vee D \leq F_{\text{cnf}} \vee D \leq q \vee D$, as desired. This completes the proof of Theorem 9.17.   □

In applications, boolean functions f are usually defined as set-theoretic predicates. In this case we say that f accepts a set $S \subseteq \{1, \ldots, n\}$ and write $f(S) = 1$ if and only if f accepts its incidence vector. Let $\overline{S} = \{1, \ldots, n\} \setminus S$ denote the complement of S. We say that a set S is a

- *Positive input* for f if $f(S) = 1$;
- *Negative input* for f if $f(\overline{S}) = 0$.

Put differently, a positive (negative) input is a set of variables which, if assigned the value 1 (0), forces the function to take the value 1 (0) regardless of the values assigned to the remaining variables. The minimal (under set inclusion) positive inputs for f are called *minterms* of f. Similarly, the maximal negative inputs for f are called *maxterms* of f.

Note that one and the same set S can be both a positive and a negative input! For example, if $f(x_1, x_2, x_3)$ outputs 1 iff $x_1 + x_2 + x_3 \geq 2$, then $S = \{1, 2\}$ is both positive and negative input for f, because $f(1, 1, x_3) = 1$ and $f(0, 0, x_3) = 0$.

To re-formulate the definition of t-simplicity (Definition 9.16) in terms of positive/negative inputs, note that if C is a CNF, then $C \leq f$ means that every negative input of f must contain at least one clause of C (looked at as set of indices of its variables). Similarly, $f \leq D \vee \bigvee_{i \in I} x_i$ means that every positive input must either intersect the set I or contain at least one monomial of D. Thus, if $\mathcal{F}_1$ ($\mathcal{F}_0$) is a family of positive (negative) inputs of f, and $\#_k(\mathcal{F})$ denotes the maximum number of members of $\mathcal{F}$ containing a fixed k-element set, then Theorem 9.17 gives the following more explicit lower bound.

Theorem 9.18. *For every integers* $2 \leq r, s \leq n$, *every monotone circuit computing* f *must have size at least the minimum of*

$$\frac{|\mathcal{F}_1| - (s-1) \cdot \#_1(\mathcal{F}_1)}{(s-1)^r \cdot \#_r(\mathcal{F}_1)} \quad and \quad \frac{|\mathcal{F}_0|}{(r-1)^s \cdot \#_s(\mathcal{F}_0)}.$$

That is, a monotone boolean function requires large monotone circuits if its positive as well as negative inputs are "scattered" well enough.

9.5 Explicit Lower Bounds

In order to show that a given boolean function cannot be computed by a monotone circuit of size at most t, it is enough, by Theorem 9.17, to show that the function is not t-simple for at least one(!) choice of parameters s and r. In this section we demonstrate how this can be used to derive strong lower bounds for concrete boolean functions.

9.5.1 Detecting Triangles

We begin with the simplest example, yielding a polynomial lower bound. We will also present more "respectable" applications leading to exponential lower bounds, but this special case already demonstrates the common way of reasoning fairly well.

Let us consider a monotone boolean function Δ_n, whose input is an undirected graph on n vertices, represented by $v = \binom{n}{2}$ variables, one for each possible edge. The value of the function is 1 if and only if the graph contains a triangle (three incident vertices). Clearly, there is a monotone circuit of size $\mathcal{O}(n^3)$ computing this function: just test whether any of $\binom{n}{3}$ triangles is present in the graph. Thus, the following theorem is tight, up to a poly-logarithmic factor.

Theorem 9.19. *Any monotone circuit, detecting whether a given n-vertex graph is triangle-free, must have* $\Omega(n^3 / \log^4 n)$ *gates.*

Proof. Let t be the minimal number for which Δ_n is t-simple. By Theorem 9.17, it is enough to show that $t = \Omega(n^3 / \log^4 n)$. For this proof, we take $s := \lfloor 5 \log^2 n \rfloor$ and $r := 2$. According to the definition of t-simplicity, we have only two possibilities.

Case 1: Every positive input for Δ_n either intersects a fixed set I of s edges, or contains at least one of $L \leq t s^r = t s^2$ 2-element sets of edges $R_1, \ldots, R_L$.

As positive inputs for Δ_n we take all triangles, that is, graphs on n vertices with exactly one triangle; we have $\binom{n}{3}$ such graphs. At most $s(n-2)$ of them will have an edge in I. Each of the remaining triangles must contain one of ts^2 given pairs of edges R_i. Since two edges can lie in at most one triangle, we conclude that, in this case,

$$t \geq \frac{\binom{n}{3} - s(n-2)}{s^2} = \Omega(n^3/\log^4 n).$$

Case 2: Every negative input for Δ_n contains at least one of $t(r-1)^s = t$ sets of edges $S_1, \ldots, S_t$, each of size $|S_i| = s$.

In this case we consider the graphs $E = E_1 \cup E_2$ consisting of two disjoint non-empty cliques E_1 and E_2 (we consider graphs as sets of their edges). Each such graph E is a negative input for Δ_n, because its complement is a bipartite graph, and hence, has no triangles. The number of such graphs is a half of the number 2^n of all binary strings of length n excluding the all-0 and all1 strings. Hence, we have $2^{n-1} - 1$ such graphs, and each of them must contain at least one of the sets $S_1, \ldots, S_t$. Every of these sets of edges S_i is incident to at least $\sqrt{2s}$ vertices, and if $E \supseteq S_i$ then all these vertices must belong to one of the cliques E_1 or E_2. Thus, at most $2^{n-\sqrt{2s}} - 1$ of our negative inputs E can contain one fixed set S_i, implying that, in this case,

$$t \geq \frac{2^{n-1} - 1}{2^{n-\sqrt{2s}} - 1} \geq 2^{\sqrt{2s}-1} \geq 2^{3\log n} \geq n^3.$$

Thus, in both cases, $t = \Omega(n^3/\log^4 n)$, and we are done.   □

9.5.2 Graphs of Polynomials

Our next example is the following monotone boolean function introduced by Andreev (1985). Let $q \geq 2$ be a prime power, and set $d := \lfloor (q/\ln q)^{1/2}/2 \rfloor$. Consider boolean $q \times q$ matrices $A = (a_{i,j})$. Given such a matrix A, we are interested in whether it contains a graph of a polynomial $h : \mathrm{GF}(q) \to \mathrm{GF}(q)$, that is, whether $a_{i,h(i)} = 1$ for all rows $i \in \mathrm{GF}(q)$.

Let f_n be a monotone boolean function of $n = q^2$ variables such that $f_n(A) = 1$ iff A contains a graph of at least one polynomial over $\mathrm{GF}(q)$ of degree at most $d-1$. That is,

$$f_n(X) = \bigvee_h \bigwedge_{i \in \mathrm{GF}(q)} x_{i,h(i)},$$

where h ranges over all polynomials over $\mathrm{GF}(q)$ of degree at most $d - 1$. Since we have at most q^d such polynomials, the function f_n can be computed by a monotone boolean circuit of size at most q^{d+1}, which is at most $n^{O(d)} = 2^{O(n^{1/4}\sqrt{\ln n})}$. We will now show that this trivial upper bound is almost optimal.

Theorem 9.20. *Any monotone circuit computing the function f_n has size at least* $2^{\Omega(n^{1/4}\sqrt{\ln n})}$.

Proof. Take a minimal t for which the function f_n is t-simple. Since $n = q^2$ and (by our choice) $d = \Theta(n^{1/4}\sqrt{\ln n})$, it is enough by Theorem 9.17 to show that $t \geq q^{\Omega(d)}$. For this proof, we take $s := \lfloor d \ln q \rfloor$ and $r := d$, and consider input matrices as bipartite $q \times q$ graphs. In the proof we will use the well-known fact that no two distinct polynomials of degree at most $d - 1$ can coincide on d points. According to the definition of t-simplicity, we have only two possibilities.

Case 1: Every positive input for f_n either intersects a fixed set I of at most s edges, or contains at least one of $L \leq ts^r$ r-element sets of edges $R_1, \ldots, R_L$.

Graphs of polynomials of degree at most $d - 1$ are positive inputs for f_n. Each set of l $(1 \leq l \leq d)$ edges is contained in either 0 or precisely q^{d-l} of such graphs. Hence, at most sq^{d-1} of these graphs can contain an edge in I, and at most q^{d-r} of them can contain any of the given graphs R_i. Therefore, in this case we have

$$ t \geq \frac{q^d - sq^{d-1}}{s^r \cdot q^{d-r}} = \left(\frac{q}{s}\right)^{\Omega(r)} = q^{\Omega(d)}. $$

Case 2: Every negative input for f_n contains at least one of $K \leq tr^s$ s-element sets of edges $S_1, \ldots, S_K$.

Let E be a random bipartite graph, with each edge appearing in E independently with probability $\gamma := (2d \ln q)/q$. Since there are only q^d polynomials of degree at most $d - 1$, the probability that the complement of E will contain the graph of at least one of them does not exceed $q^d (1 - \gamma)^q \leq q^{-d}$, by our choice of γ. Hence, with probability at least $1 - q^{-d}$, the graph E is a negative input for f. On the other hand, each of the sets S_i is contained in E with probability $\gamma^{|S_i|} = \gamma^s$. Thus, in this case,

$$ t \geq \frac{1 - q^{-d}}{r^s \gamma^s} = \left(\frac{q}{2d^2 \ln q}\right)^{\Omega(s)} = 2^{\Omega(s)} = q^{\Omega(d)}, $$

where the third inequality holds for all $d \leq (q/\ln q)^{1/2}/2$.

We have proved that the function f can be t-simple only if $t \geq q^{\Omega(d)}$. By Theorem 9.17, this function cannot be computed by monotone circuits of size smaller than $q^{\Omega(d)}$. □

9.6 Circuits with Real-Valued Gates

We now consider monotone circuits where, besides boolean AND and OR gates, one may use arbitrary monotone *real-valued* functions $\varphi : \mathbb{R}^2 \to \mathbb{R}$ as gates. Such a function φ is *monotone* if $\varphi(x_1, x_2) \leq \varphi(y_1, y_2)$ whenever $x_1 \leq y_1$ and $x_2 \leq y_2$. The corresponding circuits are called *monotone real circuit.*

First lower bounds for monotone circuits with real-valued gates were proved by Pudlák et al. (1997), via an extension of Razborov's argument, and by

Haken and Cook (1999), via an extension of the "bottleneck counting" argument of Haken (1995).

As in boolean circuits, inputs for such circuits are also binary strings $x \in \{0, 1\}^n$; the output must also be a binary bit 0 or 1. But at each intermediate gate any monotone function $f : \{0, 1\}^n \to \mathbb{R}$ may be computed. Hence, unlike in boolean case, here we have uncountable number of possible gates $\varphi : \mathbb{R}^2 \to \mathbb{R}$, and one may expect that at least some monotone boolean functions can be computed much more efficiently by such circuits. Exercise 9.6 shows that this intuition is correct: so-called "slice functions" can be computed by a very small monotone circuit with real-valued gates, but easy counting shows that most slice functions cannot be computed by boolean circuits of polynomial size, even if NOT gates are allowed! Thus, monotone real circuits may be even exponentially more powerful than circuits over $\{\wedge, \vee, \neg\}$.

It is therefore somewhat surprising that the (simple) criterion for boolean circuits (Theorem 9.17) remains true also for circuits with real-valued gates. The only difference from the boolean case is that now in the definition of t-simplicity we take slightly larger CNFs and DNFs, which does not greatly change the asymptotic values of the resulting lower bounds.

We say that a monotone boolean function f is *weakly t-simple* if the conditions in Definition 9.16 hold with (a) replaced by

(a′) $|C| \leq t \cdot (2r)^{s+1}$ and $|D| \leq t \cdot (2s)^{r+1}$

That is, the only difference from the definition of t-simplicity is a slightly larger upper bound on the number of clauses in C and monomials in D.

Theorem 9.21. (Criterion for Real Circuits) *Let f be a monotone boolean function. If f can be computed by a monotone real circuit of size t then f is weakly t-simple.*

Proof. The theorem was first proved in (Jukna 1999) using finite limits. A much simpler proof, which we present below, is due to Avi Wigderson. The argument is similar to that in the boolean case (Theorem 9.21). We only have to show how to construct the approximators for real-valued gates. The idea is to consider *thresholds* of real gates and approximate the thresholded values. For a real-valued function $f : \{0, 1\}^n \to \mathbb{R}$ and a real number a, let $f^{(a)}$ denote the boolean function that outputs 1 if $f(x) \geq a$, and outputs 0 otherwise.

Now let $\varphi : \mathbb{R}^2 \to \mathbb{R}$ be a gate at which the function $f(x)$ is computed, and let $g(x)$ and $h(x)$ be functions $g, h : \{0, 1\}^n \to \mathbb{R}$ computed at the inputs of this gate. A simple (but crucial) observation is that then

$$\varphi\big(g(x), h(x)\big) \geq a \iff \exists b, c : g(x) \geq b, \ h(x) \geq c \text{ and } \varphi(b, c) \geq a.$$

The ($\Rightarrow$) direction is trivial: just take $b = g(x)$ and $c = h(x)$. The other direction ($\Leftarrow$) follows from the monotonicity of φ: $\varphi(g(x), h(x)) \geq \varphi(b, c) \geq a$.

Together with the fact that $f^{(a)}(x) = 1$ iff $\varphi(g(x), h(x)) \geq a$, this allows us to express each threshold function $f^{(a)}$ of a gate $f = \varphi(g, h)$ from the thresholds of its input gates as:

$$f^{(a)} = \bigvee_{\varphi(b,c)\geq a} (g^{(b)} \wedge h^{(c)}) \tag{9.4}$$

as well as

$$f^{(a)} = \bigwedge_{\varphi(b,c)<a} (g^{(b)} \vee h^{(c)}). \tag{9.5}$$

It is convenient to think these expressions as an infinite AND and an infinite OR, respectively. However, since the number of settings $x \in \{0,1\}^n$ for input variables is finite, the real gates take only finite number of possible values, and we therefore only need finite expressions.

Fix a pair $1 \leq s, r < n$ of integer parameters. As before, every threshold $f^{(a)}$ is approximated by two functions: an s-CNF $f_{\mathrm{cnf}}^{(a)}$ (*left* approximator) and an r-DNF $f_{\mathrm{dnf}}^{(a)}$ (*right* approximator). The approximators for the thresholds of the input variables are 0, 1, or the variable itself, depending on the value of the threshold; they can always be represented by at most one literal and thus never fail.

Now let $f = \varphi(g, h)$ be an intermediate gate with two input gates g and h, and suppose that, for all (finitely many!) reals b, c, the left and right approximators for threshold functions $g^{(b)}$ and $h^{(c)}$ of its input gates are already constructed.

To construct the *left* approximator $f_{\mathrm{cnf}}^{(a)}$ of $f^{(a)}$ from the approximators of its two input gates g and h, we first consider the representation

$$f^{(a)} = \bigvee_{\varphi(b,c)\geq a} (g_{\mathrm{dnf}}^{(b)} \wedge h_{\mathrm{dnf}}^{(c)}).$$

Since the monomials in the r-DNFs $g_{\mathrm{dnf}}^{(b)}$ and $h_{\mathrm{dnf}}^{(c)}$ have length at most r, all the subexpressions $g_{\mathrm{dnf}}^{(b)} \wedge h_{\mathrm{dnf}}^{(c)}$ can be turned into a single $2r$-DNF D_a such that

$$D_a(x) = 1 \quad \text{iff} \quad f^{(a)}(x) = 1 \quad \text{iff} \quad f(x) \geq a. \tag{9.6}$$

After that we use the same procedure as before (that is, Lemma 9.15) to convert this DNF into an s-CNF $f_{\mathrm{cnf}}^{(a)}$. This can be done for each (of the finitely many) threshold values a, and we only need to ensure that the number of errors introduced when approximating the whole gate f does not depend on this number of thresholds.

When forming the s-CNF $f_{\mathrm{cnf}}^{(a)}$, we introduce errors as we throw away clauses that become longer than s. We want to count the number of inputs $x \in \{0,1\}^n$ such that $f^{(a)}(x) = 0$ while $f_{\mathrm{cnf}}^{(a)}(x) = 1$ for some a, that is, the union over a of the errors introduced in a gate by $f_{\mathrm{cnf}}^{(a)}$. To do this, let us list in the increasing order $a_1 < a_2 < \ldots < a_N$ all the $N \leq 2^n$ possible values $f(x)$ the gate f can output when the input vector x ranges over $\{0,1\}^n$. Hence,

$$D := D_{a_1} \vee D_{a_2} \vee \cdots \vee D_{a_N}$$

is a $2r$-DNF, and this DNF makes no error on x, that is, $D(x) = f(x)$. By (9.6), we have that

$$D_{a_1} \geq D_{a_2} \geq \cdots \geq D_{a_N} .$$

That is, every monomial of $D_{a_{i+1}}$ contains at least one monomial of D_{a_i}. Hence, if $t(D)$ denotes the family of all transversals of D, that is, the family of all subsets of variables, each of which intersects all the monomials of D, then

$$t(D_{a_1}) \subseteq t(D_{a_2}) \subseteq \cdots \subseteq t(D_{a_N}) ,$$

implying that $t(D) = t(D_{a_N})$. This means that all the clauses (= transversals), which we throw away (because they are longer than s) when forming an s-CNF f_{cnf} from the DNF D, are precisely those clauses, which we would throw away when converting the $2r$-DNF D_{a_N} into an s-CNF. Thus, by Lemma 9.15, all the errors that may appear during the construction of the left approximator f_{cnf}, can be corrected by an exact $(s + 1)$-CNF C of size $|C| \leq (2r)^{s+1}$. That is, for every input x such that $f(x) = 0$ but $f_{\mathrm{cnf}}(x) = 1$, we have that $C(x) = 0$.

A dual argument can be used to bound the number of errors introduced when constructing the right approximator f_{dnf}. Note that we cannot use the DNF (9.6) for this purpose since D is a $2r$-DNF, not an r-DNF. But we can argue as above by using the expression (9.5) instead of (9.4). Then all the introduced errors can be corrected by an exact $(r + 1)$-DNF D of size $|D| \leq (2s)^{r+1}$. The rest of the proof is the same as that of Theorem 9.17.   □

Since the definitions of t-simple functions and of weakly t-simple function are almost the same, Theorem 9.21 allows us to extend lower bounds for the monotone *boolean* circuits (we proved above) to the monotone *real* circuits. For example, the same argument as in the proof of Theorem 9.20 yields

Theorem 9.22. *Any monotone real circuit computing the polynomial function f_n has size at least $2^{\Omega(n^{1/4}\sqrt{\ln n})}$.*

Lower bounds for monotone real circuits have found intriguing applications in proof complexity. In particular, Pudlák et al. (1997) used such bounds to prove the first exponential lower bound on the length of so-called "cutting plane proofs", a proof system for solving integer programming problems. We will describe this result in Chap. 19.

The extension of the lower-bounds criterion from monotone boolean circuits to monotone real circuits shows the power of the criterion. On the other hand, monotonicity is crucial here.

Proposition 9.23. *Any boolean function of n variables can be computed using $n-1$ real monotone fanin-2 gates and one non-monotone unary gate.*

Proof. For an input vector $x \in \{0, 1\}^n$, let $\mathrm{bin}(x) = \sum_{i=1}^{n} x_i 2^{i-1}$ be the number whose binary code is x. It is easy to see that $\mathrm{bin}(x)$ can be computed by a circuit $C(x)$ using $n - 1$ real fanin-2 gates of the form $g(u, v) = u + 2v$. This can be done via the recurrence:

$$\mathrm{bin}(x) = x_1 + 2 \cdot \mathrm{bin}(x') = g(x_1, \mathrm{bin}(x')),$$

where $x' = (x_2, \ldots, x_n)$. These gates are monotone.

Now, every boolean function f defines the unique set of numbers

$$L_f = \{\mathrm{bin}(x) \mid f(x) = 1\}.$$

Hence, in order to compute f, it is enough to attach the (non-monotone) output gate testing whether $C(x) \in L_f$.   □

9.7 Criterion for Graph Properties

Fix a set V of $|V| = n$ vertices, and let $\binom{n}{2}$ be the set of all potential edges $e = \{u, v\}$ with $u \neq v \in V$ on these vertices. Assign a boolean variable x_e to each potential edge e. Then every 0-1 vector x of length $\binom{n}{2}$ defines the graph $S_x = \{e \mid x_e = 1\}$; we consider graphs as sets $S \subseteq \binom{n}{2}$ of their edges. Thus, every boolean function f of $\binom{n}{2}$ variables defines some property of n-vertex graphs.

An example of a graph property is the clique function $f_n = \mathrm{CLIQUE}(n, k)$ we have considered in Sects. 7.5 and 9.1. If applied directly, the symmetric lower-bounds criterion (Theorem 9.18) cannot yield strong lower bounds for this function. In this case, we can take as positive inputs of f_n the family $\mathcal{F}$ of all $\binom{n}{k}$ cliques on k vertices. But then we would only have that $\#_r(\mathcal{F}) \leq \binom{n-\sqrt{r}}{k-\sqrt{r}}$ because some sets S of $|S| = r$ edges may touch at most $\sqrt{r}$ vertices, with the worst case of S being a clique. Hence, the fraction

$$\frac{|\mathcal{F}|}{s^r \cdot \#_r(\mathcal{F})} \geq \frac{\binom{n}{k}}{s^r \binom{n-\sqrt{r}}{k-\sqrt{r}}} \approx \frac{n^{\sqrt{r}}}{s^r} = \left(\frac{n}{s^{\sqrt{r}}}\right)^{\sqrt{r}}$$

in this case is too small: we cannot take s and r large enough. The reason for this failure is that, so far, we only used a trivial measure of "length" for clauses and monomials—the total number of variables in them. But in the case of graph properties, variables x_e correspond to edges. Thus, clauses and monomials correspond in this case to graphs (sets of edges). Say a clause $c = \bigvee_{e \in S} x_e$ corresponds to the graph S. We therefore have more flexibility to define an appropriate notion of "length" of a monomial than just as the number of variables in it. We can, say, define the "length" of a graph S as the number $v(S)$ of vertices touched by (incident with) the edges in S, or as the number $\kappa(S)$ of connected components in S, or somehow else. It makes therefore sense to extend the lower-bounds criterion for the case of different length measures. We will now show that this can be done quite easily.

By a *legal length measure* we will mean any non-negative measure $\mu(S)$ of graphs satisfying the following conditions for some non-negative constants c, d:

$$\mu(S) \leq \mu(S \cup \{e\}) \leq \mu(S) + c \quad \text{and} \quad |S| \leq \mu(S)^d .$$

Parameter c tells us how much the measure of a graph can increase when one edge is added, and d tells us how much smaller can the measure of a graph be when compared to the total number of edges in it. For simplicity of exposition, we will only consider length measures with $c = d = 2$. For arbitrary c and d the arguments are the same, although the bounds we get are slightly worse.

Note that the length measure $\mu(S) = |S|$ (the total number of edges) we have considered in the previous sections has all these properties. The measure $\mu(S) =$ the number of vertices touched by the edges in S also has these properties. If we could use $\mu(S)$ instead of $|S|$, then only at most $\binom{n-r}{k-r}$ of k-cliques would contain a fixed graph S with $\mu(S) = r$, and the fraction

$$\frac{|\mathcal{F}|}{s^r \cdot \#_r(\mathcal{F})} \geq \frac{\binom{n}{k}}{s^r \binom{n-r}{k-r}} \approx \left(\frac{n}{s}\right)^r$$

would then already be large enough. We have therefore only to show that our lower bounds criteria can be extended to the case of arbitrary legal length measures.

Now, when some length measure of graphs is fixed, we can define the notions of k-CNF and of exact k-CNF in a similar way. By a k-CNF *relative to* μ we will now mean a monotone CNF each whose clause has μ-length at most k. In an *exact k-CNF relative to* μ we require that all clauses have μ-length at least k; and similarly for DNFs.

It is not difficult to verify that the Monotone Switching Lemma remains true for any pair of length measures for clauses and for monomials. The only difference is that now we have slightly worse upper bounds on $|D|$ and $|C|$, namely $|D| \leq s^{4r}$ and $|C| \leq r^{4s}$.

Proof. Argue as in the proof of Lemma 9.15. Regardless of which length measure for clauses we use, each clause of length s will have at most s^2 variables. Construct the "transversal tree" T in the same manner. Having a length measure μ for monomials, we now define DNFs f_{dnf} and D in the same way with the words "monomial of length" replaced by "monomial of μ-length". Namely, the DNF f_{dnf} now consists of all paths of μ-length smaller than r, and the DNF D consists of all paths whose μ-length reached the threshold r for the first time, that is, D consists of all paths p such that $\mu(p) \geq r$ but $\mu(p') < r$, where p' is the path p without its last edge. Since adding one edge can only increase the measure by an additive factor 2, every monomial in D has length (not just μ-length) at most $2r$. Since every node of T has fan-out at most s^2, this gives the desired upper bound $|D| \leq (s^2)^{2r} = s^{4r}$ on the total number of monomials in D.   □

Thus, in the case of graph properties f we have a more flexible lower-bounds criterion allowing us to choose different length measures for positive inputs (graphs accepted by f) and negative inputs (graphs whose complements are rejected by f). Let η be some length measure for negative inputs, and μ be some length measure for positive inputs.

Definition 9.24. (Approximators) By an (r, s)-*approximator* of f of size t we will mean a triple $(\mathcal{R}, \mathcal{S}, I)$ where [1]

- I is a graph of η-length $\leq s$;
- $\mathcal{R}$ is a family of $|\mathcal{R}| \leq t(2s)^{4r}$ graphs of μ-length $\geq r$, and
- $\mathcal{S}$ is a family of $|\mathcal{S}| \leq t(2r)^{4s}$ graphs of η-length $\geq s$

such that at least one of the following two conditions holds:

1. Every positive input of f either intersects the graph I or contains at least one of the graphs in $\mathcal{R}$.
2. Every negative input of f contains at least one of the graphs in $\mathcal{S}$.

Theorem 9.25. *If a monotone boolean function can be computed by a monotone real circuit of size t, then it has an (r, s)-approximator of size t for any $1 \leq r, s \leq n - 1$ and for every pair of length measures.*

The proof of this theorem is the same as that of Theorem 9.21: just use the modified version of the Monotone Switching Lemma. We leave a detailed proof as an exercise.

9.8 Clique-Like Problems

We consider graphs on a fixed set V of $|V| = n$ vertices. We have $m = \binom{n}{2}$ boolean variables, one for each potential edge. Then each boolean function $f : \{0, 1\}^m \rightarrow \{0, 1\}$ describes some graph property. A prominent NP-complete graph property is a monotone boolean function CLIQUE(n, k) which accepts a given graph on n vertices iff it contains a k-clique, that is, a subgraph on k vertices whose all vertices are pairwise adjacent. Instead of proving a lower bound on this function we will do this for a much larger class of "clique-like" functions.

An *a-coclique* is formed by assigning each vertex a color from the set $\{1, 2, \ldots, a\}$, and putting edges between those pairs of vertices with different colors. Note that no such graph can have an $(a + 1)$-clique.

Let $2 \leq a < b \leq m$ be integers. An (a, b)-*clique function* is a monotone boolean function f such that, for every graph G on m vertices,

$$
f(G) = \begin{cases} 0 & \text{if } G \text{ is an } a\text{-coclique;} \\ 1 & \text{if } G \text{ is a } b\text{-clique;} \\ \text{any value} & \text{otherwise.} \end{cases}
$$

Hence, CLIQUE(n, k) is an (a, b)-clique function with $a = k - 1$ and $b = k$.

---

[1] We take $(2r)^{4r}$ instead of just r^{4s} in order to cover also the real-valued case.

Theorem 9.26. (Jukna 1999) *Let* $32 \le a < b \le n/32$, *and let* f *be an* (a,b)-*clique function. Then the minimal number of gates in a monotone real circuit computing* f *is exponential in* $\min\{a, n/b\}^{1/4}$.

Proof. Let f be an (a, b)-clique function. We are going to apply the refined version of the lower-bounds criterion (Theorem 9.21). To do this, we must first choose appropriate length measure μ for positive inputs an a length measure η for negative inputs.

What to take as positive inputs and how to measure their length is clear. All b-cliques are positive inputs for f. A natural measure for a clique S is to take

$$\mu(S) := \text{the number of vertices touched by the edges in } S \text{ .}$$

It is clear that $\mu(S)$ is a legal length measure:

$$\mu(S \cup \{e\}) \le \mu(S) + \mu(\{e\}) = \mu(S) + 2 \quad \text{and} \quad |S| \le \binom{\mu(S)}{2} < \mu(S)^2 .$$

Our choice of negative inputs is also clear: we take all complements of a-cocliques. Each such complement G_h is defined by a coloring h of vertices in a colors: two vertices u and v are adjacent in G_h iff $h(u) = h(v)$. But what should we take as a length measure $\eta(S)$ of such graphs in this case?

Having a graph S of a given η-measure $\eta(S) = s$, we want that as few as possible a-colorings h can color the edges of S monochromatically, that is, color both endpoints of each edge $e \in S$ by the same color. If S is a tree with s vertices, then we could take the same measure $\eta(S) = \mu(S) = $ the number of vertices touched by the edges in S. Now, $G_h \supseteq S$ implies that h must assign the same color to all $s = \eta(S)$ vertices of S, and we can have at most $a \cdot a^{n-s} = a^{n-s+1}$ such colorings. Thus, if S is a connected graph then we could take $\eta(S)$ be the maximum number of edges in its spanning tree. For general (not necessarily connected) graphs we can do the same, and consider the measure:

$$\eta(S) = \text{maximum number of edges in a forest } F \subseteq S.$$

Since every tree with m edges has $m + 1$ vertices, $\eta(S)$ is just the number of vertices minus the number of connected components in S. But is η a legal length measure? The first condition $\eta(S) \le \eta(S \cup \{e\}) \le \eta(S) + c$ clearly holds with $c = 1$. But does the second condition $|S| \le \eta(S)^2$ hold? To show that it does, let m be the number of vertices touched by edges in S, and suppose that S consists of k connected components, the i-th of which has m_i vertices. We may assume that $m_i \ge 2$ for all i. Then $(m_i - 1)^2 \ge \binom{m_i}{2}$ holds for all i, and we obtain that

$$\eta(S)^2 = \left[\sum_{i=1}^{k} (m_i - 1) \right]^2 \ge \sum_{i=1}^{k} (m_i - 1)^2 \ge \sum_{i=1}^{k} \binom{m_i}{2} \ge |S|.$$

Thus, both measures $\mu(S)$ and $\eta(S)$ are legal length measures. By Theorem 9.25 it remains to choose parameters r, s and to show that our function f can have an (r, s)-approximator of size t only if t is large enough. For this purpose, we set (with foresight):

$$r := \lfloor (a/32)^{1/4} \rfloor \quad \text{and} \quad s := \lfloor (n/32b)^{1/4} \rfloor .$$

According to Definition 9.24 we have only two possibilities, depending on what of the two of its items holds.

Case 1: (Positive inputs) There exist a set I of $|I| \le s^2$ edges and a family $Q_1, \ldots, Q_L$ of $L \le t(2s)^{4r}$ r-cliques such that every b-clique must either intersect the set I or contain at least one of the cliques Q_i.

At least $\binom{n}{b} - s^2 \binom{n-2}{b-2} \ge \frac{1}{2}\binom{n}{b}$ of b-cliques must avoid a fixed set I of $|I| \le s^2$ edges. Each of these b-cliques must contain at least one of r-cliques Q_i. Since only $\binom{n-r}{b-r}$ of b-cliques can contain one clique Q_i, and we only have $L \le t(2s)^{4r}$ of the Q_i, in this case we have the lower bound

$$t \ge \frac{\frac{1}{2}\binom{n}{b}}{(2s)^{4r}\binom{n-r}{b-r}} = \left(\frac{n}{16s^4 b}\right)^{\Omega(r)} = 2^{\Omega(a^{1/4})} .$$

Case 2: (Negative inputs) Recall that negative inputs are graphs G_h corresponding to colorings h of vertices in a colors; two vertices u and v are adjacent in G_h iff $h(u) = h(v)$. Recall also that $\eta(S)$ is the maximum number $|F|$ of edges in a spanning forest $F \subseteq S$. Thus, in the second case of Definition 9.24 there must be a family $\mathcal{F}$ of $|\mathcal{F}| \le t(2r)^{4s}$ forests with $|F| \ge s$ edges in each $F \in \mathcal{F}$ such that every graph G_h contains at least one of these forests. That is, for every coloring h, there must be at least one forest $F \in \mathcal{F}$ such that $h(u) = h(v)$ for all edges of F.

Fix one forest $F \in \mathcal{F}$, and let $T_1, \ldots, T_d$ be all its connected components (trees). All vertices in each of these trees must receive the same color. Since each tree T_i has $|T_i| + 1$ vertices, the total number of vertices in the forest F is $m = \sum_{i=1}^{d}(|T_i| + 1) = |F| + d \ge s + d$. There are a^d ways for the coloring h to color the trees T_i, and at most $a^{n-m} \le a^{n-(s+d)}$ ways to color the remaining $n - m$ vertices. Thus, the number of graphs G_h containing one fixed forest $F \in \mathcal{F}$ does not exceed $a^d a^{n-(s+d)} = a^{n-s}$. Since we only have $|\mathcal{F}| \le t(2r)^{4s}$ forests in $\mathcal{F}$, in this case we have the lower bound

$$t \ge \frac{a^n}{(2r)^{4s} a^{n-s}} = \left(\frac{a}{16r^4}\right)^s = 2^{(n/b)^{1/4}} . \qquad \square$$

As mentioned above, the class of clique-like functions includes some NP-complete problems, like CLIQUE(n, k). But the class of (a, b)-clique functions is much larger—so large that it also includes some graph properties computable by non-monotone circuits of polynomial size!

A *graph function* is a function φ assigning each graph G a real number $\varphi(G)$. Such a function φ is *clique-like* if

$$\omega(G) \le \varphi(G) \le \chi(G),$$

where $\omega(G)$ is the clique number, that is, the maximum number of vertices in a complete subgraph of G, and $\chi(G)$ is the chromatic number, that is, the smallest number of colors which is enough to color the vertices of G so that no adjacent vertices receive the same color.

Although we always have that $\omega(G) \leq \chi(G)$, the gap between these two quantities can be quite large: results of Erdős (1967) imply that the maximum of $\chi(G)/\omega(G)$ over all n-vertex graphs G has the order $\Theta(n/\log^2 n)$. So, at least potentially, the class of clique-like functions is large enough. And indeed, Tardos (1987) observed that this class includes not only NP-complete problems (like the clique function) but also some problems from P.

Lemma 9.27. (Tardos 1987) *There exists an explicit monotone clique-like graph function φ which is computable in polynomial time.*

Proof. In his seminal paper on Shannon-capacity of graphs Lovász (1979a) introduced the capacity $\vartheta(G)$. The function $\varphi'(G) := \vartheta(\overline{G})$, where $\overline{G}$ denotes the complement of G, is a monotone clique-like function. Grötschel et al. (1981) gave a polynomial time approximation algorithm for ϑ. That is, given a graph G and a rational number $\epsilon > 0$ the algorithm computes, in polynomial time, a function $g(G, \epsilon)$ such that

$$\vartheta(G) \leq g(G, \epsilon) \leq \vartheta(G) + \epsilon.$$

Now, for any $0 < \epsilon < 1/2$ the function $\lfloor g(\overline{G}, \epsilon) \rfloor$ is a polynomial time computable clique-like function. This function might not be monotone. Let us therefore consider the monotone function

$$\varphi(G) = \lfloor g(\overline{G}, n^{-2}) + e(G) \cdot n^{-2} \rfloor,$$

where n is the number of vertices and $e(G)$ the number of edges in G. This is the desired monotone clique-like function computable in polynomial time.   □

Fix k to be the square root of the number n of vertices, and let f_ϕ denote the monotone boolean function of $\binom{n}{2}$ boolean variables encoding the edges of a graph on n vertices, whose values are defined by

$$f_\phi(G) = 1 \quad \text{iff} \quad \varphi(G) \geq k.$$

Observe that $f_\phi(G) = 1$ if $\omega(G) \geq k$, and $f_\phi(G) = 0$ if $\chi(G) \leq k - 1$. Thus, f_ϕ is a $(k-1, k)$-clique function. Theorem 9.26 and Lemma 9.27 immediately yield the following tradeoff between monotone real and non-monotone boolean circuits.

Theorem 9.28. *For every clique-like graph function φ, the boolean function f_ϕ can be computed by a non-monotone boolean circuit of polynomial size, but any monotone real circuit requires $2^{\Omega(n^{1/8})}$ gates.*

Thus, there are explicit monotone boolean functions, whose boolean non-monotone circuits are exponentially smaller than their monotone real circuits. We will use this theorem later in Sect. 19.4 to prove exponential lower bounds for widely used proof system—cutting plane proofs.

But what about the other direction: can every non-monotone boolean circuit computing a monotone boolean function be transformed into a monotone real circuit without an exponential blow-up in size? Using counting arguments one can give a *negative* answer (see Exercises 9.4–9.6).

9.9   What About Circuits with NOT Gates?

As we mentioned at the very beginning, no non-linear lower bounds are known for circuits using NOT gates. So, what is missing in the arguments we described in this and the previous chapters?

A possible answer is that the arguments are just too general! In order to show that no circuit with t gates can compute a given boolean function f, we have to show that no such circuit C can separate the set $f^{-1}(0)$ from $f^{-1}(1)$, that is, reject *all* vectors in $f^{-1}(0)$ and accept *all* vectors in $f^{-1}(1)$. Current arguments for monotone circuits (and formulas) do much more: there are relatively small subsets $A \subseteq f^{-1}(0)$ and $B \subseteq f^{-1}(1)$ (sets of particular negative and positive inputs) such that every monotone circuit separating A from B must be large.

To be more specific, let A be the set of all complete $(k-1)$-partite graphs on n vertices, and B be the set of all k-cliques. Hence, for *any* k-clique function f, members of A are negative inputs and members of B are positive inputs for f. We have shown that any monotone circuit separating A from B must have exponential size.

On the other hand, A *can* be separated from B by a small circuit if we allow just one NOT gate be used at the top of the circuit! Indeed, each graph in A has at least $K = \Omega(n)$ edges, whereas each graph in B (a k-clique) has only $\binom{k}{2}$ edges, which is smaller than K for $k = o(\sqrt{n})$. Hence, if $g = \neg Th_K$ is the negation of the threshold-K function, then $g(a) = 0$ for all $a \in A$, and $g(b) = 1$ for all $b \in B$. Since threshold functions have small monotone circuits (at most quadratic in the number of input variables), the resulting circuit is also small, separates A from B, and has only one NOT gate.

That is, it is not hard to separate the pair A, B by a monotone circuit—it is only hard to do this separation in the "right" direction: reject all $a \in A$, and accept all $b \in B$. This motivates the following definition.

Let f be a monotone boolean function. Say that a pair A, B with $A \subseteq f^{-1}(0)$ and $B \subseteq f^{-1}(1)$ is r-*hard* if every monotone circuit separating a 2^{-r} fraction of A from a 2^{-r} fraction of B (either in a "right" or in a "wrong" direction) must have super-polynomial size.

Exercise 9.7 shows that any r-hard pair A, B requires a super-polynomial number of gates in any circuit that separates A from B and uses up to r NOT gates. In the next chapter we will show that $r = \lceil \log(n+1) \rceil$ is a critical number of allowed NOT gates: having an r-hard pair for such an r would imply a super-polynomial lower bound for general non-monotone circuits! The best result known today is that

the clique function produces an r-hard pair for r about $\log \log n$; this was shown by
Amano and Maruoka (2005).

■ **Research Problem 9.29.** Exhibit an explicit pair A, B of disjoint subsets of
$\{0, 1\}^n$ which is r-hard for $r \gg \log \log n$.

9.10 Razborov's Method of Approximations

To describe the Method of Approximations in its full generality, it will again be
convenient to look at boolean functions $f : \{0, 1\}^X \to \{0, 1\}$ as computing set-
theoretic predicates $f : 2^X \to \{0, 1\}$. In this way we get a 1-to-1 correspondence
between boolean functions f and families $A(f) = \{S \subseteq X \mid f(S) = 1\}$ of subsets
of X with the properties $A(f \vee g) = A(f) \cup A(g)$ and $A(f \wedge g) = A(f) \cap A(g)$.
If f is monotone, then $A(f)$ is monotone with respect to set inclusion: if $E \in A(f)$
and $E \subseteq F$ then $F \in A(f)$.

Every family $\mathcal{F} \subseteq 2^X$ can be extended to a monotone family $\ulcorner\mathcal{F}\urcorner$ defined by

$$\ulcorner\mathcal{F}\urcorner := \bigcup_{F \in \mathcal{F}} \ulcorner F \urcorner, \quad \text{where} \quad \ulcorner F \urcorner := \{E \subseteq X \mid F \subseteq E\}.$$

In particular, if $F = \emptyset$ is the empty set, then $\ulcorner F \urcorner = 2^X$, whereas $\ulcorner\mathcal{F}\urcorner = \emptyset$
(empty family), if $\mathcal{F} = \emptyset$. The reason to consider monotone families is that
we only consider monotone boolean functions f, and for them we have that
$\ulcorner A(f) \urcorner = A(f)$.

Thus, each monotone circuit for a monotone boolean function f starts with the
basic monotone families $A(x_1), \ldots, A(x_n), A(1) = 2^X, A(0) = \emptyset$ corresponding to
input variables and the two constant functions, applies set-theoretic union $(\cup)$ and
intersection $(\cap)$ operations to them, and finally produces the family $A(f)$. The idea
is now to approximate the operations $\cup$ and $\cap$ by some other set-theoretic operations
$\sqcup$ and $\sqcap$. This leads to the following definition.

A collection $\mathfrak{M} \subseteq 2^X$ of monotone families with two operations $\sqcup$ (join) and $\sqcap$
(meet) is a *legitimate lattice* if it satisfies the following two conditions:

1. Families $A(x_1), \ldots, A(x_n), A(1), A(0)$ belong to $\mathfrak{M}$.
2. $\mathfrak{M}$ is a lattice with respect to set inclusion, that is, $M, N \subseteq M \sqcup N$ and $M \sqcap N \subseteq$
   M, N for all $M, N \in \mathfrak{M}$.

Note that the second condition implies that

$$M \cup N \subseteq M \sqcup N \quad \text{and} \quad M \sqcap N \subseteq M \cap N.$$

Thus, if we replace the gates $\cup$ and $\cap$ in our circuit by the lattice operations $\sqcup$ and
$\sqcap$, then some element $M \in \mathfrak{M}$ instead of the target family $A(f)$ could be computed.
To capture the errors arising at each gate, define:

$$\delta_-(M, N) := (M \sqcup N) \setminus (M \cup N),$$

$$\delta_+(M, N) := (M \cap N) \setminus (M \sqcap N).$$

Define the distance $\rho(f, \mathfrak{M})$ of a boolean function f from a lattice $\mathfrak{M}$ as the smallest number t for which there exist elements M, M_i, N_i $(1 \le i \le t)$ of the lattice $\mathfrak{M}$ such that

$$M \setminus A(f) \subseteq \delta_-(M_1, N_1) \cup \cdots \cup \delta_-(M_t, N_t),$$

$$A(f) \setminus M \subseteq \delta_+(M_1, N_1) \cup \cdots \cup \delta_+(M_t, N_t).$$

The proof of the following theorem is by easy induction on the number of gates, and we leave it as an exercise.

Theorem 9.30. *For every legitimate lattice $\mathfrak{M}$, every monotone boolean circuit computing f requires at least $\rho(f, \mathfrak{M})$ gates.*

In order to apply this theorem for a given monotone boolean function f, we have to define an appropriate legitimate lattice $\mathfrak{M}$ and show that f has a large distance from this lattice.

If we take $\mathfrak{M}$ to be a trivial lattice consisting of *all* monotone families, then $\rho(f, \mathfrak{M}) = 0$ for any monotone boolean function. So, in order to have a nontrivial distance, one has to consider some nontrivial lattices. For this, we need to achieve the following two goals:

1. Every family $M \in \mathfrak{M}$ must differ from $A(f)$ in many members.
2. The "error-families" $\delta_-(M_i, N_i)$ and $\delta_+(M_i, N_i)$ must be relatively small.

Crucial here is the second goal. Razborov achieves this goal by ensuring that each family in $\mathfrak{M}$ has relatively few minimal (w.r.t. set-inclusion) members. This, in turn, is achieved by introducing a clever "closure" operation, and by applying this operation when the union of two families in $\mathfrak{M}$ has too many minimal members.

9.10.1   Construction of Legitimate Lattices

Let $r \ge 2$ a fixed integer. Say that sets $F_1, \ldots, F_r$ imply a set F_0, and write $F_1, \ldots, F_r \vdash F_0$, if $F_i \cap F_j \subseteq F_0$ for all $1 \le i < j \le r$. We write $\mathcal{F} \vdash F$ if there exist not necessarily distinct members $F_1, \ldots, F_r$ of $\mathcal{F}$ such that $F_1, \ldots, F_r \vdash F$.

A general construction of legitimate lattices is as follows.

1. Fix an appropriate "ambient" family $\mathcal{P} \subseteq 2^X$. In the case of the clique function a natural choice is the family of all cliques on $\le s$ vertices, whereas in the case of the perfect matching function such is the family of all matchings with $\le s$ edges; s is a parameter.
2. Say that a family $\mathcal{F} \subseteq \mathcal{P}$ is *r-closed* (or just *closed*) if $\mathcal{F} \vdash F$ and $F \in \mathcal{P}$ implies $F \in \mathcal{F}$.
3. Define $\mathfrak{M} = \{\ulcorner \mathcal{A} \urcorner \mid \mathcal{A} \subseteq \mathcal{P}$ and $\mathcal{A}$ is r-closed$\}$.

Since the intersection of closed families is also closed, there is the smallest closed family containing $\mathcal{A}$, which we will denote by $\mathcal{A}^\star$.

Lemma 9.31. *For every family* $\mathcal{P} \subseteq 2^X$, $\mathfrak{M}$ *is a legitimate lattice with lattice operations given by*

$$\ulcorner\mathcal{A}\urcorner \sqcap \ulcorner\mathcal{B}\urcorner = \ulcorner\mathcal{A}\cap\mathcal{B}\urcorner \ \text{ and } \ \ulcorner\mathcal{A}\urcorner \sqcup \ulcorner\mathcal{B}\urcorner = \ulcorner(\mathcal{A}\cup\mathcal{B})^\star\urcorner.$$

Proof. First note that the condition (a) in the definition of a legitimate lattice is fulfilled: we have $A(x_i) = \ulcorner\{x_i\}\urcorner$ $A(1) = \ulcorner\mathcal{P}\urcorner$ and $A(0) = \ulcorner\emptyset\urcorner$.

Let $\mathfrak{A}$ denote the set of all r-closed families $\mathcal{A} \subseteq \mathcal{P}$. As the partially ordered with respect to the set-inclusion set, the set $\mathfrak{A}$ is a lattice with $\inf(\mathcal{A}_1,\mathcal{A}_2) = \mathcal{A}_1 \cap \mathcal{A}_2$ (intersection of two closed families is closed) and $\sup(\mathcal{A}_1,\mathcal{A}_2) = (\mathcal{A}_1 \cup \mathcal{A}_2)^\star$. The mapping $\ulcorner\urcorner : \mathfrak{A} \to \mathfrak{M}$ is a homomorphism of partially ordered under set-inclusion sets. So, to finish the proof of the lemma, it is enough to show that this mapping is in fact an isomorphism. That is, to show that $\ulcorner\mathcal{A}_1\urcorner \subseteq \ulcorner\mathcal{A}_2\urcorner$ implies $\mathcal{A}_1 \subseteq \mathcal{A}_2$.

To show this, let $\ulcorner\mathcal{A}_1\urcorner \subseteq \ulcorner\mathcal{A}_2\urcorner$ and $E_1 \in \mathcal{A}_1$. Then $E_1 \in \ulcorner\mathcal{A}_1\urcorner$, and hence, $E_1 \in \ulcorner\mathcal{A}_2\urcorner$. That is, there must exist a set $E_2 \in \mathcal{A}_2$ such that $E_2 \subseteq E_1$. But then $E_2,\ldots,E_2 \vdash E_1$, implying that $E_1 \in \mathcal{A}_2$, since $\mathcal{A}_2$ is r-closed. We have therefore shown that $\ulcorner\mathcal{A}_1\urcorner \subseteq \ulcorner\mathcal{A}_2\urcorner$ implies $\mathcal{A}_1 \subseteq \mathcal{A}_2$, as desired.												$\square$

The main property of closed families is that they cannot have too many *minimal* members with respect to set-inclusion.

A set family $\mathcal{F}$ is an *antichain* if for no distinct A, B in $\mathcal{F}$ do we have $A \subset B$. For a family $\mathcal{F}$, let $\min(\mathcal{F})$ denote the antichain consisting of all smallest members of $\mathcal{F}$ with respect to set-inclusion.

Lemma 9.32. *If $\mathcal{F}$ is r-closed and $|F| \leq s$ for all $F \in \mathcal{F}$, then $|\min(\mathcal{F})| \leq s!r^s$.*

Proof. Assume that $|\min(\mathcal{F})| > s!r^s$. Then the Sunflower Lemma (applied with $l = s$ and $p = r + 1$) gives us $r + 1$ sets $F_0, F_1, \ldots, F_r$ in $\min(\mathcal{F})$ forming a sunflower. Since $\mathcal{F}$ is an antichain, the core E of this sunflower is a proper subset of each of the F_i, and hence, also $E \subset F_0$. But $F_i \cap F_j = E$ for all $1 \leq i < j \leq r$ implies that $F_1, \ldots, F_r \vdash E$, and hence, E must be a member of $\mathcal{F}$ since $\mathcal{F}$ is r-closed. This contradicts our assumption that $F_0 \in \min(\mathcal{F})$.												$\square$

9.11 A Lower Bound for Perfect Matching

The *perfect matching function* is a monotone boolean function f_m of m^2 variables. Inputs for this function are subsets $E \subseteq K_{m,m}$ of edges of a fixed complete bipartite $m \times m$ graph $K_{m,m}$, and $f_m(E) = 1$ iff E contains a perfect matching, that is, a set of m vertex-disjoint edges. Taking a boolean variable $x_{i,j}$ for each edge of $K_{m,m}$, the function can be written as

$$f_m = \bigvee_{\sigma \in S_m} \bigwedge_{i=1}^{m} x_{i,\sigma(i)},$$

where S_m is the set of all $m!$ permutations of $1, 2, \ldots, m$. The function f_m is also known as a *logical permanent* of a boolean $m \times m$ matrix, the adjacency matrix of E. Hopcroft and Karp (1973) showed that this sequence of functions $(f_m \mid m = 1, 2, \ldots)$ can be computed by a deterministic Turing machine in time $\mathcal{O}(m^{5/2})$. Hence, f_m can be computed by a non-monotone circuit using only $\mathcal{O}(m^5)$ gates. But what about *monotone* circuits for this function?

Using his Method of Approximations, Razborov (1985b) was able to prove a super-polynomial lower bound $m^{\Omega(\log m)}$ also for this function. Fu (1998) showed that, after an appropriate modification, Razborov's proof works also for monotone real circuits.

The lattice $\mathfrak{M}_m$ with large distance $\rho(f_m, \mathfrak{M}_m)$ from f_m will depend on two parameters r and s which we will set later. Namely, let $\mathfrak{M}_m$ be the lattice constructed as above when starting with the ambient family $\mathcal{P} = \mathrm{Per}_s$, where

$$\mathrm{Per}_s = \{E \subseteq K_{m,m} \mid E \text{ is a matching and } |E| \leq s\}$$

is the set of all matchings with up to s edges. That is, each element of $M \in \mathfrak{M}_m$ is produced by taking an r-closed collection $\mathcal{A} \subseteq \mathrm{Per}_s$ of matchings, each with $\leq s$ edges, and including in $\mathfrak{M}_m$ the monotone family $M = \ulcorner \mathcal{A} \urcorner$ of all graphs (not just matchings) containing at least one matching in $\mathcal{A}$. In particular, minimal (under inclusion) members of each M are matchings of size at most s, that is, $\min(M) \subseteq \mathrm{Per}_s$.

Our goal is to prove that, for appropriately chosen parameters r and s, we have $\rho(f_m, \mathfrak{M}_m) = m^{\Omega(\log m)}$.

It will be convenient to use probabilistic language. Let E_+ be a random graph taking its values in the set of all $m!$ perfect matchings with equal probability $1/m!$. It is clear that

$$\mathrm{Prob}[f_m(\mathrm{E}_+) = 1] = 1.$$

Let h be a random two-coloring assigning each vertex of $K_{m,m}$ a value 0 or 1 independently with probability $1/2$. This coloring defines a random graph $\mathrm{E}_- = \{(u, v) \mid \mathrm{h}(u) = \mathrm{h}(v)\}$.

Lemma 9.33.
$$\mathrm{Prob}[f_m(\mathrm{E}_-) = 0] \geq 1 - \frac{2}{\sqrt{m}}.$$

Proof. Let U and V be the two parts of $K_{m,m}$; hence, $|U| = |V| = m$. The graph E_- has a perfect matching iff $\sum_{u \in U} \mathrm{h}(u) = \sum_{v \in V} \mathrm{h}(v)$. Hence,

$$\mathrm{Prob}[f_m(\mathrm{E}_-) = 1] = \mathrm{Prob}\left[\sum_{u \in U} \mathrm{h}(u) = \sum_{v \in V} \mathrm{h}(v) \right]$$

$$= \sum_{j=0}^{m} \text{Prob}\left[\sum_{u \in U} h(u) = j\right] \cdot \text{Prob}\left[\sum_{v \in V} h(v) = j\right]$$

$$\leq \max_{0 \leq j \leq m} \text{Prob}\left[\sum_{v \in V} h(v) = j\right] \leq \binom{m}{m/2} \cdot 2^{-m} \leq \frac{2}{\sqrt{m}} . \qquad \square$$

In order to show that the distance $\rho(f_m, \mathfrak{M}_m)$ is large, it is enough to show that, for every two members M_1, M_2 of the lattice $\mathfrak{M}_m$, the probabilities $\text{Prob}[E_+ \in \delta_+(M_1, M_2)]$ and $\text{Prob}[E_- \in \delta_-(M_1, M_2)]$ are small.

9.11.1 Error-Probability on Accepted Inputs

The case of E_+ is relatively simple. Recall that E_+ is a random perfect matching.

Lemma 9.34. *For any $M_1, M_2 \in \mathfrak{M}_m$ we have that*

$$\text{Prob}[E_+ \in \delta_+(M_1, M_2)] \leq (s! r^s)^2 \cdot \frac{(m - s - 1)}{m!} .$$

Proof. Let $M_1 = \ulcorner \mathcal{A}_1 \urcorner$ and $M_2 = \ulcorner \mathcal{A}_2 \urcorner$. Since for any family $\mathcal{F}$ and any two sets A, B we have that $\ulcorner \min(\mathcal{F}) \urcorner = \ulcorner \mathcal{F} \urcorner$ and $\ulcorner A \urcorner \cap \ulcorner B \urcorner = \ulcorner A \cup B \urcorner$, the error-set

$$\delta_+(M_1, M_2) = (M_1 \cap M_2) \setminus (M_1 \sqcap M_2) = (\ulcorner \mathcal{A}_1 \urcorner \cap \ulcorner \mathcal{A}_2 \urcorner) \setminus (\ulcorner \mathcal{A}_1 \cap \mathcal{A}_2 \urcorner)$$

is the union of sets $\ulcorner E_1 \cup E_2 \urcorner \setminus (\ulcorner \mathcal{A}_1 \cap \mathcal{A}_2 \urcorner)$ over all $E_1 \in \min(\mathcal{A}_1)$ and $E_2 \in \min(\mathcal{A}_2)$. Fix any two such sets E_1 and E_2, and let $E = E_1 \cup E_2$. Our goal is to upper-bound the probability $\text{Prob}[E_+ \in \ulcorner E \urcorner] = \text{Prob}[E \subseteq E_+]$. We have three possibilities.

Case 1: E is not a matching. In this case $\text{Prob}[E \subseteq E_+] = 0$.

Case 2: E is a matching and $|E| \leq s$, that is, $E \in \text{Per}_s$. Since $\mathcal{A}_1$ is closed, $E_1 \in \mathcal{A}_1$ and $E \in \text{Per}_s$ implies that $E = E_1 \cup E_2 \in \mathcal{A}_1$. Similarly, $E \in \mathcal{A}_2$. Hence $E \in \mathcal{A}_1 \cap \mathcal{A}_2$, implying that $\ulcorner E_1 \cup E_2 \urcorner \setminus (\ulcorner \mathcal{A}_1 \cap \mathcal{A}_2 \urcorner) = \emptyset$.

Case 3: E is a matching but $|E| \geq s + 1$. In this case

$$\text{Prob}[E \subseteq E_+] = \frac{(m - |E|)!}{m!} \leq \frac{(m - s - 1)!}{m!} .$$

Since, by Lemma 9.32, $|\min(\mathcal{A}_1)| \cdot |\min(\mathcal{A}_1)| \leq (s! r^s)^2$, we are done. $\qquad \square$

9.11.2 Error-Probability on Rejected Inputs

To upper bound the probability $\text{Prob}[\text{E}_- \in \delta_-(M_1, M_2)]$ requires more work. The problem is that the events $e_1 \in \text{E}_-$ and $e_2 \in \text{E}_-$ for edges e_1, e_2 are not necessarily independent. Still, the following lemma shows that the events are independent if the edges come from a fixed forest. Recall that a *forest* is a graph without cycles.

Lemma 9.35. *Let* $E = \{(u_1, v_1), \ldots, (u_p, v_p)\} \subseteq K_{m,m}$ *be a forest. Then the events* $(u_i, v_i) \in \text{E}_-$ *are independent, and each happens with probability* $1/2$.

Proof. It is enough to show that, for any subset $K \subseteq \{1, \ldots, p\}$ of indices, the event

$$(u_i, v_i) \in \text{E}_- \text{ for all } i \in K, \text{ and } (u_j, v_j) \notin \text{E}_- \text{ for all } j \notin K$$

happens with probability 2^{-p}. By the definition of E_-, this event is equivalent to the event that the values $\text{h}(u_i), \text{h}(v_i)$ satisfy the following system of linear equations over GF(2):

$$\text{h}(u_i) + \text{h}(v_i) = \chi_K + 1 \quad i = 1, \ldots, p, \tag{9.7}$$

where $\text{h}(u_i), \text{h}(v_i)$ are treated as variables, and χ_K is the characteristic function of the set K. Since E is a forest, the left-hand side of this system is linearly independent (see Exercise 9.8). Thus, the system has exactly 2^{2m-p} solutions, as desired. □

Lemma 9.36. *Let* $\mathcal{F} \subseteq \text{Per}_s$ *be a set of* $|\mathcal{F}| = r$ *pairwise disjoint matchings. Then there exits a subset* $\mathcal{F}_0 \subseteq \mathcal{F}$ *of* $|\mathcal{F}_0| \geq \sqrt{r}/s$ *matchings such that* $\cup \mathcal{F}_0$ *is a forest.*

Proof. Choose $\mathcal{F}_0 \subseteq \mathcal{F}$ such that $\cup \mathcal{F}_0$ is a forest and $|\mathcal{F}_0|$ is maximal. It is enough to show that $|\mathcal{F}_0| \geq \sqrt{r}/s$.

To show this, assume that $|\mathcal{F}_0| < \sqrt{r}/s$, and let $E_0 = \cup \mathcal{F}_0$; hence, $|E_0| < \sqrt{r}$. Let $U_0 \subseteq U$ and $V_0 \subseteq V$ be the sets of vertices incident with at least one edge of E_0. Then $|U_0| < \sqrt{r}$ and $|V_0| < \sqrt{r}$. Since $\mathcal{F}$ contains $|\mathcal{F}| = r > |U_0 \times V_0|$ matchings, at least one of these matchings E_1 must have no edge in $U_0 \times V_0$ (every edge can belong to at most one matching in $\mathcal{F}$, since these matchings are disjoint). Since E_1 is a matching and E_0 is a forest lying in $U_0 \times V_0$, the graph $E_0 \cup E_1$ is a forest as well. But $E_1 \cap E_0 = \emptyset$ implies that $E_1 \notin \mathcal{F}_0$, a contradiction with the maximality of $|\mathcal{F}_0|$. □

Now we are able to upper-bound $\text{Prob}[\text{E}_- \in \delta_-(M_1, M_2)]$. Note that the number of matchings in Per_s is

$$|\text{Per}_s| \leq \sum_{i=0}^{s} \binom{m}{i}^2 \cdot i! \leq m^s \sum_{i=0}^{s} \binom{m}{i} \leq m^{2s}.$$

Lemma 9.37. *For any* $M_1, M_2 \in \mathfrak{M}_m$ *we have that*

$$\text{Prob}[\text{E}_- \in \delta_-(M_1, M_2)] \leq (1 - 2^{-s})^{\sqrt{r}/s} \cdot m^{2s}.$$

Proof. Let $M_1 = \ulcorner \mathcal{A}_1 \urcorner$, $M_2 = \ulcorner \mathcal{A}_2 \urcorner$ and $\mathcal{A}_3 = \mathcal{A}_1 \cup \mathcal{A}_2$. Then $\delta_-(M_1, M_2) = \ulcorner \mathcal{A}_3^* \urcorner \setminus \ulcorner \mathcal{A}_3 \urcorner$, where $\mathcal{A}_3^*$ is the closure of $\mathcal{A}_3$. Hence, there is a sequence of families $\mathcal{A}_3, \mathcal{A}_4, \ldots, \mathcal{A}_p = \mathcal{A}_3^*$ such that $\mathcal{A}_{i+1} = \mathcal{A}_i \cup \{E_i\}$ with $\mathcal{A}_i \vdash E_i$ and $E_i \notin \mathcal{A}_i$. Hence, $\delta_-(M_1, M_2)$ is the union of all sets $\ulcorner E_i \urcorner \setminus \ulcorner \mathcal{A}_i \urcorner$, $i = 3, \ldots, p - 1$. Since $p \leq |\text{Per}_s| \leq m^{2s}$, it remains to show that $\mathcal{A} \subseteq \text{Per}_s$ and $\mathcal{A} \vdash E_0$ implies that

$$\text{Prob}[E_- \in \ulcorner E_0 \urcorner \setminus \ulcorner \mathcal{A} \urcorner] \leq (1 - 2^{-s})^{\sqrt{r}/s}. \tag{9.8}$$

To prove this, let $E_1, \ldots, E_r$ be matchings in $\mathcal{A}$ such that $E_1, \ldots, E_r \vdash E_0$. Hence, the sets $E_i^* := E_i \setminus E_0$ must be disjoint. If at least one of these sets is empty, then $\ulcorner E_0 \urcorner \subseteq \ulcorner \mathcal{A} \urcorner$, and the inequality (9.8) trivially holds. Otherwise, we can use Lemma 9.36 to choose a subset $\mathcal{F}_0 \subseteq \{E_1^*, \ldots, E_r^*\}$ such that $\cup \mathcal{F}_0$ is a forest and $|\mathcal{F}_0| \geq \sqrt{r}/s$. Then

$$\text{Prob}[E_- \in \ulcorner E_0 \urcorner \setminus \ulcorner \mathcal{A} \urcorner] \leq \text{Prob}[E_0 \subseteq E_- \text{ and } E_i \not\subseteq E_- \text{ for all } i = 1, \ldots, r]$$

$$\leq \text{Prob}[E_i^* \not\subseteq E_- \text{ for all } i = 1, \ldots, r]$$

$$\leq \text{Prob}[E^* \not\subseteq E_- \text{ for all } E^* \in \mathcal{F}_0].$$

By Lemma 9.35, all events $E^* \not\subseteq E_-$ for $E^* \in \mathcal{F}_0$ are independent, and

$$\text{Prob}[E^* \subseteq E_-] = 2^{-|E^*|} \geq 2^{-s}.$$

Therefore,

$$\text{Prob}[E^* \not\subseteq E_- \text{ for all } E^* \in \mathcal{F}_0] = \prod_{E^* \in \mathcal{F}_0} \text{Prob}[E^* \not\subseteq E_-] \leq (1 - 2^{-s})^{\sqrt{r}/s}.$$

This finishes the proof of (9.8), and thus of the lemma. $\square$

Theorem 9.38. (Razborov 1985b) *Every monotone circuit computing the perfect matching function f_m must have $m^{\Omega(\log m)}$ gates.*

Proof. By Theorem 9.30, it is enough to show that $\rho(f_m, \mathfrak{M}_m) = m^{\Omega(\log m)}$. For the proof we assume that m is sufficiently large, and set the parameters r and s to

$$s := \lfloor (\log m)/8 \rfloor \quad \text{and} \quad r := \lfloor m^{1/4}(\log m)^8 \rfloor.$$

Let M, M_i, N_i $(1 \leq i \leq t)$ be elements of the lattice $\mathfrak{M}_m$ such that

$$M \setminus A(f_m) \subseteq \bigcup_{i=1}^{t} \delta_-(M_i, N_i), \tag{9.9}$$

$$A(f_m) \setminus M \subseteq \bigcup_{i=1}^{t} \delta_+(M_i, N_i). \tag{9.10}$$

We consider two cases; $M = \emptyset$ and $M \neq \emptyset$.

Case 1: $M = \emptyset$. In this case, (9.10) implies that the entire set $A(f_m)$ must lie in the union of error-sets $\delta_+(M_i, N_i)$, $i = 1, \dots, t$. Since E_+ lies in $A(f_m)$ with probability 1, the sum of probabilities $\text{Prob}[E_+ \in \delta_+(M_i, N_i)]$ must be at least 1 as well. Together with Lemma 9.34, this implies that (for sufficiently large m)

$$t \geq \frac{m!}{(m - s - 1)!(s!r^s)^2} \geq (m/2)^s \cdot (sr)^{-2s}$$

$$= \left(\frac{m}{2r^2 s^2} \right)^s$$

$$\geq \left(\frac{32m}{m^{1/2}(\log m)^{18}} \right)^{(\log m)/8}$$

$$= m^{\Omega(\log m)}.$$

Case 2: $M \neq \emptyset$. By the construction of $\mathfrak{M}_m$, there exists a matching $E \in \text{Per}_s$ for which $\ulcorner E \urcorner \subseteq M$. Together with (9.9), this implies that

$$\ulcorner E \urcorner \subseteq A(f_m) \cup \delta_-(M_1, N_1) \cup \cdots \cup \delta_-(M_t, N_t).$$

We have

$$\text{Prob}[E_- \in \ulcorner E \urcorner] = 2^{-|E|} \geq 2^{-s} \qquad \text{by Lemma 9.35}$$

$$\text{Prob}[E_- \in A(f_m)] \leq 2m^{-1/2} \qquad \text{by Lemma 9.33}$$

$$\text{Prob}[E_- \in \delta_-(M_i, N_i)] \leq (1 - 2^{-s})^{\sqrt{r}/s} \cdot m^{2s} \qquad \text{by Lemma 9.37}$$

This implies that

$$t \geq (2^{-s} - 2m^{-1/2})(1 - 2^{-s})^{-\sqrt{r}/s} m^{-2s}$$

$$\geq \tfrac{1}{8} m^{-1/8} \cdot \exp \left(\frac{2^{-s}\sqrt{r}}{s} \right) \cdot m^{-2s}$$

$$\geq \tfrac{1}{8} m^{-\frac{1}{8} - \frac{1}{4}\log m} \exp \left(\frac{8m^{-1/8} \cdot m^{1/8} \cdot (\log m)^4}{\log m} \right)$$

$$= m^{\Omega(\log^3 m)}. \qquad \square$$

■ **Research Problem 9.39.**   Can the lower bound $m^{\Omega(\log m)}$ for perfect matching be improved to $2^{\Omega(m^\epsilon)}$ for a constant $\epsilon > 0$?

Exercises

9.1. A *partial b–(n,k,λ) design* is a family $\mathcal{F}$ of k-element subsets of $\{1,\ldots,n\}$ such that any b-element set is contained in at most λ of its members. We can associate with each such design $\mathcal{F}$ a monotone boolean function $f_{\mathcal{F}}$ such that $f_{\mathcal{F}}(S) = 1$ if and only if $S \supseteq F$ for at least one $F \in \mathcal{F}$. Assume that $\ln|\mathcal{F}| < k-1$ and that each element belongs to at most N members of $\mathcal{F}$. Use Theorem 9.17 to show that for every integer $a \geq 2$, every monotone circuit computing $f_{\mathcal{F}}$ has size at least

$$L := \min\left\{\frac{1}{2}\left(\frac{k}{2b\ln|\mathcal{F}|}\right)^a, \frac{|\mathcal{F}| - a\cdot N}{\lambda\cdot a^b}\right\}.$$

Hint: Take $r = a$, $s = b$ and show that under this choice of parameters, the function $f_{\mathcal{F}}$ can be t-simple only if $t \geq L$. When doing this, note that the members of $\mathcal{F}$ are positive inputs for $f_{\mathcal{F}}$. To handle the case of negative inputs, take a random subset T in which each element appears independently with probability $p = (1 + \ln|\mathcal{F}|)/k$, and show that T is not a negative input for $f_{\mathcal{F}}$ with probability at most $|\mathcal{F}|(1-p)^k \leq e^{-1}$.

9.2. Derive Theorem 9.20 from the previous exercise.

Hint: Observe that the family of all q^d graphs of polynomials of degree at most $d-1$ over $\mathrm{GF}(q)$ forms a partial b–(n,k,λ) design with parameters $n = q^2$, $k = q$ and $\lambda = q^{d-b}$.

9.3. Andreev (1987b) showed how, for any prime power $q \geq 2$ and $d \leq q$, to construct an explicit family $\mathcal{F}$ of subsets of $\{1,\ldots,n\}$ which, for every $b \leq d+1$, forms a partial b–(n,k,λ) design with parameters $n = q^3$, $k = q^2$, $\lambda = q^{2d+1-b}$ and $|\mathcal{F}| = q^{2d+1}$. Use Exercise 9.1 to show that the corresponding boolean function $f_{\mathcal{D}}$ requires monotone circuits of size exponential in $\Omega(n^{1/3-o(1)})$.

9.4. A boolean function $f(x_1,\ldots,x_n)$ is a *k-slice function* if $f(x) = 0$ for all x with $|x| < k$, and $f(x) = 1$ for all x with $|x| > k$, where $|x| = x_1 + \cdots + x_n$. Show that some slice functions require DeMorgan circuits of size $2^{\Omega(n)}$.

Hint: Take $k = n/2$ and argue as in the proof of Theorem 1.14.

9.5. (Rosenbloom 1997) Given a vector $x = (x_1,\ldots,x_n)$ in $\{0,1\}^n$, associate with it the following two integers $h_+(x) := |x|2^n + b(x)$ and $h_-(x) := |x|2^n - b(x)$, where $|x| = x_1 + \cdots + x_n$ and $b(x) = \sum_{i=1}^{n} x_i 2^{i-1}$. Prove that for any two vectors $x \neq y$,

1. if $|x| < |y|$, then $h_+(x) < h_+(y)$ and $h_-(x) < h_-(y)$;
2. if $|x| = |y|$, then $h_+(x) \leq h_+(y)$ if and only if $h_-(x) \geq h_-(y)$.

9.6. Let $f(x_1,\ldots,x_n)$ be a k-slice function, $0 \leq k \leq n$. Use the previous exercise to show that f can be computed by a circuit with $\mathcal{O}(n)$ monotone real-valued functions as gates.

Hint: As the last gate take a monotone function $\varphi : \mathbb{R}^2 \to \{0,1\}$ such that

$$\varphi(h_+(x), h_-(x)) = f(x)$$

for all inputs x of weight $|x| = k$.

9.7. Let f be a boolean function and suppose that it can be computed by a circuit of size t with at most r negations. Show that for any $A \subseteq f^{-1}(0)$ and $B \subseteq f^{-1}(1)$, there is a monotone boolean function g such that g can be computed by a monotone circuit of size at most t and either g or its negation $\neg g$ rejects a 2^{-r} fraction of inputs from A and accepts a 2^{-r} fraction of inputs from B.

Hint: Argue by induction on r. If $r \geq 1$, then consider the first negation gate and the function g which is computed at the gate immediately before this negation gate. Let $\epsilon \in \{0, 1\}$ be such that $g(a) = \epsilon$ for at least one half of the inputs $a \in A$. If also one half of the inputs $b \in B$ have $g(b) = \epsilon \oplus 1$, then either g or $\neg g$ has the property stated in the lemma. If this is not the case, try to apply the induction hypothesis.

9.8. Let G be a graph with n vertices and m edges, and let M be its $m \times n$ edge-vertex adjacency 0-1 matrix. That is, there is a 1 in the i-th row and j-th column iff the j-th vertex is an endpoint of the i-th edge. Show that the rows of M are linearly independent over $GF(2)$ if and only if G is a forest.

Hint: In any non-empty forest there are at least two vertices of degree 1. If some subset of rows sums up to zero, then the subgraph formed by the corresponding edges must have minimum degree at least 2.

9.9. A set $A \subseteq \{0, 1\}^n$ of vectors is Downward Closed if $x \in A$ and $y \leq x$ implies $y \in A$. Similarly, a set is Upward Closed if $x \in A$ and $x \leq y$ implies $y \in A$. Note that, if a boolean function $f : \{0, 1\}^n \to \{0, 1\}$ is monotone, then $f^{-1}(0)$ is Downward Closed and $f^{-1}(1)$ is Upward Closed. Prove the following result due to Kleitman (1966): if A, B are Downward Closed subsets of $\{0, 1\}^n$, then

$$|A \cap B| \geq \frac{|A| \cdot |B|}{2^n}.$$

Hint: Apply induction on n, the case $n = 0$ being trivial. For $a \in \{0, 1\}$, set $c_a = |A_a|$ and $d_a = |B_a|$, where $A_a = \{(x_1, \ldots, x_{n-1}) \mid (x_1, \ldots, x_{n-1}, a) \in A\}$. Apply induction to show that $|A \cap B| \geq (c_0 d_0 + c_1 d_1)/2^{n-1}$ and use the equality $c_0 d_0 + c_1 d_1 = (c_0 + c_1)(d_0 + d_1) + (c_0 - c_1)(d_0 - d_1)$ together with $A_1 \subseteq A_0$ and $B_1 \subseteq B_0$.

9.10. Show that Kleitman's theorem (Exercise 9.9) implies the following: Let A, B be upward closed and C downward closed subsets of $\{0, 1\}^n$. Then

$$|A \cap B| \geq \frac{|A| \cdot |B|}{2^n} \quad \text{and} \quad |A \cap C| \leq \frac{|A| \cdot |C|}{2^n}.$$

Hint: For the first inequality, apply Kleitman's theorem to the complements of A and B. For the second inequality, take $B := \{0, 1\}^n \setminus C$, and apply the first inequality to the pair A, B to get $|A| - |A \cap C| = |A \cap B| \geq 2^{-n}|A|(2^n - |C|)$.

9.11. Let $f : 2^{[n]} \to \{0, 1\}$ be a monotone boolean function, and let $\mathcal{F}$ be the family of all subsets $S \subseteq [n]$ that are both positive and negative inputs of f, that is $f(S) = 1$ and $f(\overline{S}) = 0$. Show that $|\mathcal{F}| \leq |f^{-1}(0)| \cdot |f^{-1}(1)|/2^n$.

9.12. (Flower Lemma, Håstad et al. 1995) A *blocking set* of a family $\mathcal{F}$ is a set which intersects all the members of $\mathcal{F}$; the minimum number of elements in a blocking set is the *blocking number* of $\mathcal{F}$ and is denoted by $\tau(\mathcal{F})$; if $\emptyset \in \mathcal{F}$ then we set $\tau(\mathcal{F}) = 0$. A *restriction* of a family $\mathcal{F}$ onto a set Y is the family $\mathcal{F}_Y := \{S \setminus Y \mid S \in \mathcal{F}, S \supseteq Y\}$. A *flower* with k petals and a core Y is a family $\mathcal{F}$ such that $\tau(\mathcal{F}_Y) \geq k$. Note that every sunflower is a flower with the same number of petals, but not every flower is a sunflower (give an example). Prove the following "flower lemma":

Let $\mathcal{F}$ be a family of sets each of cardinality s, and $k \geq 1$ and integer. If $|\mathcal{F}| > (k-1)^s$ then $\mathcal{F}$ contains a flower with k petals.

Hint: Induction on s. If $\tau(\mathcal{F}) \geq k$ then the family $\mathcal{F}$ itself is a flower with at least $(k-1)^s + 1 \geq k$ petals (and an empty core). Otherwise, some set of size $k-1$ intersects all the members of $\mathcal{F}$, and hence, at least $|\mathcal{F}|/(k-1)$ of the members must contain some point x.

9.13. Let f be a monotone boolean function of n variables, and suppose that all its maxterms have length at most t. Show that then for every $s = 1, \ldots, n$ the function f has at most t^s minterms of length s.

Hint: Let $\mathcal{F}$ be the family of all minterms of f of length s. Every maxterm must intersect all the minterms in $\mathcal{F}$. Assume that $|\mathcal{F}| > t^s$ and apply the Flower Lemma to get a contradiction with the previous sentence.

9.14. Use Exercise 9.13 to give an alternate proof of the Monotone Switching Lemma.

Chapter 10
The Mystery of Negations

The main difficulty in proving nontrivial lower bounds on the size of circuits using AND, OR and NOT is the presence of NOT gates—we already know how to prove even exponential lower bounds for monotone functions if no NOT gates are allowed. The effect of such gates on circuit size remains to a large extent a mystery. It is therefore worth describing what we actually know about this mystery. Among the basic questions concerning the role of NOT gates are the following:

1. For what monotone boolean functions are NOT gates useless, that is, do not lead to much more efficient circuits?
2. Given a function f, what is the minimum number $M(f)$ of NOT gates in a circuit computing f? Note that $M(f) = 0$ if f is monotone.
3. Given a circuit, to what extent can we decrease the number of NOT gates in it without a substantial increase in circuit size? In particular, how much can the size of a circuit increase when trying to compute f using the smallest possible number $M(f)$ of NOT gates?
4. Suppose that a function f of n variables *can* be computed by a circuit of size polynomial of n, but every circuit with $M(f)$ negations computing f requires super-polynomial size. What, then, is the minimal number of negations sufficient to compute f in polynomial size? In other words, how many NOT gates do we need in order to achieve super-polynomial savings in circuit size?

In this chapter we answer these questions.

10.1 When are NOT Gates Useless?

Let us consider circuits with AND, OR and NOT gates. Recall that such a circuit is a DeMorgan circuit if NOT gates are only applied to input variables. A circuit is monotone if it has no negated inputs.

S. Jukna, *Boolean Function Complexity*, Algorithms and Combinatorics 27,
DOI 10.1007/978-3-642-24508-4_10, © Springer-Verlag Berlin Heidelberg 2012

As we have already mentioned, current methods are not able to prove lower bounds larger than $5n$ for general circuits. On the other hand, we know how to prove even exponential lower bounds for monotone circuits where we have no NOT gates at all. Even better, there is a large class of monotone boolean functions for which NOT gates are almost useless, that is, monotone circuits for such functions are almost as efficient as non-monotone ones. These are the so-called "slice functions".

10.1.1   Slice Functions

A boolean function $f(x)$ is a *k-slice function* if $f(x) = 0$ when $|x| < k$, and $f(x) = 1$ when $|x| > k$; here and throughout, $|x| = x_1 + \ldots + x_n$ is the number of 1s in x (see Fig. 10.1). Note that slice functions are monotone! They are, however, nontrivial only on the k-th slice of the binary n-cube $\{0,1\}^n$. Note also that, for *every* boolean function f, the function g defined by

$$g = f \wedge \mathrm{Th}_k \vee \mathrm{Th}_{k+1}$$

is a k-slice function. Here, as before, $\mathrm{Th}_k(x)$ is the threshold-k function which accepts a given vector x iff $|x| \geq k$. An important property of slice functions is that NOT gates are almost useless when computing them. This is because we can replace each negated input in a circuit for a k-slice function f by a small monotone circuit computing a threshold function. The idea, due to Berkowitz (1982), is to consider threshold functions $\mathrm{Th}_k(x - x_i)$ where

$$x - x_i := (x_1, \ldots, x_{i-1}, x_{i+1}, \ldots, x_n)$$

is the vector x with its i-th component removed. A simple (but crucial) observation is that, for all input vectors $x \in \{0,1\}^n$ with exactly k ones, $\mathrm{Th}_k(x - x_i)$ is the negation of the i-th bit x_i:

$$\mathrm{Th}_k(x - x_i) = \neg x_i \text{ for all vectors } x \text{ with } |x| = k. \qquad (10.1)$$

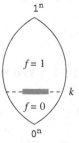

Fig. 10.1 A k-slice function. For inputs with $< k$ ones it takes value 0, for inputs with $> k$ ones it takes value 1, and is only non-trivial on inputs with exactly k ones

Indeed, if $x_1 + \cdots + x_n = k$, then $x_1 + \cdots + x_{i-1} + x_{i+1} + \cdots + x_n \geq k$ if and only if $x_i = 0$. It is known that all these n threshold functions $\mathrm{Th}_k(x - x_i)$, $i = 1, \ldots, n$ can be simultaneously computed by a monotone circuit of size $\mathcal{O}(n \log^2 n)$. This was proved by Mike Paterson (unpublished), Wegener (1985a), and Valiant and Vazirani (1986). Hence, if we replace all n negated inputs in a (non-monotone) circuit

$$f(x_1, \ldots, x_n) = F(x_1, \ldots, x_n, \neg x_1, \ldots, \neg x_n)$$

for a k-slice function f by outputs of this circuit, we obtain a *monotone* circuit

$$F_+(x_1, \ldots, x_n) = F(x_1, \ldots, x_n, \mathrm{Th}_k(x - x_1), \ldots, \mathrm{Th}_k(x - x_n)).$$

It is not difficult to verify that F_+ also computes f. That $F_+(x) = F(x)$ for all inputs x with $|x| = k$ ones follows from (10.1). To show that the same holds for all remaining input vectors, observe that

$$F(x, 0, \ldots, 0) \leq F(x, \neg x_1, \ldots, \neg x_n) \leq F(x, 1, \ldots, 1).$$

This holds because the circuit F itself is monotone, that is, has only AND and OR gates (negations are only on inputs). Since f is a k-slice function, $|x| < k$ implies $f(x) = 0$ *independent* of whether $x_i = 0$ or $x_i = 1$. Hence, on such input vectors,

$$F_+(x_1, \ldots, x_n) = F(x_1, \ldots, x_n, 0, \ldots, 0) \leq f(x_1, \ldots, x_n) = 0.$$

The case of input vectors with more than k ones is dual.

What we have just proved is the following:

Theorem 10.1. *If f is a slice function of n variables, then any non-monotone DeMorgan circuit for f can be transformed to a monotone circuit by adding at most $\mathcal{O}(n \log^2 n)$ gates.*

Thus, any lower bound $\omega(n \log^2 n)$ on the monotone(!) complexity of a slice function would yield superlinear lower bound on their non-monotone complexity. Unfortunately, existing methods for monotone circuits (and formulas) do not work for slice functions. The obstacle is that either the set of positive or the set of negative inputs of a slice function is not "scattered" well enough. For the lower-bounds criterion (Theorem 9.17) to work, we need that the number of positive (as well as negative) inputs of f containing a fixed r-element set is relatively small. Now, if f is a k-slice function with, say, $k \leq n/2$, then the only interesting negative inputs are $(n - k)$-element sets, corresponding to the vectors on the k-th slice of the n-cube on which the function takes value 0. But then up to $2^{(n-k)-r} \geq 2^{n/2-r}$ such inputs may share r common elements.

When trying to understand the monotone complexity of k-slice functions, it is important to first understand the case $k = 2$. This leads to so-called "graph complexity", a notion we have already described in Sect. 1.7 and which we will apply in Sect. 11.6 for depth-3 circuits.

10.1.2 Negated Inputs as New Variables

There is yet another bridge between monotone and non-monotone complexities. Namely, Lipton (2010) observed that it is possible to slightly modify *any* boolean function f of n variables to obtain a *monotone* boolean function g_f of $2n$ variables so that $C(f) \geq C_+(g_f) - 4n$, where $C(f)$ is the minimum size of a DeMorgan circuit computing f, and $C_\uparrow(g_f)$ the minimum size of a monotone circuit computing g_f. To show this, let $f(x)$ be any boolean function of n variables. Take a set y of new n variables and define a boolean function $g_f(x, y)$ by

$$g_f(x, y) := [f(x) \wedge \alpha(x, y)] \vee \beta(x, y),$$

where
$$\alpha(x, y) = \bigwedge_{i=1}^{n}(x_i \vee y_i) \text{ and } \beta(x, y) = \bigvee_{i=1}^{n}(x_i \wedge y_i).$$

That is, $\alpha(x, y) = 1$ iff $x \vee y = \mathbf{1}$ and $\beta(x, y) = 1$ iff $x \wedge y \neq \mathbf{0}$ (component-wise OR and AND).

Claim 10.2. For any boolean function f, g_f is a monotone function.

Proof. If $g(x, y) = g_f(x, y)$ is not monotone, there must be vectors a, b so that $g(a, b) = 1$ and changing some bit from 0 to 1 makes $g = 0$. Clearly, $\beta(a, b) = 0$; otherwise, after the change β would still output 1. Since $g(a, b) = 1$ it must be the case that $\alpha(a, b) = 1$. But then after the change β must be equal to 1, a contradiction. □

Claim 10.3. For any boolean function f,

$$g_f(x_1, \ldots, x_n, \neg x_1, \ldots, \neg x_n) = f(x_1, \ldots, x_n).$$

Proof. Let y be the vector $y = (\neg x_1, \ldots, \neg x_n)$. Then, by definition, $\alpha(x, y) = 1$ and $\beta(x, y) = 0$. □

Theorem 10.4. (Lipton 2010, Theorem 11.1) *For any boolean function f of n variables,* $C(f) \leq C_+(g_f) \leq C(f) + 4n$.

Proof. The first inequality $C(f) \leq C_+(g_f)$ follows from Claim 10.3. Now suppose that f has a circuit $F(x_1, \ldots, x_n, \neg x_1, \ldots, \neg x_n)$ of size L. This is a monotone circuit with fanin-2 AND and OR gates; inputs are variables and their negations. Replace the negated inputs $\neg x_1, \ldots, \neg x_n$ by new variables $y = (y_1, \ldots, y_n)$, extend the circuit by adding an AND with a circuit monotone computing $\alpha(x, y)$ and adding an OR with a circuit monotone computing $\beta(x, y)$. Let $F'(x, y)$ the resulting monotone circuit:

$$F'(x, y) = [F(x, y) \wedge \alpha(x, y)] \vee \beta(x, y).$$

It is clear that F' has size at most $L + 4n$. We claim that F' is the desired monotone circuit, that is, $g_f(x, y) = F'(x, y)$.

Suppose that F' is different from g_f for some values of the inputs x and y. Then, clearly, $\beta(x, y) = 0$; otherwise, they would agree. Also $\alpha(x, y)$ must equal 1; again, if not, the two values could not disagree. We now claim that for each k, $x_k = \neg y_k$. Suppose that this was false. Then, let $x_k = y_k$ for some k. Clearly, the common value cannot be 1 since $\beta = 0$. Also the common value cannot be 0 since $\alpha = 1$. This proves that for each k, $x_k = \neg y_k$. But then

$$f(x) = F(x_1, \ldots, x_n, \neg x_1, \ldots, \neg x_n) = F'(x, y).$$

Since, by Claim 10.3,

$$f(x) = g_f(x_1, \ldots, x_n, \neg x_1, \ldots, \neg x_n) = g_f(x, y),$$

we have that $g_f(x, y) = F'(x, y)$. This is a contradiction with our assumption that g_f and F' differ on input (x, y). $\qquad\square$

10.2 Markov's Theorem

We now consider circuits over $\{\wedge, \vee, \neg\}$ that are not necessarily DeMorgan circuits. That is, now inputs of NOT gates may be arbitrary gates, not just input variables. The *inversion complexity*, $I(f)$, of a boolean function f is the minimum number of NOT gates in any such circuit that computes f. It is clear that $I(f) \leq n$ for every boolean function f of n variables: just take a DeMorgan circuit.

More than 50 years ago, Markov (1957) made an amazing observation that *every* boolean (and even multi-output) function of n variables can be computed by a circuit with only about $\log n$ negations! Moreover, this number of negations is in general necessary! To state and prove this classical result, we need a concept of the "decrease" of functions.

For two binary vectors $x = (x_1, \ldots, x_n)$ and $y = (y_1, \ldots, y_n)$ we write, as before, $x \leq y$ if $x_i \leq y_i$ for all i. We also write $x < y$ if $x \leq y$ and $x_i < y_i$ for at least one i. A boolean function $f : \{0, 1\}^n \to \{0, 1\}$ is *monotone* if $x \leq y$ implies $f(x) \leq f(y)$. A *chain* in the binary n-cube is an increasing sequence $Y = \{y^1 < y^2 < \ldots < y^k\}$ of vectors in $\{0, 1\}^n$. Note that no chain can contain more than $n + 1$ vectors. A typical chain of this length consists of vectors $1^i 0^{n-i}$, $i = 0, 1, \ldots, n$.

Given such a chain, we look at how many times our boolean function f changes its value from 1 to 0 along this chain, and call this number the decrease of f on this chain. That is, we count how many times the following event happens along the chain: $x < y$ but $f(x) > f(y)$. Formally, say that i is a *jump position* (or a *jump-down* position) of f along a chain $Y = \{y^1 < y^2 < \ldots < y^k\}$, if $f(y^i) = 1$ and $f(y^{i+1}) = 0$. The number of all jump-down positions is the *decrease* $d_Y(f)$

of f on the chain Y. The *decrease* $d(f)$ of f is the maximum of $d_Y(f)$ over all chains Y.

Note that we only count the jump-down positions from 1 to 0: those j for which $f(y^j) = 0$ and $f(y^{j+1}) = 1$ (the jump-up positions) do not contribute to $d_Y(f)$. In particular, we have that $d(f) \leq n/2$ for every boolean function f of n variables, and $d(f) = 0$ for all monotone functions.

Theorem 10.5. (Markov 1957) *For every boolean function f,*

$$d(f) = 2^{I(f)} - 1.$$

That is, the minimum number $I(f)$ of negations that are enough to compute f is equal(!) to the length of the binary code of $d(f)$, $I(f) = \lceil \log(d(f) + 1) \rceil$.

Remark 10.6. The same result also holds for circuit computing *sets* F of boolean functions. We can view each subset F of $|F| = m$ functions as an operator $F : \{0, 1\}^n \to \{0, 1\}^m$. In this case, the decrease of F along a chain $Y = \{y^1 < y^2 < \ldots < y^k\}$ is the number vectors y^i such that $F(y^i) \not\leq F(y^{i+1})$.

We prove the lower and upper bounds on $I(f)$ separately.

Lemma 10.7. (Upper bound) $d(f) \leq 2^{I(f)} - 1$.

Proof. We use induction on $I(f)$. The basis case $I(f) = 0$ is obvious because then f is monotone, and $d(f) = 0$. For the induction step, assume that $d(g) \leq 2^{I(g)} - 1$ holds for all boolean functions with $I(g) \leq I(f) - 1$. In any circuit for f (containing at least one negation) there is a NOT gate whose input does not depend on the outputs of any other NOT gates. So, f may be decomposed as $f(x) = g(\neg h(x), x)$, where $I(g) = I(f) - 1$ and $I(h) = 0$ (h is a monotone function).

Fix a chain $Y = \{y^1 < y^2 < \ldots < y^k\}$ for which $d_Y(f) = d(f)$. By monotonicity of h, there is an $1 \leq l \leq k$ such that h outputs 0 on all vectors in $Y_0 := \{y^1 < y^2 < \ldots < y^l\}$, and outputs 1 on all remaining vectors in $Y_1 := Y \setminus Y_0$. Now let $g_i(x) := g(\neg i, x)$ for $i = 0, 1$. Then $I(g_i) \leq I(g) \leq I(f) - 1$, so by the induction hypothesis, $d_X(g_i) \leq 2^{I(g_i)} - 1$ for every chain X. In particular,

$$d_{Y_0}(f) = d_{Y_0}(g_0) \leq 2^{I(g_0)} - 1 \leq 2^{I(f)-1} - 1,$$
$$d_{Y_1}(f) = d_{Y_1}(g_1) \leq 2^{I(g_1)} - 1 \leq 2^{I(f)-1} - 1.$$

It follows that $d(f) = d_Y(f) \leq d_{Y_0}(f) + d_{Y_1}(f) + 1 \leq 2^{I(f)} - 1$.   □

Lemma 10.8. (Lower bound) $d(f) \geq 2^{I(f)} - 1$.

Proof. We have to prove that $I(f) \leq M(f)$ where $M(f) := \lceil \log(d(f) + 1) \rceil$ is the length of the binary code of $d(f)$. We will do this by induction on $M(f)$. The base case $M(f) = 0$ is again obvious, because then $d(f) = 0$, so f is monotone and $I(f) = 0$.

Fig. 10.2 Chain Y_0 ends in x, and chain Y_1 starts with x

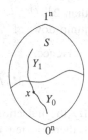

For the induction step, suppose $I(g) \leq M(g)$ holds for all boolean functions g such that $M(g) \leq M(f) - 1$. Let S be the set of all vectors $x \in \{0, 1\}^n$ such that $d_Y(f) < 2^{M(f)-1}$ for every chain Y *starting* with x:

$$S = \{x : d_Y(f) < 2^{M(f)-1} \text{ for any chain } Y \text{ starting in } x\}.$$

Note that the set S is upwards closed: if $x \in S$ and $x \leq y$, then $y \in S$. This holds because each chain starting in y can be extended to a chain starting in x.

Claim 10.9. For every chain Y *ending* in a vector outside the set S we also have $d_Y(f) < 2^{M(f)-1}$.

Proof. Assume that there is a chain Y_0 ending in a vector $x \notin S$ and such that $d_{Y_0}(f) \geq 2^{M(f)-1}$ (Fig. 10.2). The fact that x does not belong to S means that there must be a chain Y_1 starting in x for which $d_{Y_1}(f) \geq 2^{M(f)-1}$. But then the decrease $d_{Y_0 \cup Y_1}(f)$ of f on the combined chain $Y_0 \cup Y_1$ is

$$d_{Y_0 \cup Y_1}(f) = d_{Y_0}(f) + d_{Y_1}(f) \geq 2^{M(f)} = 2^{\lceil \log(d(f)+1) \rceil} > d(f),$$

contradicting the definition of $d(f)$. □

Consider now two functions f_0 and f_1 defined as follows:

$$f_0(x) = \begin{cases} f(x) & \text{if } x \in S, \\ 0 & \text{if } x \notin S, \end{cases} \quad \text{and} \quad f_1(x) = \begin{cases} 1 & \text{if } x \in S, \\ f(x) & \text{if } x \notin S. \end{cases} \tag{10.2}$$

Claim 10.10. Both $d(f_0)$ and $d(f_1)$ are strictly smaller than $2^{M(f)-1}$.

Proof. We show this for f_0 (the argument for f_1 is similar). Let Y be a chain for which $d_Y(f_0) = d(f_0)$. Let x be a vector which Y starts in and y be a vector which Y ends in. If $x \in S$ or $y \notin S$, then $d(f_0) < 2^{M(f)-1}$ by Claim 10.9 and definition of S. So, assume that $x \notin S$ and $y \in S$. Since the set S is upwards closed, some initial part Y_0 of the chain Y lies outside S and the remaining part Y_1 lies in S. By the definition of the function f_0, it is constant 0 on Y_0, and coincides with f on Y_1. By the definition of the set S, we have that the decrease of f_0 on Y_1 is smaller than $2^{M(f)-1} - 1$. Since $f_0(z) = 0$ for all $z \in Y_0$, there cannot be any additional jump-down of f_0 along the entire chain $Y = Y_0 \cup Y_1$. $\qquad\square$

Hence,

$$M(f_i) = \lceil \log(d(f_i) + 1) \rceil \le \lceil \log 2^{M(f)-1} \rceil = M(f) - 1.$$

By the induction hypothesis, $I(f_i) \le M(f_i) \le M(f) - 1$ for both $i = 0, 1$. It therefore remains to show that

$$I(f) \le 1 + \max \left\{ I(f_0), I(f_1) \right\}. \tag{10.3}$$

For this, we need one auxiliary result. A *connector* of two boolean functions $f_0(x)$ and $f_1(x)$ of n variables is a boolean function g of $n + 2$ variables such that $g(0, 1, x) = f_0(x)$ and $g(1, 0, x) = f_1(x)$.

Claim 10.11. Every pair of functions $f_0(x)$, $f_1(x)$ has a connector g such that

$$I(g) \le \max \left\{ I(f_0), I(f_1) \right\}.$$

Proof. We argue by induction on $r := \max \left\{ I(f_0), I(f_1) \right\}$. If $r = 0$ then both functions f_i are monotone, and we can take

$$g(u, v, x) = (u \wedge f_1) \vee (v \wedge f_0).$$

For the induction step, let $C_i(x)$ be a circuit with $I(f_i)$ negations computing $f_i(x)$. Replacing the first NOT gate in C_i by a new variable ξ we obtain a circuit $C_i'(\xi, x)$ on $n + 1$ variables which contains one NOT gate fewer. Let $f_i'(\xi, x)$ be the function computed by this circuit; hence, $I(f_i') \le r - 1$. Moreover, if $h_i(x)$ is the monotone function computed immediately before the first NOT gate in C_i, then

$$f_0(x) = f_0'(\neg h_0(x), x) \quad \text{and} \quad f_1(x) = f_1'(\neg h_1(x), x). \tag{10.4}$$

By the induction hypothesis, there is a boolean function $g'(u, v, \xi, x)$ (the connector of the pair f_0', f_1') such that $I(g') \le \max \left\{ I(f_0'), I(f_1') \right\} \le r - 1$,

$$g'(0, 1, \xi, x) = f_0'(\xi, x) \quad \text{and} \quad g'(1, 0, \xi, x) = f_1'(\xi, x).$$

We now replace the variable ξ in $g'(u, v, \xi, x)$ by the function

$$Z(u, v, x) := \neg \big((u \wedge h_1(x)) \vee (v \wedge h_0(x)) \big).$$

Since $Z(0, 1, x) = \neg h_0(x)$ and $Z(1, 0, x) = \neg h_1(x)$, (10.4) implies that the obtained function $g(u, v, x)$ is a connector of f_0 and f_1. Since the functions h_0 and h_1 are monotone, we have $I(g) \le 1 + I(g') \le r$, as desired. $\qquad\square$

We now can complete the proof of Lemma 10.8 as follows. Let $\chi_S(x)$ be the characteristic function of S, that is, $\chi_S(x) = 1$ for $x \in S$, and $\chi_S(x) = 0$ for $x \notin S$. Let g be a connector of f_0 and f_1 guaranteed by Claim 10.11. By the definition of the functions f_0 and f_1, we then have that our original function $f(x)$ can be computed as

$$f(x) = g(\neg\chi_S(x), \chi_S(x), x).$$

Indeed, if $x \in S$ then $f(x) = f_0(x) = g(0, 1, x) = g(\neg\chi_S(x), \chi_S(x), x)$, and similarly for all vectors $x \notin S$. Since the set S is upwards closed, its characteristic function $\chi_S(x)$ is monotone, and hence, requires no NOT gates. Thus, Claim 10.11 implies

$$I(f) \leq 1 + I(g) \leq 1 + \max\{I(f_0), I(f_1)\}.$$

This completes the proof of (10.3), and thus the proof of Lemma 10.8. $\qquad\square$

Remark 10.12. (Nondeterministic circuits) A *nondeterministic circuit* is a circuit $C(x, y)$ whose input variables are partitioned into two groups: "actual" inputs $x_1, \ldots, x_n$, and "guess" inputs $y_1, \ldots, y_m$. A circuit computes a boolean function $f(x)$ in a natural way: $f(x) = 1$ iff $C(x, y) = 1$ for at least one $y \in \{0, 1\}^m$. Let $I(f, m)$ denote the minimum number r such that f can be computed by a nondeterministic circuit over $\{\wedge, \vee, \neg\}$ with r negations and at most m guess inputs. Nondeterministic circuits were (apparently) first introduced by Karchmer and Wigderson (1993b), where several tight *combinatorial* characterizations of such circuits are presented. Morizumi (2009b) extended Markov's theorem by showing (see Exercises 10.4 and 10.5) that

$$I(f, m) = \lceil \log_2(d(f)/2^m + 1) \rceil.$$

Remark 10.13. (Symmetric functions) For circuits computing *symmetric* boolean functions, Tanaka et al. (1996) established the following structural result. Let f be a symmetric boolean function of n variables. Suppose that $d(f) = m$, where $m = 2^r - 1$ for some integer r. For $a \in \{0, 1\}^n$, let $d_f(a)$ denote the maximum decrease of f along a chain *ending* in a; note that this number depends only on the number of ones in a. Consider an arbitrary circuit $G = (g_1, g_2, \ldots, g_t)$ computing f and using r negations. For $i = 1, \ldots, r$, let h_i be the boolean function computed at the input of the i-th NOT gate. Then, for every $a \in \{0, 1\}^n$, the 0-1 sequence $(h_1(a), \ldots, h_r(a))$ is the binary representation of $d_f(a)$.

10.3 Formulas Require Exponentially More NOT Gates

We now consider *formulas*, that is, circuits with AND, OR and NOT gates whose fanout in a circuit is 1. The only difference from the general circuits (over the same basis) considered in the previous section is that now the underlying graph of a circuit is a tree, not an arbitrary directed acyclic graph. It is intuitive that requiring fanin

1 should restrict the power of circuits. And indeed, we will now show that the minimal number of NOT gates in formulas must be exponentially larger than in circuits.

Define the *inversion complexity*, $I_F(f)$, of a boolean function f in the class of formulas as the minimum number of NOT gates contained in a formula computing f.

By Markov's theorem, the minimum number of NOT gates in a *circuit* for f is about[1] $\log d(f)$, where $d(f)$ is the decrease of f. In the case of *formulas* we have:

Theorem 10.14. (Nechiporuk 1962) *For every boolean function f, we have*

$$I_F(f) = d(f).$$

This result was apparently not known in the West, and it was independently proved by Morizumi (2009). We again prove the lower and upper bounds on $I_F(f)$ separately.

Lemma 10.15. (Lower bound) $I_F(f) \geq d(f)$.

Proof. Let C be a formula computing f, and let $I_F(C)$ be the number of NOT gates in it. Fix a chain $Y = \{y^1 < y^2 < \ldots < y^k\}$ for which $d_Y(f) = d(f)$. Our goal is to show that $d_Y(f) \leq I_F(C)$. This follows by induction on the leafsize of C using the following three inequalities: $d_Y(f \wedge g) \leq d_Y(f) + d_Y(g)$, $d_Y(f \vee g) \leq d_Y(f) + d_Y(g)$ and $d_Y(\neg f) \leq d_Y(f) + 1$.

The first two inequalities are trivial because every jump-down position of $f \wedge g$ as well as for $f \vee g$ must be a jump-down position of at least one of the functions f and g. To see the third inequality, observe that each jump-down position for $\neg f$ is a jump-up position for f. Hence, $d_Y(\neg f) - d_Y(f) = f(y^k) - f(y^1)$, and $d_Y(\neg f) \leq d_Y(f) + 1$ follows.   □

Lemma 10.16. (Upper bound) $I_F(f) \leq d(f)$.

Proof. The original proof by Nechiporuk (1962) is somewhat complicated because he describes an explicit formula. But, as observed by Morizumi 2009, one can also argue by induction on $d(f)$, as in the proof of Lemma 10.7. The base case $d(f) = 0$ is trivial, since then f is monotone and $I_F(f) = 0$.

For the induction step, suppose that $d(f) \geq 1$, and $I_F(f') \leq d(f')$ for all boolean functions f' such that $d(f') \leq d(f) - 1$. Let S be the set of all vectors $x \in \{0, 1\}^n$ such that $d_Y(f) = 0$ for every chain Y *starting* with x:

$$S = \{x : d_Y(f) = 0 \text{ for any chain } Y \text{ starting in } x\}.$$

Note that the set S is upwards closed: if $x \in S$ and $x \leq y$ then $y \in S$. This holds because each chain starting in y can be extended to a chain starting in x.

---

[1] As before, all logarithms are to the basis of two.

As in the proof of Markov's theorem, consider two functions f_0 and f_1 defined by Eq. 10.2. Let also χ_S be the characteristic function of the set S itself, that is, $\chi_S(x) = 1$ for $x \in S$, and $\chi_S(x) = 0$ for $x \notin S$. It is easy to see that

$$f = f_0 \vee (f_1 \wedge \neg\chi_S).$$

Claim 10.17. $d(f_0) = d(\chi_S) = 0$ and $d(f_1) \leq d(f) - 1$.

Proof. Since the set S is upwards closed, its characteristic function χ_S is monotone, implying that $d(\chi_S) = 0$. That $d(f_0) = 0$ follows from the fact that f_0 cannot take value 1 on a chain Y until Y enters the set S.

To show that $d(f_1) \leq d(f) - 1$, assume that $d(f_1) \geq d(f)$. Since we only count the number of changes of values of f on a chain from 1 to 0 (not from 0 to 1), the maximum $d(f_1) = \max_X d_X(f_1)$ is achieved on a chain X *ending* in a vector y such that $f_1(y) = 0$. Since $d_Y(f) = 0$ for all chains Y *starting* with some vector in S, there must be a chain X which *ends* in some vector $y \notin S$ and for which $d_X(f_1) \geq d(f)$ holds. On the other hand, the fact that y is not in S implies that there must be a chain Y *starting* in y such that $d_Y(f) \geq 1$. But then for the combined chain $X \cup Y$ we have that $d_{X \cup Y}(f) = d_X(f) + d_Y(f) \geq d(f) + 1$, contradicting the definition of $d(f)$. $\square$

By Claim 10.17 and the induction hypothesis, we have that $I_F(f_0) = 0$, $I_F(\chi_S) = 0$ and $I_F(f_1) \leq d(f) - 1$. Hence, the desired upper bound follows:

$$I_F(f) \leq I_F(f_0) + I_F(f_1) + I_F(\chi_S) + 1 \leq d(f).$$ $\square$

10.4 Fischer's Theorem

According to Markov's theorem, every boolean function of n variables *can* be computed by a circuit with at most $M(n) = \lceil \log(n + 1) \rceil$ NOT gates. The next important step was made by Fischer (1974): restricting the number of negations to $M(n)$ entails only a polynomial blowup in circuit size!

Theorem 10.18. (Fischer 1974) *If a function on n variables can be computed by a circuit over $\{\wedge, \vee, \neg\}$ of size t, then it can be computed by a circuit of size at most $2t + \mathcal{O}(n^2 \log^2 n)$ using at most $\lceil \log(n + 1) \rceil$ NOT gates.*

Proof. It is easy to show that every circuit of size t can be transformed to a circuit of size at most $2t$ such that all negations are placed only on the input variables. Hence, it is enough to show how to compute the (multi-output) function

$$\mathrm{INV}_n(x_1, \ldots, x_n) := (\neg x_1, \ldots, \neg x_n)$$

by a circuit of size $\mathcal{O}(n^2 \log^2 n)$ using $M(n) := \lceil \log(n+1) \rceil$ negations; the function INV_n is also known as an *inverter*.

We already know (see Eq. 10.1) that, on inputs $x \in \{0, 1\}^n$ with exactly k ones, the negation $\neg x_i$ of its i-th bit can be computed as $\neg x_i = \mathrm{Th}_k(x - x_i)$, where

$$\mathrm{Th}_k(x - x_i) := \mathrm{Th}_k(x_1, \ldots, x_{i-1}, x_{i+1}, \ldots, x_n) \, .$$

Using this observation, we can also simulate the behavior of $\neg x_i$ on *all* inputs. For each $i = 1, 2, \ldots, n$ consider the function

$$f_i(x) := \bigwedge_{k=1}^{n} \left(\neg \mathrm{Th}_k(x) \vee \mathrm{Th}_k(x - x_i) \right) \, .$$

Claim 10.19. For any $x \in \{0, 1\}^n$ and any $1 \le i \le n$, we have that $f_i(x) = \neg x_i$.

Proof. Take an arbitrary vector $a \in \{0, 1\}^n$. If $\neg x_i(a) = 1$ then $a_i = 0$, implying that in this case $\mathrm{Th}_k(a) = \mathrm{Th}_k(a - a_i)$ for all $k = 1, \ldots, n$, and hence, $f_i(a) = 1$. If $\neg x_i(a) = 0$ then $a_i = 1$. So, for $k = |a|$, we then have $\mathrm{Th}_k(a) = 1$ and $\mathrm{Th}_k(a - a_i) = 0$, implying that $f_i(a) = 0$.   □

It can be shown (we will not do this) that all the functions $\mathrm{Th}_k(x)$ and $\mathrm{Th}_k(x - x_i)$ ($0 \le k \le n, 1 \le i \le n$) can be computed by a *monotone* circuit of size $\mathcal{O}(n^2 \log^2 n)$. Hence, it remains to compute the function

$$\neg T(x) := \left(\neg \mathrm{Th}_1(x), \neg \mathrm{Th}_2(x), \ldots, \neg \mathrm{Th}_n(x) \right)$$

using at most $M(n) = \lceil \log(n + 1) \rceil$ negations. To do this, we first take a monotone circuit $C_1(x)$ computing the function

$$T(x) := \left(\mathrm{Th}_1(x), \mathrm{Th}_2(x), \ldots, \mathrm{Th}_n(x) \right) \, .$$

Observe that the outputs of this circuit belong to the set A_{sort} of all inputs $y \in \{0, 1\}^n$ whose bits are *sorted* in decreasing order $y_1 \ge y_2 \ge \ldots \ge y_n$. That is, A_{sort} consists of $n + 1$ strings of the form $1^i 0^{n-i}$, $i = 0, 1, \ldots, n$. Using only $M(n)$ negations it is possible to construct a circuit $C_2(y)$ of size $\mathcal{O}(n)$ which computes $\mathrm{INV}_n(y)$ correctly on all inputs in A_{sort} (Exercise 10.3). Thus, the circuit $C(x) = C_2(C_1(x))$ computes $\neg T(x)$, as desired.   □

Remark 10.20. The additive term $\mathcal{O}(n^2 \log^2 n)$ in Theorem 10.18 has been subsequently improved. Namely, Tanaka and Nishino (1994) proved that INV_n can be computed by a circuit with $r = \lceil \log_2(n + 1) \rceil$ NOT gates and using at most $\mathcal{O}(n \log^2 n)$ gates in total; the depth of the constructed circuit is $\mathcal{O}(\log^2 n)$. Later, Beals et al. (1998) improved the size to $\mathcal{O}(n \log n)$; the depth of their circuit is $\mathcal{O}(\log n)$. By increasing the depth to $\mathcal{O}(\log^{1+o(1)} n)$ and allowing $\mathcal{O}(\log^{1+o(1)} n)$ negations, Morizumi and Suzuki (2010) were able to reduce the size to $\mathcal{O}(n)$.

10.5 How Many Negations are Enough to Prove P ≠ NP?

In order to prove the well known conjecture that $P \neq NP$, it would be enough to prove that some functions $f : \{0, 1\}^n \to \{0, 1\}^n$ in NP cannot be computed by circuits of polynomial (in n) size. By the results of Markov and Fischer, it would be enough to prove a "weaker" result. Namely, let

$P^{(r)} =$ class of all sequences of functions $f : \{0, 1\}^n \to \{0, 1\}^n$ computable by
polynomial-size circuits with at most r NOT gates.

Let CLIQUE be the monotone boolean function of $\binom{n}{2}$ variables which accepts a given input graph on n vertices iff it contains a clique on $n/2$ vertices (see Sect. 1.5). Since $P \neq NP$ if CLIQUE $\notin P$, Markov–Fischer results imply that:

If CLIQUE $\notin P^{(r)}$ for $r = \lceil \log(n + 1) \rceil$, then $P \neq NP$.

The breakthrough result of Razborov (1985a), see Theorem 9.1, states that

$$\text{CLIQUE} \notin P^{(r)} \text{ for } r = 0.$$

Amano and Maruoka (2005) showed that essentially the same argument yields a stronger result:

$$\text{CLIQUE} \notin P^{(r)} \text{ even for } r = (1/6) \log \log n.$$

At first glance, this development looks like a promising way to prove that $P \neq NP$: just extend the bound to circuits with a larger and larger number r of allowed NOT gates. But how large must the number r of allowed NOT gates be in order to yield the conclusion $P \neq NP$? This question motivates the following parameter for functions f:

$$R(f) = \min\{r : f \notin P^{(r)} \text{ implies } f \notin P\}.$$

By the results of Markov and Fischer, for any f, we have that

$$0 \leq R(f) \leq \lceil \log(n + 1) \rceil$$

holds for every function f of n variables. This parameter is most interesting for *monotone* functions since they need no NOT gates at all, if we don't care about the circuit size. We already know that $R(f) = 0$ for a large class of monotone boolean functions f, namely for slice functions. But no specific slice function f with $f \notin P^{(0)}$ is known.

On the other hand, if it were the case that $R(f) \leq (1/6) \log \log n$ for every monotone function f, then we would already have that CLIQUE $\notin P$, and hence, that $P \neq NP$. Unfortunately, it was shown in Jukna (2004) that there *are* monotone functions f in P for which $R(f)$ is near to Markov's $\log n$-border.

Theorem 10.21. *There is an explicit monotone function* $f : \{0, 1\}^n \to \{0, 1\}^n$ *such that* $f \in P$ *but* $f \notin P^{(r)}$ *unless* $r \geq \log n - \mathcal{O}(\log \log n)$.

Proof. The proof idea is to take a monotone boolean function $g : \{0, 1\}^n \to \{0, 1\}$ which is *feasible* (that is, belongs to P), and consider a monotone multi-output function $f : \{0, 1\}^{kn} \to \{0, 1\}^k$ computing $k = 2^r$ copies of g on *disjoint* sets of variables. We call such a function f a k-*fold extension* of g. We then show that, if g requires *monotone* circuits of exponential size, then f requires circuits of super-polynomial size, even if up to r NOT gates are allowed.

Claim 10.22. Let f be a monotone boolean function, and k be a power of 2. If the k-fold extension of f can be computed by a circuit with $\log k$ NOT gates, then f can be computed by a *monotone* circuit of the same size.

Proof. It is enough to prove the lemma for $k = 2$ (we can then iterate the argument). Thus, take a circuit with one NOT gate computing two copies $f_0 = f(Y_0)$ and $f_1 = f(Y_1)$ of the monotone function $f(X)$ on disjoint sets of variables. Let g be the monotone(!) boolean function computed at the input to the (unique) NOT gate.

We have only two possibilities: either some minterm of g lies entirely in Y_1, or not. In the first case we assign the constant 1 to all the variables in Y_1, whereas in the second case we assign the constant 0 to all the variables in Y_0. As the function g is monotone, in both cases it turns into a constant function (1 in the first case, and 0 in the second), and the subsequent NOT gate can be eliminated. But since $Y_0 \cap Y_1 = \emptyset$, the setting $Y_\epsilon \mapsto \epsilon$ does not affect the function $f_{1-\epsilon}$. Hence, we obtain a circuit which contains no NOT gates and computes either f_0 or f_1, and hence, also $f(X)$ after renaming the input variables.   □

To finish the proof of Theorem 10.21, we will make use of an explicit monotone boolean clique-like function T_m in m variables considered by Tardos (1987). In Sect. 9.8 we have shown (see Theorem 9.28) that this function is feasible—can be computed by a non-monotone circuit of size $m^{\mathcal{O}(1)}$—but every monotone circuit computing it requires size is exponential in $\Omega(m^{1/16})$.

Let $n = km$ where $k = 2^r$ and $r = \lfloor \log n - 32 \log \log n \rfloor$; hence, k is about $n/(\log n)^{32}$. Consider the k-fold extension f_n of T_m. Then f_n can be computed by a (non-monotone) circuit of size at most $k \cdot m^{\mathcal{O}(1)}$, which is, of course, polynomial in n. Hence, $f_n \in P$. On the other hand, Claim 10.22 and Theorem 9.28 imply that every circuit with at most r NOT gates computing f_n must have size exponential in $m^{1/16} \approx (n/k)^{1/16} = (\log n)^{32/16} = (\log n)^2$. Thus, $f_n \notin P^{(r)}$ unless $r \geq \log n - 32 \log \log n$.   □

The definition of Tardos' function T_m is somewhat complicated. Much more explicit is the logical permanent function $f_m(x) = \bigvee_{\sigma \in S_m} \bigwedge_{i=1}^m x_{i,\sigma(i)}$ of m^2 variables, where S_m is the set of all $m!$ permutations of $1, 2, \dots, m$. This function also belongs to P, but requires monotone circuits of size $m^{\Omega(\log m)}$ (see Theorem 9.38).

For the k-fold extensions f_n of this function the same argument yields $R(f_n) = \Omega(\log n)$.

The message of Theorem 10.21 is that, in the context of the P vs. NP problem, it is important to understand the role of NOT gates when their number r is very close indeed to the Markov–Fisher upper bound of $\log n$.

The function f_n in Theorem 10.21 has many output bits. It would be interesting to prove a similar result for a *boolean* (that is, single output) function.

■ **Research Problem 10.23.** Find an explicit sequence of monotone *boolean* (one-output) functions f_n such that $R(f_n) = \Omega(\log n)$.

Exercises

10.1. Show that $I(\text{INV}_n) = \lceil \log_2(n + 1) \rceil$.

10.2. Show that $I(\neg \oplus_n) = \lfloor \log_2(n + 1) \rfloor$, where $\oplus_n(x) = x_1 \oplus x_2 \oplus \cdots \oplus x_n$.

10.3. Let $n = 2^r - 1$, and consider the set A_{sort} of all $n + 1$ vectors $x \in \{0, 1\}^n$ whose bits are *sorted* in decreasing order $x_1 \geq x_2 \geq \ldots \geq x_n$. Construct a circuit C_n of size $\mathcal{O}(n)$ which has at most r NOT gates and computes the inverter $\text{INV}_n(x)$ for all inputs $x \in A_{sort}$.

Hint: Let $x = (x_1, \ldots, x_n) \in A_{sort}$. Take the middle bit x_m ($m = n/2$) and show that the output of C_n can be obtained from the output of $C_{n/2}$ and the output of $\neg x_m$. For this, observe that $\neg x_m = 1$ implies $\neg x_j = \neg x_m$ for all $j > m$, whereas $\neg x_m = 0$ implies $\neg x_j = \neg x_m$ for all $j < m$.

10.4. Let, as before, $I(f, m)$ denote the minimum number r such that f can be computed by a *nondeterministic* circuit over $\{\wedge, \vee, \neg\}$ with r negations and at most m guess inputs (see Remark 10.12). Prove that

$$I(f, m) \geq \lceil \log_2(d(f)/2^m + 1) \rceil.$$

Hint: Induction on m, the basis case being Markov's theorem. For the induction step, let $C(x, y)$ be a nondeterministic circuit with m guess bits $y = (y_1, \ldots, y_m)$ and r negations computing f. Take a chain $X = \{x^1 < x^2 < \ldots, x^k\}$ with $d_X(f) = d(f)$, and let $I \subseteq \{1, \ldots, k\}$ be the set of jump-down positions of f along this chain. For each $i \in I$ there must exist a setting $y^i \in \{0, 1\}^m$ of values to the guess bits such that $C(x^i, y^i) = 1$, whereas $C(x^{i+1}, y) = 0$ for *all* settings y. Look at the last, m-th position y_m^i of vectors $y^i, i \in I$. If $y_m^i = 0$ for at least half of these vectors, then fix the last guess bit of $C(x, y)$ to 0; otherwise, fix it to 1. Let f' be a boolean function computed by the resulting nondeterministic circuit with $m - 1$ guessing bits. Argue that $d(f') \geq d_{f'}(X) \geq \lceil d(f)/2 \rceil$, and use the induction hypothesis.

10.5. Prove that $I(f, m) \leq \max\{I(f) - m, 0\} + 1 \leq \lceil \log_2(d(f)/2^m + 1) \rceil + 1$.

Hint: By Markov's theorem, there is a *deterministic* circuit C which computes f and contains $I(f)$ NOT gates $N_1, \ldots, N_{I(f)}$. Let i_k and o_k be the input and the output of N_k, respectively. Let z be the output of C. Use m nondeterministic guess inputs $y_1, \ldots, y_m$ to guess the outputs of the first m NOT gates and one additional NOT gate to guarantee correctness of the guess. Compute $z \wedge (\wedge_{k=1}^{m}(i_k \vee o_k)) \wedge (\neg \vee_{k=1}^{m}(i_k \vee o_k))$ as the output of the new (nondeterministic) circuit. Show that $C'(x, y) \neq 0$ only if $i_k = \neg y_k$ for all $k = 1, \ldots, m$. In this case, $z = f$ since $o_k = y_k = \neg i_k$ for all k.

Chapter Notes

Besides the results described above, the question about the power of NOT gates was considered by many authors. In particular, Okolnishnikova (1982) and Ajtai and Gurevich (1987) showed that there exists monotone functions that can be computed by polynomial-size, constant-depth circuits with unbounded-fanin gates, but cannot be computed by *monotone*, polynomial-size, constant depth circuits. Moreover, it was shown by Santha and Wilson (1993) that in the class of constant-depth circuits we need much more than $\lceil \log(n + 1) \rceil$ negations: there is a (multi-output) function computable in constant depth that cannot be computed in constant depth with $o(n/\log^{1+\epsilon} n)$ negations. (Note that this result does not contradict with the Markov–Fischer upper bound: their simulation requires logarithmic depth.) Another line of research was to restrict the *use* of NOT gates. For circuits of logarithmic depth, a lower bound $R(f) = \Omega(n)$ was proved by Raz and Wigderson (1989) under the restriction that all the negations are placed on the input variables: there is an explicit monotone function (corresponding to the connectivity problem for graphs) that can be computed by polynomial-size, depth $O(\log^2 n)$ circuits, but cannot be computed by polynomial-size, depth $k \log n$ circuits using only $o(n/2^k)$ negated variables.

Part IV
Bounded Depth Circuits

Chapter 11
Depth-3 Circuits

We consider boolean circuits with unbounded-fanin AND and OR gates. Inputs are variables and their negations. Conjunctive and disjunctive normal forms are such circuits of depth two, and exponential lower bounds for them are easy to prove. For example, any depth-2 circuit computing the parity function $x_1 \oplus x_2 \oplus \cdots \oplus x_n$ must have 2^{n-1} gates. The situation with depth-3 circuits is much more complicated—this is the first nontrivial case.

11.1 Why is Depth 3 Interesting?

A Π_3 *circuit* is a circuit of depth 3 whose gates are arranged in three layers: AND gate at the top of the circuit (this is the output gate), OR gates on the next (middle) layer, and AND gates on the bottom (next to the inputs) layer (see Fig. 11.1). Inputs are variables and their negations. As before, a circuit is a formula if each its gate has fanout at most 1. Thus, Π_3 formulas are just ANDs of DNFs. A Σ_3 *circuit* is defined dually by interchanging the OR and AND gates. Thus, a Σ_3 formula is just an OR of DNFs. The *size* of a circuit is the total number of gates in it.

There are several methods for proving strong lower bounds for depth-3 circuits, and even for depth-d circuits with an arbitrary constant d. We will discuss these methods in this and the next chapter. For depth d the obtained lower bounds are exponential in $n^{1/(d-1)}$; for depth 3 this is exponential in $\sqrt{n}$. However, these results do not solve the problem completely, because most boolean functions require much larger circuits.

Namely, let $C_d(n)$ be the Shannon function for depth-d circuits, that is, the smallest number t such that every boolean function can be computed by a depth-d circuit containing t gates. Let also $L_d(n)$ be the Shannon function for depth-d *formulas*; here all gates have fanout at most 1, and we count the leaves, not the gates. It can be easily shown that $L_2(n) = n2^n$ (see Exercise 11.1). Lupanov (1961, 1977) proved that, for boolean functions of maximal circuit complexity, depth-3 circuits are already as powerful as circuits of any fixed depth: for $d \geq 3$ we have

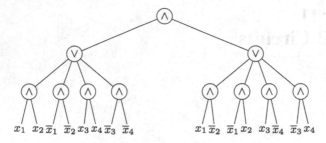

Fig. 11.1 A Π_3 formula for parity $x_1 \oplus x_2 \oplus x_3 \oplus x_4$ of $n = 4$ variables

$$L_d(n) \sim L_3(n) \sim 2^n / \log n \quad \text{and} \quad C_d(n) \sim C_3(n) \sim 2^n / n.$$

Thus, there exist boolean functions that require far more than $2^{\sqrt{n}}$ gates in depth-d circuits, and it would be interesting to exhibit an *explicit* function requiring more than this number of gates.

Even to find an explicit boolean function f of n variables such that any depth-3 circuit for f requires $2^{g(n)}$ gates, for some $g(n) = \omega(n/\log\log n)$, is an important open problem. Namely, this would give the first super-linear lower bound on the size of log-depth circuits with NOT and fanin-2 AND and OR gates, thus resolving an old problem in circuit complexity. We explain this implication next.

A *binary circuit* is a circuit in which all boolean functions of at most two variables can be used as gates. Let $\mathrm{NC}^1_{\mathrm{lin}}$ denote[1] the set of all boolean functions $f_n(x_1, \ldots, x_n)$ for which there exist constants $c_1, c_2 > 0$ such that f_n can be computed by a binary circuit of depth $c_1 \log n$ and size $c_2 n$.

Fix an arbitrarily small constant $\epsilon > 0$, and let $\Sigma_3(f_n)$ denote the smallest number t such that f_n can be written as a sum of t CNFs each with at most $2^{n^{\epsilon}}$ clauses. Note that the top gate is now a *sum* gate (over the reals), not just an OR gate. Thus, what we obtain is a restricted version of a Σ_3 circuit: the circuit is a formula (all gates have fanout 1) and, for every input vector, at most one AND gate on the middle layer is allowed to output 1. Formally $\Sigma_3(f_n)$ depends on ϵ; however, we suppress this for notational convenience.

An important consequence of Lemma 1.4 is that any log-depth circuit of linear size can be reduced to a Σ_3 circuit of moderate fanin of middle layer gates and not too large fanin of the top gate.

Lemma 11.1. (Valiant 1983) *If $f_n \in \mathrm{NC}^1_{\mathrm{lin}}$ then $\log \Sigma_3(f_n) = \mathcal{O}(n/\log\log n)$.*

---

[1] Usually, the nonuniform class NC^k denotes the class of all boolean functions computable by binary circuits of depth $d = \mathcal{O}(\log^k n)$ and polynomial size. Thus, superscript "1" tells us that we are dealing with log-depth circuits, and subscript "lin" tells that we only allow linear size. The acronym "NC" stands for "Nick's Class" and was suggested by Stephen Cook after Nick Pippenger for his research on circuits with polylogarithmic depth and polynomial size.

Proof. The idea is to decompose a given circuit into subcircuits of depth $d \leq \epsilon \log n$; by Lemma 1.4, this can be done by removing a relatively small number of wires. Since each gate has fanin at most 2, each subcircuit can depend on at most $2^d = n^\epsilon$ its inputs. We can thus write each subcircuit as a CNF with at most 2^{n^ϵ} clauses. It then remains to combine these CNFs into a depth-3 formula computing the original function.

To be more precise, take a circuit C of depth $c_1 \log n$ with $c_2 n$ fanin-2 gates. Hence, the circuit has at most $S \leq 2c_2 n$ wires. We are going to apply Lemma 1.4 which states the following. Let $d = 2^k$ and $1 \leq r \leq k$ be integers. In any directed graph with S edges and depth d it is possible to remove rS/k edges so that the depth of the resulting graph does not exceed $d/2^r$.

Now apply this lemma with k about $\log(c_1 \log n)$ and r about $\log(c_1/\epsilon)$ (a constant). This gives us a set E with $|E| \leq Sr/k = \mathcal{O}(n/\log\log n)$ wires whose removal leaves us with a circuit of depth at most $d = 2^{-r} \cdot c_1 \log n = \epsilon \log n$.

Take a set of new variables $y = (y_e \mid e \in E)$, one for each cut wire. For each such wire $e = (u, v) \in E$, let C_e be the subcircuit of C whose output gate is u. Each subcircuit C_e depends on some x-variables (inputs of the original circuit) and some y-variables (variables attached to removed wires). Moreover, each subcircuit C_e depends on at most $2^d = n^\epsilon$ variables because each of these subcircuits has depth at most $\epsilon \log n$, and each gate has fanin at most 2. Hence, the test $y_e = C_e(x, y)$ can be written as a CNF $\phi_e(x, y)$ with at most $2^{2^d} = 2^{n^\epsilon}$ clauses. Consider the CNF

$$\psi(x, y) = \phi_0(x, y) \wedge \bigwedge_{e \in E} \phi_e(x, y),$$

where ϕ_0 is the CNF of the last subcircuit, rooted in the output gate of the whole circuit. The CNF ψ has $(|E| + 1)2^{n^\epsilon}$ clauses, and for every assignment $\alpha = (\alpha_e \mid e \in E)$ in $\{0, 1\}^{|E|}$, we have that $\psi(x, \alpha) = 1$ iff $C(x) = 1$ and the computation of C on input x is consistent with the values assigned to cut wires by α. Since the computation of C on a given vector x cannot be consistent with two assignments $\alpha_1 \neq \alpha_2$, the function computed by our circuit C can be written as a sum $C(x) = \sum_\alpha \psi(x, \alpha)$, over all $\alpha \in \{0, 1\}^{|E|}$, of $s = 2^{|E|}$ CNFs, each with at most $(|E|+1)2^{n^\epsilon}$ clauses. $\square$

■ **Research Problem 11.2.** Exhibit an explicit boolean function of n variables requiring depth-3 circuits of size $2^{\omega(n/\log\log n)}$.

The best we can do so far are lower bounds of the form $2^{\Omega(\sqrt{n})}$. The only known strongly exponential lower bounds were obtained by Paturi et al. (2000) under the restriction that the bottom OR gates have fanin 2, that is, when the circuit is just an OR of 2-CNFs. Currently known lower-bound techniques seem incapable of providing a lower bound better than $2^{\min\{k, n/k\}}$ on the number of gates, where k is the bottom fanin.

11.2   An Easy Lower Bound for Parity

A binary vector is *odd* if it has an odd number of 1s; otherwise the vector is *even*. A parity function is a boolean function $x_1 \oplus x_2 \oplus \cdots \oplus x_n$ which accepts all odd vectors and rejects all even vectors. Recall that a formula is a circuit in which all gates have fanout at most 1. The top fanin is the fanin of the output gate.

Theorem 11.3. (Tsai 2001) *Every Π_3 formula of top fanin t computing $x_1 \oplus x_2 \oplus \cdots \oplus x_n$ requires at least $t2^{(n-1)/t}$ AND gates on the bottom layer.*

Proof. Let s_i be the fanin of the i-th OR gate on the middle layer. The ANDs at bottom layer can be labeled with (i,j) for $1 \le i \le t$ and $1 \le j \le s_i$ (see Fig. 11.1). Let $h_{i,j}$ denote the (i,j)-th AND. Then the circuit computes the function $\bigwedge_{i=1}^{t} \bigvee_{j=1}^{s_i} h_{i,j}$. By the distributive rule $x \wedge (y \vee z) = (x \wedge y) \vee (x \wedge z)$, this is an OR of ANDs of the form $H = h_{1,j_1} \wedge h_{2,j_2} \wedge \cdots \wedge h_{t,j_t}$. We call these "big" ANDs H the *monomials* produced by the circuit. We claim that: *each monomial H accepts at most one odd vector.* To show this, say that a variable x_i is *seen* by a gate, if either x_i or $\overline{x}_i$ is an input to this gate.

Case 1: Each of n variables is seen by at least one of $h_{1,j_1}, h_{2,j_2}, \ldots, h_{t,j_t}$. In this case, H is a (possibly inconsistent) product of *all* n variables, and hence, can accept at most one vector.

Case 2: Some variable x_i is seen by *none* of the gates $h_{1,j_1}, h_{2,j_2}, \ldots, h_{t,j_t}$. We claim that in this case $H^{-1}(1) = \emptyset$. Indeed, if the set $H^{-1}(1)$ of accepted inputs is nonempty, that is, if the monomial H contains no variable together with its negation, then $H^{-1}(1)$ must contain a pair of two vectors that only differ in the i-th position. But this is impossible, since one of these two vectors must be even, and the circuit would wrongly accept it.

Hence, we have $s_1 s_2 \cdots s_t$ monomials H, and each of them can accept at most one odd vector. Since we have 2^{n-1} odd vectors, this implies $s_1 s_2 \cdots s_t \ge 2^{n-1}$. Since our circuit is a formula, the total number of AND gates on the bottom layer is $s_1 + \cdots + s_t$. Using the arithmetic-geometric mean inequality, we can conclude that $s_1 + \cdots + s_t \ge t(s_1 s_2 \cdots s_t)^{1/t} \ge t2^{(n-1)/t}$. $\square$

11.3   The Method of Finite Limits

The above argument only gives nontrivial lower bounds for circuits with small top fanin, much smaller than n. We now describe another, less-trivial argument which works for circuits with arbitrary top fanin. This approach, suggested by Håstad et al. (1995), is based on so-called "finite limits".

Definition 11.4. (Finite limits) A vector $y \in \{0,1\}^n$ is a *k-limit* for a set of vectors $B \subseteq \{0,1\}^n$ if for every k-element subset $S \subseteq \{1,\ldots,n\}$ of positions there exists a vector $x \in B$ such that

$$y \neq x \text{ but } y_i = x_i \text{ for all } i \in S.$$

This concept, suggested by Sipser (1985), captures the following "information bottleneck": if y does not belong to B but is a k-limit for B then the fact that $y \notin B$ cannot be detected by looking at k or fewer bits of y. If a k-limit y for B satisfies the stronger condition

$$y \neq x \text{ and } y \leq x \text{ but } y_i = x_i \text{ for all } i \in S,$$

then we call y a *lower k-limit* for B.

The following lemma reduces the lower bounds problem for depth-3 circuits to a purely combinatorial problem about finite limits. We say that a circuit C *separates* a pair $A, B \subseteq \{0, 1\}^n$, $A \cap B = \emptyset$ if

$$C(x) = \begin{cases} 1 & \text{for } x \in A, \\ 0 & \text{for } x \in B. \end{cases}$$

We also say that a circuit has *bottom neg-fanin* k if each gate on the bottom (next to the inputs) level at most k *negated* input-variables as inputs; the total number of inputs to the gate may be arbitrary.

Lemma 11.5. (Limits and circuit size) *If every $1/\ell$ fraction of vectors in B has a lower k-limit in A, then every Π_3 circuit of bottom neg-fanin k separating (A, B) must have top fanin larger than ℓ.*

Proof. Suppose, for the sake of contradiction, that (A, B) can still be separated by a Π_3 circuit of bottom neg-fanin and top fanin ℓ. Since the last (top) gate is an AND gate, some of the OR gates g on the middle layer must separate a pair (A, B') for some $B' \subseteq B$ of size $|B'| \geq |B|/\ell$. By our assumption, the set A must contain a vector a which is a lower k-limit for the set B'. Hence,

$$g(a) = 1, \text{ but } g(b) = 0 \text{ for all } b \in B'.$$

To obtain the desired contradiction, we will show that the gate g, and hence, the whole circuit C, is forced to (incorrectly) reject the limit a.

Take an arbitrary AND gate h on the bottom layer feeding in g, and let S be the corresponding set of negated input variables to h:

$$h(x) = \bigwedge_{i \in S} \overline{x}_i \wedge \bigwedge_{j \in T} x_j.$$

Since $|S| \leq k$ and since a is a lower k-limit for B', we know that there must exist a vector $b = b_S$ in B' such that $a \leq b$ and $a_i = b_i$ for all $i \in S$. Since g is an OR gate and since g must reject all vectors in B', we also know that $h(b) = 0$. If some *negated* variable feeding into h computes 0 on b, then it does the same on a

(since a coincides with b on all positions in S), and hence $h(a) = 0$. Otherwise, the 0 is produced on b by some *non-negated* variable. Since $a \leq b$, this variable must produce 0 on a as well, and hence $h(a) = 0$. Since this holds for every AND gate h feeding into the OR gate g, this implies that the gate g must (incorrectly) reject a, a contradiction.   □

In order to show that a given boolean function f cannot be computed by a Π_3 circuit with fewer than ℓ gates, we can now argue as follows:

1. Assume that f can be computed by such a circuit.
2. Assign some variables of f to constants in order to reduce the bottom fanin of the circuit to k.
3. Choose appropriate subsets $A \subseteq f^{-1}(1)$ and $B \subseteq f^{-1}(0)$, and show that every subset $B' \subseteq B$ of size $|B'| \geq |B|/\ell$ has a lower k-limit $y \in A$.
4. Apply Lemma 11.5 to get a contradiction.

The bottom fanin can be reduced using the following simple fact.

Proposition 11.6. *Let $\mathcal{F}$ be a family of subsets of $[n]$, each of cardinality more than k. If*

$$|\mathcal{F}| \leq \left(\frac{n}{m}\right)^{k/2} \tag{11.1}$$

then some subset T of $[n]$ of size $|T| = n - m$ intersects all members of $\mathcal{F}$.

Proof. We construct the desired set T via the following "greedy" procedure. Since each set in $\mathcal{F}$ has more than k elements, and since we only have n elements in total, at least one element x_1 must belong to at least k/n fraction of sets in $\mathcal{F}$. Include such an element x_1 in T, remove all sets from $\mathcal{F}$ containing x_1 (hence, at most a $(1 - k/n)$ fraction of sets in $\mathcal{F}$ remains), and repeat the procedure with the remaining sub-family of $\mathcal{F}$, etc. Our goal is to show that, if the number $|\mathcal{F}|$ of sets in the original family $\mathcal{F}$ satisfied (11.1), then after $n - m$ steps all the sets of $\mathcal{F}$ will be removed.

The sub-family resulting after $n - m$ steps has at most $\alpha|\mathcal{F}|$ sets, where

$$\alpha = \left(1 - \frac{k}{n}\right)\left(1 - \frac{k}{n-1}\right)\cdots\left(1 - \frac{k}{m+1}\right).$$

Using the estimate $1 + x \leq e^x$ and known estimates $H_n = \ln n + \gamma_n$ on harmonic series $H_n = 1 + 1/2 + /3 + \cdots + 1/n$ with $\frac{1}{2} < \gamma_n < \frac{2}{3}$, we obtain that

$$\alpha \leq \exp\left(-\frac{k}{n} - \frac{k}{n-1} - \cdots - \frac{k}{m+1}\right)$$

$$= e^{-k(H_n - H_m)} \leq e^{-k(\ln n - \ln m - 1/6)}$$

$$\leq \left(\frac{n}{2m}\right)^{-k} \leq \left(\frac{n}{m}\right)^{-k/2}.$$

Thus, after $n - m$ steps no sets will remain, as desired.   □

Given a Π_3 circuit we will want to set a small subset of the variables to 1 so that the resulting circuit has neg-fanin k. We will use Proposition 11.6 to find which variables to set to 1.

The next task—forcing a k-limit—depends on which boolean function we are dealing with. To demonstrate how this can be done, let us consider the Majority function $\mathrm{Maj}_n(x_1, \ldots, x_n)$, which accepts an input vector x iff it contains at least as many 1s as 0s.

11.4   A Lower Bound for Majority

To handle the case of the majority function, we need the following "limit lemma" for threshold functions.

Lemma 11.7. *Let* $B \subseteq \{0,1\}^n$ *be a set of vectors, each with exactly* r *ones. If* $|B| > k^r$ *then there is a lower k-limit y for B with fewer than r ones.*

Proof. We use induction on the number r of ones in vectors of B. If $r = 1$ and $|B| \geq k + 1$ then $\mathbf{0} = (0, \ldots, 0)$ is the desired k-limit for B. Suppose now that the lemma holds for all $r' \leq r - 1$ and prove it for r. So, take a set B of $|B| > k^r$ vectors each with r ones. If $\mathbf{0}$ is a k-limit for B, then we are done.

Otherwise, by the definition of a k-limit, there must be a set of k coordinates such that every vector in B has at least one 1 among these coordinates. Hence, at least a $1/k$ fraction of vectors in B must have a 1 in some, say i-th, coordinate. Replace in all these vectors the i-th 1 by 0, and let B' be the resulting set of vectors. Since each vector in B has exactly $r - 1$ ones and we have $|B'| \geq |B|/k > k^{r-1}$ vectors in total, the induction hypothesis gives us a vector y with fewer than $r - 1$ ones which is a lower k-limit for B'. The i-th coordinate of y is 0. Replacing this coordinate by 1 we obtain a vector y' with at most $r - 1$ ones; y' is the desired lower k-limit for B.   □

Theorem 11.8. (Håstad et al. 1995) *Any depth-3 circuit computing the majority function* Maj_n *has size at least* $2^{\Omega(\sqrt{n})}$.

Proof. Let ℓ be the minimal size of a depth-3 circuit computing $\neg\mathrm{Maj}_n$, the negation of majority, and hence, the minimal size of a depth-3 circuit computing Maj_n itself. Since $\neg\mathrm{Maj}_n$ is self-dual (that is, complementing the output and all inputs does not change the function), we can w.l.o.g. assume that we have a Π_3 circuit.

Let $k \leq n$ and $r \leq n/2$ be parameters (to be specified later). Set $m := n/2 + r$ and assume that the size ℓ of our circuit satisfies the inequality

$$\ell \leq \left(\frac{n}{m}\right)^{k/2}. \tag{11.2}$$

For each bottom AND gate that has $\geq k + 1$ negated variables, that set of variables will be a set in the family $\mathcal{F}$. Since there are at most ℓ such gates, $|\mathcal{F}| \leq \ell$. Hence, by Proposition 11.6, there will be $|T| = n - m$ variables that intersect all the members of $\mathcal{F}$. If we set all the variables in T to 1, this will kill off (evaluate to 0) all bottom AND gates that had $\geq k + 1$ negated variables. So, the resulting Π_3 circuit has bottom neg-fanin at most k (that is, at most k negated variables enter each bottom AND gate). This circuit computes a boolean function $f : \{0, 1\}^m \to \{0, 1\}$ of m variables such that $f(x) = 1$ iff x has fewer than $n/2 - (n - m) = r$ ones. Hence, the new circuit separates the pair (A, B) of sets

$$A = \{\text{all vectors in } \{0, 1\}^m \text{ with fewer than } r \text{ ones}\}$$

and
$$B = \{\text{all vectors in } \{0, 1\}^m \text{ with precisely } r \text{ ones}\}.$$

Since the new circuit has size at most ℓ and its bottom neg-fanin is at most k, Lemma 11.5 implies that no $1/\ell$ fraction of vectors in B can have a lower k-limit in A. Together with Lemma 11.7, this implies that $|B|/\ell = \binom{m}{r}/\ell$ cannot be larger than k^r. Hence,

$$\ell \geq \binom{m}{r} \cdot k^{-r} \geq \left(\frac{m}{kr}\right)^r.$$

By our assumption (11.2), this lower bound holds for any parameters k, r and $m = n/2 + r$ satisfying

$$\left(\frac{m}{kr}\right)^r \leq \left(\frac{n}{m}\right)^{k/2}. \tag{11.3}$$

To ensure this, we can take, say, $k = 2\sqrt{m}$ and $r = \sqrt{m}/4$. Under this choice, (11.3) is fulfilled, and we obtain the desired lower bound

$$\ell \geq \left(\frac{m}{kr}\right)^r = 2^{\Omega(r)} = 2^{\Omega(\sqrt{n})}. \qquad \square$$

Remark 11.9. It would be interesting to extend the lower bounds argument based on finite limits to circuits of larger depths. For this, however, we need stronger "limit lemmas" than, say, Lemma 11.7. To see the difficulty, recall that in the case of depth-3 circuits it was enough to show (see Lemma 11.5) that every sufficiently large subset B of $f^{-1}(1)$ has a k-limit in the entire set $f^{-1}(0)$ of rejected input vectors. In the case of depth-4 circuits we need a stronger statement that every sufficiently large subset B of $f^{-1}(1)$ has a k-limit in *every* sufficiently large subset A of $f^{-1}(0)$.

11.5 NP $\neq$ co-NP for Depth-3 Circuits

In this section we will exhibit a boolean function f of n variables such that f has a Σ_3 circuit of size $\mathcal{O}(n)$ but its complement $\neg f$ requires Σ_3 circuits of size $2^{\Omega(\sqrt{n})}$.

Since Σ_3 circuits have an OR gate on the top, they constitute a nondeterministic model of computation: guess a CNF and evaluate it. For an added thrill, we can consider the "depth-3 version" of the class NP to consist of all boolean functions computable by Σ_3 circuits of polynomial size. In these terms, we are going to prove that NP ≠ co-NP in the class of depth-3 circuits.

Note that we cannot take the majority function Maj_n for this purpose just because it is self-dual:

$$\neg\text{Maj}_n(\neg x_1, \ldots, \neg x_n) = \text{Maj}_n(x_1, \ldots, x_n).$$

Hence, by Theorem 11.8, both Maj_n and $\neg\text{Maj}_n$ require Σ_3 circuits of exponential size. We therefore must use another function.

So, let $f_{s,m}$ be the boolean function with $n = 2sm$ variables defined by

$$f_{s,m}(x, y) = \bigvee_{i=1}^{s} \bigwedge_{j=1}^{m} (\neg x_{i,j} \vee \neg y_{i,j}). \tag{11.4}$$

This is an important function, known as the *Iterated Disjointness function*. function. The function takes two sequences $x = (x_1, \ldots, x_s)$ and $y = (y_1, \ldots, y_s)$ of subsets of $[m] = \{1, \ldots, m\}$, and accepts the pair (x, y) iff $x_i \cap y_i = \emptyset$ for at least one $i \in [s]$.

It is clear (from its definition) that $f_{s,m}$ can be computed by a Σ_3 circuit of size $1 + s(m + 1) = \mathcal{O}(n)$. We shall show that for $s = m = \sqrt{n}$, this function requires Π_3 circuits of size $2^{\Omega(\sqrt{n})}$, implying that any Σ_3 circuit for its negation $\neg f_{s,m}$ requires this size.

Theorem 11.10. (Håstad et al. 1995) *Any Π_3 circuit computing $S_{\sqrt{n},\sqrt{n}}$ has size at least $2^{\Omega(\sqrt{n})}$.*

By a result of Klawe et al. (1984), the function $S_{\sqrt{n},\sqrt{n}}$ has Π_3 circuits of size $2^{O(\sqrt{n})}$, thus the bound is optimal.

Proof. We first prove a lower bound for the subfunction $g_{s,m}(x) = f_{s,m}(x, 1)$ of $f_{s,m}(x, y)$ obtained by setting all y-variables to 1. $\qquad\qquad\square$

Claim 11.11. *If $g_{s,m}(x) = \bigvee_{i=1}^{s} \bigwedge_{j=1}^{m} \neg x_{i,j}$ is computed by a Π_3 circuit of size ℓ and bottom neg-fanin k, then $\ell \geq (m/k)^s$.*

Proof. Any circuit for $g_{s,m}$ must separate the pair (A, B) where $A \subseteq \{0, 1\}^{sm}$ is the set of all vectors with at most $s - 1$ ones and B is the set of m^s vectors with exactly s ones killing all ANDs in f. Now assume that $\ell < (m/k)^s$. Then, by Lemma 11.5, no subset $B' \subseteq B$ of size $|B'| \geq |B|/\ell > m^s/(m/k)^s = k^s$ can have a lower k-limit in A, a contradiction with Lemma 11.7. $\qquad\qquad\square$

The theorem now follows from the following claim by taking $s = m = \sqrt{n}$ and $k = \sqrt{n}/2$.

Claim 11.12. For any $k \leq sm$, any Π_3 circuit computing $f_{s,m}$ has size at least $\min\{2^k, (m/k)^s\}$.

Proof. Take a Π_3 circuit computing $f_{s,m}(x, y)$; let ℓ be its size, and assume that $\ell \leq 2^k$. We claim that then there exists a setting of constants to variables such that the resulting circuit has bottom fanin k and computes $f_{s,m}$. Together with Claim 11.11, this claim implies that either $\ell \geq 2^k$ or $\ell \geq (m/k)^s$, and we are done. So it remains to prove the claim.

The most natural way is to randomly set one variable from each pair $x_{i,j}, y_{i,j}$ to 1. Any such setting will leave us with a circuit computing $g_{s,m}$. It remains therefore to show that at least one of such settings will leave no bottom AND gate with more than k negated inputs.

If a bottom AND gate contains both $x_{i,j}$ and $y_{i,j}$ negatively for some i, j then it is always reduced to 0. Otherwise, such an AND gate with $> k$ negated inputs is *not* reduced to 0 with probability $\leq 2^{-(k+1)}$. Since we have at most $\ell \leq 2^k$ such AND gates, the probability that some of them will not be reduced to 0 does not exceed $\ell \cdot 2^{-(k+1)} \leq 1/2$. This, in particular, means that such a setting of constants exists. $\square$

Remark 11.13. Note that every Π_3 circuit is also a Σ_4 circuit (with the top OR gate missing). The tradeoff between Σ_3 and Σ_4 circuits was proved earlier by Håstad (1989), who gave a lower bound $2^{\Omega(n^{1/6}/\sqrt{\log n})}$ for the size of Σ_3 circuits computing a function which has a Σ_4 circuit of size $\mathcal{O}(n)$. A tradeoff between Σ_3 and Π_3 formulas was established by Shumilina (1987) using the threshold-2 function Th_2^n. This function has a Σ_3 formula of leafsize (number of input literals) $\mathcal{O}(n \log n)$ (see Exercise 11.7), but requires Π_3 formulas of leafsize $\Omega(n^{3/2})$. Actually, Shumilina (1987) established an *exact* complexity of Th_2^n in the class of Π_3 formulas. Namely, the minimum leafsize of a Π_3 formula computing Th_2^n is $2n(k-1) + m(n-k^2)$ if $n \leq km$, and is $2n(m-1) - k(m^2 - n)$ if $n \geq km$, where $k = \lfloor \sqrt{n} \rfloor$ and $m = \lceil \sqrt{n} \rceil$.

Remark 11.14. Recently, Razborov and Sherstov (2010) showed that the Iterated Disjointness function (11.4) is hard in yet another respect: if $A = (a_{x,y})$ in an $n \times n$ ± 1 matrix with $n = m^3$ and $a_{x,y} = 1 - 2 \cdot f_{m,m^2}(x, y)$, then A has signum rank $2^{\Omega(n^{1/3})}$ (cf. Sect. 4.11). Recall that the *signum rank* of a real matrix A with no zero entries is the least rank of a matrix $B = (b_{x,y})$ such that $a_{x,y} \cdot b_{x,y} > 0$ for all x, y. This result resolved an old problem about the power of probabilistic unbounded-error communication complexity; we have already mentioned this in Sect. 4.11.

The strongest known lower bounds for depth-3 circuits computing explicit boolean functions of n variables have the form $2^{\Omega(\sqrt{n})}$. We have seen how such a lower bound can be derived for the majority function. To break this "square root barrier" is an important open problem. It is especially interesting in view of possible consequences for log-depth circuits (see Lemma 11.1).

■ **Research Problem 11.15.** Prove an explicit lower bound $2^{\omega(\sqrt{n})}$ for Σ_3 circuits.

To get such lower bounds, one could try to use the graph-theoretic approach introduced in Sect. 1.7.

11.6 Graph Theoretic Lower Bounds

The idea of graph complexity (introduced in Sect. 1.7) is to reduce the lower bounds problem for boolean functions to that for bipartite graphs. Given a graph $G = (V, E)$, we associate a boolean variable x_v with each of its vertices $v \in V$. A boolean circuit $F(x)$ of these variables *represents* the graph G if it accepts all edges and rejects all non-edges. On other subsets of vertices the circuit F may output arbitrary values. That is, we only require that

the function F acts correctly on input vectors with exactly two ones!

On inputs $x \in \{0, 1\}^{|V|}$ with $\sum_{v \in V} x_v \neq 2$ the function may output any value in $\{0, 1\}$. Only if x contains exactly two 1s in, say, positions u and v, the circuit must output $F(x) = 1$ if and only if u and v are adjacent in G. In particular, every graph $G = (V, E)$ on $|V| = n$ vertices can be represented by the following monotone Σ_3 formula with at most $2n$ gates:

$$F(x) = \bigvee_{u \in S} x_u \wedge \Big(\bigvee_{v:uv \in E} x_v \Big), \tag{11.5}$$

where $S \subseteq V$ is an arbitrary vertex-cover of G, that is, a set of vertices such that every edge of G has is endpoint in S.

As already mentioned in Sect. 1.7, our motivation to consider graph representation is that even moderate lower bounds for the monotone complexity of graphs imply strong lower bounds for the non-monotone circuit complexity of boolean functions.

If G is a bipartite $n \times n$ graph with $n = 2^m$, then we can identify its vertices with vectors in $\{0, 1\}^m$ and consider the *characteristic function* f_G of G defined by: $f_G(u, v) = 1$ iff u and v are adjacent in G.

Now fix an arbitrary model $\mathfrak{M}$ of circuits. A *bottom* gate is a gate whose inputs are only literals (variables and their negations). We only require that bottom gates are either ORs or Parities of their inputs; the remaining (non-bottom) gates may be arbitrary, depending on the circuit model one deals with. For a boolean function f and a graph G, let $L_{\mathfrak{M}}(f)$ denote the smallest size of a circuit in $\mathfrak{M}$ computing f, and $L_{\mathfrak{M}}^+(G)$ the smallest size of a positive circuit in $\mathfrak{M}$ representing G; being positive here means having no negated literals as inputs.

The Magnification Lemma (Lemma 1.32 in Sect. 1.7) immediately yields:

Proposition 11.16. *For every bipartite graph G, $L_{\mathfrak{M}}(f_G) \geq L_{\mathfrak{M}}^+(G)$.*

Thus, any lower bound $L_{\mathfrak{M}}^{+}(G) \geq n^{\alpha}$ for an $n \times n$ graph G immediately gives a lower bound $L_{\mathfrak{M}}(f_G) \geq 2^{\alpha m}$ for its characteristic function f_G; recall that f_G has only $2m$ variables. Hence, any lower bound $L_{\mathfrak{M}}(G) \geq n^{\alpha}$ for an explicit graph G with $\alpha = \omega(\ln \ln n / \ln n)$ would give us a super-polynomial lower bound for an explicit boolean function!

To start with, let us consider the simplest model—that of CNFs. Each such circuit is an AND of ORs of input literals. In the case of graphs, we only need to consider *monotone* CNFs

$$F(x) = \left(\bigvee_{v \in S_1} x_v \right) \wedge \left(\bigvee_{v \in S_2} x_v \right) \wedge \cdots \wedge \left(\bigvee_{v \in S_r} x_v \right).$$

Such a CNF rejects a pair $\{u, v\}$ of vertices iff at least one of the complements $I_i = \overline{S}_i$ covers this pair, that is, contains both endpoints u and v. Hence, F represents a graph G iff $I_1, \ldots, I_r$ are independent sets of G whose union covers all non-edges of G. Thus, if $\mathrm{cnf}(G)$ denotes the minimum number of clauses in a monotone CNF representing the graph G, and if A_G is the adjacency matrix of G, then

$$\mathrm{cnf}(G) = \mathrm{Cov}(\overline{A}_G), \tag{11.6}$$

where $\overline{A}$ is the complement of A, and $\mathrm{Cov}(A)$ is the smallest number of all-1 submatrices of A covering all its ones. This immediately yields strong lower bounds for many explicit graphs. For example, if G is a (bipartite) complement of an n-to-n matching, then $\overline{A}_G$ is an identity matrix, implying that $\mathrm{cnf}(G) \geq n$. By Proposition 11.16, this implies that the boolean function $f(x, y)$ of $2m$ variables, defined by $f(x, y) = 1$ iff $x \neq y$, requires CNFs with at least 2^m clauses.

Of course, such a lower bound for CNFs is far from being interesting: we already know that, say, the parity of $2m$ variables needs even 2^{2m-1} clauses. Still, strong lower bounds for CNF-size of graphs could imply impressive lower bounds for boolean functions, if we could prove such bounds for graph *properties*.

To illustrate this, let us consider bipartite $K_{2,2}$-free graphs, that is, bipartite graphs without four-cycles. It is conjectured that $K_{2,2}$-freeness of graphs makes them hard for CNFs. The following problem is just Problem 4.9 (in the Exercises of Chap. 4) re-stated in graph-theoretic terms.

■ **Research Problem 11.17.** If G is a $K_{2,2}$-free graph of average degree D, does then $\mathrm{cnf}(G) \geq D^{\Omega(1)}$?

Together with Proposition 11.16, a positive answer to this problem would resolve Problem 11.2. Let us see why this is true.

As we have already mentioned in Sect. 7.2, explicit constructions of dense graphs G_n without four-cycles are known (see Examples 7.4 and 7.5). These graphs are d-regular bipartite $n \times n$ graphs with $d = \Theta(\sqrt{n})$. Now suppose that G_n can be represented by a monotone Σ_3 circuit of size s. Then some subgraph H of G_n with at least $\epsilon n^{3/2}/s$ edges can be represented by a monotone CNF with at most s clauses. The graph H is still $K_{2,2}$-free and has average degree $D \geq \epsilon \sqrt{n}/s$. A positive answer to Problem 11.17 would imply that $s \geq D^\delta \geq (\epsilon\sqrt{n}/s)^\delta$, from which a lower bound $s \geq (\epsilon\sqrt{n})^{\delta/2} = n^{\Omega(1)}$ on the size s follows.

Thus, an affirmative answer to Problem 11.17 would resolve Problem 11.2, and hence, give us an explicit boolean function of m variables that cannot be computed by DeMorgan circuits whose depth is logarithmic in m and size is linear in m. To exhibit such a boolean function has been an open problem in circuit complexity for more than 30 years.

Note that $K_{2,2}$-freeness in Problem 11.17 is not crucial: one can consider any *hereditary* property of graphs, that is, a property which cannot be destroyed by removing edges.

Finally, let us mention that graphs of small degree *have* small circuits, and hence, are "bad candidates" for a strong lower bound. Namely, we have already proved in Sect. 4.2 (see Lemma 4.5) that $\mathrm{Cov}(A) = \mathcal{O}(d \ln |A|)$ for every boolean matrix A, where $|A|$ is the total number of ones in A, and d is the maximal number of zeros in a line (row or column) of B. Thus, (11.6) implies that $\mathrm{cnf}(G) = \mathcal{O}(d \log n)$ for every bipartite $n \times n$ graph G of maximum degree d.

11.7   Depth-2 Circuits and Ramsey Graphs

The problem with using the graph-theoretic approach in boolean function complexity is that "combinatorially complicated" graphs are not necessarily "computationally complicated". To illustrate this, let us consider Ramsey-type graphs.

A graph over vertex set V is said to be a *t-Ramsey graph* if for every $S \subseteq V$ satisfying $|S| = t$, the graph induced by S is neither empty, nor complete. A bipartite graph over vertex sets L and R, $|L| = |R| = n$, is said to be a *t-Ramsey graph* if for every $S \subseteq L$ and $T \subseteq R$ satisfying $|S| = |T| = t$, the bipartite graph induced by S and T is neither empty, nor complete. That is, neither the graph nor its complement contains a complete bipartite $t \times t$ graph $K_{t,t}$. We call a graph (bipartite or not) just a *Ramsey graph* if it is t-Ramsey for $t = 2 \log n$.

A celebrated result of Erdős from 1947 shows that non-bipartite Ramsey graphs exist. Irving (1978) proved that bipartite t-Ramsey graphs exist already for $t = \mathcal{O}(\log n / \log \log n)$. But constructing *explicit* Ramsey graphs is a notoriously hard problem.

The best known explicit construction of non-bipartite t-Ramsey graphs due to Frankl and Wilson only achieves $t = \exp(\sqrt{\log n \log \log n})$. In the bipartite case, even going from $t = n^{1/2}$ to $t = n^\delta$ for an arbitrary constant $\delta > 0$ was only recently obtained by Pudlák and Rödl (2004), Barak et al. (2010), and Ben-Sasson and Zewi (2010). Moreover, these constructions are not explicit in the way that would satisfy Erdős. The constructions are algorithmic: when given a pair of vertices, the algorithm runs in time polynomial in the length of its input and answers whether this pair is an edge of the constructed graph.

In view of these difficulties to construct Ramsey graphs, such graphs might be a promising place to look for explicit functions that require large circuits, but alas, they are not: there *exist* bipartite Ramsey $n \times n$ graphs that can be represented as a Parity of just $2 \log n$ ORs of variables.

To show this, we consider depth-2 circuits whose output gate is a parity gate and bottom (next to the inputs) gates are OR gates; inputs are variables (no negated inputs are allowed). Such a circuit has the form

$$F(x) = \bigoplus_{i=1}^{r} \bigvee_{v \in I_i} x_v \qquad (11.7)$$

Let $r(G)$ denote the smallest number r of OR gates in such a circuit representing the graph G. By letting $S_u = \{i \mid u \notin I_i\}$, we see that vertices u and v are adjacent in G iff $r - |S_u \cap S_v|$ is odd. Thus, the adjacency matrix A_G of G can be represented as a boolean matrix of scalar products of vectors of length r over GF(2). This implies that

$$r(G) \text{ is at least the rank of } A_G \text{ over GF(2)-1.}$$

The *Sylvester graph* is a bipartite $m \times m$ graph H_m with $m = 2^r$ whose vertices are vectors in GF(2)^r. Two vertices are adjacent in H_m iff their scalar product over GF(2) is equal to 1. Note that H_m can be represented by a very small circuit of the form (11.7): $r(H_m) \leq r = \log m$. We will show that, nevertheless, H_m contains a large induced subgraph that is Ramsey.

Theorem 11.18. *There exist bipartite $n \times n$ Ramsey graphs H such that $r(H) \leq 2 \log n$.*

Proof. Let $\mathbb{F} = $ GF(2) and r be a sufficiently large even integer. With every subset $S \subseteq \mathbb{F}^r$ we associate a bipartite graph $H_S \subseteq S \times S$ such that two vertices $u \in S$ and $v \in S$ are adjacent if and only if $\langle u, v \rangle = 1$, where $\langle u, v \rangle$ is the scalar product over $\mathbb{F}$. Thus, $H_m = H_S$ with $S = \mathbb{F}^r$ and $m = 2^r$.

We are now going to show that H_m contains an induced $n \times n$ subgraph H_S with $n = \sqrt{m}$ which is a Ramsey graph. The fact that H_S is an *induced* subgraph implies that (11.7) is also a representation of H_S: just set to 0 all variables x_v with $v \notin S$. Thus, $r(H_S) \leq r(H_m) \leq \log m = 2 \log n$. To prove that such a subgraph exists, we first establish one Ramsey type property of graphs H_S for arbitrary subsets $S \subseteq \mathbb{F}^r$.

Lemma 11.19. (Pudlák and Rödl 2004) *Suppose every vector space $V \subseteq \mathbb{F}^r$ of dimension $\lfloor (r+1)/2 \rfloor$ intersects S in fewer than t elements. Then neither H_S nor the bipartite complement $\overline{H}_S$ contains $K_{t,t}$.*

Proof. The proof is based on the observation that any copy of $K_{t,t}$ in H_S would give us a pair of subsets X and Y of S of size t such that $\langle u, v \rangle = 1$ for all $u \in X$ and $v \in Y$. Viewing the vectors in X as the rows of the coefficient matrix and the vectors in Y as unknowns, we obtain that the sum $\dim(X') + \dim(Y')$ of the dimensions of vector spaces X' and Y', spanned by X and by Y, cannot exceed $r + 1$. Hence, at least one of these dimensions is at most $(r+1)/2$, implying that either $|X' \cap S| < t$

or $|Y' \cap S| < t$. However, this is impossible because both X' and Y' contain subsets X and Y of S of size t. $\qquad\square$

It remains therefore to show that a subset $S \subseteq \mathbb{F}^r$ of size $|S| = 2^{r/2} = \sqrt{m}$ satisfying the condition of Lemma 11.19 exists. We show this by probabilistic arguments.

Let $m = 2^r$ and let $S \subseteq \mathbb{F}^r$ be a random subset where each vector $u \in \mathbb{F}^r$ is included in S independently with probability $p = 2^{1-r/2} = 2/\sqrt{m}$. By Chernoff's inequality, $|S| \geq pm/2 = 2^{r/2}$ with probability at least $1 - e^{-\Omega(pm)} = 1 - o(1)$.

Now let $V \subseteq \mathbb{F}^r$ be a subspace of $\mathbb{F}^r$ of dimension $\lfloor (r+1)/2 \rfloor = r/2$ (remember that r is even). Then $|V| = 2^{r/2} = \sqrt{m}$ and we may expect $p|V| = 2$ elements in $|S \cap V|$. By Chernoff's inequality, $\mathrm{Prob}[|S \cap V| \geq 2c] \leq 2^{-2c}$ holds for any $c > 2e$. The number of vector spaces in $\mathbb{F}^r$ of dimension $r/2$ does not exceed $\binom{r}{r/2} \leq 2^r/\sqrt{r}$. We can therefore take $c = r/2$ and conclude that the set S intersects some $r/2$-dimensional vector space V in $2c = r$ or more elements with probability at most $2^{r-(\log r)/2-r} = r^{-1/2} = o(1)$. Hence, with probability $1-o(1)$ the set S has cardinality at least $2^{r/2}$ and $|S \cap V| < r$ for every $r/2$-dimensional vector space V. Fix such a set S' and take an arbitrary subset $S \subseteq S'$ of cardinality $|S| = 2^{r/2}$. By Lemma 11.19, neither H_S nor $\overline{H}_S$ contains a copy of $K_{r,r}$. $\qquad\square$

We have seen that some "combinatorially complicated" graphs can be represented by very small circuits, even in depth 2. On the other hand, some "combinatorially simple" graphs require large circuits of this type (11.7). This follows from our observation above that $r(G)$ is just the rank of the adjacency matrix A_G of G over GF(2). In particular, if M is an n-matching (a set of n vertex-disjoint edges), then A_M is a permutation matrix with exactly one 1 in each row and column. Since A_M has full rank, we obtain that $r(M) \geq n - 1$.

Remark 11.20. Arora et al. (2009) considered a related question, albeit one outside the graph complexity framework we've been considering. Suppose the characteristic function f_G of a graph G has "low" circuit complexity. What can then be said about the properties of the graph G itself? Let AC^0 be the family of all graphs G on n vertices ($n = 1, 2, \ldots$) whose characteristic functions f_G can be computed by a constant-depth circuit with a polynomial (in the number of variables of f_G) number of NOT and unbounded-fanin AND and OR gates. They observed that Håstad's theorem (Theorem 12.2 in the next chapter) implies that *none* of the graphs in AC^0 is t-Ramsey for t smaller than $\exp(\log n/\mathrm{poly}(\log\log n))$. On the other hand, they show that AC^0 contains good expanders, and that many algorithmic problems on AC^0-graphs are no easier to solve than on general graphs.

11.8 Depth-3 Circuits and Signum Rank

In this section we consider Σ_3 *formulas*, that is, Σ_3 circuits with fanout-1 gates. By the *size* of such a formula we will now mean the number of OR gates on the bottom (next to the inputs) level. We already know that every bipartite $n \times n$ graph can be represented by a Σ_3 formula of size n using formula (11.5). Our goal is to relate the size of Σ_3 formulas representing bipartite graphs to the signum rank of their adjacency matrices. Recall that the *signum rank*, $\operatorname{signrk}(A)$, of a boolean matrix A is the minimum rank, $\operatorname{rk}(M)$, of a real matrix M such that $M[x, y] < 0$ if $A[x, y] = 0$, and $M[x, y] > 0$ if $A[x, y] = 1$.

Theorem 11.21. (Lokam 2003) *Let G be a bipartite $n \times n$ graph, and A its 0-1 adjacency matrix. If G can be represented by a monotone Σ_3 formula of size S, then*

$$\operatorname{signrk}(A) \leq 2^{\mathcal{O}(S^{1/3} \log^{5/3} S)}.$$

Recall that an OR of variables represents a union of stars, that is, the bipartite complement of a complete bipartite graph. Adjacency matrices of complete bipartite graphs are primitive matrices, that is, boolean matrices of rank 1. Hence, if a graph G can be represented by a monotone Σ_3 formula of size S, then its adjacency matrix A can be written as

$$A = \bigvee_{i=1}^{t} \bigwedge_{j=1}^{d_i} \overline{R}_{ij}, \tag{11.8}$$

where $d_1 + \cdots + d_t \leq S$ and $\overline{R}_{ij} = J - R_{ij}$ are complements of primitive matrices R_{ij}; as usually, J stands for the all-1 matrix, and boolean operations on matrices are computed component-wise. Our goal is to upper-bound the signum rank of such matrices A in terms of S. For this, we need one result concerning representations of boolean functions by real polynomials.

Lemma 11.22. *If a boolean matrix $H = \bigvee_{i=1}^{d} R_i$ is an OR of d primitive matrices then, for every $k > 2$, there exists a real matrix M such that $|M[x, y] - H[x, y]| \leq 1/k$ for all entries (x, y), and*

$$\operatorname{rk}(M) \leq d^{\mathcal{O}(\sqrt{d} \log k)}.$$

Proof. Let $f(z_1, \ldots, z_d) = \bigvee_{i=1}^{d} z_i$. By Lemma 2.6, there exists a real polynomial $p(z_1, \ldots, z_d)$ of degree $r \leq c \sqrt{n} \ln k$ approximating $f(z)$ with the factor $1/k$, that is, $|p(z) - f(z)| \leq 1/k$ for all $z \in \{0, 1\}^d$. Syntactically substitute the matrix R_i for z_i in this polynomial, but interpret the product $z_i \cdot z_j$ as an entry-wise product $R_i \circ R_j$ of matrices. Thus, a monomial $z_{i_1} z_{i_2} \cdots z_{i_t}$ is replaced by the rank-1 boolean matrix $R_{i_1} \circ R_{i_2} \circ \cdots \circ R_{i_t}$. The matrix obtained by computing the polynomial $p(R_1, \ldots, R_d)$ in this way gives us the desired matrix M. Since M is a linear combination of rank-1 matrices, one for each of at most $m = \sum_{i=0}^{r} \binom{d}{i} \leq (ed/r)^r \leq d^{\mathcal{O}(r)}$ possible monomials of p, it follows that the rank of M is at most the number m of these monomials, as desired. $\qquad \square$

Now let A be a boolean matrix of the form (11.8), and let $d = \max_i d_i$.

Lemma 11.23. *There exist real $n \times n$ matrices B and C of ranks*

$$\text{rk}(B) \leq \exp(\sqrt{d} \log d \log t) \ and \ \text{rk}(C) \leq \prod_{i=1}^{t} d_i$$

such that

(i) $B[x, y] \leq -1/6$ *if* $A[x, y] = 0$, *and* $B[x, y] \geq +1/6$ *if* $A[x, y] = 1$;
(ii) $C[x, y] \geq 1$ *if* $A[x, y] = 0$, *and* $C[x, y] = 0$ *if* $A[x, y] = 1$.

Proof. The matrix A is an OR $A = \bigvee_{i=1}^{t} A_i$ of t matrices, each of which has the form $A_i = \bigwedge_{j=1}^{d} \overline{R}_{ij}$. Since $\overline{A}_i = \bigvee_{j=1}^{d_i} R_{ij}$, we can apply Lemma 11.22 with $k = 3t$ to each matrix $\overline{A}_i$ and obtain a real matrix M_i that approximates A_i with the factor $1/(3t)$. Since $A_i = J - \overline{A}_i$, the matrix $N_i = J - M_i$ approximates A_i with the same factor. Consider now the matrix

$$B = N_1 + \cdots + N_t - \frac{1}{2} \cdot J.$$

Let us verify that this matrix has the desired property (i):

- If $A[x, y] = 0$ then $A_i[x, y] = 0$ for all i; hence, $|N_i[x, y]| \leq 1/(3t)$ for all $i = 1, \ldots, t$, implying that $B[x, y] \leq 1/3 - 1/2 = -1/6$.
- If $A[x, y] = 1$ then $A_i[x, y] = 1$ for at least one i; for this i, we have that $N_i[x, y] \geq 1 - 1/(3t)$. Since $N_i[x, y] \geq -1/(3t)$ for all i, we obtain that $B[x, y] \geq 1 - \sum_{i=1}^{t}(1/3t) - 1/2 = 1/6$.

Hence, B satisfies (i). Since $N_i = J - M_i$ and $\text{rk}(M_i) \leq \mathcal{O}(\sqrt{d_i} \log d_i)$, the subadditivity of rank yields $\text{rk}(B) \leq \exp(\sqrt{d} \log d \log t)$.

To construct the matrix C, recall (again) that our matrix A is an OR $A = \bigvee_{i=1}^{t} A_i$ of t matrices, each of which has the form $A_i = \bigwedge_{j=1}^{d_i} \overline{R}_{ij}$. Define $C_i = \sum_{j=1}^{d_i} R_{ij}$, and let C be the component-wise product of the matrices $C_1, \ldots, C_t$, that is,

$$C[x, y] = \prod_{i=1}^{t} \sum_{j=1}^{d_i} R_{ij}[x, y].$$

If $A[x, y] = 0$ then $\forall i \ \exists j : R_{ij}[x, y] = 1$, implying that $C[x, y] \geq 1$. If $A[x, y] = 1$ then $\exists i \ \forall j : R_{ij}[x, y] = 0$, implying that $C[x, y] = 0$. Hence, C satisfies (ii). Since the rank of a component-wise product of two matrices does not exceed the product of their ranks, we obtain that $\text{rk}(C) \leq \prod_{i=1}^{t} \text{rk}(C_i) \leq \prod_{i=1}^{t} d_i$. $\square$

Proof of Theorem 11.21. We want to upper-bound the signum rank of a boolean matrix A of the form

$$A = \bigvee_{i=1}^{t} \bigwedge_{j=1}^{d_i} \overline{R}_{ij},$$

where $d_1 + \cdots + d_t \leq S$. Some of the fanins d_i may be small, some may be large. Large ANDs are "bad" because then our upper bound on the rank, given by Lemma 11.23 is "too large". The good news, however, is that we cannot have too

many large ANDs since the total sum $\sum_{i=1}^{t} d_i$ is upper bounded by S. We therefore
take a threshold D (to be specified later) and split these ANDs into "small" and
"large" subsets $I = \{i \mid d_i \leq D\}$ and $J = \{i \mid d_i > D\}$. Consider the
corresponding matrices

$$A_s = \bigvee_{i \in I} \bigwedge_{j=1}^{d_i} \overline{R}_{ij} \quad \text{and} \quad A_l = \bigvee_{i \in J} \bigwedge_{j=1}^{d_i} \overline{R}_{ij}.$$

We first apply Lemma 11.23 to A_s to get a matrix B such that

$$A_s[x, y] = 1 \Rightarrow B[x, y] \geq 1/6, \text{ and } A_s[x, y] = 0 \Rightarrow B[x, y] \leq -1/6. \quad (11.9)$$

Furthermore, we have that $\text{rk}(B) \leq \exp(\sqrt{D} \log D \log t)$.
   We then apply Lemma 11.23 to A_l to get a matrix C such that

$$A_l[x, y] = 1 \Rightarrow C[x, y] = 0, \text{ and } A_l[x, y] = 0 \Rightarrow C[x, y] \geq 1. \quad (11.10)$$

Since $\sum_{i=1}^{t} d_i \leq S$, we have that $|J| \leq \sum_{i=1}^{t} d_i / D \leq S/D$. Using the arithmetic-
geometric mean inequality $\left(\prod_{i=1}^{n} x_i \right)^{1/n} \leq \frac{1}{n} \sum_{i=1}^{n} x_i$ we can estimate the rank of
C as follows:

$$\text{rk}(C) \leq \prod_{i \in J} d_i \qquad\qquad \text{by Lemma 11.23}$$

$$\leq \left(\frac{1}{|J|} \sum_{i \in J} d_i \right)^{|J|} \qquad \text{arithmetic-geometric mean inequality}$$

$$\leq \exp(|J| \log S)$$

$$\leq \exp((S/D) \log S) \qquad \text{since } |J| \leq S/D.$$

Now define the matrix M by

$$M[x, y] = B[x, y] \cdot C[x, y] + \frac{1}{12}.$$

Using (11.9) and (11.10), it is easy to verify that $M[x, y] \leq -1/12$ if $A[x, y] = 0$,
and $M[x, y] \geq 1/12$ if $A[x, y] = 1$. Thus, $\text{signrk}(A) \leq \text{rk}(M)$. Since the rank of a
component-wise product of two matrices does not exceed the product of their ranks,
we obtain that $\text{rk}(M) \leq \text{rk}(B) \cdot \text{rk}(C) + 1$, which is at most exponential in

$$\sqrt{D} \log D \log t + \frac{S}{D} \log S.$$

By setting $D = (S/ \log S)^{2/3}$, this is at most exponential in $S^{1/3} \log^{5/3} S$, as
desired. $\square$

The adjacency matrix A of the Sylvester $n \times n$ graph H_n is a Hadamard matrix, and we already know (see Corollary 4.43) that its signum rank is at least $\Omega(\sqrt{n})$. Together with Theorem 11.21, this implies

Corollary 11.24. *Every monotone Σ_3 formula representing the Sylvester graph S_n must have size at least* $\log^{3-o(1)} n$.

Thus, using the signum rank one can derive nontrivial lower bounds on the depth-3 representation complexity of graphs. The result of Razborov and Sherstov (2008) mentioned in Remark 11.14 implies, however, that the signum rank alone cannot lead to lower bounds substantially larger than $\Omega(\log^3 n)$.

11.9 Depth-3 Circuits with Parity Gates

In this section we will use graph-theoretic arguments to prove *truly* exponential lower bounds for modified Σ_3 circuits, where all gates on the bottom level are Parity gates (not OR gates). A lower bound for a boolean function f of n variables is *truly exponential* if it has the form 2^{cn} for a constant $c > 0$.

A $\Sigma_3^\oplus$ circuit is a Σ_3 circuits with the OR gates on the bottom (next to the inputs) layer replaced by Parity gates. Hence, at each AND gate on the middle layer the characteristic function of some affine subspace over GF(2) is computed. The fanin of the top OR gate tells us how many affine subspaces lying within $f^{-1}(1)$ do we need to cover the whole set $f^{-1}(1)$.

Let $\Sigma_3^\oplus(G)$ denote the smallest top fanin of a $\Sigma_3^\oplus$ representing the graph G. For a boolean function f, let $\Sigma_3^\oplus(f)$ denote the smallest top fanin of a $\Sigma_3^\oplus$ circuit computing f. Note that $\Sigma_3^\oplus(G) \leq n$ for every bipartite $n \times n$ graph $G = (V_1 \cup V_2, E)$ because G can be represented by a formula of the form

$$F(x) = \bigvee_{u \in V_1} x_u \wedge \Big(\bigoplus_{v \in V_2 : uv \in E} x_v \Big).$$

Our starting point is the following immediate consequence of Proposition 11.16: For every bipartite graph G, $\Sigma_3^\oplus(f_G) \geq \Sigma_3^\oplus(G)$. Hence, if $\Sigma_3^\oplus(G) \geq n^\epsilon$, then $\Sigma_3^\oplus(f_G) \geq 2^{\epsilon m}$, and we have a truly exponential lower bound for f_G; recall that f_G is a boolean function of $2m$ variables.

We are going to prove a general lower bound: any dense graph without large complete subgraphs requires large top fanin of $\Sigma_3^\oplus$ circuits. This immediately yields exponential lower bounds for many explicit boolean functions.

A graph is $K_{a,b}$-*free* if it does not contain a complete $a \times b$ subgraph. For a graph G, by $|G|$ we will denote the number of edges in it. It turns out that every dense enough graph without large complete subgraphs requires large $\Sigma_3^\oplus$ circuits; this was observed in (Jukna 2006).

Theorem 11.25. *If an $n \times n$ graph G is $K_{a,b}$-free, then every $\Sigma_3^{\oplus}$ circuit representing G must have top fanin at least $|G|/(a+b)n$.*

To prove the theorem, we first give a combinatorial characterization of the top fanin of $\Sigma_3^{\oplus}$ circuits representing bipartite graphs (Lemma 11.26), and then a general lower bound on this characteristics (Lemma 11.27).

A *fat matching* is a union of vertex-disjoint bipartite cliques (these cliques need not to cover all vertices). Note that a matching (a set of vertex-disjoint edges) is also a fat matching. A *fat covering* of a graph G is a family of fat matchings such that each of these fat matchings is a subgraph of G and every edge of G is an edge of at least one member of the family. Let $\mathrm{fat}(G)$ denote the minimum number of fat matchings in a fat covering of G.

Theorem 11.25 is a direct consequence of the following two lemmas.

Lemma 11.26. *For every bipartite graph G, $\mathrm{fat}(G) = \Sigma_3^{\oplus}(G)$.*

Proof. Let U and V be the color classes of G, and let $g = \bigoplus_{v \in A \cup B} x_v$ with $A \subseteq U$ and $B \subseteq V$ be a gate on the bottom level of a $\Sigma_3^{\oplus}$ circuit representing G. Since g is a parity gate, it accepts a pair uv of vertices $u \in U$, $v \in V$ iff either $u \in A$ and $v \notin B$, or $u \notin A$ and $v \in B$. Thus, g represents a fat matching $(A \times \overline{B}) \cup (\overline{A} \times B)$ where $\overline{A} = U \setminus A$ and $\overline{B} = V \setminus B$ (see Fig. 11.2c). Since the intersection of two fat matchings is again a fat matching (show this!), each AND gate on the middle level represents a fat matching. Hence, if the circuit has top fanin s, then the OR gate on the top represents a union of these s fat matchings, implying that $s \geq \mathrm{fat}(G)$.

To show $\Sigma_3^{\oplus}(G) \leq \mathrm{fat}(G)$, let $M = A_1 \times B_1 \cup \cdots \cup A_r \times B_r$ be a fat matching. Let A be the union of the A_i, and B the union of the B_i. We claim that the following AND of Parity gates represents M:

$$F = \bigoplus_{u \in A} x_u \wedge \bigoplus_{v \in B} x_v \wedge \bigoplus_{w \in A_1 \cup \overline{B}_1} x_w \wedge \cdots \wedge \bigoplus_{w \in A_r \cup \overline{B}_r} x_w.$$

Indeed, if a pair $e = uv$ of vertices belongs to M, say, $u \in A_1$ and $v \in B_1$, then the first three sums accept uv because $u \in A_1$ and $v \notin \overline{B}_1$. Moreover, the mutual disjointness of the A_i as well as of the B_i implies that $u \notin A_i$ and $v \in B_1 \subseteq \overline{B}_i$ for all $i = 2, \ldots, r$. Hence, each of the last sums accepts the pair uv as well. To prove

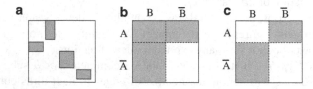

Fig. 11.2 (a) An adjacency matrix of a fat matching, (b) the adjacency matrix of a graph represented by an OR gate $g = \bigvee_{v \in A \cup B} x_v$, and (c) the adjacency matrix of a graph represented by a parity gate $g = \bigoplus_{v \in A \cup B} x_v$

the other direction, suppose that a pair uv of vertices is accepted by F. The last r sums ensure that, for each $i = 1, \ldots, r$, one of the following must hold: (a) $u \in A_i$ and $v \in B_i$; (b) $u \notin A_i$ and $v \notin B_i$. The first two sums of F ensure that (b) cannot happen for all i. Hence, (a) must happen for some i, implying that uv belongs to M. $\qquad\square$

Lemma 11.27. *Let G be a bipartite $n \times n$ graph. If G is $K_{a,b}$-free then* fat(G) *is a at least $|G|/(a+b)n$.*

Proof. Let $H = \bigcup_{i=1}^{t} A_i \times B_i$ be a fat matching, and suppose that $H \subseteq G$. By the definition of a fat matching, the sets $A_1, \ldots, A_t$, as well as the sets $B_1, \ldots, B_t$ are mutually disjoint. Moreover, since G contains no copy of $K_{a,b}$, we have that $|A_i| < a$ or $|B_i| < b$ for all i. Hence, if we set $I = \{i : |A_i| < a\}$, then

$$|H| = \sum_{i=1}^{t} |A_i \times B_i| = \sum_{i=1}^{t} |A_i| \cdot |B_i| \le \sum_{i \in I} a \cdot |B_i| + \sum_{i \notin I} |A_i| \cdot b \le (a+b)n.$$

Thus, no fat matching $H \subseteq G$ can cover more than $(a+b)n$ edges of G, implying that we need at least $|G|/(a+b)n$ fat matchings to cover all edges of G. $\qquad\square$

Remark 11.28. Note that Theorem 11.25 remains true if we allow not only parity gates be used at a bottom level, but also when allow any boolean functions that take value 1 on inputs with exactly one 1, and take value 0 on inputs with no ones at all.

There are many explicit bipartite graphs which are dense enough and do not have large complete bipartite subgraphs. By Theorem 11.25 and Proposition 11.16, each of these graphs G immediately give us an explicit boolean function f_G requiring $\Sigma_3^{\oplus}$ circuits of truly exponential size.

To give an example, consider the *disjointness function*. This is a boolean function $DISJ_{2m}$ in $2m$ variables such that

$$DISJ_{2m}(y_1, \ldots, y_m, z_1, \ldots, z_m) = 1 \text{ if and only if } \sum_{i=1}^{m} y_i z_i = 0.$$

Note that this function has a trivial Π_2 circuit (a CNF) of size $\mathcal{O}(m)$.

Theorem 11.29. *Every $\Sigma_3^{\oplus}$ circuit for $DISJ_{2m}$ has top fanin at least $2^{0.08m}$.*

Proof. The graph G_f of the function $f = DISJ_{2m}$ is a bipartite graph $G_m \subseteq U \times V$ where U and V consist of all $n = 2^m$ subsets of $[m] = \{1, \ldots, m\}$, and $uv \in G_m$ iff $u \cap v = \emptyset$. The graph G_m can contain a complete bipartite $a \times b$ subgraph $A \times B \ne \emptyset$ only if $a \le 2^k$ and $b \le 2^{m-k}$ for some $0 \le k \le m$, because then

$$\left(\bigcup_{u \in A} x_u \right) \cap \left(\bigcup_{v \in B} x_v \right) = \emptyset.$$

In particular, G_m can contain a copy of $K_{a,a}$ only if $a \le 2^{m/2} = \sqrt{n}$. Since G_m has

Fig. 11.3 Schematic
description of discriminators:
$\Delta_F(A)$ is large in case (**a**),
and $\Delta_F(A) = 0$ in case (**b**)

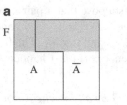
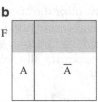

$$|G_m| = \sum_{u \in U} d(u) = \sum_{u \in U} 2^{m-|u|} = \sum_{i=0}^{m} \binom{m}{i} 2^{m-i} = 3^m \geq n^{1.58}$$

edges, Theorem 11.25 yields that any $\Sigma_3^{\oplus}$ circuit representing G_m—and hence, any $\Sigma_3^{\oplus}$ circuit computing $DISJ_{2m}$—must have top fanin at least

$$\frac{|G_m|}{2an} \geq \frac{n^{1.58}}{n^{1.5}} = n^{0.08} = 2^{0.08m}. \qquad \qquad \square$$

We now consider a generalization of $\Sigma_3^{\oplus}$ circuits, where we allow to use an arbitrary *threshold* gate, instead of an OR gate, on the top. To analyze such circuits, we will use the so-called "discriminator lemma" for threshold gates.

Let $\mathcal{F}$ be a family of subsets of a finite set X. A family $F_1, \ldots, F_t$ of members of $\mathcal{F}$ is a *threshold cover* of a given set $A \subseteq X$, if there exists a number $0 \leq k \leq t$ such that, for every $x \in X$, we have that $x \in A$ if and only if x belongs to at least k of the F_i. Let $\text{thr}_{\mathcal{F}}(A)$ denote the minimum number t of members of $\mathcal{F}$ in a threshold cover of A.

To lower bound $\text{thr}_{\mathcal{F}}(A)$ the following measure turns out to be useful:

$$\Delta_{\mathcal{F}}(A) = \max_{F \in \mathcal{F}} \Delta_F(A),$$

where

$$\Delta_F(A) = \left| \frac{|A \cap F|}{|A|} - \frac{|\overline{A} \cap F|}{|\overline{A}|} \right|.$$

If $\Delta_{\mathcal{F}}(A)$ is small, this means that every member F of $\mathcal{F}$ is split between the set A and its complement $\overline{A}$ in a "balanced" manner: the portion of $F \cap A$ in A is almost the same as the portion of $F \cap \overline{A}$ in $\overline{A}$ (see Fig. 11.3). The following lemma is a special case of a more general lemma proved by Hajnal et al. (1993); see the next section.

Lemma 11.30. (Discriminator Lemma) $\text{thr}_{\mathcal{F}}(A) \geq 1/\Delta_{\mathcal{F}}(A)$.

Proof. Let $F_1, \ldots, F_t \in \mathcal{F}$ be a threshold-k covering of A, that is, $x \in A$ iff x belongs to at least k of the F_i's. Our goal is to show that $\Delta_{\mathcal{F}}(A) \geq 1/t$.

Since every element of A belongs to at least k of the sets $A \cap F_i$, the average size of these sets must be at least k. Since no element of $\overline{A}$ belongs to more than $k - 1$

of the sets $\overline{A} \cap F_i$, the average size of these sets must be at most $k - 1$. Hence,

$$1 \leq \sum_{i=1}^{t} \frac{|A \cap F_i|}{|A|} - \sum_{i=1}^{t} \frac{|\overline{A} \cap F_i|}{|\overline{A}|} \leq t \cdot \max_{1 \leq i \leq t} \left| \frac{|A \cap F_i|}{|A|} - \frac{|\overline{A} \cap F_i|}{|\overline{A}|} \right|. \qquad \square$$

Now we are able to prove strong lower bounds on the size of $\Sigma_3^{\oplus}$ circuits with an arbitrary threshold gate on the top.

A *Hadamard matrix* of order n is an $n \times n$ matrix with entries ± 1 and with row vectors mutually orthogonal over the reals. A graph associated with a Hadamard matrix M (or just a Hadamard graph) of order n is a bipartite $n \times n$ graph where two vertices u and v are adjacent if and only if $M[u, v] = +1$.

Theorem 11.31. *Any $\Sigma_3^{\oplus}$ circuit which has an arbitrary threshold gate on the top and represents an $n \times n$ Hadamard graph must have top fanin $\Omega(\sqrt{n})$.*

Proof. Let A be an $n \times n$ Hadamard graph. Take an arbitrary $\Sigma_3^{\oplus}$ circuit which has an arbitrary threshold gate on the top and represents A. Let s be the fanin of this threshold gate, and let $\mathcal{F}$ be the set of all fat matchings participating in the representation. Then, by Lemma 11.26, $s \geq \text{thr}_{\mathcal{F}}(A)$. To prove $s = \Omega(\sqrt{n})$ it is enough, by Lemma 11.30, to show that for every fat matching $F = \bigcup_{i=1}^{\ell} S_i \times R_i$,

$$\left| \frac{|A \cap F|}{|A|} - \frac{|\overline{A} \cap F|}{|\overline{A}|} \right| = \mathcal{O}(n^{-1/2}).$$

Since both the graph A and its bipartite complement $\overline{A}$ have $\Theta(n^2)$ edges, it is enough to show that $\left| |A \cap F| - |\overline{A} \cap F| \right| \leq n^{3/2}$. By Lindsey's Lemma (see Appendix A for the proof), the absolute value of the sum of all entries in any $a \times b$ submatrix of an $n \times n$ Hadamard matrix does not exceed $\sqrt{abn}$. Thus, the absolute value of the difference

$$|A \cap (S_i \times R_i)| - |\overline{A} \cap (S_i \times R_i)|$$

does not exceed $\sqrt{s_i r_i n}$, where $s_i = |S_i|$ and $r_i = |R_i|$. Since both sums $\sum_{i=1}^{\ell} s_i$ and $\sum_{i=1}^{\ell} r_i$ are at most n, we obtain

$$\left| |A \cap F| - |\overline{A} \cap F| \right| = \left| \sum_{i=1}^{\ell} |A \cap (S_i \times R_i)| - \sum_{i=1}^{\ell} |\overline{A} \cap (S_i \times R_i)| \right|$$

$$\leq \sum_{i=1}^{\ell} \sqrt{s_i r_i n} \leq \sqrt{n} \sum_{i=1}^{\ell} \frac{s_i + r_i}{2} \leq n^{3/2}. \qquad \square$$

Recall that the *inner product function* is a boolean function of $2m$ variables defined by

$$IP_m(x_1,\ldots,x_m,y_1,\ldots,y_m) = \sum_{i=1}^m x_i y_i \bmod 2.$$

Since the graph G_f of $f = IP_{2m}$ is a Hadamard $n \times n$ graph with $n = 2^m$, Theorem 11.31 immediately yields

Corollary 11.32. *Any* $\Sigma_3^\oplus$ *circuit which has an arbitrary threshold gate on the top and computes* IP_{2m} *must have top fanin* $\Omega(2^{m/2})$.

11.10 Threshold Circuits

A boolean function $f(x_1,\ldots,x_n)$ is a *real threshold function* if there exist real numbers $w_0, w_1, \ldots, w_n$ such that for every $x \in \{0,1\}^n$, $f(x) = 1$ if and only if $w_1 x_1 + \cdots + w_n x_n \geq w_0$. The absolute values $|w_i|$ are called *weights*. Since some finite amount of precision is always sufficient, it is easy to see that we can assume that the weights are integers. Let $W(n)$ be the smallest number W such that every possible real threshold function of n variables can be realized using integer weights of magnitude $\leq W$. It is well-known (see, for example, Muroga 1971) that

$$W(n) \leq 2^{\frac{1}{2}n\log n - n(1+o(1))}.$$

Since, as shown by Yajima and Ibaraki (1965), and by Smith (1966), there are at least $2^{n(n-1)/2}$ distinct real threshold functions, we also have that $W(n) = 2^{\Omega(n)}$. A much tighter lower bound

$$W(n) \geq 2^{\frac{1}{2}n\log n - n(2+o(1))}$$

on the weight was obtained by Håstad (1994) when n is a power of 2, and by Alon and Vũ (1997) for arbitrary n.

A *threshold circuit* is a circuit using arbitrary real threshold functions as gates. For a boolean function $f(x_1,\ldots,x_n)$, let $T(f)$ denote the smallest number of gates, and $T_w(f)$ the smallest sum of weights in a threshold circuit computing f. Let also $T(n)$ and $T_w(n)$ denote the corresponding Shannon functions. Zacharova (1963) proved that

$$T_w(n) \sim \frac{2^n}{n}.$$

By improving earlier estimates of Nechiporuk (1964, 1965), Lupanov (1973) established the asymptotic for the number of gates:

$$T(n) \sim 2\sqrt{\frac{2^n}{n}}.$$

In the class of partially defined functions that are defined on N input vectors, he also proved that for $N \to \infty$,

$$T(n, N) \sim 2\sqrt{\frac{N}{\log N}}.$$

Finally, for the Shannon function $S(n)$ restricted to symmetric boolean functions, he proved that

$$S(n) \sim 2\sqrt{\frac{n}{\log n}}.$$

The lower bound here follows from the lower bound on $T(n)$ because every boolean function $f(x_1, \ldots, x_n)$ of n variables can be considered as a symmetric boolean function F of $2^n - 1$ variables: just assign weight 2^{i-1} to x_i; then

$$f(x) = F(\underbrace{x_1}_{2^0}, \underbrace{x_2, x_2}_{2^1}, \underbrace{x_3, x_3, x_3, x_3}_{2^2}, \ldots, \underbrace{x_n, \ldots, x_n}_{2^{n-1}}).$$

This map from inputs in $\{0, 1\}^n$ to integers $\{0, \ldots, 2^n - 1\}$ is clearly injective. So, if any symmetric function of 2^n variables can be computed using M gates, then any boolean function of n variables can be computed using M gates. This implies that $T(n) \leq S(2^n)$.

To show the upper bound, take a threshold circuit $C(x) = y$ computing the binary representation $y \in \{0, 1\}^m$ of the sum $\sum_{i=1}^{n} x_i$, with $m = \lceil \log_2(n + 1) \rceil$. Since one can design such a circuit with only m gates (see Exercise 11.9), this implies $S(n) \leq m + T(m)$.

11.10.1   General Threshold Circuits

The problem of proving *explicit* lower bounds for threshold circuits is even harder than, say, for DeMorgan circuits. In the latter model any boolean function essentially depending on all its n variables requires at least $n - 1$ gates. For example, if we define the *exact-k function* by $\text{Ex}_k^n(x) = 1$ iff $x_1 + \cdots + x_n = k$, then this function depends on all its variables but, as shown in Fig. 11.4, this function can be computed by a threshold circuit using only three gates! Thus, even proving non-constant lower bounds on $T(f)$ is a non-trivial task.

In the case of unrestricted threshold circuits the strongest remains the lower bound $T(IP_n) \geq n/2$ proved by Gröger and Turán (1991) for the inner product function $IP_n(x, y) = x_1 y_1 + \cdots + x_n y_n \bmod 2$.

In the proof of this bound, by a *rectangle* we will mean a Cartesian product $R = X \times Y$ of two subsets of vectors $X, Y \subseteq \{0, 1\}^n$; its *dimension* is $\dim(R) = \min\{|X|, |Y|\}$. A boolean function on such a rectangle is a mapping $f : X \times Y \to$

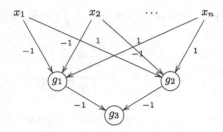

Fig. 11.4 A general form of a threshold circuit of size 3 computing any of the functions $f_k(x)=1$ iff $x_1 + \cdots + x_n = k$, for a given integer $0 < k < n$. The first gate $g_1(x)$ outputs 1 iff $(-1)x_1 + \cdots (-1)x_n \geq -k + 1$, and the second $g_2(x)$ outputs 1 iff $x_1 + \cdots + x_n \geq k + 1$. The last gate $g_3(y_1, y_2)$ outputs 1 iff $-y_1 - y_2 \geq 0$, that is, iff $y_1 = y_2 = 0$

$\{0, 1\}$. A function $f(x, y)$ is *monochromatic* on a subset $S \subseteq X \times Y$ if it takes the same value on all inputs $(x, y) \in S$.

The weakness of real threshold functions is captured by the following lemma.

Lemma 11.33. *Let $f(x, y)$ be a boolean function on a rectangle of dimension d. If f is a real threshold function, then it is monochromatic on a subrectangle of dimension at least $\lfloor d/2 \rfloor$.*

Proof. Let $f : X \times Y \to \{0, 1\}$ where $X, Y \subseteq \{0, 1\}^n$ and $d = |X| \leq |Y|$. If f is a real threshold function then there exist real numbers $a_1, \ldots, a_n, b_1, \ldots, b_n$ and c such that, for every $(x, y) \in X \times Y$, $f(x, y) = 1$ if and only if $a \cdot x + b \cdot y \geq c$, where $a \cdot x = \sum_{i=1}^{n} a_i x_i$ and $b \cdot y = \sum_{i=1}^{n} b_i y_i$.

Order the elements of X according to the value of $a \cdot x$, and elements in Y according to the value of $b \cdot y$, resolving ties arbitrarily. Let $1 \leq t \leq d$ be the smallest number such that the t-th elements $x^t \in X$ and $y^t \in Y$ satisfy $a \cdot x^t + b \cdot y^t < c$. Then $f(x, y)$ is constant 0 on the rectangle $\{x^1, \ldots, x^t\} \times \{y^1, \ldots, y^t\}$, and is constant 1 on the rectangle $\{x^{t+1}, \ldots, x^d\} \times \{y^{t+1}, \ldots, y^d\}$. Since one of these rectangles must have dimension at least $\lfloor d/2 \rfloor$, we are done.   □

For a boolean function $f : X \times Y \to \{0, 1\}$, let $\mathrm{mono}(f)$ denote the maximal dimension of a subrectangle $X' \times Y' \subseteq X \times Y$ on which f takes the same value.

Theorem 11.34. (Gröger and Turán 1991) *If $f(x, y)$ is a boolean function of $2n$ variables, then any threshold circuit computing f must have at least $n - \log \mathrm{mono}(f)$ gates.*

Proof. Let $g_1, g_2, \ldots, g_t$ be a threshold circuit computing f. By Lemma 11.33, there is a rectangle $R_1 \subseteq \{0, 1\}^n \times \{0, 1\}^n$ of dimension 2^{n-1} such that the function computed at the first gate g_1 takes the same value $c_1 \in \{0, 1\}$ on all inputs $(x, y) \in R_1$. Replace the gate g_1 by the constant c_1. The resulting circuit still computes f correctly on all inputs in R_1. The first gate in this new circuit is the gate g_2 of the original circuit one of which input is apparently set to the constant c_1. In any case, this is a real threshold function on R_1, and Lemma 11.33 again gives us a rectangle

$R_2 \subseteq R_1$ of dimension 2^{n-2} such that the function computed at the first gate g_2 takes the same value $c_2 \in \{0, 1\}$ on all inputs $(x, y) \in R_2$.

Arguing in this way, we obtain that the original circuit must output the same value on some rectangle R_t of dimension 2^{n-t}. This implies that $\text{mono}(f) \geq 2^{n-t}$, from which the desired lower bound $t \geq n - \log \text{mono}(f)$ on the number of gates follows. □

Corollary 11.35. $T(IP_n) \geq n/2$.

Proof. Consider a $2^n \times 2^n$ matrix H defined by $H[x, y] = (-1)^{IP_n(x,y)}$. For every $x \neq 0$, we have that $IP_n(x, y) = 1$ for exactly half of vectors y. Hence, H is a Hadamard matrix (every two rows are orthogonal over the reals). By Lindsey's Lemma (see Appendix A for the proof), H can contain an $k \times k$ monochromatic submatrix only if $k^2 \leq \sqrt{k \cdot k \cdot 2^n}$, that is, only if $k \leq 2^{n/2}$. Thus, $\text{mono}(IP_n) \leq 2^{n/2}$, and it remains to apply Theorem 11.34. □

Remark 11.36. Impagliazzo et al. (1997) considered the question of how the number of *wires* in threshold circuits depends on their depth. Let $W_d(f)$ denote the smallest number of wires in a general threshold circuit of depth d computing f. For the parity function $\oplus_n(x) = x_1 \oplus x_2 \oplus \cdots \oplus x_n$, they proved a lower bound $W_d(\oplus_n) \geq n^{1+\theta(d)}$ where $\theta(d) = c(1+\sqrt{2})^{-d}$ and $c > 0$ is a constant independent on n and d. They also proved that any depth-d threshold circuit computing $\oplus_n(x)$ must have at least $(n/2)^{1/2(d-1)}$ gates. Similar results for threshold circuits with polynomially bounded weights were proved earlier by Paturi and Saks (1994).

11.10.2 Threshold Circuits of Depth Two

Super-polynomial lower bounds are only known for depth-2 and depth-3 threshold circuits under various additional restrictions. The *weight* of a threshold circuit $C(x)$ of n variables is the maximal absolute value of weights occurring in gates of C. We say that C is a *bounded-weight* circuit if its weight is at most some polynomial in n. A circuit is *unweighted* if each its weight is either 0 or 1. Note that unweighted threshold circuits are just unbounded fanin circuits with (boolean) threshold functions as gates.

Goldmann and Håstad (1992) proved that any threshold circuit of depth d can be simulated by a bounded-weight circuit of depth $d + 1$ with only a polynomial increase of size. But, so far, no exponential lower bound is known for depth-3 circuits, even for unweighted circuits. Even threshold circuits of depth 2 are hard to deal with: here exponential lower bounds are only known when weights are bounded.

It can be shown (see Exercise 11.11) that the inner product function IP_n can be computed by an unweighted threshold circuit of depth 3 using $\mathcal{O}(n)$ gates. In the class of depth-2 threshold circuits, exponential lower bounds for IP_n are known when either the weights of the top (output) gate or the weights of the bottom (next to the inputs) gates are bounded. In the first case the proof is based on so-called "Discriminator Lemma", and is based on known lower bounds on the signum rank in the second case.

Let $f(x)$ be a boolean function of n variables, and $A, B \subseteq \{0, 1\}^n$ be disjoint sets. Let P_A, (resp. P_B) denote the uniform probability distribution on A (resp. B). Hence, $P_A(f(x) = 1) = |\{x \in A \mid f(x) = 1\}|/|A|$, and similarly for P_B. Then f is an ϵ-*discriminator* for A and B if

$$|P_A(f(x) = 1) - P_B(f(x) = 1)| \geq \epsilon.$$

In particular, f is a 1-discriminator for A and B if f separates these two sets in that $f(a) \neq f(b)$ for all $(a, b) \in A \times B$. If $A = g^{-1}(1)$ and $B = g^{-1}(0)$ for some boolean function g, and if $|A| = |B|$, then f being an ϵ-discriminator for A and B means that f coincides with g on a fraction $(1 + \epsilon)/2$ of all inputs.

A *threshold combination* of boolean functions $f_1, \ldots, f_m : \{0, 1\}^n \to \{0, 1\}$ of total weight α is a boolean function of the form

$$T_\alpha^m(x) = 1 \quad \text{if and only if} \quad \sum_{i=1}^m a_i f_i(x) \geq a_0$$

where $a_0, a_1, \ldots, a_m$ are integers, and $\sum_{i=1}^m |a_i| = \alpha$.

The following lemma is an extension of Lemma 11.30 to threshold gates with arbitrary weights.

Discriminator Lemma. If a threshold combination of $f_1, \ldots, f_m$ of total weight accepts all vectors in A and rejects all vectors in B, then some f_i is a $(1/\alpha)$-discriminator for A and B.

Proof. Let the random variable $f_i^A(x)$ (resp. $f_i^B(x)$) be the output of f_i when x is distributed uniformly on A (resp. B). Then $\sum_{i=1}^m a_i f_i^A(x) \geq a_0$ and $\sum_{i=1}^m a_i f_i^B(x) \leq a_0 - 1$. Taking expectations and rearranging, we obtain

$$1 \leq \sum_{i=1}^n a_i (\mathrm{E}\left[f_i^A(x)\right] - \mathrm{E}\left[f_i^B(x)\right])$$

$$\leq \alpha \cdot \max_{1 \leq i \leq m} |P_A(f_i(x) = 1) - P_B(f_i(x) = 1)|. \qquad \square$$

Theorem 11.37. (Hajnal et al. 1993) *If the weights of the top gate in a threshold circuit of depth 2 computing IP_n are at most $2^{o(n^{1/3})}$, then the top gate must have fanin at least $2^{\Omega(n^{1/3})}$.*

Proof. Take a depth-2 threshold circuit computing $IP_n(x, y)$. Assume that all weights of the top (output) gate are at most $2^{o(n^{1/3})}$. Let m be the fanin of the output gate. Our goal is to show that $m \geq 2^{\Omega(n^{1/3})}$.

Fig. 11.5 The $N \times N$ matrix of $f(x, y)$ with $N = 2^n$ is divided into $N^{2/3}$ consecutive squares of size $N^{2/3} \times N^{2/3}$. There are at most $2 \cdot N^{1/3}$ squares containing both a 0 and a 1. The 1-entries not in these squares can be covered by $N^{1/3}$ rectangles of height $N^{2/3}$ and width $\leq N$

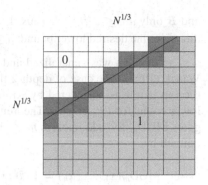

Let α be the sum of weights of the output gates; hence, $\alpha \leq m \cdot 2^{o(n^{1/3})}$. By the Discriminator Lemma, some of the bottom gates $f(x, y)$ must be a $(1/\alpha)$-discriminator for

$$A = \{(x, y) \mid IP_n(x, y) = 1\} \text{ and } B = \{(x, y) \mid IP_n(x, y) = 0\}.$$

The gate f is of the form

$$f(x, y) = 1 \text{ if and only if } \sum_{i=1}^{n} a_i x_i + \sum_{i=1}^{n} b_i y_i \geq c$$

for some integers a_i, b_i and c. The $2^n \times 2^n$ matrix H with $H[x, y] = (-1)^{IP_n(x,y)}$ is a Hadamard matrix. By Lindsey's Lemma (see Appendix A for the proof), the absolute value of the sum of all entries in any $a \times b$ submatrix of an $N \times N$ Hadamard matrix does not exceed $\sqrt{abN}$.

Now consider the set $F = \{(x, y) \mid f(x, y) = 1\}$ of inputs accepted by the gate f. We will view this set as a $2^n \times 2^n$ matrix where the rows are indexed by vectors x in increasing order of $\sum_{i=1}^{n} a_i x_i$, and columns are indexed by vectors y in increasing order of $\sum_{i=1}^{n} b_i y_i$; the entry (x, y) of the matrix is $f(x, y)$. In this matrix every entry either to the right of, or below an entry which is 1, is also equal 1.

Divide the matrix into $2^{2n/3}$ consecutive squares of size $2^{2n/3} \times 2^{2n/3}$ (Fig. 11.5). There are at most $2 \cdot 2^{n/3}$ squares containing both a 0 and a 1. The 1-entries not in these squares can be covered by at most $2^{n/3}$ rectangles of height $2^{2n/3}$ and width $\leq 2^n$. Thus, using the Lindsey Lemma, the absolute value of the difference $|A \cap F| - |B \cap F|$ does not exceed

$$2 \cdot 2^{n/3} \cdot 2^{4n/3} + 2^{n/3} \cdot \sqrt{2^{2n/3} \cdot 2^n \cdot 2^n} = 3 \cdot 2^{5n/3}.$$

In our case the sets A and B are of about the same size: $|A| = 2^{2n-1} - 2^{n-1}$ and $|B| = 2^{2n-1} + 2^{n-1}$. Hence, the bottom gate f can be an ϵ-discriminator for A

and B only if $\epsilon \leq 2^{-\Omega(n^{1/3})}$. Thus, $1/\alpha \leq 2^{-\Omega(n^{1/3})}$. Together with $\alpha \leq m \cdot 2^{o(n^{1/3})}$, this gives the desired lower bound $m \geq 2^{\Omega(n^{1/3})}$ on the fanin m of the top gate.   □

The theorem was generalized and extended by Krause (1996), and Krause and Waack (1995). The case of depth-2 threshold circuits with unrestricted weights of the top gate was considered by Krause and Pudlák (1997). They proved that the lower bound for IP_n given in Theorem 11.37 also holds in this case, if only modular gates MOD_m for odd integers $m \geq 3$ are used at the bottom level. Such gates are defined by:

$$\mathrm{MOD}_m(x_1, \dots, x_n) = 1 \quad \text{if and only if} \quad x_1 + \dots + x_n = 0 \bmod m.$$

Forster et al. (2001) considered real threshold circuits when the weights of the top gate are arbitrary and bottom gates are threshold functions with bounded but exponentially large weights.

Theorem 11.38. (Forster et al. 2001) *If the weights of the bottom gates of a threshold circuit of depth 2 computing IP_n do not exceed $2^{n/3}$, then the top gate must have fanin at least $2^{\Omega(n)}$.*

Proof. We will view boolean functions $f(x, y)$ of $2n$ variables as boolean $2^n \times 2^n$ matrices of their values. By the rank, $\mathrm{rk}(f)$, of a boolean function $f(x, y)$ we will mean the real rank of its matrix. In particular, the matrix of IP_n is (the boolean version of) a $2^n \times 2^n$ Hadamard matrix H_{2^n}. By the result of Forster (2002), we know that this matrix has signum rank at least $2^{n/2}$ (see Corollary 4.43). Thus, to prove the theorem, it is enough to show that every depth-2 threshold circuit of small top-fanin and bounded weight of bottom gates has small signum rank. We first show that small-weight threshold functions (that is, our bottom gates) have small rank.

Claim 11.39. *If $f(x, y)$ is a real threshold function with integer weights not exceeding W, then $\mathrm{rk}(f) \leq 2nW + 1$.*

Proof. Since $f(x, y)$ is a real threshold function with integer weights not exceeding W, there are integers a_i, b_i and c of absolute value at most W such that

$$f(x, y) = 1 \quad \text{if and only if} \quad \sum_{i=1}^{n} a_i x_i + \sum_{i=1}^{n} b_i y_i \geq c.$$

Let F be the $2^n \times 2^n$ matrix of $f(x, y)$. For each integer u between $-nW$ and nW, let F_u be the submatrix of F formed by all rows x with $\sum_{i=1}^{n} a x_i = u$. Since these submatrices are disjoint, and each of them has rank at most 1, the subadditivity of rank implies that the rank of F is at most the number $2nW + 1$ of these submatrices.   □

Since, by the assumption of the theorem, all bottom gates have weights $W \leq 2^{n/3}$, and since the matrix of IP_n has signum rank at least $2^{n/2}$, the theorem follows directly from the following claim.

Claim 11.40. If a boolean function $f(x, y)$ is computed by a depth-2 threshold circuit with top-fanin s, then f has signum rank at most $\mathcal{O}(snW)$ where W is the maximum weight of a bottom gate.

Proof. Let the top gate have weights $w_1, \ldots, w_s$ and threshold w_0. Hence,

$$f(x, y) = 1 \text{ if and only if } \sum_{i=1}^{s} w_i f_i(x, y) - w_0 \geq 0,$$

where $f_i(x, y)$ is a threshold function computed at the i-th bottom gate. Hence, if $F, F_1, \ldots, F_s$ are the corresponding $2^n \times 2^n$ matrices, then the value of F only depends on the signum of the matrix

$$M = w_1 F_1 + \cdots + w_s F_s - w_0 J,$$

where J is the all-1 $2^n \times 2^n$ matrix. By Claim 11.39, $rk(F_i) \leq 2nW + 1$ for all $i = 1, \ldots, s$. Thus, the signum rank of F does not exceed

$$rk(M) \leq 1 + \sum_{i=1}^{s} rk(F_i) \leq 1 + s(2nW + 1) = \mathcal{O}(snW),$$

as claimed. □

■ **Research Problem 11.41.** Prove an exponential lower bound for unrestricted depth-2 threshold circuits.

11.10.3 Threshold Circuits of Depth Three

In the case of depth 3, even the power of *unweighted* threshold circuits remains unclear. In view of the fact that we do not have lower bounds for unbounded weight threshold circuits of depth-2, this is not surprising: Goldmann et al. (1992) proved that every depth-d circuit with unbounded weights *can* be simulated by a depth-$(d + 1)$ circuit consisting of majority gates—the increase in size is only polynomial; see the survey of Razborov (1992d) for more information. Note that such (majority) circuits constitute a subclass of unweighted threshold circuits.

Unweighted threshold circuits of depth 3 are also important due to the following impressive result of Yao (1990). He showed that every boolean function in ACC^0 is computable by depth-3 threshold circuits of "moderate" size and with only AND gates on the bottom level. An ACC *circuit* (an alternating circuit with counting gates) of depth-d is a circuit formed by d alternating levels of unbounded-fanin AND and OR and arbitrary modular functions MOD_m as gates. The class ACC^0 consists of all (sequences of) boolean functions computable by constant-depth ACC circuits of size polynomial in the number of variables; we will consider such circuits in the next chapter.

Theorem 11.42. (Yao 1990) *If* $f \in ACC^0$ *and* f *has* n *variables, then* f *can be computed by an unweighted depth-3 threshold circuit of size* $2^{(\log n)^{\mathcal{O}(1)}}$ *and AND gates of fanin at most* $(\log n)^{\mathcal{O}(1)}$ *at the bottom.*

We omit the somewhat technical proof of this important result.

Unfortunately, so far we cannot prove large lower bounds for unweighted threshold circuits of depth 3. We only can do this under additional restrictions of the circuit structure. Below we will prove the largest known lower bound for depth-3 threshold circuits with AND gates on the bottom. Unfortunately, the bound is "only" super-polynomial, and does not imply lower bounds for constant-depth ACC circuits.

Recall that the generalized inner product function is defined as

$$ GIP_{n,s}(x) = \bigoplus_{i=1}^{n} \bigwedge_{j=1}^{s} x_{ij} . $$

Håstad and Goldmann (1991) combined Lemma 12.34 with the Discriminator Lemma to prove the following lower bound for depth-3 threshold circuits with restricted bottom fanin.

Theorem 11.43. (Håstad and Goldmann 1991) *Any depth-3 threshold circuit which computes* $GIP_{n,s}$ *and has bottom fanin at most* $r < s$, *must be of size* $\exp(\Omega(n/r4^r))$.

In particular, any depth-3 threshold circuit which computes $GIP_{n,\log n}$ and has bottom fanin at most $(\log n)/3$, must be of size $\exp(n^{\Omega(1)})$. Now consider the following boolean function

$$ f_n(x) = \bigoplus_{i=1}^{n} \bigwedge_{j=1}^{\log n} \bigoplus_{k=1}^{n} x_{ijk} . $$

Theorem 11.44. (Razborov and Wigderson 1993) *Any unweighted threshold circuit of depth-3 which computes* $f_n(x)$ *and has unbounded-fanin AND gates at the bottom, must be of size* $n^{\Omega(\log n)}$.

Proof. Let C be an unweighted depth-3 threshold circuit computing $f_n(x)$. Suppose that C has only AND gates on the bottom (next to the inputs) level. The strategy of the proof is to hit C with a random restriction in order to reduce the bottom fanin. Then we apply Theorem 11.43 to the resulting sub-circuit.

Set $p := (2 \ln n)/n$. Let ρ be the random restriction which assigns each variable independently to $*$ with probability p, and to $0, 1$ with probabilities $(1 - p)/2$. Given a boolean function g of n variables and a restriction ρ, we will denote by g_ρ the function we get by doing the substitutions prescribed by ρ.

Let K be a monomial, that is, a conjunction of literals. Denote by $|K|$ the number of literals in K. We are going to show that for each K we have

$$\text{Prob}[|K_\rho| \geq \tfrac{1}{3}\log n] \leq n^{-\Omega(\log n)}. \tag{11.11}$$

To show this, consider two cases.

Case 1: $|K| \leq (\log n)^2$. In this case we have

$$\text{Prob}[|K_\rho| \geq \tfrac{1}{3}\log n] \leq \binom{(\log n)^2}{\tfrac{1}{3}\log n} \cdot p^{\tfrac{1}{3}\log n} \leq \mathcal{O}(p\log n)^{(\log n)/3} \leq n^{-\Omega(\log n)}.$$

Case 2: $|K| \geq (\log n)^2$. In this case we have

$$\text{Prob}[|K_\rho| \geq \tfrac{1}{3}\log n] \leq \text{Prob}[K_\rho \not\equiv 0] = \left(\frac{1+p}{2}\right)^{|K|} \leq n^{-\Omega(\log n)}.$$

Now, when we have (11.11), the reduction to Theorem 11.43 becomes easy. Namely, if our original circuit C had size at most $n^{\epsilon \log n}$ for a sufficiently small $\epsilon > 0$ then, by (11.11), the probability that C_ρ has an AND gate on the bottom level of fanin larger than $\tfrac{1}{3}\log n$ would tend to 0. On the other hand, we have $n \log n$ sums $g(x) = \bigoplus_{k=1}^{n} x_{ijk}$ in f_n, and the probability that some of them will be evaluated by ρ to a constant, is also at most

$$(1-p)^n n \log n \leq e^{-pn} n \log n = e^{-2\ln n} n \log n = \frac{\log n}{n} \to 0.$$

So there exists an assignment ρ such that both of these events happen. That is, after this assignment ρ we are left with a depth-3 threshold circuit C' which has bottom fanin at most $\tfrac{1}{3}\log n$ and computes a subfunction f_n' of f_n where none of the sums $g(x)$ is set to a constant. By setting (if necessary) some more variables to constants, we will obtain a circuit of bottom fanin at most $\tfrac{1}{3}\log n$ computing $GIP_{n,\log n}$. By Theorem 11.43, this is only possible if size(C'), and hence also size(C), is at least $\exp(n^{\Omega(1)})$, contradicting our assumption that size$(C) \leq n^{\epsilon \log n}$.

□

The reason why Theorem 11.44 does not imply large lower bounds for ACC circuits[2] is that Yao's reduction (mentioned above) requires much larger lower bounds, namely bounds of the form $\exp((\log n)^\alpha)$ for $\alpha \to \infty$.

In Theorem 11.44 bottom gates are required to be AND gates. But, as mentioned by Razborov and Wigderson (1993), Johan Håstad observed that the same argument actually gives super-polynomial lower bounds also for unweighted depth-3 circuits whose bottom gates are arbitrary boolean functions of restricted fanin. The restriction is that, if the computed function has N variables, then the fanin of bottom gates

---

[2]And can not imply since, by its definition, the function f_n itself can be computed by a small ACC circuit of depth 3.

cannot exceed $N^{1-\epsilon}$, for an arbitrarily small but fixed constant $\epsilon > 0$. To see this, fix a constant $\epsilon > 0$. Let $n = N^{\epsilon/2}$, $m = N/n \log n$, and consider the following function of N variables:

$$f_N(x) = \bigoplus_{i=1}^{n} \bigwedge_{j=1}^{\log n} \bigoplus_{k=1}^{m} x_{ijk}.$$

Theorem 11.45. *Any unweighted depth-3 threshold circuit which computes f_N and has fanin at most s at the bottom must be of size at least the minimum of $(m/s)^{\Omega(\log n)}$ and $2^{n^{\Omega(1)}}$.*

Proof. We follow the proof of Theorem 11.44 with $p := (2 \ln n)/m$. The same analysis as in Case 1 shows that for any function of at most s variables, the probability that f_ρ depends on at least $(\log n)/3$ variables does not exceed $(s/m)^{\Omega(\log n)}$. The rest is the same.   □

Thus, any unweighted depth-3 threshold circuit computing f_N and having fanin at most $N^{1-\epsilon}$ at the bottom, must be of size $N^{\Omega(\log N)}$.

Exercises

11.1. Show that every boolean function of n variables can be written as a DNF with at most $n2^{n-1}$ literals, and that boolean functions requiring this number of literals exist.

Hint: For the upper bound use induction on n, and use the parity function for the lower bound.

11.2. For a bipartite graph G, let (as before) cnf(G) denote the smallest number of clauses in a monotone CNF representing G. Define the *intersection dimension* int(G) of G as the smallest number r for which it is possible to assign each vertex v a subset $S_v \subseteq \{1, \ldots, r\}$ such that u and v are adjacent in G iff $S_u \cap S_v = \emptyset$. Prove that cnf$(G) = $ int(G).

Hint: Given a monotone CNF $C_1 \wedge \cdots \wedge C_r$, let $S_u = \{i \mid x_u \notin C_i\}$.

11.3. Show that a bipartite graph can be represented by a monotone Σ_3 circuit with top fanin s and middle fanin r iff it is possible to assign each vertex v a boolean $s \times r$ matrix A_v such that u and v are adjacent in G iff the product-matrix $A_u \cdot A_v^T$ (over the reals) has at least one 0 on the diagonal.

Hint: Previous exercise.

11.4. Show that almost all bipartite $n \times n$ graphs require monotone Σ_3 circuits of size $\Omega(\sqrt{n})$.

Hint: Previous exercise.

11.5. ■ **Research Problem.** Prove or disprove: there exists a bipartite $2^m \times 2^m$ graph G such that G can be represented by a monotone Σ_3 circuit of size $2^{\text{polylog}(m)}$, but its bipartite complement $\overline{G}$ cannot be represented by a monotone Σ_3 circuit of such size.

Comment: Note that here G need not be explicit—a mere existence would be enough! This would separate the second level of the communication complexity hierarchy introduced by Babai et al. (1986), and thus solve an old problem in communication complexity.

11.6. (Khasin 1969) Let $n = kr$ and consider colorings of $c : [n] \to [k]$ of the set $[n] = \{1, \ldots, n\}$ by k colors. Say that such a coloring is *balanced* if each color is used for the same number r of points. Given a k-element set of points, say that it is *differently colored* if no two of its points get the same color. Prove that there exist $\ell = \mathcal{O}(ke^k \log n)$ balanced colorings such that every k-element subset of $[n]$ is differently colored by at least one of them.

Hint: Consider independent copies $c_1, \ldots, c_\ell$ of a balanced coloring c selected at random from the set of all $n!/(r!)^k$ such colorings. Show that for every k-element subset S of $[n]$, c colors S differently with probability $p = r^k \cdot \binom{n}{k}^{-1}$. Use the union bound to show that, with probability at least $1 - \binom{n}{k}(1 - p)^\ell$, every k-element subset S will be colored differently by at least one of $c_1, \ldots, c_\ell$. Recall that $r = n/k$ and show that this probability is nonzero for $\ell = \mathcal{O}(ke^k \log n)$.

11.7. Consider the *k-threshold function* $\text{Th}_k^n(x_1, \ldots, x_n)$ which outputs 1 if and only if $x_1 + \cdots + x_n \geq k$. Use the previous exercise to show that Th_k^n can be computed by a monotone Σ_3 formula of size $\mathcal{O}(ke^k n \log n)$.

Hint: Each balanced k-coloring c of $\{1, \ldots, n\}$ gives us a CNF formula $F_c = \bigwedge_{i=1}^k \bigvee_{c(j)=i} x_j$. Use the previous exercise to combine them into an Or-And-Or formula for Th_k^n.

11.8. (Parity decision trees) A $\oplus$-decision tree of n variables $x_1, \ldots, x_n$ is a binary tree whose internal nodes are labeled by subsets $S \subseteq [n]$ and whose leaves have labels from $\{0, 1\}$. If a node has label S then the test performed at that node is to examine the parity $\bigoplus_{i \in S} x_i$. If the result is 0, one descends into the left subtree, whereas if the result is 1, one descends into the right subtree. The label of the leaf so reached is the value of the function (on that particular input). Let $DISJ_{2n}(x, y)$ be a boolean function of $2n$ variables defined by $DISJ_{2n}(x, y) = 1$ iff $x_i y_i = 0$ for all $i = 1, \ldots, n$. Show that any $\oplus$-decision tree for $DISJ_{2n}$ requires $2^{\Omega(n)}$ leaves.

Hint: Transform the decision tree into a $\Sigma_3^{\oplus}$ circuit.

11.9. Let $m = \lceil \log_2(n+1) \rceil$, and consider the function $\text{Sum}_{n,m} : \{0, 1\}^n \to \{0, 1\}^m$ which, given a vector $x \in \{0, 1\}^n$ outputs the binary code of the sum $x_1 + x_2 + \cdots + x_n$. Show that $T(\text{Sum}_{n,m}) = m$.

Hint: Let $y = \text{Sum}_n(x)$. Then $y_m = 1$ iff $x_1 + \cdots + x_n \geq 2^{n-1}$. Also $y_{m-1} = 1$ iff $x_1 + \cdots + x_n - y_m 2^{n-1} \geq 2^{n-2}$, etc.

11.10. Show that every symmetric boolean function of n variables can be computed by an unweighted depth-2 threshold circuit using $2n + 3$ gates.

Hint: Let f be a symmetric boolean function defined by $S = \{s_1, \ldots, s_k\} \subseteq \{0, 1, \ldots, n\}$, that is, $f(x) = 1$ iff $\sum_{i=1}^n x_i \in S$. As an output gate take the threshold function Th_{k+1}^{2k}. As its inputs take the outputs of $\text{Th}_{s_i}^n(x)$ and $\text{Th}_{\leq s_i}^n(x)$ for $i = 1, \ldots, k$; here $\text{Th}_{\leq s_i}^n(x)$ outputs 1 iff $\sum_{i=1}^n x_i \leq s_i$.

11.11. Recall that the *inner product function* is a boolean function of $2n$ variables defined by $IP_n(x, y) = \sum_{i=1}^{n} x_i y_i \bmod 2$. Show that this function can be computed by an unweighted threshold circuit of depth 3 using $\mathcal{O}(n)$ gates.
Hint: Exercise 11.10.

Chapter 12
Large-Depth Circuits

We now consider circuits of depth $d \geq 3$. Out of attempts to prove lower bounds for such circuits, two powerful methods emerged.

The first is a "depth reduction" argument: One tries to reduce the depth one layer at a time, until a circuit of depth 2 (or depth 1) remains. The key is the so-called Switching Lemma, which allows us to replace CNFs on the first two layers by DNFs, thus reducing the depth by 1. This is achieved by randomly setting some variables to constants. If the total number of gates in a circuit is not large enough and the initial circuit depth is small enough, then we will end with a circuit computing a constant function, although a fair number of variables were not set to constants. For functions like the Parity function, this yields the desired contradiction.

The second major tool is a version of Razborov's Method of Approximations, which we have already seen applied to monotone circuits. Given a bounded-depth circuit for a boolean function $f(x)$, one uses this circuit to construct a polynomial $p(x)$ of low degree which differs from $f(x)$ on relatively few inputs, if the circuit does not have too many gates. This immediately implies a lower bound on the circuit size of any boolean function, like the Majority function, which cannot be approximated well by low-degree polynomials.

12.1 Håstad's Switching Lemma

Recall that a boolean function is a t-CNF function if it can be written as an AND of an arbitrary number of clauses, each being an OR of at most t literals (variables and negated variables). Dually, a boolean function is an s-DNF if it can be written as an OR of an arbitrary number of monomials, each being an AND of at most s literals.

In the "depth reduction" argument, an important step is to be able to transform t-CNF into s-DNF, with s as small as possible. If we just multiply the clauses we can get very long monomials, much longer than s. So the function itself may not be an s-DNF. In this case, we can try to assign constants 0 and 1 to some variables and "kill off" all long monomials (that is, evaluate them to 0). If we set some variable

x_i, say, to 1, then two things will happen: the literal $\neg x_i$ gets value 0 and disappears from all clauses, and all clauses containing the literal x_i disappear (they get value 1).

Of course, if we set all variables to constants, then we would be done: no monomials at all would remain. The question becomes interesting if we must leave some fairly large number of variables unassigned. This question is answered by the so-called switching lemma.

Recall that a *restriction* is a map ρ of the set of variables $X = \{x_1, \ldots, x_n\}$ to the set $\{0, 1, *\}$. The restriction ρ can be applied to a function $f(x_1, \ldots, x_n)$, then we get the function f_ρ (called a *subfunction* of f) where the variables are set according to ρ, and $\rho(x_i) = *$ means that x_i is left unassigned. Note that f_ρ is a function of the variables x_i for which $\rho(x_i) = *$. We can then apply another restriction π of the remaining variables to obtain a subfunction $f_{\rho\pi}$ of f_ρ, etc.

Suppose that p is a real number between 0 and 1. A *p-random restriction* assigns each variable x_i a value in $\{0, 1, *\}$ independently with probabilities

$$\mathrm{Prob}[\rho(x_i) = *] = p$$

and

$$\mathrm{Prob}[\rho(x_i) = 0] = \mathrm{Prob}[\rho(x_i) = 1] = \frac{1-p}{2}.$$

Thus, on average, such a restriction leaves a p fraction of variables unassigned. We will sometimes abbreviate the notation and write $\mathrm{Prob}[*]$ rather than $\mathrm{Prob}[\rho(x_i) = *]$. Note that the probability that more than s variables remain unassigned does not exceed $\binom{n}{s} p^s \leq (epn/s)^s$. This, in particular, is an upper bound on the probability that f_ρ cannot be written as an s-CNF.

The Switching Lemma is a substantial improvement of this trivial observation: if f is a t-CNF, then f_ρ will not be an s-DNF with probability at most $(8pt)^s$. Important here is that this "error probability" does not depend on the total number of variables. In fact, we have an even stronger statement (see Exercise 1.7 for *why* this statement is stronger): f_ρ will have a minterm longer than s with at most this probability. Recall that a *minterm* of a boolean function f is a minimal (under inclusion) subset of its variables such that the function can be converted into the constant-1 function by fixing these variables to constants 0 and 1 is some way. Let $\min(f)$ denote the length of the longest minterm of f.

Switching Lemma. Let f be a t-CNF, and let ρ be a p-random restriction. Then $\mathrm{Prob}[\min(f_\rho) > s] \leq (8pt)^s$.

This version[1] of the Switching Lemma is due to Håstad (1986, 1989). Somewhat weaker versions of this lemma were proved earlier by Ajtai et al. (1983), Furst et al. (1984), and Yao (1985). All these proofs used probabilistic arguments. A novel, non-probabilistic proof was later found by Razborov (1995), and we present it in the next section. Actually, his argument also yields an upper bound

---

[1] With a smaller constant 5 instead of 8.

$\text{Prob}[D(f_\rho)] > s \leq (8pt)^s$, where $D(f)$ is the minimum depth of a decision tree computing f; see, for example, a survey of Beame (1994) for details.

12.2 Razborov's Proof of the Switching Lemma

Throughout this section, let s and ℓ be integers with $1 \leq s \leq \ell \leq n$, where n is the total number of variables. We denote by $\mathcal{R}^\ell$ the set of all restrictions leaving exactly ℓ variables unassigned. Hence,

$$|\mathcal{R}^\ell| = \binom{n}{\ell} 2^{n-\ell}.$$

Recall that a *minterm* of f is a restriction $\rho : [n] \to \{0, 1, *\}$ such that $f_\rho \equiv 1$ and which is minimal in the sense that unspecifying any single value $\rho(i) \in \{0, 1\}$ already violates this property. The *support* of ρ is the set of all bits i with $\rho(i) \neq *$, and the *length* of ρ is the size of its support. Let $\min(f)$ be the length of the longest minterm of f, and let

$$\text{Bad}_f(\ell, s) := \{\rho \in \mathcal{R}^\ell \mid \min(f_\rho) > s\}.$$

In particular, $\text{Bad}_f(\ell, s)$ contains all restrictions $\rho \in \mathcal{R}^\ell$ for which f_ρ cannot be written as an s-DNF.

Lemma 12.1. (Razborov 1995) *If f is a t-CNF then*

$$|\text{Bad}_f(\ell, s)| \leq |\mathcal{R}^{\ell-s}| \cdot (2t)^s.$$

Before giving the proof this lemma, let us show that it indeed implies the Switching Lemma. To see this, take a random restriction ρ in $\mathcal{R}^\ell$ for $\ell = pn$, where n is the total number of variables. Then, by this lemma, for every $p \leq 1/2$, the probability that f_ρ cannot be written as an s-DNF is at most

$$\frac{|\text{Bad}_f(\ell, s)|}{|\mathcal{R}^\ell|} \leq \frac{\binom{n}{\ell-s} 2^{n-\ell+s} (2t)^s}{\binom{n}{\ell} 2^{n-\ell}} \leq \left(\frac{\ell}{n-\ell}\right)^s (4t)^s = \left(\frac{4tp}{1-p}\right)^s \leq (8pt)^s.$$

Proof of Lemma 12.1. A cute idea of this proof (which itself is relatively simple) is to use the following "coding principle":

> In order to prove that some set A cannot be very large, try to construct a mapping Code : $A \to B$ of A to some set B which is a priori known to be small, and *give a way to retrieve* each element $a \in A$ from its code Code(a). Then Code is injective, implying that $|A| \leq |B|$.

Let F be a t-CNF formula for f. Fix an order of its clauses and fix an order of literals in each clause. We want to upper-bound the number $|\mathrm{Bad}_f(\ell, s)|$ of restrictions $\rho \in \mathcal{R}^\ell$ that are "bad" for f, that is, for which the subfunction f_ρ contains a minterm of length $> s$. Our goal is to use the underlying CNF formula F to construct an encoding

$$\text{Code} : \mathrm{Bad}_f(\ell, s) \to \mathcal{R}^{\ell - s} \times S \quad \text{with} \quad S \subseteq \{0, 1\}^{ts + s} \quad \text{and} \quad |S| \leq (2t)^s$$

such that, knowing the formula F, we can reconstruct a restriction ρ from $\text{Code}(\rho)$. By the coding principle, we will be then done.

To construct the desired encoding, fix a bad restriction $\rho \in \mathrm{Bad}_f(\ell, s)$. We know that the subfunction f_ρ must contain a minterm π' of length $s' \geq s + 1$. By unspecifying an arbitrary subset of $s' - s$ variables, we truncate π' to π so that π has length exactly s. After ρ is applied to F, some clauses get the value 1 and disappear, while some literals get the value 0 and disappear from the remaining clauses. Moreover, ρ cannot set any clause to 0, for otherwise we would have $f_\rho \equiv 0$. Further, we have that $f_{\rho\pi}$ cannot be constant because π' was a minterm of f_ρ.

Consider the first clause C_1 of F which is not set to 1 by ρ but is set to 1 by $\rho\pi$. Let π_1 be the portion of π that assigns values to variables in C_1. Let also $\overline{\pi}_1$ be the uniquely determined restriction which has the same support as π_1 and does *not* set the clause C_1 to 1. That is, $\overline{\pi}_1$ evaluates all the literals "touched" by π_1 to 0.

Define the string $\mathbf{a}_1 \in \{0, 1\}^t$ based on the fixed ordering of the variables in clause C_1 by letting the j-th component of $\mathbf{a}_1$ be 1 if and only if the j-th variable in C_1 is set by π_1 (and hence, also by $\overline{\pi}_1$). That is, $\mathbf{a}_1$ is just the characteristic vector of the (common) support of restrictions π_1 and $\overline{\pi}_1$. Note that there must be at least one 1 in $\mathbf{a}_1$ (the support of π_1 cannot be empty). Here is a typical example:

$$
\begin{array}{cccccc}
C_1 = & x_3 \lor & \neg x_4 \lor & x_6 \lor & x_7 \lor & x_{12} \\
\pi_1 = & * & 1 & * & 1 & 0 \\
\overline{\pi}_1 = & * & 1 & * & 0 & 0 \\
\mathbf{a}_1 = & 0 & 1 & 0 & 1 & 1
\end{array}
$$

The main property of the string $\mathbf{a}_1$ is that *knowing C_1 and $\mathbf{a}_1$ we can reconstruct $\overline{\pi}_1$*: string $\mathbf{a}_1$ tells us what literals of C_1 must be set by the restriction $\overline{\pi}_1$, and the property that C_1 does not evaluate to 1 allows us to infer the restriction itself.

Now, if $\pi_1 \neq \pi$, we repeat the above argument with $\pi \setminus \pi_1$ in place of π, $\rho\pi_1$ in place of ρ and find a clause C_2 which is the first clause not set to 1 by $\rho\pi_1$. Based on this we generate π_2, $\overline{\pi}_2$ and $\mathbf{a}_2$ as before. Continuing in this way we get a sequence of clauses $C_1, C_2, \ldots$. Each C_i contains some variable that was not in C_j for $j < i$, so we must stop after we have identified $m \leq s$ clauses. Hence, $\pi = \pi_1 \pi_2 \ldots \pi_m$.

Let $\mathbf{b} \in \{0, 1\}^s$ be a vector that indicates for each variable set by π (which are the same as those set by $\overline{\pi} = \overline{\pi}_1 \overline{\pi}_2 \ldots \overline{\pi}_m$) whether it is set to the same value to which $\overline{\pi}$ sets it. (Recall that π_i must set at least one literal of C_i to 1 and may set some of them to 0, whereas $\overline{\pi}_i$ sets all these literals to 0.) We encode the restriction ρ by the string

$$\mathrm{Code}(\rho) := \langle \rho\overline{\pi}_1\overline{\pi}_2 \ldots \overline{\pi}_m, \mathbf{a}_1, \ldots, \mathbf{a}_m, \mathbf{b} \rangle.$$

Our goal is to show that: (1) the mapping $\rho \mapsto \mathrm{Code}(\rho)$ is injective, and (2) its range is not too large. To achieve the first goal, it is enough to show how to reconstruct ρ uniquely, given $\mathrm{Code}(\rho)$.

First note that it is easy to reconstruct $\overline{\pi}_1$. Identify the first clause of F which is not set to 1 by $\rho\overline{\pi}_1\overline{\pi}_2 \ldots \overline{\pi}_m$. Since none of the $\overline{\pi}_i$ sets a clause to 1, this must be clause C_1. Now use $\mathbf{a}_1$ to identify the variables of C_1 that are set by $\overline{\pi}_1$, and use $\mathbf{b}$ to identify how π_1 would set these variables. Thus we have reconstructed the sub-restrictions π_1 and $\overline{\pi}_1$. Knowing these sub-restrictions and the entire restriction $\rho\overline{\pi}_1\overline{\pi}_2 \ldots \overline{\pi}_m$ we can reconstruct the restriction $\rho\pi_1\overline{\pi}_2 \ldots \overline{\pi}_m$.

Now we can identify C_2: it is the first clause of F which is not set to 1 by $\rho\pi_1\overline{\pi}_2 \ldots \overline{\pi}_m$. Then we use $\mathbf{a}_2$ to identify the variables of C_2 set by $\overline{\pi}_2$, and use $\mathbf{b}$ to identify how π_2 would set these variables. Continuing in this way, we can reconstruct the restriction $\pi_1\pi_2 \ldots \pi_m$ and thus the original restriction ρ.

To finish the proof of the lemma, it is enough to upper-bound the range of the mapping Code. First, observe that restrictions $\rho\overline{\pi}_1\overline{\pi}_2 \ldots \overline{\pi}_m$ belong to $\mathcal{R}^{\ell-s}$. Hence, the number of such restrictions does not exceed $|\mathcal{R}^{\ell-s}|$. The number of strings $\mathbf{b} \in \{0,1\}^s$ is clearly at most 2^s. Finally, each $(\mathbf{a}_1, \ldots, \mathbf{a}_m)$ is a string in $\{0,1\}^{mt}$ with the property that each substring $\mathbf{a}_j \in \{0,1\}^t$ has at least one 1 and the total number of 1s in all $\mathbf{a}_j$ is s. The number of such strings $(\mathbf{a}_1, \ldots, \mathbf{a}_m)$ with k_i ones in $\mathbf{a}_i$ is

$$\prod_{i=1}^m \binom{t}{k_i} \leq \prod_{i=1}^m t^{k_i} = t^{\sum_{i=1}^m k_i} = t^s.$$

The number of positive integers $k_1, \ldots, k_m$ such that $k_1 + \cdots + k_m = s$ is $\binom{s-1}{m-1} \leq 2^s$ (show this!). Thus the range of $\mathrm{Code}(\rho)$ does not exceed $|\mathcal{R}^{\ell-s}| \times (2t)^s$, as desired. $\quad\square$

12.3 Parity and Majority Are Not in AC^0

An *alternating circuit* of depth d, or an AC circuit, is a circuit formed by d alternating levels of unbounded-fanin AND and OR gates; inputs are variables and their negations. The class[2] AC^0 consists of all (sequences of) boolean functions computable by constant-depth alternating circuits of size polynomial in the number of variables. We will now use the Switching Lemma to show that the parity function

$$\mathrm{Parity}_n(x) = x_1 + x_2 + \cdots + x_n \bmod 2$$

---

[2]Usually, AC^k denotes the class of all boolean functions computable by AC circuits of depth $d = \mathcal{O}(\log^k n)$. Thus, superscript "0" tells us that we are dealing with constant-depth circuits.

cannot be computed by such circuits of polynomial size; that is,

$$\text{Parity}_n \notin AC^0.$$

It can be shown that, for every $d \geq 2$, the parity function *can* be computed by depth-$(d + 1)$ circuits of size $2^{\mathcal{O}(n^{1/d})}$ (see Exercise 12.1). In particular, if we allow circuit depth to be about $\log n / \log \log n$ then $\text{Parity}_n(x)$ can be computed using only $\mathcal{O}(n^2 / \log n)$ gates.

Theorem 12.3 below shows that the upper bound $2^{\mathcal{O}(n^{1/d})}$ is almost optimal. The theorem itself is a direct consequence of a fact that every function in AC^0 can be reduced to a constant function by setting relatively few variables to constants.

Let $R(f)$ denote the minimal number r such that f can be made constant by fixing r variables to constants 0 and 1. The larger $R(f)$ is, the more "robust" the function is. For example, $R(f) = 1$ if f is an OR or an AND of literals, whereas $R(f) = n$ if f is the parity of n variables.

The following theorem states that functions computable by small circuits of constant depth are not robust enough.

Theorem 12.2. *If a boolean function f of n variables can be computed by a depth-$(d + 1)$ alternating circuit of size S, then*

$$R(f) \leq n - \frac{n}{c_d (\log S)^{d-1}} + 2 \log S ,$$

where $c_d > 0$ is a constant depending only on d.

This gives the following lower bound on the size S:

$$2 \log S \geq \frac{n}{c_d (\log S)^{d-1}} - (n - R(f)). \tag{12.1}$$

Proof. Fix a depth-$(d + 1)$ circuit of size S computing f. Our first goal is to reduce the fanin of gates on the first (next to the inputs) layer. Suppose that they are OR gates (a symmetric argument applies if they are AND gates).

We think of each OR gate on the bottom layer as a 1-DNF. We apply the Switching Lemma with $t = 1$, $s = 2 \log S$ and $\text{Prob}[*] = p := 1/16$, and deduce that after a random restriction each of the these 1-DNFs becomes an s-CNF (in fact, a single clause of length $\leq s$) with probability at least

$$1 - (8pt)^s = 1 - 2^{-s} = 1 - S^{-2}.$$

Since we have at most S of the 1-DNFs, this in particular implies that there is a restriction that makes all these 1-DNFs expressible as an OR of at most s input literals. We apply such a restriction, and what we obtain is a circuit of depth $d + 1$ such that each bottom gate has fanin at most $k := 2 \log S$ and the circuit still computes a subfunction of f on $n_1 = n/32$ variables.

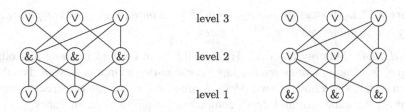

input variables

Fig. 12.1 After the Switching Lemma is applied to the CNFs computed at the first two levels (from the *bottom*), the levels 2 and 3 can be collapsed into one level

We now apply the Switching Lemma to the first two bottom layers of the obtained circuit with $\text{Prob}[*] = q := 1/(16k)$ and both s and t equal to k. We get that, for each AND gate on layer 2, after the restriction the gate can be replaced by a k-DNF with probability at least

$$1 - (8qk)^k = 1 - 2^{-k} = 1 - S^{-2}.$$

Hence, there is a restriction for which this is true for all the (at most) S gates at layer 2. We apply this restriction, replace each layer-2 gate with a k-DNF, and use associativity to collapse the OR gate of each DNF into an OR gates of the second layer of the original circuit. In this way we collapse layer 2 with layer 3 (see Fig. 12.1).

Now we have a circuit of depth d that computes a subfunction of f on

$$qn_1 = \frac{n}{16^2 k} = \frac{n}{ck} \qquad (c = 256)$$

variables, and such that every bottom gate has fanin at most k.

If we repeat the same argument another $d - 2$ times, we will eventually end up with a circuit of depth 2 such that the fanin of the bottom gates is at most $k = 2 \log S$ and the circuit computes a subfunction of f on

$$m := \frac{n}{(ck)^{d-1}} = \frac{n}{c_d (\log S)^{d-1}}$$

variables, where c_d is a constant depending only on d. Every k-DNF (as well as every k-CNF) can be evaluated to a constant 1 (resp., constant 0) by setting at most $k = 2 \log S$ variables to constants. We therefore have that the original function f can be made constant by fixing $n - m + k$ variables of f implying that $R(f) < n - m + k$, as desired. $\qquad\square$

Since $R(\text{Parity}_n) = n$, (12.1) immediately yields the following lower bound for the parity function.

Theorem 12.3. (Håstad 1986) *Any depth-*$(d+1)$ *alternating circuit computing the parity of* n *variables requires* $2^{\Omega(n^{1/d})}$ *gates.*

Remark 12.4. (Majority function) Theorem 12.3 can be used to show that other "simple" boolean functions require large constant-depth circuits as well. For this, it is enough to show how, from a depth-d circuit for f, we can construct a circuit of almost the same size and depth computing the parity function $\text{Parity}_n(x)$. To illustrate this, let us consider the majority function $\text{Maj}_n(x)$ of an even number n of variables.

Note that, when directly applied, the lower bound (12.1) will only give a constant lower bound on for Maj_n, because $R(\text{Maj}_n) \le n/2$. To obtain a larger lower bound, we take an arbitrary depth-d circuit computing Maj_n. Let S be its size. Having such a circuit, we can compute all functions

$$E_k^{n/2}(x) := \text{Th}_k^{n/2}(x) \wedge \neg\text{Th}_{k+1}^{n/2}(x)$$

for all odd $k \le n/2$ using depth-d circuits of size $\mathcal{O}(nS)$. Now, $\text{Parity}_{n/2}(x)$ is just an OR of these functions, implying that $\text{Parity}_{n/2}$ can be computed by a depth-$(d+1)$ circuit of size $\mathcal{O}(nS)$. Thus, Theorem 12.3 implies that Maj_n requires depth-d circuits of size $S = 2^{\Omega(n^{1/d})}$.

Remark 12.5. We know that the majority function does not belong to AC^0. This function outputs value 1 on all inputs of weight $n/2$, and outputs value 0 on inputs of weight smaller than $n/2$; the weight of a binary vector is the number of ones in it. But what if we will only require that the circuit outputs 0 on inputs of "small" weight, and outputs 1 on inputs of "large" weight; on other inputs the circuit can output any value. It turns out that this relaxed problem is actually much easier.

A $1/4$-*approximate selector* is any boolean function whose value is 0 if the number of ones in the input is at most $n/4$, 1 if the number of ones is at least $3n/4$, and can be either 0 or 1 otherwise. Such a function provides a rough estimate of the number of ones and is extremely useful in parallel computation. It was shown by Ajtai et al. (1983), and Ajtai and Ben-Or (1984) that there exist polynomial-size, constant-depth circuits that compute a $1/4$-approximate selector function.

Remark 12.6. By Theorem 12.2, any polynomial-size circuit of constant depth can be made to output a constant by fixing $n - n/\text{polylog}(n)$ inputs. That is,

$$f \in \text{AC}^0 \text{ implies } R(f) \le n - \frac{n}{\text{polylog}(n)}.$$

This yields a superpolynomial size lower bound for a constant depth circuit computing any function that cannot be made constant by setting $n - n/\text{polylog}(n)$ input bits. However, it does not say anything about functions which *can* be made constant by setting this many bits. Can one prove at least super-linear lower bounds for such functions? Motivated by this question, Chaudhuri and Radhakrishnan (1996) used a direct combinatorial argument to prove that if a boolean function

f of n variables can be computed by a depth-k alternating circuit of size S, then

$$S = \Omega\left(R(f)^{1+\frac{1}{4k}}\right).$$

Now, if f is any $1/4$-approximate selector, then $R(f) \geq n/4$, and we obtain a super-linear lower bound $S = \Omega(n^{1+1/4^k})$ on the size S of any depth-k circuit computing f. They also proved that a $1/4$-approximate selectors computable by a depth-k circuit of size $S = \Omega(n^{1+1/2^k})$ exist.

Using similar arguments as in the proof of Theorem 12.2, the following general lower bound for bounded-depth circuits can be derived.

Let $C(f)$ denote the minimum number k such that f can be written as a k-DNF *and* as a k-CNF. The number $C(f)$ is also known as the *certificate complexity* of f.

Theorem 12.7. *Let f be a boolean function computable by a depth-d circuit of size S, and let ρ be a p-random restriction with $p = 16^{-d} k^{-d+1}$. Then*

$$\mathrm{Prob}[C(f_\rho) > k] \leq S \cdot 2^{-k}.$$

Proof. (Due to Linial et al. 1993) We view the restriction ρ as obtained by first having a restriction with $\mathrm{Prob}[*] = 1/16$, and then $d - 1$ consecutive restrictions each with $\mathrm{Prob}[*] = 1/(16k)$.

With high probability, after the first restriction, at the bottom level of the circuit all fanins are at most k. To see this, consider two cases for each gate at the bottom level of the original circuit.

Case 1: The original fanin is $\geq 2k$. In this case, the probabilitysss that the gate was not eliminated by ρ, that is, that no input to this gate got assigned a 0 if this is an AND gate, or no input got assigned a 1 if this is an OR gate is at most $((1 + p)/2)^{2k} < (0.6)^{2k} < 2^{-k}$.

Case 2: The original fanin is $\leq 2k$. In this case, the probability that at least k inputs got assigned a $*$ is at most $\binom{2k}{k} p^k \leq (2e)^k (1/16)^k < 2^{-k}$.

Thus the probability of failure at the first stage is at most $m_1 2^{-k}$, where m_1 is the number of gates at the bottom level.

We now apply $d - 2$ more restrictions with $\mathrm{Prob}[*] = 1/(16k)$. After each of these, we use the Switching Lemma (with $t = k$) to convert the lower two levels from CNF to DNF (or vice versa), and collapse the second and third levels (from the bottom) to one level, reducing the depth by one. For each gate at distance two from the inputs, the probability that it has a minterm (respectively, maxterm) of size larger than k is bounded by $(8pk)^s = 2^{-k}$. The probability that *some* of these gates has a minterm (respectively, maxterm) of size larger than k is no more than $m_i 2^{-k}$, where m_i is the number of gates of level i of the original circuit.

After these $d - 2$ stages we are left with a CNF (or DNF) formula of bottom fanin at most k. We now apply the last restriction with $\mathrm{Prob}[*] = 1/(16k)$ and, by the

Switching Lemma, get a function f_ρ with $\min(f_\rho) \le k$. The probability of failure at this stage is at most $(8pk)^k = 2^{-k}$.

To compute the total probability of failure, we observe that each gate of the original circuit contributed a 2^{-k} probability of failure exactly once.   □

12.3.1 Majority of AC^0 Circuits

Aspnes et al. (1994) proved that the parity function remains hard to compute by shallow circuits, even if a majority gate is allowed as an output gate. By an *output-majority circuit* we will mean an unbounded fanin circuit over $\{\wedge, \vee, \neg\}$ whose last (output) gate is a majority gate Maj; recall that this function outputs 1 iff at least half of its inputs are 1s.

Theorem 12.8. (Aspnes et al. 1994) *Every majority-output circuit of depth $d + 1$ for* $\text{Parity}_n(x)$ *requires size at least* $2^{\Omega(n^{1/4d})}$.

Proof. We first show that AC-circuits of small depth and small size can be approximated well enough by low-degree polynomials. To spare parenthesis, we will often say that a function $g(n)$ "is at most about" $h(n)$ if $g(n) = \mathcal{O}(h(n))$.

Claim 12.9. For any $\epsilon > 0$ and any boolean function f computed by an AC-circuit of depth d and size S, there exists a real multivariate polynomial of degree at most about $((\log(S/\epsilon) \log S)^d$ which computes f for all but at most $\epsilon 2^n$ input vectors.

Proof. Set $r := \lfloor \log(S/\epsilon) \rfloor$. Consider the distribution of the inputs of each gate when the inputs of the circuit are chosen uniformly at random. By Lemma 2.8, using S as an upper bound on the number of inputs to a gate g, there exists some polynomial of degree at most about $r \log S$ which computes the value of the gate with probability at least $1 - 2^{-r} \ge 1 - \epsilon/S$ when the inputs to the circuit are generated uniformly. The composition of these polynomials is a polynomial of degree at most about $(r \log S)^d$ which computes f with probability at least $1 - S(\epsilon/S) = 1 - \epsilon$, that is, which computes f for all but at most $\epsilon 2^n$ inputs.   □

We now add a majority gate on the top of the circuit and obtain the following analog of Claim 12.9. Recall from Sect. 2.7 that a real polynomial $p(x)$ *signum-represents* a given boolean function $f(x)$ if $p(x) > 0$ for all $x \in f^{-1}(1)$, and $p(x) < 0$ for all $x \in f^{-1}(0)$.

Claim 12.10. For any $\epsilon > 0$ and any boolean function computed by a majority-output circuit of depth $d + 1$ and size S, there exists a real multivariate polynomial of degree at most about $((\log(S^2/\epsilon) \log S)^d$ which signum-represents f for all but at most $\epsilon 2^n$ input vectors.

Proof. Suppose that the majority gate has k inputs; hence, $k \le S$. For the subcircuit generating the i-th input, use Claim 12.9 to obtain a polynomial p_i of degree at most about $(\log(kS/\epsilon) \log S)^d$ which computes that input for all but $\epsilon 2^n/k$ inputs. Then

$\sum_{i=1}^{k} p_i - k/2$ is a polynomial of degree at most about $(\log(S^2/\epsilon) \log S)^d$ which signum-represents f for all but at most $\epsilon 2^n$ inputs.   □

We now turn to the actual proof of Theorem 12.8. Suppose the size of the circuit is S. Then by Claim 12.10 (applied with, say, $\epsilon = 1/4$), there exists a real polynomial p of degree k at most about $(\log(4S^2) \log S)^d = (\log S)^{2d}$ which signum-represents Parity(x) for all but at most $2^n/4$ input vectors. But by Corollary 2.12 the degree of p must be $k = \Omega(\sqrt{n})$, and thus $\log S = \Omega(n^{1/4d})$. □

Remark 12.11. Recently, there was an interesting development concerning the size of constant-depth circuits computing the clique function CLIQUE(n, k). Recall that this function has $\binom{n}{2}$ variables x_{ij}, one for each potential edge in a graph on a fixed set of n vertices; the function outputs 1 iff the associated graph contains a clique (complete subgraph) on some k vertices. It is easy to see that this function can be computed by a depth-2 circuit (a monotone DNF) of size $\binom{n}{k} \le n^k$. Rossman (2008) showed that, for every fixed integer k, any depth-d circuit computing CLIQUE(n, k) must have $c_d n^{k/4}$ gates. Note that only the multiplicative constant in the lower bound depends on the circuit depth.

12.3.2   Parity is Even Hard to Approximate

We have seen that shallow circuits of small size cannot compute the parity function Parity(x) *exactly*. But perhaps such circuits can at least *approximate* this function? Note that a trivial circuit, which outputs 0 on all input vectors, agrees with Parity(x) on a $1/2$ fraction of all 2^n inputs. It turns out that this is almost the best shallow circuits can do.

Theorem 12.12. (Håstad 1986) *Every alternating depth-d circuit of size $S \le 2^{o(n^{1/d})}$ can agree with* Parity(x) *on at most a $1/2 + 1/S$ fraction of input vectors.*

Proof. We follow the simplified proof due to Klivans and Vadhan as described in the survey of Viola (2009). Let $F(x)$ be a depth-d circuit over $\{\wedge, \vee, \neg\}$ of size $S \le 2^{o(n^{1/d})}$, and suppose that it agrees with Parity(x) on more than a $1/2 + 1/S$ fraction of input vectors. Hence,

$$\mathrm{Prob}[F(x) = \mathrm{Parity}(x)] \ge \frac{1}{2} + \frac{1}{S},$$

where x is a random vector uniformly distributed in $\{0, 1\}^n$. We will use $F(x)$ to design a majority-output circuit of depth d and size $2^{o(n^{1/d})}$ which computes Parity(x) on all inputs, contradicting Theorem 12.8. For this purpose, associate with each vector $a \in \{0, 1\}^n$ the circuit

$$F_a(x) := F(a \oplus x) \oplus \mathrm{Parity}(a).$$

Each circuit F_a has the same depth and size as the original circuit F. Moreover, for any fixed $x \in \{0, 1\}^n$, we have that

$$F_a(x) \oplus \text{Parity}(x) = F(a \oplus x) \oplus \text{Parity}(a) \oplus \text{Parity}(x)$$
$$= F(a \oplus x) \oplus \text{Parity}(a \oplus x).$$

Thus for every fixed $x \in \{0, 1\}^n$, we have that

$$\text{Prob}[F_a(x) = \text{Parity}(x)] \geq \frac{1}{2} + \epsilon \quad \text{with} \quad \epsilon = \frac{1}{S},$$

where this time a is a random vector uniformly distributed in $\{0, 1\}^n$. Now pick $m := nS^2$ independent copies $a_1, \ldots, a_m$ of a, and consider the (random) output-majority circuit

$$\overline{F}(x) := \text{Maj}(F_{a_1}(x), \ldots, F_{a_m}(x)).$$

By the Threshold Trick (Lemma 1.5; see also the proof of Theorem 1.6), the probability that this circuit makes an error on at least one of all 2^n possible inputs is at most $2^n e^{-2\epsilon^2 m} = 2^n e^{-2n} \ll 1$. Therefore, there must be a setting of the random inputs which gives the correct answer for all inputs. The obtained circuit is no longer probabilistic, and its size is at most $mS + 1$, which is polynomial in S, the size of the original circuit F.   □

12.4   Constant-Depth Circuits and Average Sensitivity

We will call two boolean vectors (assignments) $x, y \in \{0, 1\}^n$ *neighbors* and write $x \sim y$, if x and y differ in exactly one position. The *sensitivity*, s(f, x), of a boolean function f on an assignment $x \in \{0, 1\}^n$ is the number of neighbors of x on which f takes different value than $f(x)$:

$$\text{s}(f, x) = |\{y : y \sim x \text{ and } f(y) \neq f(x)|$$

Sensitivity is an important measure of boolean functions: as shown by Khrapchenko (1971), high sensitivity implies large formula size (see Theorem 6.27). Namely, if we set for $a \in \{0, 1\}$,

$$d_a(f) := \frac{1}{|f^{-1}(a)|} \sum_{x \in f^{-1}(a)} \text{s}(f, x),$$

then Theorem 6.27 implies that every DeMorgan formula computing f must have at least $d_0(f) \cdot d_1(f)$ leaves.

We are now going to show that high "average sensitivity" forces large circuits of constant depth as well. The *average sensitivity*, as(f), of f is the expected sensitivity of f on a random assignment:

$$\text{as}(f) := 2^{-n} \sum_x \text{s}(f, x).$$

For example, if $f(x) = \text{Parity}(x)$ is the parity of n variables, then $\text{s}(f, x) = n$ for every assignment x, implying that as$(f) = n$. We have already seen that Parity(x) requires constant-depth circuits of exponential size. It turns out that *any* boolean function of large average sensitivity requires large constant-depth circuits.

Theorem 12.13. (Boppana 1997) *For every boolean function f, every depth-d circuit computing f must have a number of gates that is exponential in* as$(f)^{1/(d-1)}$.

A somewhat weaker lower bound $2^{\Omega(\text{as}(f)^{1/d})}$ was proved earlier by Linial et al. (1993) using Fourier transforms of boolean functions. Actually, they proved a more general result: for every $t \geq 0$, every depth-d circuit computing f requires at least

$$2^{\Omega(t^{1/d})} \sum_{S \subseteq [n] : |S| > t} \widehat{f}(S)^2$$

gates, where $\widehat{f}(S)$ is the S th Fourier coefficient of f.

As an immediate consequence, we obtain that all functions in AC^0 have small average sensitivity.

Corollary 12.14. *If a boolean function f can be computed by an alternating depth-d circuit of size S, then* as$(f) = \mathcal{O}(\log^{d-1} S)$.

We split the proof of Theorem 12.13 into a sequence of lemmas, each of which might be of independent interest. We first show that depth-2 circuits have low average sensitivity.

Lemma 12.15. *If f is a k-DNF or a k-CNF function, then* as$(f) \leq 2k$.

Proof. Let f be a k-DNF function. (The k-CNF case is dual.) Take an assignment x for which $f(x) = 1$. Then there must be a monomial M of length at most k such that $M(x) = 1$. Hence,

$$\text{s}(f, x) = |\{y \mid x \sim y, f(y) = 0\}| \leq |\{y \mid x \sim y, M(y) = 0\}| \leq k.$$

Since $|f^{-1}(1)| \leq 2^n$, Exercise 2.29 yields

$$\text{as}(f) = \frac{1}{2^{n-1}} \sum_{x \in f^{-1}(1)} \text{s}(f, x) \leq \frac{k}{2^{n-1}} 2^n = 2k. \qquad \square$$

Let Q_n be the n-dimensional binary hypercube. Recall that this is a graph whose vertices are vectors in $\{0,1\}^n$. Two vectors $x, y \in \{0,1\}^n$ are adjacent in Q_n, written as $x \sim y$, if x and y differ in exactly one position. This is a regular graph of degree n with 2^n vertices and $n2^{n-1}$ edges. A *random edge* is a random variable (x, y) that is uniformly distributed over the set Q_n of all edges.

Lemma 12.16. *If (x, y) is a random edge in Q_n, then*

$$\mathrm{as}(f) = n \cdot \mathrm{Prob}[f(x) \neq f(y)].$$

Proof. Every boolean function f of n variables defines a bipartite graph G_f with parts $f^{-1}(0)$ and $f^{-1}(1)$, where $(x, y) \in f^{-1}(0) \times f^{-1}(1)$ is an edge of G_f iff $y = x \oplus e_i$ for some $i \in [n]$. Every vertex of Q_n has degree n. By Euler's theorem, the sum of degrees of all vertices in any graph is equal to two times the number of edges. Thus, there are $n2^n/2 = n2^{n-1}$ edges in Q_n. By Exercise 2.28, $\mathrm{as}(f) = |G_f|/2^{n-1}$. Hence, (x, y) belongs to G_f with probability

$$\mathrm{Prob}[f(x) \neq f(y)] = |G_f|/n2^{n-1} = (\mathrm{as}(f)2^{n-1})/n2^{n-1} = \mathrm{as}(f)/n. \qquad \square$$

Lemma 12.17. *(Sensitivity of restrictions) If f is a boolean function, and ρ is a p-random restriction, then $\mathrm{E}[\mathrm{as}(f_\rho)] = p \cdot \mathrm{as}(f)$.*

Proof. Let $x \sim y$ be a random edge of Q_n, independent of ρ. Let i be the unique position for which $x_i \neq y_i$. Let ρx denote the assignment in $\{0,1\}^n$ whose j-th position $(j = 1, \ldots, n)$ is x_j if ρ assigns $*$ to this position (that is, $\rho_j = *$), and otherwise is the bit 0 or 1 assigned by ρ to that position (that is, $\rho_j = 0$ or $\rho_j = 1$). If for example, $x = (1, 1, 0, 1, 1)$ and $\rho = (0, *, 1, *, 0)$ then $\rho x = (0, 1, 1, 1, 0)$.

Observe that $\rho x = \rho y$ if $\rho_i \neq *$, and $\rho x \sim \rho y$ if $\rho_i = *$. Let (x', y') be the conditional random variable $(\rho x, \rho y)$ conditioned on the event $\rho_i = *$. By the implication $\rho_i = * \Rightarrow \rho x \sim \rho y$, the pair (x', y') is always an edge of Q_n. Furthermore, (x', y') is a random edge of Q_n, because we favored no particular edge. By Lemma 12.16,

$$\mathrm{E}[\mathrm{as}(f_\rho)] = \mathrm{E}[n \cdot \mathrm{Prob}[f(\rho x) \neq f(\rho y)]|\rho]$$

$$= n \cdot \mathrm{E}[\mathrm{Prob}[f(\rho x) \neq f(\rho y)]|\rho]$$

$$= n \cdot \mathrm{Prob}[f_\rho(x) \neq f_\rho(y)]$$

$$= n \cdot \mathrm{Prob}[f(\rho x) \neq f(\rho y)]$$

$$= n \cdot \mathrm{Prob}[\rho x \neq \rho y \wedge f(\rho x) \neq f(\rho y)]$$

$$= n \cdot \mathrm{Prob}[\rho_i = * \wedge f(\rho x) \neq f(\rho y)]$$

$$= n \cdot \mathrm{Prob}[\rho_i = *] \cdot \mathrm{Prob}[f(\rho x) \neq f(\rho y)|\rho_i = *]$$

$$= n \cdot p \cdot \mathrm{Prob}[f(\rho x) \neq f(\rho y)|\rho_i = *]$$

$$= n \cdot p \cdot \text{Prob}[f(x') \neq f(y')]$$
$$= p \cdot \text{as}(f). \qquad \square$$

As the next step in our proof of Theorem 12.13, we give a version of the Switching Lemma that applies to a collection of functions. Let F be a nonempty, finite set of boolean functions. The *DNF complexity*, dnf(F), of F is the minimum number k such that every function $f \in F$ can be written as a k-DNF. The *CNF complexity* cnf(F) is defined similarly. If ρ is a restriction, define $F_\rho = \{f_\rho \mid f \in F\}$.

Lemma 12.18. *Let F be a nonempty, finite set of boolean functions, and let ρ be a p-random restriction. If $16p \cdot \text{cnf}(F) \leq 1$ then $\text{E}[\text{dnf}(F_\rho)] \leq \log(4|F|)$.*

Proof. Let t be a non-negative integer, and let $D := \text{dnf}(F_\rho)$. By the Switching Lemma and our assumption $16p \cdot \text{cnf}(F) \leq 1$, we have

$$\text{Prob}[D > t] \leq \sum_{f \in F} \text{Prob}[\text{dnf}(f_\rho) > t] \leq \sum_{f \in F} (8p \cdot \text{cnf}(f))^t \leq |F|/2^t.$$

Let l be an integer such that $2^l < |F| \leq 2^{l+1}$. Then

$$\text{E}[D] = \sum_{t=1}^{\infty} \text{Prob}[D \geq t]$$

$$= \sum_{t=1}^{l} \text{Prob}[D \geq t] + \sum_{t=l+1}^{\infty} \text{Prob}[D \geq t]$$

$$\leq l + \sum_{t=l+1}^{\infty} \text{Prob}[D \geq t] \leq l + \sum_{t=l+1}^{\infty} |F|/2^t$$

$$= l + |F|/2^l \leq \log|F| + 2 = \log(4|F|). \qquad \square$$

By a (d, w, k)-*circuit* we will mean an unbounded-fanin circuit such that: (1) its depth is precisely $d + 1$, (2) every gate on the bottom (next to inputs) level has fanin at most k, and (3) each of the remaining levels has at most w gates.

Lemma 12.19. *If f is a boolean function that is computable by a (d, w, k)-circuit, then $\text{as}(f) \leq 2k[16 \log(4w)]^{d-1}$.*

Proof. The proof is by induction on d. In the base case $d = 1$ we have either a k-DNF or a k-CNF, and the result follows from Lemma 12.15. For the induction step, assume that $d \geq 2$. If $k = 0$ then f is a constant function, and hence has average sensitivity 0. So assume that $k \geq 1$, and take a (d, w, k)-circuit that computes f; by duality, we may assume that the bottom level consists of OR gates. Let F be the set of functions computed by the gates at the next level; hence, $|F| \leq w$ and cnf$(F) \leq k$.

Let ρ be a p-random restriction, where $p = 1/16k$. Set $l := \mathrm{dnf}(F_\rho)$. By merging levels 2 and 3, we see that f_ρ is computable by a $(d - 1, w, l)$-circuit. So we may apply the induction hypothesis with $f := f_\rho$ and $k := l$. Then

$$
\begin{aligned}
\mathrm{as}(f) &= \frac{1}{p}\mathrm{E}[\mathrm{as}(f_\rho)] && \text{by Lemma 12.17} \\
&= 16k \cdot \mathrm{E}[\mathrm{as}(f_\rho)] && \text{since } p = 1/16k \\
&\le 16k \cdot \mathrm{E}\left[2l(16\log(4w))^{d-2}\right] && \text{by the induction hypothesis} \\
&= 32k \cdot (16\log(4w))^{d-2} \cdot \mathrm{E}[l] \\
&\le 32k \cdot (16\log(4w))^{d-2} \cdot \log(4w) && \text{by Lemma 12.18} \\
&= 2k(16\log(4w))^{d-1}. && \square
\end{aligned}
$$

Proof of Theorem 12.13. We can now finish the proof of Theorem 12.13 as follows. Let C be a circuit of depth $d + 1$ and size w that computes f. Let C' be C with a "dummy" level of gates inserted between inputs and the bottom level, whose gates have fanin 1. Then C' is a $(d + 1, w, 1)$-circuit that computes f. Now apply Lemma 12.19. $\square$

O'Donnell and Wimmer (2007) used Boppana's theorem to show that the majority function (like the parity function) is even hard to approximate by AC^0 circuits. Namely, they showed that if a boolean function $f(x)$ coincides with the majority function $\mathrm{Maj}_n(x)$ on all but an ϵ fraction of all 2^n input vectors, where $\omega(1/\sqrt{n}) \le \epsilon \le 1/2 - \omega(1/\sqrt{n})$, then $\mathrm{as}(f) = \Omega(\mathrm{as}(\mathrm{Maj}_n))$. Using the fact that $\mathrm{as}(\mathrm{Maj}_n) = \Theta(\sqrt{n})$ (see Exercise 2.32), Theorem 12.13 implies

Theorem 12.20. (O'Donnell–Wimmer 2007) *For any constant $\epsilon > 0$, every alternating depth-d circuit of size $2^{o(n^{1/(2d-2)})}$ can agree with $\mathrm{Maj}_n(x)$ on at most a $1 - \epsilon$ fraction of input vectors.*

12.5   Circuits with Parity Gates

We already know that the Parity function cannot be computed by constant depth circuits using a polynomial number of unbounded-fanin AND and OR gates. It is natural, then, to extend our circuit model and allow Parity functions to be used as gates as well. What functions remain difficult to compute in this model? We will show that the Majority function Maj_n, which accepts an input vector of length n iff it has at least as many 1s as 0s, is one such function.

In the case of monotone circuits, we obtained high lower bounds on circuit size by approximating them with CNFs and DNFs. In the case of non-monotone circuits of bounded depth, similar bounds can be obtained via particular approximations by *polynomials.* (We have already used approximation by polynomials in the proof of Theorem 12.8.) In order to show that a given boolean function f requires large circuits we then argue as follows:

1. Show that functions, computable by small circuits, can be approximated by low-degree polynomials.
2. Prove that the function f is hard to approximate by low-degree polynomials.

To achieve the first goal, one takes a small circuit computing the given boolean function f. Each subcircuit computes some boolean function. One tries to inductively assign each gate a low-degree polynomial over GF(2) which approximates the function computed at that gate well enough. This is done in a bottom-up manner. The main problem is to approximate OR and AND gates without a big blow-up in the degree of resulting polynomials.

The AND of n variables $x_1, \ldots, x_n$ is represented by the polynomial consisting of just one monomial $\prod_{i=1}^{n} x_i$, and the OR is represented by the polynomial $1 - \prod_{i=1}^{n}(1 - x_i)$. These polynomials correctly compute AND and OR on all input vectors x, but the degree of these polynomials is n. Fortunately, Lemma 2.7 shows that the degree *can* be substantially reduced if we allow errors: for every integer $r \geq 1$, and every prime power $q \geq 2$, there exists a multivariate polynomial $p(x)$ of n variables and of degree at most $r(q-1)$ over GF(q) such that $\text{dist}(p, \text{OR}_n) \leq 2^{n-r}$. Here, as before, $\text{dist}(f, g)$ is the distance between two function $f, g : \mathbb{F}^n \rightarrow \mathbb{F}$ defined as the number

$$\text{dist}(p, f) = |\{x \in \{0, 1\}^n \mid p(x) \neq f(x)\}|$$

of boolean inputs x on which the polynomial p outputs a wrong value. Using this result we can upper bound the degree of polynomials approximating small size circuits.

Lemma 12.21. *Let $f(x)$ be a boolean function of n variables, and suppose that f can be computed by a depth-d circuit over the basis $\{\wedge, \vee, \oplus\}$ using M gates. Then, for every integer $r \geq 1$, there exists a polynomial $p(x)$ of degree at most r^d over GF(2) such that*

$$\text{dist}(f, p) \leq M \cdot 2^{n-r}.$$

Proof. Our goal is to show that, if a boolean function f can be computed by a small-depth circuit with a small number of AND, OR and Parity gates, then f can be approximated well enough by a low-degree polynomial. This is done in a bottom-up manner. Input literals x_i and $1 - x_i$ themselves are polynomials of degree 1, and need not be approximated. Also, since the degree is not increased by computing the sum $\bigoplus_{i=1}^{t} p_i(x)$, parity gates do not have to be approximated either.

To approximate OR and AND gates we apply Lemma 2.7 with $q = 2$. By this lemma, for any OR $f(x) = \bigvee_{i=1}^{t} p_i(x)$ of polynomials of degree at most h over GF(2) there exists a polynomial $p(x)$ of degree at most hr over GF(2) such that $p(x)$ and $f(x)$ disagree on at most 2^{n-r} input vectors $x \in \{0, 1\}^n$. Thus the functions computed by the gates at the i-th level will be approximated by polynomials of degree at most r^i. Since we have only d levels, the function f computed at the top gate will be approximated by a polynomial $p(x)$ of degree at most r^d. Since, by Lemma 2.7, at each of M gates we have introduced at most 2^{n-r} errors, $p(x)$ can differ from $f(x)$ on at most $M \cdot 2^{n-r}$ inputs.   □

Our next goal is to show that the majority function is hard to approximate by low-degree polynomials. We will show this not for Majority function itself but rather to a closely related function, the threshold-k function Th_k^n. This function is 1 when at least k of the inputs are 1. Note that each such function is a subfunction of the Majority function in $2n$ variables: just set some $n-k$ variables to 1 and some k of the remaining variables to 0. It is therefore enough to prove a high lower bound on Th_k^n for at least one threshold value $1 \leq k \leq n$. We will consider $k = \lceil (n+h+1)/2 \rceil$ for an appropriate h.

Lemma 12.22. *Let $n/2 \leq k \leq n$. Then*

$$\mathrm{dist}(p, \mathrm{Th}_k^n) \geq \binom{n}{k}$$

for every polynomial $p(x_1, \ldots, x_n)$ of degree $d \leq 2k - n - 1$ over GF(2).

Proof. (Due to Lovász et al. 1995) Let $p(x)$ be a polynomial of degree $d \leq 2k - n - 1$ over GF(2), and let $U = \{x \mid p(x) \neq \mathrm{Th}_k^n(x)\}$ denote the set of all vectors where $p(x)$ differs from Th_k^n. Let A denote the set of all 0–1 vectors of length n containing exactly k ones.

Consider the 0–1 matrix $M = (m_{a,u})$ whose rows are indexed by the members of A, columns are indexed by the members of U, and

$$m_{a,u} = 1 \text{ if and only if } a \geq u.$$

Our goal is to prove that the columns of M span the whole linear space; since the dimension of this space is $|A| = \binom{n}{k}$, this will mean that we must have $|U| \geq \binom{n}{k}$ columns.

The fact that the columns of M span the whole linear space follows directly from the following claim saying that every unit vector lies in the span:

Claim 12.23. *If $a \in A$ and $U_a = \{u \in U \mid m_{a,u} = 1\}$ then, for every $b \in A$,*

$$\sum_{u \in U_a} m_{b,u} = \begin{cases} 1 & \text{if } b = a; \\ 0 & \text{if } b \neq a. \end{cases}$$

To prove the claim, observe that by the definition of U_a, we have (all sums are over GF(2)):

$$\sum_{u \in U_a} m_{b,u} = \sum_{\substack{u \in U \\ u \leq a \wedge b}} 1 = \sum_{x \leq a \wedge b} \left(\mathrm{Th}_k^n(x) + p(x) \right) = \sum_{x \leq a \wedge b} \mathrm{Th}_k^n(x) + \sum_{x \leq a \wedge b} p(x)$$

where $a \wedge b$ denotes the componentwise AND of vectors a and b. The second term
of this last expression is 0, since $a \wedge b$ has at least $n - 2(n - k) = 2k - n \geq d + 1$
ones (see Exercise 12.2). The first term is also 0 except if $a = b$.

This completes the proof of the claim, and thus the proof of the lemma.   □

Theorem 12.24. (Razborov 1987) *Every unbounded-fanin depth-d circuit over*
$\{\wedge, \vee, \oplus\}$ *computing* Maj_n *requires* $2^{\Omega(n^{1/2d})}$ *gates.*

Proof. Since every threshold function Th_n^k is a subfunction of the Majority function
in $2n$ variables, it is enough to prove such a lower bound for a depth-d circuit
computing a k-threshold function Th_k^n for some $n/2 \leq k \leq n$ (to be specified later).
Take such a circuit of size M computing Th_k^n. Lemmas 12.22 and 12.21 imply that

$$M \geq \binom{n}{k} 2^{r-n}. \tag{12.2}$$

Taking $r = \lfloor n^{1/(2d)} \rfloor$ and $k = \lceil (n + r^d + 1)/2 \rceil = \lceil (n + \sqrt{n} + 1)/2 \rceil$, and using
the estimate $\binom{n}{k} = \Theta(2^n/\sqrt{n})$ valid for all $k = n/2 \pm \Theta(\sqrt{n})$, the right hand side
of (12.2) becomes $2^{\Omega(r)}$, and we are done.   □

12.6   Circuits with Modular Gates

An AC[p] circuit of depth d is an AC circuit of depth d where, besides AND, OR
and NOT gates, the *counting* gates MOD_p defined by

$$\mathrm{MOD}_p(x_1, \ldots, x_m) = 1 \quad \text{iff} \quad x_1 + \ldots + x_m = 0 \bmod p$$

can be used. In particular, an AC[2] circuit is an AC^0 circuit where unbounded fanin
parity functions are also allowed to be used as gates. The class $\mathrm{AC}^0[p]$ consists of
all (sequences of) boolean functions computable by constant-depth AC[p] circuits
of polynomial (in the number of variables) size. The class ACC^0 is the union of the
classes $\mathrm{AC}^0[p]$ for $p = 2, 3, 4, \ldots$.

It is clear that $\mathrm{AC}^0 \subset \mathrm{ACC}^0$. Moreover, this inclusion is strict, because a single
MOD_2 gate computes the (negation of the) parity function which, as we know, does
not belong to AC^0. Thus, MOD_2 cannot be computed by a constant depth circuit
with polynomial number of NOT, AND and OR gates. But what if, besides NOT,
AND and OR gates, we allow some modular gates MOD_p, for some $p \geq 3$, be used
as gates—can MOD_2 then be computed more easily? It turns out that the use of
gates MOD_p, where $p \geq 3$ is a prime power, does not help to compute MOD_2 much
more efficiently. We will show this for the special case $p = 3$: Parity $\notin \mathrm{AC}^0[3]$.

The general proof idea will be the same as in Sect. 12.5: show that functions
computable by small circuits with MOD_3 gates can be approximated by low-degree
polynomials over GF(3), and prove that the parity function is hard to approximate
by such polynomials.

Lemma 12.25. *Let $f(x)$ be a boolean function of n variables, and suppose that f can be computed by an* AC[3] *circuit of depth d using M gates. Then, for every integer $r \geq 1$, there exists a polynomial $p(x)$ of degree at most $(2r)^d$ over* GF(3) *such that* $\mathrm{dist}(f, p) \leq M \cdot 2^{n-r}$.

The proof of this lemma is exactly the same as that of Lemma 12.21. The only difference is that now we work in the larger field GF(q) for $q = 3$. This results in a slightly worse upper bound $r(q - 1) = 2r$ (instead of r) on the degree of polynomials approximating OR and AND gates.

To apply Lemma 12.25 to the parity function, we have to show that this function cannot be approximated well enough by small degree polynomials over GF(3).

Lemma 12.26. *There is a constant $c > 0$ such that* $\mathrm{dist}(p, \mathrm{Parity}) \geq c2^n$ *for any polynomial of degree at most $\sqrt{n}$ over* GF(3).

Proof. For this proof, we represent boolean values by 1 and -1 rather than 0 and 1. Namely, we replace each boolean variable x_i by a new variable $y_i = 1 - 2x_i$. Hence, $y_i = 1$ if $x_i = 0$, and $y_i = -1$ if $x_i = 1$. The parity function then turns to the product of the y_i:

$$\bigoplus_{i=1}^n x_i = 1 \quad \text{if and only if} \quad \prod_{i=1}^n y_i = -1.$$

Suppose that $p(y)$ is a polynomial over GF(3) $= \{-1, 0, +1\}$ of degree at most $\sqrt{n}$.

We need to show that this polynomial differs from $\prod_{i=1}^n y_i$ on at least a fraction c of the vectors in $\{1, -1\}^n$, for some constant $c > 0$. For this, let A be the set of all vectors $a \in \{1, -1\}^n$ such that $p(a) = \prod_{i=1}^n a_i$. We wish to show that A is "small", that is, has size at most $(1 - c)2^n$ for an absolute constant $c > 0$. We will do this by upper bounding the number $|F|$ functions in the set F of all functions $f : A \to \{-1, 0, 1\}$: since $|F| = 3^{|A|}$ we may bound the size of A by showing that $|F|$ is small.

We claim that every function in F can be represented as a multilinear polynomial over GF(3) of degree at most $(n + \sqrt{n})/2$. We can represent each function $f \in F$ as a polynomial

$$q_f(y) = \sum_{a \in A} f(a) \prod_{i=1}^n (-a_i y_i - 1).$$

This polynomial agrees with f on all $y \in \{1, -1\}^n$, but its degree can be as large as n. We can, however, use the fact that $y_i^2 = 1$ for $y_i \in \{1, -1\}$ and replace each monomial $M = \prod_{i \in S} y_i$ of q_f with $|S| > (n + \sqrt{n})/2$ by a monomial $M' = \prod_{i \notin S} y_i \cdot p(y)$. Since for every $y \in \{1, -1\}^n$,

$$\prod_{i \notin S} y_i \cdot \prod_{i=1}^n y_i = \prod_{i \in S} y_i \cdot \prod_{i \notin S} y_i^2 = \prod_{i \in S} y_i ,$$

we have that $M'(y) = M(y)$ for all $y \in A$, and

$$\text{degree}(M') \leq (n - |S|) + \text{degree}(p) \leq \frac{n - \sqrt{n}}{2} + \sqrt{n} = \frac{n + \sqrt{n}}{2}.$$

Thus every function in F can be represented as a multilinear polynomial over GF(3) of degree at most $(n + \sqrt{n})/2$.

The number of multilinear monomials of degree at most $(n + \sqrt{n})/2$ is

$$N = \sum_{i=0}^{\frac{n + \sqrt{n}}{2}} \binom{n}{i} \leq (1 - c)2^n$$

for a constant $c > 0$ and large n. Since, $|F| \leq 3^N$, we conclude that

$$|A| = \log_3 |F| \leq N \leq (1-c)2^n. \qquad \square$$

Combining the two lemmas above we obtain the following

Theorem 12.27. (Smolensky 1987) *Any* AC[3] *circuit of depth* d *computing the parity function requires* $2^{\Omega(n^{1/2d})}$ *gates.*

Proof. Let M be the minimum size of an AC[3] circuit of depth-d computing the parity function Parity of n variables. Taking $r = n^{1/2d}/2$ in Lemma 12.25, we obtain that there must exist a polynomial $p(x)$ of degree at most $\sqrt{n}$ over GF(3) such that
$$\text{dist}(p, \text{Parity}) \leq M \cdot 2^{n - n^{1/2d}/2}.$$
But Lemma 12.26 implies that $\text{dist}(p, \text{Parity}) \geq c2^n$, and the desired lower bound on M follows. $\qquad \square$

Thus, Parity $\notin$ AC^0[3]. More generally, by results of Razborov (1987) and Smolensky (1987), the function MOD$_m$ does not belong to AC^0[p] for prime p unless m is a power of p. Since MOD$_m$ is a symmetric function, and since every symmetric boolean function can be computed by an unweighted threshold circuit using a linear number of gates (even in depth 2; see Exercise 11.10), this also implies that Maj$_n \notin$ AC^0[p] for every prime number p. Much less is known about the power of AC^0[p] circuits where p is a *composite* number. In particular, even the case $p = 6$ remains unclear.

Recently, Williams (2011) showed that so-called NEXP-complete functions do not belong to ACC^0 (we shortly discuss this result in Sect. 20.4). One such boolean function, called *succinct 3-SAT*, is defined as follows. Given a binary string x of length n, the function interprets it as a code of a DeMorgan circuit C_x of $n^{1/10}$ input variables. Then the function uses the circuit C_x to produce the string of all its $2^{n^{1/10}}$ possible outputs, interprets this huge string as a code of a (also huge) 3-CNF F_x, and accepts the initial string x iff the CNF F_x is satisfiable.

Fig. 12.2 A symmetric
(r, s)-circuit. The output gate
is a symmetric boolean
function of s variables, and its
inputs are ANDs of r literals
each

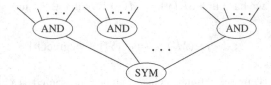

But the ACC^0 circuits are suspected to be of much weaker power. In particular, it is conjectured that even the Majority function does not belong to ACC^0.

■ **Research Problem 12.28.** Prove or disprove that Majority $\notin \mathrm{ACC}^0$.

An interesting and useful property of ACC^0 circuits (which was also used by Williams (2011)) is that they can be relatively efficiently simulated by depth-2 circuits, as we will see next.

ACC^0 *and Symmetric Depth-2 Circuits*

A depth-2 *symmetric* (r, s)-*circuit* is a circuit of the form $g(K_1, \ldots, K_s)$, where $g : \{0, 1\}^s \to \{0, 1\}$ is a symmetric boolean function, and each K_i is an AND of at most r literals (see Fig. 12.2). Since the negation of a symmetric function is also symmetric, one can also take ORs instead of ANDs.

Let SYM be the class of all (sequences of) boolean function of n variables that can be computed by a symmetric $(r, 2^r)$-circuit with $r \leq (\log n)^{\mathcal{O}(1)}$. This class is surprisingly rich. Allender (1989) showed that $\mathrm{AC}^0 \subseteq$ SYM. Then Yao (1990) showed that ACC^0 is contained in the probabilistic version of SYM. Finally, Beigel and Tarui (1994) showed that $\mathrm{ACC}^0 \subseteq$ SYM. Furthermore, Allender and Gore (1994) showed that the corresponding symmetric circuit can be efficiently constructed.

Theorem 12.29. (Yao 1990; Beigel–Tarui 1994) $\mathrm{ACC}^0 \subseteq$ SYM.

The full proof of this important result is somewhat technical, and we omit it. Theorem 12.29 allows us to extract some special properties of functions in ACC^0. In particular, it implies that these functions can be "easily separated" by low-degree polynomials, and that the corresponding to these functions bipartite graphs have small "intersection dimension".

ACC^0 *and Low-Degree Polynomials*

By a multilinear ± 1 polynomial of weight w we will mean polynomial

$$p(x_1, \ldots, x_n) = \sum_{S \subseteq [n]} \lambda_S \prod_{i \in S} x_i$$

over the reals with coefficients $\lambda_S \in \{-1, 0, +1\}$ such that $|\{S : \lambda_S \neq 0\}| = w$. A polynomial $p(x)$ *separates* a boolean function f if $p(x) \neq p(y)$ for all $x \in f^{-1}(1)$ and $y \in f^{-1}(0)$. Say that a boolean function f of n variables is *easy* to separate if it can be separated by a polynomial of poly-logarithmic in n degree d and weight $w \leq 2^d$. In particular, all symmetric functions are easy to separate: they all are separated by the polynomial $p(x) = x_1 + x_2 + \cdots + x_n$.

Lemma 12.30. *Every function in ACC^0 is easy to separate.*

Proof. Let $f(x)$ be a boolean function of n variables, and suppose that $f \in \text{ACC}^0$. By Theorem 12.29, there exists a constant c such that f can be computed by a symmetric (r, s)-circuit with $r \leq (\log n)^c$ and $s \leq 2^{(\log n)^c}$. That is, there exist monomials $g_1, \ldots, g_s$ and a subset $T \subseteq \{0, 1, \ldots, s\}$ such that $f(x) = 1$ iff the number of the g_i accepting x belongs to T. This, in particular, means that

$$\sum_{i=1}^{s} g_i(x) \neq \sum_{i=1}^{s} g_i(y) \quad \text{for all } x \in f^{-1}(1) \text{ and } y \in f^{-1}(0).$$

Now, each monomial g_i has a form $g_i(x) = \bigwedge_{i \in I} x_i \bigwedge_{j \in J} \overline{x}_j$ with $I \cap J = \emptyset$ and $|I \cup J| \leq r$. On 0–1 vectors x this monomial outputs the same values as the polynomial $p_i(x) = \prod_{i \in I} x_i \prod_{j \in J} (1 - x_j)$. This is a ± 1 polynomial of weight at most $2^r \leq s$. Hence, $p(x) := \sum_{i=1}^{s} p_i(x)$ is a ± 1 polynomial of weight at most $s^2 \leq 2^{2(\log n)^c}$, implying that f is easy to separate, as desired. □

■ **Research Problem 12.31.** Exhibit an explicit boolean function which is not easy to separate by a polynomial.

ACC^0 *and Graph Complexity*

Let $G = (V_1 \cup V_2, E)$ be a bipartite $n \times n$ graph, and $L \subseteq \{0, 1, \ldots\}$ a subset of natural numbers. An *intersection representation* relative to L is an assignment $v \mapsto S_v$ of subsets $S_v \subseteq \{1, \ldots, t\}$ such that two vertices $u \in V_1$ and $v \in V_2$ are adjacent in G iff $|S_u \cap S_v| \in L$. The smallest number t for which such a representation exists is the *intersection dimension* of G *relative to L*, and is denoted by $\text{idim}_L(G)$. The *intersection dimension*, $\text{idim}(G)$, of G is the minimum of $\text{idim}_L(G)$ over all sets L.

If $n = 2^m$, then we can encode the vertices by binary vectors of length m, and associate with G a boolean function f_G of $2m$ variables (the characteristic function of G) such that $f_G(u, v) = 1$ iff $uv \in E$.

Easy counting shows that graphs with $\text{idim}(G) = \Omega(n)$ exist. On the other hand, graphs whose characteristic functions belong to ACC^0 have small intersection dimension.

Lemma 12.32. *If $f_G \in \text{ACC}^0$ then $\text{idim}(G) \leq 2^{(\log \log n)^{O(1)}}$.*

362   12 Large-Depth Circuits

Proof. If $f_G \in \text{ACC}^0$, then Theorem 12.29 implies that f_G can be computed by a symmetric $(r, 2^r)$-circuit with r at most about $(\log m)^c$ for a constant c. In particular, f_G can be computed by a depth-2 circuit with a symmetric gate g of fanin $s = 2^r$ on the top, and ORs of literals on the bottom level. By the Magnification Lemma (Lemma 1.32 in Sect. 1.7), the graph G is then represented by a depth-2 circuit with g on the top level, and OR gates $\vee_{v \in I_1} x_v, \ldots, \vee_{v \in I_s} x_v$ on the bottom level. Since the gate g is symmetric, there exists a subset $L \subseteq \{0, 1, \ldots, s\}$ such that g rejects an input vector iff the number of 1s in this vector belongs to L. Hence, the sets $S_v = \{i \mid v \notin I_i\}$ give us an intersection representation of G, implying that $\text{idim}(G) \le s \le 2^{(\log m)^c} = 2^{(\log \log n)^c}$. □

Problem 12.33. Exhibit an explicit bipartite $n \times n$ graph of intersection dimension at least $2^{(\log \log n)^{\omega(1)}}$.

By Lemma 12.32, the characteristic function of any such graph lies outside ACC^0. Actually, by the results of Green et al. (1995), it would be enough to prove such a lower bound on $\text{idim}_L(G)$ for the specific set L consisting of all natural numbers whose binary representations have bit 1 in the middle. Such sets (also called *middle-bit* set) consist of disjoint intervals of consecutive numbers.

ACC⁰ *and Communication Complexity*

For a boolean function f of $n = km$ variables we fix a partition of its variables in k blocks, each of size m, and let $c_k(f)$ denote the k-party communication complexity of f with respect to this partition (see Sect. 5.4).

Symmetric depth-2 circuits are related to communication games via the following simple observation: if $[n] = \{1, , \ldots, n\}$ is partitioned into disjoint blocks of size n/k then every $(k-1)$-element subset of $[n]$ must be disjoint with at least one of the blocks. This holds because, if a set has an element in each block, then it must have at least k elements in total. We now can imagine that the i-th player can see all bits except those in the i-th block. Then every $(k-1)$-element set is seen by at least one of the players.

Lemma 12.34. (Håstad–Goldmann 1991) *Every symmetric depth-2 circuit which computes f and has bottom fanin $k-1$ must have top fanin at least $2^{c_k(f)/k}$.*

Proof. Fix a symmetric $(k-1, s)$-circuit computing f. We will show that $c_k(f) \le k \log_2 s$. By the observation we just made, each bottom gate can be evaluated by at least one of the k players. Partition the bottom gates among the players such that all the gates assigned to a player can be evaluated by that player. Now each of the first $k-1$ player broadcasts to the k-th players the *number* of her gates that evaluate to 1. This can be done using at most $(k-1)\log_2 s$ bits. Finally, the k-th player can evaluate the top gate and tell the one bit result to the others. The total number of bits used is $1 + (k-1)\log_2 s \le k \log_2 s$. □

Note that there was nothing special in having only AND gates on the bottom: the lemma remains true also when arbitrary boolean functions of fanin at most $k - 1$ are used as bottom gates.

Recall that the *generalized inner product function* GIP(x) is a boolean function of kn variables, arranged in an $n \times k$ matrix $x = (x_{ij})$, and is defined by:

$$\mathrm{GIP}_{n,k}(x) = \bigoplus_{i=1}^{n} \bigwedge_{j=1}^{k} x_{ij}.$$

Note that this function belongs to ACC^0, and hence also to SYM. We have already proved that the k-party communication complexity of GIP$_{n,k}$ is $\Omega(n4^{-k})$ (see Theorem 5.15). Together with Lemma 12.34, this immediately yields the following corollary.

Corollary 12.35. *Every depth-2 symmetric circuit of bottom fanin $k - 1$ computing* GIP$_{n,k}$ *must have top fanin* $2^{\Omega(n/k4^k)}$.

Remark 12.36. In order to use Theorem 12.29 and Lemma 12.34 to show that a given boolean function f of n variables does not belong to ACC^0, we need non-trivial lower bounds on the k-party communication complexity $c_k(f)$ of f in the case when $k \geq (\log n)^{\omega(1)}$, that is, when the number k of players is larger than poly-logarithmic in n. Unfortunately, known lower bound on $c_k(f)$ (some of which we presented in Chap. 5) are trivial already for $k = \log n$.

12.7 Circuits with Symmetric Gates

Recall that a boolean function $g(y_1, \ldots, y_t)$ is *symmetric* if its value only depends on the value of the sum $y_1 + \cdots + y_t$ of the values of its input bits. In other words, g is symmetric if the is a function $F : \{0, 1, \ldots, n\} \to \{0, 1\}$ such that $g(y_1, \ldots, y_t) = F(y_1 + \cdots + y_t)$. A *symmetric circuit* is a circuit whose gates are arbitrary symmetric boolean functions. The *size* of such a circuit is the number of wires in it. For a boolean function f, let S$_d(f)$ denote the minimum number of wires in a symmetric depth-d circuit computing f, and let S(f) denote the minimum number of wires in an arbitrary symmetric circuit computing f.

For the corresponding Shannon function S(n), Grinchuk (1996) proved that S$(n) \sim 2^n/n$, and S$(n) \sim 2^n/\log_2 n$ in the case of formulas (symmetric circuits with fanout-1 gates). But proving high lower bounds for *explicit* boolean functions even in the class of depth-2 symmetric circuits is a difficult task. By Theorem 12.29, a lower bound S$_2(f) \geq 2^{(\log n)^{\omega(1)}}$ implies $f \notin$ ACC^0. Actually, even proving strong lower bounds for symmetric circuits of depth 1 (where only one symmetric boolean function can be used as gate) is not a trivial task.

As we have mentioned in Sect. 11.10, every boolean function $f(x) = f(x_1,\ldots,x_n)$ can be computed by a symmetric circuit consisting of just one symmetric gate g:

$$f(x) = g(\underbrace{x_1}_{2^0}, \underbrace{x_2, x_2}_{2^1}, \underbrace{x_3, x_3, x_3, x_3}_{2^2}, \ldots, \underbrace{x_n, \ldots, x_n}_{2^{n-1}}).$$

This shows that $S_1(f) \le 2^0 + 2^1 + \cdots + 2^{n-1} = 2^n - 1$ for every boolean function f of n variables. On the other hand, direct counting shows that boolean functions f of n variables requiring

$$S_1(f) \ge 2^n - n^2$$

exists. To show this, let $L(n,m)$ denote the number of distinct boolean functions f of n variables such that $S_1(f) \le m$. For each such function f, there is a function $F : \{0,1,\ldots,m\} \to \{0,1\}$ and non-negative integers $\lambda_1,\ldots,\lambda_n$ such that $\lambda_1 + \cdots + \lambda_n \le m$ and $f(x) = F(\lambda_1 x_1 + \cdots + \lambda_n x_n)$. For every integer $0 \le r \le m$, the equation $\lambda_1 + \cdots + \lambda_n = r$ has $\binom{n+r-1}{r}$ non-negative integer solutions $\lambda_1,\ldots,\lambda_n$: the number of such solutions is exactly the number of possibilities to interrupt a sequence of $n + r - 1$ ones by $r - 1$ zeros. Thus, the inequality $\lambda_1 + \cdots + \lambda_n \le m$ has $\sum_{r=0}^m \binom{n+r-1}{r} = \binom{n+m}{n}$ solutions. Since there are only 2^{m+1} functions $F : \{0,1,\ldots,m\} \to \{0,1\}$, we obtain that

$$L(n,m) \le 2^{m+1}\binom{n+m}{n} \le \frac{(n+m)^n}{n!},$$

which is smaller than the total number 2^{2^n} of boolean functions, as long as $n \ge 5$ and $m < 2^n - n^2$.

Smolensky (1990) used algebraic arguments to show that the function

$$f(x_1,\ldots,x_n) = (x_1 \oplus x_2) \wedge (x_3 \oplus x_4) \wedge \cdots \wedge (x_{n-1} \oplus x_n)$$

requires $S_1(f) \ge 2^{n/2}$. Then Grinchuk (1996) exhibited another boolean function requiring an almost-maximal number of wires to realize by one symmetric gate. His function is defined by:

$$f_n(x_1,\ldots,x_n) = x_1 x_2 \vee \overline{x}_1 x_2 x_3 \vee \overline{x}_1 \overline{x}_2 x_3 x_4 \vee \cdots \vee \overline{x}_1 \ldots \overline{x}_{n-2} x_{n-1} x_n.$$

That is, the value $f(a)$ on a given input $a = (a_1,\ldots,a_n)$ is the value of that bit which occurs after the first (from the left) occurrence of a 1 in a.

Theorem 12.37. (Grinchuk 1996) $S_1(f_n) \ge 2^{n-1} - 1$.

Proof. Let $\lambda = S_1(f_n)$. Then there exist nonnegative integers $\lambda_1,\ldots,\lambda_n$ such that $\lambda_1 + \cdots + \lambda_n = \lambda$, and a function $F : \{0,1,\ldots,\lambda\} \to \{0,1\}$ such that $f_n(x) =$

$F(l(x))$, where $l(x) := \lambda_1 x_1 + \cdots + \lambda_n x_n$. Say that two input vectors $a, b \in \{0, 1\}^n$ are *equivalent* if $l(a) = l(b)$.

Claim 12.38. No two vectors a and b with $a_1 = b_1 = 0$ are equivalent.

Proof. Assume that $a = (0, a_2, \ldots, a_n)$ and $b = (0, b_2, \ldots, b_n)$ are equivalent, that is, $l(a) = l(b)$. Let i be the first position (from the left) in which a and b differ for the first time. Assume w.l.o.g. that $a_i = 0$ and $b_i = 1$. Now consider the vectors $a' = (0, \ldots, 0, 1, 0, a_{i+1}, \ldots, a_n)$ and $b' = (0, \ldots, 0, 1, 1, b_{i+1}, \ldots, b_n)$. Since $l(a) - l(a') = l(b) - l(b')$, we have that $l(a') = l(b')$. But $f_n(a') = 0 \neq 1 = f_n(b')$, a contradiction. $\qquad\square$

Since the values of $l(x)$ are integers between 0 and λ, the claim implies that $S_1(f_n) = \lambda$ must be at least $2^{n-1} - 1$, as desired. $\qquad\square$

12.8 Rigid Matrices Require Large Circuits

We now consider boolean circuits computing boolean matrices, that is, 0–1 matrices of some fixed dimension. Inputs are *primitive matrices*, that is, boolean matrices of rank 1. Each of these matrices consist of an all-1 submatrix, and has zeros elsewhere.

Boolean operations on matrices are computed component-wise. For example, $A \vee B$ is a boolean matrix whose entries are ORs $A[i, j] \vee B[i, j]$ of the corresponding entries of A and B. Thus each such circuit computes some matrix. As before, the *depth* of a circuit is the length of a longest path from an input to an output gate. The *size* is the number of gates. Our goal is to show that matrices of high "rigidity" require large circuits.

> Note that, given a boolean matrix A, we want to compute (or produce) the matrix A *itself* starting form primitive matrices and using component-wise boolean operations, like AND, OR, NOT or Parity. Just like, given a boolean function, we want to compute it starting from "primitive" boolean functions (variables and their negations). A related (albeit different) question (which we will consider in the next chapter) is to compute the linear transformation $y = Ax$ given by the matrix A.

The *rigidity*, $\mathrm{Rig}_M(r)$, of a boolean matrix M over $\mathrm{GF}(2)$ is the smallest number of entries of A that must be changed in order to reduce its rank over $\mathrm{GF}(2)$ down to r. That is,

$$\mathrm{Rig}_M(r) = \min\{|B| : \mathrm{rk}(M \oplus B) \leq r\},$$

where $|B|$ is the total number of 1s in B. It is known that matrices M of rigidity at least about $(n - r)^2 / \ln n$ exist (see Proposition 13.35 in the next chapter). However, the largest known lower bound on the rigidity of an *explicit* matrix remains the lower bound of about $(n^2/r) \ln(n/r)$ due to Friedman (1993).

In the following theorem, M is an arbitrary boolean $n \times n$ matrix, $d \geq 2$ an integer, and $f(r) := (\ln r)^{1/(d-1)}$.

Theorem 12.39. (Razborov 1989b) *If* $\text{Rig}_M (r) \geq n^2/2^{f(r)}$ *then every depth-d circuit over* $\{\wedge, \vee, \neg, \oplus\}$ *computing* M *must have* $2^{\Omega(f(r))}$ *gates.*

Proof. We will again use the approximation method. This time we will approximate matrices, computed at intermediate gates of the circuit, by matrices of small rank. Set $s := 2 \cdot f(r)$ where $f(r) := (\ln r)^{1/(d-1)}$. Note that for constant r the theorem is obvious, so we assume that r and s are large enough.

Suppose we have an unbounded-fanin circuit over $\{\wedge, \vee, \neg, \oplus\}$ of depth at most d and size ℓ computing the matrix M. We have to show that $\text{Rig}_M (r) \geq n^2/2^{f(r)}$ implies $\ell = 2^{\Omega(f(r))}$.

At each gate of the circuit some boolean matrix A is computed. We inductively assign to each gate on the i-th layer an *approximator*, which is a boolean matrix A' of rank

$$\text{rk}(A') \leq \ell^{i s^i}. \tag{12.3}$$

As before, the assignments are done inductively, first to the inputs, then working up to the output. Each assignment introduces some *errors*, that is, positions the approximator A' differs from the matrix obtained by applying the true operator at that gate to the approximators of the gates feeding into it. Our goal is to assign approximators in such a way that

$$\text{at most } n^2/2^s \text{ errors are introduced at each gate.} \tag{12.4}$$

We first show that (12.3) and (12.4) already imply the theorem. To see this, let M' be the approximator of the matrix computed at the last gate. We have two cases: either the rank of M' is large or not.

Case 1: $\text{rk}(M') \geq r$. In this case, (12.3) implies that $r \leq \ell^S$, where $S := \mathcal{O}(s^d)$ is about

$$(\ln r)^{d/(d-1)} = (\ln r)^{1-1/(d-1)} = (\ln r)/f(r).$$

Hence, $\ln \ell \geq (\ln r)/S = \Omega(f(r))$, as desired.

Case 2: $\text{rk}(M') \leq r$. In this case, our assumption $\text{Rig}_M (r) \geq n^2/2^{f(r)}$ implies that $|M \oplus M'| \geq n^2/2^{f(r)}$. On the other hand, (12.4) implies that

$$|M \oplus M'| \leq \ell \cdot n^2/2^s = \ell \cdot n^2/2^{2f(r)}.$$

Comparing these two estimates, the desired lower bound $\ell \geq 2^{f(r)}$ follows.

It therefore remains to show how to assign approximators satisfying (12.3) and (12.4). Approximators of input matrices are matrices themselves. Recall that these matrices have rank ≤ 1.

If the gate is a NOT gate and the unique gate feeding into it has an approximator A', then we assign this gate the approximator $\neg A'$. Since $\text{rk}(\neg A') \leq \text{rk}(A') + 1$, the rank condition (12.3) is fulfilled.

If the gate is a parity gate, then let its approximator be just the sum modulo 2 of the approximators of all its $m \leq \ell$ inputs gates. The rank condition (12.3) is fulfilled by the subadditivity of rank.

So far we have introduced no errors at all. The source of errors are, however, AND and OR gates. For these gates we use the following approximation lemma.

Lemma 12.40. *Let $s \geq 1$ be an integer. If $A = \bigvee_{i=1}^{h} A_i$ is an OR of boolean $n \times n$ matrices, each of rank at most r, then there is a boolean matrix C such that $\mathrm{rk}(C) \leq 1 + (1 + hr)^s$ and $|A \oplus C| \leq n^2/2^s$.*

Proof. Let $\mathcal{L}$ be the linear space of boolean matrices over GF(2) generated by $A_1, \ldots, A_h$. By the subadditivity of rank, we have that $\mathrm{rk}(B) \leq hr$ for every $B \in \mathcal{L}$. Take a matrix $B = (b_{ij})$ in $\mathcal{L}$ at random. That is,

$$B = \lambda_1 A_1 \oplus \lambda_2 A_2 \oplus \cdots \oplus \lambda_h A_h,$$

where the λ_i are independent uniformly distributed 0–1 random variables. Let $A = (a_{ij})$ be the OR of matrices $A_1, \ldots, A_h$. If $a_{ij} = 0$ then clearly $b_{ij} = 0$. If $a_{ij} = 1$ then the (i, j)-th entry of at least one of the matrices $A_1, \ldots, A_h$ is 1, and hence,

$$b_{ij} = \lambda_1 A_1[i, j] \oplus \lambda_2 A_2[i, j] \oplus \cdots \lambda_h A_h[i, j]$$

equals 0 with probability $1/2$. That is, $\mathrm{Prob}[b_{ij} = 0] = 0$ if $a_{ij} = 0$, and $\mathrm{Prob}[b_{ij} = 0] = 1/2$ if $a_{ij} = 1$. Thus if we let $C = (c_{ij})$ to be the OR of s independent copies of B, then $\mathrm{Prob}[c_{ij} = 0] = 1$ if $a_{ij} = 0$, and $\mathrm{Prob}[c_{ij} = 0] \leq 2^{-s}$ if $a_{ij} = 1$. That is, the expected number of positions, where C deviates from A, does not exceed $n^2/2^s$.

Thus there exists a matrix C of the form $C = \bigvee_{k=1}^{s} B_k$ such that $|A \oplus C| \leq n^2/2^s$ and $\mathrm{rk}(B_i) \leq hr$ for each i. Using the rule $x \vee y = (x \oplus 1) \wedge (y \oplus 1) \oplus 1$, this OR can be written as an all-1 matrix plus an AND of s matrices, each of which has rank at most $1 + hr$. Since the AND of matrices is a component-wise product of their entries, and component-wise product is bilinear in the space of rows of matrices, this implies that $\mathrm{rk}(A \wedge B) \leq \mathrm{rk}(A) \cdot \mathrm{rk}(B)$. Since we have an AND of s matrices each of rank at most $1 + hr$, this gives the desired upper bound $\mathrm{rk}(C) \leq 1 + (1 + hr)^s$ on the rank of C. □

Now, if the gate is an OR gate at the i-th layer of our circuit, and if it has h inputs, then Lemma 12.40, applied with $r = \ell^{(i-1)s^{i-1}}$ and $h = \ell$, yields the desired approximator satisfying (12.3). The case of an AND gate reduces to that of OR gates by DeMorgan rules. □

Theorem 12.39 has several interesting consequences. Babai et al. (1986) introduced the communication complexity analogue PH^{cc} of the complexity class PH, and proved that PH^{cc} coincides with the class of boolean $n \times n$ matrices M whose constant depth circuit complexity over the basis $\{\wedge, \vee\}$ does not exceed $\exp\left((\ln \ln n)^{\mathcal{O}(1)}\right)$. Theorem 12.39 immediately implies that, if

$$\mathrm{Rig}_M\,(r) \geq \frac{n^2}{\exp(\ln r)^{o(1)}} \quad \text{for } r \geq 2^{(\ln \ln n)^{\omega(1)}},$$

then $M \notin PH^{cc}$. That is, the class PH^{cc} does not contain highly rigid matrices.

Razborov (1988) used probabilistic arguments to show that unbounded-fanin circuits over $\{\wedge, \oplus, 1\}$ of small depth can efficiently compute some combinatorially "complicated" matrices, sharing many extremal properties of random matrices. Together with Theorem 12.39 this implies that matrices of low rigidity can share many properties of random matrices.

Chapter Notes

Apparently, the first non-trivial lower bound for small-depth circuits was proved by Lupanov (1970). From the current point of view, this result is not very impressive. But this was the first attempt to say something non-trivial about small-depth circuits, and the result is tight. He considered the following boolean function

$$f_n(x, y) = \bigvee_{i=1}^{n} x_i \wedge \bigwedge_{j=i}^{n} y_i$$

and proved that, for $d \geq 1$, the smallest number of leaves in a monotone depth-$(d+1)$ *formula* for f_n is asymptotically equal to $c_d n^{1+1/d}$ where $c_d = (1-1/(d+1))(d!)^{1/d}$. The function f_n naturally arises when computing the sum of two n-bit numbers.

The first superpolynomial lower bound for AC^0 circuits was proved by Tkachov (1980). He considered circuits of depth 3; however, in contrast with current conventions, he also let negation gates count towards the depth. Thus his class is smaller than what we call depth 3 now and contains the depth 3 monotone circuits. Unfortunately this result was published in some unimportant proceedings with small distribution and remained almost unknown outside the Soviet Union. His method is based on a clever counting.

The most successful method, the method of random restrictions, was introduced by Furst et al. (1984), and Ajtai et al. (1983). This resulted in superpolynomial lower bounds $n^{\Omega(\log n)}$ for AC^0 circuits.

In the attempt to prove exponential lower bounds, the simpler case of *monotone* small-depth circuits was studied. Valiant (1983) proved that the clique problem needs exponential size circuits when the depth is restricted to 3. Boppana (1986) proved that depth-d monotone circuits computing the Majority function require size about $2^{n^{1/(d-1)}}$, while Klawe et al. (1984) obtained similar lower bounds for the Iterated Disjointness function we have considered in Sect. 11.5.

The first breakthrough in proving exponential lower bounds without restricting circuits to be monotone was obtained by Yao (1985) who proved that depth-d

circuits computing the Parity function require about $2^{n^{1/4d}}$ gates. Finally, Håstad (1986) improved this to an almost optimal lower bound of about $2^{n^{1/(d-1)}}$; see Theorem 12.3.

Exercises

12.1. Show that, for every integer $d \geq 3$, the parity function Parity(x) of n variables can be computed by depth-$(d + 1)$ circuits over $\{\wedge, \vee, \neg\}$ of size $2^{\mathcal{O}(n^{1/d})}$.
Hint: Consider a circuit of depth d consisting of parity gates of fanin $r = n^{1/d}$. Each such gate can be replaced by a CNF as well as by a DNF of size $2^r = 2^{n^{1/d}}$. When doing this, the depth increases till $2d$. To reduce the depth, use the associativity of OR to collapse consecutive layers of OR gates into a single layer; the same with AND gates.

12.2. Let $h = \prod_{i \in S} x_i$ be a monomial of degree $d = |S| \leq n - 1$, and let a be a 0–1 vector with at least $d + 1$ ones. Show that, over GF(2), $\sum_{b \leq a} h(b) = 0$. *Hint*: There are only two possibilities: either $a_i = 1$ for all $i \in S$, or not.

12.3. Let $p = 1/\sqrt{n}$, and consider a p-random restriction ρ on n variables.

(a) Let C be a clause, and $t > 0$ an integer. Show that C_ρ will depend on more than t variables with probability at most $n^{-t/3}$.

> *Hint*: Consider two cases depending on whether: (1) C contains more than $m := t \log n$ literals, or (2) C contains at most m literals. Show that in the first case C_ρ will be non-constant with probability at most $((1 + p)/2)^m$, whereas in the second case C_ρ will contain at least t variables with probability at most $\binom{m}{t} p^t$. Show that both of these bounds are at most $n^{-t/3}$ if n is large enough.

(b) Prove the following weaker version of the Switching Lemma: for every integer constants $t, k \geq 1$ there is a constant $s = s(t, k)$ with the following property: if F is a t-CNF on n variables, then

$$\text{Prob}[F_\rho \text{ depends on} \geq s \text{ variables}] \leq n^{-k}.$$

> *Hint*: Argue by induction on t. Use the previous exercise for the base case $b(1, k) = 3k$. For the induction step, take a maximal set of clauses in F whose sets of variables are pairwise disjoint, and let Y be the union of these variable sets. Hence, each clause of F has at least one variable in Y. Consider two cases depending on whether $|Y| \geq k 2^t \log n$ or not. If $|Y| \geq k 2^t \log n$, then use the disjointness of clauses determining Y to show that F_ρ becomes constant with probability at least $1 - n^{-k}$. In the case when $|Y| \leq k 2^t \log n$ show that, for every i, the probability that more than i variables in Y will remain unassigned is at most $n^{-i/3}$; cf. part (a). Take $i = 4k$, set these $4k$ free variables of Y to constants in all possible ways to obtain a $(t - 1)$-CNF F', and apply induction hypothesis to F'.

Chapter 13
Circuits with Arbitrary Gates

In this chapter we consider unbounded-fanin circuits with *arbitrary* boolean functions as gates. The *size* of such a circuit is defined as the total number of wires (rather than gates) it has. Of course, then every single boolean function f of n variables can be computed by a circuit of size n: just take one gate—the function f itself. The problem, however, becomes nontrivial if instead of one function, we want to *simultaneously* compute m boolean functions $f_1, \ldots, f_m$ on the same set of n variables $x_1, \ldots, x_n$, that is, to compute an (n, m)-*operator* $f : \{0, 1\}^n \to \{0, 1\}^m$. Note that in this case the phenomenon which causes complexity of circuits is *information transfer* instead of *information processing* as in the case of circuits computing a single function.

As before, a circuit computing a given (n, m)-operator can be imagined as a directed acyclic graph with n input nodes corresponding to the variables $x_1, \ldots, x_n$, m output nodes corresponding to the boolean functions $f_1, \ldots, f_m$ to be computed, and each non-input node computing an arbitrary boolean function of its inputs.

Note that we cannot expect larger than quadratic lower bounds for general circuits: every operator $f : \{0, 1\}^n \to \{0, 1\}^n$ can be computed using at most n^2 wires, even in depth 1. On the other hand, using counting arguments, it can be shown that operators requiring $\Omega(n^2)$ wires in any circuit with general gates exist (see Exercise 13.1).

In this chapter we will concentrate on general circuits of depth 2—the first nontrivial case. These circuits are powerful, and their study was strongly motivated by work of Valiant (1977), who showed that any operator with $\omega(n^2/\log\log n)$ depth-2 wire complexity also cannot be computed by linear-size, logarithmic-depth boolean circuits of fanin 2 (see Lemma 13.32 below).

There are known superlinear lower bounds of $\Omega(n \ln^{3/2} n)$ for depth-2 circuits. Many superlinear lower bound proofs are given by algebraic arguments (based on the matrix rigidity) or given by graphic theoretic arguments based on various superconcentration properties of graphs: Pippenger (1977, 1982), Dolev et al. (1983), Pudlák and Savický (1993), Pudlák (1994), Alon and Pudlák (1994), Pudlák et al. (1997), Radhakrishnan and Ta-Shma (2000), Raz and Shpilka (2003), Gál et al. (2011).

S. Jukna, *Boolean Function Complexity*, Algorithms and Combinatorics 27,   371
DOI 10.1007/978-3-642-24508-4_13, © Springer-Verlag Berlin Heidelberg 2012

The advantage of arguments based on superconcentrators is that they generally provide rich addition to the structural information about how the circuits for a given operator must look like. The disadvantage of these arguments is only numerical: even for depth-2 circuits, these arguments cannot lead to larger than $\Omega(n \ln^2 n)$ lower bounds on the number of wires, as proven by Radhakrishnan and Ta-Shma (2000).

For depth-2 circuits, larger lower bounds of $\Omega(n^{3/2})$ were recently proved using a much simpler information theoretic argument, and we present it below. The argument itself is reminiscent of Nechiporuk's argument for formulas (Theorem 6.16): an operator requires many wires if the number of its distinct "sub-operators" is large.

13.1 Entropy and the Number of Wires

As mentioned above, counting arguments yield that *most* operators $f : \{0, 1\}^n \to \{0, 1\}^n$ require about n^2 wires in any circuit using arbitrary boolean functions as gates (see Exercise 13.1). But where are these "hard" operators? In particular, what is the complexity of often used operators like matrix product or cyclic convolution (corresponding to product of polynomials)? What we need are lower bounds for *specific* operators. That is, we want to understand what properties of operators make them hard to compute. In this section we will show that high "entropy" of operators is one of these properties.

An operator $f : \{0, 1\}^n \to \{0, 1\}^m$ maps binary strings of length n to binary strings of length m. Each such operator can be viewed as a sequence $f = (f_1, \ldots, f_m)$ of m (not necessarily distinct) boolean functions $f_i : \{0, 1\}^n \to \{0, 1\}$, each on the same set of n variables. The *range* of f is the set

$$\mathrm{Range}(f) = \{ f(a) \mid a \in \{0, 1\}^n \} \subseteq \{0, 1\}^m$$

of distinct values taken by f. Define the *entropy*, $\mathrm{E}(f)$, of an operator f as the logarithm base 2 of the number of distinct values taken by f. That is,

$$\mathrm{E}(f) := \log_2 |\mathrm{Range}(f)|.$$

It is clear that for any operator $f = (f_1, \ldots, f_m) : \{0, 1\}^n \to \{0, 1\}^m$, we have that $\mathrm{E}(f) \leq \min\{n, m\}$, just because $|\mathrm{Range}(f)| \leq \min\{2^n, 2^m\}$. We will use the following properties of entropy:

(P1) $\mathrm{E}(f) \leq |\{f_1, \ldots, f_m\}|$. That is, $\mathrm{E}(f)$ cannot exceed the number of *distinct* boolean functions in f. This holds because only different functions can produce different values.

(P2) $\mathrm{E}(f) \geq r$ if we have r distinct single variables among the functions $f_1, \ldots, f_m$. This is because the operator f must take at a minimum of 2^r distinct values on r distinct variables.

(P3) $E(f) \leq E(g)$ if every function f_i of f can be computed as some boolean function applied to the functions of operator g. Indeed, in this case $g(a) = g(b)$ implies $f(a) = f(b)$. Hence, f cannot take more distinct values than g.

(P4) Suppose that there is a subset $S \subseteq [n]$ such that from the value $f(x)$ one can always infer the values of all input bits x_i with $i \in S$. Then $E(f) \geq |S|$. This is a direct consequence of (P2) and (P3).

Properties (P1) and (P3) imply that, if a depth-2 circuit for an operator f has no direct input–output wires, then there must be at least $E(f)$ nodes on the middle layer. To lower bound the number of *wires*, we introduce the concept of an "augmented operator".

Given an operator $f = (f_1, \dots, f_m)$, a subset $I \subseteq [n]$ of its inputs and a subset $J \subseteq [m]$ of its outputs, define the *augmented operator* $f_{I,J}$ of f as the operator

$$f_{I,J} = (f_j^i \mid i \in I, \, j \in J)$$

consisting of $|I| \cdot |J|$ (not necessarily distinct) boolean functions f_j^i with $i \in I$ and $j \in J$, where

$f_j^i : \{0,1\}^{n-|I|} \to \{0,1\}$ is a subfunction of f_j obtained by setting the i-th variable to 1 and all remaining variables in I to 0.

Note that $f_{I,J}$ has as its domain the bits $\{x_l \mid l \notin I\}$. Thus, $f_{I,J}$ maps binary strings of length $n - |I|$ ($|I|$ variables are fixed) to binary strings of length $|I| \cdot |J|$ (the number of augmented functions f_j^i we have). In particular, by (P1), we always have that $E(f_{I,J}) \leq |I| \cdot |J|$. On the other hand, the entropy of an augmented operator $f_{I,J}$ may be much larger than that of the operator f itself. To see this, consider the following $(n, 1)$-operator (the inner product function): $f(x, y) = x_1 y_1 + x_2 y_2 + \cdots + x_n y_n \bmod 2$. Then $E(f) = 1$ because f takes only two values. But if we let I to correspond to the x-variables, and $J = \{1\}$ (we have only one output function) then, for every $i \in I$,

$$f^i(x, y) = 0 \cdot y_1 + \cdots + 1 \cdot y_i + \cdots + 0 \cdot y_n = y_i.$$

Thus, the augmented operator in this case is $f_{I,J}(x, y) = (y_1, \dots, y_n)$, implying that $E(f_{I,J}) \geq n$.

Now take an arbitrary circuit computing an operator f. Let I be some set of its input nodes, and J some subset of its output nodes. Define a J-*cut* to be any set V of non-input nodes such that every path from an input node to a node in J goes through at least one node in V. Say that a node $v \in V$ *can see* an input if there is a path from that input to v. The *weight* of a node v relative to a given set I of inputs is the number of inputs in I seen by that node. The weight of a set V of nodes relative to I, denoted by $w_I(V)$, is the sum of weights of all nodes in V relative to I. That is,

$$w_I(V) = \sum_{v \in V} |I(v)|,$$

where $I(v)$ is the set of all inputs in I from which there is a path to v. In other words, $w_I(V)$ is the sum, over all nodes $v \in V$, of the number of input variables in I on which the function g_v computed at v can depend.

The following lemma, which was implicitly used by Cherukhin (2005) and was made explicit in terms of entropy in Jukna (2010), states that this number of variables must be large if the augmented operator has high entropy.

Lemma 13.1. *If V is a J-cut then $|V| + w_I(V) \geq \mathrm{E}(f_{I,J})$.*

Proof. For a node $v \in V$, let g_v be a boolean function computed at this node. For $i \in I$, let g_v^i be the subfunction of g_v obtained by setting $x_i = 1$ and $x_j = 0$ for all $j \in I \setminus \{i\}$. Let also g_v^0 be obtained from g_v by setting to 0 *all* variables x_i with $i \in I$. Consider the operator $g = (g_v^i : v \in V, i \in I)$. A simple but crucial observation is:

> If there is no path from the i-th input to v, then g_v cannot depend on the i-th input variable, implying that $g_v^i = g_v^0$.

In particular, this implies that $g_v^i = g_v^0$ for all $i \notin I(v)$. Thus, for each node $v \in V$, the function g_v computed at this node constitutes at most $1 + |I(v)|$ distinct functions to the operator g: the function g_v^0 and at most $|I(v)|$ distinct functions g_v^i with $i \in I(v)$. We have therefore shown that the total number of distinct boolean functions in g does not exceed $|V| + w_I(V)$. By (P1), this implies that $\mathrm{E}(g) \leq |V| + w_I(V)$.

To finish the proof, observe that, since V is a J-cut, all functions f_j with $j \in J$ must be computable from the set of functions g_v with $v \in V$. Hence, the augmented operator $f_{I,J}$ must also be computable from g. In particular, if on some two input vectors, the operator g takes the same value, then the operator $f_{I,J}$ is forced to take the same value, as well. Together with (P3) this implies $\mathrm{E}(f_{I,J}) \leq \mathrm{E}(g)$, and hence, also $\mathrm{E}(f_{I,J}) \leq |V| + w_I(V)$. $\square$

13.2 Entropy and Depth-Two Circuits

In this section we apply Lemma 13.1 to depth-2 circuits with general gates computing operators $f : \{0,1\}^n \to \{0,1\}^m$. We will assume that there are no direct wires from input to output nodes: this can be easily achieved by adding n new nodes of fanin 0 on the middle layer labeled with input variables. Thus, a depth-2 circuit for f consists of three layers. The first (input) layer contains n nodes $1, \ldots, n$ corresponding to input variables $x_1, \ldots, x_n$, the middle layer consists of some number of nodes, each computing some boolean function of its inputs, and the third (output) layer consists of m nodes $1, \ldots, m$ corresponding to m components $f_1, \ldots, f_m$ of f; at each output node an arbitrary boolean function can be computed as well.

Our goal is to prove a general lower bound on the smallest number $s_2(f)$ of wires in a depth-2 circuit computing a given operator f. To do this, take an arbitrary depth-2 circuit computing a given operator f. Let I be a subset of input nodes and J a subset of output nodes. Let also Wires(I, J) denote the number of wires leaving I plus the number of wires entering J.

Lemma 13.2. *In a depth-2 circuit computing an operator f, for any subset I of inputs and any subset J of outputs, we have that*

$$\text{Wires}(I, J) \geq \text{E}(f_{I,J}).$$

Proof. Let V be the set of all nodes on the middle layer from which there is a wire to a node in J. In particular, V is a J-cut, and $|V|$ is at most the number of wires entering the nodes in J. On the other hand, for every node $v \in V$, $|I(v)|$ is at most the number of wires starting in I and entering v. Hence, $w_I(V) = \sum_{v \in V} |I(v)|$ is at most the total number of wires leaving nodes in I. Lemma 13.1 implies that Wires$(I, J) \geq |V| + w_I(V) \geq \text{E}(f_{I,J})$. $\quad\square$

Remark 13.3. Note that our lower bound on the number of wires going from V to J is very "pessimistic": we lower bound this number just by the number $|V|$ of the starting nodes of these wires, as if these nodes had fanout 1. Here, apparently, is some space for an improvement.

Definition 13.4. (Entropy of operators) The *entropy*, $\text{E}(f)$, of an operator f is the maximum of $\text{E}(f_{I_1,J_1}) + \text{E}(f_{I_2,J_2}) + \cdots + \text{E}(f_{I_p,J_p})$ over all sequences $I_1, \ldots, I_p$ of disjoint subsets of inputs, and all sequences $J_1, \ldots, J_p$ of disjoint subsets of outputs.

Since no wire can leave more than one input node, and no wire can enter more than one output node, Lemma 13.2 immediately yields the following lower bound for depth-2 circuits.

Theorem 13.5. *For every operator f, $s_2(f) \geq \text{E}(f)$.*

Remark 13.6. Taking *disjoint* subsets of inputs and outputs in Theorem 13.5 is not crucial. It is enough to require that no element belongs to more than k of the sets $I_1, \ldots, I_p$, and no element belongs to more than k of the sets $J_1, \ldots, J_p$. The argument taking disjoint subsets utilizes $k = 1$. Now, if $d(i)$ is the number of wires leaving the input i, then the sum

$$\sum_{t=1}^{p} \sum_{i \in I_t} d(i) = \sum_{i=1}^{n} \sum_{t:i \in I_t} d(i) \leq k \sum_{i=1}^{n} d(i)$$

is at most k times larger than the total number $\sum_{i=1}^{n} d(i)$ of wires leaving the inputs. Since the same also holds for the number of wires entering the output nodes, Lemma 13.2 implies

$$s_2(f) \geq \frac{1}{k} \Big[\text{E}(f_{I_1,J_1}) + \text{E}(f_{I_2,J_2}) + \cdots + \text{E}(f_{I_p,J_p}) \Big].$$

13.3   Matrix Product Is Hard in Depth Two

Let $n = m^2$. The operator $f = \text{Mult}_n(X, Y)$ of matrix product takes two m-by-m matrices X and Y as inputs, and produces their product $Z = X \cdot Y$. Since Z is just a sequence of m^2 scalar products, each of $2m$ variables (row of X times a column of Y), all these scalar products can be computed by depth-1 circuit using $2m \cdot m^2 = 2n^{3/2}$ wires.

Raz and Shpilka (2003) proved that Mult_n cannot be computed by a circuit of *any* constant depth using $\mathcal{O}(n)$ wires. For depth-2 circuits their lower bound on the number of wires has the form $\Omega(n \ln n)$ for both finite and infinite rings. However, the difference between $n \ln n$ and $n^{3/2}$ is still large. Jukna (2010) provided a nearly matching lower bound for the Mult_n operator over the field GF(2) using the entropy argument.

Theorem 13.7. *Any depth-2 circuit for* $\text{Mult}_n(X, Y)$ *requires at least* $n^{3/2}$ *wires.*

Proof. It is enough to observe that if we take I to be the i-th row of the first input matrix X, and J to be the i-th row of the output matrix Z, then the augmented operator $f_{I,J}$ contains all $m^2 = n$ single variables of Y among its boolean functions. Indeed, if we set $x_{ij} = 1$ and all other entries of X to 0, then the product $E_{ij} \cdot Y$ of Y with the resulting boolean matrix E_{ij} is just the j-th row of Y (see Fig. 13.1). When doing this for all $j = 1, \ldots, m$, we obtain all $n = m^2$ variables $Y = \{y_{ij}\}$ among the functions in $f_{I,J}$. By the property (P2) of the entropy function, we then have that $\text{E}(f_{I,J}) \geq n$.

Since we have $m = n^{1/2}$ rows, we have m sets $I_1, \ldots, I_m$ of inputs and m sets $J_1, \ldots, J_m$ of outputs. Since the I_i's and J_i's are disjoint, Theorem 13.5 implies that every depth-2 circuit computing $f(X, Y) = X \cdot Y$ must have at least $\sum_{i=1}^{m} \text{E}(f_{I_i, J_i}) \geq mn = n^{3/2}$ wires. $\square$

Remark 13.8. Note that the entropy of the matrix product operator $f(X, Y) = X \cdot Y$ is large only for this special "row-wise" partition $I_1, \ldots, I_m$ and $J_1, \ldots, J_m$ of inputs and outputs, where I_i is the i-th row of the input matrix X, and J_i is the i-th row of the output matrix $Z = X \cdot Y$. In particular, $\text{E}(f_{I_i, J_j}) \leq |J_j| = m = \sqrt{n}$ for $i \neq j$, because in this case the assignments of constants to the i-th row of X does not affect the results computed at the j-th row of Z, which are m scalar products of m columns of Y with the j-th row of X.

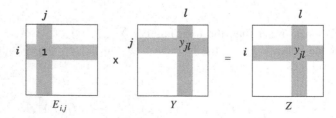

Fig. 13.1 For every fixed i and j, the product $E_{ij} \cdot Y$ gives the j-th row of Y

Remark 13.9. (Cyclic convolution) There are, however, "more complicated" operators whose entropy remains large under *any* partitions of inputs and outputs. Consider, for example, the operator of *cyclic convolution* $f = \text{Conv}(x, y)$. This operator takes two boolean vectors $x = (x_0, \ldots, x_{n-1})$ and $y = (y_0, \ldots, y_{n-1})$ as inputs and outputs the vector $z = (z_0, \ldots, z_{n-1})$, where $z_j = \sum_{i=0}^{n-1} x_i y_{i+j} \bmod 2$ and $i + j$ is taken modulo n. In other words, the j-th output z_j is the scalar product of vector x with the cyclic shift of vector y by j positions to the left. It can be shown (see Exercise 13.2) that, if $n = pq$ then for every partition of x-variables into p consecutive intervals $\mathcal{I} = \{I_1, \ldots, I_p\}$ of length q, there exists a partition of the output vector z into q disjoint sets $\mathcal{J} = \{J_1, \ldots, J_q\}$ such that $\text{E}(f_{I,J}) \geq n$ for *all* $I \in \mathcal{I}$ and $J \in \mathcal{J}$.

Remark 13.10. (Limitations) How large can entropy of operators be? Recall that in the definition of the entropy of operators we first split the inputs into p blocks $I_1, \ldots, I_p$ of some sizes $a_1, \ldots, a_p$, and the outputs into p blocks $J_1, \ldots, J_p$ of some sizes $b_1, \ldots b_p$. Then we just take the sum

$$\text{E}(f) = \text{E}(f_{I_1, J_1}) + \text{E}(f_{I_2, J_2}) + \cdots + \text{E}(f_{I_p, J_p})$$

of the entropies of the corresponding (to these blocks) augmented operators. Say that a partition is *balanced* if $a_1 \leq a_2 \leq \ldots \leq a_p$ and $b_1 \geq b_2 \geq \ldots \geq b_p$. Note that the partition (into the rows) which we used for the matrix product is balanced—there all a_i's and b_i's were equal.

Since each of the sets $\{f_j^i \mid i \in I_t, j \in J_t\}$ can have at most $|I_t \times J_t| = a_t b_t$ functions, the entropy of this set cannot exceed $a_t b_t$. If the partition is balanced, then Chebyshev's inequality (see Hardy et al. 1952, Theorem 43, page 43) yields

$$\text{E}(f) \leq \sum_{t=1}^{p} a_t b_t \leq \frac{1}{p} \left(\sum_{t=1}^{p} a_t \right) \left(\sum_{t=1}^{p} b_t \right) \leq \frac{nm}{p}.$$

On the other hand, we have a trivial upper bound $\text{E}(f) \leq pn$. Substituting $p \geq \text{E}(f)/n$ in the previous inequality, we obtain that $\text{E}(f) \leq n\sqrt{m}$. Thus, at least with respect to balanced partitions, the entropy of any (n, m)-operator does not exceed $n\sqrt{m}$. In particular, for such partitions, matrix multiplication has the largest possible entropy $\Theta(n^{3/2})$ among all (n, n)-operators.

13.3.1 Restricted Matrix Product Is Easy in Depth Three

As observed by Drucker (2011), the proof of Theorem 13.7 actually gives a lower bound $n^{3/2}$ for the following restricted version of matrix product: $\text{mult}_n(X, Y) = X \cdot Y$ if X contains exactly one 1-entry, and $\text{mult}_n(X, Y) = 0$ (all-0 matrix) otherwise. Yet it can be shown that the operator mult_n *can* be computed by a depth-3 circuit using only a *linear* number of wires, as we will see next. This shows a separation between depth-2 and depth-3 circuits.

Theorem 13.11. (Drucker 2011) *The operator* mult_n *can be computed by a depth-3 circuits using* $\mathcal{O}(n)$ *wires.*

Proof. Let $n = m^2$. Given two boolean $m \times m$ matrices X and Y, we want to detect whether X is "good" (contains exactly one 1-entry) and, if this is the case, to compute the matrix $Z = X \cdot Y$. To detect whether X is good, we just put one "security gate" s on the first (next to inputs) layer. This gate takes all X-variables as inputs and outputs 1 if X has exactly one 1-entry, and 0 otherwise.

The goal of the remaining operator computed at the first layer is to determine the unique 1-entry in a good matrix X. Our goal is to do this using $\mathcal{O}(m)$ instead of m^2 gates. For this, we associate with each position $(i, j) \in [m]^2$ a distinct 2-element subset $S_{i,j}$ of $[3m]$; this can be done since $\binom{3m}{2} > m^2$. Then we put on the first layer $3m$ "hash gates" $h_1, \ldots, h_{3m}$. The (i, j)-th entry x_{ij} is wired to gates h_p and h_q where $S_{i,j} = \{p, q\}$. Each hash gate computes the sum modulo 2 of its inputs.

Observe that if we are promised that X contains exactly one 1-entry in, say, position (i, j), then this position can be determined from the value of the operator $H(X) = (h_1(X), \ldots, h_{3m}(X))$. In fact, this value is just the characteristic vector $1_{(i,j)} \in \{0, 1\}^{3m}$ of the set $S_{i,j}$.

Next we put on the second layer "row gates" $r_1, \ldots, r_m$ and "column gates" $c_1, \ldots, c_m$. Each row gate r_k takes $h_1, \ldots, h_{3m}$ and s as inputs. We define $r_k = 1$ iff $s = 1$ and the hash gates output $1_{(k,j)}$ for some $j \in [m]$. The column gate c_l takes $h_1, \ldots, h_{3m}$, and the l-th column of Y as inputs. We define $c_l = y_{j,l}$ if the hash gates output $1_{(i,j)}$ for some $i \in [m]$, and $c_l = 0$ otherwise. Finally, we compute the entries $z_{k,l}$ of the product matrix on the last, third layer by letting $z_{k,l} := r_k \cdot c_l$.

We argue that this circuit computes mult. First suppose that X does not have exactly one 1-entry. Then $s(X) = 0$, so all row gates output 0 and $z_{k,l} = 0$ for all (k, l), as required. Next suppose $x_{i,j} = 1$ while all other entries of X are 0. Then $H(X) = 1_{(i,j)}$ and $s(X) = 1$. It follows that $r_k = 1$ iff $k = i$, whereas $(c_1, \ldots, c_m)$ is the j-th row $(y_{j,1}, \ldots, y_{j,m})$ of Y. Thus, $z_{k,l} = r_k \cdot c_l$ is 0 for $k \neq i$, and is $y_{j,l}$ for $k = i$. This is precisely the (k, l)-entry of $\text{mult}_n(X, Y)$.

It remains to count the wires. Each X-variable is connected to 2 hash-gates, for $2m^2$ wires in total leading to hash gates. The security gate s has m^2 inputs. Each row and column gate has at most $2m$ inputs, for a total of at most $(2m)^2$ wires entering the second layer. Finally, each output gate $z_{k,l}$ has 2 inputs, so the total number of wires in the circuit is $\mathcal{O}(m^2) = \mathcal{O}(n)$ as desired.   □

13.4   Larger-Depth Circuits

For a function $f : \mathbb{N} \to \mathbb{N}$ such that $1 \leq f(n) < n$ define

$$f^*(n) := \min\{k \mid \underbrace{f(f(\cdots f(n) \cdots))}_{k \text{ times}} \leq 1\}.$$

Now define the functions $\lambda_d(n)$:

$$\lambda_1(n) := \lfloor \sqrt{n} \rfloor, \quad \lambda_2(n) := \lceil \log_2 n \rceil \quad \text{and} \quad \lambda_d(n) := \lambda_{d-2}^*(n) \quad \text{for } d \geq 3.$$

In particular, $\lambda_3(n) = \Theta(\ln \ln n)$. Since $\log \log n$ is the two-fold composition of $\log n$, we obtain that $\lambda_5(n) = \Theta(\lambda_4(n))$. By induction, $\lambda_{2d+1}(n) = \Theta(\lambda_{2d}(n))$ for all $d \geq 2$. These slowly growing functions arose when dealing with so-called "superconcentrators". Although we will not use them later, let us summarize some results about superconcentrators.

An n-superconcentrator is a directed graph with n input and n output nodes, such that for every $r < n$, any r input nodes may be connected to any r output nodes in some order by r vertex-disjoint directed paths. Let $c(n)$ denote the minimum number of wires in an n-superconcentrator, and $c_d(n)$ the minimum number of wires in an n-superconcentrator of depth at most d. It is clear that $c_1(n) = n^2$: one has to take the complete bipartite graph. It is therefore somewhat surprising that much fewer wires are enough if we allow just one more layer of wires.

- Pippenger (1977) proved that $c_2(n) = \mathcal{O}(n \ln^2 n)$ and $c_2(n) = \Omega(n \ln n)$.
- Alon and Pudlák (1994) improved the lower bound to $\Omega(n \ln^{3/2} n)$.
- Finally, Radhakrishnan and Ta-Shma (2000) proved an almost optimal bound $c_2(n) = \Theta(n \ln^2 n / \ln \ln n)$.
- Alon and Pudlák (1994) proved that $c_3(n) = \Theta(n \ln \ln n)$.
- For $d \geq 4$, bounds $c_d(n) = \Theta(n \lambda_d(n))$ were proved by Dolev et al. (1983), and Pudlák (1994).
- By improving an earlier upper bound $c(n) = \mathcal{O}(n)$ of Valiant (1976), Pippenger (1977) showed that $c_d(n) = \mathcal{O}(n)$ holds already when $d = \mathcal{O}(\ln n)$; moreover, his superconcentrator is quite regular: it has *constant* maximum degree. Basalygo (1981) proved that we actually have $c_d(n) \leq 36n + \mathcal{O}(\ln n)$ for $d = \mathcal{O}(\ln n)$. The depth was improved to $d = \min\{k : \lambda_k(n) \leq 1\}$ by Dolev et al. (1983).

Raz and Shpilka (2003) used superconcentrator-type properties of circuits to prove lower bounds $\Omega(n \lambda_d(n))$ for the operator of matrix product. Their proof is based on the following graph-theoretic lemma generalizing results of Dolev et al. (1983) and Pudlák and Rödl (1994). We formulate the lemma in a slightly different (weaker) form. The proof of the lemma is somewhat technical, and we omit it.

Lemma 13.12. (Raz–Shpilka 2003) *Let $d \geq 2$ be an integer. For every constant $\epsilon > 0$, there is a constant $\delta > 0$, depending only on d and ϵ, such that if a leveled directed acyclic graph of depth d has more than n vertices and fewer than $\delta n \lambda_d(n)$ edges, then there exists a set W of inputs and outputs and a set S of inner nodes such that $|W \cup S| \leq \epsilon n$, $|S| \geq \sqrt{n}$ and at most $\epsilon^2 n^2 / |S|$ input–output paths do not go through a node in $W \cup S$.*

Cherukhin (2008b) observed that an appropriate combination of this lemma with the entropy argument (described in Sect. 13.1) allows one to increase known lower bounds for depth-d circuits from $\Omega(n \lambda_d(n))$ to $\Omega(n \lambda_{d-1}(n))$. In particular, known lower bound of $\Omega(n \ln \ln n)$ for depth-3 circuits can be improved to $\Omega(n \ln n)$.

Cherukhin gave this combined argument only for one special operator—cyclic convolution. A natural question therefore is: can this combination be used to improve, say, lower bounds of Raz and Shpilka (2003) for the matrix product?

Below we give a negative answer. We first put Cherukhin's argument in a more general setting to characterize which operators' lower bounds the combined argument can improve. We discover that this is only possible for operators whose entropy remains high under highly unbalanced partitions of input and output. Cyclic convolution has this property but unfortunately matrix multiplication does not.

By an *n-operator* we will mean any operator $f : \{0,1\}^N \to \{0,1\}^m$ with $N, m \geq n$. By a (p,q)-*partition* of inputs and outputs of such an operator we will mean a partition $\mathcal{I}$ of some n inputs into $|\mathcal{I}| = p$ disjoint subsets I of size n/p, and a partition $\mathcal{J}$ of some n outputs into $|\mathcal{J}| = q$ disjoint subsets J of size n/q. (Here and though the proof we shall ignore floors and ceilings whenever appropriate as this does not affect the asymptotic nature of our result; hence me may assume n/p and n/q are integers.)

Definition. Say that an n-operator f has *strong multiscale entropy* if there exist constants $C, \gamma > 0$ such that for every integer p lying between $C\sqrt{n}$ and n, there is a $(p, n/p)$-partition $\mathcal{I}, \mathcal{J}$ of its inputs and outputs, such that $E(f_{I,J}) \geq \gamma n$ for all $I \in \mathcal{I}$ and $J \in \mathcal{J}$.

Remark 13.9 shows that the operator of cyclic convolution has strong multiscale entropy. The following theorem, due to the author of the book, was never published before.

Theorem 13.13. *Let $d \geq 3$ be a constant. Every depth-d circuit computing an operator of strong multiscale entropy must have $\Omega(n\lambda_{d-1}(n))$ wires.*

In particular, the operator of cyclic convolution requires this number of wires.

Proof. Let f be an n-operator of strong multiscale entropy, and take an arbitrary depth-d circuit computing this operator, where $d \geq 3$ is constant. Let $\epsilon > 0$ be a small enough constant; it is enough to take $\epsilon = \min\{1/C, \gamma/17\}$, where C and γ are constant from the definition of operators with strong multiscale entropy. Assume, for the sake of contradiction, that the total number of wires in the circuit is smaller than $\delta n\lambda_{d-1}(n)$, where $\delta = \delta(d, \epsilon)$ is a constant from Lemma 13.12. Our goal is to derive a contradiction with Lemma 13.2.

Let L_i be the set of nodes on the i-th layer of our circuit, $i = 0, 1, \ldots, d$. Consider the graph induced by the last $d - 1$ layers $L_1, \ldots, L_d$. When applied to this graph, which is of depth $d - 1$, Lemma 13.12 gives us a set W of inputs and outputs, and a set S of inner nodes such that $|W \cup S| \leq \epsilon n$, $|S| \geq \sqrt{n}$ and at most $\epsilon^2 n^2/|S|$ paths from the nodes in L_1 to the nodes in L_d do not go through a node in $W \cup S$. To obtain the desired contradiction, take a (p, q)-partition $\mathcal{I}, \mathcal{J}$ of inputs in L_0 and outputs in L_d with

$$p := \left\lceil \frac{|S|}{\epsilon} \right\rceil \geq \frac{\sqrt{n}}{\epsilon} \quad \text{and} \quad q := \frac{n}{p} \geq \frac{\epsilon n}{2|S|}.$$

Since f is an operator of a strong multiscale entropy, we know that $\mathrm{E}(f_{I,J}) \geq \gamma n$ must hold for every $I \in \mathcal{I}$ and $J \in \mathcal{J}$. We will use the sets W and S guaranteed by Lemma 13.12 to choose a set $I \in \mathcal{I}$ of inputs, a set $J \in \mathcal{J}$ of outputs and a J-cut V such that

$$|V| + w_I(V) \leq 16\epsilon n < \gamma n, \tag{13.1}$$

which contradicts Lemma 13.1.

By Lemma 13.12, at most $b := \epsilon^2 n^2/|S|$ of the paths from L_1 to L_d can avoid the set $W \cup S$. Recall that $|W \cup S| \leq \epsilon n$ and $|S| \geq \sqrt{n}$. Since we have $q \geq \epsilon n/2|S|$ disjoint sets $J \in \mathcal{J}$, there must exist a $J \in \mathcal{J}$ such that J has at most

$$|J \cap W| \leq \frac{2|W|}{q} \leq 4|S| \leq 4\epsilon n$$

nodes in W and at most $2b/q \leq 4\epsilon n$ of these b "bad" paths can enter J. Thus, if we take the set $L_1' \subseteq L_1$ of $|L_1'| \leq 4\epsilon n$ starting points of these "bad" paths, then the set

$$V := W' \cup S \cup (J \cap W) \text{ with } W' := L_1' \cup (L_1 \cap W)$$

is a J-cut of size $|V| \leq 10\epsilon n$ (see Fig. 13.2).

To finish the proof of (13.1), and thus the proof of the theorem, it remains to show that $w_I(V) \leq 6\epsilon n$ holds for at least one $I \in \mathcal{I}$. Since $W' \subseteq L_1$ is a set of nodes on the first (next to input) layer of our original circuit, each $I(v)$ for $v \in W'$ is just the set of *wires* going from I to v. Since all p sets in $\mathcal{I}$ are disjoint, the sum $\sum_{I \in i} w_I(W')$ cannot exceed the total number of wires which, by our assumption, is at most $\delta n \lambda_{d-1}(n) \leq n \log_2 n$. Hence, there exists an $I \in \mathcal{I}$ such that

$$w_I(W') \leq \frac{n \log_2 n}{p} \leq \epsilon \sqrt{n} \log n \leq \epsilon n$$

for all large enough n. Further, we use $|J \cap W| \leq 2|W|/q \leq 4|S|$ and a trivial estimate $w_I(U) \leq |I| \cdot |U|$, holding for any set U of nodes, to obtain that

$$w_I(S) \leq |I| \cdot |S| \leq \frac{n|S|}{p} \leq \epsilon n$$

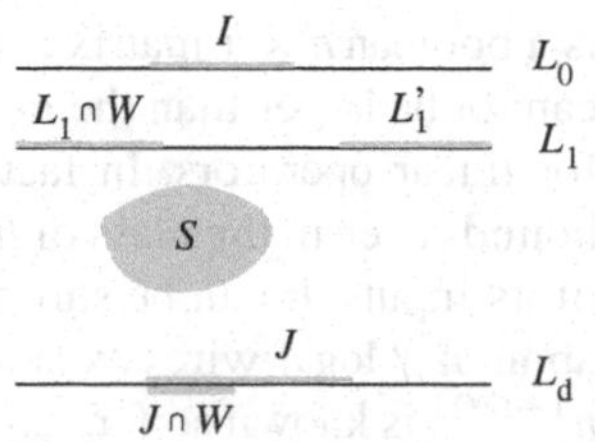

Fig. 13.2 Every path from the input layer L_0 to output nodes in J must contain a node in the set $V = L_1' \cup (L_1 \cap W) \cup S \cup (J \cap W)$. Then, V is a J-cut

and

$$w_I(J \cap W) \le |I| \cdot |J \cap W| \le \frac{n}{p} \cdot 4|S| \le 4\epsilon n.$$

Thus, $w_I(V) \le 6\epsilon n$. This completes the proof of Theorem 13.13. $\square$

Remark 13.14. A possible big imbalance of partitions resulting from the proof of Theorem 13.13 (with $p = \Omega(n)$ and $q = \mathcal{O}(1)$) arises from the trivial upper bound $w_I(S) \le |I| \cdot |S|$ given at the end of the proof. Actually, if we do not have any additional information about the set S than that given in Lemma 13.12, this imbalance cannot be avoided. In particular, we cannot exclude the possibility of the following undesired situation: from every input *at least one* path goes through at least one node in S. In this case $w_I(S) = |I| \cdot |S| = n|S|/p$. Then, in order to achieve $w_I(S) \le \epsilon n$, we would be forced to take $p = \Omega(n)$, and hence, the block length $|I| = \mathcal{O}(1)$. But since we need $\mathrm{E}(f_{I,J}) \ge \gamma n$ and since $\mathrm{E}(f_{I,J}) \le |I| \cdot |J|$ holds for all I and J, this forces $|J| = \Omega(n)$, and hence, $q = \mathcal{O}(1)$. Thus, the weakness of the entire argument is that the choice of parameters p and q is forced by the size $|S|$ of the set S guaranteed by Lemma 13.12, and we only know that $\sqrt{n} \le |S| \le \epsilon n$.

We have just seen that the combination of Lemmas 13.12 and 13.2 (supercon-centrators plus entropy) can only work for operators that have high entropy under very unbalanced partitions, where inputs are split into blocks of constant size. Remark 13.9 shows that the operator of cyclic convolution has this property. This is a lucky exception: many other important operators, like the matrix product operator, do not have this property.

Remark 13.15. An interesting question is: can the property of strong multiscale entropy alone lead to higher lower bounds than given in Theorem 13.13? Recently, Drucker (2011) gave a *negative* answer: there is an explicit operator with this property that is computable in depth d with $\mathcal{O}(n\lambda_{d-1}(n))$ wires, for $d = 2, 3$ and for even $d \ge 6$. Roughly speaking, the operator is a simplified variant of cyclic convolution.

13.5   Linear Circuits for Linear Operators

We now consider *linear* operators, that is operators of the form $f(x) = Ax$ where A is a boolean $n \times n$ matrix and computations are in GF(2). Entropy of such operators cannot be larger than the rank of A, and the entropy method does not seem to work for linear operators. In fact, for such operators it is difficult to prove high lower bounds even in the class of *linear* circuits, where each gate computes the sum mod 2 of its inputs. It can be shown that matrices A requiring linear depth-2 circuits with about $n^2/\log n$ wires exist (see Theorem 13.18 below), but no explicit lower bound $n^{1+\Omega(1)}$ is known so far.

Fig. 13.3 A matrix

$A = \begin{bmatrix} 0 & 0 & 1 & 0 & 0 & 0 \\ 0 & 1 & 0 & 0 & 1 & 1 \\ 1 & 0 & 0 & 1 & 0 & 0 \\ 1 & 1 & 0 & 1 & 1 & 1 \\ 1 & 1 & 1 & 1 & 1 & 1 \end{bmatrix}$ and a

linear depth-2 circuit
computing $y = Ax$. Each
non-input gate is the sum
mod 2 of its inputs

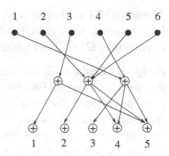

Recall that a linear circuit has n input nodes $x_1, \ldots, x_n$ and m output nodes $y_1, \ldots, y_m$. Each non-input node computes the sum mod 2 of its inputs. Thus, every such circuit computes a linear transformation $y = Ax$ for some boolean $m \times n$ matrix A (Fig. 13.3). A special property of linear circuits is that they only need to correctly compute the transformation for basis vectors x.

Namely, say that a (not necessarily linear) circuit *represents* a boolean $m \times n$ matrix $A = (a_{ij})$ if on each input vector $e_j = (0, \ldots, 0, 1, 0, \ldots, 0)$ with precisely one 1 in the j-th position, the circuit outputs the j-th column of A. That is, a circuit represents A if it correctly computes the linear operator Ax over GF(2) on all n unit vectors $e_1, \ldots, e_n$; on other input vectors x the circuit can output arbitrary values. In particular, if a circuit is linear then it represents the matrix A if for every $i \in [m]$ and $j \in [n]$,

$$a_{ij} = 1 \text{ iff the number of paths from } x_j \text{ to } y_i \text{ is odd.} \tag{13.2}$$

Proposition 13.16. *A linear circuit computes* $y = Ax$ *iff it represents* A.

Proof. If a circuit (not necessarily linear) computes $y = Ax$ for all inputs x, then it clearly represents A. For the other direction, observe that the behavior of a linear circuit on all input vectors x is completely determined by its behavior on n unit vectors: just write each input vector $x = (x_1, \ldots, x_n)$ as the sum $x = x_1 e_1 \oplus \cdots \oplus x_n e_n$ and use the linearity of gates. $\square$

It is clear that every matrix A can be represented by a depth-1 linear circuit, but the number of wires in this case is just the total number $|A|$ of ones in A. However, already in depth-2 we can have much more compact representations.

Example 13.17. Recall that a *primitive matrix* is a 0-1 matrix of rank 1, that is, a boolean matrix consisting of one all-1 submatrix and zeros elsewhere. Such a matrix R can be specified by a subset S of rows and a subset T of columns such that B has ones in all positions in $S \times T$, and has zeros elsewhere. As before, we call $|S| + |T|$ the *weight* of R. In depth-1, such a matrix requires $|S| \cdot |T|$ wires. But in depth-2, already $|S| + |T|$ wires are enough: just take one vertex on the middle layer and connect it to all inputs in S and all outputs in T.

Theorem 13.18. (Lupanov 1956) *Every boolean $n \times n$ matrix can be represented by a linear depth-2 circuit with $\mathcal{O}(n^2/\ln n)$ wires, and matrices requiring linear circuits with $\Omega(n^2/\ln n)$ wires in any depth exist.*

Proof. By Lemma 1.2, the matrix A can be decomposed into primitive matrices so that their total weight (sum of weights of primitive matrices in the decomposition) does not exceed $2n^2/\log n$. Since each primitive matrix of weight w can be represented using w wires (see Example 13.17), and since the matrices in the decomposition are disjoint (have no common 1-entries), we are done.

The lower bound follows by counting arguments: there are 2^{n^2} boolean $n \times n$ matrices and, by Lemma 1.21, at most $(9t)^t$ linear circuits with t wires.   □

Thus, if we denote by $\mathrm{lin}_2(Ax)$ the minimum number of wires in a linear depth-2 circuit computing Ax, then $n \times n$ matrices with $\mathrm{lin}_2(Ax) = \Omega(n^2/\ln n)$ exist. Nothing similar, however, is known for *general* (non-linear) circuits. Let $\mathrm{s}_2(Ax)$ denote the minimum number of wires in a depth-2 circuits with *arbitrary* boolean functions as gates computing the linear transformation $y = Ax$.

Problem 13.19. Do matrices A with $\mathrm{s}_2(Ax) = \Omega(n^2/\ln n)$ exist?

One can show that the answer is *positive* if either all gates on the output layer or all gates on the middle layer are required to be linear. The case when all output gates are linear is simple (see Exercise 13.7). The case when only middle gates are required to be linear can be proved using Kolmogorov complexity arguments (see Exercise 13.8).

Concerning *explicit* lower bounds, no lower bound $n^{1+\Omega(1)}$ is known, even in the case of linear circuits! The first nontrivial lower bound $\mathrm{lin}_2(Ax) = \Omega(n \ln n)$ for linear circuits was proved by Alon et al. (1990) using the Sylvester matrix; nearly the same lower bound holds for general (non-linear) depth-2 circuits as well (see Theorem 13.29 below). It was also stated by Pudlák and Rödl (1994) (on page 260, without proof) that the lower bound $\mathrm{lin}_2(Ax) = \Omega(n \ln^{3/2} n)$ can be derived from a well-known bound on the rigidity due to Friedman (1993). Recently, Gál et al. (2011) proved the same lower bound for *general* circuits, while computing good linear codes.

Let A be boolean $m \times n$ matrix. Say that A is *good* if there exist constants $\alpha, \beta > 0$ such that $n \geq \alpha m$ and every two vectors in $\mathrm{Range}(A) = \{Ax \mid x \in \mathrm{GF}(2)^n\}$ have Hamming distance at least βm. Thus, every good matrix is a generator matrix of a linear self-correcting code $\mathrm{Range}(A)$ of rate $n/m = \Omega(1)$ and minimal distance $\Omega(n)$.

Theorem 13.20. (Gál et al. 2011) *If A is a good $m \times n$ matrix, then any depth-2 circuit computing Ax must have $\Omega(n \ln^{3/2} n)$ wires.*

Proof. (Sketch) Consider a depth-2 circuit C computing $y = Ax$. The circuit has n input nodes, $m = \mathcal{O}(n)$ output nodes, and some number t of middle nodes $v_1, v_2, \ldots, v_t$. Suppose the middle nodes have total degrees (fanins plus fanouts) $d_1 \geq d_2 \geq \ldots \geq d_t$. Hence, $L = \sum_{i=1}^{t} d_i$ is the total number of wires in the circuit. The proof of the theorem proceeds by establishing the following inequality for any integer r between $\log n$ and $\sqrt{n}/4$ (where $\epsilon > 0$ is a constant):

$$\sum_{i \geq r} \left(\frac{d_i}{\epsilon n \sqrt{\log n}} \right)^2 \geq \frac{1}{r}. \tag{13.3}$$

The desired lower bound $\sum_i d_i = \Omega(n \cdot \log n \cdot \sqrt{\log n})$ then follows from the Monotone Sums Lemma (see Appendix A).

To prove (13.3), fix an r in the desired interval, and let P be the number of length-2 paths from an input node to an output node that only use middle nodes v_i with $i > r$. Note that $P \leq \sum_{i>r} d_i^2$. By averaging, there is a set I of $|I| \geq n/2$ input nodes such that each of them has at most $s := 2P/n$ such paths to an output node.

Each middle node v_i computes some boolean function $g_i(x)$. We now replace each v_i with $i > r$ by a constant $g_i(0)$. Let C' be the modified circuit. Let also $k \leq |I|$ be a parameter (to be fixed later), and say that an input string $x \in \{0,1\}^n$ is *legal* if it has k ones, and $x_j = 0$ for all $j \notin I$.

For every such string x, the outputs $C(x)$ and $C'(x)$ can differ in at most ks coordinates, because there are at most ks output nodes for which there is a path from an input node with value 1 through a middle node v_i with $i > r$. If ks is smaller than half of the distance of the code, we can decode any such x from the value $C'(x)$. However, the values $C'(x)$ depend only on $\leq r$ bits (so many non-fixed gates are on the middle layer). Since we have $\binom{|I|}{k}$ legal strings x, the number r of these bits must satisfy the inequality

$$r \geq \log \binom{|I|}{k} \geq k \log \frac{|I|}{k} \geq k \log \frac{n}{2k}.$$

The only constraint on k is that $ks = 2kP/n$ must be less that half of the distance of the code, and the later is $\Omega(n)$. This allows us to set $k = \Theta(n^2/P)$, yielding

$$r = \Omega \left(\frac{n^2}{P} \log \frac{P}{n} \right).$$

Using this lower bound as well as inequalities $\log n \leq r \leq \sqrt{n}/4$, one can derive that $P = \Omega((n^2/r) \log n)$. Together with $P \leq \sum_{i>r} d_i^2$, this already establishes the desired inequality (13.3). □

Gál et al. (2011) also showed that, using large distance alone, one cannot hope to substantially improve the lower bound of Theorem 13.20: there exist generator matrices A of codes of distance $\Omega(n)$ such that $\text{lin}_2(Ax) = \mathcal{O}(n(\ln n/\ln \ln n)^2)$. The existence of such generator matrices is proved using probabilistic arguments.

Actually, the authors (personal communication) can improve the lower bound in
Theorem 13.20 to an optimal bound of $\lim_2(Ax) = \Omega(n(\ln n/\ln\ln n)^2)$.

13.6   Circuits with OR Gates: Rectifier Networks

We now consider circuits where each gate is the OR of its inputs. Such a circuit for
a boolean $m \times n$ matrix $A = (a_{ij})$ has n input nodes $x_1, \ldots, x_n$ and m output nodes
$y_1, \ldots, y_m$. At the i-th output of such a circuit the OR $y_i = \bigvee_{j:a_{ij}=1} x_j$ of the input
variables is computed. That is, the circuit computes the operator $y = Ax$ over the
boolean semiring. Note that in this case a circuit represents the matrix A if for every
$i \in [m]$ and $j \in [n]$,

$$a_{ij} = 1 \text{ iff there exists a path from } x_j \text{ to } y_i. \tag{13.4}$$

Thus, we can just ignore the gates, and consider our circuit a directed acyclic graph
with this property (13.4). Such a model for matrix representation, known as *rectifier
network*, was introduced by Lupanov (1956), and was subsequently intensively
studied in the Russian literature. The *size* of such a network is again the total number
of wires.

It is easy to see that Theorem 13.18 remains true also in the case of rectifier
networks: every $n \times n$ matrix can be represented by a depth-2 rectifier network using
$\mathcal{O}(n^2/\ln n)$ wires, and matrices requiring this number of wires exist. But unlike
in the case of linear circuits, where no *explicit* lower bound larger than $n \ln^{3/2} n$
in depth-2 is known, explicit lower bounds for OR-circuits (rectifier networks) are
easier to obtain, even without depth restrictions!

A lower bound of $\Omega(n^{3/2})$ was first obtained by Nechiporuk (1969) for the
"point-line incidence" matrix defined in Example 7.4. Using a different matrix
constructed by Brown (1966), a larger lower bound $\Omega(n^{5/3})$ was later obtained by
Pippenger (1980). Similar lower bounds were also obtained by Mehlhorn (1979)
and Wegener (1980).

> It is not difficult to show that any rectifier network representing a boolean matrix A must
> contain at least $|A|/r(A)$ wires, where $|A|$ is the total number of 1-entries in A, and $r(A)$ is
> the maximum number of entries in an all-1 submatrix of A. This lower bound follows from the
> following simple observation: the sets of inputs and outputs connected by paths going through
> a fixed wire must form an all-1 submatrix of A. Since we always need at least $|A|/r(A)$ all-1
> matrices to cover all ones of A, this number of wires is necessary. Unfortunately, this lower
> bound is too weak: if, say, A is an $n \times n$ matrix, then $r(A)$ is at least the maximum number of
> 1s in a row or a column of A, implying that $|A|/r(A) \le n$. Much larger lower bounds can be
> obtained using the fact that coverings resulting from circuits must have some special properties.

Say that a boolean matrix A is (s,t)-*free* if it does not contain any $(s+1) \times (t+1)$
all-1 submatrix. A matrix is k-*free* if it is (k,k)-free.

Theorem 13.21. *If A is a boolean k-free matrix, then any rectifier network representing A must have at least $|A|/k^2$ wires.*

Proof. (Due to Pippenger 1980) Take a rectifier network F for A. For a node w in F, let s_w be the number of input nodes from which w is reachable, and t_w the number of output nodes reachable from w. Let us call a wire $e = (u, v)$ "eligible" if $s_u \leq k$ and $t_v \leq k$. If (i, j) is a 1-entry of A (that is, $a_{ij} = 1$), we say that e "covers" this entry if there is a path from the j-th input node to u, and there is a path from v to the i-th output node.

Since each eligible wire $e = (u, v)$ can cover at most $s_u \cdot t_v \leq k^2$ 1-entries of A, it remains to prove the following claim.

Claim 13.22. *Every 1-entry of A is covered by at least one eligible wire.*

To prove the claim, take a 1-entry (i, j) of A. Then there must be a path $v_0, v_1, \ldots, v_r$ in our circuit beginning in $v_0 = j$ and ending in $v_r = i$. Letting $s_l := s_{v_l}$ be the number of inputs from which v_l is reachable, and $t_l := t_{v_l}$ denote the number of outputs reachable from v_l, we have that

$$s_1 \leq s_2 \leq \ldots \leq s_r \text{ and } t_1 \geq t_2 \geq \ldots \geq t_r.$$

Let p be the largest number for which $s_p \leq k$, and q the smallest number for which $t_q \leq k$. If $q \leq p + 1$, then the wire $e = (v_p, v_{p+1})$ covering the entry (i, j) is eligible, and we are done. So assume for the sake of contradiction that $q \geq p + 2$. By the definition of positions p and q we have that $s_{p+1} > k$ and $t_{p+1} > k$. But then at least $k + 1$ inputs are connected with at least $k + 1$ outputs going through the node v_{p+1}, contradicting the k-freeness of A. This completes the proof of the claim, and thus the proof of the theorem.   □

We already know explicit constructions of 1-free $n \times n$ matrices A with $|A| = \Omega(n^{3/2})$ ones; see Examples 7.4 and 7.5. For these matrices, Theorem 13.21 yields that any OR-circuit representing them requires $\Omega(n^{3/2})$ wires.

Example 13.23. (Brown's construction) The following construction of dense 2-free matrices is due to Brown (1966). Let p be an odd prime and let d be a non-zero element of $\mathbb{Z}_p = \{0, 1, \ldots, p - 1\}$ (the field of integers modulo p) such that d is a quadratic non-residue modulo p if $p \equiv 1$ modulo 4, and a quadratic residue modulo p if $p \equiv 3$ modulo 4. Let A be a boolean $n \times n$ matrix with $n = p^3$ whose rows and columns correspond to all triples of elements in $\mathbb{Z}_p$. The entry of A corresponding to a row (a_1, a_2, a_3) and column (b_1, b_2, b_3) is 1 iff the sum $(a_1 - b_1)^2 + (a_2 - b_2)^2 + (a_3 - b_3)^2$ modulo p is equal to d. Brown showed that this matrix has $|A| = p^4(p - 1) = \Omega(n^{5/3})$ ones, and is 2-free.

Thus, Brown matrices require OR-circuits with $\Omega(n^{5/3})$ wires. Subsequent constructions of dense square-free matrices have lead to even higher lower bounds.

Example 13.24. (Norm graphs) Let q be a prime-power, $t \geq 2$ an integer, and consider the field $GF(q^t)$ with q^t elements. The norm of an element a of this field is defined as the element

$$N(a) := a \cdot a^q \cdots a^{q^{t-1}} = a^{(q^t-1)/(q-1)}$$

of this field. Now let $n = q^t$, and construct a bipartite $n \times n$ graph with vertices in each part being elements of $\mathrm{GF}(q^t)$. Two vertices a and b are adjacent iff $N(a+b)$ = 1. It is known that the number of solutions in $\mathrm{GF}(q^t)$ of the equation $N(x) = 1$ is $(q^t-1)/(q-1)$; this and other basic facts about finite fields can be found in the book by Lidl and Niederreiter (1986). Hence, each vertex of this graph has degree $d = (q^t - 1)/(q - 1)$, implying that the total number of edges is $dq^t \geq q^{2t-1} = n^{2-1/t}$. Kollár et al. (1996) proved that, for any t distinct elements $a_1, \ldots, a_t$ of $\mathrm{GF}(q^t)$, the system of equations $N(a_1+x) = 1, N(a_2+x) = 1, \ldots, N(a_t+x) = 1$ has at most $t!$ solutions $x \in \mathrm{GF}(q^t)$. This immediately implies that the constructed graph has no copy of a complete bipartite $t \times (t! + 1)$ graph, and hence, the adjacency matrix of this graph is $(t - 1, t!)$-free. Explicit matrices with slightly worse parameters were constructed earlier by Andreev (1986).

For the adjacency matrices of norm graphs, Theorem 13.21 yields almost maximal lower bounds $\Omega(n^{2-\epsilon})$ for an arbitrarily small constant $\epsilon > 0$. Nothing similar, however, is known for linear circuits, even in depth 2.

■ **Research Problem 13.25.** Can a k-free matrix constructed in Example 7.4 or in Example 7.5 be represented by a *linear* depth-2 circuits using fewer than $n^{1+\Omega(1)}$ wires?

Remark 13.26. In this problem it is important that we only consider circuits of depth 2. Gashkov and Sergeev (2010) showed that 1-free $n \times n$ matrix A with $|A| = \Omega(n^{3/2})$ ones, constructed in Example 7.4, as well as Brown's matrix and some other dense square-free matrices, *can* be represented by linear circuits of depth $\mathcal{O}(\ln n)$ using only $\mathcal{O}(n \ln n \ln \ln n)$ wires. Their construction is based not on the *properties* of matrices (their square-freeness) but rather on their *construction*. Actually, the authors can show (personal communication) that, for every integer $k \geq 1$, the resulting circuits of depth $2k - 1$ have only $\mathcal{O}(n^{1+1/k})$ wires.

13.6.1 Circuits with OR and AND Gates

One may ask whether the number of wires in an OR-circuit can be substantially decreased if one also allows AND gates? As shown by Nechiporuk (1969), Pippenger (1980), and Mehlhorn (1979), at least for k-free matrices this is not the case: the number of wires can only be decreased by a factor at most $1/k$ (see Exercise 13.13).

   Since the number of wires in a monotone circuit with unbounded fanin AND and OR gates is proportional to the number of gates in a standard monotone circuit (with fanin-2 AND and OR gates), we obtain the following lower bound on the monotone complexity of monotone operators f. Let $\mathrm{C}_{\wedge,\vee}(f)$ denote the smallest number of gates in a fanin-2 monotone circuit computing f. For a boolean $m \times n$ matrix $A = (a_{ij})$, let $f_A : \{0,1\}^n \rightarrow \{0,1\}^m$ denote the monotone operator $f_A(x) = Ax$ over the boolean semiring $(\wedge, \vee, 0, 1)$. That is, the i-th component of f_A is just the disjunction $a_{i1}x_1 \vee a_{i2}x_2 \vee \cdots \vee a_{in}x_n$.

Theorem 13.27. *If A is a k-free matrix, then* $C_{\wedge,\vee}(f_A) = \Omega(|A|/k^3)$.

Thus, constructions described above give us explicit $n \times n$ matrices A with $C_{\wedge,\vee}(f_A) = \Omega(n^{2-\epsilon})$. Similar (almost optimal) lower bounds can be obtained for yet another important operator, the so-called "boolean cyclic convolution".

A boolean $n \times n$ matrix A is *circulant* if its i-th row for $i = 1, \ldots, n-1$ is the cyclic shift of the first row $(a_0, a_1, \ldots, a_{n-1})$ by i positions to the left:

$$A = \begin{pmatrix} a_0 & a_1 & \ldots & a_{n-2} & a_{n-1} \\ a_1 & a_2 & \cdot\cdot & \cdot\cdot & a_0 \\ \vdots & \vdots & \cdot\cdot & \cdot\cdot & \vdots \\ a_{n-2} & a_{n-1} & a_0 & & a_{n-3} \\ a_{n-1} & a_0 & \ldots & a_{n-3} & a_{n-2} \end{pmatrix}. \tag{13.5}$$

Note that if $S = \{i : a_i = 1\} \subseteq \mathbb{Z}_n = \{0, 1, \ldots, n-1\}$ is the set of all positions of 1s in the first row of A, then A is k-free if and only if the set S is k-*sparse* in the following sense: $I + J \not\subseteq S$ for every pair of subsets $I, J \subseteq \mathbb{Z}_n$ of size $|I| = |J| = k + 1$; here $I + J$ is the set of all possible sums $i + j$ modulo n with $i \in I$ and $j \in J$.

Grinchuk (1988) used probabilistic arguments to show that k-sparse subsets $S \subseteq \mathbb{Z}_n$ of size $|S| = \Omega(n^{1-\sqrt{3/k}}/k^4)$ exist. This implies the existence of circulant k-free $n \times n$ matrices A with $|A| = \Omega(n^{2-\sqrt{3/k}}/k^4)$ ones. Grinchuk and Sergeev (2011) recently improved this to $|A| = \Omega(n^{2-3/k}/k^3)$. In particular, for $k = \Theta(\log n)$ we have that $|A| = \Omega(n^2/\log^3 n)$.

Circulant matrices are related to the operator of cyclic convolution which we have already considered above; see Remark 13.9. The operator of *boolean cyclic convolution* $z = f_n(x, y)$ is a boolean operator of $2n$ variables defined in a similar way:

$$z_i = \bigvee_{j=0}^{n-1} x_i y_{i+j \bmod n} \quad \text{for} \quad i = 0, 1, \ldots, n-1.$$

That is, $f_n(x, y) = Yx$ is a result of a matrix-vector product over the boolean semiring, where Y is the circulant matrix of the form (13.5) induced by the vector $y = (y_0, \ldots, y_{n-1})$ of the last n variables. Since *every* circulant matrix can be obtained from the matrix Y by substituting constants to the last n variables $y_0, \ldots, y_{n-1}$ of $f_n(x, y)$, the result of Grinchuk and Sergeev (together with Theorem 13.27) implies that $C_{\wedge,\vee}(f_n) = \Omega(n^2/\log^6 n)$.

Gashkov and Sergeev (2011) gave a general construction showing how sparse subsets of *vectors* can be transformed into sparse subsets of *numbers*. In particular, they show that the mapping $\psi : \mathbb{Z}_n^t \to \mathbb{Z}_{(2n-1)^t}$ given by

$$\psi(a_0, \ldots, a_{t-1}) = \sum_{i=0}^{t-1} a_i (2n-1)^i$$

translates every k-sparse subset $S \subseteq \mathbb{Z}_n^t$ of vectors into a k-sparse subset $\psi(S) \subseteq \mathbb{Z}_{(2n-1)^t}$ of numbers. Together with the construction of norm graphs given in Example 13.24, this yields *explicit* k-free circulant $n \times n$ matrices A with $|A| = \Omega(n^{2-1/t}/2^t)$ ones, where $k = t!$.

13.6.2 Asymptotic Bounds

Finally, let us mention some results about the asymptotic behavior of the Shannon function of rectifier networks. Let $B(n)$ denote the minimal number of wires which is enough to represent any boolean $n \times n$ matrix by a rectifier network. Let $B_r(n)$ denote the analogous number in the class of networks of depth at most r. Let also $B_r(n, \alpha)$ denote the minimal number of wires which is enough to represent any boolean $n \times n$ matrix with αn^2 ones. Finally, let $\lambda_x := -x \log_2 x - (1-x) \log_2(1-x)$ be the binary entropy function.

Lupanov (1956) proved that

$$B_2(n) \sim \frac{n^2}{\log n}.$$

Nechiporuk (1969a) proved that the asymptotic is achieved at depth 3:

$$B(n) \sim B_3(n) \sim \frac{n^2}{2 \log n}.$$

He also proved that

$$B_2(n, \alpha) \sim \lambda_\alpha \frac{n^2}{\log n} \quad \text{and} \quad B_3(n, \alpha) \sim \lambda_\alpha \frac{n^2}{2 \log n}$$

as long as $\log n = o(\lambda_\alpha n)$ and $-\log \min(\alpha, 1 - \alpha) = o(\log n)$. For the minimal number $B(m, n)$ of wires which is enough to represent any boolean $m \times n$ matrix, Orlov (1970) proved that

$$B_2(k \log n, n) \sim (k + 1)n$$

holds for every positive integer k, and

$$B(m,n) \sim B_2(m,n) \sim 2^{m+1} + n$$

holds as long as $n \geq 2(2^m - m - 1)$.

In all these estimates, the upper bounds were obtained by constructing networks with a special property that every input is connected with every output by *at most one* path. Thus, the same asymptotic equalities also hold for linear circuits.

13.7 Non-linear Circuits for Linear Operators

A positive answer to Problem 13.19 would mean that, for at least some matrices A, using non-linear gates cannot help significantly to compute the linear transformation $y = Ax$. In this section we will show that non-linear gates *can* help much if we only want to represent the matrix A, that is, to correctly compute $y = Ax$ only for vectors x with exactly one 1.

Theorem 13.18, together with Proposition 13.16, implies that, in the class of *linear* circuits, some matrices A require $\Omega(n^2/\ln n)$ wires to represent them. We will now show that in the class of *general* circuits the situation is entirely different; this was observed in Jukna (2010).

Theorem 13.28. *Every boolean $n \times n$ matrix A can be represented by a depth-2 circuit with $\mathcal{O}(n \ln n)$ wires.*

Proof. We construct the desired depth-2 circuit representing $A = (a_{ij})$ as follows. Let m be the smallest even integer such that $\binom{m}{m/2} \geq n$; hence $m = O(\ln n)$. Take m middle nodes $V = \{v_1, \ldots, v_m\}$. To each input variable x_j assign its *own* subset $S_j \subseteq V$ of $|S_j| = m/2$ middle nodes; hence, $S_{j_1} \subseteq S_{j_2}$ iff $j_1 = j_2$. Join x_j with all nodes in S_j. Finally, connect each $v \in V$ with all output nodes. The total number of wires is then $n(m/2) + nm = \mathcal{O}(n \ln n)$.

Now we assign gates to the nodes. If v is a node on the middle layer connected to inputs $x_{j_1}, \ldots, x_{j_k}$, then assign to v the gate $g_v = x_{j_1} \oplus \cdots \oplus x_{j_k}$. To the i-th output node we assign the gate

$$\phi_i = a_{i1}h_1 \oplus a_{i2}h_2 \oplus \cdots \oplus a_{in}h_n, \text{ where } h_k = \prod_{v \in S_k} g_v.$$

Then

$$h_k(e_j) = 1 \text{ iff } g_v(e_j) = 1 \text{ for all } v \in S_k$$
$$\text{iff } x_j \text{ is connected to all nodes in } S_k$$
$$\text{iff } S_k \subseteq S_j$$
$$\text{iff } k = j.$$

Hence, $h_j(e_j) = 1$ and $h_k(e_j) = 0$ for all $k \neq j$. Thus, if $f_i(x)$ is the function computed at the i-th output gate then, for all $j = 1, \dots, n$, we have that

$$f_i(e_j) = \phi_i(e_j) = a_{i1} \cdot 0 \oplus \cdots \oplus a_{ij} \cdot 1 \oplus \cdots \oplus a_{in} \cdot 0 = a_{ij},$$

as desired. $\square$

We now show that the upper bound $n \ln n$ in Theorem 13.28 is almost optimal. To do this, we will use the Sunflower Lemma proved in Sect. 9.1. Recall that a *sunflower* with k petals is a family $S_1, \dots, S_k$ of k finite sets, each two of which share precisely the same set of common elements, called the *core* of the sunflower. That is, there is a set C (the core of the sunflower) such that $S_i \cap S_j = C$ for all $1 \leq i < j \leq k$. The Sunflower Lemma states that every family of more than $s!(k-1)^s$ sets, each of which has cardinality at most s, contains a sunflower with k petals.

For a matrix A, let $\mathrm{dist}(A)$ denote the smallest Hamming distance between the columns of A. Alon et al. (1990) proved that every matrix A requires $\Omega(d \cdot \ln n / \ln \ln n)$ wires to be presented by a linear depth-2 circuits, where $d = \mathrm{dist}(A)$. The next theorem extends this result to general (non-linear) circuits.

Theorem 13.29. *Every depth-2 circuit representing a boolean $n \times n$ matrix with* $\mathrm{dist}(A) = d$ *requires at least* $\Omega(d \cdot \ln n / \ln \ln n)$ *wires.*

Proof. Fix a minimal depth-2 circuit with arbitrary gates representing a given matrix A. Without loss of generality, we may assume that there are no direct wires from inputs to outputs: this can be easily achieved by adding at most n new wires. Let $x_1, \dots, x_n$ be its input nodes, and $S_1, \dots, S_n$ be sets of their neighbors on the middle layer. Let $f_1, \dots, f_n$ be the functions computed at the output nodes. Since the circuit represents A, we must have that $f_i(e_j) = a_{ij}$ for all $1 \leq i, j \leq n$.

Let L_1 be the number of wires leaving the input nodes, and L_2 the number of wires entering the output nodes. Hence, $L_1 = \sum_{i=1}^{n} |S_i|$, and $L_1 + L_2$ is the total number of wires. Set $m := c \ln n / \ln \ln n$ for a sufficiently small constant $c > 0$. If we have $L_1 > mn$ wires leaving the input nodes, then we are done. So, assume that $L_1 \leq mn$. Our goal is to show that we must have $L_2 \geq m \cdot \mathrm{dist}(A)$ wires entering the output nodes.

Our assumption $\sum_{i=1}^{n} |S_i| \leq mn$ implies that at least $n/2$ of the sets S_i must be of size at most $s = 2m$. Hence, if the constant c in the definition of m is small enough then, by the Sunflower Lemma, these sets must contain a sunflower with $k = 2m$ petals. Having such a sunflower with a core C, we can pair its members arbitrarily $(S_{p_1}, S_{q_1}), \dots, (S_{p_m}, S_{q_m})$; hence, $S_{p_i} \cap S_{q_i} = C$ for all $i = 1, \dots, m$. Important for us will only be that the symmetric differences

$$S_{p_i} \oplus S_{q_i} = (S_{p_i} \setminus S_{q_i}) \cup (S_{p_i} \setminus S_{q_i}) = (S_{p_i} \cup S_{q_i}) \setminus C$$

of these pairs of sets are pairwise disjoint. Hence, we have m pairwise disjoint subsets $S_{p_i} \oplus S_{q_i}$ of nodes on the middle layer, and we only have to show that each of these sets has at least $\mathrm{dist}(A)$ outgoing wires: then $L_2 \geq m \cdot \mathrm{dist}(A)$.

Fix one of the pairs (S_p, S_q). Since the circuit represents the matrix A, the value $f(e_j)$ of the computed operator $f = (f_1, \ldots, f_n)$ on the j-th unit vector must be the j-th column of A. Since the Hamming distance between the p-th and the q-th columns of A must be at least d, there must exist a set I of $|I| \geq \text{dist}(A)$ rows such that

$$f_i(e_p) \neq f_i(e_q) \text{ for all } i \in I. \tag{13.6}$$

Claim 13.30. Every output f_i with $i \in I$ must be adjacent to at least one node in $S_p \oplus S_q$.

Proof. Let V be the set of all nodes on the middle layer. For a node $v \in V$, let $g_v(x_1, \ldots, x_n)$ be the boolean function computed at this node. Let $\mathbf{0}$ denote the all-0 vector. Observe that, if a wire (j, v) is present, then the values $g_v(e_j)$ and $g_v(\mathbf{0})$ must be different: would they be the same, then we could remove the wire (j, v) and replace g_v by a new boolean function g_v' obtained from g_v by fixing the j-th variable x_j of g_v to 0. The behavior of this new gate would then be the same on all unit vectors. But then we would have one wire fewer, contradicting the minimality of our circuit.

This observation implies that $g_v(e_p) = g_v(e_q)$ for all $v \notin S_p \oplus S_q$. Indeed, if $v \notin S_p \cup S_q$, then neither the wire (p, v) nor the wire (q, v) is present, implying that $g_v(e_p) = g_v(\mathbf{0}) = g(e_q)$. If $v \in S_p \cap S_q$, then both wires (p, v) and (q, v) must be present, and the above observation implies that $g_v(e_p) \neq g_v(\mathbf{0})$ as well as $g_v(e_q) \neq g_v(\mathbf{0})$. Hence, in this case we also have that $g_v(e_p) = g_v(e_q)$, just because g_v can take only two values. Thus the behavior of every gate g_v with $v \notin S_p \oplus S_q$ is the same on both unit vector e_p and e_q.

To finish the proof of Claim 13.30, take the boolean function f_i computed at the i-th output gate with $i \in I$. If there were no wire from a node in $S_p \oplus S_q$ to this output gate, then f_i would also be forced to take the same value on both unit vector e_p and e_q, contradicting (13.6). □

By Claim 13.30, for each of m pairs (S_{p_i}, S_{q_i}) of subsets of nodes on the middle layer, there must be at least $|I| \geq \text{dist}(A)$ wires going from the vertices in $S_{p_i} \oplus S_{q_i}$ to the output layer. Since the sets $S_{p_i} \oplus S_{q_i}$, $i = 1, \ldots, m$, are pairwise disjoint, the total number of wires going from the middle layer to the output layer must be at least $m \cdot \text{dist}(A)$, as desired. □

There are explicit boolean $n \times n$ matrices H_n (so-called Sylvester matrices) such that $\text{dist}(H_n) \geq n/2$ but, still, the entire linear transformation $y = H_n x$ can be computed by a *linear* depth-2 circuit with $n \log n$ wires (Exercise 13.6). Thus, the lower bound in Theorem 13.29 is almost tight.

Remark 13.31. Drucker (2011) has recently shown that the lower bound in Theorem 13.29 is, in fact, tight: the factor $1/\ln\ln n$ *cannot* be removed. He uses particular combinatorial designs to construct a boolean $n \times n$ matrix A such that $\text{dist}(A) = \Omega(n)$ but A can be represented, and even $f_A(x) = Ax$ can be computed by a linear depth-2 circuit using only $\mathcal{O}(n \ln n / \ln\ln n)$ wires. He also shows that there exists a matrix with $\text{dist}(A) = \Omega(n)$ such that Ax can be computed by a linear depth-3 circuit using only $\mathcal{O}(n)$ wires. Thus, large distance between columns alone cannot force a large number of wires.

13.8   Relation to Circuits of Logarithmic Depth

A depth-2 circuit of *width* r has n boolean variables $x_1, \ldots, x_n$ as input nodes, r arbitrary boolean functions $h_1, \ldots, h_r$ as gates on the middle layer, and arbitrary boolean functions $g_1, \ldots, g_n$ as gates on the output layer. Direct input-output wires, connecting input variables with output gates, are now allowed! Such a circuit computes an operator $f = (f_1, \ldots, f_n) : \text{GF}(2)^n \to \text{GF}(2)^n$ if, for every $i = 1, \ldots, n$,

$$ f_i(x) = g_i(x, h_1(x), \ldots, h_r(x)). $$

The *degree* of such a circuit is the maximum, over all output gates g_i, of the number of wires going directly from input variables $x_1, \ldots, x_n$ to the gate g_i. That is, we ignore the wires incident with the gates on the middle layer. Let $\text{Deg}_r(f)$ denote the smallest degree of a depth-2 circuit of width r computing f.

It is clear that $\text{Deg}_n(f) = 0$: just put the functions $f_1, \ldots, f_n$ on the middle layer. Hence, this parameter is only nontrivial for $r < n$. Especially interesting is the case when $r = \mathcal{O}(n/\ln\ln n)$:

Lemma 13.32. *If* $\text{Deg}_r(f) = n^{\Omega(1)}$ *for* $r = \mathcal{O}(n/\ln\ln n)$, *then* f *cannot be computed by a circuit of depth* $\mathcal{O}(\ln n)$ *using* $\mathcal{O}(n)$ *fanin-2 gates.*

Proof. Suppose that $f = (f_1, \ldots, f_n)$ can be computed by a circuit Φ of depth $\mathcal{O}(\ln n)$ using $\mathcal{O}(n)$ fanin-2 gates. By Valiant's lemma (Lemma 1.4), for an arbitrarily small constant $\epsilon > 0$, any such circuit can be reduced to a circuit of depth at most $\epsilon \log n$ by removing a set of at most $r = \mathcal{O}(n/\log\log n)$ edges.

Put on the middle layer all the r boolean functions computed at these (removed) edges, and connect each middle node with all inputs as well as with all outputs. Because a subcircuit of Φ computing each f_i has depth at most $\epsilon \log n$, each such subcircuit can depend on at most $2^{\epsilon \log n} = n^\epsilon$ original input variables. By joining the i-th output node with all these n^ϵ inputs we obtain a desired depth-2 circuit of degree at most n^ϵ computing our operator f. Since this holds for arbitrarily small constant $\epsilon > 0$, we are done. $\square$

The highest known lower bound for an explicit operator f, proved by Pudlák et al. (1997) has the form $\text{Deg}_r(f) = \Omega((n/r)\ln(n/r))$, and is too weak to have a consequence for log-depth circuits.

A natural question therefore was to improve the lower bound on the degree, at least for *linear* circuits, that is, for depth-2 circuits whose middle gates as well as output gates are linear boolean functions over GF(2). Such circuits compute linear operators $f_A(x) = Ax$ for some matrix A over GF(2). By Lemma 13.32, this would give a super-linear lower bound for log-depth circuits over $\{\oplus, 1\}$. (Yes, even over this basis no super-linear lower bound is known so far!)

This last question attracted the attention of many researchers because of its relation to a purely algebraic characteristic of the underlying matrix A—its rigidity. Recall that the *rigidity*, $\text{Rig}_A(r)$, of a matrix A over GF(2) is the smallest number of entries of A that must be changed in order to reduce its rank over GF(2) until r. That is,

$$\text{Rig}_A(r) = \min\{|B| : \text{rk}(A \oplus B) \leq r\},$$

where $|B|$ is the number of ones in B. For a linear operator $f_A(x) = Ax$ over GF(2), let Lin-$\deg_r(f_A)$ denote the minimum degree of a *linear* depth-2 circuit of width r computing f_A.

Proposition 13.33. *Let A be a boolean $n \times n$ matrix, $\text{Rig}_A(r)$ its rigidity and $f_A(x) = Ax$ the corresponding linear operator over GF(2). Then*

$$\text{Lin-}\deg_r(f_A) \geq \text{Rig}_A(r)/n.$$

Proof. Fix a depth-2 circuit Φ of width r computing f_A. If we set all direct input–output wires to 0, then the resulting degree-0 circuit will compute some linear transformation $A'x$. The operator $y = A'x$ takes $2^{\text{rk}(A')}$ different values. Hence, the operator $H : \text{GF}(2)^n \to \text{GF}(2)^r$ computed by r boolean functions on the middle layer of Φ must take at least so many different values, as well. This implies that the width r must be large enough to fulfill $2^r \geq 2^{\text{rk}(A')}$, from which $\text{rk}(A') \leq r$ follows. On the other hand, A' differs from A in at most dn entries, where d is the degree of the original circuit Φ. Hence, $\text{Rig}_A(r) \leq dn$ from which $d \geq \text{Rig}_A(r)/n$ follows. $\square$

Problem 13.34. Exhibit an explicit boolean $n \times n$ matrix A of rigidity $\text{Rig}_A(r) \geq n^{1+\epsilon}$ for $r = \mathcal{O}(n/\ln\ln n)$.

By Lemma 13.32 and Proposition 13.33, this would give us a linear operator $f_A(x) = Ax$ which cannot be computed by log-depth circuit over $\{\oplus, 1\}$ using a linear number of parity gates. Motivated by its connection to proving lower bounds for log-depth circuits, matrix rigidity (over different fields) was considered by many authors.

It is clear that $\text{Rig}_A(r) \leq (n - r)^2$ for any $n \times n$ matrix A: just take an arbitrary $r \times r$ submatrix A' of A and set to 0 all entries outside A. Valiant (1977) proved that $n \times n$ matrices A with $\text{Rig}_A(r) = (n - r)^2$ exist if the underlying field is infinite. For finite fields the lower bound is only slightly worse.

Proposition 13.35. *There exist $n \times n$ matrices A over GF(2) such that, for all $r < n - \sqrt{2n} + \log n$,*

$$\text{Rig}_A(r) \geq \frac{(n - r)^2 - 2n - \log n}{\log(2n^2)}.$$

Proof. Direct counting. Recall that the rigidity $\mathrm{Rig}_A(r)$ of A over $\mathrm{GF}(2)$ is the smallest number $|B|$ of nonzero entries in a boolean matrix B such that $\mathrm{rk}(A \oplus B) \leq r$. For $|B| = s$ there are at most $\binom{n^2}{s} \leq n^{2s}$ possibilities to choose s nonzero entries of B, and at most $\binom{n}{r}^2 \leq 2^{2n}$ possibilities to choose a nonsingular $r \times r$ minor of $A \oplus B$. Assuming that s is strictly smaller than the lower bound on $\mathrm{Rig}_A(r)$, given in the proposition, it can be verified that the number of possible matrices A with $\mathrm{Rig}_A(r) \leq s$ is upper bounded by $2^{n^2}/n$, which is smaller than the total number 2^{n^2} of such matrices. $\square$

The problem, however, is to exhibit an explicit matrix A of large rigidity. The problem is particularly difficult if we require A to be a *boolean* matrix or at least a matrix with relatively few different entries. What we need are explicit matrices A with $\mathrm{Rig}_A(r) \geq n^2/r^{1-\delta}$ for a constant $\delta > 0$; this would already solve Problem 13.34.

We now show that it is this "$-\delta$" which makes the problem difficult: explicit $n \times n$ ± 1 matrices A of rigidity $\mathrm{Rig}_A(r) = \Omega(n^2/r)$ over the reals are easy to present.

Let $n = 2^m$. The $n \times n$ Sylvester ± 1-matrix $S_n = (s_{ij})$ by labeling the rows and columns by m-bit vectors $x, y \in \mathrm{GF}(2)^m$ and letting $s_{ij} = (-1)^{\langle x, y \rangle}$. Hence,

$$S_2 = \begin{bmatrix} +1 & +1 \\ +1 & -1 \end{bmatrix}, \quad S_4 = \begin{bmatrix} +1 & +1 & +1 & +1 \\ +1 & -1 & +1 & -1 \\ +1 & +1 & -1 & -1 \\ +1 & -1 & -1 & +1 \end{bmatrix} \text{ and } S_{2n} = \begin{bmatrix} S_n & S_n \\ S_n & \overline{S}_n \end{bmatrix},$$

where $\overline{S}_n$ is the matrix obtained from S_n by flipping all $+1$'s to -1's and all -1's to $+1$'s. The rigidity of these matrices over the reals is $n^2/4r$.

Theorem 13.36. *If $r \leq n/2$ is a power of 2 then* $\mathrm{Rig}_{S_n}(r) \geq n^2/4r$.

Proof. (Due to Midrijanis 2005) Divide S_n uniformly into $(n/2r)^2$ submatrices of size $2r \times 2r$. One can easily verify that these submatrices each have full rank over the reals. So we need to change at least r elements of each submatrix to reduce each of their ranks to r, a necessary condition to reducing the rank of S_n to r. The total number of changes is then at least $r \cdot (n/2r)^2 = n^2/4r$. $\square$

This proof works for any matrix whose submatrices have full rank. Consider the $n \times n$ matrix $B = (b_{ij})$ where $b_{ij} = 1$ if $i \equiv j \mod 2r$, and $b_{ij} = 0$ otherwise. By the same proof $\mathrm{Rig}_B(r) \geq n^2/4r$ even though the rank of B is only $2r$.

In fact, it was observed by many authors that *any* Hadamard matrix has rigidity $\Omega(n^2/r)$. Recall that a *Hadamard matrix* of order n is an $n \times n$ matrix with entries ± 1 and with row vectors mutually orthogonal over the reals. It is easy to verify that the Sylvester matrix S_n constructed above has this property. It follows from the definition that a Hadamard matrix H of order n satisfies $HH^T = nI_n$, where I_n is the $n \times n$ identity matrix. Hence, the eigenvalues of H are all $\pm\sqrt{n}$

Theorem 13.37. *Let H be an $n \times n$ Hadamard matrix. If $r \leq n/2$ then* $\mathrm{Rig}_H(r) \geq n^2/4r$.

Proof. (Due to De Wolf (2006)) Let R be the minimum number of changes that brought the rank of H down to r. By a simple averaging argument, we can find $2r$ rows of H that contain a total of at most $2rR/n$ changes. If $n \le 2rR/n$, then $R \ge n^2/2r$ and we are done. Hence, we can assume that $n - 2rR/n > 0$.

Consider the $n-2rR/n$ columns that contain no changes in the above set of rows. We thus get a $2r \times (n - 2rR/n)$ submatrix B that contains no changes and hence is a submatrix of H. By definition of R, this submatrix must have rank at most r. But every $a \times b$ submatrix of H must have rank at least ab/n (see Lemma A.4 in Appendix A). Thus, we get $r \ge \mathrm{rk}(B) \ge 2r(n - 2rR/n)/n$. Rearranging this inequality, we get $R \ge n^2/4r$. $\square$

These bounds on rigidity are, however, still too weak to have consequences for log-depth circuits over $\{\oplus, 1\}$.

The best known lower bounds for the rigidity of explicit $n \times n$ matrices over a finite field is due to Friedman (1993) and have the form $\Omega((n^2/r) \ln(n/r))$. As shown by Shokrollahi et al. (1997), such bounds can also be obtained using the following combinatorial fact. The fact itself is an almost direct consequence from well-known bounds for the Zarankiewicz problem.

Lemma 13.38. *Let $\log^2 n \le r \le n/2$ and let n be sufficiently large. If in an $n \times n$ matrix fewer than*

$$\frac{n^2}{4r} \log \frac{n}{r-1} \tag{13.7}$$

entries are marked, then there exists an $r \times r$ submatrix with no marked entries.

Proof. Let $M = (m_{ij})$ be an arbitrary $n \times n$ matrix, some of whose entries are marked. Let $A = (a_{ij})$ be a 0-1 matrix with $a_{ij} = 1$ iff m_{ij} has *not* been marked. Let R be the number of marked entries in M. Obviously $|A| = n^2 - R$. Define

$$\mu(n,r) := n(n - r + 1) \left(1 - \left(\frac{r-1}{n}\right)^{1/r}\right).$$

It is well known (see, for example Bollobás (1978), page 310) that if A has

$$|A| > (r - 1)^{1/r}(n - r + 1)n^{1-1/r} + (r - 1)n = n^2 - \mu(n,r)$$

ones, then A contains an $r \times r$ all-1 submatrix. Hence, this condition is satisfied if only $R < \mu(n,r)$ entries were marked. It remains to show that $\mu(n,r)$ is at most (13.7) as long as $\log^2 n \le r \le n/2$. As $n(n-r+1) \ge n^2/2$ for $r \le n/2$, it suffices to show that

$$1 - \left(\frac{r-1}{n}\right)^{1/r} \ge \frac{1}{2r} \log \frac{n}{r-1}$$

holds for $r \geq \log^2 n$. Setting $K := n/(r-1)$, this inequality is equivalent to

$$\left(1 - \frac{\log K}{2r}\right)^{r/\log K} \geq \left(\frac{1}{K}\right)^{1/\log K} = \frac{1}{2}.$$

This holds because for large n the left-hand side converges to $1/\sqrt{e} > 1/2$. □

As observed by Lokam (2009), the lemma cannot be substantially improved. This can be shown using the following result of Stein (1974) and Lovász (1975), which itself has already found many other applications.

Theorem 13.39. (Lovász–Stein theorem) *Let A be a boolean $N \times M$ matrix. Suppose that each row of A has at least v ones and each column at most a ones. Then A contains an $N \times K$ submatrix C with no all-0 rows and such that $K \leq N/a + (M/v) \ln a \leq (M/v)(1 + \ln a)$.*

For us, the following consequence of this theorem will be important. Let $\mathcal{F}$ be a finite family of subsets of some finite set X. The *blocking number* $\tau(\mathcal{F})$ of $\mathcal{F}$ is the smallest size $|T|$ of a set $T \subseteq X$ intersecting all members of $\mathcal{F}$.

Corollary 13.40. *If each member of $\mathcal{F}$ has at least v elements, and each point $x \in X$ belongs to at most a of the sets in $\mathcal{F}$, then*

$$\tau(\mathcal{F}) \leq \frac{|X|}{v}(1 + \ln a).$$

Proof. Let A be the incidence matrix of $\mathcal{F}$. That is, the rows of A correspond to the members F of $\mathcal{F}$, and columns to the points x in the underlying set X. The (F, x)-th entry of A is 1 iff $x \in F$. By the assumption of the theorem, each row has at least v ones, and each column has at most a ones. Theorem 13.39 implies that there must be a subset T of $|T| \leq |X|(1 + \ln a)/v$ columns such that every row of A has at least one 1 in these columns. Thus, the set T must intersect every member of $\mathcal{F}$, and we are done. □

Theorem 13.41. (Lokam 2009) *In every $n \times n$ matrix it is possible to mark at most $\mathcal{O}((n^2/r) \ln(n/r))$ entries so that every $r \times r$ matrix will contain at least one marked entry.*

Proof. Given an $n \times n$ matrix M, let $\mathcal{F}$ be the family of all its $r \times r$ submatrices. The underlying set X in our case is the set of all $|X| = n^2$ entries in M, and each entry belongs to exactly $a = \binom{n-1}{r-1}^2$ members of $\mathcal{F}$. Moreover, each member of $\mathcal{F}$ has $v = r^2$ elements. Corollary 13.40 gives us a set T of

$$|T| \leq \frac{|X|}{v}(1 + \ln a) = \frac{n^2}{r^2}\left(1 + \ln \binom{n-1}{r-1}^2\right) = \mathcal{O}\left(\frac{n^2}{r^2} \ln \frac{n}{r}\right)$$

entries of M intersecting every $r \times r$ submatrix of M. □

More results on various versions of rigidity can be found in the survey of Lokam (2009).

Exercises

13.1. Consider circuits of arbitrary depth with all boolean functions allowed as gates. Prove that operators $f : \{0, 1\}^n \to \{0, 1\}^n$ requiring $\Omega(n^2)$ wires in such circuits exist.

Hint: Show that: (1) in an optimal circuit no gate has fanin larger than n, (2) if there are L wires in a circuit, then at most $n/2$ gates can have fanin larger than $2L/n$.

13.2. Let $\mathrm{Conv}(x, y)$ be the operator of cyclic convolution defined in Sect. 13.3. Let $n = pq$. Show that for any partition of the input vector $x = (x_0, \ldots, x_{n-1})$ into p consecutive intervals $I_1, \ldots, I_p$ of length q, there exists a partition of the output vector $z = (z_0, \ldots, z_{n-1})$ into q disjoint sets $J_1, \ldots, J_q$ such that $\mathrm{E}(\mathrm{Conv}_{I_i, J_j}) \geq n$ for *all* i and j. *Hint*: Consider residue classes modulo p.

13.3. Recall that the rank of an $n \times n$ matrix A over some field is the smallest number r such that A can be written as a product $A = B \cdot C$ of an $n \times r$ matrix B and an $r \times n$ matrix C. For a boolean matrix A, let $|A|$ be the number of 1s in A. Define the *weighted rank* of A by:

$$\mathrm{Rk}(A) = \min\{|B| + |C| : A = B \cdot C\}.$$

That is, now we are interested not in the dimension of the matrices B and C but rather in the total number of 1s in them. Prove that $\mathrm{lin}_2(A) = \mathrm{Rk}(A)$, that is, the smallest number of wires in a linear depth-2 circuit representing a matrix A is equal to the weighted rank of A.

Hint: Take the adjacency matrices of the bipartite graphs formed by the first and the second level of wires.

13.4. Prove that $\mathrm{lin}_2(Ax) = L$ if and only if there exist primitive matrices $B_1, \ldots, B_t$ (that is, boolean matrices of rank 1) of dimensions $r_1 \times s_1, \ldots, r_t \times s_t$ such that $A = \bigoplus_{i=1}^t B_i$ and $\sum_{i=1}^t (r_i + s_i) = L$.

Hint: Each matrix B_k is uniquely described by a set I_k of its rows and a set J_k of its columns containing at least one 1. For each k, put a node v_k on the middle layer and connect it with all inputs in J_k and all outputs in I_k.

13.5. Let $\Delta_n = (d_{ij})$ be a triangular boolean $n \times n$ matrix, that is, $d_{ij} = 1$ iff $i \leq j$. Let n be a power of two. Show that $\mathrm{lin}_2(\Delta_n x) = O(n \log n)$. *Hint*: Show that $\mathrm{lin}_2(\Delta_n x) \leq n + 2 \cdot \mathrm{lin}_2(\Delta_{n/2} x)$.

Comment: Pudlák and Vavrín (1991) showed that the rigidity of Δ_n is $\Theta(n^2/r)$ for all $r = o(n)$.

13.6. Let $n = 2^r$ and consider a boolean $n \times n$ matrix H_n whose rows and columns are indexed by vectors in $\mathrm{GF}(2)^r$, and $H_n[x, y]$ is the scalar product of x and y over $\mathrm{GF}(2)$. Such matrices are known as $(0, 1)$-*Sylvester matrices*. Show that:

(a) H_n can be defined inductively by

$$H_2 = \begin{bmatrix} 0 & 0 \\ 0 & 1 \end{bmatrix}, \quad H_4 = \begin{bmatrix} 0 & 0 & 0 & 0 \\ 0 & 1 & 0 & 1 \\ 0 & 0 & 1 & 1 \\ 0 & 1 & 1 & 0 \end{bmatrix} \text{ and } H_{2n} = \begin{bmatrix} H_n & H_n \\ H_n & \overline{H}_n \end{bmatrix},$$

where $\overline{H}_n$ is the matrix obtained from H_n by flipping all its entries.
(b) $\mathrm{lin}_2(H_n x) \le 2n \log n$. *Hint*: Exercise 13.3.
(c) The matrix H_n cannot be represented by a general depth-2 circuit using fewer than $\Omega(n \ln n / \ln \ln n)$ wires. *Hint*: Show that $\mathrm{dist}(H_n) \ge n/2$ and apply Theorem 13.29.

13.7. Show that Problem 13.19 has an affirmative answer if all output gates are required to be linear: if a depth-2 circuit Φ computes a linear operator $f_A(x) = Ax$ and has linear gates on the output layer, then Φ can be transformed into an equivalent *linear* circuit of the same size and width. *Hint*: Replace the operator H, computed at the middle layer, by a linear operator $H'(x) := \sum_{i=1}^n x_i H(e_i) \mod 2$.

13.8. We consider depth-2 circuits whose middle gates are linear (output gates may be arbitrary). Let A be a boolean $n \times n$ matrix. Show that, if the linear transformation Ax can be computed by such a circuit using L wires, then the matrix A can be encoded using $\mathcal{O}(L \log n)$ bits. That is, there exists a binary string ξ of length $\mathcal{O}(L \log n)$ such that the matrix A can be reconstructed from ξ. Use this to conclude that some matrices will require $\Omega(n^2 / \log n)$ wires.

Hint: At the middle layer a linear transformation is computed. Use the fact that every linear space is uniquely described by any of its bases. To get the last conclusion, show that some matrices cannot be encoded using substantially fewer than n^2 bits.

13.9. Recall that a boolean function f is *symmetric* if there is a set T of natural numbers (called also the *type* of f) such that f accepts a binary vector x iff the number of 1s in x belongs to T. A *symmetric* depth-2 circuit is a depth-2 circuit with parity gates on the middle layer, and symmetric boolean functions of the same type on the output layer. Let $\mathrm{sym}_T(A)$ denote the smallest number of nodes on the middle layer of a symmetric depth-2 circuit of type T representing a boolean matrix $A = (a_{ij})$. (Actually, on the middle layer, we can allow any gates g such that $g(\mathbf{0}) = 0$ and $g(x) = 1$ for every vector x with exactly one 1.) Let also $\mathrm{sym}(A)$ be the minimum of $\mathrm{sym}_T(A)$ over all types $T \subseteq \{0, 1, \ldots\}$. Show that:

(a) $\mathrm{sym}_T(A) =$ smallest number r for which it is possible to assign to each row/column i a subset $S_i \subseteq \{1, \ldots, r\}$ such that $a_{ij} = 1$ iff $|S_i \cap S_j| \in T$.
(b) Show that $\mathrm{sym}(A) = \Omega(n)$ for almost all $n \times n$ matrices A.

13.10. (Threshold vs. parity types) Let $n = 2^m$, and consider an $n \times n$ Sylvester matrix H_n. Recall that its rows and columns are labeled by vectors in $\mathrm{GF}(2)^m$, and the entries of H_n are the scalar products of these vectors over $\mathrm{GF}(2)$. Hence, $\mathrm{sym}_T(H_n) \leq m$ for the type T consisting of all odd natural numbers. A type $T \subseteq \{0, 1, \ldots\}$ is a threshold-k type if $T = \{k, k+1, \ldots\}$. Prove that $\mathrm{sym}_T(H_n) = \Omega(\sqrt{n})$ for any threshold type T.

Hint: Let $r = \mathrm{sym}_T(H_n)$, and consider an assignment $i \mapsto S_i$ of subsets $S \subseteq \{1, \ldots, r\}$ to rows/columns of $H_n = (h_{ij})$ such that $h_{ij} = 1$ iff $|S_i \cap S_j| \geq k$. Take $E = \{(i, j) : h_{ij} = 1\}$ and consider the family $\mathcal{F} = \{F_1, \ldots, F_r\}$ with $F_k = \{(i, j) : k \in S_i \cap S_j\}$. Show that $r \geq \mathrm{thr}_{\mathcal{F}}(E)$, where $\mathrm{thr}_{\mathcal{F}}(E)$ is the threshold cover number of E dealt with in the Discriminator Lemma (Lemma 11.30). Then use Lindsey's Lemma (proved in Appendix A) to show that $\mathrm{disc}_{\mathcal{F}}(E) = \mathcal{O}(n^{-1/2})$.

13.11. ■ **Research Problem.** Exhibit an explicit boolean $n \times n$ matrix A such that $\mathrm{sym}(A) \geq 2^{(\log \log n)^{\alpha}}$ for some $\alpha(n) \to \infty$.

Comment: This is a reformulation of Research Problem 12.33 in terms of matrices: just consider bipartite graphs as their adjacency matrices.

13.12. ■ **Research Problem.** Say that $T \subseteq \{0, 1, \ldots\}$ is an *interval type* if $T = \{a, a+1, \ldots, b\}$ for some non-negative integers $a \leq b$. Let $\mathrm{sym}_{\mathrm{int}}(A)$ denote the minimum of $\mathrm{sym}_T(A)$ over all interval types T. Exhibit an explicit boolean $n \times n$ matrix A such that $\mathrm{sym}_{\mathrm{int}}(A)$ is larger than $2^{(\log \log n)^c}$ for any constant c. *Comment*: This would be a major step towards resolving the previous problem.

13.13. Let $A = (a_{ij})$ be a boolean $m \times n$ matrix, and consider the operator $f(x) = Ax$ over the boolean semiring. Thus, the operator computes m disjunctions $f_i(x) = \bigvee_{j : a_{ij} = 1} x_j$, $i = 1, \ldots, m$. Suppose that A is k-free, and that $f(x) = Ax$ can be computed by a monotone circuit using L fanin-2 AND and OR gates. Show that then $f(x)$ can also be computed by a circuit containing at most $k \cdot L$ OR gates and no AND gates at all.

Hint: Take the *last* AND gate $g = h_1 \wedge h_2$ in the circuit. Let s be the number of variables among $x_1, \ldots, x_n$ that imply g, that is, the number of terms of length 1 in the disjunctive normal form of g. Let also t be the number of functions among $f_1, \ldots, f_m$ that are implied by g. Argue that s and t cannot be both larger than k. If $s \leq k$, then replace the AND gate g by a circuit computing the OR of the corresponding s variables. If $t \leq k$, then replace g by the constant 0, and let $f'_1, \ldots, f'_m$ be the functions computed after this replacement. Show that, for each of the i with $f'_i \neq f_i$, either $f_i = f'_i \vee h_1$ or $f_i = f'_i \vee h_2$ must hold.

13.14. ■ **Research Problem.** Prove or disprove: if a linear operator $f_A(x) = Ax$ can be computed by a depth-2 circuit of degree d and width w, then f_A can also be computed by a *linear* depth-2 circuit of degree $\mathcal{O}(d)$ and width $\mathcal{O}(w)$.

13.15. An *extension* of a partial m-by-n matrix M with entries in $\{0, 1, *\}$ is obtained by setting all $*$-entries to constants 0 and 1. Let $\mathrm{mr}(M)$ denote the smallest possible rank of an extension of M over $\mathrm{GF}(2)$. Let F be a depth-2 circuit computing a linear operator $f_A(x) = Ax$ over $\mathrm{GF}(2)$. Say that the (i, j)-th entry of A is *seen* by the circuit, if there is a direct wire from x_j to the i-th output gate. Replace all entries of A seen by the circuit with $*$'s, and let A_F be the resulting

$(0, 1, *)$-matrix. Note that the original matrix A is one of the extensions of A_F; hence, $\mathrm{rk}(A) \geq \mathrm{mr}(A_F)$. Prove the following:

(a) If the circuit F is linear, then $\mathrm{width}(F) \geq \mathrm{mr}(A_F)$.

> *Hint*: Every assignment of constants to direct input–output wires leads to a depth-2 circuit of degree $d = 0$ computing a linear operator Bx, where B is an extension of A_F. Argue that the operator $H : \mathrm{GF}(2)^n \rightarrow \mathrm{GF}(2)^w$ computed by $w = \mathrm{width}(F)$ boolean functions on the middle layer of F must take at least $2^{\mathrm{rk}(B)}$ different values.

(b) Every depth-2 circuit F computing a linear operator can be transformed into an equivalent *linear* depth-2 circuit of the same degree and width at most $\mathrm{mr}(A_F)$.

> *Hint*: Let B be an extension of A_F of rank $r = \mathrm{mr}(A_F)$. Take any r linearly independent rows of B and put on the middle layer r scalar products of the input vector x with these rows. Show that the resulting linear circuit computes the same linear operator.

13.16. ■ **Research Problem.** Let M be a partial m-by-n matrix with entries in $\{0, 1, *\}$. An operator $f = (f_1, \ldots, f_m) : \mathrm{GF}(2)^n \rightarrow \mathrm{GF}(2)^m$ is *consistent* with M if the i-th coordinate f_i of f can only depend on variables corresponding to $*$-entries in the i-th row of M. Let $\mathrm{Sol}(M)$ denote the maximum, over all extensions A of M and all operators f consistent with M, of the number of solutions of the system of equalities $Ax = f(x)$ over $\mathrm{GF}(2)$. Prove or disprove that there exists a constant $\epsilon > 0$ such that $\mathrm{Sol}(M) \leq 2^{n - \epsilon \cdot \mathrm{mr}(M)}$.

Comment: Some partial results towards this problem were obtained by Jukna and Schnitger (2011). Together with Exercise 13.15, an affirmative answer would give an affirmative answer to the research problem stated in Exercise 13.14.

Part V
Tracking Programs

Chapter 14
Decision Trees

A decision tree is an algorithm for computing a function of an unknown input. Each node of the tree is labeled by a variable and the branches from that node are labeled by the possible values of the variable. The leaves are labeled by the output of the function. The process starts at the root, knowing nothing, works down the tree, choosing to learn the values of some of the variables based on those already known and eventually reaches a decision. The decision tree complexity of a function is the minimum depth of a decision tree that computes that function.

14.1 Adversary Arguments

Let $f : \{0, 1\}^n \to \{0, 1\}$ be a boolean function. A deterministic *decision tree* for f is a binary tree whose internal nodes have labels from $x_1, \ldots, x_n$ and whose leaves have labels from $\{0, 1\}$. If a node has label x_i then the test performed at that node is to examine the i-th bit of the input. If the result is 0, one descends into the left subtree, whereas if the result is 1, one descends into the right subtree. The label of the leaf so reached is the value of the function on that particular input (Fig. 14.1).

The *depth* of a decision tree is the number of edges in a longest path from the root to a leaf, or equivalently, the maximum number of bits tested on such a path.

Let $D(f)$ denote the minimum depth of a decision tree computing f.

When trying to prove that every decision tree for a given boolean function requires large depth, one possibility is to use the so-called *adversary argument*. The idea is that an all-powerful malicious adversary pretends to choose a "hard" input for the solver (a decision tree). When the solver wants to look at a bit, the adversary sets that bit to whatever value will make the solver do the most work. If the solver does not look at enough bits before terminating, then there will be several different inputs, each consistent with the bits already seen, that should result in different outputs. Whatever the solver outputs, the adversary can "reveal" an input that has all the examined bits but contradicts the solver's output, and then claim that was the input which he was using all along. Since the only information the solver has

Fig. 14.1 A decision tree of
depth 3. It accepts only three
inputs $(0, 0, 0)$, $(1, 0, 0)$ and
$(0, 1, 1)$

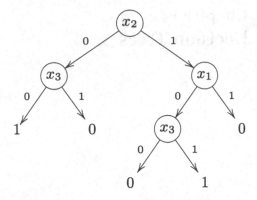

is the set of bits it examined, the algorithm cannot distinguish between a malicious
adversary and an honest user who actually chooses an input in advance and answers
all queries truthfully.

Let us demonstrate this on decision trees for the following problem on graphs:
Given an n-vertex graph G as input, we would like to know how many pairs of
its vertices a decision tree might have to inspect in order to determine whether G
is connected. To express this problem as a boolean function, associate a boolean
variable x_e with each possible pair $e = \{u, v\}$ of vertices $u \neq v$. Then, each
assignment x of $\binom{n}{2}$ boolean values to these variables gives us a graph G_x with
the edge-set $E = \{e \mid x_e = 1\}$. The graph connectivity function f_n is defined by:

$$f_n(x) = 1 \text{ if and only if } G_x \text{ is connected.}$$

Proposition 14.1. $D(f_n) \geq n^2/4 \geq \frac{1}{2}\binom{n}{2}$.

Proof. Fix a partition $V = V_1 \cup V_2$ of the vertex-set into two equal size parts
$|V_1| = |V_2| = n/2$. Imagine an adversary that constructs a graph, edge by edge,
in response to the queries of a solver (decision tree). If the decision tree queries an
edge e whose both endpoints lie in V_1 or both lie in V_2, then the adversary replies
with "$x_e = 1$" (the edge is present). On crossing edges $e \in V_1 \times V_2$ the adversary
replies with "$x_e = 0$". Thus, the graph constructed along that path is disconnected:
it consists of two vertex disjoint cliques. But the decision tree cannot detect this
unless it has already queried all $n^2/4$ crossing edges.   □

Theorem 14.2. $D(f_n) = \binom{n}{2}$.

Proof. The adversary now maintains two graphs, Y and M ("yes" and "maybe"),
both on all n vertices. The graph Y contains all the edges that the solver knows are
definitely in the input graph. The graph M contains all the edges that the solver (the
decision tree) thinks might be in the input graph, or in other words, all the edges of
Y plus all the unexamined edges. Initially, Y is empty and M is complete. Suppose
now that the solver asks whether an edge e is in the input graph or not. The adversary
strategy is:

- If $M \setminus \{e\}$ is connected then answer "$x_e = 0$" and remove e from M.
- If $M \setminus \{e\}$ is not connected then answer "$x_e = 1$" and add e to Y.

Notice that, at each step, Y is a subgraph of M, and M is connected: if the removal of an edge e from M would result in a disconnected graph, the adversary adds e to Y, and hence, keeps that edge in M. Further, if M has a cycle, then none of its edges can belong to Y: deleting any edge in a cycle cannot disconnect a graph. This, in particular, implies that Y is acyclic. We claim that, if $Y \neq M$ then Y is disconnected. To show this, assume that Y is connected. The only connected acyclic graph is a spanning tree, that is, a tree on the entire set of vertices. Thus, Y is a spanning tree and some edge e is in M but not in Y. But then there is a cycle in M that contains e, all of whose other edges are in Y. This violates the fact (we just established) that no cycle in M can have an edge in Y.

Now, if the solver terminates before examining all $\binom{n}{2}$ edges, then there is at least one edge in M which is not in Y. Since the solver cannot distinguish between M and Y, even though M is connected and Y is not, the solver cannot give the correct output for both graphs. Thus, in order to be correct, any algorithm must examine every edge.   □

14.2   P = NP ∩ co-NP for Decision Tree Depth

Given an input $a = (a_1, \dots, a_n)$ from $\{0, 1\}^n$, we would like to know whether $f(a) = 1$ or $f(a) = 0$. How many bits of a must we see in order to answer this question? It is clear that seeing $D(f)$ bits is always enough: just look at those bits of a which are tested along the (unique) path from the root to a leaf.

In a deterministic decision tree all the tests are made in a prescribed order independent of individual inputs. Can we do better if we relax this and allow for each input a to choose its own smallest set of bits to be tested? This question leads to a notion of "nondeterministic" decision tree.

A *nondeterministic decision tree* for a boolean function $f(x_1, \dots, x_n)$ is a (not necessarily binary) tree each whose edge is labeled by a literal (a variable or a negated variable). One literal can label several edges leaving one and the same node. Such a tree T computes f in a nondeterministic manner: $T(a) = 1$ if and only if there exists a path from a root to a leaf such that all literals along this path are consistent with the input a, that is, are evaluated to 1 by this input. Let $C_1(f)$ denote the smallest depth of a nondeterministic tree computing f, and define the dual measure by $C_0(f) := C_1(\neg f)$. It is not difficult to verify that

$$C_1(f) = \min\{k \mid f \text{ can be written as a } k\text{-DNF}\}$$

and

$$C_0(f) = \min\{k \mid f \text{ can be written as a } k\text{-CNF}\} = C_1(\neg f).$$

It is important to note that $C_1(f)$ is *not* the length of a longest minterm of f. Recall that a *minterm* of a boolean function f is a minimal under inclusion subset of its variables such that the function can be made constant-0 function by fixing these variables to constants 0 and 1 in some way. Let $\min(f)$ denote the length of the longest minterm of f. Exercise 1.7 shows that there are boolean functions f of $n + \log n$ variables such that

$$\min(f) \geq n \quad \text{but} \quad C_1(f) \leq 1 + \log n.$$

How do the depths of nondeterministic and deterministic trees are connected? It is clear that $\max\{C_0(f), C_1(f)\} \leq D(f)$, that is, for every input a, seeing its $D(f)$ bits is enough to determine the value $f(a)$, be it 0 or 1. Is this upper bound optimal? The following example shows that this may not be the case: there are boolean functions f for which

$$\max\{C_0(f), C_1(f)\} \leq \sqrt{D(f)}.$$

For example, for the monotone boolean function $f(X)$ on $n = m^2$ boolean variables defined by

$$f = \bigwedge_{i=1}^{m} \bigvee_{j=1}^{m} x_{ij} \tag{14.1}$$

we have $C_0(f) = C_1(f) = m$ but $D(f) = m^2$ (see Exercise 14.1), implying that $D(f) = C_0(f) \cdot C_1(f)$.

It turns out that the example given above is, in fact, the worst case. Namely, the following theorem has been re-discovered by many authors in different contexts: Blum and Impagliazzo (1987), Tardos (1989), and Hartmanis and Hemachandra (1991).

Theorem 14.3. *For every boolean function f, $D(f) \leq C_0(f) \cdot C_1(f)$.*

Proof. Induction on the number of variables n. If $n = 1$ then the inequality is trivial. For the induction step, let (say) $f(0, \dots, 0) = 0$; then some set Y of $k \leq C_0(f)$ variables can be chosen such that by fixing their value to 0, the function is 0 independently of the other variables. We can assume w.l.o.g. that the set $Y = \{x_1, \dots, x_k\}$ of the first k variables has this property.

Take a complete deterministic decision tree T_0 of depth k on these k variables. Each of its leaves corresponds to the unique input $a = (a_1, \dots, a_k)$ in $\{0,1\}^k$ reaching this leaf. Replace such a leaf by a minimal depth deterministic decision tree T_a for the subfunction

$$f_a := f(a_1, \dots, a_k, x_{k+1}, \dots, x_n).$$

Obviously, $C_0(f_a) \leq C_0(f)$ and $C_1(f_a) \leq C_1(f)$. We claim that the latter inequality can be strengthened:

$$C_1(f_a) \leq C_1(f) - 1. \tag{14.2}$$

To prove this, take an arbitrary input $(a_{k+1}, \ldots, a_n)$ of f_a which is accepted by f_a. Together with the bits $(a_1, \ldots, a_k)$, this gives an input of the whole function f with $f(a_1, \ldots, a_n) = 1$. According to the definition of the quantity $C_1(f)$, there must be a set $Z = \{x_{i_1}, \ldots, x_{i_m}\}$ of $m \leq C_1(f)$ variables such that fixing them to the corresponding values $x_{i_1} = a_{i_1}, \ldots, x_{i_m} = a_{i_m}$, the value of f becomes 1 independently of the other variables. A simple (but crucial) observation is that

$$Y \cap Z \neq \emptyset. \tag{14.3}$$

Indeed, if $Y \cap Z = \emptyset$ then the value of $f(0, \ldots, 0, a_{k+1}, \ldots, a_n)$ should be 0 because fixing the variables in Y to 0 forces f to be 0, but should be 1, because fixing the variables in Z to the corresponding values of a_i forces f to be 1, a contradiction.

By (14.3), only $|Z \setminus Y| \leq m - 1$ of the bits of $(a_{k+1}, \ldots, a_n)$ must be fixed to force the subfunction f_a to obtain the constant function 1. This completes the proof of (14.2).

Applying the induction hypothesis to each of the subfunctions f_a with $a \in \{0, 1\}^k$, we obtain $D(f_a) \leq C_0(f_a) \cdot C_1(f_a) \leq C_0(f)(C_1(f) - 1)$. Altogether,

$$D(f) \leq k + \max_a D(f_a) \leq C_0(f) + C_0(f)(C_1(f) - 1) = C_0(f)C_1(f). \qquad \square$$

14.3 Certificates, Sensitivity and Block Sensitivity

Let $f : \{0, 1\}^n \to \{0, 1\}$ be a boolean function, and $a \in \{0, 1\}^n$. An f-*certificate* of a is a subset $S \subseteq \{1, \ldots, n\}$ such that $f(b) = f(a)$ for all vectors $b \in \{0, 1\}^n$ such that $b_i = a_i$ for all $i \in S$. That is, the value $f(a)$ can be determined by looking at only bits of a in the set S.

Certificates are related to monomials and clauses as follows. Associate with each vector $a \in \{0, 1\}^n$ and each subset $S \subseteq [n]$ the monomial

$$M_{S,a} := \bigwedge_{i \in S} x_i^{a_i},$$

where $x_i^1 = x_i$ and $x_i^0 = \neg x_i$, as well as the clause

$$C_{S,a} := \bigvee_{i \in S} x_i^{1-a_i}.$$

If $f(a) = 1$, then S is an f-certificate for a iff $M_{S,a}(x) \leq f(x)$ for all $x \in \{0, 1\}^n$. If $f(a) = 0$, then S is an f-certificate for a iff $C_{S,a}(x) \geq f(x)$ for all $x \in \{0, 1\}^n$.

By $C(f, a)$ we denote the minimum size of an f-certificate for a. The *certificate complexity* of f is $C(f) = \max_a C(f, a)$. Therefore, considering what was mentioned before,

$$C(f) = \max\{C_1(f), C_0(f)\}$$

$$= \min\{k \mid f \text{ can be written as a } k\text{-DNF and as a } k\text{-CNF}\}.$$

Theorem 14.3 gives the following relation between the decision tree depth of boolean functions and their certificate complexity:

$$C(f) \leq D(f) \leq C(f)^2.$$

A similar relation also exists between certificate complexity and another important measure of boolean functions—their sensitivity and "block sensitivity".

Recall that the *sensitivity* of a boolean function $f : \{0, 1\}^n \rightarrow \{0, 1\}$ on $a \in \{0, 1\}^n$ is defined as the number of distance-1 neighbors b of a such that $f(b) \neq f(a)$. For example, if $f(x) = x_1 \vee x_2 \vee \cdots \vee x_n$, then $s(f, \mathbf{0}) = n$ but $s(f, a) = 0$ for every vector a with at least two 1s. The *sensitivity* (or *maximum sensitivity*) of f is defined as $s(f) = \max_a s(f, a)$.

Study of sensitivity of boolean functions originated from Cook and Dwork (1982) and Reischuk (1982). They showed an $\Omega(\log s(f))$ lower bound on the number of steps required to compute a boolean function f on a so-called "consecutive read exclusive write parallel random access machine" (CREW RAM). Such a machine is a collection of synchronized processors computing in parallel with access to a shared memory with no write conflicts. Nisan (1989) then found a way to modify the definition of sensitivity to characterize the minimum number of steps required to compute a function on a CREW PRAM. For this purpose, he introduced a related notion called "block sensitivity".

A natural generalization of sensitivity is to flip blocks of bits rather than single bits. To formalize this, we use the following notation. For a vector $a \in \{0, 1\}^n$ and a subset $S \subseteq [n]$ of its bit-positions, let a^S denote the vector a, with *all* bits a_i with $i \in S$ flipped to opposite values. That is, a^S differs from a exactly on the bits in S. For example, if $a = (0, 1, 1, 0, 1)$ and $S = \{1, 3, 4\}$, then $a^S = (1, 1, 0, 1, 1)$. In particular, if $S = \{i\}$ then $a^S = a \oplus e_i$.

We say that f is *sensitive* to S on a if $f(a^S) \neq f(a)$. The *block sensitivity* of f on a, denoted $\mathrm{bs}(f, a)$, is the largest number t for which there exist t disjoint sets (blocks) $S_1, \ldots, S_t \subseteq [n]$ such that f is sensitive on a to each of these sets, that is, $f(a^{S_i}) \neq f(a)$ for all $i = 1, \ldots, t$. The *block sensitivity* of a boolean function f is $\mathrm{bs}(f) = \max_a \mathrm{bs}(f, a)$.

It is clear that $s(f, a) \leq \mathrm{bs}(f, a)$: this follows by considering the partition where every S_i is a singleton. Moreover, we also have that $\mathrm{bs}(f, a) \leq C(f, a)$: any certificate of a must include at least one variable from each set to which f is sensitive on this input a. Hence

$$s(f) \leq bs(f) \leq C(f) \leq D(f).$$

It can be shown (see Exercise 14.4) that

$$s(f) = bs(f) = C(f) \text{ for every } monotone \; f.$$

It can also be shown (Exercise 14.6) that

$$s(f) \geq \frac{n+1}{2} \text{ for every } symmetric \; f \text{ of } n \text{ variables.}$$

The biggest known gap between $s(f)$ and $bs(f)$ is quadratic.

Lemma 14.4. (Rubinstein 1995) *There are boolean functions f with*

$$bs(f) \geq s(f)^2/2.$$

Proof. For an even m, let g be a boolean function of m variables such that $g(x) = 1$ if and only if $x_{2i-1} = x_{2i} = 1$ for some $1 \leq i \leq m/2$ and $x_j = 0$ for all other positions j. For example, if $m = 8$, then

$$g^{-1}(1) = \begin{bmatrix} 1\,1\,0\,0\,0\,0\,0\,0 \\ 0\,0\,1\,1\,0\,0\,0\,0 \\ 0\,0\,0\,0\,1\,1\,0\,0 \\ 0\,0\,0\,0\,0\,0\,1\,1 \end{bmatrix}.$$

By taking blocks $S_i = \{2i - 1, 2i\}$ for $i = 1, \ldots, m/2$, we see that $bs(g) \geq bs(g; 0) = m/2$. Moreover, $s(g, a) = 1$ for all $a \in g^{-1}(0)$, and $s(g, b) = m$ for all $b \in g^{-1}(1)$.

Now let $f = g_1 \vee \cdots \vee g_m$ be an OR of m copies of g on disjoint sets of variables; hence, f has $n = m^2$ variables. Then

$$bs(f) \geq bs(f; 0) = m \cdot bs(g; 0) \geq m^2/2 = n/2.$$

On the other hand, for every $a \in f^{-1}(0)$ we have that $s(f, a) \leq \sum_{i=1}^{m} s(g, a) \leq m$, and for every $b \in f^{-1}(1)$ we have that $s(f, b) \leq \max_j s(g_j, b) \leq m$; this last inequality holds because $f(b) = 1$ and $f(b^i) = 0$ implies that $g_j(b) = 1$ for *exactly one* of the g_j. Hence, $s(f) \leq m = \sqrt{n}$.   □

It remains unknown whether there is a polynomial relationship between sensitivity and block sensitivity.

■ **Research Problem 14.5.** Do there exist constants c, d such that $bs(f) \leq c \cdot s(f)^d$ holds for all boolean functions f?

More information about this problem, known as the *Sensitivity Conjecture*, can be found in a survey of Hatami et al. (2011).

14.3.1 Block Sensitivity Versus Certificate Complexity

That $C(f)$ can be at most quadratic than $\mathrm{bs}(f)$ was shown by Nisan (1989).

Theorem 14.6. (Nisan 1989) *For every boolean function f,*

$$C(f) \leq \mathrm{s}(f) \cdot \mathrm{bs}(f) \leq \mathrm{bs}(f)^2.$$

Proof. Take an arbitrary input $a \in \{0,1\}^n$. Our goal is to show that $C(f,a) \leq \mathrm{s}(f) \cdot \mathrm{bs}(f)$. First we show that minimal blocks to which a function f is sensitive cannot have more than $\mathrm{s}(f)$ variables. Let $S \subseteq [n]$ a minimal with respect to set-inclusion subset such that $f(a^S) \neq f(a)$.

Claim 14.7. $|S| \leq \mathrm{s}(f)$.

Proof. If we flip one of the S-variables in a^S, then the function value must flip from $f(a^S)$ to $f(a)$, for otherwise S would not be minimal. So, every S-variable is sensitive for f on input a^S, implying that $\mathrm{s}(f) \geq |S|$.   □

Now let $S_1, \dots, S_t$ be disjoint minimal sets of variables that achieve the block sensitivity $t = \mathrm{bs}(f,a) \leq \mathrm{bs}(f)$. Consider the set $S = S_1 \cup \dots \cup S_t$. By the previous claim, we have that $|S_i| \leq \mathrm{s}(f)$ for all i. Hence, $|S| \leq \mathrm{s}(f) \cdot t \leq \mathrm{s}(f) \cdot \mathrm{bs}(f)$, and it remains to show that S is an f-certificate of a.

If S is not an f-certificate of a, then let $b \in \{0,1\}^n$ be an input that coincides with a on S, and $f(b) \neq f(a)$. Let S_{t+1} be the set of positions on which b differs from a; hence, $b = a^{S_{t+1}}$. Now f is sensitive to S_{t+1} on a and S_{t+1} is disjoint from $S_1, \dots, S_t$ which contradicts $t = \mathrm{bs}(f,a)$. Hence, S is a an f-certificate of a, as claimed.   □

14.3.2 Block Sensitivity Versus Depth

Since $D(f) \leq C(f)^2$, Theorem 14.6 gives an upper bound $D(f) \leq \mathrm{bs}(f)^4$ on the decision tree depth. A better upper bound $D(f) \leq \mathrm{bs}(f)^3$ can be obtained from the following result.

Theorem 14.8. (Beals et al. 2001) *If a boolean function f can be written as a k-DNF or a k-CNF, then $D(f) \leq k \cdot \mathrm{bs}(f)$.*

The inequality is tight because, if $f(x) = x_1 \vee x_2 \vee \dots \vee x_n$, then $D(f) = n$, f is a 1-DNF and $\mathrm{bs}(f) = n$.

Proof. Suppose that a boolean function f of n variables can be written as a k-DNF (the case of k-CNF is dual). Let $t := \mathrm{bs}(f)$. We will describe an algorithm which, given an input vector $a \in \{0,1\}^n$, queries at most $k \cdot t$ bits of a to compute the value $f(a)$.

Stage 1: Repeat the following at most $t = \mathrm{bs}(f)$ times:

- Pick a monomial M consistent with the values of all queries made so far.
- If there is no such monomial, then return value 0 and stop.
- Otherwise, query *all* not yet queried variables of M and assign them values corresponding to the bits of a.
- If all these values agree with a then return value 1 and stop.

Stage 2: If Stage 1 does not stop after performing it t times, then pick a vector $b \in \{0, 1\}^n$ consistent with all queries made so far and return value $f(b)$.

The nondeterministic "pick" can easily be made deterministic by choosing the first monomial M and the first vector y in some fixed in advance order. Since the algorithm runs for at most $t = \mathrm{bs}(f)$ steps and each step queries at most k variables, at most $t \cdot k$ variables are queried in total.

It remains to show that the algorithm always returns the right answer. If it returns an answer in Stage 1, this is either because no monomial is consistent with a (and hence $f(a)$ must be 0) or because a is found to agree with a particular monomial (and hence $f(a)$ must be 1). In both cases the algorithm gives the right answer.

Now consider the case where the algorithm returns an answer only in Stage 2. We will show that $f(b) = f(a)$ for all vectors $b \in \{0, 1\}^n$ that are consistent with the path constructed by vector a in step (1). Suppose not. Then there are consistent vectors b and c with $f(b) = 0$ and $f(c) = 1$. On input a, the algorithm has queried all variables of a sequence of $t = \mathrm{bs}(f)$ monomials $M_1, \ldots, M_t$, and both vectors b and c coincide with a on all these variables. Moreover, since $f(c) = 1$, there also must be a monomial M_{t+1} consistent with c. We will derive from these monomials disjoint non-empty sets $S_1, \ldots, S_{t+1}$ of variables such that f is sensitive to each S_i on input b. This will imply that $\mathrm{bs}(f, b) \geq t + 1 = \mathrm{bs}(f) + 1$, contradicting the definition of $\mathrm{bs}(f)$.

For every $i = 1, \ldots, t + 1$, define S_i as the set of variables in the monomial M_i that are inconsistent with the corresponding bits of b (and hence, also of a). Clearly, each S_i is non-empty because $f(b) = 0$. Note that b^{S_i} is already consistent with M_i, so $f(b^{S_i}) = 1$, which shows that f is sensitive to each S_i on b. To obtain the desired contradiction, it remains therefore to show that all the S_i are pairwise disjoint.

To show this, take a variable $x_k \in S_i$ and assume that $x_k \in S_j$ for some $j > i$. Assume w.l.o.g. that $a_k = 1$; hence, also $b_k = 1$. Then both monomials M_i and M_j must contain the same literal $\neg x_k$. This already implies that $j \neq t + 1$, because $c_k = a_k$ and $M_{t+1}(c) = 1$. So, $i < j \leq t$, meaning that M_j has been chosen *after* all variables of M_i, including x_k, where queried. But this is impossible because M_j is not consistent with the (already queried) value a_k. Thus, no two of the sets S_i can share a common variable, as desired. $\qquad\square$

Since every boolean function f can be written as a k-DNF and as a k-CNF with $k = C(f)$, and since $C(f) \leq s(f) \cdot \mathrm{bs}(f)$, we obtain the following

Corollary 14.9. $D(f) \leq s(f) \cdot bs(f)^2 \leq bs(f)^3$.

■ **Research Problem 14.10.** Is $D(f) = \mathcal{O}(bs(f)^2)$?

14.3.3 Sensitivity and Degree of Polynomials

We now relate block sensitivity of a boolean function f with the degree of real polynomials representing f. Recall that a multilinear polynomial $p : \mathbb{R}^n \to \mathbb{R}$ *represents* a boolean function $f : \{0, 1\}^n \to \{0, 1\}$ if $p(a) = f(a)$ for all $a \in \{0, 1\}^n$. We already know (see Sect. 2) that every function $f : \{0, 1\}^n \to \mathbb{R}$ has a unique representation as a multilinear polynomial over $\mathbb{R}$:

$$f(x) = \sum_{a \in \{0,1\}^n} f(a) \prod_{i:a_i=1} x_i \prod_{j:a_j=0} (1 - x_j) = \sum_{S \subseteq [n]} c_S \prod_{i \in S} x_i.$$

The *degree*, $\deg(f)$, of a boolean function f is the degree of the unique multilinear real polynomial p that represents f. The AND of n variables $x_1, \ldots, x_n$ is represented by the polynomial consisting of just one monomial $\prod_{i=1}^{n} x_i$, and the OR is represented by the polynomial $1 - \prod_{i=1}^{n} (1 - x_i)$. Hence, both of these functions have degree n.

Besides that the degree is an interesting algebraic parameter of boolean functions, it can be used to lower-bound the depth of decision trees: it can be easily shown (see Exercise 14.5) that $D(f) \geq \deg(f)$.

How is the degree related to sensitivity? On the one hand, there are boolean functions f such that

$$s(f) = bs(f) = C(f) = \sqrt{n} \text{ but } \deg(f) = n.$$

Take, for example, the function f of $n = m^2$ variables defined by (14.1): since the degree of each AND as well as of each OR of m variables is m, we have that $\deg(f) = m^2 = n$.

On the other hand, $\deg(f)$ may also be significantly smaller than $C(f)$, and hence, than $s(f)$ and $bs(f)$. To see this, consider the boolean function $f : \{0, 1\}^n \to \{0, 1\}$ constructed in the proof of Lemma 2.5. This function has degree $\deg(f) \leq n^{0.631\cdots}$ and maximal sensitivity $s(f, \mathbf{0}) = n$. Hence,

$$s(f) = bs(f) = C(f) = n \text{ but } \deg(f) \leq n^{0.631\cdots}.$$

Our goal is now to show that $\deg(f)$ can be no more than quadratically smaller than $bs(f)$. This shows that the gap of the last example is close to optimal.

Theorem 14.11. (Nisan–Szegedy 1994) *For every boolean function f,*

$$\deg(f) \geq \sqrt{bs(f)/2}.$$

Proof. Let $f(x)$ be a boolean function of n variables, and let $q : \mathbb{R}^n \to \mathbb{R}$ be the multilinear polynomial of degree d representing f; hence $q(x) = f(x)$ for all $x \in \{0,1\}^n$. By Lemma 2.4, we know that every boolean function f of n variables, which rejects the all-0 vector and accepts all n vectors with exactly one 1, has $\deg(f) \geq \sqrt{n/2}$. It is therefore enough to construct a multilinear polynomial p of $t = \mathrm{bs}(f)$ variables satisfying the conditions of this lemma.

Let $t = \mathrm{bs}(f)$, and $a \in \{0,1\}^n$ be an input such that $\mathrm{bs}(f,a) = \mathrm{bs}(f)$. We assume without loss of generality that $f(a) = 0$. Let also $S_1, \ldots, S_t$ be the corresponding disjoint subsets of $[n]$ such that $f(a^{S_i}) = 1$ for all $i = 1, \ldots, t$. We transform $q(x_1, \ldots, x_n)$ into a multilinear polynomial $p(y_1, \ldots, y_t)$ of t new variables by replacing every variable x_j in p as follows

$$x_j := \begin{cases} y_i & \text{if } a_j = 0 \text{ and } j \in S_i, \\ 1 - y_i & \text{if } a_j = 1 \text{ and } j \in S_i, \\ a_j & \text{if } j \notin S_1 \cup \cdots \cup S_t. \end{cases}$$

That is, for $y \in \{0,1\}^t$ we have that

$$p(y) = q(a \oplus y_1 S_1 \oplus y_2 S_2 \oplus \cdots \oplus y_y S_t)$$

where

$$y_i S_i = (0, \ldots, 0, \overbrace{y_1, \ldots, y_i}^{S_i}, 0, \ldots, 0).$$

It is clear that p is a multilinear polynomial of degree at most d, and $p(y)$ takes values in $\{0,1\}$ for all $y \in \{0,1\}^t$, since $p(x)$ does this for all $x \in \{0,1\}^n$. Moreover, we have that $p(0) = q(a) = f(a) = 0$, and

$$p(e_i) = q(a^{S_i}) = f(a^{S_i}) = 1$$

for all unit vectors $e_i \in \{0,1\}^t$, $i = 1, \ldots, t$. We can therefore apply Lemma 2.4 and conclude that $d = \deg(q) \geq \deg(p) \geq \sqrt{t/2} = \sqrt{\mathrm{bs}(f)/2}$. □

Together with Corollary 14.9, Theorem 14.11 gives the following relation between the depth of decision trees and the degree of boolean functions.

Corollary 14.12. (Nisan–Szegedy 1994) $D(f) \leq 8 \cdot \deg(f)^6$.

This upper bound was improved by Nisan and Smolensky (unpublished); see the survey of Buhrman and de Wolf (2002).

Theorem 14.13.

$$D(f) \leq \deg(f)^2 \cdot \mathrm{bs}(f) \leq 2 \cdot \deg(f)^4.$$

Proof. (Due to Nisan and Smolensky) Let f be a boolean function of n variables. By a *maxonomial* of f we will mean a monomial with maximal degree in the multilinear polynomial representing f.

Claim 14.14. For every maxonomial M of f, there is a set S of variables in M such that $f(0^S) \neq f(0)$.

Proof. Obtain a subfunction g of f by setting all variables outside M to 0. This g cannot be constant 0 or 1, because its unique polynomial representation (as obtained from p) contains M. Thus there must be some subset S of the variables in M that makes $g(0^S) \neq g(0)$ and hence $f(0^S) \neq f(0)$. $\square$

Claim 14.15. There exists a set of $\deg(f) \cdot \mathrm{bs}(f)$ variables that intersects each maxonomial of f.

Proof. Greedily take all variables in maxonomials of f, as long as there is a maxonomial that is still disjoint from those taken so far. By Claim 14.14, each maxonomial contains a sensitive block on $\mathbf{0}$. Since there can be at most $\mathrm{bs}(f)$ disjoint sensitive blocks, this procedure can go on for at most $\mathrm{bs}(f)$ maxonomials. Since each maxonomial of f contains only $\deg(f)$ variables, the claim follows. $\square$

We can now finish the proof of the theorem as follows. Let $a \in \{0, 1\}^n$ be an arbitrary input vector. By Claim 14.15, there is a set of $\deg(f) \cdot \mathrm{bs}(f)$ variables that intersects each maxonomial of f. Query all these variables. This induces a restriction g of f on the remaining variables, such that $\mathrm{bs}(g) \leq \mathrm{bs}(f)$ and $\deg(g) < \deg(f)$ (because the degree of each maxonomial in the representation of f drops by at least one). Repeating this inductively for at most $\deg(f)$ times, we reach a constant function and learn the value $f(a)$. This algorithm uses at most $\deg(f)^2 \cdot \mathrm{bs}(f)$ queries, hence $D(f) \leq \deg(f)^2 \cdot \mathrm{bs}(f)$ where, by Theorem 14.11, $\mathrm{bs}(f) \leq 2 \cdot \deg(f)^2$. $\square$

It is conjectured that $\deg(f) = \Theta(\mathrm{s}(f)^2)$.

■ **Research Problem 14.16.** Do there exist constants c, d such that $\deg(f) \leq c \cdot \mathrm{s}(f)^d$ holds for all boolean functions f?

14.4 Sensitivity and Subgraphs of the n-Cube

Problem 14.16 is related to the following problem about the maximum degree of induced subgraphs of the n-dimensional binary hypercube Q_n. For a set $S \subseteq \{0, 1\}^n$ of its vertices, let $Q_n[S]$ denote the subgraph of Q_n induced by S. That is, S is the vertex-set of $Q_n[S]$, and edges of $Q_n[S]$ are all edges of Q_n connecting two vertices in S.

Let $\Delta(n)$ denote the maximal number D such that, for every subset $S \subseteq \{0, 1\}^n$ of size $|S| \neq 2^{n-1}$, the maximum degree of $Q_n[S]$ or $Q_n[\overline{S}]$ is at least D. The condition $|S| \neq 2^{n-1}$ is necessary because if we take $S = \{x \mid \sum_{i=1}^{n} x_i \text{ is even}\}$, then the subgraph of Q_n induced by S is empty, that is, has zero degree. It is known that $\Delta(n) < \sqrt{n} + 1$.

Example 14.17. (Chung et al. 1988). For simplicity, assume that $n = m^2$ is a square number. Look at vectors $x \in \{0, 1\}^n$ as $m \times m$ matrices. Say that x is *even* (resp., *odd*) if it has an even (resp., odd) number of ones. Let S_0 be the set of all even matrices containing at least one all-1 row, and let S_1 be the set of all odd matrices containing no all-1 rows. Consider the subgraph G of Q_n induced by S, and let $d_S(x)$ denote the degree of x in this graph. It can be shown that $d_S(x) \leq m = \sqrt{n}$ for all $x \in S$ (and the same also holds for the subgraph induced by the complement $\overline{S}$).

Indeed, by flipping one 1 to 0 we can destroy only one all-1 row. So, if a vector x belongs to S_0, then only its neighbors in S_1 can be those corresponding to the positions of the unique all-1 row of x. Since each row has only m positions, this implies that $d_S(x) \leq m = \sqrt{n}$. If we take a vector $x \in S_1$ then (again) all its neighbors lying in S must belong to S_0, that is, must have an all-1 row. Since flipping one 0 to 1 we can produce only one all-1 row, the only neighbors of x in S are those corresponding to "almost all-1" rows of x, that is rows with exactly one 0. Since we only have m rows, this again implies that $d_S(x) \leq m = \sqrt{n}$.

On the other hand, Gotsman and Linial (1992) showed that $\deg(f) \leq s(f)^d$ as long as $\Delta(n) \geq n^{1/d}$; in fact, they show that these two inequalities are equivalent. Thus, to solve Problem 14.16 it would be enough to prove or disprove that $\Delta(n) \geq n^\epsilon$ for a constant $\epsilon > 0$.

Theorem 14.18. (Gostman–Linial 1992) *The following are equivalent for any monotone function $h : \mathbb{N} \to \mathbb{R}$:*

(a) $\Delta(n) \geq h(n)$.
(b) *For any boolean function f, $s(f) \geq h(\deg(f))$.*

Proof. It will be convenient to switch to the ± 1 notation and consider boolean functions as colorings $f : Q_n \to \{-1, +1\}$ of the hypercube Q_n whose vertices are vectors in $\{0, 1\}^n$, and two vertices are adjacent if and only if they differ in exactly one coordinate. This transformation can be done via mapping $a \in \{0, 1\}$ to $(-1)^a \in \{-1, +1\}$. The degree, $\deg(f)$, of f is then the maximum size $|I|$ of a subset $I \subseteq [n]$ for which the Fourier coefficient

$$\widehat{f}(I) = \sum_{x \in Q_n} f(x) \prod_{i \in I} x_i$$

is nonzero; recall from Sect. 2.3 that all these coefficients are real numbers between -1 and $+1$. Associate with a subgraph G of Q_n induced by a set of vertices $S \subseteq Q_n$ a boolean function $g : Q_n \to \{-1, +1\}$ such that $g(x) = 1$ if and only if $x \in S$. Let $d_G(x)$ denote the degree of x in the graph G. Observe that

$$d_G(x) = n - s(g, x) \text{ for all } x \in S \tag{14.4}$$

and the same holds in the subgraph induced by $\overline{S}$. Let $\mathrm{E}\,[g] = 2^{-n} \sum_x g(x)$ be the average of g on C_n. Observe that

$$\mathrm{E}\,[g] = 0 \text{ iff } |S| = 2^{n-1}. \tag{14.5}$$

By (14.4) and (14.5), (a) and (b) are equivalent to the following:

(A) For any boolean function g, $\mathrm{E}\,[g] \neq 0$ implies $s(g, x) \leq n - h(n)$ for some $x \in Q_n$.
(B) For any boolean function f, $s(f) < h(n)$ implies $\deg(f) < n$.

To see the equivalence of (A) and (B), define

$$g(x) = f(x) \prod_{i=1}^{n} x_i .$$

Since $\prod_{i=1}^{n} x_i$ is the parity function in the ± 1 notation, and since the parity function is sensitive to all n variables, we have that $s(g, x) = n - s(f, x)$ for all $x \in \{-1, +1\}^n$, and

$$\widehat{g}(I) = 2^{-n} \sum_x g(x) \prod_{i \in I} x_i = 2^{-n} \sum_x f(x) \prod_{i \notin I} x_i = \widehat{f}([n] \setminus I)$$

for all $I \subseteq [n]$. In particular, $\mathrm{E}\,[g] = \widehat{g}(\emptyset) = \widehat{f}([n])$, where $\widehat{f}([n])$ is the highest order Fourier coefficient in the representation of f as a polynomial

(A) $\Rightarrow$ (B): Assume that $\deg(f) = n$, that is, $\widehat{f}([n]) \neq 0$. This is equivalent to $\mathrm{E}\,[g] \neq 0$. By (A), there exists a vector x such that $s(g, x) \leq n - h(n)$, that is, $s(f, x) \geq h(n)$, contradicting the premise $s(f) < h(n)$ of (B).

(B) $\Rightarrow$ (A): Assume that $s(g, x) > n - h(n)$ for all x. This implies that $s(f) < h(n)$. By (B), we have that $\deg(f) < n$, which is equivalent to $\mathrm{E}\,[g] = \widehat{g}(\emptyset) = \widehat{f}([n])=0$, contradicting the premise $\mathrm{E}\,[g] \neq 0$ of (A). $\square$

Sherstov (2010) has recently proved the following lower bound on $\deg(f)$ based on rank. Define the *AND-rank* of a boolean function $f : \{0, 1\}^n \to \{0, 1\}$ as the rank over $\mathbb{R}$ of the $2^n \times 2^n$ boolean matrix M whose entries are given by $M[x, y] := f(x \wedge y)$, where $x \wedge y = (x_1 \wedge y_1, \ldots, x_n \wedge y_n)$. The *OR-rank* is defined similarly. Let $R(f)$ be the maximum of the AND-rank and the OR-rank of f. Then $\deg(f) = \Omega(\log R(f))$.

14.5   Evasive Boolean Functions

To prove that some boolean function f requires decision trees of large depth, it is useful to imagine the situation as a game between Alice and Bob. This time the players are not cooperative: Alice acts as an "adversary". Bob knows the function

$f : \{0, 1\}^n \to \{0, 1\}$ but does not know the actual input vector $x \in \{0, 1\}^n$. He can ask Alice what the i-th bit of x is. Then what the j-th bit is, and so on. He stops when he definitely knows the answer "$f(x) = 0$" or "$f(x) = 1$". Alice's goal is to inductively construct (depending on what bits Bob has already asked about) an input x on which Bob is forced to make many queries. That is, Alice tries to construct an "evasive" path forcing Bob to make his tree deep. This is the adversary argument we described in Sect. 14.1.

We now demonstrate this argument on symmetric functions. Recall that a boolean function is *symmetric* if every permutation of its variables leaves its value unchanged. That is, a boolean function is symmetric if and only if its value depends only on *how many* of its variables (not on *which* of them) are 0 or 1.

A boolean function f of n variables is called *evasive* if it has maximal possible depth, that is, if $D(f) = n$.

Lemma 14.19. *Every non-constant symmetric boolean function is evasive.*

Proof. Let $f : \{0, 1\}^n \to \{0, 1\}$ be the symmetric boolean function in question. Since f is not constant, there is a k with $1 \le k \le n$ such that if $k - 1$ variables have value 1, then the function has value 0, but if k variables are 1 then the function's value is 1 (or the other way round).

Using this, we can propose the following strategy for Alice. She thinks of a 0–1 sequence of length n and Bob can ask the values of each bit. Alice answers 1 on the first $k - 1$ questions and 0 on every question that follows. Thus, after $n - 1$ questions, Bob cannot know whether the number of 1s is $k - 1$ or k, that is, he cannot know the value of the function. $\qquad\square$

Every boolean function f of n variables splits the n-cube $\{0, 1\}^n$ into two disjoint blocks $f^{-1}(0)$ and $f^{-1}(1)$. Since the number 2^n of vectors in the n-cube is even, the sizes of these blocks must be both even or both must be odd. It turns out that all boolean functions with odd block size are evasive.

Lemma 14.20. *If $|f^{-1}(0)|$ is odd then f is evasive.*

Proof. Consider an arbitrary deterministic decision tree that computes the function f. Let v be an arbitrary node in this tree. If the depth of v is d, then exactly 2^{n-d} of the possible inputs lead to v. In particular, any node whose depth is at most $n - 1$ is reached by an even number of possible inputs. On the other hand, each input reaches exactly one leaf. Thus, if $|f^{-1}(0)|$ is odd, there must be a leaf which is reached by a single input x with $f(x) = 0$; this leaf has depth n. $\qquad\square$

Symmetric functions are very special; the following class is significantly more general. Call a boolean function of n variables *weakly symmetric* if for all pairs x_i, x_j of variables, there is a permutation of the variables that takes x_i into x_j but does not change the value of the function. For example, the function

$$(x_1 \wedge x_2) \vee (x_2 \wedge x_3) \vee \cdots \vee (x_{n-1} \wedge x_n) \vee (x_n \wedge x_1)$$

is weakly symmetric but not symmetric (check this!).

Theorem 14.21. (Rivest–Vuillemin 1976) *Let n be a prime power. If $f : \{0,1\}^n \to \{0,1\}$ is weakly symmetric, and $f(\mathbf{0}) \neq f(\mathbf{1})$, then f is evasive.*

Proof. Every permutation $\pi : [n] \to [n]$ on the input coordinates induces a permutation $\widehat{\pi} : \{0,1\}^n \to \{0,1\}^n$ on the set of possible input vectors:

$$\widehat{\pi}(x_1,\ldots,x_n) = (x_{\pi(1)},\ldots,x_{\pi(n)}).$$

Let Γ be the set of all permutation π that leave the value of the function unchanged, that is,

$$\Gamma = \{\pi \mid f(\widehat{\pi}(x)) = f(x) \text{ for all vectors } x\}.$$

It can be easily verified that Γ forms a group. Moreover, since the function f is weakly symmetric, this group is *transitive*, that is, for any pair of ground elements i and j, there is a permutation $\pi \in \Gamma$ such that $\pi(i) = j$.

We define the *orbit* of a vector $x \in \{0,1\}^n$ to be the set of images of x under permutations in Γ:

$$\text{orbit}(x) = \{\widehat{\pi}(x) \mid \pi \in \Gamma\}.$$

Claim 14.22. For any vector x except $\mathbf{0}$ or $\mathbf{1}$, the size $|\text{orbit}(x)|$ is divisible by n.

Proof. Since $x \neq \mathbf{0}$ and $x \neq \mathbf{1}$, the orbit of x has more than one element. Let $|x|$ denote the number of 1s in x. Then

$$\sum_{y \in \text{orbit}(x)} |y| = \sum_{y \in \text{orbit}(x)} \sum_{i=1}^n y_i = \sum_{i=1}^n \sum_{y \in \text{orbit}(x)} y_i.$$

Since Γ is transitive, for every i, there must be a permutation $\pi \in \Gamma$ such that $\pi(i) = 1$. Thus the last summand does not actually depend on i, implying that

$$\sum_{y \in \text{orbit}(x)} |y| = n \cdot \sum_{y \in \text{orbit}(x)} y_1.$$

On the other hand, since all vectors in the orbit have the same number of 1s, we have

$$\sum_{y \in \text{orbit}(x)} |y| = |\text{orbit}(x)| \cdot |x|.$$

Thus, $|\text{orbit}(x)| \cdot |x|$ is divisible by n. On the other hand, $0 < |x| < n$ implies that $|x|$ is not divisible by n. Since n is prime power, Euclid's theorem implies that $|\text{orbit}(x)|$ must be divisible by n. $\square$

By Lemma 14.20, the function f is evasive if

$$S := \sum_{x \in f^{-1}(0)} (-1)^{|x|} \neq 0.$$

If $f(x) = 0$, then the orbit of x contributes

$$\sum_{y \in \text{orbit}(x)} (-1)^{|y|} = |\text{orbit}(x)| \cdot (-1)^{|x|}$$

to this sum, since all vectors in orbit(x) have the same number of 1s. By Claim 14.22, this is a multiple of n, except for the cases $x = \mathbf{0}$ and $x = \mathbf{1}$. Since exactly one of the vectors $\mathbf{0}$ and $\mathbf{1}$ is in $f^{-1}(0)$, the sum S is either one more or one less than a multiple of n. In either case, $S \neq 0$, so f must be evasive. $\quad\square$

14.6 Decision Trees for Search Problems

So far we have considered decision trees solving *decision problems*. That is, for each input the decision tree must give an answer "yes" (1) or "no" (0). For example, if $n = \binom{v}{2}$ then each input $x \in \{0,1\}^n$ can be interpreted as a graph G on v vertices, where $x_e = 1$ means that the edge e is present in G, and $x_e = 0$ means that the edge e is not present in G. There are a lot of decision problem for graphs. Is the graph connected? Has the graph a clique of size k? Is the graph colorable by k colors?

But decision alone is often not what we actually need. Knowing the answer "the graph has a triangle", we would like to find any of these triangles. Given an unsatisfiable CNF and an assignment to its variables, we would like to find a clause which is not satisfied.

In general, a search problem is specified by n boolean variables and a collection W of "witnesses". In addition, this collection must have the property that every assignment to the n variables is associated with at least one witness. That is, a *search problem* is specified by a relation $F \subseteq \{0,1\}^n \times W$ such that, for every $x \in \{0,1\}^n$ there exists at least one $w \in W$ such that $(x, w) \in F$. The problem itself is:

Given an $x \in \{0,1\}^n$, find a witness $w \in W$ such that $(x, w) \in F$.

With every boolean function $f : \{0,1\}^n \to \{0,1\}$ we can associate the relation $F \subseteq \{0,1\}^n \times W$, where $W = \{0,1\}$ and $(x,w) \in F$ if and only if $f(x) = w$. Hence, decision problems (=boolean functions) are special case of search problems.

Example 14.23. Consider the graphs G_x on v vertices, encoded by binary strings $x \in \{0,1\}^n$ of length $n = \binom{v}{2}$, one bit for each potential edge. As a set W of witnesses we can take some special element λ and the set of all triangles. Define the relation F by: $(x, w) \in F$ if $w = \lambda$ and graph G_x is triangle-free, or $w \neq \lambda$ and w is a triangle in G_x. Then the search problem is, given an input $x \in \{0,1\}^n$, either to answer "no triangle" if G_x is triangle-free, or to find a triangle in G_x.

Given a bipartite graph $G = (U \cup V, E)$, define the search problem Degree(G) in the following way. We have $|E|$ variables x_e, one for each edge $e \in E$. Each assignment $x \in \{0, 1\}^E$ to these variables is interpreted as a subgraph G_x of G, defined by those edges e for which $x_e = 1$, that is, $G_x = \{e \in E \mid x_e = 1\}$. The search problem Degree(G) is:

Given an input vector x, find a vertex whose degree in G_x is not one.

It is clear that such a vertex always exist, as long as the sides of the graph are not equal. Thus, as long as $|U| \neq |V|$, Degree(G) is a valid search problem. Note also that Degree(G) can be solved by a *nondeterministic* decision tree of depth at most d, where d is the maximum degree of G. For this, it is enough to guess a vertex of degree $\neq 1$ and check the incident edges of this vertex.

We will now show that *deterministic* decision trees must have much larger depth. For this, we take a bipartite $(2n) \times n$ graph $G = (U \cup V, E)$ of maximum degree d. Suppose that G has the following *expansion property*:

Every subset $S \subseteq U$ of $|S| \leq n/4$ vertices has at least $2|S|$ neighbors in V.

Such graphs exist for $d = \mathcal{O}(1)$ and infinitely many n's, and can be efficiently constructed using known expander graphs. The following theorem, as well as separations between deterministic, nondeterministic and randomized decision trees for search problems, were proved by Lovász et al. (1995).

Theorem 14.24. (Lovász et al. 1995) *Let G be a bipartite $2n \times n$ graph of maximum degree d. If G has the expansion property, then every deterministic decision tree for* Degree(G) *requires depth $\Omega(n/d)$.*

Proof. We use an adversary argument. At each step, Bob (a deterministic decision tree) queries some edge $e \in E$. Based on what edges Bob has queried so far, Alice (the adversary) answers either "$x_e = 1$" (the edge e is present) or "$x_e = 0$" (the edge e is not present) in the subgraph. That is, Alice constructs a subgraph of G step-by-step depending on what edges (pairs of vertices) Bob has queried so far. We will show that Alice can cause Bob to probe $\Omega(n/d)$ edges of G. The adversary will be limited to produce (at the end) a subgraph of G in which all vertices in U have degree at most 1 and all vertices in V have degree exactly 1. Hence, the answer is a vertex in U.

To describe the adversary strategy we need some definitions. For step i (after i edges were already probed), let G_i be the subgraph of G obtained by removing all edges $e \in E$ such that:

- The edge e was already probed and was rejected;
- The edge e was not probed yet but e is adjacent in G with at least one already probed and accepted edge.

That is, G_i contains all edges of G that are still possible for the adversary to use in her final subgraph without violating the above limitations.

A set $S \subseteq U$ cannot be matched to V in G_i if it has fewer than $|S|$ neighbors in G_i. Let $S(G_i)$ denote a minimum cardinality unmatchable set in G_i. By the above limitation on the adversary, at step i the subgraph G_i contains a (partial) matching from U to V. Bob cannot know the answer as long as there is no isolated vertex in G_i. Such a vertex itself is a minimum unmatchable set of size 1.

Initially, since the graph G has an expansion property, we have that $|S(G)| > n/4$. Thus, Alice's strategy is to make sure that the minimum unmatchable set size does not decrease too fast.

To describe her strategy, suppose that an edge $e = (u, v)$ is probed in step i (after i edges were already probed). In order to give an answer "$x_e = 1$" or "$x_e = 0$", Alice first constructs two sets of vertices:

- $S^0(e)$ = the minimum unmatchable set that would occur in G_{i+1} if Alice answered "$x_e = 0$".
- $S^1(e)$ = the minimum unmatchable set that would occur in G_{i+1} if Alice answered "$x_e = 1$".

Alice then chooses the answer on e so as to make $S(G_{i+1})$ the larger of $S^0(e)$ and $S^1(e)$. The heart of the argument is the following claim

Claim 14.25. $|S(G_{i+1})| \geq \frac{1}{2}|S(G_i)|$.

Proof. Assume e is asked in step $i + 1$. By the above strategy,

$$|S(G_{i+1})| = \max\{|S^0(e)|, |S^1(e)|\}.$$

Consider the set $S = S^0(e) \cup S^1(e)$. This set cannot be matched into V in G_i, for otherwise either $S^0(e)$ or $S^1(e)$ would be matchable after the decision about e is made. Thus, S contains an unmatchable set for step i of cardinality no more than $|S^0(e) \cup S^1(e)| \leq 2 \cdot \max\{|S^0(e)|, |S^1(e)|\} = 2 \cdot |S(G_{i+1})|$. $\qquad\Box$

We can now complete the proof of the theorem by the following argument. During the game between Alice and Bob, a sequence $S_0, S_1, \ldots, S_t$ of minimum unmatchable sets $S_i = S(G_i)$ of vertices in U is constructed. At the beginning $|S_0| > n/4$, and $|S_t| = 1$ at the end. Moreover, by Claim 14.25, we have that the cardinality of the S_i does not decrease by more than a factor of 2. It must therefore be a step i at which $n/16 \leq |S_i| \leq n/8$ and S_i has fewer than $|S_i|$ neighbors in the i-th subgraph G_i of G. However, by the expansion property of G, the set S_i has had at least $2|S_i|$ neighbors in the original graph G. Since at each step and for any set, the number of its neighbors can drop down by at most a factor of $1/d$, it follows that at least $|S_i|/d = \Omega(n/d)$ edges were probed up to step i. $\qquad\Box$

In a *randomized* decision tree, Bob (the decision tree) is allowed to flip a not necessarily fair coin at each step to decide which variable to test next. These random flips are "for free": only queries of variables contribute to the depth of the tree. The complexity measure in this case is the expected depth of the tree under the worst case input. Equivalently, a random decision tree can be defined as a probability distribution over all deterministic decision trees.

We now show that using random flips the search problem Degree(G) can be solved much more efficiently.

Theorem 14.26. *For every bipartite* $2n \times n$ *graph* G *of maximum degree* d, *the search problem* $\mathrm{Degree}(G)$ *can be solved by a randomized decision tree of expected depth* $\mathcal{O}(d^2)$.

Proof. Let $G = (U \cup V, E)$ be a bipartite $2n \times n$ graph of maximum degree d. Consider the following random decision tree. Pick at random a vertex $u \in U$ and independently a vertex $v \in V$, query *all* edges that are incident to each of the two vertices, that is, at most $2d$ edges are being checked. If u or v produce a witness stop, otherwise repeat this process until done.

Claim 14.27. In each iteration, the probability that a witness is discovered is at least $1/(d + 1)$.

Proof. Let H be a subgraph of G determined before the i-th iteration. That is, H consists of all edges e of G, the query to which was answered as "$x_e = 1$". If there are more than $2nd/(d + 1)$ edges in H, then at least $n/(d + 1)$ of the n vertices in V are of degree at least 2. In this case the fact that $v \in V$ is chosen at random proves the claim. If, on the other hand, H has less then $2nd/(d + 1)$ edges, then at least $2n/(d + 1)$ of the vertices in U are of degree 0 in H. Thus, the fact that $u \in U$ is chosen at random proves the claim in this case.   □

We get that the expected number of iterations is $d + 1$, in each of them at most $2d$ edges are probed which yields the desired upper bound on the expected depth of the tree.   □

14.7   Linear Decision Trees

In standard decision trees we have considered so far, at each node a test on a single variable is made. If the result is 0, one descends into the left subtree, whereas if the result is 1, one descends into the right subtree. The label of the leaf so reached is the value of the function (on that particular input). We will now consider decision trees where more general test are allowed.

A *real threshold function* is a boolean function $f : \{0, 1\}^n \to \{0, 1\}$ for which there exists real numbers $a_1, \dots, a_n, b$ such that, for every vector $x \in \{0, 1\}^n$,

$$f(x) = 1 \text{ if and only if } \textstyle\sum_{i=1}^{n} a_i x_i \geq b;$$

the sum here is over the reals. In a *linear decision tree*, at each node a real threshold function on the entire input vector x is evaluated. If the result is 0 one descends into the left subtree, whereas if the result is 1, one descends into the right subtree. Note that the decision trees considered above correspond to simplest threshold functions $f(x) = 1$ iff $x_i \geq 1$.

By a *rectangle* we will mean a cartesian product $R = X \times Y$ of two subsets of vectors $X, Y \subseteq \{0, 1\}^n$; its *dimension* is $\dim(R) = \min\{|X|, |Y|\}$. A boolean function on such a rectangle is a mapping $f : X \times Y \to \{0, 1\}$. A function $f(x, y)$

is *monochromatic* on a subset $S \subseteq X \times Y$ if it takes the same value on all inputs $(x, y) \in S$. For a boolean function $f : X \times Y \to \{0, 1\}$, let mono$(f)$ denote the maximal dimension of a subrectangle $X' \times Y' \subseteq X \times Y$ on which f takes the same value.

We already know (see Lemma 11.33) that if $f(x, y)$ is a real threshold function defined on a rectangle of dimension k, then mono$(f) \geq k/2$.

Theorem 14.28. (Gröger–Turán 1991) *If $f(x, y)$ is a boolean function of $2n$ variables, then any linear decision tree computing f must have depth at least $n - \log \operatorname{mono}(f)$.*

Proof. Let $X = Y = \{0, 1\}^n$ and $f : X \times Y \to \{0, 1\}$. Consider a linear decision tree T computing $f(x, y)$. We define a sequence of rectangles $R_i \subseteq X \times Y$ such that $\dim(R_i) \geq 2^{n-i}$ and inputs in R_i follow the same path in T. To do this, let $R_0 := X \times Y$, assume that R_i is defined and let v_i be the node of T where the inputs in R_i arrive after i test are evaluated (thus v_0 is the root). Assume that v_i is *not* a leaf, and let f_i be the linear test made in v_i. Apply Lemma 11.33 to $g_i : R_i \to \{0, 1\}$ to get a subrectangle $R_{i+1} \subseteq R_i$ of dimension $(2^{n-i})/2 = 2^{n-(i+1)}$ on which f_i is constant. Clearly all inputs in R_{i+1} follow the same path of length $i + 1$ in T.

Now assume that v_i is a leaf. Then f must be constant on R_i, implying that $2^{n-i} \leq \dim(R_i) \leq \operatorname{mono}(f)$, from which the lower bound $2^i \geq 2^n/\operatorname{mono}(f)$, and hence, also the desired lower bound $i \geq n - \log \operatorname{mono}(f)$ on the depth of the tree follows. $\qquad\square$

The *inner product function* is a boolean function $IP_n(x, y)$ of $2n$ variables defined by: $IP_n(x, y) = 1$ if and only if $\sum_{i=1}^{n} x_i y_i \bmod 2 = 1$. That is, $IP_n(x, y) = 1$ if and only if the vectors x and y share an odd number of common 1-coordinates. Note that IP_n is a boolean function on the rectangle $X \times Y$ of dimension $N = 2^n$ with $X = Y = \{0, 1\}^n$.

We already know (see the proof of Corollary 11.35) that mono$(IP_n) \leq 2^{n/2}$. Together with Theorem 14.28, this implies that IP_n requires deep linear decision trees.

Corollary 14.29. *Every linear decision tree computing the inner product function IP_n requires depth at least $n/2$.*

Gröger and Turán (1991) also proved a similar lower bound of $\Omega(n)$ for *randomized* linear decision trees.

14.8 Element Distinctness and Turán's Theorem

Let D be some finite domain. We consider general decision trees computing functions $f : D^n \to \{0, 1\}$. We have n variables $x_1, \ldots, x_n$ taking their values in D. At each node of a decision tree an arbitrary function $g : D^2 \to \{0, 1\}$ may be computed. If the result of the test $g(x_i, x_j)$ is 0 one descends into the left subtree, whereas if the result is 1, one descends into the right subtree.

The *element distinctness function* over D is the function $\mathrm{ED}_n : D^n \to \{0, 1\}$ such that, for every input string $a \in D^n$, $\mathrm{ED}_n(a) = 1$ iff $a_i \neq a_j$ for all positions $i \neq j$. We restrict our domain to $D = [n] := \{1, 2, \ldots, n\}$. Note that in this case, ED_n accepts a string $a \in [n]^n$ iff a is a permutation of $[n]$.

Theorem 14.30. (Boppana 1994) *Any general decision tree computing* ED_n *over the domain* $D = \{1, \ldots, n\}$ *must have depth at least* $\Omega(n \sqrt{\log n})$.

Proof. Given a decision tree T for ED_n and one of its leaves l, define the *computation graph* G_l as follows. The vertex set is $[n]$, and an edge is placed between i and j iff on the paths to l there is a node at which a test $g(x_i, x_j)$ or a test $g(x_j, x_i)$ for some function g is made. Let $\alpha(G_l)$ denote the size of a largest independent set in G_l, and let $A = \{a \in [n]^n \mid \mathrm{ED}_n(a) = 1\}$ be the set of all accepted inputs.

Claim 14.31. *If* l *is a 1-leaf then at most* $n!/\alpha(G_l)!$ *inputs* $a \in A$ *can reach* l.

Proof. Fix a largest independent set S in G_l, of size $s = \alpha(G_l)$. Say that two inputs $a, b \in A$ are *equivalent* if $a_i = b_i$ for all $i \notin S$. Notice that this equivalence relation partitions the set A into

$$n(n-1) \cdots (n - |\overline{S}| + 1) = n(n-1) \cdots (s+1) = \frac{n!}{s!}$$

equivalence classes, one for each setting of distinct values outside S. Therefore, it suffices to show that no two equivalent inputs reach the same leaf l.

Assume, for the sake of contradiction, that some two equivalent inputs $a \neq b \in A$ reach l. Let $k \in S$ be a position for which $a_k \neq b_k$. Consider the input c that equals a on all positions except the k-th one, and equals b on the k-th position, that is, $c_k = b_k$ and $c_i = a_i$ for all $i \neq k$. The input c must be rejected by our decision tree, since some two of its positions must be equal (both a and b were permutations of $[n]$). To obtain the desired contradiction, we will now show that input c reaches the leaf l too.

To show this, note that since c differs from a on only position k, the only place the computation on c can diverge from that on a is at a node at which a test $g(x_i, x_k)$ or $g(x_k, x_i)$ is made. Since the set S is independent, and since $k \in S$, the position i must be outside S. But by the definition of c, and since a and b are equivalent ($a_i = b_i$ for all $i \notin S$), it follows that

$$(c_i, c_k) = (a_i, b_k) = (b_i, b_k).$$

In other words, the computation on c follows the same direction as b does. Since b reaches the leaf l, c will reach l too. This gives the desired contradiction. □

To finish the proof of the theorem, we will use the following celebrated theorem of Turán (1941) which states that a sparse graph contains a large independent set (see Exercise 14.11 for the proof): if G is a graph with n vertices and m edges, then

$$\alpha(G) \geq \frac{n^2}{2m+n}. \tag{14.6}$$

Now let T be a decision tree for the element distinctness function ED_n, and let h be its depth. Then, for every leaf l of T, the computation graph G_l has at most h edges. By Claim 14.31 and by Turán's theorem, every leaf l of T can be reached by at most $n!/s!$ inputs $a \in A$, where $s := n^2/(2h+n)$. Since there are at most 2^h leaves in T, it follows that T can accept at most $2^h n!/s!$ inputs $a \in A$. Since there are $|A| = n!$ such inputs in total, we obtain the inequality $2^h n!/s! \geq |A| = n!$, and hence, the inequality $2^h \geq s!$. Using Stirling's formula $s! \geq (s/e)^s$, and taking logarithms, we obtain the inequality

$$h \geq s \log(s/e) = \frac{n^2}{2h+n} \log \frac{n^2}{(2h+n)e}.$$

Solving for h, we find that $h = \Omega(n\sqrt{\log n})$, as desired. $\square$

14.9   P ≠ NP ∩ co-NP for Decision Tree Size

The *size* of a decision tree is the number of all its leaves. Let $\mathrm{Size}(f)$ denote the minimum size of a *deterministic* decision tree computing f. The minimum size of a *nondeterministic* decision tree for f is denoted by $\mathrm{dnf}(f)$. Note that $\mathrm{dnf}(f)$ is just the minimal number of monomials in a DNF of f. That is, $\mathrm{dnf}(f)$ is the minimal number t such that f can be written as an Or of t monomials.

We already know that P = NP ∩ co-NP for decision trees if we consider their *depth* as complexity measure. In this section we will show that the situation changes drastically if we consider their size (the total number of nodes) instead of the depth: in this case we have P ≠ NP ∩ co-NP. Namely, there are explicit boolean functions f such that *both* f and its negation $\neg f$ have nondeterministic decision trees of size N, whereas the size of any deterministic decision tree for f is $N^{\Omega(\log N)}$.

Let f be a boolean function, and suppose we know that $\mathrm{dnf}(f)$ is small. Is then the decision tree also small? The following examples show that it may be *not* the case:

$$f = \bigvee_{i=1}^{m} \bigwedge_{j=1}^{m} x_{ij}.$$

It can be shown that $\mathrm{Size}(f) \geq 2^{\mathrm{dnf}(f)}$ (Exercise 14.9). This shows that P ≠ NP for decision tree size. Well, this function has very small DNF (of size m) but the DNF of its negation

$$\neg f = \bigwedge_{i=1}^{m} \bigvee_{j=1}^{m} \neg x_{ij}$$

is huge—it has m^m monomials. It is therefore natural to ask what happens if *both* the function f *and* its negation $\neg f$ have small DNFs? Put differently, does $P = NP \cap$ co-NP for decision trees if we consider the *size* as their complexity measure? Below we answer this question negatively.

The sum $N(f) := \mathrm{dnf}(f) + \mathrm{dnf}(\neg f)$ will be called the *weight* of f. It is clear that $N(f) \le \mathrm{Size}(f)$ (just because every decision tree represents *both* the function and its negation). But what about the other direction: is $\mathrm{Size}(f)$ polynomial in $N(f)$? For a long time only a *quasi-polynomial* relation was known: the decision tree size of any boolean function is *quasi-polynomial* in its weight.

Theorem 14.32. (Ehrenfeucht and Haussler 1989) *Let f be a boolean function of n variables and $N = \mathrm{dnf}(f) + \mathrm{dnf}(\neg f)$ be its weight. Then* $\mathrm{Size}(f) \le n^{\mathcal{O}(\log^2 N)}$.

Proof. (Due to Petr Savický) The idea is to apply the following simple "greedy" strategy: given DNFs for f and $\neg f$, let the decision tree *always* test the "most popular" literal first.

Assume, we have DNFs for both f and $\neg f$, and let N be the total number of monomials in these two DNFs. Since the disjunction of these two DNFs is a tautology (that is, outputs 1 on all inputs), there must exist a monomial of length at most $\log N$, just because monomial of length k accepts only 2^{n-k} of the inputs.

Select one of such monomials and denote its length by k. The selected monomial belongs to one of the two DNFs. By the cross-intersection property of monomials (see Exercise 14.3), every monomial in the other DNF contains at least one literal which is contradictory to at least one literal in the selected monomial. Hence, there is a literal in the selected monomial, which is contradictory to at least a $1/k$-portion of the monomials in the other DNF. Thus, if we evaluate this literal to 1, then all these monomials will get the value 0 and so will disappear from the DNF.

Test this variable first and apply this strategy recursively to both restrictions which arise. By the observation we just made, for each node v, at least one of its two successors is such that at least one of the two DNFs in it decreases by a factor of $1 - 1/k$. Let us call the corresponding outgoing edge(s) *decreasing*. Now, if v is a node (not a leaf) such that the path from the source to v contains s decreasing edges, at least one of the two initial DNFs was decreased at least $s/2$ times, and each time it was decreased by a factor of $1 - 1/k \ge 1 - 1/\log N$. If s were at least $2\log^2 N$ then at least one of the DNFs at v would have only

$$N \left(1 - \frac{1}{\log N} \right)^{s/2} < N \cdot e^{-s/(2\log N)} \le N \cdot e^{-\log N} = N^{1 - \log e} < 1$$

monomials, which is impossible (because v is not a leaf). Thus, *every* path to a leaf has at most n edges, and among them at most $s := 2\log^2 N$ can be decreasing. Recall that for *every* node at least one of the out-going edges was decreasing.

Assume w.l.o.g. that every node has *exactly* one decreasing edge (if there were two, we simply ignore one of them). Mark decreasing edges by 1 and the remaining edges by 0. Then every leaf corresponds to a 0–1 string of length at most n with at most s ones. The number of such strings (and hence, the total number of leaves) does not exceed $L(n, s/2)$, where $L(n, t)$ denotes the maximal possible number of leaves in a decision tree of depth n such that *every* path from the root to a leaf has at most t 1-edges.

It remains to estimate $L(n, t)$ for $t = s/2$. Clearly, we have the following recurrence:

$$L(n, t) \leq L(n - 1, t) + L(n - 1, t - 1) \text{ with } L(0, t) = L(n, 0) = 1. \quad (14.7)$$

By induction on n and t, it can be shown that

$$L(n, t) \leq \sum_{i=0}^{t} \binom{n}{i} \leq \left(\frac{ne}{t}\right)^t.$$

Indeed, using the identity $\binom{n-1}{k} + \binom{n-1}{k-1} = \binom{n}{k}$, the induction hypothesis together with the recurrence (14.7) yields:

$$L(n, t) \leq L(n - 1, t) + L(n - 1, t - 1) \leq \sum_{i=0}^{t} \binom{n-1}{i} + \sum_{i=0}^{t-1} \binom{n-1}{i}$$

$$= 1 + \sum_{i=1}^{t} \left[\binom{n-1}{i} + \binom{n-1}{i-1}\right] = 1 + \sum_{i=1}^{t} \binom{n}{i} = \sum_{i=0}^{t} \binom{n}{i}.$$

Thus,

$$\text{Size}(f) \leq L(n, s/2) = L(n, \log^2 N) \leq \left(\frac{n}{\log^2 N}\right)^{O(\log^2 N)}. \qquad \square$$

■ **Research Problem 14.33.** Is it possible to improve the upper bound $\text{Size}(f) \leq n^{O(\log^2 N)}$ in Theorem 14.32 to $\text{Size}(f) \leq 2^{O(\log^2 N)}$?

In the next section we will exhibit explicit boolean functions f requiring deterministic decision trees of size $N^{\Omega(\sqrt{\log N})}$ (iterated majority function) and even $N^{\Omega(\log N)}$ (iterated NAND function), where $N = \text{dnf}(f) + \text{dnf}(\neg f)$ its the weight of f. Namely, Jukna et al. (1999) proved the following lower bound.

Theorem 14.34. *There are explicit boolean functions f such that both f and $\neg f$ have DNFs of size N, but any deterministic decision tree for f has size $N^{\Omega(\log N)}$.*

That is, for the size of decision trees we have that P ≠ NP ∩ co-NP. The rest of this section is devoted to the proof of this theorem. For this purpose, we will use an argument which has many applications in engineering. The argument is based on harmonic analysis of boolean functions, and is known as the "spectral argument".

14.9.1   Spectral Lower Bound

We will use Fourier transforms over the field $\mathbb{R}$ of real numbers (see Sect. 2). For this, it will be convenient to switch to $(-1, +1)$-notation, that is, to consider boolean functions as mappings from $\{-1, +1\}^n$ to $\{-1, +1\}$, where the correspondence $1 \rightarrow -1$ and $0 \rightarrow +1$ is assumed. To convert from the standard $0/1$ representation to the Fourier ± 1 representation map $x \mapsto 1 - 2x = (-1)^x$. To convert from the Fourier ± 1 representation to the standard $0/1$ representation map $x \mapsto (1-x)/2$; hence, $\neg x$ maps to $(1+x)/2$.

Example 14.35. Suppose $n = 3$ and f is the Majority function Maj_3. So, in the ± 1 notation we have that $\text{Maj}_3(1, 1, 1) = 1$, $\text{Maj}_3(1, 1, -1) = 1, \ldots$, $\text{Maj}_3(-1, -1, -1) = -1$. Denoting $x = (x_1, x_2, x_3)$, we can write

$$\text{Maj}_3(x) = \left(\frac{1+x_1}{2}\right)\left(\frac{1+x_2}{2}\right)\left(\frac{1+x_3}{2}\right) \cdot (+1)$$

$$+ \left(\frac{1+x_1}{2}\right)\left(\frac{1+x_2}{2}\right)\left(\frac{1-x_3}{2}\right) \cdot (+1)$$

$$+ \quad \cdots$$

$$+ \left(\frac{1-x_1}{2}\right)\left(\frac{1-x_2}{2}\right)\left(\frac{1-x_3}{2}\right) \cdot (-1).$$

If we actually expand out all of the products, tremendous cancelation occurs and we get

$$\text{Maj}_3(x) = \frac{1}{2}x_1 + \frac{1}{2}x_2 + \frac{1}{2}x_3 - \frac{1}{2}x_1 x_2 x_3. \tag{14.8}$$

We could do a similar interpolate-expand-simplify procedure even for a function $f : \{-1, 1\}^n \rightarrow \mathbb{R}$, just by multiplying each x-interpolator by the desired value $f(x)$. Note that after expanding and simplifying, the resulting polynomial will always be *multilinear*, that is, have no variables squared, cubed, etc. In general, a multilinear polynomial over variables $x_1, \ldots, x_n$ has 2^n terms, one for each monomial

$$\chi_S(x) := \prod_{i \in S} x_i \,,$$

where $S \subseteq [n] = \{1, \ldots, n\}$; for $S = \emptyset$ this monomial is constant 1. Hence, every function $f : \{-1, 1\}^n \rightarrow \mathbb{R}$ can be expressed (in fact, even uniquely) as a multilinear polynomial

$$f(x) = \sum_{S \subseteq [n]} c_S \cdot \chi_S(x) \,, \tag{14.9}$$

where each c_S is a real number. The S-th coefficient c_S is call the S-th Fourier coefficient of f, and is usually denoted by $\widehat{f}(S)$. It can be computed as

$$\widehat{f}(S) = 2^{-n} \sum_x f(x)\chi_S(x).$$

Proposition 14.36. *If f does not depend on the i-th variable, then $\widehat{f}(S) = 0$ for every S with $i \in S$.*

Proof. For a vector $x \in \{-1,1\}^n$ and a coordinate $i \in [n]$, let $x^{(i)}$ denote the vector x with its i-th coordinate x_i replaced by $-x_i$. If $i \in S$, then we have that $f(x^{(i)}) = f(x)$ but $\chi_S(x^{(i)}) = -\chi_S(x)$, implying that $\sum_x f(x)\chi_S(x) = 0$. $\square$

This proposition allows us to compute Fourier coefficients of arithmetic combination of some functions with disjoint sets of variables.

Proposition 14.37. *Let $S = S_1 \cup S_2$ be a partition of S into two disjoint nonempty blocks. Let $g, h : \{-1,1\}^S \to \{-1,1\}$ be functions such that g only depends on variables x_i with $i \in S_1$, and h only depends on variables x_i with $i \in S_2$. Then*

$$\widehat{f}(S) = \begin{cases} 0 & \text{if } f = g + h; \\ \widehat{g}(S_1) \cdot \widehat{h}(S_2) & \text{if } f = g \cdot h. \end{cases}$$

We leave the proof of this as an exercise.

The following general lower bound on the size of decision trees is a combination of Lemma 4 in Linial et al. (1993) with Lemma 5.1 of Kushilevitz and Mansour (1991).

Lemma 14.38. (Spectral Lower Bound) *For every boolean function f of n variables and every subset of indices $S \subseteq \{1, \ldots, n\}$ we have the bound*

$$\text{Size}(f) \geq 2^{|S|} \cdot \sum_{T \supseteq S} |\widehat{f}(T)|. \tag{14.10}$$

Proof. Take a decision tree for f of size $\text{Size}(f)$. For a leaf l, let $\text{val}(l) \in \{-1, +1\}$ be its label (recall that we are in ± 1-notation), and let I_l be the set of indices of those variables, which are tested on the path to l. Let $B_l \subseteq \{-1, +1\}^n$ be the set of all the inputs that reach leaf l; hence, $|B_l| = 2^{n-|I_l|}$.

Since each input reaches a *unique* leaf, the sets B_l are pairwise disjoint. Hence, for every $T \subseteq [n]$,

$$\widehat{f}(T) = 2^{-n} \sum_x f(x) \cdot \chi_T(x) = 2^{-n} \sum_l \sum_{x \in B_l} f(x) \cdot \chi_T(x) = \sum_l \text{val}(l) \cdot \Delta(T, l),$$

where

$$\Delta(T,l) := 2^{-n} \sum_{x \in B_l} \chi_T(x).$$

Now, if $T \not\subseteq I_l$, that is, if some variable x_i with $i \in T$ is not tested along the path from the root to the leaf l, then $\chi_T(x) = +1$ for exactly half of the inputs $x \in B_l$, and hence, $\Delta(T,l) = 0$. If $T \subseteq I_l$ then the value of χ_T is fixed on B_l to either $+1$ or -1, and so,

$$|\Delta(T,l)| = 2^{-n} \cdot |B_l| = 2^{-|I_l|}.$$

Thus, in both cases, $|\Delta(T,l)| \le 2^{-|I_l|}$. Since for any $S \subseteq [n]$ there are only $2^{|I_l|-|S|}$ sets T satisfying $S \subseteq T \subseteq I_l$, we conclude that

$$\sum_{T:T \supseteq S} |\hat{f}(T)| \le \sum_{T:T \supseteq S} \sum_l |\Delta(T,l)| = \sum_l \sum_{T:T \supseteq S} |\Delta(T,l)|$$

$$\le \sum_l 2^{-|S|} = 2^{-|S|} \cdot \text{Size}(f),$$

and the desired bound (14.10) follows.   □

We are going to apply Lemma 14.38 for $S = [n]$ to the Iterated Majority function and for $S = \emptyset$ to the Iterated NAND function.

14.9.2 Explicit Lower Bounds

Recall that our goal is to exhibit a boolean function f which requires decision tree of size super-polynomial in its weight $N = \text{dnf}(f) + \text{dnf}(\neg f)$. For this purpose, we take the Iterated Majority function which is defined as follows.

The majority of three boolean variables is given by

$$\text{Maj}_3(x_1, x_2, x_3) = x_1 x_2 \vee x_1 x_3 \vee x_2 x_3.$$

In Example 14.35 we have shown that in the $(-1, +1)$-representation (that is, when the correspondence $1 \to -1$ and $0 \to +1$ is assumed) we have that

$$\text{Maj}_3(x_1, x_2, x_3) = \frac{1}{2}x_1 + \frac{1}{2}x_2 + \frac{1}{2}x_3 - \frac{1}{2}x_1 x_2 x_3.$$

Now consider the monotone function F_h in $n = 3^h$ variables which is defined by the balanced read-once formula of height h in which every gate is Maj_3, the majority of three variables. That is, $F_0 = x$, $F_1 = \text{Maj}_3(x_1, x_2, x_3)$ and for $h \ge 2$,

$$F_h = \text{Maj}_3(F_{h-1}^{(1)}, F_{h-1}^{(2)}, F_{h-1}^{(3)}) \tag{14.11}$$

where $F_{h-1}^{(v)}$ are three copies of F_{h-1} with disjoint(!) sets of variables.

Theorem 14.39. *Let $f = F_h$ be the iterated majority function and $N = \mathrm{dnf}(f) + \mathrm{dnf}(\neg f)$ be its weight. Then $\mathrm{Size}(F_h) \geq N^{\Omega(\sqrt{\log N})}$.*

Proof. It can be shown (Exercise 14.14) that the function $F_h(x_1, x_2, \ldots, x_n)$ has $n = 3^h = 2^{c \cdot h}$ variables, where $c = \log 3 > 3/2$, and has weight

$$N = 2 \cdot 3^{2^h - 1} = 2^{\Theta(2^h)} = 2^{\Theta(n^{2/3})}.$$

Since $2^{\Omega(n)} \geq 2^{\Omega(\log^{3/2} N)} = N^{\Omega(\sqrt{\log N})}$, it is enough to prove the lower bound $\mathrm{Size}(F_h) \geq 2^{\Omega(n)}$. To prove this, we will apply Lemma 14.38 with $S = [n] = \{1, \ldots, n\}$. Letting

$$a_h := \left| \widehat{F}_h([n]) \right|$$

denote the absolute value of the leading Fourier coefficient of F_h, this lemma yields $\mathrm{Size}(F_h) \geq a_h \cdot 2^n$. It remains therefore to prove an appropriate lower bound on a_h. We proceed by induction on h.

Clearly, $a_0 = 1$, since F_0 is a variable (cf. Exercise 14.13), and $a_1 = 1/2$ by the above representation of Maj_3.

For the inductive step recall that in the $(-1, +1)$-representation,

$$\mathrm{Maj}_3(x_1, x_2, x_3) = \frac{1}{2}\left(\sum_{i=1}^{3} x_i - \prod_{i=1}^{3} x_i \right).$$

Thus,

$$F_h = \frac{1}{2} \sum_{v=1}^{3} F_{h-1}^{(v)} - \frac{1}{2} \prod_{v=1}^{3} F_{h-1}^{(v)}.$$

By Proposition 14.37, the first summand does not contribute to $\widehat{F}_h([n])$ and we obtain that

$$a_h = \frac{1}{2} a_{h-1}^3.$$

Together with the condition $a_0 = 1$, this recursion resolves to

$$a_h = 2^{-3^0} \cdot a_{h-1}^3 = 2^{-3^0 - 3^1} \cdot a_{h-2}^3 = 2^{-3^0 - 3^1 - 3^2} \cdot a_{h-3}^3 = \ldots = 2^{-\Delta},$$

where

$$\Delta = 3^0 + 3^1 + 3^2 + \cdots + 3^{h-1} = \frac{3^h - 1}{3 - 1} = (3^h - 1)/2 = (n - 1)/2.$$

Thus

$$\mathrm{Size}(F_h) \geq a_h \cdot 2^n \geq 2^{-(n-1)/2} \cdot 2^n = 2^{(n+1)/2},$$

as desired. $\square$

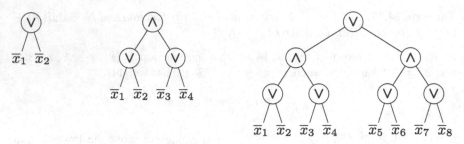

Fig. 14.2 Iterated NAND functions G_1, G_2 and G_3

We now present a boolean function almost matching the upper bound in Theorem 14.32.

The *iterated NAND function* is a boolean function G_h in $n = 2^h$ variables which is computed by the balanced read-once formula of height h in which every gate is NAND, the negated AND operation $NAND(x, y) = \neg(x \wedge y) = \neg x \vee \neg y$. Up to complementation of the inputs this is equivalent to a monotone read-once formula with alternating levels of AND and OR gates (see Fig. 14.2).

Theorem 14.40. *Let $f = G_h$ be the iterated majority function and $N = \mathrm{dnf}(f) + \mathrm{dnf}(\neg f)$ be its weight. Then* $\mathrm{Size}(f) \geq N^{\Omega(\log N)}$.

Proof. We have that $\mathrm{dnf}(G_0) = \mathrm{dnf}(\neg G_0) = 1$ (since G_0 is a single variable), and it is easy to see that for every $h \geq 1$ we have $\mathrm{dnf}(G_h) \leq 2 \cdot \mathrm{dnf}(\neg G_{h-1})$ and $\mathrm{dnf}(\neg G_h) \leq \mathrm{dnf}(G_{h-1})^2$. By induction on h one obtains $\mathrm{dnf}(G_h) \leq 2^{2^{(h+1)/2}-1}$ and $\mathrm{dnf}(\neg G_h) \leq 2^{2^{(h/2)+1}-2}$. Since $n = 2^h$, we have $N \leq 2^{2^{(h/2)+1}} = 4^{\sqrt{n}}$, and our statement boils down to showing $\mathrm{Size}(G_h) \geq 2^{\Omega(n)}$.

Let us say that a Fourier coefficient $\widehat{G}_h(S)$ is *dense* if for every subtree of height 2, S contains the index of at least one of the four variables in that subtree. We are going to calculate exactly the sum of absolute values of dense coefficients. Denote this sum by c_h. Note that in the $(-1, +1)$-representation, we have $NAND(x, y) = (xy - x - y - 1)/2$. Hence,

$$G_h = \frac{1}{2}\left(G_{h-1}^{(1)} \cdot G_{h-1}^{(2)} - G_{h-1}^{(1)} - G_{h-1}^{(2)} - 1\right), \qquad (14.12)$$

where $G_{h-1}^{(1)}$, $G_{h-1}^{(2)}$ are two copies of G_{h-1} with disjoint sets of variables.

In order to compute c_2, we use the following transformation. Let $f_1 = G_1^{(1)} + 1/2$ and $f_2 = G_1^{(2)} + 1/2$. Then it follows from (14.12) that

$$G_2 = \frac{1}{2}f_1 f_2 - \frac{3}{4}f_1 - \frac{3}{4}f_2 + \frac{1}{8}.$$

Since each monomial in f_1 and f_2 contains at least one variable and the sets of variables of f_1 and f_2 are disjoint, there are no common monomials in the four terms in the above expression of G_2. Hence, it is easy to calculate the sum of the absolute values of the coefficients in the non-constant monomials, which is $c_2 = 1/2 \cdot r_1 \cdot r_2 + 3/4 \cdot (r_1 + r_2) = 27/8 = 3.375$, where $r_1 = r_2 = 3/2$ is the sum of the absolute values of the coefficients in f_1 and f_2.

In order to compute c_h for $h > 2$, we use (14.12) directly. Only the first term $G_{h-1}^{(1)} \cdot G_{h-1}^{(2)}$ in this equation can contribute to dense coefficients, and its individual contributions do not cancel each other. Hence, we have the recursion

$$c_h = \frac{1}{2} c_{h-1}^2.$$

This resolves to $c_h = 2(c_2/2)^{2^{h-2}}$ which is $2^{\Omega(n)}$ since $c_2 > 2$. The proof is now completed by applying Lemma 14.38 (this time with $S = \emptyset$). $\square$

Exercises

14.1. Consider the following function $f(X)$ on $n = m^2$ boolean variables:

$$f = \bigwedge_{i=1}^{m} \bigvee_{j=1}^{m} x_{ij}. \tag{14.13}$$

Show that for this function f we have that $C_0(f) = C_1(f) = m$ but $D(f) = m^2$.
Hint: Take an arbitrary deterministic decision tree for f and construct a path from the root by the following "adversary" rule. Suppose we have reached a node v labeled by x_{ij}. Then follow the outgoing edge marked by 1 if and only if *all* the variables x_{il} with $l \neq j$ were already tested before we reached the node v.

14.2. Let $f : \{0,1\}^n \to \{0,1\}$ be a boolean function, and let $k = k(f)$ be the largest natural number such that $|f^{-1}(0)|$ is divisible by 2^k. Show that $D(f) \geq n - k(f)$.
Hint: The number of inputs $x \in f^{-1}(0)$ leading to a given leaf of depth d is either 0 or 2^{n-d}.

14.3. Let D_1 be a DNF of a boolean function f, and D_2 be a DNF of its negation $\neg f$. Show the following *cross-intersection property*: if K is a monomial in D_1, then every monomial in D_2 contains at least one literal which is contradictory to at least one literal in K.

14.4. Show that $s(f) = bs(f) = C(f)$ for every *monotone* boolean function f.
Hint: Since $s(f) \leq bs(f) \leq C(f)$ holds for any boolean function f, it is enough to show that $C(f) \leq s(f)$. Take a vector a with $C(f,a) = C(f)$, assume that $f(a) = 0$. Let S be an f-certificate of a of smallest size, and let b be the vector a with all bits outside S set to 1. Show that f is sensitive on b to each bit $i \in S$.

14.5. Show that for every boolean function f, $\deg(f) \leq D(f)$.

Hint: The tests along paths to 1-leaves define a multilinear polynomial.

14.6. Recall that a boolean function $f(x)$ of n variables is *symmetric* if its value only depends on the number $|x| := x_1 + \cdots + x_n$ of ones in x. That is, there exists a subset $L \subseteq \{0, 1, \ldots, n\}$ such that $f(x) = 1$ if and only if $|x| \in L$. Show that symmetric functions have high sensitivity: $s(f) \geq (n + 1)/2$ holds for every non-constant symmetric boolean function f of n variables.

Hint: We can assume (why?) that there is a non-negative integer $k \leq (n - 1)/2$ such that $k \notin L$ and $k + 1 \in L$. Take a vector a with $|a| = k$ ones, and argue that $s(f, a) \geq n - k$.

14.7. Recall that the *average sensitivity*, $as(f)$, of a boolean function f is the expected sensitivity of f on a random assignment: $as(f) := 2^{-n} \sum_a s(f, a)$. Let T be a decision tree, and let p_a be the length of the unique path in it consistent with a. The *average depth* of T is $2^{-n} \sum_a p_a$. Show that the average depth of any decision tree for f is at least $as(f)$.

14.8. Let $L(f)$ denote the minimum leafsize of a DeMorgan formula computing f. Let $as(f)$ be the average sensitivity of f, and let $p = |f^{-1}(1)|/2^n$. Show that

$$L(f) \geq \frac{as(f)^2}{4p(1 - p)}.$$

Hint: Theorem 6.27 and Exercise 2.28.

14.9. Show that, for the boolean function f defined by (14.13), we have that $Size(f) \geq 2^{dnf(f)}$.

Hint: Observe that all the minterms and maxterms of f have length m. Show that every such function requires a decision tree of size at least 2^m.

14.10. Let $G = ([n], E)$ be a graph on n vertices and let d_i denote the degree of the i-th vertex. Prove that

$$\alpha(G) \geq \sum_{i=1}^{n} \frac{1}{d_i + 1}. \tag{14.14}$$

Hint: Let $\pi : [n] \to [n]$ be a random permutation taking its values uniformly and independently with probability $1/n!$. Let A_i be the event that $\pi(j) > \pi(i)$ for all d_i neighbors j of i. Show that

$$\text{Prob}[A_i] = \binom{n}{d_i + 1} \frac{d_i!(n - d_i - 1)!}{n!} = \frac{1}{d_i + 1}.$$

Let U be the (random) set of those vertices i for which A_i holds. Show that $E[|U|] = \sum_{i=1}^{n} 1/(d_i + 1)$. Fix such a set U, and show that for every edge $\{i, j\}$ of G, either $\pi(i) < \pi(j)$ or $\pi(j) < \pi(i)$.

14.11. (Turán's theorem) Derive from Exercise 14.10 the Turán theorem: If a graph G has n vertices and $nk/2$ edges, then $\alpha(G) \geq n/(k + 1)$. Show that this is equivalent to (14.6).

Hint: Fixing the total number of edges, the sum $\sum_{i=1}^{n} 1/(d_i + 1)$ is minimized when the d_i's are as nearly equal as possible. By Euler's theorem $\sum_{i=1}^{n} d_i$ is exactly two times the number of edges in G.

14.12. Let f be a monotone boolean function. Suppose that every minterm of f has length $\geq s$ and every maxterm has length $\geq r$. Show that any decision tree computing f must have at least $\binom{s+r}{s}$ leaves. *Hint*: $\binom{n}{k} = \binom{n-1}{k} + \binom{n-1}{k-1}$.

14.13. Let $f = x_i$ be a single variable. Show that $\widehat{f}(\{i\}) = 1$.

14.14. Show that the iterated majority function F_h, defined by (14.11), has $n = 3^h$ variables and its weight is $2 \cdot 3^{2^h - 1}$.

Hint: Observe that: (1) dnf$(F_0) = 1$ and dnf$(F_h) = 3 \cdot$ dnf$(F_{h-1})^2$, and (2) the minimal DNF of the negation $\neg F_h$ coincides with the DNF of F_h with all the variables negated.

14.15. A $\vee$-*decision tree* is a generalization of a deterministic decision tree, where at each node an OR $g(x) = \bigvee_{i \in S} x_i$ of some subset S of variables can be tested. Hence, decision trees correspond to the case when $|S| = 1$. Consider the threshold-k function $\text{Th}_k^n(x_1, \ldots, x_n) = 1$ if and only if $x_1 + \cdots + x_n \geq k$. Show that any $\vee$-decision tree for Th_k^n requires at least $\binom{n}{k-1}$ leaves.

Hint: Look at Th_k^n as accepting/rejecting subsets of $[n]$. Suppose that some two different $(k-1)$-element subsets $A, B \subseteq [n]$ reach the same leaf. Show that then also the set $C = A \cup B$ will reach that leaf.

14.16. Consider the search problem for a given relation $F \subseteq \{0, 1\}^n \times W$ (see Sect. 14.6). Our goal is to give a game-theoretic lower bound on the minimum *size*, Size(F), of a decision tree solving the search problem for F. There are two players, Prover and Delayer. Given an input vector $x \in \{0, 1\}^n$, the goal of the Prover is to find a witness $w \in W$ such that $(x, w) \in F$. The goal of Delayer is to delay this happening as long as possible. The game proceeds in rounds. In each round, the Prover suggests a variable x_i to be set in this round, and Delayer either chooses a value 0 or 1 for x_i or leaves the choice to the Prover. In this last case, the Delayer scores one point, but the Prover can then choose the value of x_i. The game is over when a desired witness w is found. Let Score(F) denote the maximal number of points the Delayer can earn in this game independent of what strategy the Prover uses. Prove that Size$(F) \geq 2^{\text{Score}(F)}$.

Hint: Prove the converse direction: if the search problem for F can be solved by a decision tree of size S, then the Prover has a strategy under which the Delayer can earn at most $\log S$ points. If the Delayer defers the choice to the Prover, then let the Prover use the "take the smaller sub-tree" strategy.

Chapter 15
General Branching Programs

A branching program is a generalization of a decision tree where the underlying graph can be an arbitrary directed acyclic graph. The model of branching programs is one of the most fundamental *sequential* (in contrast to *parallel*, as circuits or formulas) model of computations. This model captures in a natural way the deterministic space whereas nondeterministic branching programs do the same for the nondeterministic mode of computation.

In this chapter we first establish almost quadratic lower bounds for general branching programs and show that counting programs are not much weaker than nondeterministic ones. Then we prove a surprising result of Barrington that branching programs of constant(!) width are not much weaker than DeMorgan formulas. Finally, we establish some width versus length bounds for oblivious programs.

15.1 Nechiporuk's Lower Bounds

The best we can do so far for unrestricted programs is a quadratic lower bound $\Omega(n^2/\log^2 n)$ for deterministic programs, and $\Omega(n^{3/2}/\log n)$ for nondeterministic programs. These bounds can be shown by counting arguments due to Nechiporuk (1966): just compare the number of subfunctions with the number of distinct subprograms.

Let $BP(f)$ denote the number of *nodes* in a deterministic branching program, $S(f)$ the number of *contacts* (labeled wires) in a switching network, and $NBP(f)$ the number of *contacts* in a nondeterministic branching program computing f. Recall that a switching network is a nondeterministic branching programs whose underlying graph is undirected; in this case unlabeled wires are redundant.

Let $f(X)$ be a boolean function depending on all its variables. Let $Y_1, \ldots, Y_m$ be disjoint subsets of the variable set X. For every $i \in [m]$, let $c_i(f)$ be the number of distinct subfunctions of f on the variables Y_i obtained by fixing the remaining variables to constants in all possible ways.

S. Jukna, *Boolean Function Complexity*, Algorithms and Combinatorics 27,   439
DOI 10.1007/978-3-642-24508-4_15, © Springer-Verlag Berlin Heidelberg 2012

Theorem 15.1. (Nechiporuk 1966) *There exists a constant $\epsilon > 0$ such that*

$$\text{BP}(f) \geq \text{S}(f) \geq \epsilon \sum_{i=1}^{m} \frac{\log c_i(f)}{\log \log c_i(f)}$$

and

$$\text{NBP}(f) \geq \frac{1}{4} \sum_{i=1}^{m} \sqrt{\log c_i(f)}.$$

The lower bound on $\text{S}(f)$ was proved by Nechiporuk (1966); his argument was then extended to $\text{NBP}(f)$ by Pavel Pudlák (unpublished).

Proof. Let $N(r, h)$ denote the number of switching networks of r variables and with h contacts. We already known (see Lemma 1.21) that the number of graphs with h edges does not exceed $(9h)^h$. Every switching network with h contacts is obtained from one of these graphs by labeling its edges by literals. Since such a labeling can be done in at most $(2r)^h$ ways, we obtain $N(r, h) \leq (9h)^h (2r)^h \leq (18rh)^h$.

Now take a partition $Y_1, \ldots, Y_m$ of variables of f, and take a minimal switching network computing f. For each $i \in [m]$, each setting of constants to variables outside Y_i yields an induced subnetwork whose contacts are the contacts of the original network labeled by variables from Y_i. Say there are h_i such contacts. Then the obtained subnetworks can compute at most $N(|Y_i|, h_i)$ different boolean functions. Since we have $c_i(f)$ such subfunctions, this implies $(18|Y_i| h_i)^{h_i} \geq c_i(f)$, where $|Y_i| \leq h_i$ because f depends on all its variables. We thus obtain that h_i must be a constant times $\log c_i(f) / \log \log c_i(f)$. Since $\text{S}(f) = h_1 + \cdots + h_m$, we are done.

To prove the lower bounds on $\text{NBP}(f)$, let $G(V, E)$ be a nondeterministic branching program computing f. Any fixing of the variables outside Y_i to constants results in a reduced branching program for the resulting subfunction. Let $E_i \subseteq E$ be the set of wires whose labels are literals of variables from Y_i, and let V_i be the set of vertices touched by these wires. Then without loss of generality the reduced program uses only the vertices V_i, on which we have the contacts E_i and perhaps some extra unlabeled wires (switches) that resulted from fixing values. That is, each reduced program is obtained by drawing some additional (to the contacts in E_i) unlabeled wires between the nodes in V_i. Thus, there are at most $2^{|V_i|^2}$ different possible programs, and as $|V_i| \leq 2|E_i|$ and the size of our program is $\sum_{i=1}^{m} |E_i|$, the desired lower bound on $\text{NBP}(f)$ follow.   □

Remark 15.2. Note that the same lower bound as for $\text{NBP}(f)$ holds for nondeterministic branching programs with *any* acceptance mode. In particular, this bound holds for parity branching programs.

Recall that the *element distinctness function* ED_n is a boolean function of $n = 2m \log m$ variables divided into m consecutive blocks with $2 \log m$ variables in each of them. Each of these blocks encode a number in $[m^2]$. The function accepts an input $x \in \{0, 1\}^n$ if and only if all these numbers are distinct.

We have already shown that ED_n has $2^{\Omega(n)}$ subfunctions on each of these $m = \Omega(n/\log n)$ blocks (see Lemma 1.1). Thus, Theorem 15.1 immediately yields

Theorem 15.3. (Nechiporuk 1966) *The element distinctness function* ED_n *requires deterministic branching programs and switching networks of size* $\Omega(n^2/\log^2 n)$, *and nondeterministic branching programs of size* $\Omega(n^{3/2}/\log n)$.

Using a similar argument as in Remark 6.18, one can show that Theorem 15.1 cannot yield larger than $\Omega(n^2/\log^2 n)$ lower bounds on $BP(f)$. Beame and McKenzie (2011) showed that Theorem 15.1 cannot yield larger than $\Omega(n^{3/2}/\log n)$ lower bounds on $NBP(f)$.

To prove this last claim, take an arbitrary partition of the n variables into disjoint subsets $Y_1, \ldots, Y_m$, and consider the sum $s(f) := \sum_{i=1}^{m} \sqrt{\log c_i(f)}$. Using the fact that $c_i(f) \leq \min\{2^{2^{|Y_i|}}, 2^{n-|Y_i|}\}$, the sum $s(f)$ is at most $\sum_{i=1}^{m} h(y_i)$, where $y_i = |Y_i|$ and $h(x) := \min\{2^{x/2}, \sqrt{n-x}\}$. Clearly, $h(x) = 2^{x/2}$ for $x \leq \log(n/2)$, and hence, $h(x) + h(y) \leq h(x+y)$ if $x + y \leq \log(n/2)$. We can therefore assume that at most one y_i is smaller than $t := \frac{1}{2}\log(n/2)$. Such a small y_i has $h(y_i) \leq 2^{t/2} < n^{1/4}$. There are at most n/t larger y_i, and each of them has $h(y_i) \leq \sqrt{n - y_i} \leq \sqrt{n}$. Hence,

$$s(f) \leq \sum_{i=1}^{m} h(y_i) \leq (n/t) \cdot \sqrt{n} + n^{1/4} = \mathcal{O}(n^{3/2}/\log n). \qquad \Box$$

15.1.1 Lower Bounds for Symmetric Functions

If f is a symmetric boolean function of n variables, then f can have at most $n - |Y| + 1$ distinct subfunctions on any set Y of variables. Thus, Nechiporuk's method cannot yield superlinear lower bounds for symmetric functions. A series of lower bounds were proved for such functions using Ramsey-type arguments. Let $BP(f)$ denote the minimal size of a deterministic, and $\oplus BP(f)$ the minimal size of a parity branching program.

- Pudlák (1984) proved that $BP(Maj_n) = \Omega(n \ln \ln n / \ln \ln \ln n)$.
- Babai et al. (1990) improved this to $BP(Maj_n) = \Omega(n \ln n / \ln \ln n)$.
- For switching networks, an intermediate model whose size lies between $BP(f)$ and $NBP(f)$, Grinchuk (1987,1989) proved than Maj_n requires $\Omega(\alpha n)$ contacts where $\alpha = \alpha(n)$ is an extremely slowly growing function.
- In the most powerful model of nondeterministic branching programs, Razborov (1990b) proved $NBP(Maj_n) = \Omega(\beta n)$ where $\beta = \beta(n)$ is also an extremely slowly (but faster than α) growing function.
- Karchmer and Wigderson (1993a) proved $\oplus BP(Maj_n) = \Omega(\beta n)$.

The function β in these lower bounds has the form $\beta(n) = \log\log\log^* n$, where $\log^* n$ is the maximal natural number r such that $t(r) \le n$, where $t(0) = 1$ and $t(x+1) = 2^{t(x)}$.

All these proofs employ quite nontrivial combinatorics, mainly Ramsey theory. The obtained (barely non-linear) lower bounds might be seen as "too weak". The point however is that one cannot expect larger than quadratic bounds for such functions: Lupanov (1965b) proved that, for every sequence (f_n) of symmetric boolean functions, $\text{BP}(f_n) = \mathcal{O}(n^2/\log n)$ and $\text{NBP}(f_n) = \mathcal{O}(n^{3/2})$. Krasulina (1987, 1988) proved that $\text{S}(\text{Maj}_n) = \mathcal{O}(\frac{1}{p}n\ln^4 n)$, where $p = (\ln\ln n)^2$. This was later improved by Sinha and Thathachar (1997) to $\text{BP}(\text{Maj}_n) = \mathcal{O}(\frac{1}{p}n\ln^3 n)$ where $p = (\ln\ln n)(\ln\ln\ln n)$.

Actually, the above mentioned papers of Grinchuk and Razborov give more. Namely, it is possible to completely characterize symmetric boolean functions having size $\mathcal{O}(n)$ in each of the three basic models (deterministic branching programs, switching networks and nondeterministic branching programs) and, moreover, this characterization is the same for all three models. They show that such functions are exactly the symmetric boolean functions f_n with the following property: $f_n(x) = f_n(y)$ whenever there is a constant $T > 0$ such that $T \le |x|, |y| \le n - T$ and $|x| - |y|$ is divisible by T; here, as before, $|x|$ is the number of ones in x.

15.2 Branching Programs Over Large Domains

One can define branching programs also for functions $f : D^n \to \{0,1\}$ over larger domains D than $\{0,1\}$. In this case, instead of just tests $x_i = 0$ and $x_i = 1$ the program is allowed to make tests $x_i = d$ for elements $d \in D$. Different edges leaving the same node may make the same test—this is why a program is nondeterministic. As before, an input $a \in D^n$ is accepted if and only if all the tests along at least one path from the source node s to the target node t are passed by a, that is, if all the tests $x_i = d$ made along at least s-t path are consistent with the input string a in that $a_i = d$. A *switching network* is a nondeterministic branching program whose underlying graph is undirected. Edges are labeled by tests $x_i = d$, and are called *contacts*. Note that in this case unlabeled edges are redundant since we can always contract them. We are going to prove a lower bound on the number of contacts which is about $|D|$ times the binary length $n\log|D|$ of the input.

As our domain D we take $\mathbb{Z}_q := \{0,1,\ldots,q-1\}$. The counting modulo q function $\text{Mod}_q^n : \mathbb{Z}_q^n \to \{0,1\}$ is defined by:

$$\text{Mod}_q^n(x) = 0 \text{ if and only if } x_1 + \cdots + x_n = 0 \pmod q.$$

That is, Mod_q^n *rejects* the input vector if the sum of its components is divisible by q. In particular, Mod_2^n is the parity function.

Cardot (1952) proved that every switching network computing Mod_2^n must have $4n-4$ contacts, and this bound is optimal. In contrast, the number of contacts needed to compute this functions in the class of parallel-serial networks (that are equivalent to DeMorgan formulas) lies between n^2 and $(9/8)n^2$ (see Sect. 6.8).

Theorem 15.4. (Rychkov 2009) *The function* Mod_q^n *can be computed by a switching network with* q^2n *contacts, and every switching network computing* Mod_q^n *must have* $\Omega(qn\log q)$ *contacts.*

Proof. To prove the upper bound, take a graph $G = (V, E)$ whose set of vertices is divided into disjoint subsets $V_0, V_1, \ldots, V_n$ each of size q. The set of edges is defined as $E = E_1 \cup \cdots \cup E_n$ where $E_i = V_{i-1} \times V_i$; hence, $|E| = q^2n$. Identify the vertices in each V_i with members of $\mathbb{Z}_q = \{0, 1, \ldots, q-1\}$. Label the edge joining $u \in V_{i-1}$ with $v \in V_i$ by the test $x_i = (v - u) \bmod q$. As the source node take the vertex v_0 in V_0 numbered by 0. Observe that for every input vector $a \in \mathbb{Z}_q^n$ and for every $1 \leq i \leq n$, the set of contacts in $E_i i$ consistent with a forms a perfect matching between V_{i-1} and V_i. Thus, there is exactly one path $p = (v_0, v_1, \ldots, v_n)$ from the source node v_0 to a node v_n in V_n which is consistent with a. Moreover, we have that $v_i - v_{i-1} = a_i \bmod q$ for all $i = 1, \ldots, n$. Thus,

$$v_n = v_n - v_0 = (v_1 - v_0) + (v_2 - v_1) + (v_3 - v_2) + \cdots + (v_n - v_{n-1})$$

$$= a_1 + a_2 + a_3 + \cdots + a_n \bmod q.$$

Thus, if we remove from V_n the node numbered by 0 and glue all remaining vertices in V_n to one target node, the obtained switching network will compute Mod_q^n.

To prove the lower bound, take a switching network $F(x)$ computing Mod_q^n. We claim that for every pair $1 \leq i < j \leq n$, the network must contain $\Omega(q\log q)$ contacts labeled by tests on variables x_i and x_j. This already implies the desired lower bound $\Omega(qn\log q)$ on the total number of contacts. By symmetry, it is enough to prove the claim for $i = 1$ and $j = 2$. The idea is to use Hansel's result (see Exercise 1.12) that any monotone switching network computing the threshold-2 function Th_2^m must have $\Omega(m\log m)$ contacts.

From F we obtain a network F' depending only on x_1 and x_2 as follows. First, remove from F all contacts labeled by tests $x_i = a$ for $i \geq 3$ and $a \neq 0$. Then contract all contacts labeled by tests $x_i = 0$ for $i \geq 3$. The resulting network computes the function $F'(x_1, x_2) = F(x_1, x_2, 0, \ldots, 0)$. That is, F' accepts a pair (a, b) of integers if and only if $a + b \notin \{0, q\}$. Thus, for every input (a, b) with $a + b \notin \{0, q\}$, there must be a consistent path p in F' from the source node s to the target node t such that along p

(i) Only tests $x_1 = a$ and $x_2 = b$ are made, and

(ii) Each test $x_1 = a$ and $x_2 = b$ is made at least once: if, say, only tests $x_1 = a$ were made along p, then F' would be forced to wrongly accept the input $(a, q - a)$.

On the other hand, no s-t path can be consistent with the input $(0,0)$ or any of the inputs $(a, q - a)$ with $a \in \mathbb{Z}_q$.

Take q new variables $y_1, \ldots, y_q$ and relabel the contacts of F' as follows. For every $a \in \mathbb{Z}_q, a \neq 0$, replace all tests $x_1 = a$ and $x_2 = q - a$ by the test $y_a = 1$. Further, replace all tests $x_1 = 0$ and $x_2 = 0$ by the test $y_q = 1$. Let $F''(y_1, \ldots, y_q)$ be the obtained monotone boolean switching network. We claim that this network computes the threshold-2 function $\text{Th}_2^q(y_1, \ldots, y_q)$.

To show this, take an arbitrary vector $y \in \{0, 1\}^q$ with exactly two 1s in positions a and b with $1 \leq a < b \leq q$. We have to show that y must be accepted by F''.

If $b < q$ and $a + b \neq q$, then property (i) ensures that y is accepted by F''. So, assume that $b < q$ and $a + b = q$. Since $a < b$, we have that either $a + a$ or $b + b$ must not be equal to q. Assume that $a + a \neq q$ (the case $b + b \neq q$ is similar). Then F' must accept the input (a, a). By property (i), there must exist a consistent s-t path in F' along which only tests $x_1 = a$ and $x_2 = a$ are made. But by our construction, these tests are replaced by tests $y_a = 1$ and $y_{q-a} = 1$, that is, by tests $y_a = 1$ and $y_b = 1$. Hence, F'' accepts y also in this case. Finally, let $b = q$. Then a lies between 1 and $q - 1$, and hence, the input $(a, 0)$ must be accepted by F'. The corresponding path in F'' has only tests $y_a = 1$ and $y_q = 1$, implying that y is accepted by F''.

It remains to show that F'' rejects every vector $y \in \{0, 1\}^q$ with exactly one 1 in, say, position a. For the sake of contradiction, assume that F'' accepts such a vector y. Then F'' must contain an s-t path along which only the variable y_a is tested. By our construction, the corresponding path in F' can only make tests $x_1 = a$ or $x_2 = q - a$. But then (ii) implies that F' must wrongly accept the input $(a, q - a)$.

Thus, we have shown that F'' is a monotone switching network computing the threshold-2 function Th_2^q, and we already know (see Exercise 1.12) that any such network must have $\Omega(q \log q)$ contacts. $\square$

An important subclass of switching networks is that of parallel-serial switching networks (or π-schemes). We know (see Proposition 1.10) that such networks are equivalent to DeMorgan formulas. DeMorgan formulas have literals as inputs and use AND and OR operations as gates. If we will give "generalized literals" as inputs, then such a formula will compute a function $f : D^n \to \{0, 1\}$. As a generalized literal we take a function $x_i^d : D^n \to \{0, 1\}$ such that for every $a \in D^n$, $x_i^d(a) = 1$ if and only if $a_i = d$. Let $L(f)$ denote the minimum leafsize of a (generalized) DeMorgan formula computing f.

Since formulas are special switching networks, Theorem 15.4 implies that $L(\text{Mod}_q^n) = \Omega(qn \log q)$. By extending the argument of Khrapchenko (see Sect. 6.7) to functions over larger domains, Rychkov (2009) improved this lower bound to

$$L(\text{Mod}_q^n) \geq (q - 1)n^2.$$

But what about the complexity of the negation $\neg\text{Mod}_q^n$ of Mod_q^n? Note that for boolean functions $f : D^n \to \{0, 1\}$ with $D = \{0, 1\}$, we always have that $L(\neg f) = L(f)$. This, however, does not hold for larger domains D. In this case, $L(x_i^d) = 1$ but $L(\neg x_i^d) = |D| - 1$ because $\neg x_i^d$ is an OR of x_i^k over all $k \in D, k \neq d$. This last observation implies that $L(\neg f) \geq L(f)/(|D| - 1)$. For the negation $\neg \mathrm{Mod}_q^n$ of Mod_q^n, this yields the lower bound $L(\neg \mathrm{Mod}_q^n) \geq n^2$. For domains D larger than the number n of variables, Rychkov (2010) improved this to $L(\neg \mathrm{Mod}_q^n) \geq 2(|D| - 1)n$.

15.3 Counting Versus Nondeterminism

Our goal is to show that, at the cost of a slight increase of size, every nondeterministic branching program can be simulated by a parity branching program. That is, in the model of branching programs nondeterminism is not much more powerful than counting. But perhaps more interesting than the result itself is its proof: it uses in a nontrivial manner an interesting fact that random weighting of elements will almost surely isolate exactly one member of a family.

Let X be some set of n points, and $\mathcal{F}$ be a family of subsets of X. Assign a weight $w(x)$ to each point $x \in X$, and define the weight of a set E to be $w(E) = \sum_{x \in E} w(x)$. It may happen that several sets of $\mathcal{F}$ will have the minimal weight. If this is not the case, that is, if $\min_{E \in \mathcal{F}} w(E)$ is achieved by the unique $E \in \mathcal{F}$, then we say that w is *isolating for* $\mathcal{F}$.

Let X be a set with n elements, and let $w : X \to \{1, \ldots, N\}$ be a random function. Each $w(x)$ is independently and uniformly chosen over the range.

Lemma 15.5. (Mulmuley et al. 1987) *For every family $\mathcal{F}$ of subsets of X,* $\mathrm{Prob}[w \text{ is isolating for } \mathcal{F}] \geq 1 - \frac{n}{N}$.

Proof. (Due to Spencer 1995) For a point $x \in X$, set

$$\alpha(x) = \min_{E \in \mathcal{F}; x \notin E} w(E) - \min_{E \in \mathcal{F}; x \in E} w(E \setminus \{x\}).$$

A crucial observation is that the evaluation of $\alpha(x)$ does not require knowledge of $w(x)$. As $w(x)$ is selected uniformly from $\{1, \ldots, N\}$, we have that $w(x) = \alpha(x)$ with probability at most $1/N$, and hence, the probability that $w(x) = \alpha(x)$ *for some* $x \in X$ does not exceed n/N. But if w had two sets $A, B \in \mathcal{F}$ of minimal weight $w(A) = w(B)$ and $x \in A \setminus B$, then

$$\min_{E \in \mathcal{F}; x \notin E} w(E) = w(B) \quad \text{and} \quad \min_{E \in \mathcal{F}; x \in E} w(E \setminus \{x\}) = w(A) - w(x),$$

so $w(x) = \alpha(x)$. Thus, if w is *not* isolating for $\mathcal{F}$ then $w(x) = \alpha(x)$ for some $x \in X$, and we have already established that the last event can happen with probability at most n/N. □

Theorem 15.6. (Wigderson 1994) *There is a constant c such that for every boolean function f of n variables, we have that $\oplus \mathrm{BP}(f) \leq cn \cdot \mathrm{NBP}(f)^5$.*

Fig. 15.1 A fragment of the
graph G_w^l for $l = 4$,
$w(e_1) = 2$ and $w(e_2) = 1$

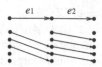

Proof. Let $G = (V, E)$ be a directed graph and $w : E \to \{1, \dots, 2|E|\}$ a weight
function on its edges. The weight of an s-t path is the sum of weights of its edges; a
path is *lightest* if its weight is minimal. Let $d_w(G)$ denote the weight of the shortest
s-t path in G; hence,

$$d_w(G) \le M := 2|V| \cdot |E|.$$

Having a weight function w and an integer l, define the (unweighted, layered)
version $G_w^l = (V', E')$ of G as follows. Replace every vertex $u \in V$ by $l + 1$
new vertices $u_0, u_1, \dots, u_l$ in V' (that is, V' consists of $l + 1$ copies of V, arranged
in layers). For every edge $e = (u, v)$ in E and every $0 \le i \le l - w(e)$ put an edge
$(u_i, v_{i+w(e)})$ in E' (see Fig. 15.1); hence, $|V'| \le (1 + l)|V|$ and $|E'| \le (l + 1)|E|$.

Claim 15.7. *If G has no s-t path, then for every w and l, G_w^l has no s_0-t_l path. If G
has an s-t path and $l = d_w(G)$, then G_w^l has an s_0-t_l path. Moreover, the later path
is unique if the lightest s-t path in G is unique.*

Proof. Let $P = (e_1, e_2, \dots, e_k)$ be an s-t path in G. The first node of this path is s.
In the new graph G_w^l the first node is s_0 and, following the path P in this new graph,
at the i-th edge e_i we move by $w(e_i)$ vertices down (in the next, $(i + 1)$-th layer of
nodes). Hence, P can produce an s_0-t_l path in G_w^l if and only if $\sum_{i=1}^k w(e_i) \le l$.
That is, a graph G_w^l has an s_0-t_l if and only if G has an s-t path *and* $\sum_{i=1}^k w(e_i) \le l$.
For $l = d_w(G)$, only lightest paths can fulfill this last condition.   □

Now let $G = (V, E)$ be a nondeterministic branching program computing a
given boolean function $f(x_1, \dots, x_n)$. Say that a weight function w is *good* for an
input $a \in \{0, 1\}^n$ if either $G(a)$ has no s-t paths or the lightest s-t path in $G(a)$ is
unique.

For each input $a \in \{0, 1\}^n$, taking the family $\mathcal{F}_a$ to be all s-t paths in the graph
$G(a)$, the isolation lemma (Lemma 15.5) implies that at least one-half of all weight
functions w are good for a. By a standard counting argument, there exists a set W of
$|W| \le 1 + \log(2^n) = n + 1$ weight functions such that, for every input a, at least one
$w \in W$ is good for a. If w is good for a, then the graph $G_w^l(a)$ with $l = d_w(G(a))$
has the properties stated in Claim 15.7. For different inputs a, the corresponding
values of l may be different, but they all lie in the interval $1, \dots, M$. Thus, there
exist $m \le (n + 1) \cdot M$ nondeterministic branching programs $H_1, \dots, H_m$ (with each
$H_j = G_w^l$ for some $w \in W$ and $1 \le l \le M$) such that, for every input $a \in \{0, 1\}^n$,
the following holds:

(i) If $|G(a)| = 0$, then $|H_j(a)| = 0$ for all j;
(ii) If $|G(a)| > 0$, then $|H_j(a)| = 1$ for at least one j.

Fig. 15.2 Construction of the
parity branching program H

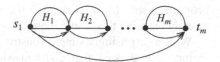

Let s_j, t_j be the specified vertices in H_j, $j = 1, \ldots, m$. We construct the desired parity branching program H as follows: to each H_j add the unlabeled edge (s_j, t_j), identify t_j and s_{j+1} for every $j < m$, and add the unlabeled edge (s_1, t_m) (see Fig. 15.2).

It is easy to see that, for every input $a \in \{0, 1\}^n$, $|H(a)| = 1 \bmod 2$ if and only if $|G(a)| > 0$. Indeed, if $|G(a)| = 0$, then by (i), $H(a)$ has precisely two s_1-t_m paths (formed by added unlabeled edges). On the other hand, if $|G(a)| > 0$, then by (ii), at least one $H_j(a)$ has precisely one s_j-t_j path, implying that the total number of s_1-t_m paths in $H(a)$ is odd. Thus, H is a parity branching program computing the same boolean function f. Since $l \leq M$ and $m \leq nM$ with $M = 2|V| \cdot |E| \leq 2|E|^2$, the size of (the number of edges in) H is at most $m(l + 1)|E| = \mathcal{O}(n|E|^5)$. □

15.4   A Surprise: Barrington's Theorem

The *length* of a branching program is the number of edges in a longest path (be it consistent or not)[1]. If the nodes are arranged into a sequence of levels with edges going only from one level to the next, then the *width* is the number of nodes of the largest level.

Intuitively, the smaller the width is, the fewer information can be transferred through each level of a program. So, if the width is small (say constant), then boolean functions with a lot of dependency between parts of its input variables "should" require very long branching programs.

In this section we will show that this intuition is wrong! Even programs of width 5 are not much longer than DeMorgan formulas. We will also give lower bounds on the length of width-restricted branching programs.

We first consider branching programs of constant(!) width. At first glance, it seems that such a drastic width restriction might be very crucial: if the width is bounded by some constant then, when going from one level to the next, we can keep only a constant amount of information about what we have done before. It was therefore conjectured by many researchers that, due to this "information bottleneck", even such function as the Majority function $\mathrm{Maj}(x_1, \ldots, x_n)$ should require very long branching programs, if their width is constant. A trivial branching

---

[1]A path is *inconsistent* if it contains two contradicting queries $x_i = 0$ and $x_i = 1$ on the same variable.

program would try to remember the number of 1s among the bits, which were already read; but this would require non-constant width of about $\log n$.

With a surprisingly simple construction, Barrington (1989) disproved this conjecture. He showed that constant-width branching programs are unexpectedly powerful: they are almost as powerful as DeMorgan formulas!

Theorem 15.8. (Barrington 1989) *If a boolean function can be computed by a DeMorgan formula of a polynomial size, then it can also be computed by an oblivious width-5 branching program of polynomial length.*

Before we present his proof, let us explain the intuition behind it.

The (wrong) intuition above (that large width is necessary to keep the collected information) relies on viewing the computation by a program as a sequential process which *gradually* collects information about the input vector. This intuition is borrowed from our view at how Turing machine works. But branching programs constitute a *nonuniform* model of computation. Each input $a \in \{0,1\}^n$ defines (at once!) a subprogram of our branching program in a natural way. Thus, for a program to correctly compute a given boolean function f, it is enough that subprograms produced by inputs in $f^{-1}(0)$ are different from those produced by inputs in $f^{-1}(1)$. It then remains to show how the program can detect this difference.

To be more specific, assume that our branching program is an oblivious program of width w and length l. That is, the nodes are arranged into a w by l array. All l levels have w nodes, and all nodes at a given level are labeled by the same variable. Moreover, at each level the 0-edges and the 1-edges going to the next level form two mappings from $[w] = \{1, \ldots, w\}$ to $[w]$. If that level is labeled by a variable x_i, then one of these mappings is given by edges corresponding to tests $x_i = 1$, and the other to tests $x_i = 0$ (see Fig. 15.3). The length $|P|$ of such a program P is the number of levels in it.

Thus, we can view such a program as a sequence of instructions $\langle i, \sigma, \tau \rangle$ where x_i is the variable tested at the corresponding level, and $\sigma, \tau : [w] \to [w]$ are the two mappings corresponding to whether $x_i = 0$ or $x_1 = 1$. Every input $a \in \{0,1\}^n$ gives a sequence of l mappings, and let $P(a) : [w] \to [w]$ be their superposition. Now, for the program to correctly compute a given boolean function f, it is enough that $P(a) \neq P(b)$ for all $a \in f^{-1}(0)$ and $b \in f^{-1}(1)$.

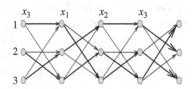

Fig. 15.3 On input vector $x = (0, 1, 1)$ this program outputs the identity permutation $P(x) = e$, whereas on input $x = (0, 1, 0)$ it produces a cyclic permutation $P(x) = \begin{pmatrix} 1 & 2 & 3 \\ 3 & 1 & 2 \end{pmatrix}$. *Bold arrows* correspond to tests $x_i = 1$, the remaining ones to tests $x_i = 0$

A basic question is: how, starting from programs for boolean functions f and g, can we build programs for $\neg f$, $f \wedge g$ and $f \vee g$? Barrington showed that this can be easily done if we restrict programs and only allow cyclic permutations σ, τ : $[w] \rightarrow [w]$ be used. A permutation is *cyclic* if it is composed of a single cycle on all its elements. For example,

$$\sigma = \begin{pmatrix} 1\,2\,3\,4\,5 \\ 3\,1\,5\,2\,4 \end{pmatrix} = \begin{pmatrix} 1\,3\,5\,4\,2 \\ 3\,5\,4\,2\,1 \end{pmatrix}$$

is a cyclic permutation, which we will denote as

$$1 \rightarrow 3 \rightarrow 5 \rightarrow 4 \rightarrow 2 \rightarrow 1 \quad \text{or shortly as} \quad \sigma = (13542).$$

The nodes of a width-w *permuting branching program* P of length ℓ are arranged into a $w \times \ell$ array, where two permutations of $[w]$ are computed between any of its two levels. Any input vector $x \in \{0, 1\}^n$ yields a permutation $P(x)$ which is the composition of the selected permutations at each level. For a boolean function f and a permutation σ, say that branching program P σ-*computes* f if for every input x,

$$P(x) = \begin{cases} \sigma & \text{if } f(x) = 1, \\ e & \text{if } f(x) = 0, \end{cases}$$

where e is the identity permutation.

The following three simple claims accumulate the basic properties of permuting branching programs that use only cyclic permutations.

Let σ and τ be *cyclic* permutations, f and g boolean functions, P and Q permuting branching programs.

Claim 15.9. (Changing output) If P σ-computes f, then there is a permuting branching program of the same size τ-computing f.

Proof. Since σ and τ are both cyclic permutations, we may write $\tau = \theta \sigma \theta^{-1}$ for some permutation θ. Then simply reorder the left and right nodes of P according to θ to obtain the τ-computing branching program P':

$$\text{if } P(x) = \sigma_1 \sigma_2 \cdots \sigma_t = \sigma \text{ then } P'(x) = \theta \sigma_1 \sigma_2 \cdots \sigma_t \theta^{-1} = \theta \sigma \theta^{-1} = \tau.$$

That is, we replace the permutation σ_1 computed at the first layer by the permutation $\theta \sigma_1$, and the permutation σ_t computed at the last layer by the permutation $\sigma_t \theta^{-1}$. $\qquad \square$

Claim 15.10. (Negation) If P σ-computes f then there is a permuting branching program of the same size σ-computing $\neg f$.

Proof. Use the previous lemma to obtain a branching program P' which σ^{-1}-computes f. Hence, $P'(x) = \sigma^{-1}$ if $f(x) = 1$, and $P'(x) = e$ if $f(x) = 0$. Then reorder the final level by σ so that the resulting program P'' σ-computes $\neg f$:

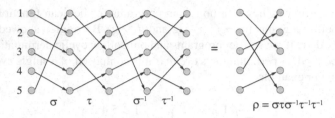

Fig. 15.4 Cyclic permutations $\sigma = (12345)$, $\tau = (13542)$ of $\{1, 2, 3, 4, 5\}$ and their commutator $\rho = \sigma\tau\sigma^{-1}\tau^{-1} = (13254)$

$$\text{if } P'(x) = \sigma_1\sigma_2\cdots\sigma_t \text{ then } P''(x) = \sigma_1\sigma_2\cdots\sigma_t\sigma.$$

In this way, $P''(x)$ outputs e if $P'(x) = \sigma^{-1}$, that is, if $f(x) = 1$; otherwise, $P''(x)$ outputs σ.   □

Claim 15.11. (Computing AND) *If P σ-computes f and Q τ-computes g, then there is a permuting branching program of length $2(|P| + |Q|)$ which $\sigma\tau\sigma^{-1}\tau^{-1}$-computes $f \wedge g$.*

Proof. Use Lemma 15.9 to get a program σ^{-1}-computing f and τ^{-1}-computing g. Then compose these four programs in the order $\sigma, \tau, \sigma^{-1}, \tau^{-1}$. This has the desired effect because replacing either σ or τ by e in $\sigma\tau\sigma^{-1}\tau^{-1}$ yields e.   □

The next claim is the only place where the value $w = 5$ is important; neither $w = 3$ nor $w = 4$ suites for this purpose.

Claim 15.12. *There are cyclic permutations σ and τ of $\{1, 2, 3, 4, 5\}$ such that their commutator $\rho = \sigma\tau\sigma^{-1}\tau^{-1}$ is cyclic.*

Proof. See Fig. 15.4.   □

By w-PBP we will mean a permuting branching program of width w all whose permutations are *cyclic* permutations of $[w]$.

Theorem 15.13. *If a boolean function can be computed by a DeMorgan circuit of depth d, then it can be computed by a 5-PBP of length 4^d.*

In particular, if a boolean function f can be computed by a DeMorgan formula of polynomial leafsize, then f can also be computed by a 5-PBP of polynomial length!

Proof. By induction on the depth d. If $d = 0$, the whole circuit for f is either a variable x_i or its negation $\neg x_i$, and f can be easily computed by a one-instruction program.

Now suppose that $d \geq 1$. By Claim 15.10, we can assume that $f = g \wedge h$, where g and h have formulas of depth $d - 1$, and thus (by induction hypothesis) 5-PBPs G and H of length at most 4^{d-1}.

Let σ and τ be the permutations from Claim 15.12. By Claim 15.9, we may assume that G σ-computes g and H τ-computes h. By Claim 15.11, there is a 5-PBP of length at most $2(\text{size}(G) + \text{size}(H)) \le 4^d$ which $\sigma\tau\sigma^{-1}\tau^{-1}$-computes f. Since, by Claim 15.12, the permutation $\sigma\tau\sigma^{-1}\tau^{-1}$ is cyclic, we are done. □

To derive Theorem 15.8 from Theorem 15.13 it is enough to show that any w-PBP of length l can be transformed into a width-w branching program of length l.

To show this, let $P(x)$ be a w-PBP σ-computing a boolean function $f(x)$. Hence, $P(x) = \sigma$ if $f(x) = 1$, and $P(x) = e$ (the identity function) if $f(x) = 0$. Choose an $i \in [w]$ such that $\sigma(i) \ne i$. Declare the i-th node on the first level as the start node of our branching program, and in the final level we let node $\sigma(i)$ be the 1-leaf, and all other nodes on this level be the 0-leafs.

15.5 Oblivious Branching Programs

Every boolean function of n variables can be computed by a trivial branching program of length $l = n$ and width $w = 2^n$: just take a decision tree. But what if we restrict the width w—how long then the program must be? To answer this question we use communication complexity arguments.

A branching program is *oblivious* if the nodes are arranged into a sequence of levels with edges going only from one level to the next, and at all nodes of each level the same variable is tested. As before, the *width* of such a program is the number of nodes of the largest level.

An *s-overlapping protocol* for a boolean function $f : \{0,1\}^n \to \{0,1\}$ is a deterministic communication protocol between two players under a restriction that each player cannot see s variables seen by the other player. The remaining $n - 2s$ variables are seen by both players! Also, before the game begins, the players (knowing the function f) are allowed to decide what s variables should be their "private" variables (not seen by the other player).

Let $c_s(f)$ denote the maximum number of bits communicated by a best deterministic s-overlapping protocol for f on the worst case input. The larger the number $n - 2s$ of common variables is, the easier is the game. Hence, $s \le t$ implies that $c_s(f) \le c_t(f)$.

Theorem 15.14. *Suppose that a boolean function* $f : \{0,1\}^n \to \{0,1\}$ *can be computed by an oblivious branching program of width w and length l. If $l \le 0.1n \log n$ then, for every $n^{0.6}/4 \le s \le n/2$,*

$$c_s(f) = \mathcal{O}\Big(\frac{l \log w}{n}\Big).$$

Because the branching program is oblivious, we can think of its labels as forming a string z of length l over the alphabet $[n]$. To obtain a communication protocol from the program, we need the following combinatorial result.

Let z be a string over an alphabet $X = \{x_1, \ldots, x_n\}$. A *substring* of z is a sequence of its consecutive letters. Given two sets $A, B \subseteq X$ of letters, say that a string z has an (r, A, B)-partition if z can be partitioned into r substrings $z = z_1 z_2 \ldots z_r$ such that none of the substrings z_i contains letters from both sets A and B. For example, if $X = \{1, 2, 3, 4, 5, 6\}$, $A = \{1, 3\}$, $B = \{2, 4\}$ and

$$z = 1\,6\,2\,3\,2\,6\,4\,1\,5\,3\,1\,2\,4\,6\,5\,1\,3\,2$$

then we have the following $(8, A, B)$-partition of Z:

$$z = \overbrace{1\,6}^{A}\ \underbrace{2}_{B}\ \overbrace{3}^{}\ \overbrace{2\,6\,4}^{A}\,1\,5\,3\,1\,\overbrace{2\,4\,6\,5}^{A}\ \underbrace{1\,3}_{B}\ \underbrace{2}_{B}$$

Lemma 15.15. (Alon and Maass 1988) *Let $A, B \subseteq X$ be two disjoint sets of size $|A| = |B| = m$. Let z be a string over X such that each $a \in A$ appears in z at most k_A times and each $b \in B$ appears in z at most k_B times. Then there are $A' \subseteq A$ and $B' \subseteq B$ of size at least $m/2^k$ such that z has a (k, A', B')-partition, where $k = k_A + k_B$.*

Proof. Induction on k. If $k = 1$, then either k_A or k_B is 0, and we can take $A' = A$ and $B' = B$. For the induction step, assume w.l.o.g. that each letter appears in z at least once (otherwise, extend z with the missing letters in an arbitrary way).

Examine the letters of z one by one until we reach a location where we already have seen $m/2$ letters of one of A and B but fewer than $m/2$ of the other; such a location must exist since $A \cap B = \emptyset$. Denote the prefix by z' and the rest of z by z''. Let it was A whose $m/2$ letters appear in z' (the case when it is B is dual). Let $A^* = \{a \in A : a \in z'\}$ be those letters of A that appear in z', and $B^* = \{b \in B : b \notin z'\}$ be those letters of B that *do not* appear in z'. It follows that $|A^*|, |B^*| \geq m/2$.

Now consider the suffix z''. Each letter of A^* appears in z'' at most $k_A - 1$ times, since each of them already appeared in z' at least once. Hence, we can apply the induction hypothesis to the string z'' for sets A^* and B^*, and obtain subsets $A' \subseteq A^*$ and $B' \subseteq B^*$ such that z'' has a $(k - 1, A', B')$-partition with

$$|A'| \geq |A^*|/2^{k-1} \geq m/2^k \text{ and } |B'| \geq |B^*|/2^{k-1} \geq m/2^k.$$

Since the prefix z' can only contain letters of A' but none of B', the entire string $z = z'z''$ also has a (k, A', B')-partition. $\square$

Proof of Theorem 15.14. Now let z be the string over $X = \{x_1, \ldots, x_n\}$ of length l corresponding to the labels of our branching program. Observe that at least $n/2$ variables must appear at most $2l/n$ times, for otherwise the length of the string would be larger than $(n/2)(2l/n) = l$. Partition these variables into two sets A and B each of size $n/4$ in an arbitrary way. By Lemma 15.15 with $m = n/4$, $k_A = k_B = 2l/n$ and $k = 4l/n$, there are disjoint sets of variables A' and B'

such that $|A'|, |B'| \geq n/(4 \cdot 2^k)$ and z is a (k, A', B')-partition. Moreover, since $l \leq 0.1n \log n$, we have that

$$k = \frac{4l}{n} \leq \frac{0.4n \log n}{n} = 0.4 \log n.$$

Hence,

$$|A'|, |B'| \geq n/(4 \cdot 2^k) \geq n^{0.6}/4.$$

Since the sequence z of variables, tested along the l levels of the program, has a (k, A', B')-partition, it is possible to split z into k substrings $z = z_1 z_2 \cdots z_k$ such that no substring z_i contains variables from both subsets A' and B'. Hence, if we give all variables in A' to Alice, all variables in B' to Bob and the rest to both players, the players can determine the value of our function by communicating according to the underlying branching program.

To carry out the simulation, the players need to tell each other, at the end of each of k blocks, the name of the node in the next level from which the simulation should proceed; for this $\log w$ bits are sufficient. Hence, the obtained protocol communicates $\mathcal{O}(k \cdot \log w)$ bits, that is, $\mathcal{O}((l \log w)/n)$ bits in total. The protocol is s-overlapping with $s \geq \min\{|A'|, |B'|\} \geq n^{0.6}/4$. ☐

Thus, to obtain a large tradeoff between the width and the length of oblivious branching programs, we need boolean functions of large overlapping communication complexity. We will now show that such are characteristic functions of good codes.

A linear (n, m, t)-code is a linear subspace $C \subseteq GF(2)^n$ of dimension $n - m$ such that the Hamming distance between any two vectors in C is at least $2t + 1$. A characteristic function of C is a boolean function $f : \{0, 1\}^n \to \{0, 1\}$ such that $f(x) = 1$ if and only if $x \in C$.

Lemma 15.16. *If f is the characteristic function of a linear (n, m, t)-code, then* $c_s(f) \geq 2t \log(s/t) - m$.

Proof. Take an arbitrary s-overlapping protocol for $f(X)$. Let $A \subseteq X$ be the set of variables seen only by Alice, and $B \subseteq X$ be the set of variables seen only by Bob. Hence, $|A|, |B| \geq s$, and at most $r := n - 2s$ variables are seen by both players. We can assume w.l.o.g. that $|A| = |B| = s$ (the fewer "forbidden" bits we have, the easier the life of the players is). Since there are only 2^r possible settings α of constants to these (common) variables, at least one of these settings gives us a subfunction f_α of f in $2s$ variables which is the characteristic function of some linear $(n - r, m - r, t)$-code C.

After this setting, our protocol turns to a usual communication protocol for the matrix $M = \{f_\alpha(x, y)\}$. From Sect. 4.2 we know that this last protocol must communicate at least $\log \text{Cov}(M)$ bits, where $\text{Cov}(M)$ is the smallest number of (not necessarily disjoint) all-1 submatrices of M covering all its 1s. (In fact, $\log \text{Cov}(M)$ is a lower bound even for *nondeterministic* communication complexity of M, but we will not need this now.)

Claim 15.17. Every row and every column of M has at most $2^s \binom{s}{t}^{-1}$ ones.

Proof. Fix one row $x \in \{0,1\}^s$ of M (the case of columns is the same). Since the Hamming distance between any two vectors in C is at least $2t + 1$, we have that any two vectors $y \neq y' \in \{0,1\}^s$ of M such that $M[x,y] = M[x,y'] = 1$ must also be at Hamming distance at least $2t + 1$. Hence, no Hamming ball of radius t over a vector y with $M[x,y] = 1$ can contain another vector y' with $M[x,y'] = 1$. Since each such ball has $\sum_{i=0}^{t} \binom{s}{i} > \binom{s}{t}$ vectors, this implies that each row and each column of M can have at most $2^s \binom{s}{t}^{-1}$ ones. $\square$

The matrix has $|M| = 2^{n-m-r} = 2^{2s-m}$ ones and, by Claim 15.17, every all-1 submatrix of M has at most $2^{2s} \binom{s}{t}^{-2}$ ones, the desired lower bound on $\mathrm{Cov}(M)$, and hence, on $c_s(f)$ follows:

$$\mathrm{Cov}(M) \geq \frac{2^{2s-m}}{2^{2s} \binom{s}{t}^{-2}} = \binom{s}{t}^2 2^{-m} \geq \left(\frac{s}{t}\right)^{2t} \cdot 2^{-m}.$$ $\square$

Consider Bose-Chaudhury codes (BCH-codes). These are linear (n, m, t)-codes C with $m \leq t \log(n+1)$. Such codes can be constructed for any n such that $n+1$ is a power of 2, and for every $t < n/2$. Since the parity-check matrix of C has m rows, the characteristic function f_C of C is just an AND of m negations of parity functions, and hence, can be computed by an oblivious branching program of constant width and length $l = mn = \mathcal{O}(tn \log n)$. If, however, we require the length be smaller than $n \log n$, then an exponential width is required.

Corollary 15.18. *Let $0 < \epsilon \leq 0.01$ be a constant, and C be a BCH-code of minimal distance $2t + 1$ with $t = n^{0.09}$. Then any oblivious branching program for f_C of length $l = \mathcal{O}(n \log n)$ must have width $w = 2^{\Omega(n^{0.09})}$.*

Proof. We apply Lemma 15.16 with $s = n^{0.6}$. Since $(s/t)^2 = n^{1+\Omega(1)}$, we obtain that

$$c_s(f_C) \geq 2t \log(s/t) - m \geq t \log(s/t)^2 - t \log(n+1) = \Omega(t \log n).$$

Hence, Theorem 15.14 implies that $l \log w = \Omega(tn \log n) = \Omega(n^{1.09} \log n)$. $\square$

Exercises

15.1. Recall that the Majority function Maj of n variables accepts an input vector if and only if the number of 1s in it is at least $n/2$. Show that any *constant-width* branching program for Maj must have length $l = \Omega(n \log n)$.
Hint: Show that $c_s(\mathrm{Maj}) = \Omega(\log s)$ and use Theorem 15.14.

15.2. (Due to Razborov 1990b) Our goal is to give an exact(!) combinatorial characterization of $\text{NBP}(f)$ for any boolean function f. Let $U := f^{-1}(0)$ and $V := f^{-1}(1)$. Denote by $\mathcal{F}$ the set of all nontrivial monotone functions $F : 2^U \to \{0, 1\}$, that is, $F(\emptyset) = 0$, $F(U) = 1$, and $F(B) = 1$ as long as $F(A) = 1$ for some $A \subseteq B$. Given $1 \le i \le n$, $\epsilon \in \{0, 1\}$, and $A \subseteq U$, define

$$\delta_{i,\epsilon}(A) := \{(F, v) \in \mathcal{F} \times V : v_i = \epsilon, F(A) = 1, F(A \cap X_i^\epsilon) = 0\},$$

where X_i^ϵ is the set of all vectors in $\{0, 1\}^n$ with ϵ in their i-th position. Let Δ be the collection of all such sets $\delta_{i,\epsilon}(A)$; this is a huge collection, $|\Delta| = 2n2^{2^{|U|}}$. Let $\text{Cov}(f)$ denote the smallest number of members of Δ whose union covers the whole set $\mathcal{F} \times V$.

(a) Show that $\text{Cov}(f) \le \text{NBP}(f)$.

Hint: Take an arbitrary nondeterministic branching program P computing f. Let s be its source, and t its target nodes. For every node w of P, let f_w be the boolean function computed by the subprogram with source s and target node w. Hence, $f_s = 1$ and $f_t = f$. Associate with every contact $e = (w, w')$ labeled by a literal x_i^ϵ the set $\delta(e) := \delta_{i,\epsilon}(U \cap f_w^{-1}(1))$. Show that the sets $\delta(e)$ cover the whole set $\mathcal{F} \times V$. For this, take a point $(F, v) \in \mathcal{F} \times V$ and consider an arbitrary path accepting the vector v. Since F is nontrivial, $F(U) = 1$ and $F(\emptyset) = 0$. Hence, there must be some contact $e = (w, w')$ on that path for which $F(U \cap f_w^{-1}(1)) = 1$ and $F(U \cap f_{w'}^{-1}(1)) = 0$. Show that $(F, v) \in \delta(e)$.

(b) Show that $\text{NBP}(f) \le \text{Cov}(f)$.

Hint: Let $\Delta_0 \subseteq \Delta$ covers the whole set $\mathcal{F} \times V$, and $|\Delta_0| = \text{Cov}(f)$. As nodes take all subsets $A \subseteq U$. For each pair $A \subseteq B$ of theses subsets include a non-labeled edge (A, B). For every set $\delta_{i,\epsilon}(A)$ include an edge $(A, A \cap X_i^\epsilon)$ labeled by x_i^ϵ. (Recall that we only count *contacts*, that is, labeled edges in $\text{NBP}(f)$.) Set $s := U$ as the source node, and $t := \emptyset$ as the target node. Show that the resulting program computes f. For this, take a vector $v \in V = f^{-1}(1)$ and show that there must be a path from $s = U$ to $t = \emptyset$ whose labels are consistent with v.

Chapter 16
Bounded Replication

Since, so far, we are unable to prove exponential lower bounds for general branching programs, it is natural to try to do this for restricted programs. We have seen that restricting the width of a program does not decrease their power too much: the resulting class of programs is almost as powerful at that of (unrestricted) formulas. Another possibility is to restrict the "length" of a program, that is, the length of a longest computation path. A path from the source to a sink is a *computation path*, if the tests made along its wires are passed by at least one input vector, that is, if the path does not contain two contradictory tests "is $x_i = 1$?" and "is $x_i = 0$?" on the same variable x_i. In a *read-k times* branching program it is required that along every computation path, each variable is tested at most k times. We will consider such programs in the next chapter.

In this chapter we restrict the number of variables that can be queried more than once during a computation. Namely, define the *replication number* of a branching program as the smallest number R such that along every computation path at most R variables are tested more than once. Sets of variables re-tested along different computations may be different! Also, the (up to R) re-tested variables may be re-tested arbitrarily often. Thus, restricted replication does not mean restricted length of computations—it may still be arbitrarily long. Finally, note that the restriction is only on *computation* paths: we have no restrictions on inconsistent paths. Branching programs with replication number R are also known as $(1, +R)$-*programs*.[1]

Note that for *every* branching program of n variables we have $0 \leq R \leq n$. Moreover, every boolean function f of n variables can be computed by a branching program with $R = 0$: just take a decision tree. However, the size S of such (trivial) branching programs is then exponential for most functions. It is therefore interesting to understand whether S can be substantially reduced by allowing larger values of R.

The goal is to prove exponential lower bounds on the size of branching programs of as large replication number R as possible. An ultimate goal is to do this for

---

[1] The meaning of this notation is that we have a read-one branching program with up to R exceptions along each computation.

S. Jukna, *Boolean Function Complexity*, Algorithms and Combinatorics 27,
DOI 10.1007/978-3-642-24508-4_16, © Springer-Verlag Berlin Heidelberg 2012

$R = n$: then we would have an exponential lower bound for *unrestricted* branching programs.

In this chapter we will come quite "close" to this goal by exhibiting boolean functions f (based on expander graphs) with the following property: there is a constant $\epsilon > 0$ such that every branching program computing f must either have replication number $R > \epsilon n$ or must have exponential size.

16.1 Read-Once Programs: No Replications

To "warm up", we start with *read-once* branching programs (1-BP), that is, programs along each path of which no variable can be tested more than once. This corresponds to programs with replication $R = 0$.

Read-once programs constitute just a small generalization of decision trees. Indeed, it is not difficult to see that the minimal size of a read-once program for a function f is precisely the minimal number of *non-isomorphic subtrees* in any decision tree for f.

Since subtrees correspond to subfunctions, it seems intuitive that the number of non-isomorphic subtrees (and hence, the size of the corresponding read-once program) must be large, if f has many different subfunctions. This motivates the following definition.

Say that a boolean function f is *m-mixed* if for every subset of m variables and for every two distinct assignments $a \neq b$ of constants to these variables, the obtained subfunctions f_a and f_b of f are distinct, that is, there exists an assignment c to the remaining variables such that $f_a(c) \neq f_b(c)$.

Lemma 16.1. (Folklore fact) *If f is an m-mixed boolean function, then every deterministic read-once branching program computing f must have at least $2^m - 1$ nodes.*

Proof. Let P be a deterministic read-once branching program computing f. Our goal is to show that the initial part of P must be a complete binary tree of depth $m - 1$. For this, it is enough to show that no two initial paths (starting in the source node) of length $m - 1$ can meet in a node. For the sake of contradiction, assume that some two paths p and q of length $m - 1$ meet in some node v.

Claim 16.2. *The sets of variables tested along p and q are the same.*

Proof. Assume that some variable x is tested along p but not along q. Let Y be the set of variables tested along q; hence, $x \notin Y$. The path q defines an assignment to the variables in Y. Extend this assignment to two assignments a and b by setting the variable x to 0 and to 1. In this way we obtain two distinct assignments to the same set of $|Y \cup \{x\}| = m$ variables. Since both of these extended assignments remain consistent with the path q, and since, due to the read-once property, the variable x cannot be tested along any path starting in v, the subfunctions of f defined by these assignments must be the same, contradicting the m-mixedness of f. □

By Claim 16.2, the paths p and q define two assignments on the same set Y of m variables. Moreover, these assignments are different since the computations on them split before they meet. But the read-once property again implies that these two assignments define the same subfunction of f, contradicting the m-mixedness of f. □

There are many natural boolean functions of n variables that are m-mixed for m about $\sqrt{n}$. We now describe one function which is m-mixed for $m = n - \mathcal{O}(\sqrt{n})$. For this, we use the following important number-theoretic result of Dias de Silva and Hamidoune (1994). Let p be a prime, and $A \subseteq \mathbb{Z}_p$. For an integer $1 \leq h \leq |A|$, let $\bigoplus_h A$ denote the set of all elements $b \in \mathbb{Z}_p$ that can be represented as sums $b = a_1 + a_2 + \cdots + a_h$ modulo p of h distinct elements $a_1, \ldots, a_h$ of A.

Theorem 16.3. $|\bigoplus_h A| \geq \min\{p, h|A| - h^2 + 1\}$.

Let n be a square of a natural number, and $n < p \leq 2n$ a prime number. Take $h = 2\sqrt{n}$ and $k = 2h$. Then $hk - h^2 + 1 = h^2 + 1 = 4n + 1 > p$. Hence, for every k-element subset A of $\mathbb{Z}_p$, we can obtain *each* element of $\mathbb{Z}_p$ as a sum of h elements in A. In particular, this holds for every k-element subset A of $[n] = \{1, \ldots, n\}$. Hence, if we define a mapping $s : \{0, 1\}^n \to \mathbb{Z}_p$ by

$$s(x) = x_1 + 2x_2 + 3x_3 + \cdots + nx_n \bmod p,$$

then we have the following interesting property.

Lemma 16.4. *For every index* $r \in [n]$, *every partial assignment with at least* $4\sqrt{n}$ *unspecified entries can be extended to an assignment* $x \in \{0, 1\}^n$ *such that* $s(x) = r$.

Proof. Let $y \in \{0, 1\}^I$ be a partial assignment, $A = [n] \setminus I$ and $s = \sum_{i \in I} iy_i \bmod p$. Assume that $|A| \geq 4\sqrt{n}$. By Theorem 16.3 and its discussion, there exist a subset $B \subseteq A$ of $|B| = 2\sqrt{n}$ elements of A whose sum is equal to $r - s$ modulo p. Hence, if we set $x_k = y_k$ for all $k \in I$, $x_k = 1$ for all $k \in B$, and $x_k = 0$ for all $k \in A \setminus B$, then $s(x) = s + (r - s) = r$, as desired. □

The *weighted sum* function is a boolean function $w_n(x)$ defined by:

$$w_n(x_1, \ldots, x_n) = \begin{cases} x_{s(x)} & \text{if } s(x) \in \{1, \ldots, n\} \\ x_1 & \text{otherwise.} \end{cases}$$

Theorem 16.5. (Savický and Žák 1996) *The function* $w_n(x)$ *is* m-mixed for $m = n - \mathcal{O}(\sqrt{n})$.

Proof. Let $f(x) = w_n(x)$, and $m = n - 4\sqrt{n} - 2$. Take an arbitrary subset $I \subseteq [n]$ of size $|I| = m$, and let $J = [n] \setminus I$ be its complement. Take any two distinct assignments $x, y \in \{0, 1\}^I$. Our goal is to find an assignment $z \in \{0, 1\}^J$ such that $f(x, z) \neq f(y, z)$. When doing this we will use a simple fact that, modulo p, the

weighted sum of the vector (y, z) is the weighted sum of the vector (x, z) plus the
difference

$$\Delta = \sum_{i \in I} i y_i - \sum_{i \in I} i x_i \bmod p.$$

We start with the simpler case where $\Delta = 0$. Fix a position $r \in I$ for which
$x_r \neq y_r$. Since $|J| \geq 4\sqrt{n}$, Lemma 16.4 gives us an assignment $z \in \{0, 1\}^J$ such
that $s(x, z) = r$. Since $\Delta = 0$, we also have that $s(y, z) = s(x, z) + \Delta = r$. Hence,
in this case we have that $f(x, z) = x_r \neq y_r = f(y, z)$, as desired.

In the following, we can assume that $\Delta \neq 0$. Fix an arbitrary $j \in J \setminus \{1\}$,
and let r be rest of $j + \Delta$ modulo p, if this rest lies in $\{1, \ldots, r\}$, and set $r := 1$
otherwise. In any case, we have that $r \neq j$ because $j \neq 1$ and $\Delta \neq 0$. To define
the desired assignment $z \in \{0, 1\}^J$, we consider two possible cases. If $r \in J$ then
set $z_j := 0$ and $z_r := 1$. If $r \in I$ then set $z_j := 1 - y_r$. In both cases we still
have at least $4\sqrt{n}$ unspecified bits in J which, by Lemma 16.4, can be set in such
way that $s(x, z) = j$; hence, $s(y, z) = s(x, z) + \Delta = j + \Delta = r$. Now, if
$r \in J$ then we have that $f(x, z) = z_j = 0 \neq 1 = z_r = f(y, z)$. If $r \in I$
then $f(x, z) = z_j = 1 - y_r \neq y_r = f(y, z)$. Thus, in both cases we have that
$f(x, z) \neq f(y, z)$, as desired. $\square$

Corollary 16.6. *Every deterministic read-once branching program computing the
weighted sum function $w_n(x)$ must have size at least $2^{n - \mathcal{O}(\sqrt{n})}$.*

16.2 P $\neq$ NP $\cap$ co-NP for Read-Once Programs

We now consider *nondeterministic* branching programs. Call such a program *read-
once* (or a 1-NBP) if along any path from the source node to the target node
every variable appears at most once. Note that this is a "syntactic" restriction: such
a program cannot contain any inconsistent paths, that is, paths along which two
contradictory tests "is $x_i = 1$?" and "is $x_i = 0$?" on the same variable x_i are made.
Recall that every boolean function of n variables can be computed by a 1-NBP of
size at most $2\sqrt{2^n}$; see (1.9).

Just as we did it for the size of decision trees, we can ask the P versus NP$\cap$co-NP
question for their (slight) generalization—read-once programs. We will show that
here we also have P $\neq$ NP $\cap$ co-NP.

Namely, we will exhibit a boolean function f of n variables (the "pointer
function") such that both f and $\neg f$ have nondeterministic read-once branching
programs of polynomial size but any deterministic read-once program for f must
have exponential size.

The *pointer function* $\pi_n(x_1, \ldots, x_n)$ is defined as follows. Let s and k be such
that $k s^2 = n$ and $k = \log n$. Arrange the n indices $1, \ldots, n$ of the variables into a
$k \times s^2$ matrix, split the i-th row $(1 \leq i \leq k)$ into s blocks $B_{i1}, B_{i2}, \ldots, B_{is}$ of size
s each, and let y_i be the OR of ANDs of variables in these blocks:

$$y_i = \bigvee_{j=1}^{s} \left(\bigwedge_{l \in B_{ij}} x_l \right) \qquad i = 1, \ldots, k, \tag{16.1}$$

Then define the pointer function by

$$\pi_n(x_1, \ldots, x_n) := x_{\mathrm{bin}(y)+1},$$

where $\mathrm{bin}(y) = 2^{k-1}y_1 + 2^{k-2}y_2 + \cdots + 2y_{k-1} + y_k$ is the number whose binary code is the vector $y = (y_1, \ldots, y_k)$.

Example 16.7. Here is an example of the pointer function of $n = 8$ variables with $k = s = 2$:

$$\underbrace{\overbrace{(x_1 \vee x_2)}^{y_1} \wedge (x_3 \vee x_4)}_{B_{11} \qquad B_{12}} \underbrace{\overbrace{(x_5 \vee x_6)}^{y_2} \wedge (x_7 \vee x_8)}_{B_{21} \qquad B_{22}}$$

On input $a = (1, 0, \ldots, 0)$ we have $y_1 = y_2 = 0$, and hence, $\mathrm{bin}(y) = 0$, implying that $\pi(a) = a_{\mathrm{bin}(y)+1} = a_1 = 1$. On input $a = (1, 0, 0, 1, 0, 0, 1, 0)$ we have $y_1 = 1$, $y_2 = 0$, and hence, $\mathrm{bin}(y) = 2$, implying that $\pi(a) = a_{\mathrm{bin}(y)+1} = a_3 = 0$.

Theorem 16.8. *Both π_n and $\neg\pi_n$ have 1-NBPs of size $\mathcal{O}(n)$ whereas any 1-BP computing π_n must have size at least $2^{s-1} = \exp\left(\Omega(n/\log n)^{1/2}\right)$.*

Proof. We first prove the upper bound. On input vector $x = (x_1, \ldots, x_n)$ in $\{0, 1\}^n$, the desired 1-NBP first guesses a binary string $a = (a_1, \ldots, a_k) \in \{0, 1\}^k$. After that it remains to test if the values $y_1 = a_1, \ldots, y_k = a_k$ satisfy the equalities (16.1) and if the corresponding (to the string a) variable $x_l := x_{\mathrm{bin}(a)} + 1$ has the value 1 (or 0 in the case of $\neg\pi_n$). It is clear that the resulting program is read-once, except that the variable x_l could be tested two times: once in the program P_i making that of the tests (16.1) for which $l \in B_{i1} \cup \ldots \cup B_{is}$, and then once more at the end of a computation. A simple (but crucial) observation is that we can safely replace the variable x_l in that program P_i by the constant 1 (or by 0, in the case of $\neg\pi_n$), so that the whole program is read-once.

We now prove the lower bound. By Lemma 16.1 it is enough to show that the function π_n is m-mixed for $m = s - 1$. To show this, take any two different assignments a and b of constants to a set of m variables in X. Since m is strictly smaller than s, we have that: (i) every block B_{ij} has at least one unspecified variable, and (ii) in every row, at least one block consists entirely of unspecified variables. This means that (regardless of the values of a and b) we can arrange the rest so that the resulting string $(y_1, \ldots, y_k)$ points to a bit x_l where the assignments a and b differ. □

16.3 Branching Programs Without Null-Paths

A *null-path* in a nondeterministic branching program (NBP) is a path from the source to a sink node containing an edge labeled by a variable x_i and an edge labeled by its negation $\neg x_i$. Hence, such a path has "zero conductivity": it cannot

Fig. 16.1 A function f
majorities A if it does not
accept a vector $x \notin A$ with at
most m ones. The function
isolates A if it additionally
rejects all inputs $x \notin A$ with
at most $2m$ ones

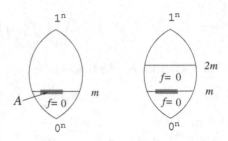

be consistent with any input vector. Although such paths seem to be "useless" (they cannot accept any vector), their presence may exponentially reduce the size (total number of edges) of the program; compare Corollary 16.11 with Proposition 16.13 below. Actually, the presence of such "redundant" paths, just like the presence of NOT gates in circuits, is exactly what makes it so difficult to analyze general branching programs. So, let us look what happens if we forbid null-paths. Let us call such programs *null-path-free* programs. Note that every *deterministic* null-path-free program is just a read-once program. So, the restriction is only interesting for *nondeterministic* branching programs and switching networks. It is clear that every 1-NBP is null-path-free but not every null-path-free NBP must be read-once.

Theorem 16.8 shows that 1-NBP may be exponentially more powerful than their deterministic counterparts, 1-BP. Thus, it is harder to prove high lower bounds even for 1-NBP. Still, we have a general lower-bounds criterion for null-path-free NBP.

For a 0-1 vector x, let $|x|$ denote the number of ones in it. We say that a set $A \subseteq \{0, 1\}^n$ is m-uniform, if $|x| = m$ for all $x \in A$. A set A is *uniform* if it is m-uniform for some $0 \le m \le n$. We also say (see Fig. 16.1) that a boolean function $f : \{0, 1\}^n \to \{0, 1\}$

- *Majorities A* if for every $x \in \{0, 1\}^n$ with $|x| \le m$, $f(x) = 1$ implies $x \in A$;
- *Isolates A* if for every $x \in \{0, 1\}^n$ with $|x| \le 2m$, $f(x) = 1$ implies $x \in A$.

Define the *k-th degree*, $d_k(A)$, of A as the maximum number of vectors in A, all of which have 1s on some fixed set of k coordinates. That is,

$$d_k(A) := \max_{|I|=k} \big| \{x \in A : x_i = 1 \text{ for all } i \in I\} \big|.$$

Define also $d(A) = \min d_k(A) \cdot d_{m-k}(A)$, where the minimum is taken over all k, $1 \le k \le m$. The following general lower bounds were proved by Jukna and Razborov (1998).

Theorem 16.9. *Let $A \subseteq \{0, 1\}^n$ be a uniform set of vectors. Then every 1-NBP majorizing A as well as every null-path-free NBP isolating A requires at least $|A|/d(A)$ nodes.*

Proof. We first consider the case of 1-NBP Let m be the number of ones in the vectors of A, and let P be a 1-NBP majorizing A. Hence, P computes some boolean function $f(x)$ such that for every vector $x \in \{0, 1\}^n$ with $|x| \le m$ ones, $f(x) = 1$ iff $x \in A$. Let $1 \le k \le m$ be an integer for which $d(A) = d_k(A) \cdot d_{m-k}(A)$.

For each input $a \in A$, fix an accepting path consistent with a. Since a has m 1-bits, and no vector with a smaller number of 1-bits can be accepted, all the m 1-bits of a must be tested along this path. Split this path into two segments (p_a, q_a), where p_a is an initial segment of the path accepting a along which exactly k 1-bits of a are tested. We denote the corresponding set of bits by I_a, and let J_a denote the set of remaining $m - k$ 1-bits of a. For a node v of P, let A_v denote the set of all inputs $a \in A$ such that v is the terminal node of p_a. We are going to finish the proof by showing that $|A_v| \le d_k(A)d_{m-k}(A)$ for every node v.

Fix some node v of P, and let $\mathcal{I} = \{I_a : a \in A_v\}$, $\mathcal{J} = \{J_b : b \in A_v\}$. Since our program is read-once, we have that $I \cap J = \emptyset$ for all $I \in \mathcal{I}$ and $J \in \mathcal{J}$. Take now an arbitrary pair $I \in \mathcal{I}$, $J \in \mathcal{J}$, and denote by $c_{I,J}$ the input defined by $c_{I,J}(i) = 1$ iff $i \in I \cup J$.

Claim 16.10. For every $I \in \mathcal{I}$ and $J \in \mathcal{J}$, the combined input $c_{I,J}$ belongs to A.

Proof. Choose some $a, b \in A_v$ such that $I = I_a$, $J = J_b$. Since I and J are disjoint, the path (p_a, q_b) is consistent with the input $c_{I,J}$. Hence, this input is accepted because the path leads to an accepting sink. But since $|I| + |J| = m$ and m is the smallest number of 1s in an accepted input, this is possible only when this combined input $c_{I,J}$ belongs to A. $\square$

With this claim in mind, we fix an arbitrary $J \subset \mathcal{J}$ and notice that $\{c_{I,J} : I \in \mathcal{I}\}$ is a set of different inputs from A, all of which have 1s on J. Hence, $|\mathcal{I}| \le d_{|J|}(A) \le d_{m-k}(A)$ (provided $J \ne \emptyset$). Similarly, $|\mathcal{J}| \le d_k(A)$ which implies $|\mathcal{I}| \cdot |\mathcal{J}| \le d_k(A) \cdot d_{m-k}(A)$. Finally, every $a \in A_v$ is uniquely determined by the pair (I_a, J_a), therefore $|A_v| \le |\mathcal{I}| \cdot |\mathcal{J}|$. This completes the proof of the desired inequality $|A_v| \le d_k(A)d_{m-k}(A)$, and thus, the proof of the theorem in the case of 1-NBP.

If our program P is null-path-free but not necessarily a 1-NBP, then along an accepting path, some variables may be tested many times. This, in particular, means that the sets I and J may not be disjoint. Still, if we require that our program not only majorities the set A but also isolates it, then Claim 16.10 applies also in this case. Indeed, as before, the path (p_a, q_b) is consistent with the combined input $c_{I,J}$, meaning that this input must be accepted by the program. But since this input has $|c_{I,J}| = |I \cup J| \le |I| + |J| \le k + m \le 2m$ ones, and since the program cannot accept any input with $\le 2m$ ones lying outside A, this implies that $c_{I,J}$ must belong to A. The rest of the proof is the same as in the case of 1-NBP. $\square$

The *perfect matching function* is a monotone boolean function PM_n of m^2 variables. Inputs for this function are subsets $E \subseteq K_{m,m}$ of edges of a fixed complete bipartite $m \times m$ graph $K_{m,m}$, and $f_m(E) = 1$ iff E contains a perfect matching, that is, a set of m vertex-disjoint edges. Taking a boolean variable $x_{i,j}$ for each edge of $K_{m,m}$, the function can be written as

$$PM_n(x) = \bigvee_{\sigma \in S_m} \bigwedge_{i=1}^{m} x_{i,\sigma(i)} ,$$

where S_m is the set of all $m!$ permutations of $1, 2, \ldots, m$.

The *exact perfect matching* function EPM_n accepts a graph E iff E *is* a perfect matching. That is, EPM_n takes a boolean $n \times n$ matrix as an input, and outputs 1 iff this is a permutation matrix, that is, each row and each column has exactly one 1.

In Sect. 9.11 we have shown that PM_n requires monotone circuits of super-polynomial size. Now we show that it also requires 1-NBP of exponential size.

Corollary 16.11. *Every* 1-NBP *computing* PM_n *as well as any null-path-free* NBP *computing* EPM_n *must have size* $2^{\Omega(n)}$.

Proof. Let A be the set of all $|A| = n!$ permutation matrices; hence, A is m-uniform with $m = n$. Since only $(n - k)!$ perfect matchings can share k edges in common, we have that $d_k(A) = (n - k)!$. In particular, taking $k = n/2$, we obtain that $d(A) \leq (n/2)! \cdot (n/2)!$. Observe that every program computing PM_n majorities A, and every program computing EPM_n must isolate A. Thus, Theorem 16.9 yields the desired lower bound $n!/d(A) \geq \binom{n}{n/2}$. $\square$

To better understand the role of null-paths, we have to first solve the following problem. Say that a nondeterministic branching program is *weakly read-once* if along any *consistent* s-t path no variable is tested more than once. That is, we now put no restrictions on inconsistent paths: only consistent paths are required to be read-once.

The following problem is one of the "easiest" questions about branching programs, but it still remains open!

■ **Research Problem 16.12.** Prove an exponential lower bound for weakly read-once nondeterministic branching programs.

That such programs may be much more powerful than 1-NBPs shows the following observation made in Jukna (1995).

Proposition 16.13. *The function* EPM_n *can be computed by a weakly read-once nondeterministic branching program of size* $\mathcal{O}(n^3)$.

Proof. To test that a given square 0-1 matrix is a permutation matrix, it is enough to test whether every row has at least one 1, and every column has at least $n - 1$ zeros. These two tests can be made by two nondeterministic branching programs P_1 and P_2 designed using the formulas

$$P_1(X) = \bigwedge_{i=1}^{n} \bigvee_{j=1}^{n} x_{i,j} \quad \text{and} \quad P_2(X) = \bigwedge_{j=1}^{n} \bigvee_{k=1}^{n} \bigwedge_{\substack{i=1 \\ i \neq k}}^{n} \neg x_{i,j} .$$

Let $P = P_1 \wedge P_2$ be the AND of these two programs, that is, the sink-node of P_1 is the source-node of P_2. The entire program has size $\mathcal{O}(n^3)$. It remains to verify that P is read-once. But this is obvious because all the contacts in P_1 are positive whereas all contacts in P_2 are negative; so every s-t path in the whole program P is either inconsistent or is read-once. $\square$

Thus, the presence of "redundant" paths—those consistent with *none* of the input vectors—may exponentially decrease the size of branching programs! To understand the actual role of such paths is one of the main problems in circuit complexity.

16.4 Parity Branching Programs

We already know how to prove exponential lower bounds for "syntactic" read-once branching programs, where along *any* input-output path each variable is tested at most once. In this case we can prove high lower bounds for deterministic as well as for nondeterministic programs. The situation with *parity* branching programs ($\oplus$-BP) is, however, much worse: here no high lower bounds are known even for syntactic read-once programs. Recall that a $\oplus$-BP is just a nondeterministic branching program with a counting acceptance mode: it accepts a given input vector a if the number of input-output paths consistent with a is odd. In a *read-once* $\oplus$-BP no variable can be re-tested along any path.

■ **Research Problem 16.14.** Prove an exponential lower bound for read-once parity branching programs.

So far, exponential lower bounds for such programs are only known under the additional restriction that the program is *oblivious*. The nodes are partitioned into at most n levels so that edges go only from one level to the next, all the edges of one level are labeled by contacts of one and the same variable, and different levels have different variables.

To prove exponential lower bounds for oblivious read-once $\oplus$-BPs, we will employ one specific property of linear codes—their "universality".

Recall that a linear code is just a set of vectors $C \subseteq \{0, 1\}^n$ which forms a linear subspace of $GF(2)^n$. The *minimal distance* of a code C is a minimal Hamming distance between any pair of distinct vectors in C. It is well known (and easy to show) that the minimal distance of C is exactly the minimum weight of (that is, the number of 1s in) a nonzero vector from C. The *dual* of C is the set $C^{\perp}$ of all those 0-1 vectors that are orthogonal over $GF(2)$ to all vectors in C, that is, the set of all vectors whose scalar product over $GF(2)$ with every vector in C is equal 0.

Say that a set of vectors $C \subseteq \{0, 1\}^n$ is k-*universal* if for any subset of k coordinates $I \subseteq \{1, \dots, n\}$ the projection of vectors from C onto this set I gives the whole cube $\{0, 1\}^I$. A nice property of linear codes is that their duals are universal.

Proposition 16.15. *The dual of every linear code of minimal distance $k + 1$ is k-universal.*

Proof. Let C be a linear code of minimal distance $k + 1$, and take a set $I \subseteq \{1, \ldots, n\}$ with $|I| \le k$. The set of all projections of vectors in the vector space $W = C^{\perp}$ onto I forms a linear subspace. If this subspace were proper, then some nonzero vector x, whose support $S_x = \{i : x_i = 1\}$ lies in I, would belong to the orthogonal complement $W^{\perp} = C$ of W. But this would mean that C has minimum distance at most $|S_x| \le |I| \le k$, a contradiction. $\Box$

A *characteristic function* of a set $C \subseteq \{0, 1\}^n$ is a boolean function f_C such that $f_C(x) = 1$ iff $x \in C$.

Theorem 16.16. *Let $C \subseteq \{0, 1\}^n$ be a linear code with minimal distance d_1, and let d_2 be the minimal distance of the dual code $C^{\perp}$. Then every oblivious read-once $\oplus$-BP computing the characteristic function f_C of C has size at least $2^{\min\{d_1, d_2\} - 1}$.*

Proof. Let P be an oblivious read-once $\oplus$-BP computing f, set $k := \min\{d_1, d_2\} - 1$, and let $I \subseteq \{1, \ldots, n\}$ be the set of bits tested on the first $k = |I|$ levels of P. Every assignment $a : I \to \{0, 1\}$ (treated for this purpose as a restriction) defines a subfunction f_a of f of $n - |I|$ variables which is obtained from f by setting x_i to $a(i)$ for all $i \in I$. Let $\mathcal{F}$ be the subspace of the 2^{n-k}-dimensional space of all boolean functions on $n - k$ variables, generated by the subfunctions f_a of f with $a : I \to \{0, 1\}$. It is not difficult to see that $size(P) \ge \dim(\mathcal{F})$.

Indeed, if $v_1, \ldots, v_r$ are the nodes at the k-th level of P, then for every assignment $a : I \to \{0, 1\}$, the subfunction f_a is a linear combination of the functions computed by the sub-programs with source-nodes $v_1, \ldots, v_r$: $f_a(b) = 1$ iff the number of accepting paths in $P(a, b)$ is odd. Hence, we need at least $r \ge \dim(\mathcal{F})$ such functions to get all the subfunctions in $\mathcal{F}$.

Now we can finish the proof as follows. Since the dual of C has distance $d_2 \ge k + 1$, we have by Proposition 16.15, that the code C itself is k-universal. This, in particular, means that for every assignment $a : I \to \{0, 1\}$ there is an assignment $x_a : \overline{I} \to \{0, 1\}$ such that $(a, x_a) \in C$. Moreover, since C has distance $d_1 > k = |I|$, we have that $(b, x_a) \notin C$ for every other assignment $b : I \to \{0, 1\}, b \ne a$. Thus, if we describe the subfunctions $f_a, a : I \to \{0, 1\}$, as rows of a $2^k \times 2^{n-k}$ matrix, then this matrix contains a diagonal $2^k \times 2^k$ submatrix with entries $f(a, x)$ such that $f(a, x) = 1$ iff $x = x_a$. So, the matrix has row-rank equal 2^k, which means that the subfunctions in $\mathcal{F}$ are linearly independent (over any field, including GF(2)). Thus, $size(P) \ge \dim(\mathcal{F}) = |\mathcal{F}| \ge 2^k$, as desired. $\Box$

To give an explicit lower bound, recall that the r-th order binary Reed–Muller code $R(r, m)$ of length $n = 2^m$ is the set of graphs (sequences of values) of all multilineal polynomials in m variables over GF(2) of degree at most r. We have $\sum_{i=0}^{r} \binom{m}{i} \approx m^r$ possible monomials, and each subset of these monomials gives us a polynomial. We list the vectors in $\{0, 1\}^m$ as $a_1, \ldots, a_n$, and associate with each polynomial $p(z_1, \ldots, z_m)$ a code-word $\big(p(a_1), \ldots, p(a_n)\big)$ in $\{0, 1\}^n$. This code is linear and has minimal distance 2^{m-r}.

Corollary 16.17. *Let $n = 2^m$ and $r = \lfloor m/2 \rfloor$. Then every oblivious read-once $\oplus$-BP computing the characteristic function of the Reed-Muller code $R(r, m)$ has size at least $2^{\Omega(\sqrt{n})}$.*

Proof. It is known (see, for example, MacWilliams and Sloane 1997, p. 374) that the dual of $R(r, m)$ is the code $R(m - r - 1, t)$. Hence, in the notation of Theorem 16.16 we have that $d_1 = 2^{m-r} \geq \Omega(\sqrt{n})$ and $d_2 = 2^{r+1} \geq \Omega(\sqrt{n})$. The desired bound follows. $\qquad\square$

For other explicit codes, as BCH-codes, the lower bound can be increased to $2^{\Omega(n)}$; see Example 16.22 below.

16.5 Linear Codes Require Large Replication

Recall that the *replication number* of a program is the minimal number R such that along every computation path, at most R variables are tested more than once. The sets of variables re-tested along different computations may be different. We will now prove exponential lower bounds for deterministic branching programs with replication number $R = \epsilon n$ for a constant $\epsilon > 0$. Recall that $R = n$ is the maximal possible value corresponding to unrestricted branching programs.

But before we start, let us first show that testing just 1 bit twice can help a lot! For this, let us again consider the pointer function π_n, introduced in Sect. 16.2. We already know (see Theorem 16.8) that any deterministic branching program of replication number $R = 0$ (that is, a read-once program) for this function must have exponential size. We now show that allowing to re-test just 1 bit along each path reduces the size drastically.

Proposition 16.18. *The pointer function π_n can be computed by a deterministic branching program of size $\mathcal{O}(n^2/\log n)$ and replication number $R = 1$.*

Proof. For each $i = 1, \ldots, k$, let P_i be an obvious 1-BP of size $s^2 = n/k \leq n/\log n$ computing the function $y_i = \bigvee_{j=1}^{s} \left(\bigwedge_{x \in B_{ij}} x \right)$. Arrange these programs into a binary tree of height k: the first level consists of P_1, the second consists of two copies of P_2 having the 0 and 1 leaves of P_1 as its start nodes, and the i-th one consists of 2^{i-1} copies of P_i. In this way we obtain a read-once program of size $\mathcal{O}(2^k n/k) = \mathcal{O}(n^2/\log n)$. This program has $2^k = n$ leaves, each labeled by the corresponding string $a - (a_1, \ldots, a_k)$ of values of $(y_1, \ldots, y_k)$, and hence, by the corresponding index $l = \text{bin}(a)$. Replace each such leaf by a size-1 branching program testing the corresponding variable x_{l+1}. The resulting program has replication number $R = 1$, computes π_n and has the desired size. $\qquad\square$

We are now going to show that some explicit boolean functions require large replication number R, growing with the number n of variables. We will present two entirely different lower bounds arguments for $(1, +R)$-branching programs. The first one, presented in this section, is numerically weaker—works only for

$R = o(n/\log n)$—but is (apparently) more instructive. Moreover, it works for important objects—characteristic functions of linear codes. A different argument, presented in the next section, gives exponential lower bounds for programs of almost maximal replication $R = \Omega(n)$, but the functions for which it works are no longer as "simple"—they are quadratic functions of good expander graphs.

The following theorem gives a general lower-bounds criterion for $(1, +R)$-branching programs: a function is hard to compute by such programs if the accepted inputs lie far away from each other, and if the function cannot be made constant 0 by fixing few variables to 0 and 1. Namely, say that a boolean function is

- *d-rare* if any two accepted inputs differ in at least d bits;
- *m-robust* if it is not possible to make the function be constant 0 by fixing fewer that m variables.

The following general lower bound was proved by Jukna and Razborov (1998) using earlier results of Zák (1995) and Savický and Zák (1997).

Theorem 16.19. *Let $0 \le d, m, R \le n$ be arbitrary integers. Every $(1, +R)$-branching program computing a d-rare and m-robust function must have size at least $2^{(\min\{d,\, m/(R+1)\}-1)/2}$.*

The idea behind the proof of this fact is the following. If all computations are long (of length at least m) and the program is not too large, a lot of computation paths must split and join again. At that node where they join again, some information about the inputs leading to this node is lost. If too much information is lost and not too many (at most R) variables may be re-tested once again, it is not possible to compute the correct value of the function.

The intuition about the "loss of information" is captured by the following notion of "forgetting pairs" of inputs. Given a branching program P and a partial input $a : [n] \to \{0, 1, *\}$, $comp(a)$ is the path in P consistent with a until we reach a node where the first test of $*$ is made. For two partial inputs a and b, let $D(a, b)$ be the set of all bits where they both are defined and have different values. The *support* $S(a)$ of a partial input a is the set of all specified bits, that is, bits i for which $a(i) \ne *$. A composition $b = a_1 a_2 \cdots a_s$ of partial inputs $a_1, a_2, \ldots, a_s$, whose supports are pairwise disjoint, is a (partial) input defined by $b(i) = a_j(i)$ for $i \in S(a_j)$. The *size* $|P|$ of a branching program P is the number of nodes.

Let a, b be (partial) inputs with $S(a) = S(b)$. Given a branching program P, the pair a, b is called a *forgetting pair* (for P) if there exists a node w such that w belongs to both $comp(a)$ and $comp(b)$, and both computations read all the variables with indices in $D(a, b)$ at least once before reaching w (Fig. 16.2). Thus, all the bits

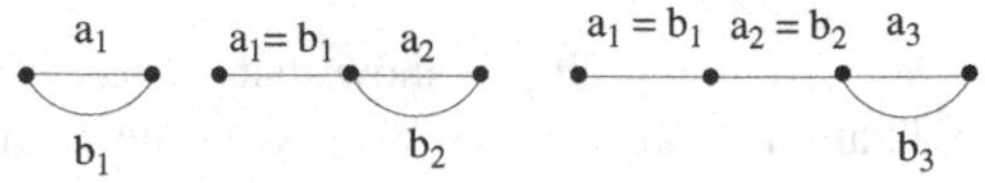

Fig. 16.2 Forgetting pairs a_1 and b_1, a_1a_2 and a_1b_2, $a_1a_2a_3$ and $a_1a_2b_3$

from $D(a, b)$ are "forgotten" when the computations reach the node w. To increase the number of these forgotten bits, we need the following definition.

Say that a sequence (a_j, b_j), $j = 1, \ldots, s$ of pairs of partial inputs $a_j \neq b_j$ forms an (s, l)-*chain* if $S(a_j) = S(b_j) = I_j$, $|I_j| \leq l$, the I_j are pairwise disjoint and, for all $j = 1, \ldots, s$, the inputs $a_1 \cdots a_{j-1} a_j$ and $a_1 \cdots a_{j-1} b_j$ form a forgetting pair.

Lemma 16.20. *Let P be a branching program in which every computation reads at least m different variables. If $s \leq m/(2 \log |P| + 1)$ and $l \leq 2 \log |P| + 1$, then P has an (s, l)-chain.*

Proof. Given a branching program P, one can get a forgetting pair by following all the computations until $r := \lfloor \log |P| \rfloor + 1$ different bits are tested along each of them. Since $|P| < 2^r$, at least two of these paths must first split and then join in some node. Take the corresponding partial inputs a_1' and b_1' and extend them to a_1 and b_1 such that $S(a_1) = S(b_1) = S(a_1') \cup S(b_1')$ and $D(a_1, b_1) \subseteq S(a_1') \cap S(b_1')$. In this way we get a forgetting pair of inputs $a_1 \neq b_1$ both of which are defined on the same set of at most $|S(a_1') \cup S(b_1')| \leq 2r - 1$ bits. We can now repeat the argument for the program P_{a_1} obtained by setting all variables x_i with $i \in S(a_1)$ to the corresponding values of a_1, and obtain next forgetting pair of inputs $a_1 a_2$ and $a_1 b_2$, etc. We can continue this procedure for s steps until $s(2r - 1) \leq s(2 \log |P| + 1)$ does not exceed the minimum number m of different variables tested on a computation of P. □

Proof of Theorem 16.19. Suppose that some branching $(1, +R)$-program P computes a d-rare and m-robust function and has size $|P|$ smaller than stated in Theorem 16.19, that is, assume that

$$2 \log |P| + 1 < \min\{d,\ m/(R + 1)\}.$$

We can assume w.l.o.g. that $d \geq 2$ (otherwise the bound becomes trivial), and this implies that every 1-term of f has size $n \geq m$. Hence, in order to force f to either 0 or 1 we must specify at least m positions, implying that every computation of P must read at least m different variables. Since $2 \log |P| + 1 < m/(R + 1)$, we can apply Lemma 16.20 with $s := R + 1$ and obtain that P must contain an (s, l)-chain with $s = R + 1$ and $l \leq 2 \log |P| + 1 < \min\{d,\ m/(R + 1)\}$. That is, we can find $R + 1$ pairwise disjoint sets I_i of size

$$|I_i| \leq 2 \log |P| + 1 < \min\{d,\ m/(R + 1)\} \tag{16.2}$$

and pairs $a_i \neq b_i$ of assignments on these sets such that all partial inputs $a_1 \cdots a_{i-1} a_i$ and $a_1 \cdots a_{i-1} b_i$ form forgetting pairs in P.

By (16.2), the partial input $a_1 \cdots a_{R+1}$ specifies strictly fewer than m variables. Since f is m-robust, $a_1 \cdots a_{R+1}$ can be extended to a totally defined input a such that $f(a) = 1$.

As the sets $I_1, \ldots, I_{R+1}$ are non-empty and pairwise disjoint, and at most R variables can be re-tested along any computation, there must exist j such that all variables with indices from I_j are tested *at most once* along $comp(a)$. Now, let w be the node that corresponds to the forgetting pair $a_1 \cdots a_{j-1} a_j$ and $a_1 \cdots a_{j-1} b_j$. The node w is on $comp(a)$. All variables with indices from $D(a_j, b_j) \subseteq I_j$ are already tested along $comp(a)$ before w, hence no such variable is tested after w, and the computation on the input c obtained from a by replacing a_j with b_j can not diverge from $comp(a)$ after the node w. Therefore, $f(c) = f(a) = 1$. But (16.2) implies that $|I_j| < d$, contradicting the d-rareness of f. This completes the proof of Theorem 16.19. $\qquad\qquad\qquad\qquad\qquad\qquad\qquad\qquad\qquad\qquad\qquad\qquad\qquad\quad$ □

This theorem is especially useful for (characteristic functions of) linear codes, that is, for linear subspaces of $GF(2)^n$. It is clear that the characteristic function f_C of a linear code C is d-rare if and only if the minimal distance of C is at least d. Also, Proposition 16.15 implies that f_C is m-robust if and only if the minimal distance of its dual $C^\perp$ is at least m. Hence, Theorem 16.19 implies the following lower bound for characteristic functions of linear codes.

Theorem 16.21. *Let C be a linear code with minimal distance d_1, and let d_2 be the minimal distance of the dual code $C^\perp$. Then every $(1, +R)$-branching program computing the characteristic function of C has size exponential in $\min\{d_1, d_2/R\}$.*

This theorem yields exponential lower bounds on the size of $(1, +R)$-branching programs computing characteristic functions of many linear codes.

Example 16.22. (BCH-codes) Let $n = 2^\ell - 1$, and let $C \subseteq \{0, 1\}^n$ be a BCH-code with designed distance $\delta = 2t + 1$, where $t \leq \sqrt{n}/4$, and let f_C be its characteristic function. Let d_2 be the minimal distance of its dual $C^\perp$. The Carliz–Uchiyama bound (see, e.g., MacWilliams and Sloane 1997, p. 280) says that $d_2 \geq 2^{\ell-1} - (t - 1)2^{\ell/2}$ which is $\Omega(n)$ due to our assumption on t. Since the minimal distance d_1 of a BCH-code is always at least its designed distance δ, we get from Theorem 16.21 that every $(1, +R)$-branching program computing f_C has size exponential in $\min\{t, n/R\}$. In particular, if $t = \omega(\log n)$ then every such program must have super-polynomial size as long as $R = o(n/\log n)$.

16.6 Expanders Require Almost Maximal Replication

We are now going to prove exponential lower bound on the size of branching programs with almost maximal replication number $R = \Omega(n)$. This was done in Jukna (2008). The functions for which we prove such a bound will be quadratic functions of a specially chosen graph, the so-called *Ramanujan graph*. Let $G = (V, E)$ be an undirected graph on $V = \{1, \ldots, n\}$. The quadratic function of G over $GF(2)$ is a boolean function

$$f_G(x_1, \ldots, x_n) = \sum_{\{i,j\} \in E} x_i x_j \bmod 2.$$

That is, given an input vector $a \in \{0, 1\}^n$, we remove all vertices i with $a_i = 0$, and count the number of the surviving edges modulo 2.

It is clear that f_G can be computed by an unrestricted branching program (with replication $R = n$) of size $\mathcal{O}(n^2)$. We will show that good expanding properties of the graph G imply that every branching program computing $f_G \wedge (x_1 \oplus x_2 \oplus \cdots \oplus x_n \oplus 1)$ must have either replication number $R > \epsilon n$ for a constant $\epsilon > 0$ or must have exponential size.

But first we will prove a general theorem telling us what properties of boolean functions do actually force the replication number of their branching programs to be large.

A boolean function $r(x_1, \ldots, x_n)$ is a *rectangular function* if there is a balanced partition of its variables into two parts such that r can be written as an AND of two boolean functions, each depending on variables in only one part of the partition. A set $R \subseteq \{0, 1\}^n$ of vectors is a *combinatorial rectangle* (or just a *rectangle*) if $R = r^{-1}(1)$ for some rectangular function r. So, each combinatorial rectangle has a form $R = R_0 \times R_1$ where $R_0 \subseteq \{0, 1\}^{I_0}$ and $R_1 \subseteq \{0, 1\}^{I_1}$ for some partition $[n] = I_0 \cup I_1$ of $[n] = \{1, \ldots, n\}$ into two disjoint parts I_0 and I_1 whose sizes differ by at most 1.

The *rectangle number*, $\rho(f)$, of a boolean function f is the maximum size of a rectangle lying in $f^{-1}(1)$:

$$\rho(f) = \max\{|R| : R \text{ is a rectangle and } f(x) = 1 \text{ for all } x \in R\}.$$

Finally, we say that a boolean function f of n variables is:

- *Sensitive* if any two accepted vectors differ in at least 2 bits;
- *Dense* if $|f^{-1}(1)| \geq 2^{n-o(n)}$, and
- *Rectangle-free* if $\rho(f) \leq 2^{n-\Omega(n)}$.

Theorem 16.23. *There is a constant $\epsilon > 0$ with the following property: if f is a sensitive, dense and rectangle-free boolean function of n variables, then any deterministic branching program computing f with the replication number $R \leq \epsilon n$ must have $2^{\Omega(n)}$ nodes.*

Proof. Let f be a sensitive and dense boolean function of n variables. Suppose also that the function f is rectangle-free, that is, $f^{-1}(1)$ does not contain a rectangle of size larger than $2^{n-\delta n}$, for some constant $\delta > 0$. Take an arbitrary deterministic branching program computing f with replication number $R \leq \epsilon n$, where $\epsilon > 0$ is a sufficiently small constant to be specified later; this constant will only depend on the constant δ. Our goal is to prove that the program must have at least $2^{\Omega(n)}$ nodes.

For an input $a \in \{0, 1\}^n$ accepted by f, let $comp(a)$ denote the (accepting) computation path on a. Since the function f is sensitive, all n bits are tested at least once along each of these paths. Split every path $comp(a)$ into two parts $comp(a) = (p_a, q_a)$, where p_a is an initial segment of $comp(a)$ along which $n/2$ different bits are tested. Hence, the remaining part q_a can test at most $n/2 + R$ different bits. (Note that we only count the number of tests of *different* bits—the total number of tests along $comp(a)$ may be much larger than $n + R$.) Let S be the number of nodes in our program.

Viewing segments p_a and q_a as monomials (ANDs of literals), we obtain that f can be written as an OR of at most S ANDs $P \wedge Q$ of DNFs P and Q satisfying the following three conditions:

(i) All monomials have length at least $n/2$ and at most $n/2 + R$. This holds by the choice of segments p_a and q_a.

(ii) Any two monomials in each of the DNFs are inconsistent, that is, one contains a variable and the other contains its negation. This holds because the program is deterministic: the paths must split before they meet.

(iii) For all monomials $p \in P$ and $q \in Q$, either $pq = 0$ (the monomials are inconsistent) or $|X(p) \cap X(q)| \leq R$ and $|X(p) \cup X(q)| = n$, where $X(p)$ is the set of variables in a monomial p. This holds because the program has replication number R and the function f is sensitive.

Now fix one AND $P \wedge Q$ for which the set B of accepted vectors is the largest one; hence, the program must have at least $|f^{-1}(1)|/|B| \geq 2^{n-o(n)}/|B|$ nodes, and it remains to show that the set B cannot be too large, namely that

$$|B| \leq 2^{n-\Omega(n)}.$$

We do this by showing that otherwise the set B, and hence, also the set $f^{-1}(1)$, would contain a large rectangle in contradiction with the rectangle-freeness of f. When doing this we only use the fact that all vectors of B must be accepted by an AND of DNFs satisfying the properties (i)–(iii) above.

By (iii) we know that every vector $a \in B$ must be accepted by some pair of monomials $p \in P$ and $q \in Q$ such that $|X(p) \cap X(q)| \leq R$. A (potential) problem, however, is that for different vectors a the corresponding monomials p and q may share *different* variables in common. This may prohibit their combination into a rectangle (see Example 16.25 below). To get rid of this problem, we just fix a set Y of $|Y| \leq R$ variables for which the set $A \subseteq B$ of all vectors in B accepted by pairs of monomials with $X(p) \cap X(q) = Y$ is the largest one. Since $R \leq \epsilon n$, we have that

$$|A| \geq |B| \Big/ \sum_{i=0}^{R} \binom{n}{i} \geq |B| \cdot 2^{-n \cdot H(\epsilon)},$$

where $H(x) = -x \log_2 x - (1 - x) \log_2(1 - x)$ is the binary entropy function.

Claim 16.24. The set A contains a rectangle C of size

$$|C| \geq \frac{|A|^2}{9 \cdot 2^{n+R}}.$$

Assuming the claim, we can finish the proof of the theorem as follows. By the rectangle-freeness of f, we know that $|C| \leq 2^{n-\delta n}$ for a constant $\delta > 0$. By

Claim 16.24, we know that

$$|A| \leq 3 \cdot 2^{(n+R)/2} |C| \leq 3 \cdot 2^{(1+\epsilon)n/2+(1-\delta)n}.$$

Hence, if $R \leq \epsilon n$ for a constant $\epsilon > 0$ satisfying $\epsilon + 2H(\epsilon) < 2\delta$, then

$$|B| \leq |A| \cdot 2^{H(\epsilon)n} \leq 3 \cdot 2^{n-(2\delta-\epsilon-2H(\epsilon))n/2} \leq 2^{n-\Omega(n)}.$$

It remains therefore to prove Claim 16.24.

Each monomial of length at most k accepts at least a 2^{-k} fraction of all vectors from $\{0,1\}^n$. Hence, there can be at most 2^k mutually inconsistent monomials of length at most k. By (i) and (ii), this implies that

$$|P| \leq 2^{n/2} \quad \text{and} \quad |Q| \leq 2^{n/2+R}. \tag{16.3}$$

For each monomial $p \in P \cup Q$, let $A_p := \{a \in A : p(a) = 1\}$ be the set of all vectors in A accepted by p; we call these vectors *extensions* of p. Note that, by the definition of the set A, $a \in A_p$ if $pq(a) = 1$ for some monomial $q \in Q$ such that $X(p) \cap X(q) = Y$.

Since, by (ii), the monomials in P are mutually inconsistent, no two of them can have a common extension. Since every vector from A is an extension of at least one monomial $p \in P$, the sets A_p with $p \in P$ form a partition of A into $|P|$ disjoint blocks. The average size of a block in this partition is $|A|/|P|$. Say that a monomial $p \in P$ is *rich* if the corresponding block A_p contains $|A_p| \geq \frac{1}{3}|A|/|P|$ vectors. Similarly for monomials in Q. By averaging, at least two-thirds of vectors in A must be extensions of rich monomials in P. Since the same also holds for monomials in Q, at least one vector $x \in A$ must be an extension of some rich monomial $p \in P$ and, at the same time, of some rich monomial $q \in Q$.

Let y be the projection of x onto $Y = X(p) \cap X(q)$. Since all variables in Y are tested in both monomials p and q, all the vectors in A_p and in A_q coincide with y on Y. Consider the set of vectors $C = C_1 \times \{y\} \times C_2$, where C_1 is the set of projections of vectors in A_q onto the set of variables $X \setminus X(q)$, and C_2 is the set of projections of A_p onto the set of variables $X \setminus X(p)$. Since both monomials p and q have at least $n/2$ variables, the set C is a rectangle of size

$$|C| = |C_1| \cdot |C_2| = |A_p| \cdot |A_q| \geq \frac{|A|}{3|P|} \cdot \frac{|A|}{3|Q|} \geq \frac{1}{9} \frac{|A|}{2^{n/2}} \cdot \frac{|A|}{2^{n/2+R}} = \frac{1}{9} \frac{|A|^2}{2^{n+R}}.$$

Hence, it remains to verify that $C \subseteq A$, i.e., that all vectors $c \in C$ are accepted by $P \wedge Q$.

The vector x belongs to C and has the form $x = (x_1, y, x_2)$ with $x_i \in C_i$. Now take an arbitrary vector $c = (c_1, y, c_2)$ in C. The vector (x_1, y, c_2) belongs to A_p. Hence, there must be a monomial $q' \in Q$ such that $X(p) \cap X(q') = Y$ and pq' accepts this vector. Since all bits of x_1 are tested in p and none of them belongs to Y, none of these bits is tested in q'. Hence, q' must accept also the vector $c = (c_1, y, c_2)$. Similarly, using the fact that (c_1, y, x_2) belongs to A_q, we can conclude that the vector $c = (c_1, y, c_2)$ is accepted by some monomial $p' \in P$. Thus, the vector c is accepted by the monomial $p'q'$, and hence, by $P \wedge Q$.

This completes the proof of Claim 16.24, and thus, the proof of Theorem 16.28. $\quad\square$

In the last step of the proof of Theorem 16.23 it was important that every vector from A is accepted by a pair of monomials sharing the *same* set of variables Y; otherwise, the rectangle C need not lie within the set A.

Example 16.25. Take $P = \{x_1, \neg x_1\}$ and $Q = \{x_2, x_1 \neg x_2\}$ with $p = x_1$ and $q = x_2$. The AND $P \wedge Q$ accepts the set of vectors $A = \{11, 01, 10\}$. The projection of $A_q = \{11, 01\}$ onto $X \setminus X(q) = \{x_1\}$ is $C_1 = \{0, 1\}$, and the projection of $A_p = \{11, 10\}$ onto $X \setminus X(p) = \{x_2\}$ is also $C_2 = \{0, 1\}$. But $C = C_1 \times C_2 \nsubseteq A$, because 00 does not belong to A.

In the proof of Theorem 16.23 it was also important that the branching program was *deterministic*: this resulted in the property (ii) above which, in its turn, gave upper bounds (16.3) on the total number of monomials. In the case of nondeterministic branching programs we do not necessarily have this property, and in this case no exponential lower bounds are known even for $R = 1$ (cf. Research Problem 16.12).

16.6.1   Quadratic Functions of Expanders are Hard

To apply Theorem 16.23, we need an explicit boolean function which is sensitive, dense and rectangle-free. Note that the first two conditions—being sensitive and dense—are easy to ensure. A more difficult task is to ensure rectangle-freeness. The problem here is that f must be rectangle-free under *any* balanced partition of its variables. For this purpose, we consider quadratic functions of graphs. Recall that a quadratic function of a graph $G = ([n], E)$ is a boolean function

$$f_G(x_1, \ldots, x_n) = \sum_{\{i,j\} \in E} x_i x_j \bmod 2.$$

What properties of the underlying graph G do ensure that its quadratic function f_G is rectangle-free? We will show that if G has a large "matching number", then f_G is rectangle free.

Define the *matching number* $m(G)$ as the largest number m such that, for every balanced partition of vertices of G, at least m crossing edges form an induced matching. (An edge is crossing if it joins a vertex in one part of the partition with a vertex in the other part. Being an induced matching means that no two endpoints of any two edges of the matching are joined by a crossing edge.)

Fig. 16.3 After the setting to
0 all variables outside the
induced matching, the
function
$f_G = \bigoplus_{\{i,j\} \in E} x_i y_j$ turns to
the inner product function
$IP_m = x_1 y_1 \oplus \cdots \oplus x_m y_m$

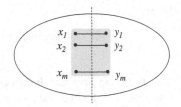

Lemma 16.26. *For every graph G on n vertices, $\rho(f_G) \leq 2^{n - m(G)}$.*

Proof. Fix an arbitrary balanced partition of the vertices of G into two parts. The partition corresponds to a partition (x, y) of the variables of f_G. Let $r = r_1(x) \wedge r_2(y)$ be an arbitrary rectangle function with respect to this partition, and suppose that $r \leq f$. Our goal is to show that r can accept at most $2^{n - m(G)}$ vectors.

By the definition of $m(G)$, some set $M = \{x_1 y_1, \ldots, x_m y_m\}$ of $m = m(G)$ crossing edges $x_i y_i$ forms an induced matching of G. We set to 0 all variables corresponding to vertices outside the matching M (see Fig. 16.3).

Since M is an *induced* subgraph of G, the obtained subfunction of f_G is just the inner product function

$$IP_m(x_1, \ldots, x_m, y_1, \ldots, y_m) = \sum_{i=1}^{m} x_i y_i \bmod 2.$$

The obtained subfunction $r' = r_1'(x_1, \ldots, x_m) \wedge r_2'(y_1, \ldots, y_m)$ of the rectangle function $r = r_1 \wedge r_2$ is also a rectangle function such that $r'(a) \leq IP_m(a)$ for all $a \in \{0, 1\}^{2m}$. Since r' was obtained from r by setting to 0 at most $n - 2m$ variables, we have that $|r^{-1}(1)| \leq |B| \cdot 2^{n-2m}$ where $B = \{a : r'(a) = 1\}$. Hence, it remains to show that $|B| \leq 2^m$. For this, let H be a $2^m \times 2^m$ matrix defined by

$$H[x, y] = (-1)^{IP_m(x,y) \oplus 1}.$$

Since, for every $x \neq 0$, $IP_m(x, y) = 1$ for exactly half of vectors y, this matrix is a Hadamard matrix. Since our set $B \subseteq \{0, 1\}^m \times \{0, 1\}^m$ lies within $IP_m^{-1}(1)$, it corresponds to an all-1 submatrix of H. By the Lindsey Lemma (see Appendix A), $|B| \leq 2^m$.   □

By Lemma 16.26, we need graphs G such that, for any balanced partition of their vertices, many crossing edges form an induced matching. To ensure this property, it is enough that the graph is "mixed enough".

Say that a graph is *s-mixed* if every two disjoint s-element subsets of its vertices are joined by at least one edge.

Lemma 16.27. *If an n-vertex graph G of maximum degree d is s-mixed, then*

$$m(G) \geq \frac{n - 2s}{4(d + 1)}.$$

Hence,

$$\log \rho(f_G) \le n - \frac{n}{4(d+1)} + \frac{s}{2(d+1)}.$$

Proof. Fix an arbitrary balanced partition of the vertices of G into two parts. To construct the desired induced matching, formed by crossing edges, we repeatedly take a crossing edge and remove it together with all its neighbors. In each step we remove at most $2d+1$ vertices. If the graph is s-mixed, then the procedure will run for m steps as long as $\lfloor n/2 \rfloor - (2d+1)m$ is at least s. □

Fix a prime power $q \ge 2^6$, and let f_G be the quadratic function of a Ramanujan graph $G = RG(n,q)$ of degree $q+1$. By Corollary 5.24, Ramanujan graphs $RG(n,q)$ are s-mixed for $s = 2n/\sqrt{q}$. Consider the function

$$f_n = f_G \wedge (x_1 \oplus \cdots \oplus x_n \oplus 1).$$

That is, given an input vector $a \in \{0,1\}^n$, we remove all vertices i with $a_i = 0$, and let $f_n(a) = 1$ iff the number of 1s in a is even and the number of surviving edges is odd.

Theorem 16.28. *There is a constant $\epsilon > 0$ such that any deterministic branching program computing f_n with the replication number $R \le \epsilon n$ requires size $2^{\Omega(n)}$.*

Proof. Our function f_n is a polynomial of degree at most 3 over GF(2). Moreover, f_n is nonzero because $f_n(a) = 1$ for an input vector $a \in \{0,1\}^n$ with precisely two 1s corresponding to the endpoints of some edge of G. Thus, Exercise 16.9 implies that f_n accepts at least 2^{n-3} vectors, and hence, is a dense function. Since $q \ge 2^6$, the graph G is s-mixed for $s = n/4$. Thus, Lemma 16.27 implies that the quadratic function f_G, and hence, also the function f_n is rectangle-free. Finally, the presence of the parity function in the definition of f_n ensures that f_n is a sensitive function.

Since f_n is sensitive, dense and rectangle-free, Theorem 16.23 implies that there is a constant $\epsilon > 0$ such that any deterministic branching program computing g_n with the replication number $\le \epsilon n$ must have size $2^{\Omega(n)}$. □

Theorem 16.28 also gives an exponential lower bound for programs working in bounded time. Say that a program works in time T if every accepting computation in it has length at most T.

Corollary 16.29. *There is a constant $\epsilon > 0$ such that any deterministic branching program computing f_n and working in time $T \le (1+\epsilon)n$ requires size $2^{\Omega(n)}$.*

Proof. Since f_n is sensitive, along each accepting computation all n its variables must be tested at least once. This means that for branching programs computing sensitive functions we always have $R \le T - n$. Hence, Theorem 16.28 yields exponential lower bounds also for the class of time $(1+\epsilon)n$ branching programs for a constant $\epsilon > 0$. □

Exercises

16.1. Let $P = \{1, \ldots, n\}$ be the set of points of a projective plane $PG(2, q)$ of order q, and let $L_1, \ldots, L_n$ be the lines viewed as subsets of P; hence $n = q^2 + q + 1$. Recall that each line has exactly $q + 1$ points, every two lines intersect in exactly one point, and exactly $q + 1$ lines meet in one point. A *blocking set* is a set of points which intersects every line. The smallest blocking sets are just the lines. Show that the characteristic function $f(x) = \bigwedge_{i=1}^{n} \bigvee_{j \in L_i} x_j$ of blocking sets is m-mixed for $m = q - 1$.

Hint: Show that for every subset $I \subseteq P$ of $|I| = q - 1$ points there must be two distinct lines L_1, L_2 such that $I \cap L_1 = I \cap L_2 = \{i\}$.

16.2. A boolean function $f(X)$ is *m-stable* if, for every $Y \subseteq X$ of size $|Y| \leq m$ and for every variable $x \in Y$, there exists an assignment $c : X \setminus Y \to \{0, 1\}$ such that either $f_c(Y) = x$ or $f_c(Y) = \neg x$. That is, after the assignment c, the value of the subfunction $f_c(Y)$ depends only on that of the variable x. Show that every m-stable boolean function is also m-mixed.

16.3. For a monotone boolean function f, let $\mathrm{Min}(f)$ be the set of all its minterms. Show that a monotone boolean function is m-stable if and only if, for every $Y \subseteq X$ of size $|Y| \leq m$ and for every variable $x \in Y$, there exists a minterm $K \in \mathrm{Min}(f)$ such that, $K \cap Y = \{x\}$ and $W \setminus (K \cup Y) \neq \emptyset$ for all $W \in \mathrm{Min}(f)$ with $x \notin W$.

16.4. The *perfect matching* function is a monotone boolean function PM_n in n^2 variables, encoding the edges of a bipartite graph with parts of size n. The function computes 1 iff the input graph contains a perfect matching. Taking a boolean variable $x_{i,j}$ for each edge of $K_{n,n}$, the function can be written as

$$PM_n = \bigvee_{\sigma \in S_n} \bigwedge_{i=1}^{n} x_{i,\sigma(i)},$$

where S_n is the set of all permutations of $\{1, \ldots, n\}$. We have shown in Sect. 9.11 that this function requires monotone circuits of size $n^{\Omega(\log n)}$. Show that the function PM_n is $(n-1)$-stable.

16.5. Call a boolean function $f(X)$ *weakly m-mixed* if for every $Y \subseteq X$ of size $|Y| = m$ and any two different assignments $a, b : Y \to \{0, 1\}$, it holds that $f_a = f_b = 0$ or $f_a \neq f_b$. Define $\mathrm{Cov}(f, k)$ to be the minimal t for which there exist mutually inconsistent monomials $K_1, \ldots, K_t$ of k literals each, such that $f \leq K_1 \vee \ldots \vee K_t$. Show that every deterministic read-once branching program computing a weakly m-mixed boolean function f must have at least $\mathrm{Cov}(f, m-1)$ nodes.

Hint: Argue as in the proof of Lemma 16.1, but consider only *accepting* paths.

16.6. Recall that a boolean function f is *d-robust* if f cannot be made a constant-0 function by fixing any its $d - 1$ variables to constants. Show that, if a boolean function f of n variables is $(d+1)$-robust, then $\mathrm{Cov}(f,k)$ is at least exponential in dk/n.

Hint: Hit the inequality $f \le K_1 \vee \ldots \vee K_t$ with a restriction a assigning random 0-1 values to randomly chosen d variables. Let K be any monomial of length k. Given that exactly s variables of K are set by a, the probability that $K_a \not\equiv 0$ is 2^{-s}. Use this to show that $\mathrm{Prob}[K_a \not\equiv 0] \le 2^{-\Omega(dk/n)}$.

16.7. Show that, if a boolean function f can be computed by a 1-NBP of size S, then f can be written as an OR of at most S rectangular functions.

16.8. Recall that the clique function $\mathrm{CLIQUE}(n,k)$ has $\binom{n}{2}$ variables x_{ij}, one for each potential edge in a graph on a fixed set of n vertices; the function outputs 1 iff the associated graph contains a clique (complete subgraph) on some k vertices. Use Exercise 16.7 to show that the size of every 1-NBP computing this function must be exponential in $\min\{k, n-k\}$.

Hint: The union of any two graphs, each with at most $k-1$ non-isolated vertices, cannot contain a k-clique.

16.9. Show that every nonzero polynomial $p(x_1, \ldots, x_n)$ of degree k over $\mathrm{GF}(2)$ has at least 2^{n-k} nonzero points, that is, $|\{a: p(a) = 1\}| \ge 2^{n-k}$.

Hint: Take a monomial $\prod_{i \in I} x_i$ of p with $|I| = k$. Show that, for every setting b of constants to variables x_i with $i \notin I$, there is a setting c of constants to the remaining k variables such that $p(b, c) = 1$.

Chapter 17
Bounded Time

In this chapter we show a time-size tradeoff for nondeterministic branching programs. By the *size* of a program in this chapter we will mean the number of nodes, not just the number of labeled edges. A program computes a given function *in time* T if every accepted input has at least one accepting computation of length at most T.

Our goal is to show that some functions of n variables cannot be computed in time $T = \mathcal{O}(n)$ unless the program has size $2^{\Omega(n)}$. In the case of *deterministic* branching programs such a result was established in a celebrated paper of Ajtai (1999b). The case of *nondeterministic* branching programs is, however, much more difficult: here no exponential lower bounds are known even for $T = n$.

Let us stress that our restriction is a "semantic" one: inconsistent paths may be arbitrarily long! Even consistent paths may be arbitrarily long: we only require that for every accepted input there *exists* at least one short path consistent with this input. The "syntactic" case, where all paths (be they consistent or not) must have length at most T, is easier to deal with (see the chapter notes and Exercise 17.4).

To obtain high lower bounds for "semantic" nondeterministic programs, we consider programs computing functions $f : D^n \to \{0, 1\}$ over domains D larger than $\{0, 1\}$. In this case, instead of just tests "is $x_i = 0$?" and "is $x_i = 1$?" the program is allowed to make tests "is $x_i = d$?" for elements $d \in D$. Different edges leaving the same node may make the same test—this is why a program is nondeterministic. As before, an input $a \in D^n$ is accepted iff all the tests along at least one path from the source node to the target node are passed by a.

The exposition below is based on the papers by Ajtai (1999a), Beame, Jayram[1], and Saks (2001), and Jukna (2009b).

---

[1] Formerly Jayram S. Thathachar.

S. Jukna, *Boolean Function Complexity*, Algorithms and Combinatorics 27,
DOI 10.1007/978-3-642-24508-4_17, © Springer-Verlag Berlin Heidelberg 2012

17.1   The Rectangle Lemma

We consider nondeterministic branching programs computing a given function f :
$D^n \to \{0, 1\}$ and working in time kn where k is an arbitrarily large constant. We
want to show that any such program must be large. As in the case of programs with
bounded replication, the idea is to show that if the number of nodes is small then the
program is forced to accept all vectors of a large rectangle. Having shown this, we
construct a function f that cannot accept many vectors of any rectangle. This will
imply that any program for f working in time kn must have large size.

Let $X = \{x_1, \ldots, x_n\}$ be a set of n variables. A subset $R \subseteq D^n$ of vectors is an
m-rectangle ($m \le n/2$), if there exist two disjoint m-element subsets X_0 and X_1 of
X, subsets $R_0 \subseteq D^{|X_0|}$ and $R_1 \subseteq D^{|X_1|}$ of vectors, and a vector w in D of length
$n - 2m$ such that (after some permutation of the variables) the set R can be written
as $R = R_0 \times \{w\} \times R_1$; the vector w is then the broomstick of the rectangle. That
is, on the variables outside $X_0 \cup X_1$ all vectors in R have the same values as the
vector w. Below is an example of a 2-rectangle over the domain $D = \{0, 1, 2\}$ with
a broomstick $w = (1, 2, 0, 0, 2)$:

$$R = \{1\} \times \begin{Bmatrix} 0 \\ 1 \end{Bmatrix} \times \{2\} \times \begin{Bmatrix} 2 \\ 0 \end{Bmatrix} \times \{0\} \times \{0\} \times \begin{Bmatrix} 21 \\ 01 \end{Bmatrix} \times \{2\}.$$

With some abuse of notation we will write $R = R_0 \times \{w\} \times R_1$, meaning that this
holds after the corresponding permutation of variables:

$$R = \overbrace{\begin{pmatrix} 02 \\ 00 \\ 12 \\ 10 \end{pmatrix}}^{R_0} \times \overbrace{\{1\} \times \{2\} \times \{0\} \times \{0\} \times \{2\}}^{w} \times \overbrace{\begin{Bmatrix} 21 \\ 01 \end{Bmatrix}}^{R_1}.$$

Note that combinatorial rectangles considered in the previous chapter are
m-rectangles with $m = n/2$. That is, we now just refine this notion, and consider
rectangles, all vectors in which have a common "broomstick" (i.e., which coincide
on some fixed set of $n - 2m$ positions) (Fig. 17.1).

The main property of m-rectangles is (again) the "cut-and-paste" property: if
some m-rectangle R contains two vectors (a_0, w, a_1) and (b_0, w, b_1), then it must
also contain both combined vectors (b_0, w, a_1) and (a_0, w, b_1).

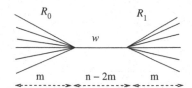

Fig. 17.1 An m-rectangle
with a broomstick w

The refined *rectangle number*, $\rho_m(f)$, of a function $f : D^n \to \{0, 1\}$ is the maximum size of an m-rectangle lying in $f^{-1}(1)$:

$$\rho_m(f) = \max\{|R| : R \text{ is an } m\text{-rectangle and } f(x) = 1 \text{ for all } x \in R\}.$$

In general, we have that $0 \le \rho_m(f) \le |D|^{2m}$ for every function $f : D^n \to \{0, 1\}$. Standard examples of functions with very small rectangle number are characteristic functions f of codes $C \subseteq D^n$ of large minimum distance d: for such functions we have that $\rho_m(f) = 1$ as long as $m \le d - 1$.

A function $f : D^n \to \{0, 1\}$ is *sensitive* if any two accepted vectors differ in at least two positions. The only property of sensitive functions we will use is that, in any branching program computing such a function, each variable must be tested at least once along any accepting computation path. The density of f is $\mu(f) = |f^{-1}(1)|/|D|^n$.

Rectangle Lemma. If a sensitive function $f : D^n \to \{0, 1\}$ can be computed by a nondeterministic branching program of size S working in time kn then, for every $m \le n/2^{k+1}$,

$$\rho_m(f) \ge \frac{1}{\Delta}|D|^{2m} \quad \text{where} \quad \Delta = \frac{(2S)^r \binom{n}{m}^2}{\mu(f)} \quad \text{and} \quad r = 8k^2 2^k.$$

We first give an application of this lemma, and then prove the lemma itself.

17.2 A Lower Bound for Code Functions

As our domain D we take a Galois field $D = \mathrm{GF}(q)$. We consider the function $g_N(Y, x)$ of $N = n^2 + n$ variables, the first n^2 of which are arranged in an $n \times n$ matrix Y. The values of the function are defined by

$$g_N(Y, x) = 1 \text{ iff the vector } x \text{ is orthogonal over } \mathrm{GF}(q) \text{ to all rows of } Y.$$

In other words, $g_N(Y, x) = 1$ iff the vector x belongs to a linear code defined by the parity-check matrix Y.

We say that a nondeterministic branching program computes $g_N(Y, x)$ in time T if for every accepted input (Y, x), there exists at least one accepting s-t path for (Y, x) along which at most T tests on x-variables are made—the first n^2 variables from Y can be tested an arbitrary number of times.

Using ideas similar to those in the proof of Theorem 15.4, it is not difficult to show that $g_N(Y, n)$ can be computed by a branching program of size $\mathcal{O}(q^2 N)$ whose every path has length at most about n^2: we only need to compute the AND of n scalar products over $\mathrm{GF}(q)$, each of at most $2n$ variables. The following theorem shows that if the time is restricted to kn for a constant k, then exponential size is necessary.

Theorem 17.1. *Let* $k \geq 1$ *be an integer. If* $q \geq 2^{3k+9}$ *then every nondeterministic branching program computing* $g_N(Y, x)$ *in time* kn *must have size exponential in* $\sqrt{N}/k^2 4^k$.

Proof. Let $d = m + 1$ where $m = n/2^{k+1}$. By the Gilbert–Varshamov bound (see, e.g., MacWilliams and Sloane 1997), linear codes C of distance d and size

$$|C| \geq \frac{q^n}{V(n, d-1)} = \frac{q^n}{V(n, m)}$$

exist, where

$$V(n, m) = \sum_{i=0}^{m} (q-1)^i \binom{n}{i} \leq dq^m \binom{n}{m}$$

is the number of vectors in a Hamming ball of radius m around a vector in $GF(q)^n$.

Let H be the parity-check matrix of such a code, and consider the function $f : GF(q)^n \to \{0, 1\}$ such that $f(x) = 1$ iff $Hx = \mathbf{0}$. That is, $f(x) = 1$ iff $x \in C$. The function $f(x) = g_N(H, x)$ is the characteristic function of the code C and is a subfunction of $g_N(Y, x)$. Hence, if the function $g_N(Y, x)$ can be computed by a nondeterministic branching program working in time kn, then the size of this program must be at least the minimum size S of a nondeterministic branching program computing $f(x)$ in time kn. To finish the proof of the theorem, it remains therefore to show that S must be exponential in m/r, where $r = 8k^2 2^k$ is from the Rectangle Lemma.

The function $f(x)$ has density $\mu(f) = |C|/q^n = 1/V(n, m)$. Hence, the Rectangle Lemma yields

$$\rho_m(f) \geq \frac{\mu(f)}{(2S)^r \binom{n}{m}^2} q^{2m} \geq \frac{q^{2m}}{(2S)^r dq^m \binom{n}{m}^3} = \frac{q^m}{(2S)^r d \binom{n}{m}^3}.$$

Recalling that $m = n/2^{k+1}$ and $q \geq 2^{3k+9}$, we obtain

$$\binom{n}{m}^3 \leq \left(\frac{en}{m}\right)^{3m} = (e^3 2^{3k+3})^m \leq (2^{3k+8})^m \leq \left(\frac{q}{2}\right)^m.$$

Hence, $(2S)^r d \geq 2^m / \rho_m(f)$. On the other hand, since the Hamming distance between any two vectors in C is at least $d = m + 1$, we have that $\rho_m(f) \leq 1$, and the desired lower bound $S = 2^{\Omega(m/r)}$ on the size of our branching program follows. $\square$

17.3   Proof of the Rectangle Lemma

We will use one purely combinatorial result which may be of independent interest.

Let $\mathcal{F} = (F_1, \ldots, F_r)$ be a sequence of not necessarily distinct subsets of some set X. By a *separator* for $\mathcal{F}$ we will mean a pair (S, T) of disjoint subsets of X such that each member of $\mathcal{F}$ is disjoint from either S or from T, that is, none of the members of $\mathcal{F}$ can contain elements from *both* sets S and T. The *size* of such a separator (S, T) is the minimum of $|S|$ and $|T|$. The *degree* d_x of a point x in $\mathcal{F}$ is the number of members of $\mathcal{F}$ that contain x. The *average degree* of $\mathcal{F}$ is

$$d = \frac{1}{|X|} \sum_{x \in X} d_x.$$

Separator Lemma. Let $\mathcal{F} = (F_1, \ldots, F_r)$ be a sequence of non-empty subsets of an n-element set, each with at most ℓ elements. Let d be the average degree of $\mathcal{F}$, and let m be the maximum size of a separator of $\mathcal{F}$.

(i)   If $\ell < n/8d$, then $m \geq n/2N$ where $N = \binom{r}{0} + \binom{r}{1} + \cdots + \binom{r}{2d}$.
(ii) If $\ell \leq n/8d\,2^d$, then $m \geq n/2^{d+1}$.

We postpone the proof of this lemma and proceed with the actual proof of the Rectangle Lemma.

For each input $a \subseteq f^{-1}(1)$, fix one accepting computation path $comp(a)$, and split it into r sub-paths $p_1, \ldots, p_r$ of length at most $\ell = kn/r$; the length of a sub-path p_i is the number of tests made along it. That is, we have r *time segments* $1, \ldots, r$, and in the i-th of them the computation on a follows the sub-path p_i.

Say that two inputs $a, b \in f^{-1}(1)$ are equivalent if the starting nodes of the corresponding sub-paths $comp(a) = (p_1, \ldots, p_r)$ and $comp(b) = (q_1, \ldots, q_r)$ coincide. Since we have at most S nodes in the program, the number of possible equivalence classes does not exceed S^r. Fix some largest equivalence class $A \subseteq f^{-1}(1)$; hence,

$$|A| \geq |f^{-1}(1)|/S^r.$$

We say that a pair of disjoint subsets of variables X_0 and X_1 is *good* for a set of vectors B if there is a coloring of time segments $1, \ldots, r$ in red and blue such that, along each computation $comp(a) = (p_1, \ldots, p_r)$ on a vector $a \in B$, the variables from X_0 are tested only in red and those from X_1 only in blue sub-paths.

Claim 17.2. For every vector $a \in f^{-1}(1)$, at least one pair of disjoint m-element subsets of variables with $m \geq n/2^{k+1}$ is good for a.

Proof. For a variable $x \in X$, let d_x be the number of sub-paths $comp(a) = (p_1, \ldots, p_r)$ along which this variable is tested. Since the computed function $f(X)$ is sensitive and since $f(a) = 1$, we know that each variable $x \in X$ is tested at least once along $comp(a)$. Since the program computes $f(X)$ in time kn, we also know that at most kn tests can be made along the whole computation $comp(a)$.

Hence, $\sum_{x \in X} d_x \leq kn$, implying that average number $\sum_{x \in X} d_x/n$ of tests made on a single variable does not exceed k. Finally, since $r = 8k^2 2^k$ we know that each sub-path can make at most

$$\ell = \frac{kn}{r} = \frac{kn}{8k^2 2^k} = \frac{n}{8k 2^k}$$

tests. We can therefore apply the Separator Lemma and obtain the desired pair X_0, X_1 of disjoint subsets of $|X_0| = |X_1| \geq n/2^{k+1}$ variables which is good for the input a. $\square$

We have only 2^r possible colorings of time intervals $1, \ldots, r$, and at most $\binom{n}{m}\binom{n-m}{m} \leq \binom{n}{m}^2$ pairs of disjoint m-element subsets of variables. Hence, by Claim 17.2, some of these pairs X_0, X_1 must be good for a subset $B \subseteq A$ of size

$$|B| \geq \frac{|A|}{2^r \binom{n}{m}^2}.$$

We can write each vector $a \in D^n$ as $a = (a_0, w, a_1)$, where a_0 is the projection of a onto X_0, a_1 is the projection of a onto X_1, and w is the projection of a onto $X \setminus (X_0 \cup X_1)$. Say that two vectors $a = (a_0, w, a_1)$ and $b = (b_0, w', b_1)$ are equivalent if $w = w'$. Since the sets of variables X_0 and X_1 are disjoint, each equivalence class is an m-rectangle.

Let $R \subseteq B$ be a largest equivalence class lying in B; hence

$$|R| \geq \frac{|B|}{|D|^{n-2m}} \geq \frac{|A|}{2^r \binom{n}{m}^2 |D|^{n-2m}} \geq \frac{|f^{-1}(1)|}{S^r 2^r \binom{n}{m}^2 |D|^{n-2m}} = \frac{1}{\Delta} |D|^{2m}.$$

So, it remains to show that all vectors of the rectangle R are accepted by the program. This is a direct consequence of the following more general claim.

Claim 17.3. If both vectors $a = (a_0, w, a_1)$ and $b = (b_0, w, b_1)$ belong to B, then both combined vectors (a_0, w, b_1) and (b_0, w, a_1) belong to A.

Proof. Let $comp(a) = (p_1, \ldots, p_r)$ and $comp(b) = (q_1, \ldots, q_r)$ be the computations on a and on b. Consider the combined vector $c = (a_0, w, b_1)$. Our goal is to show that $p_t(c) \vee q_t(c) = 1$ for all $t = 1, \ldots, r$. That is, that for each $t = 1, \ldots, r$, the combined vector c must be accepted by (must be consistent with) at least one of the sub-paths p_t or q_t.

To show this, assume that c is not accepted by p_t. Since p_t accepts the vector $a = (a_0, w, a_1)$, and this vector coincides with the combined vector $c = (a_0, w, b_1)$ on all the variables outside X_1, this means that at least one variable from X_1 must be tested along p_t. But then, by the goodness of the pair X_0, X_1, no variable from X_0 can be tested along the sub-path q_t. Since q_t accepts the vector $b = (b_0, w, b_1)$, and the combined vector $c = (a_0, w, b_1)$ coincides with this vector on all the variables outside X_0, the sub-path q_t must accept the vector c, as desired.

This completes the proof of the Rectangle Lemma. $\square$

Proof of the Separator Lemma. (i) Associate with each element $x \in X$ its *trace* $T(x) = \{i \mid x \in F_i\}$. Hence, $d_x = |T(x)|$. By double-counting,

$$\sum_{i=1}^{r} |F_i| = \sum_{x \in X} |T(x)| = dn. \tag{17.1}$$

We will concentrate on elements whose traces are not too large. Namely, say that an element $x \in X$ is *legal* if $|T(x)| \leq 2d$. It is clear that we must have at least $n/2$ legal elements, for otherwise the second sum in (17.1) would be larger than $(2d)(n/2) = dn$.

Partite the legal elements into blocks, where two elements x and y belong to the same block iff they have the same trace, that is, iff $T(x) = T(y)$. Since $|T(x)| \leq 2d$, each block in this partition is determined by a subset of $\{1, \ldots, r\}$ of size at most $2d$. So, the total number of blocks does not exceed $\sum_{i=0}^{2d} \binom{r}{i} = N$.

Say that a legal element $x \in X$ is *happy* if the (unique) block, which it belongs to, has at least $n/2N$ elements. If we will find two legal elements $x \neq y \in X$ such that both of them are happy and $T(x) \cap T(y) = \emptyset$, then we are done.

First observe that, by the same averaging argument as above, at least half of all $n/2$ legal elements must be happy (belong to large blocks); hence, at least $n/4$ elements are both legal and happy. Fix any such element x. We have only to show that there is yet another legal and happy element y which belongs to none of the $|T(x)| \leq 2d$ sets F_i containing x. For this, it is enough to observe that the total number of elements that belong to some of the sets F_i containing x is

$$\left| \bigcup_{i \in T(x)} F_i \right| \leq \sum_{i \in T(x)} |F_i| \leq |T(x)| \cdot \ell \leq 2d\ell$$

which, due to our assumption $\ell < n/8d$, is strictly smaller than the total number $n/4$ of legal and happy elements.

To prove the second claim (ii), we use a probabilistic argument due to Beame et al. (2001).

Color each set $F \in \mathcal{F}$ red or blue uniformly and independently with probability $1/2$. Say that an element $x \in X$ is *red* (respectively, *blue*) if every set that contains x is colored red (respectively, blue). Let S and T be, respectively, the set of red and of blue elements. Since every element of X occurs in at least one set, it follows that S and T are disjoint. Moreover, for each $F \in \mathcal{F}$, either $F \cap S$ or $F \cap T$ is empty. To complete the proof, we show that with positive probability both S and T have at least $(1 - \delta)2^{-d}n$ elements, where

$$\delta = \sqrt{\frac{\ell d 2^{d+1}}{n}}. \tag{17.2}$$

This will prove the lemma, since $\ell \leq n/8d\,2^d$ implies $\delta \leq 1/2$.

Each variable is red as well as blue with probability 2^{-d_x}. Hence, we can expect

$$\mu := \sum_{x \in X} 2^{-d_x} \geq n \left(\prod_{x \in X} 2^{-d_x} \right)^{1/n} = n 2^{-\sum_x d_x/n} \geq n 2^{-k} \qquad (17.3)$$

red variables as well as at least $n 2^{-k}$ blue variables. The first inequality here follows from the arithmetic-geometric mean inequality

$$\frac{1}{n} \sum_{i=1}^{n} a_i \geq \left(\prod_{i=1}^{n} a_i \right)^{1/n} .$$

We will now use Chebyshev's inequality to show that at least one coloring must produce at least $m \geq (1 - \delta)n/2^k$ red elements *and* at least so many blue elements.

Let $Z = \sum_x Z_x$ where Z_x is the indicator random variable for the event "$x \in S$"; hence, $Z = |S|$ and $\mu = E[Z]$. By Chebyshev's inequality, we have that

$$\text{Prob}[|S| < (1 - \delta)\mu] \leq \frac{\text{Var}[Z]}{\delta^2 \mu^2}, \qquad (17.4)$$

where $\text{Var}[Z] = E[Z^2] - E[Z]^2$ is the variance of Z. The variance itself can be written as

$$\text{Var}[Z] = \sum_x \text{Var}[Z_x] + \sum_{x \neq y} \text{Cov}(Z_x, Z_y), \qquad (17.5)$$

where $\text{Cov}(Z_x, Z_y) = E[Z_x Z_y] - E[Z_x] \cdot E[Z_y]$ is the covariance of Z_x and Z_y. Consider the first term in the right-hand side of (17.5). For any x, Z_x is a Bernoulli random variable, so $\text{Var}[Z_x] = E[Z_x] - E[Z_x]^2 \leq E[Z_x]$, implying that

$$\sum_x \text{Var}[Z_x] \leq E[Z] = \mu. \qquad (17.6)$$

To bound the second term in the right-hand side of (17.5), observe that if no member of $\mathcal{F}$ contains both x and y, then Z_x and Z_y are independent, implying that $\text{Cov}(Z_x, Z_y) = 0$. Thus, we are only interested in those pairs (x, y) such that *some* member of $\mathcal{F}$ contains both x and y. For any fixed x, the number of such pairs (x, y) is at most $(\ell - 1)d_x$. For each such pair,

$$\text{Cov}(Z_x, Z_y) \leq E[Z_x Z_y] \leq E[Z_x] \leq 2^{-d_x}.$$

Therefore,

$$\sum_{x \neq y} \text{Cov}(Z_x, Z_y) \leq (\ell - 1) \sum_x d_x 2^{-d_x}.$$

The last term above can be bounded using the following inequality (also due to Chebyshev): if $a_1, \ldots, a_n$ is a non-decreasing sequence and $b_1, \ldots, b_n$ a non-increasing sequence of non-negative numbers, then

$$\sum_{i=1}^{n} a_i b_i \leq \frac{1}{n} \left(\sum_{i=1}^{n} a_i \right) \left(\sum_{i=1}^{n} b_i \right).$$

Order the x's so that the sequence $\{d_x\}$ is non-decreasing. Now the second Chebyshev inequality can be applied to the sequences $\{d_x\}$ and $\{2^{-d_x}\}$. We obtain

$$\sum_{x \neq y} \mathrm{Cov}(Z_x, Z_y) \leq \frac{\ell - 1}{n} \left(\sum_x 2^{-d_x} \right) \left(\sum_x d_x \right) = d(\ell - 1)\mu \qquad (17.7)$$

because $(\sum_x d_x)/n = d$, and $\sum_x 2^{-d_x} = \mu$. Substitute the bounds (17.6) and (17.7) into (17.5). We obtain $\mathrm{Var}[Z] \leq (d(\ell - 1) + 1)\mu \leq dr\mu$, where the last inequality holds because each $x \in X$ occurs in at least one set, implying that $d = \sum_x d_x/n \geq 1$. Using this upper bound on the variance, (17.4) yields:

$$\mathrm{Prob}[|S| < (1 - \delta) \cdot \mu] < \frac{\mathrm{Var}[Z]}{\delta^2 \mu^2} \leq \frac{d\ell}{\delta^2 \mu}.$$

Substituting for δ its value as given by (17.2) and using the inequality (17.3), we obtain that the right-hand side does not exceed

$$\frac{d\ell n}{d\ell 2^{d+1}\mu} = \frac{n}{2^{d+1}\mu} \leq \frac{n}{2^{d+1}2^{-d}n} = \frac{1}{2}.$$

In a similar fashion, we obtain $\mathrm{Prob}[|T| < (1 - \delta) \cdot \mu] < 1/2$. Thus, with positive probability, both S and T have size at least $(1-\delta)\mu \geq (1-\delta)2^{-d}n$. We conclude that there is a coloring of the sets in $\mathcal{F}$ such that the induced S and T satisfy $|S|, |T| \geq n/2^{d+1}$, as desired.   □

Remark 17.4. Note that in the proof of the first claim (i) we have actually proved a stronger statement than just the existence of a large enough separator: there exist two disjoint subsets Y_0 and Y_1 of X, two disjoint subsets T_0 and T_1 of $\{1, \ldots, r\}$ such that, for both $\alpha = 0, 1$, we have that $|Y_\alpha| \geq n/2N$, $1 \leq |T_\alpha| \leq 2d$, and $T(x) = T_\alpha$ for all $x \in Y_\alpha$.

■ **Research Problem 17.5.** Prove an exponential lower bound on the size of nondeterministic *boolean* ($D = \{0, 1\}$) branching programs of n variables, all whose consistent paths have length at most n.

Note that if the computed function is sensitive, then the restriction that all consistent paths must have length at most n implies that along every consistent s-t path, each variable is tested exactly once; this is the model of weakly read-once programs we introduced in Sect. 16.3, see Problem 16.13. In the case of *deterministic* branching programs we already know how to prove exponential lower bounds on the size even when up to ϵn variables may be re-tested along each computation. Nothing similar, however, is known for boolean *nondeterministic* programs.

What makes the problem nontrivial when nondeterminism is allowed is that the restriction is only on *consistent* paths. The "syntactic" case when the restriction is on *all* paths (be they consistent or not) is much easier to analyze (see Exercise 17.3).

Chapter Notes

In a syntactic read-k program, along every path (be it consistent or not!) every variable can appear at most k times. Exponential lower bounds on the size of syntactic read-k times branching programs for boolean functions were first proved by Okolnishnikova (1991) for deterministic programs, and by Borodin et al. (1993) and Jukna (1995) for nondeterministic programs; see Exercise 17.4.

The non-syntactic case, where only *consistent* paths must be no longer than kn, remained open for a long time. In the case of *deterministic* programs, exponential lower bounds for programs working in time kn, for $k = 1+1/\log n$, were proved by Jukna and Razborov (1998). Beame et al. (2001) proved such bonds for $k = 1 + \epsilon$, where $\epsilon > 0$ is very small but constant. Such a lower bounds also follows from lower bounds on branching programs with bounded replication (see Corollary 16.29). This holds because, if the computed function is sensitive, then the replication number of the program cannot exceed the computation time minus the number of variables.

A breakthrough came with the paper of Ajtai (1999b) where he was able to modify his proof for programs over large domains to the case of the boolean domain $D = \{0, 1\}$. He achieves this by a very delicate probabilistic reasoning leading to a much sharper version of the Rectangle Lemma.

Ajtai's Rectangle Lemma. Let k be a positive integer, and $f : \{0, 1\}^n \to \{0, 1\}$ a boolean function such that $|f^{-1}(1)| \geq 2^{n - \mathcal{O}(1)}$. Suppose that f can be computed by a deterministic branching program of size $2^{o(n)}$ working in time kn. Then there exist constants $\mu, \lambda > 0$ such that $\mu^{1+0.01/k} \geq 2\lambda$, and a μn-rectangle $R = R_0 \times \{w\} \times R_1$ such that $R \subseteq f^{-1}(1)$ and $|R_0|, |R_1| \geq 2^{(\mu - \lambda)n}$.

Unfortunately, the full proof of this lemma is too long to be presented here. Using this lemma Ajtai (1999b) proves that, for every constant $k \geq 1$, every deterministic branching program computing the following boolean function

$$f(x) = 1 \text{ iff the number of pairs } i < j \text{ such that } x_i = x_j = 1 \text{ is odd}$$

in time kn must have size $2^{\Omega(n)}$. Beame et al. (2003) extended Ajtai's result to boolean *randomized* branching programs.

Exercises

17.1. Let G be a *deterministic* branching program computing some function f : $D^n \to \{0, 1\}$. For $1 \le r \le n$ let t_r denote the number of computation paths (that is, consistent paths) in G which read exactly r different input variables. Show that $\sum_{r=1}^{n} t_r |D|^{-r} = 1$.

Hint: Let $x \in D^n$ be a random input vector, and let p be a computation paths along which r different variables are tested. Show that the computation on x follows p with probability $|D|^{-r}$.

17.2. (Oblivious programs) We consider nondeterministic *oblivious* branching programs (oblivious n.b.p.) Recall that a program is oblivious if its nodes can be divided into levels such that: (a) edges from the nodes in the i-th level go to the nodes in the $(i + 1)$-th level, and (b) the edges between two consecutive levels are labeled by the tests $x_i = d$ for the same variable x_i (cf. Sect. 15.4). Prove the following version of the Rectangle Lemma for oblivious programs:

If a sensitive function $f : \{0, 1\}^n \to \{0, 1\}$ can be computed in time kn by an oblivious n.b.p. of size S, then for $r = 8k^2 2^k$ and every $m \le n/2^{k+1}$, we have that

$$S^r \ge \frac{\mu(f) 2^{2m}}{\rho_m(f)}. \tag{17.8}$$

Note that this time we do not have an "annoying" term $\binom{n}{m}^2$ in the denominator making the Rectangle Lemma useless for functions over the boolean domain $D = \{0, 1\}$.

17.3. Let $k\text{-NBP}(f)$ denote the minimum number of nodes in a syntactic k-n.b.p. computing f. Prove that for $r = 8k^2 2^k$ and every $m \le n/2^{k+1}$, we have that

$$k\text{-NBP}(f)^r \ge \frac{\mu(f) 2^{2m}}{\rho_m(f)}.$$

Hint: Show that (17.8) also holds for syntactic k-n.b.p. For this, argue as in the proof of the Rectangle Lemma to construct an equivalence class $A \subseteq f^{-1}(1)$ of size $|A| \ge |f^{-1}(1)|/S^r$. Let Y_i be the set of variables that are tested along the i-th sub-computation on at least one vector from A. Use the Separator Lemma and the fact that our program is *syntactic* k-n.b.p. to show that there must be a pair X_0, X_1 of disjoint subsets of variables, each of size at least $m = n/2^{k+1}$ such that $Y_i \cap X_0 = \emptyset$ or $Y_i \cap X_1 = \emptyset$ for each $i = 1, \dots, r$.

17.4. A Bose-Chaudhury code (BCH-code) $C = C_{n,t}$ is a linear subspace of $GF(2)^n$ such that $|C| \ge 2^n/(n + 1)^t$ and every two vectors in C differ in at least $2t + 1$ bits. Such codes can be constructed for any n such that $n + 1$ is a power of 2, and for every $t < n/2$. Let k be an arbitrarily large positive integer, and $\epsilon > 0$ a sufficiently small with respect to k constant. Let f_n be the characteristic function of a BCH code $C_{n,t}$ for $t = \lfloor \epsilon \sqrt{n} \rfloor$. Prove that the function f_n requires syntactic k-n.b.p. of size $2^{\Omega(\sqrt{n})}$.

Hint: By Claim 15.17, no code of minimal distance at least $2t + 1$ can contain an m-rectangle of size larger than $2^{2m} \binom{m}{t}^{-2}$. Use Exercise 17.3.

17.5. The goal of this exercise is to show that lower bounds on k-NBP(f) for growing k can be used to obtain lower bounds on the size of *unrestricted* branching programs. Let f be a boolean function of n variables, and let NBP(f) denote the minimum number of contacts (labeled edges) in an unrestricted(!) nondeterministic branching program computing f. Suppose that for every subset of m variables, there exists an assignment ρ of constants to these variables such that k-NBP$(f_\rho) \geq S$. Show that then

$$\mathrm{NBP}(f) \geq \max\{km,\ S\}.$$

Hint: Take an n.b.p. computing f. If we have more than m variables, each appearing as a contact more thank k times, then NBP$(f) \geq km$. If this is not the case, we can set these "popular" variables to constants. What we obtain is a k-n.b.p.

17.6. (Okolnishnikova 2009) Let $0 < \alpha < 1/2$ be a constant, and let f_n be the characteristic function of a BCH code $C_{n,t}$ with $t = n^\alpha$. Let BP(f) denote the minimum number of nodes in an unrestricted *deterministic* branching program computing f. Show that:

(a) BP$(f) = \mathcal{O}(n^{1+\alpha} \log n)$.

   Hint: The parity-check matrix of $C_{n,t}$ has $\mathcal{O}(t \log n)$ rows.

(b) NBP$(f_n) = \Omega(n \log n)$.

   Hint: Use Exercises 17.3 and 17.5.

Part VI
Fragments of Proof Complexity

Chapter 18
Resolution

Propositional proof systems operate with boolean formulas, the simplest of which are clauses, that is, ORs of literals, where each literal is either a variable x_i or its negation $\neg x_i$. A *truth-assignment* is an assignment of constants 0 and 1 to all the variables. Such an assignment *satisfies* (*falsifies*) a clause if it evaluates at least one (respectively, none) of its literals to 1. A set of clauses, that is, a CNF formula, is *satisfiable* if there is an assignment which satisfies all its clauses. The basic question is: Given an *unsatisfiable* CNF formula F, what is the size of a proof that F is indeed unsatisfiable? The *size* (or *length*) of a proof is the total number of clauses used in it.

A proof of the unsatisfiability of F starts with clauses of F (called *axioms*), at each step applies one of several (fixed in advance) simple rules of inferring new clauses from the already derived ones, and must eventually produce the empty clause $\emptyset$ which, by definition, is satisfied by none of the assignments.

For such a derivation to be a legal proof, the rules must be *sound* in the following sense: if some assignment (of constants to all variables) falsifies the derived clause, then it must falsify at least one of the clauses from which it was derived. Then the fact that $\emptyset$ was derived implies that the CNF F was indeed unsatisfiable: given any assignment $a \in \{0, 1\}^n$ we can traverse the proof going from $\emptyset$ to an axiom (a clause of F), and the soundness of the rules will give us a clause of F which is not satisfied by a.

The main goal of proof complexity is to show that some unsatisfiable CNFs require long proofs. A compelling reason to study this problem is its connection with the famous P versus NP question. It has long been known (Cook and Reckhow, 1979) that NP $=$ co-NP iff there is a propositional proof system giving rise to short (polynomial in $|F|$) proofs of unsatisfiability of all unsatisfiable CNFs F; here and throughout, $|F|$ denotes the number of clauses in F.

Thus a natural strategy to approach the NP versus co-NP problem, and hence, also the P versus NP problem is, by analogy with research in circuit complexity, to investigate more and more powerful proof systems and show that some unsatisfiable CNFs require exponentially long proofs. In this and the next chapter we will demonstrate this (currently very active) line of research on some basic proof systems, like resolution and cutting planes proofs.

The areas of circuits complexity and proof complexity look similar, at least syntactically: in proof complexity one also starts from some "simplest" objects (axioms) and applies local operations to obtain a result. But there is a big difference: the number of "objects of interest" differ drastically between the two settings. There are doubly exponentially number of boolean functions of n variables, but only exponentially many CNFs of length n. Thus a counting argument shows that some functions require circuits of exponential size (see Theorem 1.14), but no similar argument can exist to show that some CNFs require exponential size proofs. This is why even *existence* results of hard CNFs for strong proof systems are interesting in this setting as well.

The most basic proof system, called the *Frege system*, puts no restriction on the formulae manipulated by the proof. It has one derivation rule, called the cut rule: from $A \vee C$ and $B \vee \neg C$ one can derive $A \vee B$ in one step. Adding any other sound rule turns out to have little effect on the length of proofs in this system. The major open problem in proof complexity is to find any tautology that has no polynomial-size proof in the Frege system. As lower bounds for Frege are hard to obtain, we turn to subsystems of Frege which are interesting and natural. One of the simplest and most important such subsystems is called *Resolution*. This subsystem is used by most propositional, as well as first order automated theorem provers.

18.1 Resolution Refutation Proofs

The resolution proof system was introduced by Blake (1937) and has been made popular as a theorem-proving technique by Davis and Putnam (1960) and Robinson (1965). Let F be a set of clauses and suppose that F is not satisfiable. A *resolution refutation proof* (or simply, a *resolution proof*) for F is a sequence of clauses $\mathcal{R} = (C_1, \ldots, C_t)$ where $C_t = \emptyset$ is the empty clause and each intermediate clause C_i either belongs to F or is derived from some previous two clauses using the following *resolution rule*:

$$\frac{A \vee x_i \qquad B \vee \neg x_i}{A \vee B} \qquad\qquad (18.1)$$

meaning that the clause $A \vee B$ can be inferred from two clauses $A \vee x_i$ and $B \vee \neg x_i$.

In this case one also says that the variable x_i was *resolved* to derive the clause $A \vee B$; here A and B are arbitrary ORs of literals. The *size* of such a proof is equal to the total number t of clauses in the derivation. It is often useful to describe a resolution proof as a directed acyclic graph (see Fig. 18.1). If this graph is a tree, then one speaks about a *tree-like* resolution proof. For technical reasons the following "redundant" rule, the *weakening rule*, is also allowed: a clause $A \vee B$ can be inferred from A.

Observe that the resolution rule is sound: if some assignment (of constants to all variables) falsifies the derived clause $A \vee B$, then it must falsify at least one of the

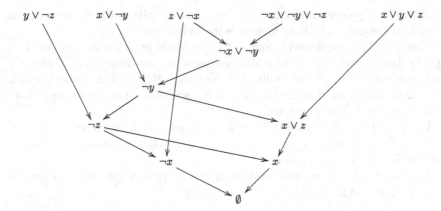

Fig. 18.1 A resolution refutation proof of an unsatisfiable CNF formula F. Leaves (fanin-0 nodes) are clauses of F, and each inner node is a clause obtained from previous ones by the resolution rule. This proof is *not* tree-like

clauses $A \vee x_i$ and $B \vee \neg x_i$ from which it was derived. It is also known (and easy to show, see Exercise 18.1) that Resolution is *complete*: every unsatisfiable set of clauses has a resolution refutation proof.

Interestingly, resolution proofs are related to the model of computation we already considered above—branching programs.

18.2   Resolution and Branching Programs

Let F be an unsatisfiable CNF formula, that is, for every input $a \in \{0, 1\}^n$ there is a clause $C \in F$ for which $C(a) = 0$. The *search problem* for F is, given a, to find such a clause; there may be several such clauses—the goal is to find at least one of them. Such a problem can be solved by a branching program with at most $n|F|$ nodes. Namely, given an assignment $a \in \{0, 1\}^n$, we can test whether all literals of the first clause in F are falsified by a. If yes, then we reach a leaf labeled by this clause. If not, then test whether all literals of the second clause in F are falsified by a, etc. Note that the resulting branching program is not read-once: if a variable appears in k clauses, then it will be retested k times.

Of course, the search problem for any unsatisfiable CNF formula can be solved by a decision tree, and hence, by a read-once branching program. But the size (total number of nodes) may be then exponential in the number n of variables.

Let $S_R(F)$ be the smallest size of a resolution refutation of F, and BP(F) the smallest size of (the number of nodes in) a deterministic branching program solving the search problem for F. It is not difficult to show that $S_R(F) \geq \text{BP}(F)$ (see the first part of the proof of Theorem 18.1 below). But the gap between these two measures may be exponential: as mentioned above, any unsatisfiable CNF F has a trivial branching program of size $n|F|$ whereas, as we will show in the next section, some CNFs require $S_R(F)$ exponential in their variables.

The first exponential lower bounds for resolution proofs were obtained long ago by Tseitin (1970) under an additional restriction that along every path every particular variable x_i can be resolved at most once. He called this model *regular* resolution. It turns out that this model exactly coincides(!) with the familiar model of read-once branching programs.

Let $1\text{-}S_R(F)$ be the smallest size of a regular resolution refutation proof for F, and $1\text{-}BP(F)$ the smallest size of a deterministic read-once branching program solving the search problem for F.

The following theorem was used implicitly by various authors and explicitly noted in Lovász et al. (1995).

Theorem 18.1. *For every unsatisfiable CNF formula F, we have*

$$S_R(F) \geq BP(F) \quad and \quad 1\text{-}S_R(F) = 1\text{-}BP(F).$$

Proof. Resolution proofs $\Rightarrow$ branching programs: To show that $S_R(F) \geq BP(F)$ and $1\text{-}S_R(F) \geq 1\text{-}BP(F)$, let $\mathcal{R}$ be a resolution refutation proof for F. Construct a branching program as follows.

- The nodes of the program are clauses C of $\mathcal{R}$.
- The source node is the last clause in $\mathcal{R}$ (the empty one), the sinks are the initial clauses from F.
- Each non-sink node C has fanout 2 and the two edges directed from C to the two clauses C_0 and C_1 from which this clause is derived by one application of the resolution rule. If the resolved variable of this inference is x_i then the edge going to the clause containing x_i is labeled by the test $x_i = 0$, and the edge going to the clause containing $\neg x_i$ is labeled by the test $x_i = 1$ (see Fig. 18.2).

It is straightforward to verify that *all* clauses on a path determined by an input $a \in \{0, 1\}^n$ are falsified by a, and hence, the last clause of F reached by this path is also falsified by a. That is, the obtained branching program solves the search problem, and is read-once if $\mathcal{R}$ was regular.

Read-once branching programs $\Rightarrow$ regular resolution proofs: It remains to prove the more interesting direction that $1\text{-}S_R(F) \leq 1\text{-}BP(F)$. Let P be a deterministic read-once branching program (1-b.p.) which solves the search problem for F. That is, for every input $a \in \{0, 1\}^n$ the (unique) computation path on a leads to a clause $C \in F$ such that $C(a) = 0$. We will associate a clause to every node of P such that P becomes a graph of a resolution refutation for F. A vertex v labeled by a variable will be associated with a clause C_v with the property that

$$C_v(a) = 0 \text{ for every input } a \in \{0, 1\}^n \text{ that reaches } v. \qquad (18.2)$$

We associate clauses inductively from the sinks backwards. If v is a sink then let C_v be the clause from F labeling this sink in the program P.

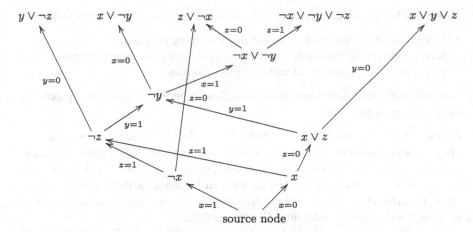

Fig. 18.2 A branching program obtained from the resolution proof given in Fig. 18.1: just reverse the direction of arcs and label them accordingly. The program is not read-once

Now assume that the node v of P corresponds to a variable x_i and has edges (v, u_0) for $x_i = 0$ and (v, u_1) for $x_i = 1$. By induction we may assume that u_0 and u_1 are labeled by clauses C_0 and C_1 satisfying (18.2).

Claim 18.2. C_0 does not contain $\neg x_i$ and C_1 does not contain x_i.

Proof. Otherwise, if C_0 contains $\neg x_i$, take an input a with $a_i = 0$ that reaches v. Such an input exists since by the read-once assumption on P, the i-th bit x_i was not asked along any path from the source to v. The input a can reach u_0 and it satisfies C_0, in contradiction to the inductive hypothesis. The proof in the case when C_1 contains x_i is similar. □

We conclude that

(i) Either $C_0 = (x_i \vee A)$ and $C_1 = (\neg x_i \vee B)$,

(ii) Or one of C_0 and C_1 does not contain $x_i, \neg x_i$ at all.

In the first case label v with $C_v = A \vee B$. In the second case label v with the clause that does not contain $x_i, \neg x_i$. (If both clauses do not contain $x_i, \neg x_i$ choose any of them.)

It is easy to see that the inductive hypothesis (18.2) holds for C_v. Indeed, if C_v were satisfied by some (partial) input a reaching the node v then, due to the read-once property, this input could be extended to two inputs a_0 and a_1 by setting the i-th bit to 0 and to 1. But $C_v(a) = 1$ implies that either $A(a) = 1$ or $B(a) = 1$ (or both). Hence, we would have that either $C_0(a_1) = 1$ or $C_1(a_1) = 1$, contradicting the inductive hypothesis (18.2).

Finally, the clause associated with the source node must be the empty clause, just because *every* input reaches it. Thus the obtained labeled digraph represents a regular resolution derivation for F (possibly with some redundant steps that correspond to the second case (ii) in the labeling above). □

We thus have a bridge between resolution refutations and branching programs:

- Every resolution refutation is a restricted branching program.
- Every *regular* resolution refutation is just a *read-once* branching program.
- Every *tree-like* resolution refutation is just a *decision tree*.

The only difference is that now these branching programs solve *search* problems, not just *decision* problems.

Remark 18.3. Note that Claim 18.2 holds for *any* deterministic branching program, not just for read-once programs: it is enough that P is a *minimal* program. Indeed, in this case a node must be reachable by (at least) two inputs a and b such that $a_i = 0$ and $b_i = 1$, for otherwise the test on the i-th bit made at the node v would be redundant. However, the fact that the branching program is read-once was essential to show that the constructed clause C_v satisfies (18.2).

To see this, let $C_0 = (x_i \vee A)$, $C_1 = (\neg x_i \vee B)$ and $C_v = A \vee B$. Suppose that the node v is reached by two inputs a and b such that $a_i = 0$ and $b_i = 1$. Assume that the bit x_i *was* tested along both paths at least once; hence, the paths must diverge after the test on x_i at the node v, that is, a cannot reach C_1, and b cannot reach C_0. Assume now that $A(a) = B(b) = 0$ but $A(b) = 1$ or $B(a) = 1$. Then $C_0(a) = 0$ and $C_1(b) = 0$ but $C_v(a) = 1$ or $C_v(b) = 1$. In the read-once case such a situation cannot occur because then every (single!) computation reaching a node v can be extended in *both* directions.

18.3   Lower Bounds for Tree-Like Resolution

Let F be an unsatisfiable CNF formula. A resolution proof for F is *tree-like* if its underlying graph is a tree. That is, tree-like resolution proof is a special case of regular resolution proofs; the corresponding branching program for the corresponding search problem is then just a decision tree. By the *size* $|T|$ of a tree-like resolution proof T we will mean the number of leaves in the corresponding decision tree. Since the search problem for any unsatisfiable CNF formula can be solved by a decision tree, Theorem 18.1 implies that any such CNF formula *has* a tree-like resolution proof, and hence, also regular resolution proof. The question, however, is: how large tree-like proofs must be?

Lower bounds on the size of tree-like resolution can be proved using the following game-theoretic argument proposed by Pudlák and Impagliazzo (2000). There are two players, Prover and Delayer. The goal of the Prover is to construct a (partial) assignment falsifying at least one clause of F. The goal of Delayer is to delay this happening as long as possible. The game proceeds in rounds. In each round

- Prover suggests a variable x_i to be set in this round, and
- Delayer either chooses a value 0 or 1 for x_i or leaves the choice to the Prover.

- In this last case, Delayer scores one point, but the Prover can then choose the value of x_i.

The game is over when one of the clauses is falsified by the obtained (partial) assignment, that is, when all the literals in the clause are assigned 0.

Pudlák and Impagliazzo observed that, if the Delayer has a strategy which scores r points, then any tree-like resolution refutation proof for F has size at least 2^r. This holds because, given a tree-like derivation, the Prover can use the following strategy: if the Delayer leaves the choice to the Prover, then the Prover chooses an assignment resulting into a *smaller* of the two subtrees.

This result can be easily extended to the case of *asymmetric* games, where the Delayer earns different number of points depending on whether the prover sets $x_i = 0$ or $x_i = 1$. As before, the game proceeds in rounds. In each round

- Prover suggests a variable x_i to be set in this round, and
- Delayer either chooses a value 0 or 1 for x_i or leaves the choice to the Prover
- The number of points earned by the Delayer is

  - 0 if Delayer chooses the value for x_i,
  - $\log_2 a$ if Prover sets x_i to 0, and
  - $\log_2 b$ if Prover sets x_i to 1.

The only requirement is that $1/a + 1/b = 1$; in this case we say that (a, b) is a legal scoring pair, and call this game the (a, b)-game. Hence, the symmetric game is one with $a = b = 2$.

Lemma 18.4. *Let F be an unsatisfiable CNF formula F. If Delayer can earn r points in some asymmetric game on F, then any tree-like resolution refutation proof for F has size at least 2^r.*

Proof. We will prove the lemma in the converse direction: if F has a tree-like resolution refutation proof T with $|T|$ leaves then, in any (a, b) game for F, the Prover has a strategy under which the Delayer can earn at most $\log |T|$ points.

Consider an arbitrary (a, b)-game on F, and let α_i be the partial assignment constructed after i rounds of the game (the i-th prefix of α). By p_i we denote the number of points that Delayer has earned after i rounds, and let T_i be the sub-tree of T which has as its root the node reached in T along the path specified by α_i. Our goal is to prove, by induction on i, that

$$|T_i| \le \frac{|T|}{2^{p_i}}. \tag{18.3}$$

The desired inequality $p_m \le \log |T|$ then follows, because T_m consists of just one clause falsified by α.

So it remains to prove the claim (18.3). At the beginning of the game ($i = 0$) we have $p_0 = 0$ and $T_0 = T$. Therefore the claim trivially holds.

For the inductive step, assume that the claim holds after i rounds and Prover asks for the value of the variable x in round $i + 1$. The variable Prover asks about is determined by T: it is the variable resolved at the root of the subtree reached after the i-th round.

It the Delayer chooses the value, then $p_{i+1} = p_i$ and (18.3) remains true after the $(i + 1)$-th round. Otherwise, let T_i^0 be the 0-subtree of T_i, and T_i^1 the 1-subtree of T_i. Since $1/a + 1/b = 1$, we have that

$$|T_i^0| + |T_i^1| = |T_i| = \frac{|T_i|}{a} + \frac{|T_i|}{b}.$$

If the Delayer defers the choice to the Prover, then the Prover can use the following "take the smaller tree" strategy: set $x = 0$ if $|T_i^0| \leq |T_i|/a$, and set $x = 1$ otherwise; in this last case we have that $|T_i^1| \leq |T_i|/b$. Thus if Prover's choice is $x = 0$, then we get

$$|T_{i+1}| \leq \frac{|T_i|}{a} \leq \frac{|T|}{a2^{p_i}} = \frac{|T|}{2^{p_i + \log a}} = \frac{|T|}{2^{p_{i+1}}},$$

as desired. Since the same holds (with a replaced by b) if Prover's choice is $x = 1$, we are done.   □

We now apply this lemma to prove that the unsatisfiable CNF corresponding to the pigeonhole principle (and even to its "weak" version) require tree-like resolution refutation proofs of exponential size.

The *weak pigeonhole principle* asserts that if $m > n$ then m pigeons cannot sit in n holes so that every pigeon is alone in its hole.[1] In terms of 0-1 matrices, this principle asserts that, if $m > n$ then no $m \times n$ 0-1 matrix can simultaneously satisfy the following two conditions:

1. Every row has at least one 1.
2. Every column has at most one 1.

To write this principle as an unsatisfiable CNF formula, we introduce boolean variables $x_{i,j}$ interpreted as:

$$x_{i,j} = 1 \text{ if and only if the } i\text{-th pigeon sits in the } j\text{-th hole.}$$

Let PHP_n^m denote the CNF consisting of the following clauses:

- Pigeon Axioms: each of the m pigeon sits in at least one of n holes:

$$x_{i,1} \vee x_{i,2} \vee \cdots \vee x_{i,n} \text{ for all } i = 1, \ldots, m.$$

---

[1] The word "weak" is used here to stress that the number m of pigeons may be arbitrarily large. The larger m is, the "more obvious" the principle is, and hence, its proof might be shorter.

- Hole Axioms: no two pigeons sit in one hole:

$$\neg x_{i_1,j} \vee \neg x_{i_2,j} \text{ for all } i_1 \neq i_2 \text{ and } j = 1,\dots,n.$$

Hence, truth assignments in this case are boolean $m \times n$ matrices α. Such a matrix can satisfy all pigeon axioms iff every row has at least one 1, whereas it can satisfy all hole axioms iff every column has at most one 1. Since $m \geq n + 1$, no assignment can satisfy pigeon axioms and hole axioms at the same time. So PHP_n^m is indeed an unsatisfiable CNF.

Theorem 18.5. (Dantchev and Riis 2001) *For any $m > n$, any tree-like resolution refutation proof of PHP_n^m has size $n^{\Omega(n)}$.*

Note that the lower bound does not depend on the number m of pigeons—it may be arbitrarily large! A smaller (but also exponential) lower bound of the form $2^{\Omega(n)}$ for an arbitrary number of pigeons was proved earlier by Buss and Pitassi (1998).

Proof. (Due to Beyersdorff et al. 2010) By Lemma 18.4, we only have to define an appropriate scoring pair (a, b) for which the Delayer has a strategy giving her many points in the (a, b) game on PHP_n^m. We first define the Delayer's strategy for an arbitrary (a, b) game, and then choose a and b so that to maximize the total score. By Lemma 18.4, it is enough to show that the Delayer can earn $\Omega(n \log n)$ points.

The goal of the Delayer is to delay an appearance of two 1s in a column and of an all-0 row as long as possible. So if Prover asks for a value of $x_{i,j}$, then the Delayer is only then forced to set it to 0 if the j-th column already has a 1. Otherwise it is beneficial for the Delayer to set $x_{i,j} = 1$ to avoid an all-0 row. But at the same time, it is beneficial for her not to set too many 1s in a row to avoid two 1s in a column. Intuitively, it would be the best for the Delayer to set just one 1 per row. Moreover, she should not wait too long: if the i-th row already has many 0s, she should try to set a 1 in it, for otherwise she could be forced (by many columns already having a 1) to set the remaining variables in this row to 0s.

To formally describe the strategy of the Delayer, let α be a partial assignment to the variables $X = \{x_{i,j} \mid i \in [m], j \in [n]\}$. For pigeon i, let $J_i(\alpha)$ be the set of "excluded free holes" for the pigeon i. These are the holes which are still free (not occupied by any pigeon) but are explicitly excluded for pigeon i by α:

$$J_i(\alpha) := \{j \in [n] \mid \alpha(x_{i,j}) = 0 \text{ and } \alpha(x_{i',j}) \neq 1 \text{ for all } i' \in [m]\}.$$

If Prover asks for a value of $x_{i,j}$, then the Delayer uses the following strategy (see Fig. 18.3):

$$\alpha(x_{i,j}) := \begin{cases} 0 & \text{if either the } i\text{-th row or the } j\text{-th column already has a 1;} \\ 1 & \text{if } |J_i(\alpha)| \geq n/2 \text{ and there is no 1 in the } j\text{-th column yet;} \\ * & \text{otherwise.} \end{cases}$$

Here $*$ means the decision is deferred to the Prover.

Fig. 18.3 The strategy of the Delayer: she sets $x_{i,j} = 1$ if $|J_i| \geq n/2$ and there is no 1 in the j-th column, and sets $x_{i,j} = 0$ if either the i-th row or the j-th column already has a 1. Otherwise, she defers the decision to the Prover

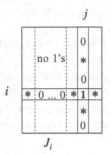 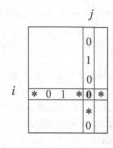

If Delayer uses this strategy, then none of the hole clauses $\neg x_{i_1,j} \vee \neg x_{i_2,j}$ from PHP_n^m will be falsified in the game. Therefore, a contradiction (a falsified clause) will be a pigeon clause $x_{i,1} \vee \cdots \vee x_{i,n}$. That is, the resulting assignment α sets all n variables in this clause to zero (pigeon i has no hole).

But after the number $|J_i(\alpha)|$ of excluded free holes for pigeon i reaches the threshold $n/2$, Delayer will not leave the choice to Prover. Instead, Delayer will try to place pigeon i into some hole. Since this hasn't happened, the Delayer was forced to set the remaining $n/2$ variables in the i-th row to 0. Since the Delayer is only then forced to set $x_{i,j}$ to 0 when the j-th column already has a 1, there must already be a 1 in each of these $n/2$ columns. Moreover, no two of these 1s can be in one row, since Delayers strategy forbids this: she always sets a "dangerous" variable (with a 1 in the same row or column) to 0. Therefore, at the end of the game at least $n/2$ variables must be set to 1, and no two of these 1s lie in one row or one column. We assume w.l.o.g. that these are the variables x_{i,j_i} for $i = 1, \ldots, n/2$. Let us check how many points Delayer earns in this game. We calculate the points separately for each pigeon $i = 1, \ldots, n/2$.

Case 1: Delayer sets x_{i,j_i} to 1. Then pigeon i was not assigned to a hole yet, and $|J_i(\alpha)| \geq n/2$. Hence, there must be a set J of $|J| \geq n/2$ 0-positions in the i-th row of α. Moreover, all these positions must be already set to 0 by the Prover (not by the Delayer) because none of the columns $j \in J$ can have a 1, by the definition of $J_i(\alpha)$. Thus before Delayer sets $\alpha(x_{i,j_i}) = 1$, she has already earned points for all $|J| \geq n/2$ previous 0-settings by the Prover. That is, in this case Delayer earns at least $(n/2) \log a$ points.

Case 2: Player sets x_{i,j_i} to 1. In this case Delayer earns $(n/2) \log b$ points.

Thus, the Delayer earns either $(n/4) \log b$ or $(n^2/8) \log a$ points. To maximize the score, we set $b = n/\log n$ and

$$a = \frac{b}{b-1} = 1 + \frac{1}{b-1} = \Omega\left(e^{1/(b-1)}\right) = 2^{\Omega\left(\frac{\log n}{n}\right)}.$$

Since $1/a + 1/b = (b-1)/b + 1/b = 1$, this is a legal scoring, and Delayer earns $\Omega(n \log n)$ points, as desired.   □

18.4 Tree-Like Versus Regular Resolution

A partial ordering of A is a binary relation $a \to b$ which is antisymmetric and transitive. That is, $a \to b$ implies $\neg(b \to a)$, and $a \to b$ and $b \to c$ implies $a \to c$. An element $a \in A$ is *minimal* if it has no predecessor, that is, if $\neg(b \to a)$ for all $b \in A$, $b \neq a$. It is clear that in each partial order there must be at least one minimal element. We consider the CNF formula GT_n expressing the negation of this property. For this we take $A = \{1, \ldots, n\}$ and associate a boolean variable x_{ij} to each pair (i, j) of elements. We interpret these variables as $x_{ij} = 1$ if and only if $i \to j$.

The CNF formula GT_n consists of three sets of clauses. The first two sets consist of all clauses $\neg x_{ij} \vee \neg x_{jk} \vee x_{ik}$ and $\neg x_{ij} \vee \neg x_{ji}$ for all distinct i, j, k. These clauses ensure that we have a partial ordering. The third set consists of n clauses

$$C_n(j) = x_{1j} \vee \cdots \vee x_{j-1,j} \vee x_{j+1,j} \vee \cdots \vee x_{n,j}, \qquad j = 1, \ldots, n$$

stating that every element j has at least one predecessor (no minimal element). In terms of graphs, the CNF formula GT_n is a negation of the property that if a directed graph is transitive and has no loops and no cycles of size two, then there must be at least one source node, that is, a node of fanin 0.

It was conjectured that GT_n requires resolution refutation proofs of exponential size. And indeed, it was shown by Bonet and Galesi (1999) that *tree-like* refutations for this CNF must be of exponential size. However, Stålmark (1996) showed that this CNF has a small *regular* resolution refutation.

Theorem 18.6. (Stålmark 1996) *The CNF formula GT_n has a regular resolution refutation of polynomial size.*

Proof. We will construct the desired refutation proof recursively. Our initial clauses (axioms) are $A(i, j, k) = \neg x_{ij} \vee \neg x_{jk} \vee x_{ik}$, $B(i, j) = \neg x_{ij} \vee \neg x_{ji}$, and the clauses $C_n(j)$ for all $j = 1, \ldots, n$. We introduce auxiliary clauses

$$C_m(j) = x_{1j} \vee \cdots \vee x_{j-1,j} \vee x_{j+1,j} \vee \cdots \vee x_{m,j}$$

for all $m = 2, \ldots, n$, stating that some element $i \in \{1, \ldots, m\}$ is smaller than j. The idea of the proof is to obtain clauses of the form $C_m(j)$ from $m = n$ down to $m = 2$ in the following way:

$$
\begin{array}{ccccc}
C_n(1) & C_n(2) & \ldots & C_n(n-1) & C_n(n) \\
C_{n-1}(1) & C_{n-1}(2) & \ldots & C_{n-1}(n-1) \\
& & \vdots & & \\
C_2(1) & C_2(2) & & &
\end{array}
$$

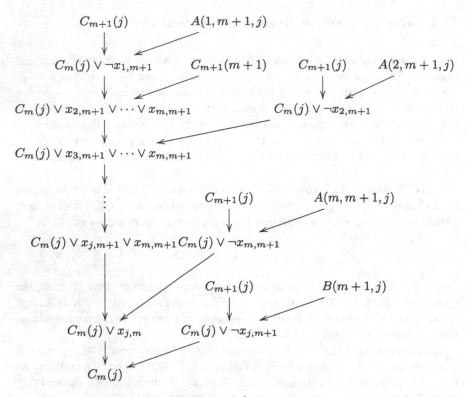

Fig. 18.4 Resolution derivation of $C_m(j)$, for $j \neq m + 1$. Note that $x_{j,m+1}$ can not be deleted in the upper part of the derivation but is removed in the last step of the derivation. Note also that the derivation is *not* tree-like: the same clause $C_{m+1}(j)$ is used many times

Note that the first (top) row corresponds to our initial CNF formula GT_n, the second to GT_{n-1}, and so on. For each m, clauses $C_m(1), \ldots, C_m(m)$ are obtained in parallel. Each $C_m(j)$ is obtained using the clauses $C_{m+1}(j)$ and $C_{m+1}(m + 1)$ derived in the previous step, and the initial clauses (axioms) $A(1, m + 1, j), A(2, m + 1, j), \ldots, A(m, m + 1, j)$ and $B(m + 1, j)$ (see Fig. 18.4). At the end we easily derive the empty clause from $C_2(1), C_2(2)$ and $B(2, 1)$.   □

Remark 18.7. Alekhnovich et al. (2007) proved that an appropriate modification GT_n' of GT_n requires regular refutations of exponential size, but has (non-regular) resolution refutations of polynomial size. More precisely, if S denotes the smallest size of a non-regular resolution refutation of GT_n', and R the smallest size of a regular resolution refutation of GT_n', then $\log R = \Omega(\sqrt[3]{S})$. Using different CNF formulas, Urquhart (2011) obtained even larger gap: $\log R = \Omega(S/\mathrm{polylog}(S))$. Thus, we have the following separations, where "$A \ll B$" stands for "proof system A is exponentially weaker that B":

tree-like resolution $\ll$ regular resolution $\ll$ general resolution.

18.5 Lower Bounds for General Resolution

General, non-tree-like resolution proofs are much harder to analyze. The first exponential lower bound for the size of such proofs was proved by Haken (1985).

Theorem 18.8. (Haken 1985) *Any resolution refutation proof of PHP_{n-1}^n requires size $2^{\Omega(n)}$.*

Proof. (Due to Beame and Pitassi 1996) The proof is by contradiction. We define an appropriate notion of a "fat" clause and show two things:

1. If PHP_{n-1}^n has a short resolution proof, then it is possible to set some variables to constants so that the resulting proof is a refutation of PHP_{m-1}^m for a large enough m, and has no fat clauses.
2. If m is large enough, then every refutation proof for PHP_{m-1}^m must have at least one fat clause.

This implies that PHP_{n-1}^n cannot have short resolution proofs.

In the case of the CNF formula PHP_{n-1}^n truth assignments α are n by $n-1$ boolean matrices. We say that a truth assignment α is *i-critical* if

- The i-th row of α is the only all-0 row, and
- Every column has exactly one 1.

Note that each such assignment α is "barely unsatisfying": it satisfies all hole axioms $\neg x_{i_1,j} \vee \neg x_{i_2,j}$ as well as the axioms of all but the i-th pigeon. That is, the only axiom it falsifies is the pigeon axiom $C_i = x_{i,1} \vee x_{i,2} \vee \cdots \vee x_{i,n-1}$. Thus an i-critical assignment corresponds to an assignment of pigeons to holes such that $n-1$ of the pigeons are mapped to $n-1$ different holes, but the i-th pigeon is mapped to no hole at all. Call an assignment α *critical* if it is i-critical for some $1 \leq i \leq m$.

The properties of critical truth assignments make it convenient to convert each clause C to a positive clause C^+ which is satisfied by precisely the same set of critical assignments as C. More precisely to produce C^+, we replace each negated literal $\neg x_{i,j}$ with the OR of all variables in the j-th column, except the i-th one:

$$X_{i,j} = x_{1,j} \vee \cdots \vee x_{i-1,j} \vee x_{i+1,j} \vee \cdots \vee x_{n,j}.$$

Note that the monotone version C^+ of every hole axiom $C = \neg x_{i_1,j} \vee \neg x_{i_2,j}$ is just the OR of *all* variables in the j-th column, and hence, is satisfied by *any* critical assignment.

Claim 18.9. For every critical truth assignment α, $C^+(\alpha) = C(\alpha)$.

Proof. Suppose there is a critical assignment α such that $C^+(\alpha) \neq C(\alpha)$. This could only happen if C contains a literal $\neg x_{i,j}$ such that $\neg x_{i,j}(\alpha) \neq X_{i,j}(\alpha)$. But this is impossible, since α has precisely one 1 in the j-th column. $\square$

Associate with each clause in a refutation of PHP_{n-1}^n the set

$$Pigeon(C) = \{i / \text{there is some } i\text{-critical assignment } \alpha \text{ such that } C(\alpha) = 0\}$$

of pigeons that are "bad" for this clause: some critical assignment of these pigeons falsifies C.

The *width*, $w(C)$, of a clause is the number of literals in it.

Claim 18.10. Every resolution refutation of PHP^n_{n-1} must have a clause C such that $w(C^+) \geq n^2/9$.

Proof. Define the *weight* of a clause C as $\mu(C) := |\text{Pigeon}(C)|$. By the definition, each hole axiom has weight 0, each pigeon axiom has weight 1, and the last (empty) clause has weight n since it is falsified by any truth assignment. Moreover, this weight measure is *subadditive*: if a clause C is derived from clauses A and B, then $\mu(C) \leq \mu(A) + \mu(B)$. This holds because every assignment (even a non-critical one) falsifying C must falsify at least one of the clauses A and B. Therefore, if C is the first clause in the proof[2] with $\mu(C) > n/3$, we must have

$$n/3 < \mu(C) \leq 2n/3. \tag{18.4}$$

Fix such a "medium heavy" clause C and let $s = \mu(C)$ be its weight. Since $n/3 < s \leq 2n/3$, it is enough to show that the positive version C^+ of this clause must have $w(C^+) \geq s(n-s)$ distinct variables.

Fix some $i \in \text{Pigeon}(C)$ and let α be an i-critical truth assignment with $C(\alpha) = 0$. For each $j \notin \text{Pigeon}(C)$, define the j-critical assignment α', obtained from α by toggling rows i and j. That is, if α maps the j-th pigeon to the k-th hole, then α' maps the i-th pigeon to this hole (see Fig. 18.5).

Now $C(\alpha') = 1$ since $j \notin \text{Pigeon}(C)$. By Claim 18.9, we have that $C^+(\alpha) = 0$ and $C^+(\alpha') = 1$. Since the assignments α, α' differ only in the variables $x_{i,k}$ and $x_{j,k}$, this can only happen when C^+ contains the variable $x_{i,k}$.

Running the same argument over all $n - s$ pigeons $j \notin \text{Pigeon}(C)$ (using the same α), it follows that C^+ must contain at least $n - s$ of the variables $x_{i,1}, \ldots, x_{i,n-1}$ corresponding to the i-th pigeon. Repeating the argument for all pigeons $i \in \text{Pigeon}(C)$ shows that C^+ contains at least $s(n-s)$ variables, as claimed. $\square$

We can now finish the proof of Theorem 18.8 as follows. Let $\mathcal{R}$ be a resolution refutation proof of PHP^n_{n-1}. Let a and $b \geq 2$ be positive constants (to be specified

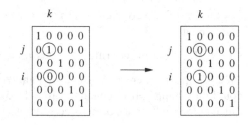

Fig. 18.5 Assignment α' is obtained from α by interchanging the i-th and j-th rows

---

[2]Recall that a proof is a *sequence* of clauses. Alternatively, one can apply Lemma 1.3.

Fig. 18.6 Setting of
constants to eliminate clauses
containing $x_{i,j}$; non-shaded
positions are not set. In this
way PHP^n_{n-1} is reduced to
PHP^{n-1}_{n-2}

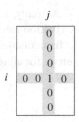

later). For the sake of contradiction, assume that

$$|\mathcal{R}| < e^{n/a}.$$

Together with $\mathcal{R}$ we consider the set $\mathcal{R}^+ = \{C^+ \mid C \in \mathcal{R}\}$ of positive versions
of clauses in $\mathcal{R}$. Say that a clause is *fat* if it contains at least n^2/b variables. Let S
be the total number of fat clauses in $\mathcal{R}^+$. Since each fat clause has at least a $1/b$
fraction of all the variables, there must be (by the pigeonhole principle!) a variable
$x_{i,j}$ which occurs in at least S/b of fat clauses in $\mathcal{R}^+$.

Set this "popular" variable to 1, and at the same time set to 0 all the variables
$x_{i,j'}$ and $x_{i',j}$ for all $j' \neq j, i' \neq i$ (see Fig. 18.6). After this setting, all the
clauses containing $x_{i,j}$ will disappear from $\mathcal{R}^+$ (they all get the value 1) and the
variables which are set to 0 will disappear from the remaining clauses.

Applying this restriction to the entire proof $\mathcal{R}$ leaves us with a refutation proof
$\mathcal{R}_1$ for PHP^{n-1}_{n-2}, where the number of fat clauses in $\mathcal{R}^+_1$ is at most $S(1 - 1/b)$.
Applying this argument iteratively $d = b \ln S < (b/a)n$ times, we are guaranteed
to have knocked out all fat clauses, because

$$S(1 - 1/b)^d < e^{\ln S - d/b} = 1.$$

Thus we are left with a refutation proof for PHP^m_{m-1}, where

$$m = n - d \geq (1 - b/a)n,$$

and where $w(C^+) < n^2/b$ for *all* its clauses. But Claim 18.10 implies that any
refutation proof of PHP^m_{m-1} must contain a clause C for which

$$n^2/b > w(C^+) \geq m^2/9 \geq (1 - b/a)^2 n^2/9.$$

To get the desired contradiction, it is enough to choose the parameters a and b so
that $(1 - b/a)^2 \geq 9/b$ which, in particular, is the case for $b = 16$ and $a = 4b = 64$.
$\square$

The reader may wonder: where in this proof did we used the fact that the clauses
in a refutation are derived using only resolution and weakening rules? The same
argument seems to work for more general derivations. And this is indeed the case:
the only important thing was that the formulas in such a derivation are clauses—this
allowed us to kill off a clause by setting just one variable to a constant.

Actually, a closer look at the proof shows that it also works for a more general derivation rule, called *semantic derivation rule*. This rule allows to derive a clause C from clauses $C_1, \ldots, C_k$ if these clauses "semantically imply" C in the following sense: for all $\alpha \in \{0, 1\}^n$,

$$C_1(\alpha) = 1, \ldots, C_k(\alpha) = 1 \quad \text{implies} \quad C_j(\alpha) = 1.$$

A *semantic proof* of an unsatisfiable CNF F is a sequence $\mathcal{R} = (C_1, \ldots, C_t)$ of clauses such that $C_t = 0$ is the empty clause and each C_j is either an axiom (belongs to F) or is obtained from k or fewer previous clauses (already derived or belonging to F) by one application of the semantic rule.

The only difference is that now instead of (18.4) we will have (cf. Lemma 1.3):

$$\frac{n}{k+1} < \mu(C) \leq \frac{kn}{k+1},$$

which results in a lower bound $w(C^+) \geq n^2/(k+1)^2$ in Claim 18.10. The rest is the same with constants $b := (k+1)^2$ and $a := (k+1)^3$. The resulting lower bound is then $e^{n/(k+1)^2}$, which is super-polynomial as long as $k \leq \sqrt{n}/\log n$.

18.6 Size Versus Width

We have already seen that "fat" clauses—those whose width (number of literals) exceeds some given threshold value—play a crucial role in trying to show that the size of a resolution proof (= the total number of lines in it) must be large. We are now going to show that this is a general phenomenon, not just an accident: if any resolution proof for an unsatisfiable CNF formula F must contain at least one fat clause, then F cannot have a short resolution proof.

The width of a clause C is just the number of literals in it. If F is a set of clauses then its width $w(F)$ is the maximum width of its clause. Recall that each resolution refutation $\mathcal{R}$ is also a set (more precisely, a sequence) of clauses. Hence, the width of a refutation is also the maximum width of a clause participating in it.

Now let F be an unsatisfiable CNF of n variables. Define its *resolution refutation width* $w_R(F)$ as the minimum width of a resolution refutation of F. The *resolution refutation size* $S_R(F)$ is, as before, the minimum number of clauses in a resolution refutation of F. That is,

$$w_R(F) = \min\{w(\mathcal{R}) : \mathcal{R} \text{ is a resolution refutation proof of } F\}$$

and

$$S_R(F) = \min\{|\mathcal{R}| : \mathcal{R} \text{ is a resolution refutation proof of } F\}.$$

Note that refutation proofs $\mathcal{R}$ achieving $w_R(F)$ and $S_R(F)$ may be different! Let also $S_T(F)$ denote the minimum number of clauses in a *tree-like* resolution refutation of F.

What is the relation between these parameters? If we use all clauses of the CNF F in its refutation, then $w_R(F) \geq w(F)$. But this is not true in general: it may happen that not all clauses of F are used in the refutation of F.

The relation $S_R(F) \leq (2n + 1)^{w_R(F)}$ between proof-size and proof-width is easy to see: since we only have $2n$ literals, the number of all possible clauses of width k does not exceed $(2n + 1)^k$. Much more interesting is the following *lower* bound on proof-size in terms of proof-width: only CNF formulas having narrow proofs can be proved in a short time!

Theorem 18.11. (Ben-Sasson and Wigderson 2001) *For any unsatisfiable k-CNF formula F of n variables,*

$$\log S_R(F) \geq \frac{(w_R(F) - k)^2}{16n} \tag{18.5}$$

and

$$\log S_T(F) \geq w_R(F) - k. \tag{18.6}$$

For the proof of this theorem we need a concept of a *restriction* of CNFs and of refutation proofs. Let F be some set of clauses (think of F as a CNF or as a refutation proof). Let x be some of its literals. If we set this literal to 0 and to 1, then we obtain two sets of clauses:

- $F_{x=0}$ is F with all clauses containing $\neg x$ removed from F (they get value 1) and literal x removed from all the remaining clauses of F (it gets value 0);
- $F_{x=1}$ is F with all clauses containing x removed from F and literal $\neg x$ removed from all the remaining clauses of F.

Note that, if F was an unsatisfiable CNF, then both CNFs $F_{x=0}$ and $F_{x=1}$ remain unsatisfiable. Moreover, if $\mathcal{R}$ was a resolution refutation proof of F and $a \in \{0, 1\}$, then $\mathcal{R}_{x=a}$ is also a resolution refutation proof of $F_{x=a}$. Indeed, if at some step in $\mathcal{R}$ a literal x is resolved using the resolution rule, then this step in $\mathcal{R}_{x=a}$ corresponds to an application of the weakening rule:

$$\frac{A \vee x \quad B \vee \neg x}{A \vee B} \quad \mapsto \quad \frac{A}{A \vee B} \quad \text{or} \quad \frac{B}{A \vee B}.$$

Lemma 18.12. *Let F be an unsatisfiable k-CNF formula. If $w_R(F_{x=1}) \leq w - 1$ and $w_R(F_{x=0}) \leq w$, then $w_R(F) \leq \max\{w, k\}$.*

Proof. The idea is to combine refutations for $F_{x=1}$ and for $F_{x=0}$ into one refutation proof for F. First we can deduce $\neg x$ from $F_{x=1}$ using clauses of width at most w. To do this, follow closely the deduction of the empty clause from $F_{x=1}$, which uses clauses of width at most $w - 1$, and add the literal $\neg x$ to every clause in that deduction. Let $\mathcal{R}$ be the resulting deduction of $\neg x$ from $F_{x=1}$. Now, from $\neg x$ and F we can deduce $F_{x=0}$ by using the resolution rule: just resolve $\neg x$ with each clause of F containing x to get $F_{x=0}$. This step does not introduce any clauses of width more than k. Finally, deduce the empty clause from $F_{x=0}$ using clauses of width at most w. $\qquad\square$

Now let W be a parameter (to be specified later), and call a clause *fat* if it has width larger than W. Set also

$$a := \left(1 - \frac{W}{2n}\right)^{-1} \geq e^{W/2n}.$$

Lemma 18.13. *If a k-CNF F has a refutation that contains fewer than a^b fat clauses then $w_R(F) \leq W + b + k$.*

Proof. We prove this by induction on b and n. The base case $b = 0$ is trivial, since then we have no fat clauses at all implying that $w_R(F) \leq \max\{W, k\} \leq W + k$.

Now assume that the claim holds for all smaller values of n and b. Take a resolution refutation $\mathcal{R}$ of F using $< a^b$ fat clauses. Since there are at most $2n$ literals and any fat clause contains at least W of them, an average literal must occur in at least a $W/2n$ fraction of fat clauses. Choose a literal x that occurs most frequently in fat clauses and set it to 1. This way we kill off (evaluate to 1) all clauses containing x. The obtained refutation $\mathcal{R}_{x=1}$ of $F_{x=1}$ has fewer than $a^b\left(1 - \frac{W}{2n}\right) = a^{b-1}$ fat clauses. By induction on b we have $w_R(F_{x=1}) \leq W + (b-1) + k$. On the other hand, since $F_{x=0}$ has one variable fewer, induction on n yields $w_R(F_{x=0}) \leq W + b + k$. The desired upper bound $w_R(F) \leq W + b + k$ now follows from Lemma 18.12. $\qquad\square$

Proof of Theorem 18.11. Choose b so that $a^b = S_R(F)$. Then

$$b = \frac{\log S_R(F)}{\log a} \leq \frac{2n \log S_R(F)}{W \log(e)} \leq \frac{4n \log S_R(F)}{W}$$

and, by Lemma 18.13,

$$w_R(F) \leq W + \frac{4n \log S_R(F)}{W} + k.$$

Choosing $W := 2\sqrt{n \log S_R(F)}$ to minimize the right-hand side yields the desired upper bound $w_R(F) \leq 4\sqrt{n \log S_R(F)} + k$. This finishes the proof of (18.5). We leave the proof of (18.6) as an exercise; hint: as the literal x to be set take the *last* literal which is resolved to get the empty clause. $\qquad\square$

Remark 18.14. That Theorem 18.11 cannot be substantially improved was shown by Bonet and Galesi (1999): there are unsatisfiable k-CNF formulas F (k being a constant) such that $S_R(F) \leq n^{\mathcal{O}(1)}$ but $w_R(F) = \Omega(\sqrt{n})$.

A general frame to prove that the proof-width $w_R(F)$, and hence, the proof-size $S_R(F)$ must be large is as follows.

1. Take an arbitrary resolution refutation proof $\mathcal{R}$ for F.
2. Define some measure $\mu(C)$ of "weight" of its clauses $C \in \mathcal{R}$ such that

   a. The weight of each axiom is small;
   b. The last (empty) clause $\emptyset$ has large weight, and
   c. The measure is subadditive: $\mu(C) \leq \mu(A) + \mu(B)$ if C is a resolvent of A and B.

3. Use the subadditivity of μ to find a clause $C \in \mathcal{R}$ of "intermediate" (large, but not too large) measure $\mu(C)$.
4. Show that any clause of intermediate μ-measure must have many literals.

To achieve these goals one usually takes $\mu(C)$ to be the smallest number of axioms in a "witness" for C. A set $\mathcal{A}$ of axioms is a *witness* for C if every assignment satisfying all axioms in $\mathcal{A}$ satisfies the clause C as well. Then one argues as follows. The minimality of $\mathcal{A}$ implies that, for any axiom $A \in \mathcal{A}$, there must exist an assignment α such that $C(\alpha) = 0$ but $B(\alpha) = 1$ for all $B \in \mathcal{A}$, $B \neq A$. Now suppose that flipping the i-th bit of α gives us an assignment α' satisfying all axioms in $\mathcal{A}$. Since $\mathcal{A}$ is a witness for C, we have that $C(\alpha') = 1$. But the assignments α and α' only differ in the i-th position, implying that the i-th variable x_i or its negation must be present in C. Note that this was the way we argued in the proof of Haken's theorem for PHP_n^{n+1}.

In the next sections we show how this idea works in other situations.

18.7 Tseitin Formulas

In this section we discuss a large class of unsatisfiable CNFs whose resolution refutation proofs have large width. These CNFs formalize the basic property of graphs: in every graph, the number of vertices of *odd* degree must be *even*. This is a direct consequence of Euler's theorem stating that the sum of degrees in any graph is two times the number of edges, and hence, is even.

Let $G = (V, E)$ be a connected graph, and $f : V \to \{0, 1\}$ an assignment of bits 0 and 1 to its vertices. Let $d(v)$ denote the degree of a vertex $v \in V$. Associate with each edge $e \subset E$ a boolean variable x_e. For each vertex $v \in V$, let A_v be a CNF formula with $2^{d(v)-1}$ clauses expressing the equality

$$L_v : \quad \bigoplus_{e:v\in e} x_e = f(v). \tag{18.7}$$

For example, the equality $x \oplus y \oplus z = 0$ is expressed by a CNF

$$(\neg x \vee \neg y \vee \neg z) \wedge (x \vee y \vee \neg z) \wedge (x \vee \neg y \vee z) \wedge (\neg x \vee y \vee z).$$

The *Tseitin formula*, $\tau(G, f)$, is the AND of all these CNF formulas A_v, $v \in V$. Tseitin (1970) used such formulas to prove the first exponential lower bound on the size of regular resolution.

Remark 18.15. If k is the maximal degree of G, then $\tau(G, f)$ is a k-CNF formula with at most $n2^{k-1}$ clauses and $nk/2$ variables. Thus, if the degree k is *constant*, then $\tau(G, f)$ is a k-CNF formula with $\mathcal{O}(n)$ clauses and $\mathcal{O}(n)$ variables.

The meaning of Tseitin's formulas is the following. The function f "charges" some of the vertices, that is, gives them value 1. Each assignment α of constants 0 and 1 to the variables x_e defines a subgraph G_α of G. Such an assignment α satisfies $\tau(G, f)$ if and only if exactly the charged vertices have odd degrees in the subgraph G_α.

It is not difficult to show that if we charge an *odd* number of vertices, that is, if $\bigoplus_{v \in V} f(v) = 1$, then $\tau(G, f)$ is not satisfiable. Indeed, otherwise the graph G would have a subgraph in which an odd number of vertices (the charged ones) have odd degree, contradicting the Euler theorem. Interestingly, the converse also holds.

Lemma 18.16. (Tseitin 1970) *For a connected graph $G = (V, E)$, the CNF $\tau(G, f)$ is satisfiable if and only if an even number of vertices are charged by f.*

Proof. Assume first that f charges an *odd* number of vertices. We have used Euler's theorem to show that then $\tau(G, f)$ is unsatisfiable. This can also be shown directly. Observe that each variable x_e with $e = \{u, v\}$ appears in exactly two equations L_u and L_v. Hence, if we sum (modulo 2) all equations in (18.7), the left hand side will be equal to 0, whereas the right hand side will be 1, a contradiction. Hence, in this case the system (18.7) is not satisfiable.

Now assume that f charges an *even* number of vertices. We have to show that then $\tau(G, f)$ is satisfiable. For this, we make the following simple observation. Now start with an all-0 assignment α. If α satisfies all equalities L_v, we are done. If α does not satisfy all equalities, then the number of unsatisfied equalities must be even (the number of vertices v with $f(v) = 1$ must be even). We take any two vertices u, v with unsatisfied equalities L_v, L_u and change all bits of α corresponding to edges on a path from u to v (such a path must exist since G is connected). The obtained assignment α' will already satisfy L_v and L_u. Moreover, by our observation,[3] we have that $L_w(\alpha') = L_w(\alpha)$ for all vertices $w \notin \{u, v\}$. Hence, the number of unsatisfied equalities decreases by two. Proceeding in this way we will eventually reach an assignment satisfying all equalities.   □

We now give a general lower bound of the resolution refutation width of unsatisfiable Tseitin formulas $\tau(G, f)$ in terms of one combinatorial characteristic of the underlying graphs $G = (V, E)$. For a subset $S \subseteq V$ of vertices, let $e(S, V \setminus S)$ denote the number of crossing edges with one endpoint lying in S and the other in

---

[3]Let $\alpha \in \{0, 1\}^E$ be an assignment, and e_1, e_2 two edges with a common endpoint v. Let α' be obtained by flipping the values of both variables x_{e_1} and x_{e_2}. Then $L_v(\alpha') = L_v(\alpha)$.

$V \setminus S$. Define the *edge expansion*, ex(G), of G as the minimum of $e(S, V \setminus S)$ over all subsets S with $n/3 \le |S| \le 2n/3$; here $n = |V|$ is the total number of vertices in G.

Theorem 18.17. (Ben-Sasson and Wigderson 2001) *Let* $G = (V, E)$ *be a connected graph, and* $f : V \to \{0, 1\}$ *satisfy* $\bigoplus_{v \in V} f(v) = 1$. *Then*

$$w_R(\tau(G, f)) \ge \text{ex}(G).$$

Proof. Fix an arbitrary resolution refutation proof $\mathcal{R}$ for $\tau(G, f)$. Recall that axioms of this proof are CNF formulas A_v corresponding to equalities (18.7). For a subset $S \subseteq V$ of vertices, let A_S be the AND of all clauses in the sets A_v, $v \in S$. Define the measure $\mu : \mathcal{R} \to \mathbb{N}$ on clauses by:

$$\mu(C) := \min\{|S| : A_S \text{ implies } C\}.$$

If C is one of the axioms, then clearly $\mu(C) = 1$. Furthermore, μ is subadditive: $\mu(C) \le \mu(A) + \mu(B)$ if C is a resolvent of A and B.

Claim 18.18. $\mu(\emptyset) = n$.

Proof. By the definition of μ, $\mu(\emptyset)$ is exactly the smallest number $|S|$ of vertices such that A_S is unsatisfiable. So it is enough to show that A_S is satisfiable for each subset $S \subseteq V$ of size $|S| < |V|$. To show this, take any vertex $v \in V \setminus S$. Consider the function $f' : V \to \{0, 1\}$ such that $f'(v) = 1 - f(v)$ and $f'(u) = f(u)$ for all $u \ne v$. Since $\bigoplus_{v \in V} f'(v) = 0$, Lemma 18.16 implies that $\tau(G, f')$ is satisfiable. Since $v \notin S$, the CNF A_S is a part of the formula $\tau(G, f')$, and hence, is satisfiable as well. Hence, $\mu(\emptyset) = n$. $\square$

By the subadditivity of μ, there must exist a clause $C \in \mathcal{R}$ such that $n/3 \le \mu(C) \le 2n/3$, an "intermediate clause" (see Lemma 1.3). Let $S \subseteq V$ be a minimal set for which A_S implies C; hence $n/3 \le |S| \le 2n/3$.

To finish the proof of the theorem, it is enough to show that $x_e \in C$ for every crossing edge $e = \{u, v\}$ with $u \in S$ and $v \in V \setminus S$. For the sake of contradiction, assume that $x_e \notin C$. By the minimality of S, there exists an assignment α which satisfies all axioms in A_S except those in A_u, and falsifies C. The assignment α', obtained from α by flipping the bit x_e, satisfies all axioms in A_S (because $v \notin S$), and hence, must satisfy the clause C. This is a contradiction because $C(\alpha) = 0$ and the new assignment α' still agrees with α for all variables of C. $\square$

Theorem 18.17 gives us a whole row of unsatisfiable k-CNF formulas $F = \tau(G, f)$ of n variables such that $k = \mathcal{O}(1)$, $|F| = \mathcal{O}(n)$ and $w_R(F) = \Omega(n)$. Together with Theorem 18.11, these CNF formulas require resolution proofs of size $2^{\Omega(n)}$. For this, it is enough that the underlying graph G has constant degree k and still has large edge extension ex(G). The existence of such graphs can be shown by simple probabilistic arguments. There are even *explicit* graphs with these properties. Such are, for example, Ramanujan graphs considered in Sect. 5.8. By the Expander Mixing Lemma (see Appendix A) these graphs have ex$(G) = \Omega(n)$.

18.8 Expanders Force Large Width

We have shown that resolution refutation proofs for Tseitin CNF formulas $\tau(G, f)$ require large width as long as the underlying graph G has good expanding properties. It turns out that a similar fact also holds for *any* unsatisfiable CNF as long as it has good expansion properties in the following sense.

Look at a CNF formula F as a *set* of its clauses. Hence, $|F|$ denotes the number of clauses in F, and $G \subseteq F$ means that the CNF G contains only clauses of F. Let var(F) denote the number of variables in F.

We say a CNF formula F is (r, c)-*expanding* if

$$\text{var}(G) \geq (1 + c)|G| \text{ for every subset } G \subseteq F \text{ of its } |G| \leq r \text{ clauses.}$$

We can associate with F a bipartite graph, where nodes on the left part are clauses of F, nodes on the right part are variables, and a clause C is joined to a variable x iff x or $\neg x$ belongs to C. Then F is (r, c)-expanding iff every subset of $s \leq r$ nodes on the left part have at least $(1 + c)s$ neighbors on the right part.

Theorem 18.19. (Ben-Sasson and Wigderson 2001) *Let F be an unsatisfiable CNF formula. If F is (r, c)-expanding, then $w_R(F) \geq cr/2$.*

We first prove three claims relating the number of clauses with the number of variables in unsatisfiable CNF formulas.

Claim 18.20. *If $|G| \leq \text{var}(G)$ for every $G \subseteq F$, then F is satisfiable.*

Proof. We will use the well-known Hall's Marriage Theorem. It states that a family of sets $\mathcal{S} = \{S_1, \ldots, S_m\}$ has a system of distinct representatives (that is, a sequence $x_1, \ldots, x_m$ of m distinct elements such that $x_i \in S_i$) iff the union of any number $1 \leq k \leq m$ of members of $\mathcal{S}$ has at least k elements.

Now assume that $|G| \leq \text{var}(G)$ for all $G \subseteq F$. Then, by Hall's theorem, we can find for each clause C of F a variable $x_C \in \text{var}(C)$ such that x_C or its negation appears in C, and for distinct clauses these variables are also distinct. We can therefore set these variables to 0 or 1 independently to make all clauses true. Hence, F is satisfiable.   □

Say that an unsatisfiable CNF formula is *minimally* unsatisfiable if removing any clause from it makes the remaining CNF satisfiable. The following claim is also known as Tarsi's Lemma.

Claim 18.21. *If F is minimally unsatisfiable, then $|F| > \text{var}(F)$.*

Proof. Since F is unsatisfiable, Claim 18.20 implies that there must be a subset of clauses $G \subseteq F$ such that $|G| > \text{var}(G)$. Let $G \subseteq F$ be a *maximal* subset of clauses with this property. If $G = F$ then we are done, so assume that $G \subset F$ and we will derive a contradiction.

Take an arbitrary sub-formula $H \subseteq F \setminus G$, and let $\text{Vars}(H)$ be the set of its variables. Due to maximality of G, $\text{Vars}(H) \setminus \text{Vars}(G)$ must have at least $|H|$ variables, for otherwise we would have that $\text{var}(G \cup H) < |G \cup H|$, a contradiction with the maximality of G.

Thus the CNF formula $F \setminus G$ satisfies the condition of Claim 18.20, and hence, can be satisfied by only setting constants to variables in $\text{Vars}(F) \setminus \text{Vars}(G)$. Since F is minimally unsatisfiable, the CNF formula G must be satisfiable using only the variables in $\text{Vars}(G)$. Altogether this gives us a truth assignment satisfying the entire formula F, a contradiction. $\qquad\Box$

As before, we say that a CNF formula F *implies* a clause A if any assignment satisfying F also satisfies A. We also say that F *minimally implies* A if the CNF formula F implies A but none of its proper subformulas (obtained by removing any clause) does this.

Claim 18.22. If F minimally implies a clause A, then $|A| > \text{var}(F) - |F|$.

Proof. Let $\text{Vars}(F) = \{x_1, \ldots, x_n\}$ and assume that $\text{Vars}(A) = \{x_1, \ldots, x_k\}$. Take a (unique) assignment $\alpha \in \{0, 1\}^k$ for which $A(\alpha) = 0$. Since F implies A, restricting F to α must yield an unsatisfiable formula F_α on variables $x_{k+1}, \ldots, x_n$. The formula F_α must also be minimally unsatisfiable because F minimally implied A. By Claim 18.21, F_α must have more than $n - k$ clauses. Hence, $|F| \geq |F_\alpha| > n - k = \text{var}(F) - |A|$, as desired. $\qquad\Box$

We now turn to the actual proof of the theorem.

Proof of Theorem 18.19. Let F be an (r, c)-expanding unsatisfiable CNF formula, and let $\mathcal{R}$ be any resolution refutation proof of F. We can assume that both numbers r and c are positive (otherwise there is nothing to prove). With each clause C in $\mathcal{R}$ associate the number

$$\mu(A) = \min\{|G| : G \subseteq F \text{ and } G \text{ implies } A\}.$$

It is clear that $\mu(A) \leq 1$ for all clauses A of F. Furthermore, μ is subadditive: $\mu(C) \leq \mu(A) + \mu(B)$ if C is a resolvent of A and B. Finally, the expansion property of F implies that $\mu(0) > r$. Indeed, by the definition, $\mu(0)$ is the smallest size $|G|$ of an unsatisfiable subformula $G \subseteq F$, and Claim 18.21 yields $|G| > \text{var}(G)$. Had we $\mu(0) \leq r$, then we would also have $|G| \leq r$ and the expansion property of F would imply $\text{var}(G) > (1 + c)|G|$, a contradiction.

Hence, the subadditivity of μ implies that the refutation $\mathcal{R}$ of F must contain a clause C such that $r/2 \leq \mu(C) < r$ (cf. Lemma 1.3). Fix some $G \subseteq F$ minimally implying C; hence, $r/2 \leq |G| = \mu(C) < r$. By the expansion of F, $\text{var}(G) \geq (1 + c)|G|$. Together with Claim 18.22 this implies $|C| > \text{var}(G) - |G| \geq c|G| \geq cr/2$, as desired. $\qquad\Box$

18.9 Matching Principles for Graphs

We already know (see Theorem 18.8) that the pigeonhole principle PHP_n^m requires resolution proof of exponential size, as long as the number m of pigeons is $m = n + 1$, where n is the number of holes. However, the larger m is, the more true the pigeonhole principle itself is, and it could be that PHP_n^m with larger number m pigeons could be refuted by much shorter resolution refutation proof. We now will use expander graphs to prove that PHP_n^m has no resolution proofs of polynomial size even if we have up to $m = n^{2-o(1)}$ pigeons.

Given a bipartite $m \times n$ graph $G = ([m], [n], E)$, we may consider the CNF formula $\mathrm{PHP}(G)$ which is an AND of the following set of axioms:

- Pigeon Axioms: $C_i = \bigvee_{(i,j) \in E} x_{i,j}$ for $i = 1, \dots, m$.
- Hole Axioms: $\neg x_{i_1,j} \vee \neg x_{i_2,j}$ for $i_1 \neq i_2 \in [m]$ and $j \in [n]$.

That is, the graph dictates what holes are offered to each pigeon, whereas hole axioms forbid (as in the case of PHP_n^m) that two pigeons sit in one hole.

Observe that, if $m > n$ and if the graph G has no isolated vertices, then the CNF formula $\mathrm{PHP}(G)$ is unsatisfiable. Indeed, every truth assignment α defines a subgraph G_α of G. Now, if α satisfies all hole axioms then G_α must be a (possibly empty) matching. But we have $m > n$ vertices of the left side. Hence, at least one of these vertices $i \in [m]$ must remain unmatched in G_α, implying that $C_i(\alpha) = 0$.

Observe also that $\mathrm{PHP}_n^m = \mathrm{PHP}(K_{m,n})$ where $K_{m,n}$ is a complete bipartite $m \times n$ graph. Moreover, if G' is a subgraph of G, then every resolution refutation for $\mathrm{PHP}(G)$ can be turned to a resolution refutation of $\mathrm{PHP}(G')$ just by setting to 0 all variables corresponding to edges of G that are not present in G'. Thus to prove a lower bound of the resolution complexity of $\mathrm{PHP}(G)$ it is enough to prove such a bound for *any* subgraph of G.

This opens plenty of possibilities to prove large lower bounds for PHP_n^m: just show that there *exists* a graph G (a subgraph of $K_{m,n}$) such that $\mathrm{PHP}(G)$ requires long resolution refutation proofs. By Theorems 18.11 and 18.19, this can be done by showing that the CNF formula $F = \mathrm{PHP}(G)$ has large expansion. This, in turn, can be achieved if the underlying graph G itself has good expansion properties.

A bipartite graph is an (r, c)-*expander* if every set of $k \leq r$ vertices on the left part has at least $(1 + c)k$ neighbors on the right part. It can be easily shown (Exercise 18.4) that if G is an (r, c)-expander then the CNF formula $\mathrm{PHP}(G)$ is (r, c)-expanding.

Using a probabilistic argument it can be shown that (r, c)-expanders with $c > 0$, $r = \Omega(n)$ and *constant* left-degree exist (Exercise 18.5). Hence, the CNF formula $F = \mathrm{PHP}(G)$ has $N = \mathcal{O}(m)$ variables and each its clause has constant width. Theorem 18.19 implies that $w_R(F) = \Omega(n)$. So by Theorem 18.11, every resolution refutation for F, and hence, for PHP_n^m must have size exponential in $w_R(F)^2/N = \Omega(n^2/m)$.

This gives super-polynomial lower bound on the size of resolution refutations of PHP_n^m for up to $m \ll n^2/\log n$ pigeons. In Sect. 18.3 we have proved that, no matter how large the number $m > n$ of pigeons is, any *tree-like* resolution refutation proof of PHP_n^m must have size $n^{\Omega(n)}$. But all attempts to overcome the "n^2 barrier" for the number of pigeons m in the case of general (not just tree-like) resolution proofs failed for many years. This was one of the most famous open problems concerning resolution proofs.

The "n^2 barrier" was finally broken by Raz (2001). He proved that, for any number $m > n$ of pigeons, the CNF PHP_n^m requires general (non-tree-like) resolution proofs of exponential size. Shortly after, Razborov (2003) found a simpler proof.

Exercises

18.1. Show that Resolution is complete: every unsatisfiable CNF formula F has a resolution refutation proof.

Hint: Show that the search problem for F can be solved by a decision tree, and use Theorem 18.1.

18.2. Show that Theorem 18.5 remains true if instead of CNF formula PHP_n^m we take its *functional* version by adding new axioms $\neg x_{i,j_1} \vee \neg x_{i,j_2}$ for all $j_1 \neq j_2$ and $i = 1, \ldots, m$. These axioms claim that no pigeon can sit in two holes.

18.3. Let F be a CNF formula and x a literal. Show that F is unsatisfiable if and only if both CNFs $F_{x=1}$ and $F_{x=0}$ are unsatisfiable.

18.4. Let G be a bipartite (r, c)-expander graph. Show that then the induced CNF formula $PHP(G)$ is (r, c)-expanding.

18.5. Show that for every constant $d \geq 5$, there exist bipartite $n \times n$ graphs of left degree d that are (r, c)-expanders for $r = n/d$ and $c = d/4 - 1$.

Hint: Construct a random graph with parts L and R, $|L| = |R| = n$, by choosing d neighbors for each vertex in L. For $S \subseteq L$ and $T \subseteq R$, let $E_{S,T}$ be the event that all neighbors of S lie within T. Argue that, $\text{Prob}[E_{S,T}] = (|T|/n)^{d|S|}$. Let E be the event that the graph is not the desired expander, i.e., that all neighbors of some subset $S \subseteq L$ of size $|S| \leq n/d$ lie within some subset $T \subseteq R$ of size $|T| < (d/4)|S|$. Use the union bound for probabilities and the estimate $\binom{n}{k} \leq (en/k)^k$ to show that $\text{Prob}[E] \leq \sum_{i=1}^{n/d} \left(\frac{e}{4}\right)^{id/2}$. Use our assumption $d \geq 5$ together with the fact that $\sum_{i=0}^{\infty} x^i = 1/(1-x)$ for any real number x with $|x| < 1$ to conclude that $\text{Prob}[E]$ is strictly smaller than 1.

18.6. Given an unsatisfiable set F of clauses, define its *boundary* ∂F to be the set of variables appearing in exactly one clause of F. Let also

$$s(F) = \min\{|G| : G \subseteq F \text{ and } G \text{ is unsatisfiable}\}.$$

Define the *expansion* of F by

$$e(F) = \max_{s \leq s(F)} \min\{|\partial G| : G \subseteq F, \ s/2 \leq |G| < s\}.$$

Prove that, for every unsatisfiable CNF F, $w_R(F) \geq e(F)$.

Hint: Take a resolution refutation proof $\mathcal{R}$ for F. Define the *witness* of a clause C in the proof to be the set $G \subseteq F$ of all those clauses in F that are used by the proof to derive C. Show that the clause C can have at most $|\partial G|$ literals (if a literal appears in an axiom $A \in F$, then the only way it can be removed from clauses derived using A is if the literal is resolved with its negation). Then, define $\mu(C)$ to be the number $|G|$ of clauses in the witness G of C in the proof. Show that: $\mu(\emptyset) \geq s(F)$, and $\mu(C) = 1$ for any clause C in F, and $\mu(C) \leq \mu(A) + \mu(B)$ if C is a resolvent of A and B.

18.7. (2-satisfiable CNFs) A CNF formula F is *k-satisfiable* if any subset of its k clauses is satisfiable. Prove the Lieberher-Specker result for 2-satisfiable CNF formulas: if F is a 2-satisfiable CNF formula then at least γ-fraction of its clauses are simultaneously satisfiable, where $\gamma = (\sqrt{5} - 1)/2 > 0.618$.

Hint: Define the probability of a literal y to be satisfied to be: a ($a > 1/2$) if y occurs in a unary clause, and $1/2$ otherwise. Observe that then the probability that a clause C is satisfied is a if C is a unary clause, and at least $1 - a^2$ otherwise (at worst, a clause will be a disjunction of two literals whose negations appear as unary clauses); verify that $a = 1 - a^2$ for $a = \gamma$.

18.8. (3-satisfiable CNFs) Given a 3-satisfiable CNF formula F of n variables, define a random assignment $\alpha = (\alpha_1, \ldots, \alpha_n) \in \{0, 1\}^n$ by the following rule:

$$\text{Prob}[\alpha_i = 1] = \begin{cases} 2/3 \text{ if } F \text{ contains a unary clause } (x_i); \\ 1/3 \text{ if } F \text{ contains a unary clause } (\neg x_i); \\ 1/2 \text{ otherwise.} \end{cases}$$

1. Why is this definition consistent?
   Hint: 3-satisfiability.
2. Show that $\text{Prob}[y(\alpha) = 1] \geq 1/3$ for each literal $y \in \{x_i, \neg x_i\}$, which appears in the formula F (independent of whether this literal forms a unary clause or not).
3. Show that the expected number of clauses of F satisfied by α is at least a $2/3$ fraction of all clauses.
   Hint: Show that each clause if satisfied by α with probability at least $2/3$. The only nontrivial case is when the clause has exactly two literals. Treat this case by keeping in mind that our formula is 3-satisfiable, and hence, cannot have three clauses of the form $(y \vee z)$, $(\neg y)$ and $(\neg z)$.

18.9. (Due to Hirsch 2000) Suppose we have a CNF formula F of n variables with *is* satisfiable. Our goal is to find a satisfying assignment. Consider the following randomized algorithm: pick an initial assignment $\alpha \in \{0, 1\}^n$ uniformly at random, and flip its bits one by one trying to satisfy all clauses. At each step, the decision on what bit of a current assignment α to flip is also random one. The algorithm first constructs a set $I \subseteq [n]$ of bits such that flipping any bit $i \in I$ increases the number of satisfied clauses. Then it chooses one of these bits at random, and flips it. If $I = \emptyset$, then the algorithm chooses one bit at random from the set of bits that do not lead to the decrease of the number of satisfied clauses. If all variables lead to such a decrease, it chooses at random a bit from $[n]$. The algorithm works in iterations, one iteration being a random choice of an initial assignment α. We are interested in how many iterations are needed to find a satisfying assignment with a constant probability.

Consider the CNF formula F which is an AND of two CNFs G and H. The first CNF G consists of $n + 1$ clauses:

$$\neg x_1 \vee x_2, \quad \neg x_2 \vee x_3, \quad \ldots, \quad \neg x_n \vee x_1 \quad \text{and} \quad \neg x_1 \vee \neg x_2.$$

The first n clauses express that in every satisfying assignment for G the values of all its bit must be equal. The last clause of G ensures that all these values must be equal to 0. Hence, $\alpha = \mathbf{0}$ is the only assignment satisfying all the $n + 1$ clauses of G. The second CNF H consists of all $n\binom{n-1}{2}$ clauses of the form $\neg x_i \vee x_j \vee x_k$ with $i \neq j \neq k$. Hence, $\alpha = \mathbf{0}$ is the unique satisfying assignment for the entire CNF $F = G \wedge H$. The clauses in H are intended for "misleading" the algorithm. Prove that, regardless of how long one iteration tends, at least $2^{\Omega(n)}$ iterations are necessary for the CNF formula F.

Hint: Show that, if c is a sufficiently large constant, then assignments α with $t := n/3 + c$ or more ones form an "insurmountable ring" around the (unique) satisfying assignment $\mathbf{0}$. Namely, if the algorithm encounters an assignment with this number of ones, then it chooses a wrong bit for flipping. That is, on such assignments α the algorithms flips some 0-bit to 1-bit, and hence, goes away from the satisfying assignment $\mathbf{0}$. When showing this, it is only important that $(k - 1)(n - k - 1) > \binom{n-k}{2} + 2$ holds for all $k \geq t$.

Chapter 19
Cutting Plane Proofs

We now turn our attention to a proof system more powerful than resolution—the so-called *cutting plane* proof system. This proof system, which can be viewed as a "geometric generalization" of resolution, originated in works on integer programming by Gomory (1963) and Chvátal (1973); as a proof system it was first considered in Cook et al. (1987) The basic idea is to use a few elementary rules to prove that a given system of linear inequalities (or "cutting planes") with integer coefficients does not have a 0-1 solution.

Why should we care about cutting planes in the context of this book? It turns out that there is an intimate methodological relation between the cutting planes system and circuit complexity: the only known exponential lower bounds for cutting plane proofs were obtained using bounds on communication complexity as well as on the size of monotone real-valued circuits.

19.1 Cutting Planes as Proofs

Let A be a matrix with integer entries, and b an integer vector. We are interested in *boolean-valued* solutions, to the system of inequalities $Ax \leq b$ (over integer arithmetic). We say that the system $Ax \leq b$ is *unsatisfiable* if it has no 0-1 solution $x \in \{0,1\}^n$. Such unsatisfiable systems are the "objects of interest" for cutting plane proof systems, taking the place of the unsatisfiable CNFs studied in resolution. Given an unsatisfiable system of inequalities, our goal is to prove its unsatisfiability using repeated applications of a few basic rules.

The cutting planes proof system provides one such set of rules: addition of inequalities, their multiplication by positive integers, and one truly powerful rule—the rounded division rule, known as *Gomory–Chvátal rule*. The idea of this last rule is that if the coefficients $a_1, \ldots, a_n$ all are multiplies of an integer $c \geq 1$ then any *integer* solution x of $a^T x \leq b$ is also a solution of $\frac{1}{c} a^T x \leq \lfloor \frac{b}{c} \rfloor$. Thus we have three rules of derivation, where $a \in \mathbb{Z}^n$, $b, c \in \mathbb{Z}$ and $c \geq 1$:

S. Jukna, *Boolean Function Complexity*, Algorithms and Combinatorics 27,
DOI 10.1007/978-3-642-24508-4_19, © Springer-Verlag Berlin Heidelberg 2012

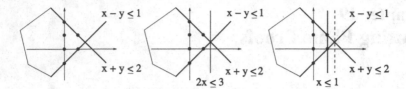

Fig. 19.1 A cutting plane derivation of $x \leq 1$ from $x - y \leq 1$ and $x + y \leq 2$. Sum rule yields $2x \leq 3$, and rounding-down gives $x \leq 1$. The half space $\{(x,y)|x \leq 3/2\}$ is replaced by the half space $\{(x,y)|x \leq 1\}$; no integer solution is lost

$$\frac{a^T x \leq b}{ca^T x \leq cb}, \qquad \frac{a_1^T x \leq b_1 \qquad a_2^T x \leq b_2}{(a_1 + a_2)^T x \leq b_1 + b_2} \quad \text{and} \quad \frac{ca^T x \leq b}{a^T x \leq \lfloor b/c \rfloor}.$$

A *cutting plane derivation* (or *CP derivation*) of an inequality $c^T x \leq d$ from a system $Ax \leq b$ is a sequence of inequalities $a_1^T x \leq b_1, \ldots, a_t^T x \leq b_t$ such that $a_t = c$, $b_t \leq d$ and each $a_i^T x \leq b_i$ is either an inequality in $Ax \leq b$ or is obtained from the previous ones by an application of one of the three rules above (Fig. 19.1). The *size* of such a derivation is the number t of inequalities in it. If the original system $Ax \leq b$ is unsatisfiable, then a *CP proof* of its unsatisfiability (or a *CP refutation*) is a derivation of a contradiction expressed as $0 \leq -1$ from $Ax \leq b$.

If the inequality is in the other direction $a^T x \geq b$, then the application of the rounded division rule yields $\frac{1}{c} a^T x \geq \lceil \frac{b}{c} \rceil$. We will often use the trivial fact that $f \geq g$ is equivalent to $-f \leq -g$. That is, we will write our systems of inequalities either as $Ax \leq b$ or as $Ax \geq b$, depending on what form is more convenient in a concrete situation.

Example 19.1. (The "triangle trick") The power of the division rule can be seen in the following simple example. Consider the system consisting of the following three inequalities: $x + y \leq 1$, $x + z \leq 1$ and $y + z \leq 1$. By adding these inequalities, we obtain $2x + 2y + 2z \leq 3$, and the division rule yields $x + y + z \leq \lfloor 3/2 \rfloor = 1$. More generally, suppose that there are three disjoint sets of indices P, Q and R and that we have the inequalities

$$\sum_{p \in P} x_p + \sum_{q \in Q} x_q \leq 1, \quad \sum_{p \in P} x_p + \sum_{r \in R} x_r \leq 1 \quad \text{and} \quad \sum_{q \in Q} x_q + \sum_{r \in R} x_r \leq 1.$$

We can derive the inequality

$$\sum_{p \in P} x_p + \sum_{q \in Q} x_q + \sum_{r \in R} x_r \leq 1$$

in three steps by adding the three inequalities together, dividing the result by two and rounding down.

Remark 19.2. Interestingly, it is enough to use the restricted rounded division rule with $c = 2$. Namely, Buss and Clote (1996) showed that any CP proof of size S can be transformed into a CP proof of size polynomial in n and S where only divisions by 2 are used.

Remark 19.3. In combinatorial optimization and in proof complexity, other types of cut rules are considered as well: various types of *Lovász–Schrijver cuts*, *lift-and-project cuts*, *split cuts*; see Exercise 19.13 for definitions of the latter two cuts. In this chapter we will only discuss proof systems based on Gomory–Chvátal cuts, as defined above.

19.2 Cutting Planes and Resolution

While cutting plane proofs apply to linear systems, not CNFs, it is not difficult to "translate" CNFs into equivalent linear systems. Thus we can study the length of refutations of CNFs in the cutting planes system, and compare this to their refutation length under resolution. We will see that resolution refutations can be naturally translated into cutting plane refutations without any increase in length, justifying our claim that the cutting planes system is a generalization of resolution.

Let us first see how to translate CNFs into linear systems. First, we replace each clause by an inequality using the translation $x_i \mapsto x_i$ and $\neg x_i \mapsto 1 - x_i$. For example, the clause $x \vee \neg y$ translates to an inequality $x + (1 - y) \geq 1$, which is the same as $x - y \geq 0$. An assignment $a = (a_1, a_2) \in \{0, 1\}^2$ satisfies the clause iff $a_1 - a_2 \geq 0$, that is, iff $a_1 = 1$ or $a_2 = 0$. In this way each unsatisfiable CNF translates to an unsatisfiable system of linear inequalities. For example, the CNF

$$(x \vee y) \wedge (\neg x \vee y) \wedge (x \vee \neg y) \wedge (\neg x \vee \neg y)$$

translates to the system

$$x + y \geq 1, \ -x + y \geq 0, \ x - y \geq 0, \ -x - y \geq -1.$$

More generally, each clause C translates to the inequality

$$\sum_i a_i x_i \geq 1 - m,$$

where m is the number of negated literals in C, and

$$a_i = \begin{cases} 1 & \text{if } x_i \in C, \\ -1 & \text{if } \neg x_i \in C, \\ 0 & \text{if neither } x_i \text{ nor } \neg x_i \text{ is in } C. \end{cases}$$

Proposition 19.4. *The cutting planes proof system can efficiently simulate resolution.*

Proof. Suppose we have a resolution refutation proof $\mathcal{R}$ of some unsatisfiable CNF. By adding a trivial derivation rule "derive $C \vee z$ from C", we can assume that each resolution inference in this proof has the form "derive C from $C \vee x_i$ and $C \vee \neg x_i$". The rule "derive $C \vee z$ from C" is known in the literature as the *weakening rule*, and it is known that this additional rule does not make the resolution system stronger.

Let $f = \sum_j a_j x_j \geq 1 - m$ be the inequality corresponding to clause C; here m is the number of negated literals in C. Then the inequality for the clause $C \vee x_i$ is $f + x_i \geq 1 - m$ (x_i comes positively in this clause), and the inequality for the clause $C \vee \neg x_i$ is $f - x_i \geq 1 - (m + 1)$. Now apply the sum-rule

$$\frac{f + x_i \geq 1 - m \qquad f - x_i \geq -m}{2f \geq 1 - 2m},$$

and then the division rule

$$\frac{2f \geq 1 - 2m}{f \geq 1 - m}$$

to obtain the inequality for the clause C.   □

We will now show that, in fact, cutting plane proofs for some CNFs may be even exponentially shorter than resolution proofs! So proving lower bounds for the former model is a more difficult task.

In a general CP proof, one derived inequality can be used many times without re-deriving it. That is, the underlying graphs of derivations may be arbitrary directed acyclic graphs (Fig. 19.2). A *tree-like* CP proof is a special case of a CP proof, where

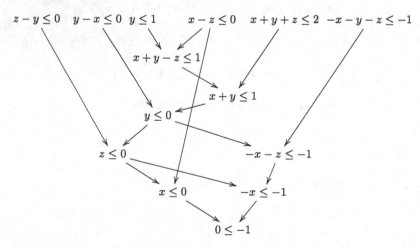

Fig. 19.2 The graph of a cutting plane proof of $0 \leq -1$. The proof is not tree-like

the underlying graph is a tree. That is, every inequality in the proof, except for the initial inequalities, is used at most once as an antecedent of an implication.

Although restricted, tree-like CP proofs are still powerful. So, for example, we already know that the CNF formula PHP_n^{n+1} formalizing the pigeonhole principle has no resolution proof of polynomial size. On the other hand, this CNF has relatively short tree-like CP proofs.

Proposition 19.5. (Cook et al. 1987) *For any $m \geq n + 1$, PHP_n^m has a tree-like cutting plane proof of size $\mathcal{O}(nm^2)$.*

Proof. When translated to the language of inequalities, the axioms for the pigeonhole principle PHP_n^m consist of the following inequalities:

- Pigeon Axioms: $x_{i1} + x_{i2} + \cdots + x_{in} \geq 1$ for all $i = 1, \ldots, m$.
- Hole Axioms: $x_{ij} + x_{kj} \leq 1$ for all $1 \leq i < k \leq m$, $j = 1, \ldots, n$.
- 0/1-axioms: $x_{ij} \geq 0$; $x_{ij} \leq 1$ for $i = 1, \ldots, m$, $j = 1, \ldots, n$.

For each j we first derive $x_{1j} + x_{2j} + \cdots + x_{mj} \leq 1$ inductively. The inequality $x_{1j} \leq 1$ is a 0/1-axiom, and inequality $x_{1j} + x_{2j} \leq 1$ is a hole axiom. Suppose we have already derived the inequality $x_{1j} + x_{2j} + \cdots + x_{kj} \leq 1$, and want to derive $x_{1j} + x_{2j} + \cdots + x_{(k+1)j} \leq 1$. Multiply the inequality

$$x_{1j} + x_{2j} + \cdots + x_{kj} \leq 1$$

by $k - 1$ and add to the result the hole axioms $x_{ij} + x_{(k+1)j} \leq 1$ for $i = 1, \ldots, k$ to get

$$kx_{1j} + kx_{2j} + \cdots + kx_{kj} + kx_{(k+1)j} \leq 2k - 1.$$

Apply division rule to get the desired inequality:

$$x_{1j} + x_{2j} + \cdots + x_{kj} + x_{(k+1)j} \leq \left\lfloor \frac{2k-1}{k} \right\rfloor = \left\lfloor 2 - \frac{1}{k} \right\rfloor = 1.$$

Now, summing these inequalities $x_{1j} + x_{2j} + \cdots + x_{mj} \leq 1$ over all holes j gives that the sum S of all variables is at most n, that is, $-S \geq -n$. On the other hand, summing pigeon inequalities $x_{i1} + x_{i2} + \cdots + x_{in} \geq 1$ over all pigeons i gives that $S \geq m$. Summing these two last inequalities gives $0 \geq m - n \geq 1$, the desired contradiction. $\qquad \square$

Remark 19.6. The size of this proof for PHP_n^m has polynomial size, but its depth is $\Omega(n)$; as usually, the *depth* of a tree is the maximum number of edges along a path from the root to a leaf. Using the "triangle trick" (Example 19.1) one can substantially reduce the depth to $\mathcal{O}(\log n)$ while keeping the size polynomial (see Exercise 19.6). Rhodes (2009) showed that this cannot be improved: *any* cutting plane proof for PHP_n^{n+1} requires depth $\Omega(\log n)$.

19.3   Lower Bounds for Tree-Like CP Proofs

We are now going to prove an exponential lower bound on the size of tree-like CP proofs using communication complexity arguments. The bound is due to Impagliazzo et al. (1994). The idea is to consider a tree-like CP proof for an unsatisfiable system of inequalities $Ax \leq b$ as a "search tree" for the following search problem: given an assignment $\alpha \in \{0, 1\}^n$ find an axiom (an inequality in our system) which is falsified by α. This step is easy: just traverse the proof starting from its last inequality $0 \leq -1$ until a "hurt" axiom (an inequality in $Ax \leq b$ falsified by α) is found. This works because of the soundness of derivation rules: if an assignment α falsifies a derived inequality then it must falsify at least one of the premises. Since the proof works with inequalities, the underlying model for the computation is a threshold decision tree.

A *threshold decision tree* is a rooted, directed tree whose vertices are labeled by threshold functions

$$f(x) = 1 \text{ if and only if } a_1 x_1 + \cdots + a_n x_n \leq b$$

with integer coefficients $a_1, \ldots, a_n, b$, and edges are labeled with either 0 or 1. The leaves of the tree are labeled with axioms (inequalities of $Ax \leq b$).

A threshold decision tree computes the search problem for a system of inequalities in the obvious way: we start at the root and evaluate the threshold function, follow the edge which is consistent with the value of the threshold function, continue until we hit a leaf and output the associated hurt axiom.

The threshold decision tree complexity of a system of inequalities is the minimum depth of any threshold decision tree for computing the search problem for this system. The threshold decision tree complexity of an unsatisfiable system is a guideline to whether the system will have efficient tree-like CP proofs.

The following lemma shows that an efficient tree-like CP proof can be converted into a small-depth threshold decision tree. Unfortunately, there is no converse to this lemma: every unsatisfiable system containing m inequalities has a trivial threshold decision tree of depth m, namely the tree obtained by evaluating the inequalities one by one.

Lemma 19.7. *If an unsatisfiable system $Ax \leq b$ has a tree-like CP proof of size S, then it has a threshold decision tree of depth $\mathcal{O}(\log S)$.*

Proof. Let T be a tree-like CP proof for the system $Ax \leq b$, and let S be the number of leaves in T. We will describe a threshold decision tree of depth $\log_{3/2} S$ which computes the search problem associated with $Ax \leq b$. The proof is by induction on the size S. Clearly if the size is 1 then the system consists of a single unsatisfiable threshold formula, so the lemma holds. Now assume that the size of T is S.

By Lemma 1.3, there must be a subtree T_0 of T rooted in some node v and such that the number $|T_0|$ of leaves in T_0 satisfies $S/3 \leq |T_0| \leq 2S/3$. Cut off this subtree T_0 from the entire tree T, assign its root (now a leaf) the always true inequality $1 \leq 1$, and let T_1 be the resulting tree. We apply the induction hypothesis to both T_0 and T_1 to obtain threshold decision trees T_0' and T_1'.

Let $c^T x \leq d$ be the inequality derived at the root v of the removed subtree T_0. In our threshold decision tree, we first query the threshold function $c^T x \leq d$. If it evaluates to 0 we proceed on the subtree T_0', otherwise, we proceed on the subtree T_1'. By the induction hypothesis, since both T_0 and T_1 have at most $2S/3$ leaves, the depth of the threshold decision tree obtained will be $1 + \log_{3/2}(2S/3) = \log_{3/2} S$.

To see that the decision tree actually computes the search function, notice that if $c^T x \leq d$ evaluates to false on a given assignment $\alpha \in \{0, 1\}^n$, then we proceed on the subproof T_0. Since the proof is sound, and the root formula of T_0 is false on α, this implies that one of the leaf formulas of T_0 must be falsified by α. Similarly, if $c^T x \leq d$ evaluates to true on α, then we proceed on T_1. Again, since the root inequality $0 \leq -1$ of T_1 is false, we will reach some leaf of T_1. This leaf cannot be the node v since $c^T x \leq d$ evaluates to true on α. Hence, we will reach one of the leaf inequalities of T_1 (axioms) falsified by α. □

Given an unsatisfiable system $Ax \leq b$, fix a partition of its variables into two parts, and consider the following Karchmer–Wigderson type communication game: for an assignment $\alpha \in \{0, 1\}^n$, Alice gets its projection onto the first part of variables, Bob gets the projection of α onto the second part, and their goal is to find an inequality falsified by α. Say that a CP proof has *bounded coefficients* if there exists a constant $k > 0$ such that the absolute values of all coefficients of inequalities used in the proof do not exceed $\mathcal{O}(n^k)$.

Lemma 19.8. *If for some partition of variables, the communication game for $Ax \leq b$ requires t bits of communication, and if we have n variables in total, then any tree-like CP proof with bounded coefficients for $Ax \leq b$ must have size $2^{\Omega(t/\log n)}$.*

Proof. Suppose that $Ax \leq b$ has a tree-like CP proof of size S, all coefficients in which are bounded. This proof gives us a threshold tree solving the search of a hurt axiom problem for $Ax \leq b$. By Lemma 19.7, this problem can be solved by a threshold tree of depth $d = \mathcal{O}(\log S)$. Since all coefficients are bounded, the communication complexity of each threshold function used in the tree does not exceed $c = \mathcal{O}(\log n)$: it is enough to send the value of a partial sum to the other player. Thus, the entire communication complexity is at most cd. But, by our assumption, t bits of communication are necessary. This implies $cd \geq t$, from which the desired lower bound $S = 2^{\Omega(t/\log n)}$ follows. □

Remark 19.9. Note that this lemma holds for very general tree-like CP proofs where, besides addition of inequalities, their multiplication by positive constants and the rounded division rule, *any* sound rule is allowed. The only restriction is that at most two (or only a constant number of) already derived inequalities can be used to derive a new inequality, and the derived inequality must have bounded coefficients.

19.3.1 Lower Bound for the Matching CNF

In Sect. 7.6 we have considered the following "find an edge" communication game FE_n for graphs on $n = 3m$ vertices:

- Alice gets a matching p consisting of m edges.
- Bob gets an $(m-1)$-element set q of vertices.
- The goal is to find an edge e such that $e \in p$ and $e \cap q = \emptyset$.

We proved (Theorem 7.26) that any deterministic communication protocol for this game requires $\Omega(n)$ bits of communication. It is therefore enough to turn this "find an edge" problem into a "find a falsified inequality" problem for an unsatisfiable system of inequalities MATCH_n of $n^{\mathcal{O}(1)}$ variables.

Assume for a moment that we already have such a system MATCH_n. Then the communication complexity of the corresponding to MATCH_n search problem is $\Omega(n)$. By Lemma 19.8, every tree-like CP proof for MATCH_n using only polynomially bounded coefficients must have size $S \geq 2^{\Omega(n/\log n)}$.

To describe the desired system of inequalities MATCH_n, we use the following encoding of matchings and subsets of vertices.

- Each m-matching $p = \{e_1, \ldots, e_m\}$ is encoded by an $m \times (3m)$ matrix $X = (x_i^j)$, where $x_i^j = 1$ iff $j \in e_i$.
- Each $(m-1)$-element subset $q = \{v_1, \ldots, v_{m-1}\}$ of $[3m]$ is encoded by an $(m-1) \times (3m)$ matrix $Y = (y_i^j)$, where $y_i^j = 1$ iff $v_i = j$.

That is, the i-th row of X specifies the i-th pair in the matching p, whereas the i-th row of Y specifies the i-th vertex in the set q. The system MATCH_n consists of three subsystems:

- $F_1(X)$ is satisfied iff p is an m-matching: every row of X has two 1s, and every column has at most one 1. Hence, $F_1(X)$ consists of inequalities:

$$\sum_{j=1}^{3m} x_i^j = 2 \qquad\qquad \text{for all rows } 1 \leq i \leq m;$$

$$\sum_{i=1}^{m} x_i^j \leq 1 \qquad\qquad \text{for all columns } 1 \leq j \leq 3m.$$

- $F_2(Y)$ is satisfied iff q is an $(m-1)$-subset of $[3m]$: every row of Y has exactly one 1, and every column has at most one 1. Hence, $F_2(Y)$ consists of inequalities:

$$\sum_{j=1}^{3m} y_i^j = 1 \qquad\qquad \text{for all rows } 1 \leq i \leq m-1;$$

$$\sum_{i=1}^{m-1} y_i^j \leq 1 \qquad \text{for all columns } 1 \leq j \leq 3m.$$

- $F_3(X, Y)$ is satisfied iff every edge $e_i = \{j_1, j_2\}$ in p has at least one endpoint in q: for all rows $1 \leq i \leq m$ of X and all pairs of columns $1 \leq j_1 \neq j_2 \leq 3m$ of Y,

$$x_i^{j_1} + x_i^{j_2} - \sum_{k=1}^{m-1}(y_k^{j_1} + y_k^{j_2}) \leq 1.$$

Every communication protocol solving the search problem for MATCH_n must in particular solve the following search problem: Given an assignment $\alpha \in \{0, 1\}^{X \cup Y}$ such that $F_1(\alpha) = F_2(\alpha) = 1$, determine i, j_1, j_2 such that

- $x_i^{j_1} = x_i^{j_2} = 1$ (the i-th row of X has 1s in columns j_1 and j_2) and
- $y_k^{j_1} = y_k^{j_2} = 0$ for all $1 \leq k \leq m - 1$ (j_1-th and j_2-th columns of Y have no 1s).

By taking $q = \{j \mid j\text{-th column of } Y \text{ has a } 1\}$, any protocol for MATCH_n can be used to solve the search problem FE_n.

By our work above, we have proved the following lower bound.

Theorem 19.10. (Impagliazzo et al. 1994) *Any tree-like CP proof for the system* MATCH_n, *all coefficients in which are polynomial in n, must have size exponential in* $n/\log n$.

In fact, a similar lower bound $2^{\Omega(n/\log^3 n)}$ for MATCH_n also holds without any restrictions on the size of coefficients used in a CP proof—being tree-like is the only restriction. For this, it is enough to observe that Theorem 7.26 about the deterministic communication complexity of the game FE_n can be extended to *randomized* protocols: $\mathfrak{c}_{1/n}(FE_n) = \Omega(n/\log n)$. It remains then to combine this lower bound with the following "randomized" version of Lemma 19.8.

Lemma 19.11. *If the search problem for* $Ax \leq b$ *has a threshold tree of depth* d, *then there exists a randomized communication protocol for this problem where* $\mathcal{O}(d \log^2 n)$ *bits are sent.*

Proof. It is enough to use two facts about threshold functions. The first (classical) fact is that any threshold function of n variables can be computed as a threshold function with weights at most $2^{\mathcal{O}(n \log n)}$ (see Muroga 1971). The second fact is that $\mathfrak{c}_{1/n}(GT_n) = O(\log^2 n)$, where $GT_n(x, y)$ is the greater-than function on two n-bit integers which outputs 1 iff $x \geq y$ (see Exercise 4.18). The rest is the same as in the proof of Lemma 19.8. $\square$

So we can prove that some systems require tree-like cutting plane proofs of exponential size. The case of general (non-tree-like) proofs is more complicated. And, in fact, general proofs may be exponentially smaller. Namely, there are unsatisfiable systems of inequalities in n variables that have CP refutation proofs of size $n^{\mathcal{O}(1)}$, but every tree-like CP proof for them must have size $2^{n^{\Omega(1)}}$. Such systems are described, for example, by Bonet et al. (2000).

19.4 Lower Bounds for General CP Proofs

The lower bound techniques in Sect. 19.3 only work for tree-like CP proofs; we have no analogue of Lemma 19.7 for general CP proofs. When trying to prove lower bounds for the size of (=number of inequalities in) general cutting plane proofs, an interesting connection with monotone circuits was discovered. The connection is via a so-called "interpolation theorem" in logic.

Craig's Interpolation Theorem for propositional logic (Craig 1957) states that for every pair of propositional formulas F, G such that $F \rightarrow G$ is a tautology (a negation of an unsatisfiable CNF), there is a formula I that $F \rightarrow I$ and $I \rightarrow G$ are tautologies, and I contains only the variables used by both F and G. Formula I is called an *interpolant* for F and G.

In this form it is a simple fact which can be proved by induction on the number of variables that occur on F but do not occur in G. But Craig proved more. He gave a way of constructing the interpolant I from a proof of $F \rightarrow G$. Thus the complexity of the interpolant is bounded in some way by the complexity of the proof. This suggests the following approach to proving lower bounds: if we can show that an implication $F \rightarrow G$ does not have a simple interpolant, then it cannot have a simple proof.

Pudlák (1997) used this approach to prove that some unsatisfiable CNFs require exponentially long cutting plane proofs. Namely, suppose that our system of inequalities $F(x, y, z)$ on three groups of variables consists of two sub-systems $A(x, y)$ and $B(y, z)$, where the inequalities in $A(x, y)$ do not have z-variables, and those in $B(y, z)$ do not have x-variables. That is, $F(x, y, z)$ is a system of inequalities $Mx \leq b$ where the coefficient matrix M has the form $M = \begin{bmatrix} * & * & 0 \\ 0 & * & * \end{bmatrix}$ where "$*$" stands for a submatrix with arbitrary coefficients. Suppose that $F(x, y, z)$ is unsatisfiable, that is, has no 0-1 solution. Then, for any truth assignment α to the y-variables, at least one of these two systems $A(x, \alpha)$ of $B(\alpha, z)$ must be unsatisfiable, for otherwise $F(x, \alpha, z)$ would be satisfiable. A so-called "interpolant" just tells us which of these two system is unsatisfiable.

An *interpolant* of F is a boolean function $I(y)$ (on the common variables y) such that for any truth assignment α to the y-variables,

- $I(\alpha) = 0$ implies $A(x, \alpha)$ is unsatisfiable, and
- $I(\alpha) = 1$ implies $B(\alpha, z)$ is unsatisfiable.

That is, given any assignment α to y-variables, the interpolant *cannot* answer "0" if $A(x, \alpha)$ is satisfiable, or answer "1" if $B(\alpha, z)$ is satisfiable. Note that this notion of interpolant makes little sense if *both* A and B are unsatisfiable: then any formula

I on the common variables will be an interpolant. An algorithm computing such a function $I(y)$ is called an *interpolating algorithm* for F.

Remark 19.12. It is not difficult to see that our definition of interpolant for unsatisfiable CNFs $A \wedge B$ agrees with Craig's definition for tautologies. The negation of $A \wedge B$ is the tautology $A \to \neg B$. Craig's interpolant I for this tautology satisfies $A \to I$ and $I \to \neg B$. Thus, if $I = 0$ then $\neg A$ must be a tautology, implying that the CNF formula A itself must be unsatisfiable. Similarly, if $I = 1$ then $\neg B$ must be a tautology, implying that the CNF formula B must be unsatisfiable.

Lemma 19.13. *Every cutting plane proof for F gives an interpolating algorithm for F running in time polynomial in the proof-size.*

Proof. Take a cutting plane proof for F. The idea is, given an assignment α to the common y-variables, to split the proof so that we get a refutation either from x-axioms $A(x, \alpha)$ or from z-axioms $B(\alpha, z)$. The only rule which can mix x-variables and z-variables in the original proof is the addition of two inequalities, yielding an inequality $f(x) + g(y) + h(z) \le D$. The strategy is (after the assignment $y \mapsto \alpha$) not to perform this rule, but rather keep two inequalities $f(x) \le D_0$ and $h(z) \le D_1$, where D_0, D_1 are integers. The sums $f(x)$ and $h(z)$ may be empty, in which case they are treated as 0. What we need is only to ensure that this pair of inequalities is at least as strong as the original inequality after the assignment α, which means that we need to ensure the property:

$$D_0 + D_1 \le D - g(\alpha). \tag{19.1}$$

To achieve this, the axioms are replaced (after the assignment $y \mapsto \alpha$) by pairs of inequalities as follows:

$$f(x) + g(y) \le a \quad \text{by a pair} \quad f(x) \le a - g(\alpha) \quad \text{and} \quad 0 \le 0;$$
$$g(y) + h(z) \le b \quad \text{by a pair} \quad 0 \le 0 \quad \text{and} \quad h(z) \le b - g(\alpha).$$

The addition rule is simulated by performing additions of the first and the second inequalities of the pairs in parallel. This clearly preserves the property (19.1) we need. The multiplication rule is simulated in a similar way.

But what about the division rule? We perform this rule also in parallel on the two inequalities in the pair. The divisibility condition is clearly satisfied, as we have the same coefficients at variables x and z as in the original inequality. Thus, the only challenge is to make sure that the property (19.1) is preserved under rounding. For this, look at an inequality $c \cdot f(x) + c \cdot h(z) \le D - c \cdot g(\alpha)$ in the proof after the assignment $y \mapsto \alpha$. By inductive assumption, we have the following inferences:

$$\frac{c \cdot f(x) \le D_0}{f(x) \le \lfloor D_0/c \rfloor} \quad \text{and} \quad \frac{c \cdot h(z) \le D_1}{h(z) \le \lfloor D_1/c \rfloor}$$

with $D_0 + D_1 \le D - c \cdot g(\alpha)$. Since $\lfloor u \rfloor + \lfloor v \rfloor \le \lfloor u + v \rfloor$, the conclusions satisfy
our property (19.1) with new values being $D_0' := \lfloor D_0/c \rfloor$, $D_1' := \lfloor D_1/c \rfloor$ and
$D' := \lfloor D/c \rfloor - g(\alpha)$ because $D_0 + D_1 \le D - c \cdot g(\alpha)$ implies

$$D_0' + D_1' = \left\lfloor \frac{D_0}{c} \right\rfloor + \left\lfloor \frac{D_1}{c} \right\rfloor \le \left\lfloor \frac{D - c \cdot g(\alpha)}{c} \right\rfloor = \left\lfloor \frac{D}{c} \right\rfloor - g(\alpha) = D'.$$

Consider now the pair corresponding to the final inequality $0 \le -1$. It is of the
form $0 \le D_0$, $0 \le D_1$ where $D_0 + D_1 \le -1$. Since D_0 and D_1 are integers, this
implies that either $D_0 \le -1$ or $D_1 \le -1$. Thus we have a proof of a contradiction
either from $A(x, \alpha)$ or from $B(\alpha, z)$ (or from both). To know which one is the case,
the algorithm may just test whether "$D_0 \le -1$" or not. Thus, by only looking at
the first inequality in each pair, the CP proof gives us an algorithm which, given an
assignment α to y-variables, computes an integer D_0 such that $D_0 \le -1$ implies
$A(x, \alpha)$ is unsatisfiable, and $D_0 \ge 0$ implies $B(\alpha, z)$ is unsatisfiable (since then
$D_1 \le -1$).   □

Having an interpolating algorithm we can turn it into a sequence of boolean
circuits. Thus if any interpolating circuit for F must be large, then F cannot have
small cutting plane proofs. This is the main idea of relating proofs with circuits.
Of course, on its own this idea is of little help: no nonlinear lower bound for circuits
computing explicit boolean functions is known. An intriguing aspect, however, is
that under some mild conditions on the form of inequalities in F, the circuits can be
made *monotone*: we only have to allow monotone real-valued functions as gates.

Namely, say that a system $A(x, y) \wedge B(y, z)$ of linear inequalities is *separated* if
in at least one of the systems A and B all y-variables appear in all inequalities with
non-negative coefficients, or all appear with non-positive coefficients.

The following theorem was proved by Pudlák (1997) based on earlier ideas of
Krajíček (1997) and Bonet et al. (1997).

Theorem 19.14. (Pudlák 1997) *If an unsatisfiable system F of linear inequalities
is separated then it has an interpolating monotone real circuit of size polynomial in
the minimal size of cutting plane proof of F.*

Proof. It is enough to turn the algorithm from Lemma 19.13 into a monotone real
circuit whose size is polynomial in the size of the underlying cutting proof of F.
Let us first realize that we only need to compute the constant D_0 (or only D_1)
corresponding to the last inequality. We shall assume w.l.o.g. that all y-variables
appear in all inequalities of $A(x, y)$ with non-negative coefficients.

Recall that, in each step, we replace an inequality $f(x) + g(\alpha) \le a$ by $f(x) \le$
D_0 with $D_0 = a - g(\alpha)$. We assume that all the coefficients of y-variables in $A(x, y)$
are non-negative. In this case, it is more convenient to talk about $-D_0 = g(\alpha) - a$;
then we do not need to multiply $g(\alpha)$ by a negative constant -1.

Thus we only need to compute successively $-D_0$ for each pair. For this, we
can use the algorithm from Lemma 19.13. Each gate will produce a new $-D_0$

from previous ones. The circuit has 0 and 1 as inputs corresponding to the truth assignment α to y-variables, but computes arbitrary integers in the inner nodes.

If $f(x) + g(y) \leq a$ is an axiom, where $g(y) = \sum c_i y_i$, then the function $\alpha \mapsto -D_0 = g(\alpha) - a$ is non-decreasing because all coefficients c_i are non-negative, by our assumption. Hence, if $\alpha \leq \beta$ are two 0-1 vectors, then $g(\alpha) \leq g(\beta)$. Thus the operations we need are:

- Addition of an integer constant,
- Multiplication by a non-negative integer constant,
- Addition,
- Division by a positive integer constant with rounding,
- To get a 0-1 output we add a threshold gate at the output gate, that is, the unary gate t defined by $t(\xi) = 1$ if $\xi \geq 1$ and $t(\xi) = 0$ otherwise.

All the operations are non-decreasing. The only inequalities where we need multiplication by negative constants are $-y_i \leq 0$ (that is, $y_i \geq 0$). These, however, can be treated as inequalities containing z-variables, that is, we can put $D_0 = 0$ for them. Thus we can make all gates non-decreasing.

As in the proof of Lemma 19.13, for each assignment α to y-variables, the input to the last gate $t(\xi)$ is an integer $\xi = -D_0$ such that $t(\xi) = 1$ implies $D_0 \leq -1$, and hence, that $A(x, \alpha)$ is unsatisfiable, and $t(\xi) = 0$ implies $D_0 \geq 0$ (D_0 is an integer), and hence, that $B(\alpha, z)$ is unsatisfiable. Thus the obtained circuit is indeed an interpolating circuit for $A(x, y) \wedge B(y, z)$.   □

Remark 19.15. As observed by Dash (2005, 2010), the same argument works not only for proof systems using the Gomory–Chvátal cuts introduced above but also for any proof system whose derivation rules ("cuts") satisfy the following three conditions:

1. If $g^T x + h^T y \leq d$ is a cut for $Ax + By \leq c$, then for any 0-1 assignment α to the y-variables, $g^T x \leq d - h^T \alpha$ is a cut for $Ax \leq c - B\alpha$.
2. If $g^T x + h^T y \leq d$ is a cut for $Ax \leq e, By \leq f$, then there are numbers r and s such that $g^T x \leq r$ is a cut for $Ax \leq e$, and $h^T y \leq s$ is a cut for $By \leq f$, and $r + s \leq d$.
3. The numbers r can be computed from A, e or the numbers s can be computed from B, f with polynomially many monotone operations.

In particular, Dash (2005, 2010) showed that so-called *lift-and-project* cuts and even *split* cuts satisfy all three conditions above; see Exercise 19.13 for definitions of these cuts.

19.4.1   The Clique-Coloring Polytope

Theorem 19.14 reduces the lower bounds task for cutting planes to that for monotone real circuits. We already know (see Theorem 9.28) that every monotone boolean function f of n variables with the following two properties requires monotone real circuits of size 2^{n^ϵ}. Inputs of f are graphs G on n vertices, encoded by $\binom{n}{2}$ boolean variables, and

- $f(G) = 1$ if G contains a k-clique,
- $f(G) = 0$ if G is $(k-1)$-colorable.

What we need is a system of linear inequalities $A(x, y) \wedge B(y, z)$ such that any interpolant $f(y)$ for this system satisfies these two conditions. That is, we only need to write the statement

a graph contains a k-clique and is $(k-1)$-colorable

as an unsatisfiable system of linear inequalities. For this, we take three groups of variables:

- y-variables $y_{i,j}$ encoding the edges: $y_{i,j} = 1$ iff the edge $\{i, j\}$ is present;
- x-variables x_i, one for each vertex, encoding cliques;
- z-variables $z_{i,c}$ encoding colorings: $z_{i,c} = 1$ iff vertex i has color c.

We want to impose the conditions:

 (i) The set of nodes $\{i : x_i = 1\}$ forms a clique of size $\geq k$.
(ii) For all $c = 1, \ldots, k-1$, the set $\{i : z_{i,c} = 1\}$ is an independent set.

The underlying graph is given by the values of y-variables. We now describe a system of inequalities Clique(x, y) corresponding to the first condition (i), and a system of inequalities Color(y, z) corresponding to the second condition (ii).

Clique(x, y): For $V = \{i : x_i = 1\}$ to form a clique, besides the inequality

$$\sum_i x_i \geq k \quad \text{or equivalently} \quad -\sum_i x_i \leq -k$$

we also need to ensure that all nodes in V are pairwise adjacent. That is, we need that $x_i = 1$ and $x_j = 1$ implies $y_{i,j} = 1$. This can be written as an inequality

$$x_i + x_j - y_{i,j} \leq 1.$$

Color(y, z): For the sets $I = \{i \mid z_{i,c} = 1\}$ to be independent sets (color classes), we first need that each vertex i receives exactly one color:

$$\sum_c z_{i,c} = 1$$

and that no two adjacent vertices $i \neq j$ receive the same color. This last condition means that $z_{i,c} = 1$ and $z_{j,c} = 1$ must imply $y_{i,j} = 0$, and this can be written as an inequality

$$y_{i,j} + z_{i,c} + z_{j,c} \leq 2.$$

We also need inequalities ensuring that the values of all variables lie between 0 and 1. To keep the same signs for y-variables, we add inequalities $-y_{ij} \leq 0$ to $\text{Clique}(x, y)$, and inequalities $y_{ij} \leq 1$ to $\text{Color}(y, z)$. Note that, after that, the y-variables occur in $\text{Clique}(x, y)$ only with negative signs, and occur with positive signs in $\text{Color}(y, z)$. Hence, the system of inequalities $F = \text{Clique}(x, y) \wedge \text{Color}(y, z)$ is separated. By Theorem 19.14, this system has an interpolating monotone real circuit $I(y)$ of size polynomial in the minimal CP proof size of F.

Let us look at what this interpolating circuit $I(y)$ does. By the definition of the interpolant, for every assignment $\alpha \in \{0, 1\}^{\binom{n}{2}}$ to y-variables:

- If the graph G_α encoded by α contains a k-clique, then $\text{Clique}(x, \alpha)$ is satisfiable, and hence, $I(\alpha)$ cannot be equal to 0. So $I(\alpha) = 1$.
- If the graph G_α is colorable by $k - 1$ colors, then $\text{Color}(\alpha, z)$ is satisfiable, and hence, $I(\alpha)$ cannot be equal to 1. So $I(\alpha) = 0$.

Thus the circuit $I(y)$ outputs 1 on graphs containing a k-clique, and outputs 0 on $(k - 1)$-colorable graphs. By Theorem 9.26, we know that, for $k = \Theta(\sqrt{n})$, the circuit $I(y)$ must have size 2^{n^ϵ} for a constant $\epsilon > 0$. This gives us

Corollary 19.16. (Pudlák 1997) *Any cutting plane derivation of the contradiction $0 \leq -1$ from $\text{Clique}(x, y) \wedge \text{Color}(y, z)$ has size exponential in $n^{\Omega(1)}$.*

The proof of Theorem 19.14 via interpolating circuits is not quite satisfying: it is not as "combinatorial" as that for resolution refutations. In particular, it is not clear how to use this theorem to show that every cutting plane algorithm solving, say, the maximum clique problem must produce many inequalities. It would be therefore nice to have a lower bounds argument for cutting plane proofs *explicitly* showing what properties of inequalities do force long derivations.

■ **Research Problem 19.17.** Find a combinatorial lower bounds argument for the size of general (non-tree-like) cutting plane proofs.

For systems $Ax \leq b$ of n variables but with a huge (exponential) number of inequalities there is yet another general argument to show that they require long cutting plane derivations. Suppose some inequality $c^T x \leq d$ holds for all integer solutions x of $Ax \leq b$. We would like to know how many applications of the cutting rules do we need to derive $c^T x \leq d$ from $Ax \leq b$.

Say that an inequality $a_i^T x \leq b_i$ of $Ax \leq b$ is *critical* with respect to $c^T x \leq d$ if the system, obtained from $Ax \leq b$ by replacing this inequality with $c^T x \geq d + 1$, has an integer solution. Intuitively, every critical inequality must be used in any derivation of $c^T x \leq d$. This intuition was made precise by Chvátal et al. (1989) as follows.

Lemma 19.18. *If a system $Ax \leq b$ has t inequalities that are critical with respect to $c^T x \leq d$, then any cutting plane derivation of $c^T x \leq d$ from $Ax \leq b$ must produce at least $(t - 1)/n$ inequalities.*

This can be used to show that cutting plane proofs of integer infeasibility of some exponentially large systems must be long (see, for example, Exercise 19.8).

19.5 Chvátal Rank

So far we were interested in the *size* of CP proofs, that is, in the total number of inequalities used in them. Another important measure is the *depth* of a CP proof, that is, the length of a longest path in the underlying derivation graph. It turns out that this measure is tightly related to a classical measure in integer linear programming—the Chvátal rank of polytopes—for which an impressive collection of lower bounds were already proved. We first introduce the Chvátal rank, relate it to the depth of CP proofs, and describe some of the most powerful lower bounds arguments for this measure.

A *rational polyhedron* is a set $P = \{x \in \mathbb{R}^n : Ax \leq b\}$ of real vectors, where A is an integer $m \times n$ matrix and $b \in \mathbb{Z}^m$ is an integer vector. In this case one says that $Ax \leq b$ *defines* P. Polyhedrons lying in a ball of finite radius are called *polytopes*. In particular, every polyhedron $P \subseteq [0, 1]^n$ is a polytope; we will mainly consider such polyhedrons.

One of the main goals of integer linear programming is, given a target linear function $c^T x$, a threshold d, and a system of linear inequalities $Ax \leq b$, to show that $c^T x \leq d$ holds for all *integer* solutions x of this system. That is, given a polyhedron P defined by $Ax \leq b$, we want to show that $c^T x \leq d$ holds in $P \cap \mathbb{Z}^n$.

Example 19.19. (Independent set problem) A set of vertices in a graph is *independent* (or *stable*) if no two of its vertices are adjacent. The cardinality of a largest independent set in a graph G is usually denoted by $\alpha(G)$. The *independent set polytope* of $G = (V, E)$ can be described by a system of inequalities:

$$x_u + x_v \leq 1 \quad \text{for all edges } \{u, v\} \in E, \tag{19.2}$$

$$x_v \geq 0 \quad \text{for all vertices } u \in V.$$

It is easy to see that a 0-1 vector x satisfies all these inequalities if and only if the set $S_x = \{v \in V : x_v = 1\}$ is an independent set in G. The independent set problem is, given a graph G and a number d, to decide whether $\alpha(G) \leq d$, that is, to decide whether $\sum_{v \in V} x_u \leq d$ holds for all 0-1 points of the independent set polytope of G. This is an NP-complete problem. The relaxed problem is, given a system of linear inequalities, to decide whether the inequality $\sum_{v \in V} x_u \leq d$ holds for all *real* solutions of this system. Known algorithms for linear programming can solve this problem in polynomial time. If the algorithm answers "Yes" then we are done because every integer solution is also a real one. The problem, however, is that when the algorithm answers "No", it might still be the case that all integer solutions satisfy our inequality. So for example, the vector $x = (1/2, \ldots, 1/2)$ lies in the independent set polytope of any n-vertex graph, and its value is $\sum_{v \in V} x_v = |V|/2$. Thus real points of this polytope say nothing about the actual size of independent sets in G.

The condition "x must be an integer vector" is hard to deal with. So one tries to add to the system $Ax \leq b$, defining the original polyhedron P, some new inequalities such that the resulting polyhedron $P_I = \{x: A'x \leq b'\}$ is the convex hull of $P \cap \mathbb{Z}^n$, that is, P_I consists of all convex linear combinations[1] of integer solutions of $Ax \leq b$. Having found P_I, it is enough to test (in polynomial time) whether the inequality $c^T x \leq d$ holds for all *real* solutions of $A'x \leq b'$. One can obtain P_I from P by iteratively shrinking the original polyhedron P using so-called "closure operations".

The *Chvátal closure* (or just a *closure*) P' of a polyhedron P is obtained by removing from P all vectors x for which there exists an inequality $c^T y < d + 1$ with $c \in \mathbb{Z}^n$ and $d \in \mathbb{Z}$ such that this inequality is valid for all $y \in P$, but $c^T x > d$. That is, P' consists of all vectors $x \in \mathbb{R}^n$ such that, for every $c \in \mathbb{Z}^n$ and $\delta \in \mathbb{R}$,

$$c^T y \leq \delta \text{ for all } y \in P \text{ implies } c^T x \leq \lfloor \delta \rfloor.$$

Intuitively, when forming a closure of P we remove all solutions x of $Ax \leq b$ that are "definitely" not integer solutions. In the literature this closure operator is also called *Chvátal-Gomory cut*.

Geometrically, the closure is obtained as an intersection of halfspaces. A *halfspace* is a set $H = \{x \in \mathbb{R}^n : c^T x \leq \delta\}$ for some integer vector $c \in \mathbb{Z}^n$ and δ a rational number. If the components of c are relatively prime integers, then the integer hull H_I of H is $H_I = \{x \in \mathbb{R}^n | c^T x \leq \lfloor \delta \rfloor\}$. The Chvátal closure P' of a polyhedron P is then the set

$$P' = \bigcap_{H:H \supseteq P} H_I,$$

where the intersection ranges over all rational halfspaces containing P. We can define the sequence of sets $P = P^{(0)} \supseteq P^{(1)} \supseteq P^{(2)} \supseteq \ldots$ by letting $P^{(i+1)}$ to be the closure of $P^{(i)}$. That is, for every $i = 0, 1, \ldots$, and every $c \in \mathbb{Z}^n$ and $d \in \mathbb{Z}$, we have that

$$\text{if } c^T x < d + 1 \text{ is valid in } P^{(i)} \text{ then } c^T x \leq d \text{ is valid in } P^{(i+1)}. \tag{19.3}$$

For a polyhedron P, let P_I denote the convex hull of $P \cap \mathbb{Z}^n$.

Theorem 19.20. (Schrijver 1980) *For every rational polyhedron P and every integer $i \geq 1$, $P^{(i)}$ is a rational polyhedron defined by a finite set of integer inequalities $a^T x \leq b$ that satisfy $a^T x < b + 1$ for any $x \in P^{(i-1)}$. Moreover, there is an integer $r \geq 0$ such that $P^{(r)} = P_I$.*

This theorem motivates the following definition. Let $P \subseteq \mathbb{R}^n$ be a polyhedron, and let $c^T x \leq d$ be an inequality valid over $P \cap \mathbb{Z}^n$.

---

[1] A convex linear combination of vectors $v_1, \ldots, v_k$ is a vector $v = \lambda_1 v_1 + \cdots + \lambda_k v_k$ where $\lambda_i \geq 0$ and $\lambda_1 + \cdots + \lambda_k = 1$. Note that if $v_1, \ldots, v_k$ are solutions of $Ax \leq b$, then v is a solution as well.

Definition 19.21. (Chvátal rank) The *Chvátal rank* of an inequality $c^T x \leq d$ relative to P is the smallest number r such that $c^T x \leq d$ is valid for all points $x \in P^{(r)}$. The Chvátal rank of a rational polyhedron P is the minimum r such that $P^{(r)} = P_I$, where P_I is the convex hull of $P \cap \mathbb{Z}^n$. The Chvátal rank of a point $w \in \mathbb{R}^n \setminus \mathbb{Z}^n$ relative to P is $\text{rk}_P(w) = \min\{r: w \notin P^{(r)}\}$, that is, the smallest number of closure operations required to remove this point w.

We are now going to show that the Chvátal rank captures the depth of cutting plane derivations with the following generalized Gomory–Chvátal cutting rule:

$$\frac{Ax \leq b}{\lambda^T Ax \leq \lfloor \lambda^T b \rfloor}, \tag{19.4}$$

where λ is non-negative vector with at most n nonzero positions (n is the number of variables), and the vector $\lambda^T A$ is an integer vector. The difference from the CP proofs we considered above is that instead of three rules of "fanin" at most 2 now we have one single rule of "fanin" at most n.

This rule (19.4) itself relies on the duality theorem of linear programming in the following form.

Theorem 19.22. (Duality theorem) *If $Ax \leq b$ has a solution and if each of its solutions satisfies a linear inequality $c^T x < d$, then there is a non-negative vector λ with at most n positive rational components such that $\lambda^T A = c^T$ and $\lambda^T b < d$.*

That is, if all real solutions of $Ax \leq b$ satisfy a given inequality, then this inequality is a linear combination of the inequalities in $Ax \leq b$. This is an almost direct consequence of a more common form of the duality theorem of linear programming stating that (if it exists) the optimal value of the linear program "maximize $c^T x$ under the constraints $Ax \leq b$" coincides with the optimal value of the dual program "minimize $\lambda^T b$ under the constraints $\lambda^T A = c^T$ and $\lambda \geq 0$". The general Gomory–Chvátal cutting rule just additionally rounds down the right-hand side.

In the cutting rule (19.4) we can assume that all components of λ lie between 0 and 1. This holds because an inequality $\lambda^T Ax \leq \lfloor \lambda^T b \rfloor$ with $\lambda \in \mathbb{R}^n$, $\lambda \geq 0$ and $\lambda^T A \in \mathbb{Z}^n$ is implied by $Ax \leq b$ and $\mu^T Ax \leq \lfloor \mu^T b \rfloor$ with the vector $\mu = \lambda - \lfloor \lambda \rfloor$ in $[0, 1]^n$:

$$\lambda^T Ax = \mu^T Ax + \lfloor \lambda \rfloor^T Ax \leq \lfloor \mu^T b \rfloor + \lfloor \lambda \rfloor^T b = \lfloor \lambda^T b \rfloor.$$

Since $\lambda^T A$ is an integer vector, its scalar product $\lambda^T Ax$ with any *integer* vector x is an integer number. Therefore we do not lose any integer solution when rounding down the right hand side. Thus every inequality derived from $Ax \leq b$ during a CP derivation is valid in $P \cap \mathbb{Z}^n$. More important is the converse proved by Chvátal (1973) and Schrijver (1980): if the polyhedron $P = \{x: Ax \leq b\}$ is rational, and if an integer inequality $c^T x \leq d$ is valid in $P \cap \mathbb{Z}^n$ then there is a CP derivation of $c^T x \leq d$ from $Ax \leq b$.

We now consider CP derivations using the generalized cutting rule (19.4). As before, each such derivation can be represented by a directed acyclic graph whose fanin-0 nodes correspond to axioms (initial inequalities) and each inner node corresponds to a derived inequality. As noted above, we may assume that each node has fanin at most n, the number of variables. The *size* of a derivation is the total number of nodes, and its *depth* is the length of (number of edges in) the longest path in it.

We are now interested in proving lower bounds on the depth of derivations. Note that, as long as we are only interested in the depth, we may restrict us to tree-like derivations, that is, derivations whose underlying graph is a tree. Our first goal is to show that the depth essentially coincides with the Chvátal rank.

19.6   Rank Versus Depth of CP Proofs

The following simple lemma shows that the Chvátal rank of an inequality is a lower bound on the depth of its CP derivation.

Lemma 19.23. *If $c^T x \leq d$ has a CP derivation from P of depth r, then the Chvátal rank of $c^T x \leq d$ relative to P is at most r.*

Proof. It is enough to show that for any application of the generalized Gomory–Chvátal cutting rule (19.4), if the system $Ax \leq b$ of premises is valid in $P^{(r)}$, then the conclusion $\lambda^T Ax \leq \lfloor \lambda^T b \rfloor$ is valid in $P^{(r+1)}$. But this is obvious by the property (19.3) because $\lambda^T Ax \leq \lambda^T b$ must also be valid in $P^{(r)}$.   □

The next lemma shows that Lemma 19.23 is almost tight.

Lemma 19.24. *Let P be a rational polyhedron, and r its Chvátal rank.*

 (i) *If $P^{(r)} \neq \emptyset$, then any integer inequality valid in $P^{(r)}$ has a CP derivation from P of depth at most $r + 1$.*
 (ii) *If $P^{(r)} = \emptyset$, then the inequality $0 \leq -1$ has a CP derivation from P of depth at most $r + 2$.*

Proof. Assume first that $P^{(r)} \neq \emptyset$, and let $c^T x \leq d$ be an integer inequality valid in $P^{(r)}$. By Theorem 19.20, $c^T x < d + 1$ is valid in $P^{(r-1)}$. To prove the first claim of the lemma it is enough to prove the following

Claim 19.25. $c^T x \leq d$ *has a CP derivation from P of depth $r + 1$.*

Proof. We argue by induction on r. In the basis case $r = 1$, we know that $c^T x < d + 1$ is valid in $P^{(0)} = P \neq \emptyset$. By Theorem 19.22, there exists a non-negative vector λ with at most n positive rational components such that $\lambda^T A = c^T$ and $\lambda^T b < d + 1$. It follows that one can derive $c^T x \leq d$ by just one application of the general Gomory–Chvátal cutting rule, because $\lfloor \lambda^T b \rfloor \leq d$.

For the inductive case $r \geq 1$, reason as follows. By Theorem 19.22, the set $P^{(r)}$ is a rational polyhedron defined by a finite system $Ax \leq b$ of integer linear inequalities $a_i^T x \le b_i$ such that $a_i^T x < b_i + 1$ holds for all $x \in P^{(r-1)}$. By the induction hypothesis, each of inequalities $a_i^T x \le b_i$ can be derived from P by a CP proof of depth r. Since, by our assumption, the inequality $c^T x < d + 1$ is valid in $P^{(r)}$, Theorem 19.22 implies that it is a linear combination of $Ax \le b$, that is, there exists a rational vector $\lambda \ge 0$ with at most n nonzero components such that $\lambda^T A = c^T$ and $\lambda^T b < d + 1$. Thus the desired inequality $c^T x \le d$ can be derived from $Ax \le b$ by just one application of the general Gomory–Chvátal cutting rule. $\square$

To finish the proof of the lemma, it remains to consider the case when $P^{(r)}$ is empty. Let $Ax \le b$ be a finite system of integer linear inequalities $a_1^T x \le b_1, \ldots, a_m^T x \le b_m$ defining $P^{(r)}$. Let $P_k^{(r)}$ be the set of all points $x \in \mathbb{R}^n$ satisfying the first k inequalities $a_1^T x \le b_1, \ldots, a_k^T x \le b_k$. Since $P^{(r)} = P_m^{(r)}$ is empty and $P_0^{(r)} = \mathbb{R}^n$, there exists a maximal k such that $P_k^{(r)}$ is not empty. For that maximal k, the inequality $a_{k+1}^T x \ge b_{k+1} + 1$ (that is, the inequality $-a_{k+1}^T x \le -b_{k+1} - 1$) holds for every $x \in P_k^{(r)}$. By Claim 19.25, there exists a proof of this inequality of depth $r + 1$. Combined with $a_{k+1}^T x \le b_{k+1}$, this gives a proof of $0 \le -1$ from P. Recall that the depth of the entire proof is at most $r + 2$, as desired. $\square$

Thus we can obtain lower bounds on the depth of CP derivations by proving lower bounds on the Chvátal rank.

19.7   Lower Bounds on Chvátal Rank

To show that the Chvátal rank of $c^T x \le d$ relative to P is larger than r, it is enough to show that $P^{(r)}$ contains a real vector w such that $c^T w > d$. This can be shown by characterizing some of the points on $P^{(i)}$ that survive in $P^{(i+1)}$. Such characterizations are usually called "protection lemmas", because they argue that certain points are protected from removal in the next round provided certain other points survived the previous round.

Let $P \subseteq \mathbb{R}^n$ be a rational polyhedron, u and v some two points in $\mathbb{R}^n$, and m a positive number. Let P' be the Chvátal closure of P.

Lemma 19.26. *If $u \in P$ and if the point $w = u - \frac{1}{m}v$ satisfies every valid over $P \cap \mathbb{Z}^n$ integer inequality $a^T x \le b$ with $a^T v < m$, then $w \in P'$.*

Proof. If $w \notin P'$ then $a^T w > b$ for some a and b such that

$$a \in \mathbb{Z}^n, b \in \mathbb{Z} \text{ and } a^T x < b + 1 \text{ for all } x \in P. \tag{19.5}$$

Hence we only need to show that $a^T w \le b$ whenever (19.5) holds. Since $a^T x < b + 1$ is valid in $x \in P$, $a^T x \le b$ is valid in $P \cap \mathbb{Z}^n$. Hence, if $a^T v < m$ then $a^T w \le b$ holds by the assumption of the lemma. So assume that $a^T v \ge m$. Since

$u \in P$ and since $a^T x < b + 1$ is valid over P, we have

$$a^T w = a^T u - \frac{1}{m} a^T v \le (b+1) - 1 = b,$$

as claimed.   □

Lemma 19.26 gives us a tool to show that some points will survive one round of Chvátal–Gomory cuts. We can easily extend it to more rounds. Let $m_1, \ldots, m_r$ be positive real numbers. Define points $w^{(0)}, w^{(1)}, \ldots$ in $\mathbb{R}^n$ inductively as follows:

$$w^{(0)} = u \quad \text{and} \quad w^{(j)} = w^{(j-1)} - \frac{1}{m_j} v.$$

Hence,

$$w^{(j)} := u - \left(\sum_{i=1}^{j} \frac{1}{m_i} \right) v. \qquad (19.6)$$

By an *m-bounded* inequality for P we will mean an integer inequality $a^T x \le b$ which is valid over $P \cap \mathbb{Z}^n$ and satisfies $a^T v < m$.

Lemma 19.27. *If $u \in P$ and if, for every $j = 1, \ldots, r$, the point $w^{(j)}$ satisfies every m_j-bounded inequality for P, then $w^{(j)} \in P^{(j)}$ for all $j = 1, \ldots, r$.*

We leave the proof as an exercise; it is the same as that of Lemma 19.26 using induction on j.

19.7.1 The Maximum Independent Set Problem

Given a graph $G = ([n], E)$, we consider the following *independent set polytope* described by the system of inequalities:

$$\sum_{i \in S} x_i \le 1 \quad \text{for all cliques } S \subseteq V \text{ in } G, \qquad (19.7)$$

$$x_i \ge 0 \quad \text{for all vertices } i \in [n].$$

We are interested in the size and depth of cutting plane derivations of $\sum_{i=1}^{n} x_i \le \alpha(G)$ from this system. Note that we now allow much more inequalities than in (19.2), so this system approximates the independent set polytope tighter than (19.2).

For a graph $G = ([n], E)$, let $\mathrm{rk}(G)$ denote the Chvátal rank of the inequality $\sum_{i=1}^{n} x_i \le \alpha(G)$ relative to the independent set polytope (19.7). By Lemma 19.24, we know that any cutting plane derivation of $\sum_{i=1}^{n} x_i \le \alpha(G)$ from (19.7) which uses generalized Gomory–Chvátal cutting rule (19.4) must have depth at least

$\mathrm{rk}(G) - 1$. The depth of a derivation is the maximum length of a path from the root to a leaf in the underlying directed acyclic graph of the derivation.

Lemma 19.28. *If an n-vertex graph G has no clique on more than k vertices, then*

$$\mathrm{rk}(G) \geq \log \frac{n}{k \cdot \alpha(G)}.$$

Proof. Let $\alpha = \alpha(G)$, and let P be the independent set polytope of G defined by (19.7). Observe that $\mathrm{rk}(G) > r$ if $P^{(r)}$ contains a vector $x = (x_1, \ldots, x_n)$ with $x_i > \alpha/n$ for all i, because then $\sum_{i=1}^{n} x_i > \alpha$. On the other hand, since G contains no clique on more than k vertices, the vector $w := (1/k, \ldots, 1/k)$ belongs to P: if S is a clique then the axiom $\sum_{i \in S} x_i \leq 1$ is satisfied by vector w because $\sum_{i \in S} w_i = |S|/k \leq 1$. Motivated by this observation, consider the vectors

$$w^{(r)} := 2^{-r} w = \left(\frac{1}{k2^r}, \ldots, \frac{1}{k2^r} \right), \qquad r = 0, 1, \ldots.$$

Claim 19.29. For every r, $w^{(r)} \in P^{(r)}$.

Note that this already implies the lemma, because $P^{(r)} = P_I$ only if $1/k2^r \leq \alpha/n$, from which $r \geq \log(n/k\alpha)$ follows.

We prove the claim by induction on r. Let $w^{(r)} \in P^{(r)}$, and let $c^T x \leq \delta$ be an inequality valid in $P^{(r)}$ with $c \in \mathbb{Z}^n$. Since $0 \in P^{(r)}$, we have that $\delta \geq 0$. If $\delta \geq 1$ then

$$c^T w^{(r+1)} = c^T (2^{-r-1} w) = \tfrac{1}{2} c^T (2^{-r} w) = \tfrac{1}{2} c^T w^{(r)} \leq \tfrac{1}{2} \delta \leq \lfloor \delta \rfloor,$$

implying that $w^{(r+1)} \in P^{(r+1)}$. Now assume that $0 \leq \delta < 1$. In this case vector c cannot have any positive components because P_I, and hence, also $P^{(r)}$ must contain all n unit vectors (zero-one vectors with exactly one 1). Thus we again have that $c^T w^{(r+1)} = c^T (2^{-r-1} w) \leq 0 = \lfloor \delta \rfloor$. $\square$

Theorem 19.30. *There exist arbitrarily large graphs G such that G has n vertices, $\alpha(G) = 2$, and $\mathrm{rk}(G) = \Omega(\log n)$.*

Proof. Erdős (1961) proved that there exist arbitrarily large n-vertex graphs G such that $\alpha(G) = 2$ and every clique in G has at most $k = \mathcal{O}(\sqrt{n} \log n)$ vertices. It remains to apply Lemma 19.28. $\square$

The following theorem substantially improves the bound of Lemma 19.28. A graph is *k-colorable* if it is possible to assign to each its vertex a number (color) from $\{1, \ldots, k\}$ so that no two adjacent vertices receive the same number.

Theorem 19.31. (Chvátal et al. 1989) *Let G be a graph with n vertices and let $k < s$ be positive integers. If every induced subgraph of G with s vertices is k-colorable, then*

$$\mathrm{rk}(G) \geq \frac{s}{k} \ln \frac{n}{k\alpha(G)}.$$

Proof. Let $\mathbf{1} = (1, \ldots, 1)^T$ be the all-1 vector of length n. Writing

$$w^{(j)} := \frac{1}{k}\left(\frac{s}{s+k}\right)^j \mathbf{1}, \qquad (19.8)$$

it is enough to show that $w^{(j)} \in P^{(j)}$ for all j: if $j < (s/k)\ln[n/k\alpha(G)]$ then, using the inequality $1 - t > e^{-t-t^2/2}$ valid for all $0 < t < 1$, we obtain that

$$\mathbf{1}^T w^{(j)} = \frac{n}{k}\left(\frac{s}{s+k}\right)^j \geq \frac{n}{k}e^{-jk/s} > \alpha(G).$$

To show that $w^{(j)} \in P^{(j)}$ for all j, we only need to verify the assumptions of Lemma 19.27 with

$$u^T := \left(\frac{1}{k}, \ldots, \frac{1}{k}\right), \quad v^T := (1, \ldots, 1) \quad \text{and} \quad m_j := s\left(\frac{s+k}{s}\right)^j.$$

Indeed, using the geometric series $1 + h + h^2 + \cdots + h^n = (1 - h^{n+1})/(1 - h)$ we have

$$\frac{1}{k} - \sum_{i=1}^{j} \frac{1}{m_i} = \frac{1}{k} - \frac{1}{s}\sum_{i=1}^{j}\left(\frac{s}{s+k}\right)^i$$

$$= \frac{1}{k} - \frac{1}{s} \cdot \frac{1 - \left(\frac{s}{s+k}\right)^{j+1}}{1 - \frac{s}{s+k}} + \frac{1}{s}$$

$$= \frac{1}{k} - \frac{s+k}{sk} + \frac{1}{k}\left(\frac{s}{s+k}\right)^j + \frac{1}{s}$$

$$= \frac{1}{k}\left(\frac{s}{s+k}\right)^j.$$

Thus with our choice of u, v and m_j the vectors in (19.8) have the desired form (19.6). Moreover, since G contains no clique with more than k vertices, we have $u \in P$: if S is a clique then the $\sum_{i \in S} u_i = |S|/k \leq 1$.

Now consider an arbitrary inequality $a^T x \leq b$ valid over $P \cap \mathbb{Z}^n$ and such that $a \in \mathbb{Z}^V$ and $a^T \mathbf{1} < m_j$. We only need to verify that $a^T w^{(j)} \leq b$.

By our assumption about the graph, the induced subgraph of G on every subset $S \subseteq V$ of $|S| \leq s$ vertices is k-colorable. That is, S can be partitioned into at most k independent sets $S = S_1 \cup S_2 \cup \cdots \cup S_k$. Hence, if we let A be the maximum of $\sum_{i \in I} a_i$ over all independent sets I of G, then

$$\sum_{i \in S} a_i \leq k \cdot \max_j \sum_{i \in S_j} a_i \leq k \cdot A.$$

Thus we have that

$$b \geq \max\{a^T x \mid x \in P \cap \mathbb{Z}^V\} = A \geq \frac{1}{k} \max \left\{ \sum_{i \in S} a_i \mid |S| \leq s \right\}. \tag{19.9}$$

We may assume that $a^T \mathbf{1} > 0$; otherwise $a^T w^{(j)} \leq 0$ and we are done, as $b \geq 0$ (the all-0 vector belongs to $P \cap \mathbb{Z}^n$). If vector a has at most s positive components, then (19.9) implies $b \geq (1/k)a^T \mathbf{1} \geq a^T w^{(j)}$. If a has at least s positive components, then (19.9) and $a^T \mathbf{1} < m_j$ imply

$$b \geq \frac{s}{k} > \frac{s}{km_j} a^T \mathbf{1} = a^T w^{(j)}. \qquad \square$$

As a consequence we obtain that some independent set polytopes have large Chvátal rank.

Theorem 19.32. *There exist arbitrarily large graphs G on n vertices with $\mathcal{O}(n)$ edges such that $\mathrm{rk}(G) = \Omega(n)$.*

Proof. Erdős (1962) proved that for every positive t there are positive integer c, a positive number δ, and arbitrarily large graphs G such that G has n vertices and cn edges, $\alpha(G) < tn$, and every induced subgraph with at most δn vertices is 3-colorable. (In fact he proved that most of n-vertex graphs with cn edges have the last two properties.) Any t smaller than $1/3$ will do for our purpose: we only need to set $k = 3$ and $s = \lfloor \delta n \rfloor$ in Theorem 19.31. $\square$

Chvátal et al. (1989) also give *upper* bounds: for every n-vertex graph G with $\alpha(G) = \alpha$, $\mathrm{rk}(G)$ is at most $n - \alpha$, if $\alpha < n$, and is at most $\alpha + 1 + (2\alpha + 1)$ $\ln[n/(2\alpha + 1)]$, if $2\alpha + 1 \leq n$.

A general upper bound on the Chvátal rank of polytopes in the 0/1-cube where found by Eisenbrand and Schulz (2003): $\mathrm{rk}(P) \leq n^2(1 + \log n)$ for every polytope $P \subseteq [0, 1]^n$; moreover, $\mathrm{rk}(P) = \mathcal{O}(n)$ if $P \cap \mathbb{Z}^n = \emptyset$.

Theorem 19.31 has several other interesting consequences. In particular, it implies that some other polytopes, including those of particular instances of the knapsack problem, also have large Chvátal rank. This can be shown by considering special rank-preserving mappings between polyhedrons.

A function $f : \mathbb{R}^s \to \mathbb{R}^t$ is *linear* if $f(x) = Cx + d$ for some integer $t \times s$ matrix C and some vector $d \in \mathbb{Z}^t$. For a polyhedron P, let $\mathrm{rk}(P)$ denote its Chvátal rank. Recall that $\mathrm{rk}(P) \geq r + 1$ if $P^{(r)} \setminus P_I \neq \emptyset$ where, as before, P_I is the convex hull of $P \cap \mathbb{Z}^n$.

Lemma 19.33. *Let $S \subseteq \mathbb{R}^s$ and $T \subseteq \mathbb{R}^t$ be polyhedrons, and $f : \mathbb{R}^s \to \mathbb{R}^t$ a one-to-one linear function. If $f(S) \subseteq T$ and $f(S \cap \mathbb{Z}^s) \supseteq T \cap \mathbb{Z}^t$, then $\mathrm{rk}(T) \geq \mathrm{rk}(S)$.*

Proof. Our first goal is to show that, for every linear mapping $f : \mathbb{R}^s \to \mathbb{R}^t$, the condition $f(S) \subseteq T$ alone implies that

$$f(S^{(i)}) \subseteq T^{(i)} \quad \text{for all} \quad i = 0, 1, \dots. \tag{19.10}$$

We show this by induction on i. The basis case $i = 0$ holds by our assumption. Now assume that (19.10) holds for some i, and consider an arbitrary point w in $S^{(i+1)}$; we need only to show that $f(w) \in T^{(i+1)}$. Since $w \in S^{(i)}$, we have $f(w) \in T^{(i)}$ by the induction hypothesis. Hence, our task reduces to showing that $a^T f(w) \leq b$ holds for every $a \in \mathbb{Z}^t$, $b \in \mathbb{Z}$ such that

$$\max\{a^T y \mid y \in T^{(i)}\} < b + 1.$$

The last inequality combined with $f(S^{(i)}) \subseteq T^{(i)}$ guarantees that

$$\max\{a^T f(x) \mid x \in S^{(i)}\} < b + 1.$$

Since $f(x) = Cx + d$ with integral matrix C and integral vector d, it follows that $\max\{a^T f(x): x \in S^{(i+1)}\} \leq b$. In particular, $a^T f(w) \leq b$ must hold because $w \in S^{(i+1)}$. This finishes the proof of (19.10).

Now let r be the Chvátal rank of S; we need to show that $\mathrm{rk}(T) \geq r$. If $r = 0$, then the desired conclusion is trivial. If $r > 0$, then there must be a point x in $S^{(r-1)} \setminus S_I$. Since $x \in S^{(r-1)}$ and $f(S^{(r-1)}) \subseteq T^{(r-1)}$, the point $f(x)$ belongs to $T^{(r-1)}$. On the other hand, since $x \notin S_I$, the fact that f is a one-to-one linear function satisfying $f(S \cap \mathbb{Z}^n) \supseteq T \cap \mathbb{Z}^n$ implies that $f(x) \notin T_I$. Hence, $f(x) \in T^{(r-1)} \setminus T_I$, implying that $\mathrm{rk}(T) \geq r$, as desired. $\square$

19.7.2 The Set-Covering Problem

Now let A be a zero-one $m \times n$ matrix, and P denote the polytope in $\mathbb{R}^n$ defined by $Ax \geq 1$ and $0 \leq x \leq 1$. The problem of minimizing a linear function over $P \cap \mathbb{Z}^n$ is known as the *set-covering problem*. This term comes from interpreting the j-th column of A as the incidence vector of a subset S_j of the ground set $[m] = \{1, \ldots, m\}$, and calling a subset $J \subseteq [n]$ a *cover* if the union of all S_j with $j \in J$ is the entire ground set.

Recall that a *vertex cover* in a graph $G = (V, E)$ is a set of its vertices containing at least one endpoint of each edge. The *vertex-cover number*, $\tau(G)$, of G is the minimum size of such a set. If we take the set E of all edges as our ground set, and let $S_j \subseteq E$ to be the set of all edges incident with the j-th vertex, then the vertex-cover problem turns to special case of the set-covering problem. On the other hand, since $\tau(G) = |V| - \alpha(G)$ holds for every graph, the vertex-cover problem is related to the independent-set problem. By formalizing this relation, we obtain the following lower bound on the rank of set-covering polytopes.

Theorem 19.34. (Set-Covering Problem) *There exist arbitrarily large $m \times n$ zero-one matrices A such that $n \leq m = \mathcal{O}(n)$, each row of A has precisely two ones, and the polytope $P = \{x: Ax \geq 1, 0 \leq x \leq 1\}$ has $\mathrm{rk}(P) = \Omega(n)$.*

Proof. Fix a graph $G = (V, E)$ guaranteed by Theorem 19.32. Let A denote the zero-one matrix whose rows correspond to edges, columns to vertices of G, and $A[e, v] = 1$ iff $v \in e$. Let S denote the independent set polytope of G defined by (19.7), and $T = \{x: Ax \geq 1, 0 \leq x \leq 1\}$. The desired conclusion now follows from Theorem 19.32 and Lemma 19.33 by taking the function $f : \mathbb{R}^V \to \mathbb{R}^V$ defined by $f(x) = 1 - x$.

The function $f(x)$ is clearly one-to-one. To verify that $f(S) \subseteq T$, take a vector $x \in S$. Since $A1 = 2 \cdot 1$ and $Ax \leq 1$, we have that

$$Af(x) = A(1 - x) = A1 - Ax = 2 \cdot 1 - Ax \geq 1,$$

It remains therefore to verify the condition $T \cap \mathbb{Z}^n \subseteq f(S \cap \mathbb{Z}^n)$. For this, take an arbitrary zero-one vector $y \in T$; hence $Ay \geq 1$. Take $x := 1 - y$; hence, $y = f(x)$. The vector x belongs to $S \cap \mathbb{Z}^n$ because

$$Ax = A(1 - y) = A1 - Ay = 2 \cdot 1 - Ay \leq 1,$$

as $Ay \geq 1$. Thus we can apply Lemma 19.33 with this function $f(x)$. □

Theorem 19.35. (Set-Partitioning Problem) *There exist arbitrarily large $m \times n$ zero-one matrices A such that $n \leq m = \mathcal{O}(n)$, each row of A has precisely three ones, and the polytope $P = \{x: Ax = 1, 0 \leq x \leq 1\}$ has* $\mathrm{rk}(P) = \Omega(n)$.

Proof. Let A be a matrix guaranteed by Theorem 19.34, and let S be the polytope defined by $Ax \geq 1$ and $0 \leq x \leq 1$. Let T denote the polytope in $\mathbb{R}^{n+m}$ defined by $Ay + z = 1, 0 \leq y \leq 1$ and $0 \leq z \leq 1$. We are going to apply Lemma 19.33 together with Theorem 19.34. For this purpose, define a linear function $f : \mathbb{R}^n \to \mathbb{R}^{n+m}$ by

$$f(x) = \begin{bmatrix} 1 - x \\ Ax - 1 \end{bmatrix} = \begin{bmatrix} y \\ z \end{bmatrix}.$$

This function is one-to-one just because $1 - x$ is such. To verify that $f(S) \subseteq T$, take a vector $x \in S$. Since $A1 = 2 \cdot 1$, we have that

$$Ay + z = A(1 - x) + (Ax - 1) = A1 - 1 = 1;$$

hence, $f(x) \in T$, as desired. It remains therefore to verify the condition $T \cap \mathbb{Z}^{n+m} \subseteq f(S \cap \mathbb{Z}^n)$. For this, take an arbitrary zero-one vector $(y, z) \in T$; hence $Ay + z = 1$. Take $x := 1 - y$. Then $x \in S \cap \mathbb{Z}^n$ because

$$Ax = A(1 - y) = A1 - Ay = 2 \cdot 1 - Ay \geq 1,$$

as $Ay = 1 - z \leq 1$. Moreover, $1 - x = y$ and

$$Ax - 1 = A(1 - y) - 1 = A1 - Ay - 1 = 1 - Ay = z$$

because $Ay + z = 1$; hence, $(y, z) = f(x)$, as desired.

Since this function $f(x)$ fulfills the conditions of Lemma 19.33, the desired lower bound $\mathrm{rk}(T) \geq \mathrm{rk}(S) = \Omega(n)$ follows from Theorem 19.34. $\qquad\square$

19.7.3 The Knapsack Problem

Let a be a vector in $\mathbb{Z}_+^n$, and let b be an integer. The problem of maximizing a linear function over all zero-one solutions of the system of linear inequalities $a^T x \leq b$, $0 \leq x \leq 1$ is known as the *zero-one knapsack problem*.

Theorem 19.36. (Knapsack Problem) *For arbitrarily large n, there exist $a \in \mathbb{Z}_+^n$ and $b \in \mathbb{Z}_+$ such that all components of a are at most 4^n, and the polytope $P = \{x : a^T x \leq b, 0 \leq x \leq 1\}$ has $\mathrm{rk}(P) = \Omega(n)$.*

Proof. By Exercise 19.7, we only need to prove that $\mathrm{rk}(T) = \Omega(n)$, where $T = \{x : a^T x = b, 0 \leq x \leq 1\}$ for particular a and b. For this purpose, take a zero-one matrix A guaranteed by Theorem 19.35. Let $a_0, a_1, \ldots, a_{m-1}$ be its rows. We know that each of these rows has precisely three ones. Define

$$a^T := \sum_{i=0}^{m-1} 4^i a_i \quad \text{and} \quad b := \sum_{i=0}^{m-1} 4^i.$$

Let T be the polytope defined by $a^T x = b$ and $0 \leq x \leq 1$ for this particular choice of a and b. Let S be the polytope defined by $Ax = 1$ and $0 \leq x \leq 1$. We are going to apply Lemma 19.33 with $f(x) = x$. Since $a^T x = b$ is a linear combination of $Ax = 1$, we have that $f(S) = S \subseteq T$. So it remains to prove that

$$T \cap \mathbb{Z}^n \subseteq S \cap \mathbb{Z}^n, \tag{19.11}$$

for then the desired conclusion will follow from Lemma 19.33 and Theorem 19.35. To prove (19.11), take an arbitrary zero-one vector $x \in T$. Then, $a^T x = b$, that is

$$\sum_{i=0}^{m-1} (a_i^T x) 4^i = \sum_{i=0}^{m-1} 4^i.$$

Since each a_i has precisely three ones, each $a_i^T x$ is one of the integers $0, 1, 2, 3$. Since the 4-ary expansion of every non-negative integer is unique, we conclude that $a_i^T x = 1$ for all $i = 0, 1, \ldots, m-1$, and so $x \in S \cap \mathbb{Z}^n$. $\qquad\square$

19.7.4   An Upper Bounds on the Proof Size

Let $\text{Size}(G)$ denote the smallest t such that there is a tree-like cutting plane derivation of $\sum_{v \in V} x_v \leq \alpha(G)$ from (19.7) using at most t applications of the generalized Gomory–Chvátal cutting rule (19.4). As before, the size of a derivation is the total number of produced inequalities.

Theorem 19.37.   (Chvátal et al. 1989) *For every graph G with n vertices,*

$$\text{Size}(G) \leq \binom{n}{\alpha(G)}.$$

Proof. Let $G = (V, E)$. For every subset W of V, let $\alpha(W)$ denote the largest number of pairwise nonadjacent vertices in W. For a vector $x \in \mathbb{R}^n$, set

$$x(W) := \sum_{u \in W} x_u.$$

Our goal is to prove by induction on $|W|$ that there is a cutting plane derivation of $x(W) \leq \alpha(W)$ from (19.7) of size at most $\binom{|W|}{\alpha(W)}$.

For this purpose, set $\alpha := \alpha(W)$. We may assume that $1 < \alpha < |W|$, for otherwise the desired conclusion is trivial. Since $\alpha < |W|$, there must be a vertex $w \in W$ such that $\alpha(W \setminus \{w\}) = \alpha$. Set $W_1 := W \setminus \{w\}$, and let W_0 denote the set of all vertices in W_1 that are not adjacent to w. Since $\alpha(W_0) \leq \alpha - 1$, there is a set W_2 such that $W_0 \subseteq W_2 \subset W_1$ and $\alpha(W_2) = \alpha - 1$.

By the induction hypothesis, there is a cutting plane derivation of $x(W_1) \leq \alpha$ from (19.7) of size at most $\binom{|W_1|}{\alpha}$ and a cutting plane derivation of $x(W_2) \leq \alpha - 1$ from (19.7) of size at most $\binom{|W_2|}{\alpha-1}$. Since $\binom{m}{k} + \binom{m}{k-1} = \binom{m+1}{k}$ holds for all integers $m \geq k \geq 1$, and since $|W_2| < |W| - 1$ and $\alpha > 1$, the total size of both of these derivations is

$$\binom{|W_1|}{\alpha} + \binom{|W_2|}{\alpha - 1} < \binom{|W| - 1}{\alpha} + \binom{|W| - 1}{\alpha - 1} = \binom{|W|}{\alpha}.$$

We claim that for every vector $x \in \mathbb{R}^n$ satisfying $x(W_1) \leq \alpha$ and $x(W_2) \leq \alpha - 1$, we have that $x(W) < \alpha + 1$. Note that this claim already finishes the induction step: one additional application of the generalized Gomory–Chvátal cutting rule (19.4) (with rounding-down) yields the desired inequality $x(W) \leq \alpha$.

To prove the claim, take an arbitrary vector $x \in \mathbb{R}^n$ satisfying $x(W_1) \leq \alpha$ and $x(W_2) \leq \alpha - 1$. Since W_2 is a proper subset of W_1, the set $U = W_1 \setminus W_2$ is not empty. If $x(U) = 0$ then $x(W) = x_w + x(W_2) \leq 1 + (\alpha - 1) = \alpha$. If $x(U) > 0$ then $x_u > 0$ for some $u \in U$. But since u is adjacent to w, the inequality $x_u + x_w \leq 1$ implies that $x_w < 1$. Since $x(W_1) \leq \alpha$, in this case we obtain that $x(W) = x_w + x(W_1) < \alpha + 1$, as desired.   □

■ **Research Problem 19.38.** Prove that there exist n-vertex graphs G for which Size(G) grows faster than every polynomial in n.

Theorem 19.37 shows that graphs with small $\alpha(G)$ are bad candidates. Exercise 19.1 shows that some graphs with large $\alpha(G)$ are also bad candidates.

19.8 General CP Proofs Cannot be Balanced

We already know how to prove large lower bounds (up to $\Omega(n)$) on the *depth* of derivations. A natural question is: can these bounds be translated to super-polynomial lower bounds on the *size* of derivations, that is, on the total number of inequalities produced in any derivation?

By the Formula Balancing Lemma (Lemma 6.1) we already know that any DeMorgan formula of leafsize S can be transformed into an equivalent DeMorgan formula of depth $\mathcal{O}(\log S)$. Hence, any lower bound t on the depth implies a lower bound $2^{\Omega(t)}$ of the leafsize. Does something similar hold for cutting plane proofs? That is, does cutting plane proofs can also be balanced?

In this section we show that general (non-tree-like) cutting plane proofs *cannot* be balanced: there are unsatisfiable systems of linear equations that have CP proofs (and even a resolution refutation proof) of polynomial size, but any such proof must have depth $\Omega(n)$.

In the proof we will use the following "protection lemma" which is based on a simple observation: any convex linear combination of solutions of a system of linear inequalities is also a solution. In particular, if two vectors y and z satisfy a linear inequality, then their average vector $x = (y + z)/2$ must satisfy this inequality as well.

Let $P \subseteq [0, 1]^n$ be a polytope, and $x \in \{0, \frac{1}{2}, 1\}^n \cap P$. A *neighbor* of x is a vector obtained from x by replacing any one of its $\frac{1}{2}$-positions by 0 or by 1. Let $N(x)$ denote the set of all neighbors of x. Hence, $|N(x)|$ is two times the number of $\frac{1}{2}$-positions in x.

Lemma 19.39. *If $N(x) \subseteq P$ then $x \in P'$.*

That is, a fractional point is protected from removal in the next round provided *all* its neighbors survived the previous round.

Proof. Let $I_x = \{i : x_i = \frac{1}{2}\}$ be the set of fractional (non-integral) positions of x. Denote by $x^{i,a}$ the vector obtained from x by replacing its i-th position by a. Thus $N(x) = \{x^{i,a} : i \in I_x, a = 0, 1\}$.

Suppose that $x \notin P'$, that is, there exist $a \in \mathbb{Z}^n$ and $b \in \mathbb{Z}$ such that $a^T y < b + 1$ is valid for all $y \in P$, but $a^T x > b$. Then $a^T x$ cannot be integer, and hence, must belong to $\frac{1}{2} + \mathbb{Z}$. Thus at least one a_i is odd, and $\sum_{j \neq i} a_j$ is even. For this i, both scalar products $a^T x^{i,0}$ and $a^T x^{i,1}$ must be integers, and hence, must be at most b. Since x is an average of $x^{i,0}$ and $x^{i,1}$, the scalar product $a^T x$ must also be at most b, a contradiction. □

Lemma 19.39 can be generalized as follows (Exercise 19.10). Say that a vector $x \in \frac{1}{2}\mathbb{Z}^n$ is a *combined average* of points in P if $I_x = \{i : x_i = \frac{1}{2}\}$ can be partitioned into sets $I_1, \ldots, I_s$ such that, for every $i = 1, \ldots, s$, we can represent x as an average of vectors in P that are 0-1 on I_i and agree with x elsewhere.

Lemma 19.40. *Let $P \subseteq \mathbb{R}^n$ be a rational polytope, and $x \in \frac{1}{2}\mathbb{Z}^n$. If x is a combined average of points in P, then $x \in P'$.*

19.8.1 Size Versus Depth of CP Proofs

Recall that a partial ordering on $[n] = \{1, \ldots, n\}$ is a binary relation $\prec$ which is antisymmetric ($i \prec j$ implies $j \nprec i$), and transitive ($i \prec j$ and $j \prec k$ implies $i \prec k$). An ordering is *total* if any two elements of $[n]$ are comparable. The unsatisfiable system of inequalities we consider is the negation of the property that every total ordering on n elements has minimal element.

More formally, we associate a variable x_{ij} to each pair (i, j) of elements in $[n]$, and consider consider the *total order polytope* defined by the following system of linear inequalities:

$$x_{ij} + x_{ji} = 1 \qquad \text{(antisymmetry and totality)}$$
$$x_{ij} + x_{jk} - x_{ik} \le 1 \qquad \text{(transitivity)}$$
$$\sum_{j:j\neq i} x_{ij} \ge 1 \qquad \text{(no maximal element } i\text{)}$$
$$0 \le x_{ij} \le 1$$

If the variables x_{ij} take only boolean values 0 and 1, interpreted as $x_{ij} = 1$ if and only if $i \prec j$, then the first two sets of inequalities ensure that $\prec$ is a total ordering, whereas the third states that there is no maximal element in this ordering.

We already know that the unsatisfiable CNF formula GT_n corresponding to this system has a resolution refutation proof of polynomial size (see Theorem 18.6). By Proposition 19.4, the system itself has a CP proof of polynomial size. We now show that any CP proof for this system must have depth $\Omega(n)$.

Theorem 19.41. (Buresh-Oppenheim et al. 2006) *The Chvátal rank of the total order polytope is $\Omega(n)$.*

Proof. We associate with a (partial) order $\prec$ a $\{0, \frac{1}{2}, 1\}$-vector $x = x_\prec$ whose component x_{ij} is $0, 1, \frac{1}{2}$ when i is smaller than, bigger than, or incomparable to j, respectively. Call a partial ordering *s-scaled* if there is a partition of $[n]$ into sets $A_1, A_2, \ldots, A_s$ such that each set A_i is totally ordered, and elements from different sets are incomparable.

Claim 19.42. If $\prec$ is s-scaled with $s > 2$, then $x_\prec$ remains after $s - 3$ rounds of Gomory–Chvátal cuts.

The claim immediately provides a lower bound of $n - 2$ for the rank of the total order polytope P since the all-$\frac{1}{2}$ vector associated with the empty order (which is n-scaled) has that rank.

We prove Claim 19.42 by induction on s. Suppose $\prec$ is 3-scaled. We need to show that $x_\prec \in P = P^{(0)}$. Transitivity inequalities clearly hold for three elements in the same A_i. A transitivity inequality that involves more than one A_i must contain at least two variables with value $\frac{1}{2}$ and therefore must be satisfied. The "no maximal element" inequalities also hold, because for every element there are at least two others to which it is not comparable, and the two associated $\frac{1}{2}$ values alone satisfy the inequality.

For a general s we let $x = x_\prec$ and $\mathrm{rk}(x) = \mathrm{rk}_P(x)$. Let $A_1, A_2, \ldots, A_s$ be the corresponding partition of $[n]$. Notice that the set $I_x = \{(i, j) : x_{ij} = \frac{1}{2}\}$ is the set of all edges connecting different components A_i and A_j of the graph when we associate $\prec$ with a graph consisting of s disjoint independent sets, and wires joining every pair of vertices from different sets. We partition the edges in I_x to $\binom{s}{2}$ sets by the components they connect and argue that x and this partition satisfy the conditions of Lemma 19.40.

For a choice of components A and B, we denote by $\prec_A$ the order which is the same as $\prec$ except all the elements of A are bigger than those of B. Similarly we define $\prec_B$. Let $y = x_{\prec_A}$ and $z = x_{\prec_B}$. Then $x = (y + z)/2$. Indeed, if i and j both belong to A or both belong to B, then $x_{ij} = y_{ij} = z_{ij} = 1$. If $i \in A$ and $j \in B$, then $x_{ij} = \frac{1}{2}$, $y_{ij} = 1$ and $z_{ij} = 0$ because $z_{ji} = 1$ and $\prec$ is antisymmetric. Since $\prec_A, \prec_B$ are $(s - 1)$-scaled we inductively have that $\mathrm{rk}(y), \mathrm{rk}(z) \geq s - 3$, and by Lemma 19.40, $\mathrm{rk}(x) \geq s - 2$. $\qquad\square$

Theorems 18.6 and 19.41 show that general (non-tree-like) cutting plane proofs *cannot* be balanced.

■ **Research Problem 19.43.** Can tree-like cutting plane proofs be balanced?

19.9 Integrality Gaps

A general approach to solve the maximum independent set problem via linear programming is, given a graph $G = (V, E)$ on n vertices, to construct a polytope $P = \{x \in \mathbb{R}^n : Ax \leq b\}$ such that: (1) the characteristic 0-1 vector of every independent set in G belongs to P, and (2) $\sum_{v \in V} x_v \leq \alpha(G)$ holds for all vectors $x \in P$. That is, the goal is to construct the convex hull of the set of all characteristic 0-1 vectors of independent sets in G. So far, we have considered constructions of such polytopes using cutting planes. But there are many other ways to construct convex hulls of integer solutions. For example, one can safely add to (19.2) an inequality $\sum_{v \in C} x_v \leq k$ for any cycle C in G with $2k + 1$ vertices, and so on. A natural question is: can any of these methods actually succeed in giving the desired convex hull? In this section we will show that no algorithm producing only "short" inequalities, that is, inequalities $a^T x \leq b$ in which at most ϵn components of a are nonzero, can succeed.

An *independent set relaxation* for G is any polytope $P = \{x : Ax \leq b\}$ containing all characteristic 0-1 vectors of independent sets in G. It is clear that the value $\max\{\sum_{v \in V} x_v : x \in P\}$ of the optimal solution of a relaxation is always at least the independence number $\alpha(G)$, but may be much larger, in general. The ratio of the value of an optimal solution of a relaxation divided by $\alpha(G)$ is called the *integrality gap* of the relaxation P. This number gives us the factor with which the polytope P can approximate the largest number of pairwise non-adjacent vertices in G; the smaller this factor is, the better is the approximation.

Each relaxation P is defined by some set of linear inequalities (constraints) $a^T x \leq b$ in variables $x_1, \ldots, x_n$. We say that P is *s-local* if in every such inequality, the coefficient vector a is nonzero for at most s coordinates, that is, if the constraint involves at most s variables. For example, the independent set polytope defined by (19.2) is 2-local.

Let $\text{Gap}_s(G)$ denote the minimum possible integrality gap over all s-local independent set relaxations for G. We are going to describe a property of graphs that make $\text{Gap}_s(G)$ large. This property is similar to that used in Theorem 19.31, but with the chromatic number replaced by the fractional chromatic number.

Let $k \geq 1$ be a positive rational number. A *fractional k-coloring* of a graph G is a sequence $I_1, \ldots, I_m$ of (not necessarily distinct) independent sets such, for every vertex, exactly the fraction $1/k$ of the I_i contain this vertex. The *fractional chromatic number* of G is

$$\chi_f(G) = \inf\{k : G \text{ has a fractional } k\text{-coloring}\}.$$

Note that every k-coloring is also a fractional k-coloring in which $m = k$ and the sets I_i (color classes) form a partition of the vertex set. Consequently, $\chi_f(G) \leq \chi(G)$. The following lower bound was proved (among other interesting results) by Arora et al. (2006).

Theorem 19.44. (Arora et al. 2006) *Let G be a graph on n vertices. If $\chi_f(H) \leq k$ for every induced subgraph H of G on s vertices, then*

$$\text{Gap}_s(G) \geq \frac{n}{k \cdot \alpha(G)}.$$

Proof. Let P be an s-local independent set relaxation for $G = (V, E)$. We will show that the all-$(1/k)$ vector $(1/k, 1/k, \ldots, 1/k)$ is feasible (belongs to P). This will clearly prove the theorem because then the value of an optimal fractional solution must be at least $\sum_{v \in V} (1/k) = n/k$.

It suffices to show that the all-$1/k$ vector is feasible for any system of constraints $Ax \leq b$ where at most s columns of A are not all-0 columns. So fix any subset S of $|S| = s$ vertices and assume that all columns of A corresponding to the vertices outside S are all-0 columns. Let H be the induced subgraph of G on S, and let $I_1, \ldots, I_m$ be a fractional k-coloring of H. Hence, every vertex of S belongs to a $1/k$ fraction of the I_j's. Each I_j is an independent set in H, and hence, also an independent set in the entire graph G. By the definition of P, the characteristic 0-1 vector $\mathbf{v}_j$ of I_j must satisfy $Ax \leq b$. Replace by $1/k$ all 0-entries of $\mathbf{v}_j$ corresponding to vertices $i \notin S$. Since A has only zeros in these columns, the resulting vectors $\mathbf{w}_j$ are still solutions of $Ax \leq b$. But then the average vector

$$\mathbf{w} = \frac{1}{m}(\mathbf{w}_1 + \mathbf{w}_2 + \cdots + \mathbf{w}_m)$$

is a solution as well. Since every vertex $i \in S$ belongs to exactly m/k of the sets $I_1, \ldots, I_m$, we have that exactly m/k of the vectors $\mathbf{w}_1, \ldots, \mathbf{w}_m$ have 1s in each coordinate $i \in S$. Thus the average vector $\mathbf{w}$ is the all-$(1/k)$ vector, as desired. $\square$

The result of Erdős mentioned in the proof of Theorem 19.32 gives us graphs G for which $\mathrm{Gap}_s(G)$ is not bounded for $s = \lfloor \delta n \rfloor$. To get larger integrality gaps, the use of *fractional* chromatic number was important. Namely, using probabilistic arguments, Arora et al. (2006) proved that, for every constants $0 < \gamma < \epsilon$, and for every large enough n, graph G_n on n vertices satisfying the conditions of Theorem 19.44 with $s = n^\gamma$ and $k \cdot \alpha(G) \leq n^\epsilon$ exist. Theorem 19.44 implies that for such graphs the integrality gap is huge: $\mathrm{Gap}_s(G_n) \geq n^{1-\epsilon}$.

Exercises

19.1. Let $G = (V, E)$ be a *bipartite* graph on n vertices, and suppose that it is regular, that is, all its vertices have the same degree. Show that the inequality $\sum_{v \in V} x_v \leq \alpha(G)$ can be derived from (19.7) by just one application of the generalized Chvátal–Gomory cut rule (19.4).
Hint: Sum all edge-inequalities.

19.2. In Remark 19.15 we claimed that Pudlák's interpolation theorem remains true for more general types of cutting plane rules. Prove this.

19.3. Let $K_n = (V, E)$ be a complete graph on n vertices; hence, $\alpha(K_n) = 1$. Let $S(n)$ be the minimum number of applications of the generalized Chvátal–Gomory cutting rule (19.4), and be the minimum depth of a CP derivation of $\sum_{v \in V} x_v \leq 1$ from the polytope defined by (19.2) with $G = K_n$.

(a) Show that $S(n) \leq n$.

    Hint: Try induction on n. Having proved $\sum_{i=1}^{n} x_i \leq 1$, add all inequalities $x_i + x_{n+1} \leq 1$ for $i = 1, \ldots, n$, then multiply $\sum_{i=1}^{n} x_i \leq 1$ by $n - 1$ and add to this inequality.

(b) Show that $D(n) \leq \log n$.

    Hint: Sum up inequalities of all $(n/2 + 1)$-cliques to show that $D(n) \leq 1 + D(n/2 + 1)$.

19.4. Let n be a power of 2, and let $D(n)$ be the minimum depth of a CP derivation of $\sum_{v \in V} x_v \leq 1$ from the polytope defined by (19.2). Show that $D(n) \leq \log n$.

19.5. In Exercise 19.3 we have derived $\sum_{i=1}^{n} x_i \leq 1$ from axioms $x_i + x_j \leq 1$ for all $1 \leq i < j \leq n$ by a CP proof of size linear in n. But this proof has depth $\Omega(n)$. Exercise 19.4 gives a proof of depth $\mathcal{O}(\log n)$, but its size is exponential. Give a CP proof which has logarithmic depth *and* polynomial size.

Hint: Use the "triangle trick" from Exercise 19.1 to design a proof whose depth $D(n)$ and size $S(n)$ satisfy recurrences $D(n) \leq 1 + D(\lceil 2n/3 \rceil)$ and $S(n) \leq 1 + 3 \cdot S(\lceil 2n/3 \rceil)$. Show that $D(n)$ is at most about $\log_{3/2} n$ and $S(n)$ is at most about $n^{\log_{3/2} 3}$.

19.6. (Rhodes 2009) Consider the polytope corresponding to the pigeonhole principle PHP_n^m as defined in the proof of Proposition 19.5. Show that this polytope has Chvátal rank $\mathcal{O}(\log n)$.

Hint: Exercise 19.5.

19.7. (Rank of faces) A *face* of a polyhedron P is the intersection $F = P \cap H$ of P with a hyperplane $H = \{x : a^T x = b\}$ such that $b = \max\{a^T x : x \in P\}$. Show that the Chvátal rank of P is at least the Chvátal rank of F.

Hint: Since $F \subseteq P$, we have $F^{(r)} \subseteq P^{(r)}$ for all integers $r \geq 0$. Show that we also have that $F \cap P_I \subseteq F_I$: every $x \in F \cap P_I$ satisfies $a^T x = b$ and is a convex combination of some points $y \in P \cap \mathbb{Z}^n$; show that then also $a^T y = b$ must hold.

19.8. Consider the following system with n variables and $n + 2^n$ inequalities:

$$\sum_{i \in I} x_i + \sum_{i \notin I} (1 - x_i) \geq 1/2 \qquad \text{for all } I \subseteq \{1, \ldots, n\};$$

$$0 \leq x_i \leq 1 \qquad \text{for all } i = 1, \ldots, n.$$

Show that this system is integer infeasible, but any CP proof of this fact requires size $\Omega(2^n/n)$.

Hint: Lemma 19.18.

19.9. Let $P \subseteq [0, 1]^n$ be the polytope defined by the system of Exercise 19.8. Let F_k denote the set of all points $x \in \{0, \frac{1}{2}, 1\}^n$ with exactly k components equal to $\frac{1}{2}$. Prove that $F_k \subseteq P^{(k-1)}$ for all $k = 1, \ldots, n$.

Hint: Try induction on k. Take a vector $y \in F_{k+1}$ and an arbitrary integer inequality $a^T x < b + 1$ valid over $P^{(k-1)}$. The task reduces to proving that $a^T y \leq b$. Show that y is a convex combination of vectors in F_k, and hence, must belong to $P^{(k-1)}$. If $a^T y$ is not an integer, then there is a position i such that $a_i \neq 0$ and $y_i = 1/2$. Consider two vectors y^0 and y^1 obtained from y be replacing its i-th component by 0 and by 1. Show that $a^T y + \frac{1}{2} \leq \max\{a^T y^0, a^T y^1\} < b + 1$.

19.10. Prove Lemma 19.40.

Hint: Argue as in the proof of Lemma 19.39 to show that there must be an i for which $\sum_{j \in I_i} a_j$ is odd. Consider the set of vectors $V \subset P$ that average to x and that differ from x exactly on I_i where they take 0-1 values. Show that $a^T v \leq b$ for all $v \in V$.

19.11. The clique-coloring polytope described in Sect. 19.4 corresponds to the following maximization problem $MP_{n,l}$ for $l = k - 1$: Maximize the number of nodes in a clique of an n-vertex graph whose chromatic number does not exceed l. Although l is a trivial solution for this problem, Corollary 19.16 show that any cutting plane proof certifying that no such graph can have a clique on more than l vertices must generate an exponential number of inequalities. That is, quick cutting plane algorithms cannot solve this maximization problem optimally. Use a lower bound on the monotone circuit size of clique like functions (Theorem 9.26) to show that such algorithms cannot even *approximate* this problem: any cutting plane proof certifying that no l-colorable graph can have a clique on $k > l$ vertices must generate an exponential in $\min\{l, n/k\}^{\Omega(1)}$ number of inequalities.

19.12. ■ **Research Problem.** Given a graph $G = (V, E)$, consider the following communication game. Alice gets a subset $A \subseteq V$, Bob gets a subset $B \subseteq V$ such that $|A \cup B| > \alpha(G)$. Hence, $A \cup B$ must contain at least one edge. The goal is to find such an edge. Does there exist n-vertex graphs G for which this game requires $\omega(\log^2 n)$ bits of communication?

Comment: By Lemma 19.8, this would imply that every tree-like CP proof with bounded coefficients for the unsatisfiability of the system (19.2) augmented with the inequality $\sum_{v \in V} x_v \geq \alpha(G) + 1$ must have super-polynomial size. This would be the first strong lower bound for a non-artificial system corresponding to an important optimization problem, the maximum independent set problem.

19.13. (Split cuts) Given a polytope $P = \{x \in \mathbb{R}^n : Ax \leq b\}$, a *cut* for P is any inequality $c^T x \leq d$ with integral coefficients such that $c^T x \leq d$ is valid in the 0-1 restriction $P \cap \{0, 1\}^n$ of P. In this case one also says that $Ax \leq b$ *implies* $c^T x \leq d$. In CP-proofs we used simplest cuts: if $a^T x \leq b$ is valid in P, and if all coordinates of a are dividable by an integer c, then $(a/c)^T x \leq \lfloor b/c \rfloor$ is valid in $P \cap \{0, 1\}^n$. There are also other types of cuts. An inequality $c^T x \leq d$ is a *lift-and-project cut* for $P = \{x \in [0, 1]^n : Ax \leq b\}$ if for some index i, $c^T x \leq d$ is satisfied by points in $P \cap \{x : x_i = 0\}$ and by points in $P \cap \{x : x_i = 1\}$. Even more powerful are so-called "split cuts". An inequality $c^T x \leq d$ is a *split cut* for P if there exist $a \in \mathbb{Z}^n$ and $b \in \mathbb{Z}$ such that $c^T x \leq d$ is satisfied in $P \cap \{x : a^T x \leq b\}$ as well as in $P \cap \{x : a^T x \geq b + 1\}$; the inequality $a^T x \leq b$ is a *witness* for this cut. In particular, any inequality valid in the whole polytope P is a (useless) split cut for P with the witness $0^T x \leq 0$.

(a) Show that each Gomory–Chvátal cut is a special case of a split cut.
(b) Consider the polytopes $P_t = \{(x_1, x_2) \in \mathbb{R}^2 : x_1 - 2tx_2, x_1 + 2tx_2 \leq 2t\}$, $t = 1, 2, \ldots$. It was observed by J. A. Bondy that every CP-proof of $x_1 \leq 0$ from $A_t x \leq b_t$ using Gomory–Chvátal cuts has size at least t, which is exponential

in the binary encoding size $\mathcal{O}(\log t)$ of P_t. Show that $x_1 \leq 0$ is a split cut for every P_t.

(c) It is known that any lift-and-project proof of the fact that the system $\sum_{i=1}^{n} x_i \geq 1/2, 0 \leq x_i \leq 1, i = 1,\ldots,n$ implies $\sum_{i=1}^{n} x_i \geq 1$ has size at least n; see Cook and Dash (2001). Show that $\sum_{i=1}^{n} x_i \geq 1$ is a split cut for the polytope P consisting of all $x \in [0,1]^n$ such that $\sum_{i=1}^{n} x_i \geq 1/2$.

Comment: Dash (2005, 2010) shows that the exponential lower bound, given in Corollary 19.16, remains valid for split cuts.

Chapter 20
Epilogue

The lower-bounds arguments we presented in this book work well for different restricted circuit classes but, so far, have not led to a non-linear lower bound for unrestricted circuits. In this concluding chapter we sketch several general results explaining this failure (the phenomenon of "natural proofs") as well as showing a possible line of further attacks (the "fusion method", indirect arguments).

20.1 Pseudo-Random Generators

When trying to prove a lower bound, we try to show that something *cannot* be computed efficiently. It turns out that this task is closely related to proving that something—namely so-called "pseudorandom generators"—*can* be efficiently computed! This object was invented by Yao (1982), Goldwasser and Micali (1984), Blum and Micali (1984), and was then studied by many authors.

Informally speaking, a pseudorandom generator is an "easy to compute" function which converts a "few" random bits to "many" pseudorandom bits that "look random" to any "small" circuit. Each of the quoted words is in fact a parameter, and we may get pseudorandom generators of different qualities according to the choice of these parameters. For example, the standard definitions are: "easy to compute" = polynomial time; "few" $= n^\epsilon$; "many" $= n$.

Definition 20.1. A function $G : \{0,1\}^l \to \{0,1\}^n$ with $l < n$ is called an (s, ϵ)-*pseudorandom generator* if for any circuit C of size s on n variables,

$$\left| \text{Prob}[C(y) = 1] - \text{Prob}[C(G(x)) = 1] \right| < \epsilon,$$

where y is chosen uniformly at random in $\{0,1\}^n$, and x in $\{0,1\}^l$.

S. Jukna, *Boolean Function Complexity*, Algorithms and Combinatorics 27,
DOI 10.1007/978-3-642-24508-4_20, © Springer-Verlag Berlin Heidelberg 2012

That is, a pseudorandom generator G stretches a short truly random seed x into a long string $G(x)$ which "fools" all circuits of size up to s: no such circuit can distinguish $G(x)$ from a truly random string y.

The quantity l is called the *seed length* and the quantity $n - l$ is called the *stretch* of the pseudorandom generator. Note that the definition is only interesting when $l < n$, for otherwise the generator can simply output the first n bits of the input, and satisfy the definition with $\epsilon = 0$ and arbitrarily large circuit size s. The larger n/l is, the stronger is the generator. Note also that, if the input x is taken in $\{0, 1\}^l$ at random, the output $G(x)$ of a generator is also a random variable in $\{0, 1\}^n$. But if $l < n$, the random variable $G(x)$ is by no means uniformly distributed over $\{0, 1\}^n$ since it can take at most 2^l values with nonzero probability.

The *existence* of $(s, 1/n)$-pseudorandom generators $G : \{0, 1\}^l \to \{0, 1\}^n$ with the seed length $l = 4 \log s$ can be proved using probabilistic arguments. For this, define G by randomly assigning strings of length n for inputs of length l, that is, $\text{Prob}[G(x) = y] = 2^{-n}$ for every $x \in \{0, 1\}^l$ and $y \in \{0, 1\}^n$. Let x be a string of length l. For a circuit C of size s with input size n, define random variable ξ_x to be $C(G(x))$. The expected value of ξ_x is precisely the fraction of strings accepted by C:

$$\text{E}[\xi_x] = \sum_{y:C(y)=1} \text{Prob}[G(x) = y] = 2^{-n}|\{y : C(y) = 1\}|.$$

We have $2^l = s^4$ such independent random variables ξ_x with the same expected value. Hence, by Chernoff's bound, the probability that the average of these variables differs from the expectation by more than $1/n$ is smaller than 2^{-s^3}. Since there are fewer than 2^{s^2} circuits of size s, most of the choices of G will be pseudorandom.

The argument above only shows the mere *existence* of pseudorandom generators with good parameters. In applications, however, we need the generators to be *constructed* in time $\ll 2^{s^2}$. It turns out that this problem (construction of good pseudorandom generators) is related to proving lower bounds on circuit size.

Definition 20.2. Let $f : \{0, 1\}^n \to \{0, 1\}$ be a boolean function. We say that f is (s, ϵ)-*hard* if for any circuit C of size s,

$$\left| \text{Prob}[C(x) = f(x)] - \frac{1}{2} \right| < \epsilon,$$

where x is chosen uniformly at random in $\{0, 1\}^n$.

The meaning of this definition is that hard functions f must be "really hard": no circuit of size s can even approximate its values, that is, any such circuit can do nothing better than just guess the value. So the function f looks random for each small circuit.

The idea of how hard boolean functions can be used to construct pseudorandom generators is well demonstrated by the following construction of a generator stretching just one bit.

Lemma 20.3. *Let f be an $(s + 1, \epsilon)$-hard boolean function of n variables. Then the function $G_f : \{0, 1\}^n \to \{0, 1\}^{n+1}$ defined by $G_f(x) := (x, f(x))$ is an (s, ϵ)-pseudorandom generator.*

Proof. The intuition is that, since f is hard, no small circuit C should be able to figure out that the last bit $f(x)$ of its input string $(x, f(x))$ is not just a random bit. By the definition of a pseudorandom generator, we want the following to hold for any circuit of size at most s on $n + 1$ variables:

$$\left| \text{Prob}[C(y) = 1] - \text{Prob}[C(G_f(x)) = 1] \right| < \epsilon,$$

where y is chosen uniformly at random in $\{0, 1\}^{n+1}$, and x in $\{0, 1\}^n$. Assume that this does not hold. Then there is a circuit C that violates this property. Without loss of generality, we may assume that

$$\text{Prob}[C(G_f(x)) = 1] - \text{Prob}[C(y) = 1] \geq \epsilon.$$

This can be done because we can take $\neg C$ if this is not the case. The above is the same as

$$\text{Prob}[C(x, f(x)) = 1] - \text{Prob}[C(x, r) = 1] \geq \epsilon,$$

where x is chosen uniformly at random in $\{0, 1\}^n$, and r is a random bit in $\{0, 1\}$ with $\text{Prob}[r = 0] = \text{Prob}[r = 1] = 1/2$. A way to interpret this inequality is to observe that when the first n input bits of C are a random string x, the circuit C is more likely to accept if the last bit is $f(x)$ than if the last bit is random. This observation suggests the following strategy in order to use C to predict $f(x)$: given an input x for which we want to compute $f(x)$, we guess a value $r \in \{0, 1\}$ and compute $C(x, r)$. If $C(x, r) = 1$ we take it as evidence that r was a good guess for $f(x)$, and output r. If $C(x, r) = 0$, we take it as evidence that r was the wrong guess for $f(x)$, and we output $1 - r$. Let $C_r(x)$ be the random circuit (with just one random bit r) we just described. We claim that

$$\Pr_{x,r}[C_r(x) = f(x)] \geq \frac{1}{2} + \epsilon.$$

Since $C_r(x) = r$ iff $C(x, r) = 1$, this can be shown by elementary calculations:

$\text{Prob}[C_r(x) = f(x)]$

$= \text{Prob}[C_r(x) = f(x)|r = f(x)] \cdot \text{Prob}[r = f(x)]$

$\quad + \text{Prob}[C_r(x) = f(x)|r \neq f(x)] \cdot \text{Prob}[r \neq f(x)]$

$= \frac{1}{2} \cdot \text{Prob}[C_r(x) = f(x)|r = f(x)] + \frac{1}{2} \cdot \text{Prob}[C_r(x) = f(x)|r \neq f(x)]$

$= \frac{1}{2} \cdot \text{Prob}[C(x, r) = 1|r = f(x)] + \frac{1}{2} \cdot \text{Prob}[C(x, r) = 0|r \neq f(x)]$

$= \frac{1}{2} \cdot \text{Prob}[C(x, r) = 1|r = f(x)] + \frac{1}{2}(1 - \text{Prob}[C(x, r) = 1|r \neq f(x)])$

$$= \tfrac{1}{2} + \text{Prob}[C(x,r) = 1 | r = f(x)]$$

$$- \tfrac{1}{2}\Big(\text{Prob}[C(x,r) = 1 | r = f(x)] + \text{Prob}[C(x,r) = 1 | r \neq f(x)]\Big)$$

$$= \tfrac{1}{2} + \Pr_{x}[C(x, f(x)) = 1] - \Pr_{x,r}[C(x,r) = 1] \geq \tfrac{1}{2} + \epsilon.$$

Thus there must be a constant $r \in \{0, 1\}$ such that $\Pr_{x}[C_r(x) = f(x)] \geq \tfrac{1}{2} + \epsilon$. Since the size of C_r is at most $s + 1$ (plus 1 could come from starting with $\neg C$ instead of C), which is a contradiction with the hardness of f. □

To push this strategy further, what we could do is to break up the input into k blocks and then apply f to them. In this way we get a generator stretching nk bits into $(n + 1)k$ pseudorandom bits. But this is not enough: for applications we need generators stretching n bits into 2^{n^ϵ} pseudorandom bits. To achieve this, we need to use intersecting blocks. But we also have to ensure that these blocks do not intersect too much. This is the main motivation for the construction of generators known as *Nisan–Wigderson generators*. The starting point of this construction is a combinatorial object known as a "partial design".

A collection of subsets $S_1, \dots, S_n$ of $[l] = \{1, \dots, l\}$ is called a *partial (m, k)-design* if $|S_i| = m$ for all i, and $|S_i \cap S_j| \leq k$ for all $i \neq j$.

Example 20.4. A standard construction of partial designs is based on the fact that no two polynomials of degree k can have more than k common roots. So let $l = m^2$ where m is a prime power, and consider the elements of $\{1, \dots, l\}$ as pairs of elements in $\text{GF}(m)$. Every polynomial $p(z)$ over $\text{GF}(m)$ gives us a subset $F_p \subseteq \{1, \dots, l\}$ defined by $F_p = \{(a, p(a)) \mid a \in \text{GF}(m)\}$. Let $\mathcal{F}$ be the family of all subsets F_p where p ranges over all polynomials of degree at most k over $\text{GF}(m)$. Then each member of $\mathcal{F}$ has exactly m elements, any two of them share at most k elements in common, and we have $|\mathcal{F}| \geq m^k$ members.

Given a partial design $S_1, \dots, S_n$ and a boolean function $f : \{0, 1\}^m \to \{0, 1\}$, the Nisan–Wigderson generator $G_f : \{0, 1\}^l \to \{0, 1\}^n$ is defined by:

$$G_f(x) = (f(x_{S_1}), f(x_{S_2}), \dots, f(x_{S_n})),$$

where x_S is the substring $(x_i \mid i \in S)$ of x. That is, the i-th bit of $G_f(x)$ is just the value of f applied to the substring of x determined by the i-th set of the design.

Using a similar argument as for the one-bit generator above, one can prove that G_f is an $(n^2/2, 2/n^2)$-secure pseudorandom generator, as long as the function f is $(n^2, 1/n^2)$-hard. Constructions of pseudorandom generators from so-called "one-way" functions were given by Håstad et al. (1999). Simpler constructions were found by Holenstein (2006). Pseudorandom generators play an important role in cryptography. Detailed treatment of pseudorandom generators can be found in the books by Goldreich (2001), and Katz and Lindell (2007).

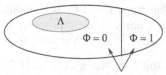

decision by circuits in Γ

A random function f should have a non-negligible chance of having the property Φ. But the value $\Phi(f)$ must be computable by a circuit in Γ taking the 2^n bits of the truth table of f as input

20.2 Natural Proofs

So far, none of existing lower-bounds arguments has been able to separate P from NP. Razborov and Rudich (1997) gave an explanation: all these proof techniques are "natural", and natural proofs cannot prove P $\neq$ NP unless good pseudorandom generators do not exist. Since the existence of such generators is widely believed, it seems very unlikely that natural proofs could show this separation.

Let B_n be the set of all 2^{2^n} boolean functions $f : \{0,1\}^n \to \{0,1\}$, and let Γ and Λ be some classes of boolean functions closed by taking subfunctions. We can think of Λ being, say, the class of all boolean functions computable by circuits of size n^2, and Γ being the class P/poly of boolean functions computable by circuits of size polynomial in n. Hence, $f \in \Lambda$ iff f can be computed by relatively small circuit (of quadratic size).

Given a specific boolean function $f_0 \in B_n$, our goal is to show that $f_0 \notin \Lambda$. A possible proof of this fact is a property $\Phi : B_n \to \{0,1\}$ of boolean functions such that $\Phi(f_0) = 1$ and $\Phi(f) = 0$ for all $f \in \Lambda$. Each such property is a witness for (or a proof of) the fact that "$f_0 \notin \Lambda$".

A Γ-*natural proof against* Λ is a property $\Phi : B_n \to \{0,1\}$ satisfying the following three conditions:

1. *Usefulness against Λ*: $\Phi(f) = 1$ implies $f \notin \Lambda$.
2. *Largeness*: $\Phi(f) = 1$ for at least $2^{-\mathcal{O}(n)}$ fraction of all functions f in B_n.
3. *Constructivity*: $\Phi \in \Gamma$. That is, when looked at as a boolean function of $N = 2^n$ variables, the property Φ itself belongs to the class Γ. The input function f here is given as its truth table (a binary string of length $N = 2^n$).

The first condition (1) is obvious: after all we want to prove that $f_0 \notin \Lambda$. If $\Lambda \neq \emptyset$, this condition also ensures that Φ cannot be trivial, that is, take value 1 on all functions. Condition (2) corresponds to our intuition that any reasonable lower bounds argument, designed for a given function f_0, should also be able to show the hardness of the hardest functions—random ones. Thus a random function f should have a non-negligible chance of having the property Φ. What makes the property "natural" is the last condition (3). That is, the requirement that the property itself can be tested by not too large circuits.

We emphasize that when property $\Phi(f)$ is computed, the input is the truth table of f, whose size is 2^n, not n. Thus a property is P/poly-natural, if it can be computed by circuits of size $2^{\mathcal{O}(n)}$, which is more than exponential in $n(!)$

Example. Let us consider the case when $\Lambda = AC^0$, the class of all boolean functions computable by constant depth circuits with polynomial number of NOT and unbounded-fanin AND and OR gates. The proof that Parity $\notin AC^0$ (Sect. 12.1) involves the following steps: (1) Show that every AC^0 circuit can be simplified to a constant by restricting at most $n - n^\epsilon$ input variables to constants, and (2) show that Parity does not have this property (Here $0 < \epsilon \le 1/2$ is a constant depending only on the depth of a circuit, and property (2) trivially holds, as long as $n - n^\epsilon \ge 1$). Thus the natural property lurking in the proof is the following:

$\Phi(f) = 1$ iff f cannot be made constant by restricting all but n^ϵ variables.

Clearly, if $\Phi(f) = 1$ then $f \notin AC^0$, so Φ is useful against AC^0. Furthermore, the number of functions that can be made constant by fixing $n - k$ variables does not exceed $2\binom{n}{k}2^{2^{n-k}} \le 2^{n2^{n-k}}$, and this is a negligible fraction of all 2^{2^n} functions. Hence, Φ has the largeness property as well. Finally, Φ is constructive in a very strong sense: given a truth table of f, the value $\Phi(f)$ can be computed by a depth-3 circuit of size $2^{\mathcal{O}(n)}$ as follows. List all $\binom{n}{k}2^{n-k} = 2^{\mathcal{O}(n)}$ restrictions of $n - k$ variables. For each one there is a circuit of depth 2 and size $2^{\mathcal{O}(n)}$ which outputs 1 iff that restriction does not leave f a constant function, that is, iff the positions in the truth sub-table, corresponding to that restriction, are not all equal. Output the AND of all these circuits. The resulting circuit has depth 3 and has size polynomial in 2^n.

Thus property Φ is AC^0-natural against AC^0.

Now we show that natural properties cannot be useful against substantially larger classes of boolean functions, like P/poly, unless good pseudorandom generators do not exist.

A *pseudorandom function generator* is a boolean function $f(x, y)$ of $n + m$ variables. By setting the y-variables at random, we obtain its random subfunction $f_y(x) = f(x, y)$ of n variables. Let $h : \{0, 1\}^n \to \{0, 1\}$ be a truly random boolean function. A generator $f(x, y)$ is *secure against Γ-attacks* if for every circuit C in Γ,

$$\left| \text{Prob}[C(f_y) = 1] - \text{Prob}[C(h) = 1] \right| < 2^{-n^2}. \tag{20.1}$$

That is, no circuit in Γ can distinguish f_y from a truly random function; here again, inputs for circuits are truth tables of boolean functions.

Theorem 20.5. *If a complexity class Λ contains a pseudorandom function generator f which is secure against Γ-attacks, then there is no Γ-natural proof against Λ.*

Proof. Suppose that a Γ-natural proof Φ against Λ exists. To get a contradiction, we will show that then the proof Φ can be used to distinguish f_y from a random function h.

Since $f(x, y)$ belongs to Λ, every subfunction $f_y(x)$ with $y \in \{0, 1\}^m$ belongs to Λ as well. The usefulness of Φ against Λ implies that $\Phi(f_y) = 0$ for all y. Hence,

$$\text{Prob}[\Phi(f_y) = 1] = 0.$$

On the other hand, the largeness of Φ implies that $\text{Prob}[\Phi(h) = 1] \geq 2^{-\mathcal{O}(n)}$. Hence,

$$\left| \text{Prob}[\Phi(f_y) = 1] - \text{Prob}[\Phi(h) = 1] \right| \geq 2^{-\mathcal{O}(n)},$$

and thus Φ is a distinguisher. But by constructivity, the boolean function Φ itself belongs to Γ, a contradiction with (20.1). $\qquad\Box$

It is known that pseudorandom *function* generators may be constructed starting from simpler objects—pseudorandom *number* generators. We have already considered these objects in the previous section (see Definition 20.1). This time we work with a particular class Γ of circuits, not with all circuits, and we take the security parameter ϵ very small, $\epsilon = 2^{-n^2}$. Namely, say that a function $g : \{0, 1\}^n \to \{0, 1\}^{2n}$ is secure against Γ-attacks if for every circuit C in Γ,

$$\left| \text{Prob}[C(y) = 1] - \text{Prob}[C(g(x)) = 1] \right| < 2^{-n^2}.$$

Here x is chosen at random from $\{0, 1\}^n$ and y is chosen at random from $\{0, 1\}^{2n}$. That is, given a random seed x, g produces a random string $y' = g(x)$ in $\{0, 1\}^{2n}$, and no circuit in Γ can distinguish this produced string y' from a truly random string y.

Starting from a pseudorandom number generator $g : \{0, 1\}^n \to \{0, 1\}^{2n}$, one can construct a pseudorandom function generator $f(x, y) : \{0, 1\}^{n+2n} \to \{0, 1\}$ as follows. Associate with g two functions $g_0, g_1 : \{0, 1\}^n \to \{0, 1\}^n$, where $g_0(x)$ is the first and $g_1(x)$ the second half of the string $g(x)$. Having a vector $y \in \{0, 1\}^{2n}$, we can define a function $S_y : \{0, 1\}^n \to \{0, 1\}^n$ which is a superposition $S_y = g_{y_n} \circ g_{y_{n-1}} \circ \cdots \circ g_{y_1}$ of these two functions g_0 and g_1 defined by the bits of y. Then just let $f(x, y)$ be the first bit of the superposition S_y applied to input x. This construction is essentially due to Goldreich et al. (1986).

It is widely believed that the class $\Lambda = \text{P/poly}$ (and even much smaller classes) contain pseudorandom number generators g_n that are secure against P/poly-attacks. It is also known that the pseudorandom function generator $f(x, y)$ constructed from g_n is then also secure against P/poly-attacks. Together with Theorem 20.5, this means that no P/poly-natural proof can lead to a super-polynomial lower bound on circuit size.

Besides natural proofs and relativization, yet another barrier towards proving $P \neq NP$—algebraization—was recently discovered by Aaronson and Wigderson (2008).

All these barriers show that it is apparently impossible (or at least very difficult) to prove that $P \neq NP$ by proving strong circuits lower bounds. This (somewhat too optimistic) hope arose from assuming that circuits are much simpler combinatorial objects than Turing machines. But any super-polynomial lower bound for a boolean function in NP would not just prove NP $\not\subseteq$ P: this would prove a much stronger separation NP $\not\subseteq$ P/poly.

There is a big difference between the classes P and P/poly: the first is "uniform" in that it requires *one* Turing machine for *all* boolean functions f_n in the sequence $(f_n \mid n = 1, 2, \ldots)$, whereas the second only requires that for each n a small circuit computing f_n exists. And indeed, non-uniformity makes the circuit model stronger than Turing machines (or equivalently, than the model of uniform circuits): there even exist functions f_n that cannot be computed by Turing machines (regardless of their running time) but do have small circuits. To see this, fix a standard enumeration $M_1, M_2, \ldots$ of Turing machines, and consider the following family of boolean functions $f_n(x)$, where $f_n(x) = 1$ iff M_n halts on the empty tape. Note that each f_n is just a constant function, equal to 0 or 1. Thus, the family of these f_n's is computable by boolean circuits of constant size, even though it cannot be computed by any Turing machine.

At the beginning of complexity theory, some people (including the great mathematician Kolmogorov) even believed that all of P could be computable by circuits of linear size. Decades passed, and this belief is still not refuted! Moreover, this belief does not contradict any reasonable complexity-theoretic conjectures, like P $\neq$ NP. To see this, assume that all boolean functions in P have circuits of size at most n^k for some fixed constant k. Using counting arguments, Kannan (1981) showed that, for every fixed constant k, already the class $\Delta_2 = \Sigma_2 \cap \Pi_2$, next after NP in the complexity hierarchy, contains boolean functions whose circuit size is $\Omega(n^k)$ (see Theorem 20.13 below). Now, if P $=$ NP then $\Delta_2 =$ P. Thus our assumption (all functions in P have small circuits) implies that P $\neq$ NP.

There are even some indications that this Kolmogorov's prediction (or something similar) could be indeed true. It is NP-hard to decide whether some subset of a given sequence of n real numbers sums up to 1. A trivial algorithm probing all subsets needs exponential time. Still, Mayer auf der Heide (1984) showed that, *for each dimension n*, this problem is solvable in time $n^4 \log n$. Another indication is given by Allender and Koucký (2010): in the class of constant-depth threshold circuits, some boolean functions cannot have circuits of polynomial size, if they do not have such circuits of size $n^{1+\epsilon}$ for an arbitrarily small constant $\epsilon > 0$. These and other results show that even circuits of small (almost linear) size may apparently accumulate unexpected power, so big that current mathematics is unable to engage such circuits. The phenomenon of Natural Proofs even gives us a "formal excuse" for our failure to engage general circuits.

The barriers themselves may not be unbreakable, at least for restricted circuit classes. Using ideas sketched in Remark 1.20, Chow (2011) showed that if one replaces the largeness condition by "$\Phi(f) = 1$ for at least $|B_n|/2^{q(n)}$ functions" where $q(n)$ is quasi-polynomial in n, then the resulting "almost-natural" proofs against P/poly exist.

Finally, the P versus NP question is *not* the main goal of the theory of circuit complexity. This theory was born long before the classes P and NP emerged, and has its own, more "prosaic" goals: prove strong lower bounds even in restricted (but interesting from both mathematical and practical points of view) circuit models—like boolean formulas, decision trees, bounded-depth circuits, time-restricted branching programs, etc. There are a lot of "barrier-free" models of circuits where no nontrivial lower bounds are known. The model of read-once(!) switching networks is a good illustration (see Problem 16.12).

20.3 The Fusion Method

In order to show, by contradiction, that a given circuit $G(x)$ is too small for computing a given boolean function $f(x)$ one could try to argue in the following way: try to combine (or "fuse") *correct* rejecting computations of G on inputs in $f^{-1}(0)$ into an *incorrect* rejecting computation on an input in $f^{-1}(1)$.

We restrict ourselves to DeMorgan circuits, that is, to circuits over $\{\wedge, \vee\}$ whose inputs are variables $x_1, \ldots, x_n$ and their negations $\neg x_1, \ldots, \neg x_n$. Such a circuit G is just a sequence $G = (g_1, \ldots, g_t)$ of gates. The first $2n$ gates are n variables followed by n their negations. Each remaining gate g_i is either and AND or an OR of some two previous gates.

For each input $a \in \{0, 1\}^n$, the computation of a circuit $G = (g_1, \ldots, g_t)$ is just a binary string $r = (r_1, \ldots, r_t)$ in $\{0, 1\}^t$ the i-th bit $r_i = g_i(a)$ of which is the output of the i-th gate g_i on input a. The main property of this string r is its "local consistency" (determined by the underlying circuit G): the first n bits are followed by n flipped bits ($r_{j+n} = \neg r_j$ for all $j = 1, \ldots, n$), and each subsequent bit r_i is

- Either the OR $r_i = r_k \vee r_l$ of two previous bits r_k and r_l, if $g_i = g_k \vee g_l$,
- Or is the AND $r_i = r_k \wedge r_l$ of two previous bits r_k and r_l, if $g_i = g_k \wedge g_l$.

In other words, we can consider a given circuit G as a sequence T_G of local tests on strings $r = (r_1, \ldots, r_t)$ in $\{0, 1\}^t$. A string $r \in \{0, 1\}^t$ is then a computation of G if and only if r passes all tests in T_C. It is clear that each computation $G(a) = (g_1(a), \ldots, g_t(a))$ on a input vector $a \in \{0, 1\}^n$ must pass all the tests. On the other hand, if r passes all tests in T_G then r is just a computation of G on input $a = (r_1, \ldots, r_n)$, and the result of this computation is the last bit r_t of r.

This suggests the following "diagonalization" argument to prove that a given function f cannot be computed by a circuit of size t:

- Show that, for every set T of $|T| \leq t$ local tests, there exists a vector r in $\{0, 1\}^t$ such that r passes all tests in T, but $r_t \neq f(r_1, \ldots, r_n)$.

There are two general ideas of how to construct such a "diagonal computation" r: the "topological approach" of Sipser (1985) and the "fusion method" first proposed by Razborov (1989a) and then put in a more general framework by Karchmer (1993). The topological approach of Sipser was demonstrated in Chap. 11 by the notion of "finite limits" used to prove lower bounds for depth-3 circuits. Now we shortly describe the fusion method.

Let $f(x_1, \ldots, x_n)$ be a given boolean function, and let $U = f^{-1}(0)$ be the set of all vectors rejected by f. We assume that $U = \{a_1, \ldots, a_m\}$ is non-empty (for otherwise we have nothing to do). We view each gate $g : \{0, 1\}^n \to \{0, 1\}$ as a (column) vector $\mathbf{g} \in \{0, 1\}^m$ of length $m = |U|$ whose jth position is the value of g when applied to the j-th vector in U. In particular, the vector $\mathbf{x}_i$ corresponding to an input variable x_i has a 1 in the j-th position iff the j-th vector of U has 1 the i-th position. Put differently, the columns $\mathbf{x}_1, \ldots, \mathbf{x}_n$ form an $m \times n$ matrix A such that $f(a) = 0$ iff a is a row of A. In this way we can consider any circuit $G = (g_1, \ldots, g_t)$ as a boolean m by t matrix M, a *computation matrix*, whose columns are the vectors $\mathbf{g}_1, \ldots, \mathbf{g}_t$:

	x_1	$\cdots$	x_n	$\neg x_1$	$\cdots$	$\neg x_n$	$\cdots$	g_i	$\cdots$	g_t
a_1	0	$\cdots$	1	1	$\cdots$	0	$\cdots$	0	$\cdots$	0
a_2	1	$\cdots$	1	0	$\cdots$	0	$\cdots$	1	$\cdots$	0
$\vdots$										$\vdots$
a_m	1	$\cdots$	0	0	$\cdots$	1	$\cdots$	0	$\cdots$	0
	$F(x_1)$	$\cdots$	$F(x_n)$	$F(\neg x_1)$	$\cdots$	$F(\neg x_n)$	$\cdots$	$F(g_i)$	$\cdots$	0

(The meaning of the last row will be clear soon when we come to the "fusion functionals".) This matrix has the following properties:

- The $(n + i)$-th column is the negation of the i-th column, for $i = 1, \ldots, n$;
- If $g_i = g_j \wedge g_k$ then the i-th column is the AND of the j-th and k-th columns;
- If $g_i = g_j \vee g_k$ then the i-th column is the OR of the j-th and k-th columns,

where here and throughout, boolean operations on boolean vectors are performed component-wise.

Each boolean function f determines the set $U = f^{-1}(0)$ of its zeros , as well as the first $2n$ columns $x_1, \ldots, x_n$ and $\neg x_1, \ldots, \neg x_n$ of a computation matrix M of any circuit for f. The remaining columns, however, are determined by the gates of a concrete circuit we are considering. To construct a "diagonal" computation we will combine columns in a new row using boolean functions $F : \{0, 1\}^m \to \{0, 1\}$ defined on the column space.

We call such a function F a *fusing functional* for f if $F(\mathbf{0}) = 0$ and $F(\neg x_i) = \neg F(x_i)$ for all $i = 1, \ldots, n$, that is, if F respects negations of "basis" columns $x_1, \ldots, x_n$. Say that a pair (a, b) of vectors in $\{0, 1\}^m$ *covers* a functional F if

$$F(a) \wedge F(b) \neq F(a \wedge b). \tag{20.2}$$

We can now introduce a combinatorial (set-covering) measure characterizing the size of circuits.

Let $\mu(f)$ be the smallest number of pairs of vectors in $\{0, 1\}^m$ satisfying the following condition: each monotone fusing functional F for f such that

$$f(F(x_1), \ldots, F(x_n)) = 1 \tag{20.3}$$

is covered by at least one of these pairs. Let $C_\wedge(f)$ be the smallest number of $\wedge$-gates in a DeMorgan circuit computing f.

Lemma 20.6. (Razborov 1989a) *For every boolean function f, we have* $C_\wedge(f) \geq \mu(f)$.

Proof. Let $U = f^{-1}(0)$, $m = |U|$ and let $G = (g_1, \ldots, g_t)$ be a circuit computing f. Fix an arbitrary monotone functional $F : \{0, 1\}^m \to \{0, 1\}$ for f satisfying (20.3). Say that a gate $g_i = g_j * g_k$ with $* \in \{\wedge, \vee\}$ *covers* F if

$$F(g_j) * F(g_k) \neq F(g_j * g_k).$$

Note that, if none of the gates in G covers F, then $r = (r_1, \ldots, r_t)$ with $r_i = F(g_i)$ would be a computation $G(a) = (g_1(a), \ldots, g_t(a))$ of our circuit G on the input

$$a := (F(x_1), \ldots, F(x_n)). \tag{20.4}$$

The fact that $F(g_t) = F(0) = 0$ would imply that this is a rejecting computation. But (20.3) implies that $f(a) = 1$, and hence, the vector a should be accepted by G, a contradiction.

Thus the functional F must be covered by at least one gate of G. It suffices therefore to show that if a $\vee$-gate covers F, then F is also covered by an $\wedge$-gate. To show this, let S be the set of all gates in G that cover F. For the sake of contradiction, assume that S contains no $\wedge$-gates. By the definition of a cover we have that for each $g_i = g_j \vee g_k$ in S, $F(g_j) \vee F(g_k) \neq F(g_j \vee g_k)$. Since F is monotone, the only possibility is that

$$F(g_j) = F(g_k) = 0 \quad \text{and} \quad F(g_j \vee g_k) = 1. \tag{20.5}$$

Let G' be a circuit identical to G except that each gate $g_i = g_j \vee g_k$ in S is replaced by the instruction $g_i' = 1 \vee 1$. We concentrate on the behavior of both circuits G and G' on the input vector defined by (20.4) defined by the functional F, and make the following two observations.

Claim 20.7. $G'(a) = 1$.

Proof. This is because we have only changed gates g_i, whose values were 0 on this input (by (20.5)). Since the circuit uses only AND and OR gates, which are monotone, we have that $G'(a) \geq G(a) = f(a) = 1$. $\quad\square$

Claim 20.8. The computation of $G'(a) = (g_1'(a), \ldots, g_t'(a))$ on input a coincides with the string $F(g_1), \ldots, F(g_t)$, that is, $g_i'(a) = F(g_i)$ for all $i = 1, \ldots, t$.

Proof. We show this by induction on the position of the gates in G'. Since the first $2n$ gates of G' are the same as in G, namely the variables $x_1, \ldots, x_n$ and their negations, the claim holds for all $i = 1, \ldots, 2n$. Take now a gate $g_i = g_j * g_k$ with $i > 2n$, and assume the claim holds for both its inputs, that is, $g_j'(a) = F(g_j)$ and $g_k'(a) = F(g_k)$.

Case 1: $g_i = g_j \wedge g_k$. Since, by our assumption, $\wedge$-gates do not cover F, we obtain: $g_i'(a) = g_j'(a) \wedge g_k'(a) = F(g_j) \wedge F(g_k) = F(g_j \wedge g_k) = F(g_i)$.

Case 2: $g_i = g_j \vee g_k$. If $g_i \notin S$, then g_i does not cover F, and the claim follows as in the previous case. If $g_i \in S$, then (20.5) holds, implying that $g_i'(a) = 1 \vee 1 = 1 = F(g_j \vee g_k) = F(g_i)$. $\quad\square$

By Claims 20.7 and 20.8, we get that $g_t'(a) = g_t(a) = f(a) = 1$ on one side, and $g_t'(a) = F(g_t) = F(0) = 0$ on the other side. Thus we have a contradiction that S can contain only $\vee$-gates, that is, that only OR gates can cover F. This means that at least one of the pairs (g_j, g_k) of vectors in $\{0, 1\}^{|U|}$, corresponding to a $\wedge$-gate $g_i = g_j \wedge g_k$ of G, will cover F in the sense of (20.2), as desired.   □

It can also be shown (we will not do this) that the lower bound in Lemma 20.6 is tight enough: $C_\wedge(f) \leq c(\mu(f) + n)^2$ for a constant c. Thus at least in principle, diagonal computations for (deterministic) circuits *can* be produced by using only monotone functionals. It turned out that different classes of fusing functionals capture different circuit models. A boolean function $F(x)$ is *self-dual* if $F(\neg x) = \neg F(x)$, and is *affine* if it is a parity (sum modulo 2) of an odd number of variables.

- Monotone functionals capture deterministic circuits as well as nondeterministic branching programs: classes P and NL.
- Monotone self-dual functionals capture nondeterministic circuits: the class NP.
- Affine functionals capture nondeterministic circuits as well as parity branching programs: classes NP and $\oplus$L.

How this happens can be found in a nice survey by Wigderson (1993) and in the literature cited therein.

20.4 Indirect Proofs

Most of the lower bound arguments described in this book go deeply into details trying to capture the behavior of circuits. The lesson given by the natural proofs phenomenon is that arguing in this way we are apparently trying to prove *much more* than actually needed. That is, when trying to prove a lower bound on "only" the total number of gates in a circuit, we are actually trying to show how any circuit for a given boolean function must work! This additional information is important in circuit design, but might be less important if we merely want to give a lower bound on the number of gates, without exploring further the work of circuits themselves. So how to prove a lower bound without proving anything else?

Actually, within the field of structural complexity, there is a line of work on circuit lower bounds where the inner structure of circuits is completely ignored. Circuits are treated as "black boxes" and the arguments are not trying to explore their structure. Instead of that one tries to obtain a lower bound using high-level arguments like counting, diagonalization and various reductions. These are "brute force" arguments and, at first glance, do not seem to work for *specific* problems.

The clue however is that many seemingly unrelated problems have essentially the same complexity. For example, in order to show that the clique function is difficult (requires circuits of super-polynomial size), it is enough to show that the class NP contains at least one difficult boolean function! It is therefore not excluded that the existence of a hard function in NP can be proved using high-level arguments, perhaps combined with some low-level ones to make reductions and diagonalization tighter.

20.4.1 Williams' Lower Bound

That a "mixture" of low and high level arguments could work was recently demonstrated by Williams (2011). He used such arguments to show that NEXP $\not\subseteq$ ACC^0; here ACC^0 is the class of all sequences of boolean functions computable by constant-depth circuits of polynomial size using AND, OR, NOT and arbitrary MOD_m gates (we considered these circuits in Chap. 12), and NEXP is the class of all sequences of boolean functions computable by nondeterministic Turing machines in exponential time.

The proof is a combination of some ideas from structural complexity (time hierarchies, compression by circuits, local checkability of computations) and algorithmic ideas (fast matrix multiplication, dynamic programming, table lookup). Using these tools, Williams first shows that the satisfiability of ACC^0 circuits (given an ACC^0 circuit with n inputs, decide whether it accepts at least one vector) can be solved in time $2^{n-n^{\Omega(1)}}$. Then he uses this to show that NEXP $\subseteq \text{ACC}^0$ would imply that every language in NEXP can be decided in subexponential nondeterministic time, contradicting the nondeterministic time hierarchy theorem. This implies that every NEXP-complete problem (sequence of boolean functions) requires ACC^0 circuits of super-polynomial size.

One NEXP-complete function is "succinct 3-SAT". Given a binary string x of length n, the function interprets it as a code of a DeMorgan circuit C_x of $n^{1/10}$ input variables. Then the function uses the circuit C_x to produce the string of all its $2^{n^{1/10}}$ possible outputs, interprets this string as a code of a 3-CNF F_x, and accepts the initial string x iff the CNF F_x is satisfiable.

Of course, this function is too "wild" to be called an "explicit" function, and the class of ACC^0 circuits is too "weak": we already know that even the Majority function requires ACC^0 circuits of exponential size if only MOD_m gates for *prime* numbers m are allowed, and it is conjectured that composite moduli cannot help much for this function. Still, this result is an excellent demonstration that high-level arguments may also yield nontrivial lower bounds.

The proof of Williams' result is somewhat technical and requires knowledge about results in structural complexity. So to demonstrate how high-level arguments work, we will describe some older (but not less impressive) results.

20.4.2 Kannan's Lower Bound

Let $\{0, 1\}^*$ denote the infinite set of all binary vectors (of arbitrary length). Subsets $L \subseteq \{0, 1\}^*$ are called *languages*. For a positive integer n, let $L_n = L \cap \{0, 1\}^*$ denote the restriction of L to strings of length exactly n. Note that there is a one-to-one correspondence between languages L and sequences $(f_n \mid n = 1, 2, \ldots)$ of boolean functions: for every string x of length $|x| = n$, let $x \in L_n$ iff $f_n(x) = 1$.

Structural complexity deals with special classes of languages, called complexity classes. The basic class is P. It consists of all languages L for which there exists a constant k such that the membership of every string $x \in \{0, 1\}^*$ in L can be decided in time $\mathcal{O}(|x|^k)$. Note a big difference between P and P/poly: for a language to belong to P/poly, it is enough that, for every n, the membership of an x of length $|x| = n$ in L_n can be detected by a circuit of size $\leq |x|^k$. That is, instead of *one* algorithm for all inputs x, it is now enough to have *different* circuits for inputs of different length. And in fact, P/poly contains even languages, the membership in which cannot be detected by *any* Turing machine, not just by one running in polynomial time.

Starting from the class P, one can define larger classes of languages by quantification. Given a class C of languages, let $\exists$C denote the class of languages L with the following property: there exists a language $L' \in$ C and a constant k such that, for every $x \in \{0, 1\}^*$, $x \in L$ iff there exist a string y of length $|y| \leq |x|^k$ such that $(x, y) \in L'$. We will write this shortly as:

$$x \in L \text{ if and only if } \exists y \ (x, y) \in L'$$

with an understanding that the quantification is only over strings y of length polynomial in $|x|$. The class $\forall$C is defined similarly with "there exists" replaced by "for all". The class co-C consists of all *complements* $\overline{L}$ of languages L in C: co-C $:= \{\overline{L} \mid L \in$ C$\}$. In this notation, we have that NP $= \exists$P. The complexity classes Σ_i and Π_i are defined inductively as follows:

$$\Sigma_0 = \Pi_0 := \text{P}, \quad \Sigma_{i+1} := \exists \Pi_i, \quad \Pi_{i+1} := \forall \Sigma_i.$$

Since co-P $=$ P, we have that $\Pi_i = $ co-Σ_i, that is, $L \in \Sigma_i$ if and only if $\overline{L} \in \Pi_i$. Note that $\Sigma_1 = \exists$P $=$ NP and $\Pi_1 = \forall$P $=$ co-NP. Let PH $:= \bigcup_{i=0}^{\infty} \Sigma_i$; "PH" stands for "polynomial hierarchy".

Intuitively, adding more and more quantifiers we can encode more and more complex languages. It is therefore believed that the PH-hierarchy is strict, that is, $\Sigma_0 \subset \Sigma_1 \subset \Sigma_2 \subset \Sigma_3 \subset \dots$. On the other hand, it is easy to show that this hierarchy would collapse if $\Sigma_k = \Pi_k$ for some k.

Proposition 20.9. *Let $k \geq 1$ be an integer. If $\Sigma_k = \Pi_k$ then, for every $m \geq k$, $\Sigma_m = \Pi_m = \Sigma_k = \Pi_k$.*

Proof. Induction on m. The claim is trivially true for $m = k$. So assume it is true for m, and prove it for $m + 1$. Take a language $A \in \Sigma_{m+1}$. By the definition, there exists a language $B \in \Pi_m$ such that $x \in A$ iff $\exists y \ (x, y) \in B$. By the induction hypothesis, $B \in \Sigma_k$. So there exists a language C in Π_{k-1} such that $(x, y) \in B$ iff $\exists z \ (x, y, z) \in C$. Thus $x \in A$ iff $\exists y \ \exists z \ (x, y, z) \in C$, implying that $A \in \Sigma_k$. $\qquad \square$

This fact, together with our belief that alternating quantification increases the class of languages, implies that some languages in NP, including the clique function, "should" require circuits of super-polynomial size.

Theorem 20.10. (Karp-Lipton 1980) *If* NP $\subseteq$ P/poly *then* $\Pi_2 = \Sigma_2$, *and hence,* PH $= \Sigma_2$.

Proof. A prominent language in NP (besides that corresponding to the clique function) not known to belong to P is the language SAT. We encode CNF formulas ϕ as binary strings $\phi \in \{0, 1\}^*$ and let $\phi \in$ SAT if ϕ is satisfiable. As with the clique-language, the language SAT is NP-complete, meaning that it is the "most complicated" language in NP. Formally, this means that, for every language $L \in$ NP, there exists a function $g : \{0, 1\}^* \to \{0, 1\}^*$ such that g is computable in polynomial time, and a string z belongs to L iff $g(z) \in$ SAT.

Now assume that NP $\subseteq$ P/poly. We are going to show that then $\Pi_2 \subseteq \Sigma_2$. This inclusion implies that Σ_2 is closed under complementation, and hence, that $\Pi_2 =$ co-$\Sigma_2 = \Sigma_2$, as desired.

The argument is roughly the following: To simulate Π_2 by Σ_2, guess a poly-size circuit F for SAT, modify F via so-called "self-reducibility" so that whenever $F(\phi) = 1$ it also produces a satisfying assignment to ϕ, then check whether all universal paths of the Π_2 computation lead to a satisfiable formula.

To be more specific, let $L \in \Pi_2$. Then there is a language $L' \in$ NP and a constant a such that $L_n = \{x \mid (x, y) \in L'$ for all y of length $|y| \leq n^a\}$. Since $L' \in$ NP, there is a polynomial-time reduction $z \mapsto \phi_z$ from L' to SAT, that is, a string z belongs to L' iff $\phi_z \in$ SAT. Since the reduction can be computed in polynomial time, there exists a constant b such that $|\phi_z| \leq |z|^b$ for all z. Hence,

$$L_n = \{x \mid \phi_{x,y} \in \text{SAT for all } y \text{ of length } |y| \leq n^a\}. \tag{20.6}$$

Since, by our assumption, SAT $\in$ P/poly, there exists a constant c and a sequence $F = (F_n \mid n = 1, 2, \ldots)$ of circuits such that size$(F_n) \leq n^c$ and for every (code of a) CNF formula $\phi \in \{0, 1\}^*$, $F(\phi) = 1$ iff ϕ is satisfiable. We can view circuits F_n as encoded by binary strings as well. The circuits F_n solve a *decision* problem (is a given CNF satisfiable or not). But the circuits can also be used to solve a *search* problem: if a CNF is satisfiable, then find a satisfying assignment; this property is called the *self-reducibility* of SAT.

Claim 20.11. There exists a polynomial-time computable function $\alpha = \alpha(\phi)$ such that $\phi \in$ SAT iff α is a satisfying assignment for ϕ.

Proof. We have circuits deciding whether a given CNF formula is satisfying of not. We can now use these circuits to construct a satisfying assignment as follows. Ask the circuit if the formula ϕ is satisfiable. If so, ask if the formula $\phi_{x_1=1}$ with the first variable x_1 set to 1 is satisfiable. If the circuit answers "yes", then we already know the first bit of a satisfying assignment, it is 1. If the circuit answers "no", then we also know the first bit of a satisfying assignment, it is 0 because the entire formula ϕ was satisfiable. Continuing in this way we can generate the a satisfying assignment after at most $|\phi|$ queries. $\square$

Using this claim, and observing that $|\phi_{x,y}| \leq (|x| + |y|)^b \leq |x|^{ab+1}$, we can rewrite (20.6) as: $x \in L_n$ iff for all y of length $|y| \leq n^a$, $\alpha(\phi_{x,y})$ is a satisfying assignment for $\phi_{x,y}$. Now we are almost done. Even though there may be no easy way to construct the circuits $F = (F_n \mid n = 1, 2, \ldots)$ for SAT, we can just try to "guess" them. Namely, $x \in L_n$ iff there *exists* a circuit of size $\leq n^{abc+c}$ with $\leq n^{ab+1}$ inputs such that for all y of length $|y| \leq n^a$, $\alpha(\phi_{x,y})$ is a satisfying assignment for $\phi_{x,y}$. Since the mapping α is computable in polynomial time, we have thus shown that L belongs to Σ_2, as desired.   □

Let Circuit$[n^k]$ denote the class of all languages L such that the membership in L can be decided by a DeMorgan circuit of size $\mathcal{O}(n^k)$. Hence,

$$\text{P/poly} = \text{Circuit}[n] \cup \text{Circuit}[n^2] \cup \text{Circuit}[n^3] \cup \cdots .$$

Lemma 20.12. (Kannan 1981) *For every constant k, $\Sigma_4 \cap \Pi_4 \not\subseteq$ Circuit$[n^k]$.*

Proof. Let F be the lexically *first* circuit on n inputs such that Size$(F) \geq n^{k+1}$, and F is minimal, that is, no circuit of smaller size is equivalent to F. By the circuit-size hierarchy theorem (Theorem 1.19), we know that such circuits with Size$(F) \leq 4n^{k+1} \leq n^{k+2}$ exist. Let L be the language computed by this sequence of circuits. It is not difficult to verify that $L \in \Sigma_4$. For this, it is enough to describe L as a Σ_4-formula. The formula will simulate a circuit F of size n^{k+2} which is not equivalent to any circuit of size n^k. The first three statements of the formula below ensure this. But we also need the formula to simulate the *same* circuit on all inputs of length n. This is accomplished by the last four statements that choose the "minimum" (in the lexical ordering) circuit F with the necessary property. In all these statements we assume that circuits themselves are encoded as binary strings. On input x of length, the desired formula accepts x if and only if

1. $\exists$ a circuit F of size at most n^{k+2} such that
2. $\forall$ circuits F_1 of size at most n^k
3. $\forall$ circuits F_2 preceding F in the lexical order
4. $\exists$ input y of length n such that $F_1(y) \neq F(y)$
5. $\exists$ circuit F_3 of size at most n^k
6. $\forall$ inputs z of length n, $F_3(z) = F_2(z)$
7. F accepts x.

Thus we have found a language $L \in \Sigma_4$ such that $L \notin$ Circuit$[n^k]$. In fact, we also have that $L \in \Pi_4$: if we replace the last expression in the Σ_4 formula above by "F rejects x", the resulting Σ_4 formula will express the complement of L.   □

Theorem 20.13. (Kannan 1981) *For every constant k, $\Sigma_2 \cap \Pi_2 \not\subseteq$ Circuit$[n^k]$.*

Proof. We consider two cases. If SAT $\notin$ Circuit$[n^k]$, then we are done because SAT $\in$ NP $\subseteq \Sigma_2 \cap \Pi_2$. If SAT $\in$ Circuit$[n^k]$, then NP $\subseteq$ P/poly and Theorem 20.10 implies that $\Sigma_4 = \Sigma_2$. In this case Lemma 20.12 gives a desired language $L \in \Sigma_2 \cap \Pi_2$ such that $L \notin$ Circuit$[n^k]$.   □

Note that Theorem 20.13 does not imply that $\Sigma_2 \cap \Pi_2 \not\subseteq P/\text{poly}$ (and hence, that $P \neq NP$) because for that to be true we would have to be able to construct a *single* language $L \in \Sigma_2 \cap \Pi_2$ such that $L \notin \text{Circuit}[n^k]$ for *every* constant k, instead of constructing a different language for each constant k.

Structural complexity has a comprehensive treatment in recent books by Goldreich (2008), and Arora and Barak (2009). The book by Lipton (2010) gives a gentle introduction to the P versus NP problem itself.

Results mentioned in this section (as well as other results in a similar fashion) show that high-level arguments may (apparently) also lead to large circuit lower bounds. Their advantage is that they avoid such barriers as the natural proofs phenomenon. Their disadvantage is that they can merely confirm our intuitive belief that such functions like Clique or SAT *do* require large circuits without telling us *why* this happens. It seems therefore reasonable to develop *both* the high-level as well as low-level arguments. To prove NP $\not\subseteq$ P/poly, we perhaps need a technique explicit enough to "get its hands dirty" by exploring some issues of how circuits work, but not sufficiently explicit to provide any general measure of gate by gate progress.

Appendix A
Mathematical Background

In this appendix we give some mathematical background that is related to topics of this book. We do not attempt to give a detailed description. Rather, we mention some definitions and basic facts used in the book. The reader is invited to open any of the standard textbooks in mathematics to fill in the details.

A.1 Basics and Notation

Some of the results are asymptotic, and we use the standard asymptotic notation: for two functions f and g, we write $f = \mathcal{O}(g)$ if $f \leq c_1 g + c_2$ for all possible values of the two functions, where c_1, c_2 are constants. We write $f = \Omega(g)$ if $g = \mathcal{O}(f)$, and $f = \Theta(g)$ if $f = \mathcal{O}(g)$ and $g = \mathcal{O}(f)$. If the ratio f/g tends to 0 as the variables of the functions tend to infinity, we write $f = o(g)$ as well as $g = \omega(f)$. Finally, $f \sim g$ denotes that $f = (1 + o(1))g$, i.e., that f/g tends to 1 when the variables tend to infinity. As customary, $\mathbb{Z}$ denotes the set of integers, $\mathbb{R}$ the set of reals, $\mathbb{Z}_n$ an additive group of integers modulo n, and $\mathrm{GF}(q)$ a finite Galois field with q elements. Such a field exists as long as q is a prime power. If $q = p$ is a prime then $\mathrm{GF}(p)$ can be viewed as the set $\{0, 1, \ldots, p - 1\}$ with addition and multiplication performed modulo p. The sum in $\mathrm{GF}(2)$ is often denoted by $\oplus$, that is, $x \oplus y$ stands for $x + y \bmod 2$. If not stated otherwise, $e = 2.718\ldots$ always denotes the base of the natural logarithm. For a positive integer n, we also use the notation $[n] = \{1, 2, \ldots, n\}$. If x is a real number, then $\lceil x \rceil$ denotes the smallest integer not smaller than x, and $\lfloor x \rfloor$ denotes the greatest integer not exceeding x.

A.2 Graphs

A *graph* on n vertices is a pair $G = (V, E)$ consisting of an n-element set V, whose members are called *vertices* (or *nodes*), and a family E of 2-element subsets of V, whose members are called *edges*. A vertex v is *incident* with an edge e if $v \in e$. The two vertices incident with an edge are its *end-vertices* or *endpoints*, and the edge *joins* its ends. Two vertices u, v of G are *adjacent*, or *neighbors*, if $\{u, v\}$ is an edge of G. The number $d(u)$ of neighbors of a vertex u is its *degree*. A graph is d-*regular* if all its vertices have the same degree.

Euler's Theorem. In every graph, the sum of degrees of all its vertices is equal to two times the number of edges.

Proof. Take the 0-1 edge-vertex adjacency matrix $A = (a_{e,v})$ whose rows are labeled by edges e and columns by vertices v, and $a_{e,v} = 1$ iff $v \in e$. The sum of ones in A is the same if we count them row-wise, and if we count them column-wise. $\square$

A *walk* of length k in G is a sequence $v_0, e_1, v_1 \ldots, e_k, v_k$ of vertices and edges such that $e_i = \{v_{i-1}, v_i\}$. A walk without repeated vertices is a *path*. A walk without repeated edges is a *trail*. A *cycle* of length k is a path $v_0, \ldots, v_k$ with $v_0 = v_k$. A (connected) *component* in a graph is a set of its vertices such that there is a path between any two of them. A graph is *connected* if it consists of one component. A *tree* is a connected graph without cycles. A *subgraph* is obtained by deleting edges and vertices. A *spanning subgraph* is obtained by deleting edges only. An *induced subgraph* is obtained by deleting vertices (together with all the edges incident to them).

A *complete graph* or *clique* is a graph in which every pair is adjacent. An *independent set* in a graph is a set of vertices with no edges between them. The greatest integer r such that G contains an independent set of size r is the *independence number* of G, and is denoted by $\alpha(G)$. The well-known Turán's theorem (see Exercise 14.11 for the proof) states that, if G is a graph with n vertices and m edges, then $\alpha(G) \geq n^2/(2m + n)$. A *bipartite $m \times n$ graph* is a graph $G = (V, E)$ whose vertex-set can be partitioned into two independent sets $V = L \cup R$ with $|L| = m$ and $|R| = n$ vertices; the two sets in such a partition are also called color classes, where L stands for the "left" class, and R for the "right" class. If $L = \{u_1, \ldots, u_m\}$ and $R = \{v_1, \ldots, v_n\}$, then the *adjacency matrix* of G (relative to this bipartition) is a boolean $m \times n$ matrix $A = (a_{ij})$ such that $a_{ij} = 1$ if and only if u_i and v_j are adjacent in G.

A *coloring* of $G = (V, E)$ is an assignment of colors to each vertex so that adjacent vertices receive different colors. In other words, this is a partition of the vertex set V into independent sets. The minimum number of colors required for doing that is the *chromatic number* of G.

The *complement* $\overline{G}$ of a graph G is a graph on the same set of vertices in which two vertices are adjacent if and only if they are non-adjacent in G. If the graph is bipartite, then its bipartite complement contains only those pairs of previously non-adjacent vertices that belong to the different parts of the bipartition. Thus a bipartite complement of a bipartite graph is again a bipartite graph with the same partition.

A.3 Linear Algebra

Informally, a *field* is a set $\mathbb{F}$ closed under addition, subtraction, multiplication and division by nonzero element; if division is not defined (or is defined not for all elements) then the set is called a *ring*. By addition and multiplication, we mean commutative and associative operations which obey distributive law. The additive identity is called zero, and the multiplicative identity is called unity. Examples of fields are the reals $\mathbb{R}$, the rationals $\mathbb{Q}$, and the set of integers modulo a prime p. We will be mostly concerned with finite fields. The cardinality of a finite field must be a power of a prime and all finite fields with the same number of elements are isomorphic. Thus for each prime power q there is essentially one field $\mathbb{F}$ with $|\mathbb{F}| = q$. This field is usually denoted as $GF(q)$ or $\mathbb{F}_q$. If p is a prime number, then the set $\mathbb{Z}_p = \{0, 1, \dots, p - 1\}$ forms a field with the arithmetic operations performed modulo p.

A *linear space* (or *vector space*) V over a field $\mathbb{F}$ is an additive Abelian group $(V, +, \mathbf{0})$ closed under (left) multiplication by elements of $\mathbb{F}$ (called *scalars*). It is required that this multiplication is distributive with respect to addition in both V, and associative with respect to multiplication in $\mathbb{F}$. Elements of V are called *vectors* or *points*. Standard examples of vector spaces are subsets $V \subseteq \mathbb{F}^n$ closed under the component-wise addition $u + v = (u_1 + v_1, \dots, u_n + v_n)$ and multiplication by scalars $\lambda v = (\lambda v_1, \dots, \lambda v_n), \lambda \in \mathbb{F}$.

Linear independence and dimension A *linear combination* of the vectors $v_1, \dots, v_m$ is a vector of the form $\lambda_1 v_1 + \dots + \lambda_m v_m$ with $\lambda_i \in \mathbb{F}$. Such a combination is a *affine combination* if $\lambda_1 + \dots + \lambda_m = 1$, and is a *convex combination* (over the reals) if additionally $\lambda_i \geq 0$ for all i. A *convex hull* of a set of vectors is the set of all their convex combinations.

A *linear subspace* of V is a nonempty subset of V, closed under linear combinations. An *affine space* is a set $S \subseteq V$ closed under affine combinations. The set of solutions to the system of equations $Ax = b$ is an affine space. Each linear subspace is also an affine space. Conversely, each affine space is a translate $U + v = \{u + v \mid u \in U\}$ of some linear subspace $U \subseteq V$.

The *span* of $v_1, \dots, v_m$, denoted by $\mathrm{Span}(\{v_1, \dots, v_m\})$, is the linear subspace formed by all linear combinations of these vectors. A vector u *linearly dependent* on the vectors $v_1, \dots, v_m$ if $u \in \mathrm{Span}(\{v_1, \dots, v_m\})$. The vectors $v_1, \dots, v_m$ are *linearly independent* if none of them is dependent on the rest. Equivalently, (in)dependence can be defined as follows. A *linear relation* among the vectors $v_1, \dots, v_m$ is a linear combination that gives the zero vector: $\lambda_1 v_1 + \dots + \lambda_m v_m = \mathbf{0}$. This relation is nontrivial if $\lambda_i \neq 0$ for at least one i. It is easy to see that the vectors $v_1, \dots, v_m$ are linearly independent if and only if no nontrivial linear relation exists between them. A *basis* of V is a set of independent vectors which spans V. A fundamental fact in linear algebra says that *any two bases of V have the same cardinality*; this number is called the *dimension* of V and is denoted by $\dim(V)$. The dimension of an affine space $S \subseteq V$ is the dimension of its corresponding linear subspace (of which S is a translate).

A further basic fact is the so-called *linear algebra bound* (see any standard linear algebra book for the proof): If $v_1, \ldots, v_k$ are linearly independent vectors in a vector space of dimension m then $k \leq m$.

Vector and matrix products   In this book we mainly consider standard vector spaces whose elements (vectors) are finite strings of elements of some fixed field $\mathbb{F}$. These strings $v \in \mathbb{F}^n$ are assumed to be in a "column" form; their transpose v^T is then a "row" vector. Both v and v^T represent the *same* vector—which of these forms should be used is only important when performing operations on vectors.

To distinguish between elements $v \in \mathbb{F}$ of the field and vectors over this field, we will sometimes write vectors as **v**. But usually it will be clear from the context when v_i denotes the i-th vector in a set of vectors and when the i-th component of some vector v.

An $m \times n$ matrix $A = (a_{ij})$ over a field $\mathbb{F}$ is a sequence of n vectors in $\mathbb{F}^m$, called columns of A. A *transpose* $A^T = (b_{ij})$ of a square $n \times m$ matrix A is the $m \times n$ matrix with $b_{ij} = a_{ji}$. A matrix is a square matrix if $m = n$.

A *scalar product* of two vectors $u^T = (u_1, \ldots, u_n)$ and $v^T = (v_1, \ldots, v_n)$ is the number

$$\langle u, v \rangle = u^T v := u_1 v_1 + \cdots + u_n v_n.$$

If $A = (a_{ij})$ is an m-by-n matrix over some field $\mathbb{F}$ and x is a vector in $\mathbb{F}^m$, then $x^T A$ is the vector in $\mathbb{F}^n$ whose j-th coordinate is the scalar product of x with the j-th *column* of A. Thus the rows of A are linearly independent if and only if $x^T A \neq \mathbf{0}$ for all $x \neq \mathbf{0}$. Similarly, if $y \in \mathbb{F}^n$, then Ay is the vector in $\mathbb{F}^m$ whose i-th coordinate is the scalar product of y with the i-th *row* of A. If A is an m-by-n matrix and B is an n-by-p matrix with columns $b_1, \ldots, b_p \in \mathbb{F}^n$, then their product is the m-by-p matrix $A \cdot B$ whose columns are $Ab_1, \ldots, Ab_p$.

Note a big difference between vector-vector products $x^T y$ and xy^T: the first is a number whereas the second is a matrix(!) whose columns are multiplies of vector x:

Orthogonality   Vectors u and v are *orthogonal* if $\langle u, v \rangle = 0$; in this case one also writes $u \perp v$. If $U \subseteq V$ is a subspace of V then the *dual* (or *orthogonal complement*) is the subspace $U^{\perp} \subseteq V$ consisting of all vectors $v \in V$ such that $\langle u, v \rangle = 0$ for all $u \in U$. The following equality connects the dimensions of two orthogonal subspaces of a finite dimensional linear space: $\dim(U) + \dim(U^{\perp}) = \dim(V)$. A consequence of this is that, for every linear subspace $U \subseteq \mathbb{R}^n$ and every vector $x \in U$, there are uniquely defined vectors $u \in U$ and $w \in U^{\perp}$ such that $x = u + w$. The vector u is then called the *projection* of x onto U.

A *Hadamard matrix* of order n is an $n \times n$ matrix with entries ± 1 and with row vectors mutually orthogonal over the reals. It follows from the definition that a Hadamard matrix H of order n satisfies $HH^T = nI_n$, where I_n is the $n \times n$ identity matrix.

Lindsey's Lemma. The absolute value of the sum of all entries in any $a \times b$ submatrix of an $n \times n$ Hadamard matrix H does not exceed $\sqrt{abn}$.

In particular, if $ab > n$ then no $a \times b$ submatrix of H is monochromatic.

Proof. Let H be an $n \times n$ Hadamard matrix, and A one of its $a \times b$ submatrices. Assume for simplicity that A consists of its first a rows and b columns. Let α be the sum of all entries of A. We want to prove that $\alpha \le \sqrt{abn}$.

Let $v_1, \ldots, v_a$ be the first a rows of H, and $y = \sum_{i=1}^{a} v_i$. If we take the vector $x = (1^b 0^{n-b})$, then $\alpha^2 = \langle x, y \rangle^2 \le \|x\|^2 \|y\|^2 = b \cdot \|y\|^2$. On the other hand, the conditions $\langle v_i, v_i \rangle = n$ and $\langle v_i, v_j \rangle = 0$ for all $i \ne j$ imply that $\|y\|^2 = \sum_{i,j=1}^{a} \langle v_i, v_j \rangle = \sum_{i=1}^{a} \langle v_i, v_i \rangle = an$. Thus $\alpha^2 \le b \cdot \|y\|^2 = abn$, as desired. $\square$

Rank The *column rank* of a matrix A is the dimension of the vector space spanned by its columns. The *row rank* of A is the dimension of the vector space spanned by its rows. One of the first nontrivial results in matrix theory asserts that the *row and column ranks are equal*; this common value is the rank of A, denoted by $\mathrm{rk}(A)$. There are several equivalent definitions of the rank of an m-by-n matrix $A = (a_{ij})$ over a given field $\mathbb{F}$:

- $\mathrm{rk}(A)$ is the smallest r such that A can be written as a sum of r rank-1 matrices, that is, there exist vectors $x_1, \ldots, x_r$ in $\mathbb{F}^m$ and $y_1, \ldots, y_r$ in $\mathbb{F}^n$ such that $A = \sum_{i=1}^{r} x_i y_i^T$.
- $\mathrm{rk}(A)$ is the smallest r such that $A = B \cdot C$ for some m-by-r matrix B and r-by-n matrix C;
- $\mathrm{rk}(A)$ is the smallest r such that A is a matrix of scalar products of vectors in $\mathbb{F}^r$: there exist vectors $u_1, \ldots, u_m$ and $v_1, \ldots, v_n$ in $\mathbb{F}^r$ such that $a_{ij} = \langle u_i, v_j \rangle$.

The following inequalities hold for the rank (if A is an m-by-n and B an n-by-k matrix):

$$\mathrm{rk}(A) - \mathrm{rk}(B) \le \mathrm{rk}(A + B) \le \mathrm{rk}(A) + \mathrm{rk}(B);$$

$$\mathrm{rk}(A) + \mathrm{rk}(B) - n \le \mathrm{rk}(AB) \le \min \{\mathrm{rk}(A), \mathrm{rk}(B)\}.$$

A componentwise product (or Hadamard product) of two m-by-n matrices $A = (a_{ij})$ and $B = (b_{ij})$ is the m-by-n matrix $A \circ B = (a_{ij} b_{ij})$.

Lemma A.1. (Rank of Hadamard product) $\mathrm{rk}(A \circ B) \le \mathrm{rk}(A) \cdot \mathrm{rk}(B)$.

Proof. Let $r = \mathrm{rk}(A)$ and $s = \mathrm{rk}(B)$. Then $A = \sum_{i=1}^{r} x_i y_i^T$ and $B = \sum_{i=1}^{s} u_i v_i^T$ for some vectors $x_i, u_i \in \mathbb{F}^m$ and $y_i, v_i \in \mathbb{F}^n$. Since $(xy^T) \circ (uv^T) = (x \circ u)(y \circ v)^T$, we can write $A \circ B$ as the sum

$$A \circ B = \sum_{i=1}^{r} \sum_{j=1}^{s} (x_i y_i^T) \circ (u_j v_j^T) = \sum_{i=1}^{r} \sum_{j=1}^{s} (x_i \circ u_j)(y_i \circ v_j)^T$$

of at most sr rank-1 matrices, implying that $\mathrm{rk}(A \circ B) \le rs = \mathrm{rk}(A) \cdot \mathrm{rk}(B)$. $\qquad$ □

Spaces of solutions If U is the space spanned by the rows of A, then the set of solutions of $Ax = \mathbf{0}$ is clearly the subspace $U^{\perp}$ of all vectors that are orthogonal to all the rows of A and, since $\dim(U) + \dim(U^{\perp}) = n$, its dimension is $n - \mathrm{rk}(A)$; the subspace $U^{\perp} = \{x \mid Ax = \mathbf{0}\}$ is also called the *kernel* of A. Thus if the underlying field is finite and $\mathbb{F}$ has $|\mathbb{F}| = q$ elements, then $Ax = \mathbf{0}$ has exactly $q^{n-\mathrm{rk}(A)}$ solutions.

Norms The *norm* (or *length*) of a vector $v = (v_1, \dots, v_n)$ in $\mathbb{R}^n$ is the number

$$\|v\| := \langle v, v \rangle^{1/2} = \left(\sum_{i=1}^{n} v_i^2 \right)^{1/2}.$$

The following basic inequality, known as the *Cauchy–Schwarz inequality*, estimates the scalar product of two vectors in terms of their norms (we have already used it in previous sections; now we will prove it):

Cauchy–Schwarz Inequality. For any real vectors $u, v \in \mathbb{R}^n$, $|\langle u, v \rangle| \le \|u\| \cdot \|v\|$ with an equality iff u and v are linearly dependent.

When expressed explicitly, this inequality turns to:

$$\left(\sum_{i=1}^{n} u_i v_i \right)^2 \le \left(\sum_{i=1}^{n} u_i^2 \right) \left(\sum_{i=1}^{n} v_i^2 \right).$$

Proof. We may assume that $u \ne \mathbf{0}$. For any constant $\lambda \in \mathbb{R}$ we have

$$0 \le \langle \lambda u - v, \lambda u - v \rangle = \langle \lambda u, \lambda u - v \rangle - \langle v, \lambda u - v \rangle = \lambda^2 \langle u, u \rangle - 2\lambda \langle u, v \rangle + \langle v, v \rangle.$$

Substituting $\lambda = \frac{\langle u, v \rangle}{\langle u, u \rangle}$ we get

$$0 \le \frac{\langle u, v \rangle^2}{\langle u, u \rangle^2} \langle u, u \rangle - 2 \frac{\langle u, v \rangle^2}{\langle u, u \rangle} + \langle v, v \rangle = \langle v, v \rangle - \frac{\langle u, v \rangle^2}{\langle u, u \rangle}$$

Rearranging the last inequality, we get $\langle u, v \rangle^2 \le \langle u, u \rangle \langle v, v \rangle = \|u\|^2 \cdot \|v\|^2$. $\qquad$ □

Of a similar vein is the following useful inequality due to Chebyshev (see Hardy et al. 1952, Theorem 43, page 43): if $a_1, \dots, a_n$ is a non-decreasing sequence and $b_1, \dots, b_n$ a non-increasing sequence of non-negative numbers, then

$$\sum_{i=1}^{n} a_i b_i \leq \frac{1}{n}\Big(\sum_{i=1}^{n} a_i\Big)\Big(\sum_{i=1}^{n} b_i\Big).$$

In some applications it is desirable to estimate the sum $\sum_{i=1}^{n} a_i$ of numbers in terms of the value $A := \sum_{i=1}^{n} a_i^2$ of the sum of *squares* of these numbers. The Cauchy–Schwarz inequality gives us an upper bound $\sum_{i=1}^{n} a_i \leq \sqrt{nA}$. If all the a_i are equal, then $\sqrt{nA}$ is also a lower bound. The situation, however, is more complicated if the numbers a_i are very different. Still, even then one can obtain a lower bound of about $\sqrt{A}\log n$, if some information is known about partial sums of the a_i. This can be derived from the following lemma due to Cherukhin (2008); somewhat weaker forms were proved earlier by Pudlák (1994) and Rychkov (1994).

A real-valued function $f(x)$ defined on an interval is called concave if, for any two points x and y in its domain and any $\lambda \in [0, 1]$, we have

$$f(\lambda x + (1 - \lambda)y) \geq \lambda f(x) + (1 - \lambda)f(y).$$

If $f(x)$ is twice-differentiable, then $f(x)$ is concave if and only if $f''(x)$ is non-positive. For example, $f(x) = \sqrt{x}$ as well as $f(x) = -x^2$ are concave functions.

Monotone Sums Lemma. Let $a_1, \ldots, a_n, b_1, \ldots, b_n$ be nonnegative real numbers such that $a_1 \geq \ldots \geq a_n$ and

$$a_r + \ldots + a_n \geq b_r + \ldots + b_n \quad \text{for all} \quad r = 1, \ldots, n. \tag{A.1}$$

Let $f(x)$ be an increasing concave function. Then

$$f(a_1) + \ldots + f(a_n) \geq f(b_1) + \ldots + f(b_n).$$

Proof. First we prove that if $x \geq y \geq \epsilon > 0$, then

$$f(x) + f(y) \geq f(x + \epsilon) + f(y - \epsilon). \tag{A.2}$$

That is, if the largest of the numbers of x, y is increased by ϵ, and the smallest one is decreased by ϵ, then the sum $f(x) + f(y)$ does not increase. To show this, set $\lambda := \epsilon/(x - y + 2\epsilon)$. Since f is concave, we obtain

$$\lambda f(y - \epsilon) + (1 - \lambda)f(x + \epsilon) \leq f(\lambda(y - \epsilon) + (1 - \lambda)(x + \epsilon))$$
$$= f(x + \epsilon - \lambda(x - y + 2\epsilon)) = f(x).$$

Similarly, $(1 - \lambda)f(y - \epsilon) + \lambda f(x + \epsilon) \leq f(y)$. Summing the last two inequalities we get (A.2).

We prove the lemma by induction on n. The base is $n = 1$. In this case the claim follows from (A.1) and the assumption that f is increasing.

We now prove the induction step ($n \geq 2$). Increase a_1 and decrease a_n by ϵ, where ϵ is the maximum possible number such that all the inequalities (A.1) are satisfied after this change. The sum $f(a_1) + \ldots + f(a_n)$ does not increase due to this change. This follows from the inequality $a_1 \geq a_n$ and the inequality (A.2). Hence, if we are able to prove the claim for the new numbers $a_1, \ldots, a_n$, the claim for the former numbers will follow.

We now prove the claim for the new numbers $a_1, \ldots, a_n$. By the maximality of ϵ, at least one of the inequalities (A.1) (except for the first one) becomes an equality. Indeed, increasing a_1 by ϵ and decreasing a_n by ϵ does not change the sum $a_1 + \ldots + a_n$. Thus the first inequality of the system (A.1) remains intact. However, the left-hand side of all subsequent inequalities decreases. Thus the maximality of ϵ implies that one of the subsequent inequalities has become an equality.

Thus, for some $k \geq 2$ we have

$$a_k + \ldots + a_n = b_k + \ldots + b_n. \tag{A.3}$$

If we subtract the equality (A.3) from the first $k - 1$ inequalities of the system (A.1), then the system (A.1) splits into two independent systems of the same type, namely one for the first $k - 1$ numbers $a_1, \ldots, a_{k-1}$; $b_1, \ldots, b_{k-1}$, and the other for the remaining $n - k + 1$ numbers $a_k, \ldots, a_n$; $b_k, \ldots, b_n$. Applying the induction hypothesis to these two systems, we get

$$f(a_1) + \ldots + f(a_{k-1}) \geq f(b_1) + \ldots + f(b_{k-1}),$$
$$f(a_k) + \ldots + f(a_n) \geq f(b_k) + \ldots + f(b_n).$$

Finally, summing the two last inequalities we get the desired claim.   □

Using the estimate $\ln n + 1/3 < H(n) < \ln n + 2/3$ for the harmonic series $H(n) := \sum_{i=1}^{n} \frac{1}{i}$, one can derive the following estimate due to Pudlák (1994).

Lemma A.2. *If* $x_1 \geq \ldots \geq x_n \geq 0$ *and* $A \geq 1$ *are real numbers, such that* $\sum_{i=r}^{n} x_i^2 \geq A/r$ *for all* $r = 1, 2, \ldots, n$, *then* $\sum_{i=1}^{n} x_i \geq \sqrt{A}(\ln n - 1)$.

Proof. Applying the Monotone Sum Lemma with $f(x) = \sqrt{x}$, $a_i = x_i^2$, $n_n = \frac{1}{n}$ and $b_i = \frac{1}{i} - \frac{1}{i+1}$ for $i = 1, \ldots, n - 1$, we obtain that $\sum_{i=1}^{n} x_i$ is at least

$$\sum_{i=1}^{n} \sqrt{b_i} = \sum_{i=1}^{n} \sqrt{\frac{A}{i(i+1)}} \geq \sqrt{A} \sum_{i=1}^{n} \frac{1}{i+1} > \sqrt{A}(\ln n - 1). \qquad \square$$

Eigenvalues A scalar $\lambda \in \mathbb{R}$ is an *eigenvalue* of a square real matrix A if the equation $Ax = \lambda x$ has a solution $x \in \mathbb{R}^n$, $x \neq \mathbf{0}$, which is the case iff the *characteristic polynomial* $p_A(z) = \det(A - zI)$ has λ as a root; here, I is a unit matrix with 1s on the diagonal, and 0s elsewhere. A nonzero vector x with

$Ax = \lambda x$ is called an *eigenvector* corresponding to the eigenvalue λ. Since p_A has degree n, we can have at n (not necessarily distinct) complex eigenvalues. If the matrix A is symmetric, that is, $A^T = A$, then all its eigenvalues are real numbers. The following are standard facts about the eigenvalues of a real symmetric $n \times n$ matrix $A = (a_{ij})$:

- A has exactly n (not necessarily distinct) real eigenvalues $\lambda_1 \geq \ldots \geq \lambda_n$.
- There exists a set of n eigenvectors $x_1, \ldots, x_n$, one for each eigenvalue, that are normalized and mutually orthogonal, that is, $\|x_i\| = 1$ and $\langle x_i, x_j \rangle = 0$ over the reals. Hence, $x_1, \ldots, x_n$ form an *orthonormal* basis of $\mathbb{R}^n$.
- The rank of A is equal to the number of its nonzero eigenvalues, including multiplicities.
- The sum of all eigenvalues $\sum_{i=1}^{n} \lambda_i$ is equal to the trace $\mathrm{tr}(A) = \sum_{i=1}^{n} a_{ii}$.
- The first two largest eigenvalues are equal to

$$\lambda_1 = \max_{x \neq 0} \frac{x^T A x}{x^T x} = \max_{\|x\|=1} x^T A x \quad \text{and} \quad \lambda_2 = \max_{x \perp 1} \frac{x^T A x}{x^T x} = \max_{x \perp 1, \|x\|=1} x^T A x,$$

where $\mathbf{1}$ is the all-1 vector, and the second equality follows since we can replace x by $x/\|x\|$, since the first maximum is over all nonzero vectors x.

The second largest eigenvalue $\lambda(G)$ of the adjacency matrix of a graph G is an important parameter telling us how much "expanding" the graph G is.

Expander Mixing Lemma. If G is a d-regular graph on n vertices and $\lambda = \lambda(G)$ is the second largest eigenvalue of its adjacency matrix, then the number $e(S, T)$ of edges between every two (not necessarily disjoint) subsets S and T of vertices satisfies

$$\left| e(S, T) - \frac{d|S| \cdot |T|}{n} \right| \leq \lambda \sqrt{|S| \cdot |T|}.$$

In particular, if $s > \lambda n / d$ then the graph is s-*mixed*, that is, for any pair of disjoint s-element subsets of vertices, there is at least one edge between these sets.

Proof. Let $\lambda_1 \geq \lambda_2 \geq \ldots \geq \lambda_n$ be the eigenvalues of the adjacency matrix M of G, and let $\mathbf{v}_1, \ldots, \mathbf{v}_n \in \mathbb{R}^n$ be the corresponding orthonormal basis of eigenvectors. That is, for each i, we have that $M\mathbf{v}_i = \lambda_i \mathbf{v}_i$, $\langle \mathbf{v}_i, \mathbf{v}_i \rangle = 1$ and $\langle \mathbf{v}_i, \mathbf{v}_j \rangle = 0$ for $j \neq i$. Here $\mathbf{v}_1$ is $\frac{1}{\sqrt{n}}$ times the all-1 vector $\mathbf{1}$. Let χ_S and χ_T be the characteristic 0-1 vectors of S and T. Expand these two vectors as linear combinations $\chi_S = \sum_{i=1}^{n} a_i \mathbf{v}_i = \langle \mathbf{a}, \chi_S \rangle$ and $\chi_T = \sum_{i=1}^{n} b_i \mathbf{v}_i = \langle \mathbf{b}, \chi_T \rangle$ of the basis vectors. Since the $\mathbf{v}_i$ are orthonormal eigenvectors,

$$e(S, T) = \chi_S^T M \chi_T = \left(\sum_{i=1}^{n} a_i \mathbf{v}_i \right)^T M \left(\sum_{i=1}^{n} b_i \mathbf{v}_i \right) = \sum_{i=1}^{n} \lambda_i a_i b_i. \tag{A.4}$$

Since the graph G is d-regular, we have $\lambda_1 = d$. The first two coefficients a_1 and b_1 are scalar products of $v_1 = \frac{1}{\sqrt{n}}\mathbf{1}$ with χ_S and χ_T; hence, $a_1 = |S|/\sqrt{n}$ and

$b_1 = |T|/\sqrt{n}$. Thus the first term $\lambda_1 a_1 b_1$ in the sum (A.4) is precisely $d|S||T|/n$. Since $\lambda = \lambda_2$ is the second largest eigenvalue, the absolute value of the sum of the remaining $n - 1$ terms in this sum does not exceed $\lambda \langle \mathbf{a}, \mathbf{b} \rangle$ which, by Cauchy–Schwarz inequality, does not exceed $\lambda \|\mathbf{a}\| \|\mathbf{b}\| = \lambda \|\chi_S\| \|\chi_T\| = \lambda \sqrt{|S||T|}$.   □

Ramanujan graphs   An n-vertex graph G is a *Ramanujan graph* if G is $(q + 1)$-regular (all vertices have the same degree $q + 1$), and $\lambda(G) \leq 2\sqrt{q}$. Explicit constructions of Ramanujan graphs on n vertices for every prime $q \equiv 1 \mod 4$ and infinitely many values of n were given in Margulis (1973), Lubotzky et al. (1988); these were later extended to the case where q is an arbitrary prime power in Morgenstern (1994) and Jordan and Livné (1997). By the Expander Mixing Lemma, for every such graph we have that

$$e(S, T) \geq \frac{q + 1}{n} |S| \cdot |T| - 2\sqrt{q|S| \cdot |T|}.$$

The spectral norm   The *spectral norm* of a matrix $A = (a_{ij})$ is defined as

$$\|A\| := \max_{x \neq 0} \frac{\|Ax\|}{\|x\|} = \max_{\|x\|=1} \|Ax\|.$$

It is also well known that

$$\|A\| = \max_{\|x\|=1, \|y\|=1} |x^T A y|.$$

The name "spectral norm" comes from the fact that

$$\|A\| = \text{ square root of the largest eigenvalue of } A^T A.$$

This holds because $x^T (A^T A) x = \langle Ax, Ax \rangle = \|Ax\|^2$. The Cauchy–Schwarz inequality implies the following useful inequality

$$x^T A y \leq \|x\| \cdot \|A\| \cdot \|y\|.$$

If $a = \max_j (|a_{1j}| + \cdots + |a_{nj}|)$ is the maximum absolute column sum of the matrix, then

$$\frac{a}{\sqrt{n}} \leq \|A\| \leq a\sqrt{n}.$$

The *Frobenius norm* of A is just the Euclidean norm $\|A\|_F := \left(\sum_{i,j} a_{ij}^2 \right)^{1/2}$ of the corresponding vector of length n^2. The following fact relates these two norms with the rank over the reals.

Lemma A.3.   (Norms and rank) *For every real matrix A,*

$$\frac{\|A\|_F}{\sqrt{\mathrm{rk}(A)}} \le \|A\| \le \|A\|_F.$$

Proof. Observe that $\|A\|_F^2$ is equal to the trace, that is, the sum of diagonal elements of the matrix $B = A^T A$. On the other hand, the trace of any real matrix is equal to the sum of its eigenvalues. Hence, $\|A\|_F^2 = \sum_{i=1}^n \lambda_i$ where $\lambda_1 \ge \ldots \ge \lambda_n$ are the eigenvalues of B. Since B has only $\mathrm{rk}(B) = \mathrm{rk}(A) = r$ nonzero eigenvalues, and since all eigenvalues of B are non-negative, the largest eigenvalue λ_1 is bounded by $\|A\|_F^2/r \le \lambda_1 \le \|A\|_F^2$. It remains to use the fact mentioned above that $\|A\| = \sqrt{\lambda_1}$. □

If H is an $n \times n$ Hadamard matrix, then $H^T H = nI_n$ where I_n is the $n \times n$ identity matrix. Hence, all n eigenvalues of $H^T H$ are equal to n, implying that H has spectral norm $\|H\| = \sqrt{n}$.

Lemma A.4. (Rank of Hadamard matrices) *Every $a \times b$ submatrix of a Hadamard $n \times n$ matrix has rank at least ab/n over the reals.*

Proof. Let H be an $n \times n$ Hadamard matrix, and A one of its $a \times b$ submatrices. Since A is a submatrix of H, we have that $\|A\| \le \|H\|$. So, the previous lemma implies that $\mathrm{rk}(A) \ge \|A\|_F^2/\|A\|^2 \ge \|A\|_F^2/\|H\|^2 = ab/n$, where the last equality follows because $\|A\|_F^2$ is precisely the number of entries in A. □

A.4 Probability Theory

A *finite probability space* consists of a finite set Ω (called a *sample space*) and a function (also called *probability distribution*) Prob $: \Omega \to [0, 1]$, such that $\sum_{x \in \Omega} \mathrm{Prob}[x] = 1$. A probability space is a representation of a random experiment, where we choose a member of Ω at random and $\mathrm{Prob}[x]$ is the probability that x is chosen. The most common probability distribution is the *uniform distribution*, which is defined as $\mathrm{Prob}[x] = 1/|\Omega|$ for each $x \in \Omega$; the corresponding sample space is then called *symmetric*.

Subsets $A \subseteq \Omega$ are called *events*. The probability of an event is defined by $\mathrm{Prob}[A] := \sum_{x \in A} \mathrm{Prob}[x]$, that is, the probability that a member of A is chosen. One of the simplest inequalities is the so-called *union bound*:

$$\mathrm{Prob}[A_1 \cup A_2 \cup \cdots \cup A_n] \le n \cdot \max_i \mathrm{Prob}[A_i].$$

This is just a weighted version of a trivial fact that $|A \cup B| \le |A| + |B|$. Be it so simple, in may situations, even this bound allows us to show that some object with desired "good" properties exists. If $A_1, \ldots, A_n$ are some "bad" events, each occurring with probability at most $p < 1/n$, then the probability that *none* of these bad events will happen is at least $1 - pn > 0$.

Conditional probability For two events A and B, the *conditional probability of A given B*, denoted Prob$[A|B]$, is the probability that one would assign to A if one knew that B occurs. Formally,

$$\text{Prob}[A|B] := \frac{\text{Prob}[A \cap B]}{\text{Prob}[B]},$$

when Prob$[B] \neq 0$. For example, if we are choosing a uniform integer from $\{1, \ldots, 6\}$, A is the event that the number is 2 and B is the event that the number is even, then Prob$[A|B] = 1/3$, whereas Prob$[B|A] = 1$.

Independent events Two events A and B are *independent* if

$$\text{Prob}[A \cap B] = \text{Prob}[A] \cdot \text{Prob}[B].$$

If $B \neq \emptyset$, this is equivalent to Prob$[A|B] = $ Prob$[A]$. It is very important to note that the "independence" has nothing to do with the "disjointness" of the events: if, say, $0 < \text{Prob}[A] < 1$, then the events A and $\overline{A}$ *are* dependent!

Random sets Let Γ be finite set, and $0 \leq p \leq 1$. A *random subset* S of Γ is obtained by flipping a coin, with probability p of success, for each element of Γ to determine whether the element is to be included in S; the distribution of S is the probability distribution on $\Omega = 2^{\Gamma}$ given by Prob$[S] = p^{|S|}(1 - p)^{|\Gamma| - |S|}$ for $S \subseteq \Gamma$. We will mainly consider the case when S is uniformly distributed, that is, when $p = 1/2$. In this case each subset $S \subseteq \Gamma$ receives the same probability Prob$[S] = 2^{-|\Gamma|}$. If $\mathcal{F}$ is a family of subsets, then its *random member* S is a uniformly distributed member; in this case, $\Omega = \mathcal{F}$ and S has the probability distribution Prob$[S] = 1/|\mathcal{F}|$. Note that, for $p = 1/2$, a random subset of Γ is just a random member of 2^{Γ}.

Random variables A *random vector* r $= (r_1, \ldots, r_n)$ in GF$(2)^n$ is obtained by flipping an unbiased 0-1 coin n times. Hence, Prob$[r = v] = 2^{-n}$ for each vector $v \in$ GF$(2)^n$. A simple, but often-used fact is that r is orthogonal to every nonzero boolean vector with the same probability $1/2$, that is,

$$\text{Prob}[\langle r, v \rangle = 0] = \frac{1}{2} \quad \text{for every } v \neq \mathbf{0} \text{ in GF}(2)^n.$$

The reason is that $v \neq \mathbf{0}$ implies that $v_i = 1$ for some position i. Hence we can partition the space GF$(2)^n$ into 2^{n-1} pairs u, u' that differ only in their i-th position. For each of these pairs, we have that $\langle v, u \rangle \neq \langle v, u' \rangle$. Hence $\langle v, u \rangle = 0$ for exactly 2^{n-1} vectors v, and r will be equal to one of these vectors with probability $2^{n-1}2^{-n} = 1/2$. This also implies that, if $u \neq v$ are two distinct vectors, then

$$\text{Prob}[\langle r, u \rangle = \langle r, v \rangle] = \frac{1}{2}.$$

Random vectors A *random variable* is a variable defined as a function $X : \Omega \to \mathbb{R}$ of the domain of a probability space. For example, if X is a uniform integer chosen from $\{1, \ldots, n\}$, then $Y := 2X$ and $Z := $ "the number of prime divisors of X" are both random variables, and so is X itself. In what follows, $\mathrm{Prob}[X = s]$ denotes the probability of the event $X^{-1}(s) = \{x \in \Omega : X(x) = s\}$. One says in this case that X takes value $s \in \mathbb{R}$ with probability $\mathrm{Prob}[X = s]$. Two random variables X and Y ar *independent* if such are the events $X^{-1}(s)$ and $X^{-1}(t)$ for all s, t in the range of X. In this book we will only consider random variables whose range is finite. It is clear that events are a special type of random variables taking only two values 0 and 1. Namely, one can identify an event $A \subseteq \Omega$ with its *indicator random variable* X_A such that $X_A(x) = 1$ if and only if $x \in A$.

Expectation and variance One of the most basic probabilistic notions is the expected value of a random variable. This is defined for any real-valued random variable X, and intuitively, it is the value that we would expect to obtain if we repeated a random experiment several times and took the average of the outcomes of X. Namely, if X takes values $s_1, \ldots, s_n$, then the *mean* or *expectation* of X is defined as the weighted average of these values:

$$E[X] := \sum_{i=1}^{n} s_i \cdot \mathrm{Prob}[X = s_i] = \sum_{x \in \Omega} X(x) \cdot \mathrm{Prob}[x].$$

In particular, if X takes each value s_i with the same probability $1/n$, then the expectation of X is just the average value $(s_1 + \cdots + s_n)/n$ of these values.

One of the most important properties of expectation is its *linearity*: If $X_1, \ldots, X_n$ are random variables and $a_1, \ldots, a_n$ real numbers, then

$$E[a_1 X_1 + \cdots + a_n X_n] = a_1 E[X_1] + \cdots + a_n E[X_n].$$

The equality follows directly from the definition $E[X]$. The power of this principle comes from there being no restrictions on the X_i's.

The *variance* of a random variable X is defined by:

$$\mathrm{Var}[X] := E\left[(X - E[X])^2\right] = E\left[X^2\right] - E[X]^2,$$

where the second equality can be easily shown using the linearity of expectation.

Probabilistic proofs of existence Many extremal problems can be defined by a pair (M, f), where M is some finite set of objects and $f : M \to \mathbb{R}$ some function assigning each object $x \in M$ its "value". For example, M could be a set of graphs, satisfying some conditions, and $f(x)$ could be the maximum size of a clique in x. Given a threshold value t, the goal is to show that an object $x \in M$ with $f(x) \geq t$ *exists*. That is, we want to show that $\max_{x \in M} f(x) \geq t$.

A general framework to solve this task is to define an appropriate probability distribution $\mathrm{Pr} : M \to [0, 1]$ and to consider the resulting probability space. In this space the target function f becomes a random variable. One tries then to show that either $\mathrm{E}\,[f] \geq t$ or $\mathrm{Prob}[f(x) \geq t] > 0$ holds. If at least one of these inequalities holds, then the existence of $x \in M$ with $f(x) \geq t$ is already shown. Indeed, were $f(x) < t$ true for *all* $x \in M$, then we would have

$$\mathrm{Prob}[f(x) \geq t] = \mathrm{Prob}[\emptyset] = 0$$

and

$$\mathrm{E}\,[f] = \sum_i i \cdot \mathrm{Prob}[f = i] < \sum_i t \cdot \mathrm{Prob}[f = i] = t.$$

The property "$\mathrm{E}\,[f] \geq t$ implies $f(x) \geq t$ for at least one $x \in M$" is sometimes called the *pigeonhole principle* of expectation: a random variable cannot always be smaller (or always greater) than its expectation.

Large deviation inequalities As such, the expectation $\mathrm{E}\,[X]$ of a random variable X is just some number: the actual values of X may lie far away from this number. The large deviation inequalities estimate the *probability* with which this happens. Let X be a random variable with finite expectation $\mathrm{E}\,[X] = \mu$ and variance $\mathrm{Var}\,[X] = \sigma^2$. Let also $a > 0$ be an arbitrary real number.

Markov's Inequality. If $X \geq 0$ then $\mathrm{Prob}[X \geq a\mu] \leq 1/a$.

Proof.

$$\mu = \mathrm{E}\,[X] = \sum_i i \cdot \mathrm{Prob}[X = i] \geq \sum_{i \geq a} a \cdot \mathrm{Prob}[X = i] = a \cdot \mathrm{Prob}[X \geq a]. \qquad \square$$

Equivalent form of Markov's inequality is: $\mathrm{Prob}[X \geq a] \leq \mu/a$. Intuitively, when $a \leq \mu$ the inequality is trivial. For $a > \mu$, it means the larger a is relative to the mean, the harder it is to have $X \geq a$.

In the case when X is not necessarily non-negative random variable, we can applying Markov's inequality to non-negative random variable $Y := (X - \mu)^2$ and obtain:

Chebyshev's Inequality. For any X, $\mathrm{Prob}[|X - \mu| \geq a\sigma] \leq 1/a^2$.

Sums of independent variables In Markov's inequality, X can be an *arbitrary* non-negative random variable. In applications, however, X is often a sum of independent random variables. In these cases, Markov's inequality can be substantially sharpened. The main observation (due to Sergei Bernstein) is that, if X is a random variable and $t > 0$, then Markov's inequality yields

$$\mathrm{Prob}[X \geq a] = \mathrm{Prob}[e^{tX} \geq e^{ta}] \leq \mathrm{E}[e^{tX}] \cdot e^{-ta}.$$

There are many resulting inequalities known under a common name "Chernoff's inequalities". We mention just one of them.

Chernoff's Inequality. Let $X_1, \ldots, X_n$ be independent random variables taking their values in the interval $[0, 1]$. Let $X = X_1 + \cdots + X_n$ and $\mu = E[X]$. Then, for every real number $a > 0$, both $\mathrm{Prob}[X \geq \mu + a]$ and $\mathrm{Prob}[X \leq \mu - a]$ are at most $e^{-a^2/2n}$.

We finish with a list of some useful (in)equalities; all numbers here are assumed to be positive.

$$1 - x \leq e^{-x} \qquad\qquad\qquad\qquad\qquad x \in \mathbb{R}$$

$$1 - x \geq e^{-x-x^2/2} \qquad\qquad\qquad\qquad 0 < x < 1$$

$$\left(1 - \frac{1}{x}\right)^x \leq e^{-1} \leq \left(1 - \frac{1}{x+1}\right)^x \qquad x \geq 1$$

$$1 - \frac{1}{x} \leq \ln x \leq x - 1 \qquad\qquad\qquad x > 0$$

$$f\left(\frac{\sum a_i x_i}{\sum a_i}\right) \leq \frac{\sum a_i f(x_i)}{\sum a_i} \qquad\qquad \text{Jensen's inequality, } f \text{ convex}$$

$$f\left(\frac{1}{n}\sum_{i=1}^{n} x_i\right) \leq \frac{1}{n}\sum_{i=1}^{n} f(x_i) \qquad\qquad \text{Jensen, special case}$$

$$\left(\prod_{i=1}^{n} x_i\right)^{1/n} \leq \frac{1}{n}\sum_{i=1}^{n} x_i \qquad\qquad \text{geometric/arithmetic mean}$$

$$\sum_{i=1}^{n} i = \frac{n(n+1)}{2} \qquad\qquad\qquad\qquad \text{arithmetic series}$$

$$\sum_{i=0}^{n} x^i = \frac{1 - x^{n+1}}{1 - x} \qquad\qquad\qquad \text{geometric series, } x \neq 1$$

$$\sum_{i=1}^{n} \frac{1}{i} = \ln n + \gamma_n \qquad\qquad\qquad \text{harmonic series, } \tfrac{1}{2} < \gamma_n < \tfrac{2}{3}$$

$$\binom{n}{k} = \text{number of } k\text{-subsets of } \{1, \ldots, n\} \quad \text{definition of } \binom{n}{k}$$

$$\binom{n}{k} = \frac{n!}{k!(n-k)!} \qquad\qquad\qquad n! = n(n-1)(n-2)\cdots 2 \cdot 1$$

$$\binom{n}{k} = \binom{n}{n-k} \qquad\qquad\qquad\qquad \text{symmetry}$$

$$\binom{n}{k} = \frac{k}{n}\binom{n-1}{k-1}$$ recurrence

$$\binom{n}{k} = \binom{n-1}{k-1} + \binom{n-1}{k}$$ Pascal triangle

$$\binom{n}{k} \geq \left(\frac{n}{k}\right)^k$$ since $(n-1)/(k-1) \geq n/k$

$$\sum_{i=0}^{k}\binom{n}{i} \leq \left(\frac{en}{k}\right)^k$$

$$\sum_{i=0}^{k}\binom{n}{i} \leq 2^{-n \cdot H(k/n)}$$ $H(\alpha) = \log_2\left(\frac{1}{\alpha}\right)^{\alpha}\left(\frac{1}{1-\alpha}\right)^{1-\alpha}$

$$\binom{cn}{k} \leq c^k\binom{n}{k}$$ $0 \leq c \leq 1$ constant

$$n! = \left(\frac{n}{e}\right)^n\sqrt{2\pi n}\,e^{\alpha_n}, \quad \alpha_n \approx \frac{1}{12n}$$ Stirling's formula

$$\binom{n}{n/2} \sim \frac{2^n}{\sqrt{\pi n/2}}$$

$$\frac{1}{2\sqrt{n}} \leq \binom{2n}{n}2^{-2n} \leq \frac{1}{\sqrt{2n}}$$

$$\binom{2n}{n+1} = \frac{n}{n+1}\binom{2n}{n}$$

$$\binom{2n}{n} = \sum_{i=0}^{n}\binom{n}{i}^2$$

$$\binom{n}{k} = (1+o(1))\frac{n^k}{k!}e^{-\frac{k^2}{2n}-\frac{k^3}{6n^2}}$$ consequence of Stirling

$$\binom{n}{k} = c\sqrt{\frac{n}{k(n-k)}}2^{nH(k/n)}$$ $1/3 \leq c \leq 1/2$

$$\binom{n}{\alpha n} = \frac{1}{\Theta(\sqrt{n})}2^{nH(\alpha)}$$ for all $0 < \alpha < 1$

References

Acronyms of conferences and their proceedings:

FOCS	=	Ann. IEEE Symp. on Foundations of Computer Science
STOC	=	Ann. ACM Symp. on the Theory of Computing
CCC	=	Ann. IEEE Conf. on Computational Complexity
ECCC	=	Electronic Colloq. on Comput. Complexity
ICALP	=	Int. Colloq. on Automata, Languages and Programming
FCT	=	Int. Symp. on Fundamentals of Computation Theory
MFCS	=	Int. Symp. on Math. Foundations of Computer Science
LNCS	=	Springer Lect. Notes in Comput. Sci.

S. Aaronson and A. Wigderson (2008): Algebrization: A new barrier in complexity theory, in 40th STOC, 731–740.

L.M. Adleman (1978): Two theorems on random Polynomial time, in 19-th FOCS, 75–83.

A. Aho, J. Ullman and M. Yannakakis (1983): On notions of information transfer in VLSI circuits, in 15th STOC, 133–139.

M. Agrawal, N. Kayal, and N. Saxena (2004): PRIMES is in P, Annals of Math. 160(2), 781–793.

M. Ajtai (1983): Σ_1^1-formulae on finite structures, Ann. Pure and Appl. Logic 24, 1–48.

M. Ajtai (1999a): Determinism versus non-determinism for linear time RAMs with memory restrictions, in 31st STOC, 632-641. Journal version in: J. Comput. Syst. Sci. 65(1) (2002) 2–37.

M. Ajtai (1999b): A non-linear time lower bound for boolean branching programs, in 40th FOCS, 60–70. Journal version in: Theory of Computing 1(1) (2005) 149–176.

M. Ajtai and M. Ben-Or (1984): A theorem on probabilistic constant depth computations, in 16th STOC, 471–474.

M. Ajtai and Y. Gurevich (1987): Monotone versus positive, J. ACM 34, 1004–1015.

M. Ajtai, J. Komlós, and E. Szemerédi (1994): An $O(n \log n)$ sorting network, in 15th STOC, 1–9.

E. Alekhnovich, J. Johannsen, T. Pitassi, and A. Urquhart (2007): An exponential separation between regular and general resolution, Theory of Computing 3(1), 81–102.

E. Allender (1989): A note on the power of threshold circuits, in 30th FOCS, 580–584.

E. Allender and V. Gore (1994): A uniform circuit lower bound for the permanent, SIAM J. Comput. 23(5), 1026–1049.

E. Allender and M. Koucký (2010): Amplifying lower bounds by means of self-reducibility, J. ACM 57(3), Article Nr. 14.

N. Alon (1986): Covering graphs by the minimum number of equivalence relations, Combinatorica 6, 201–206.

S. Jukna, *Boolean Function Complexity*, Algorithms and Combinatorics 27,
DOI 10.1007/978-3-642-24508-4, © Springer-Verlag Berlin Heidelberg 2012

N. Alon and R. Boppana (1987): The monotone circuit complexity of boolean functions, Combinatorica 7(1), 1–22.

N. Alon and W. Maass (1988): Meanders and their applications in lower bounds arguments, J. Comput. Syst. Sci. 37(2), 118–129.

N. Alon and P. Pudlák (1994): Superconcentrators of depth 2 and 3; odd levels help (rarely), J. Comput. Syst. Sci. 48, 194–202.

N. Alon and P. Seymour (1989): A counterexample to the rank-coloring conjecture, J. Graph Theory 13, 523–525.

N. Alon and Văn H. Vũ (1997): Anti-hadamard matrices, coin weighing, threshold gates, and indecomposable hypergraphs, J. Combin. Theory Ser. A 79(1), 133–160.

N. Alon, P. Frankl, and V. Rödl (1985): Geometric realizations of set systems and probabilistic communication complexity, in 26th FOCS, 277–280.

N. Alon, M. Karchmer, and A. Wigderson (1990): Linear circuits over GF(2), SIAM J. Comput. 19(6), 1064–1067.

N. Alon, Y. Matias, and M. Szegedy (1999): The space complexity of approximating the frequency moments, J. Comput. Syst. Sci. 58 (1), 137–147.

K. Amano and A. Maruoka (2005): A super-polynomial lower bound for a circuit computing the clique function with at most $(1/6) \log \log n$ negation gates, SIAM J. Comput. 35(1), 201–216.

K. Amano and J. Tarui (2008): A well-mixed function with circuit complexity 5n+ o(n): Tightness of the Lachish–Raz-type bounds, in Springer LNCS, vol. 4978, 342–350.

A.E. Andreev (1985): On a method for obtaining lower bounds for the complexity of individual monotone functions, Soviet Math. Dokl. 31(3), 530–534.

A.E. Andreev (1986): On a family of boolean matrices, Moscow Univ. Math. Bull. 41, 79–82.

A.E. Andreev (1987a): On a method for obtaining more than quadratic effective lower bounds for the complexity of π-schemes, Moscow Univ. Math. Bull. . 42(1), 63–66.

A.E. Andreev (1987b): A method for obtaining efficient lower bounds for monotone complexity, Algebra and Logics 26(1), 1–18.

A.E. Andreev (1988): The complexity of realizing partial Boolean functions by schemes consisting of functional elements, Discrete Math. and Appl. 1(3), 251–261.

A.E. Andreev (1988b): On synthesis of networks of functional elements in complete monotone bases, Mat. Voprosy Kibernetiki 1, 114–139 (in Russian).

A.E. Andreev and S. Jukna (2008): Very large cliques are easy to detect, Discrete Math. 308(16), 3717–3721.

A.E. Andreev, A.E.F. Clementi, J. D. P. Rolim (1996): Hitting properties of hard Boolean operators and their consequences on BPP, ECCC Report Nr. 55.

S. Arora and B. Barak (2009): *Computational Complexity: A Modern Approach.* Cambridge University Press.

S. Arora, D. Steurer, and A. Wigderson (2009): Towards a study of low-complexity graphs, in 36th ICALP, Springer LNCS, vol. 5555, 119–131.

S. Arora, B. Bollobás, L. Lovász, and I. Tourlakis (2006): Proving integrality gaps without knowing the linear program, Theory of Computing 2(1), 19–51.

J. Aspnes, R. Beigel, M. Furst, and S. Rudich (1994): The expressive power of voting polynomials, Combinatorica 14(2), 135–148.

S. V. Augustinovich (1980): On one approach to proving lower bounds for boolean function, in Methods of Discrete Mathematics in the Theory of Boolean Functions and Circuits vol. 35, 3–8 (in Russian).

L. Babai, P. Frankl, and J. Simon (1986): Complexity classes in communication complexity theory, in 27th FOCS, 337–347.

L. Babai, P. Pudlák, V. Rödl, and E. Szemeredi (1990): Lower bounds to the complexity of symmetric boolean functions, Theor. Comput. Sci. 74(3), 313–323.

L. Babai, N. Nisan, and M. Szegedy (1992): Multiparty protocols, pseudorandom generators for logspace, and time-space trade-offs, J. Comput. Syst. Sci. 45, 204–232.

L. Babai, A. Gál, J. Kollár, L. Rónyai, T. Szabó, and A. Wigderson (1996): Extremal bipartite graphs and super-polynomial lower bounds for monotone span programs, in 28th STOC, 603–611.

L. Babai, T. Hayes, and P. Kimmel (1998): The cost of the missing bit: communication complexity with help, in 30th STOC, 673–682.

L. Babai, A. Gál, and A. Wigderson (1999): Superpolynomial lower bounds for monotone span programs, Combinatorica 19(3), 301–319.

L. Babai, A. Gál, P. G. Kimmel, and S. V. Lokam (2003): Communication complexity of simultaneous messages, SIAM J. Comput. 33(1), 137–166.

Z. Bar-Yossef, T. S. Jayram, R. Kumar, and D. Sivakumar (2004): An information statistics approach to data stream and communication complexity, J. Comput. Syst. Sci. 68, 702–732.

A. Barak and E. Shamir (1976): On the parallel evaluation of boolean expressions, SIAM J. Computers 5(4), 678–681.

B. Barak, G. Kindler, R. Shaltiel, B. Sudakov, and A. Wigderson (2010): Simulating independence: new constructions of condensers, Ramsey graphs, dispersers and extractors, J. ACM 57(4), 2010.

D. A. Mix Barrington (1989): Bounded-width polynomial-size branching programs recognize exactly those languages in NC^1, J. Comput. Syst. Sci. 38(1), 150–164.

L. A. Basalygo (1981): Asymptotically optimal switching circuits, Prob. Info. Trans. 17, 206–211.

R. Beals, T. Nishino, and K. Tanaka (1998): On the complexity of negation-limited Boolean networks, SIAM J. Comput. 27(5), 1334–1347.

R. Beals, H. Buhrman, R. Cleve, M. Mosca, and R. de Wolf (2001): Quantum lower bounds by polynomials, J. ACM 48(4), 778–797.

P. Beame (1994): A switching lemma primer, Tech. Rep. UW-CSE-95-07-01, Dept. of Comput. Sci. and Engineering, University of Washington, November 1994.

P. Beame (2000): Proof complexity, in Computational Complexity Theory, S. Rudich and A. Wigderson (eds.), IAS/Park City Math. Series (AMS), vol. 10, 199–246.

P. Beame and J. Lawry (1992): Randomized versus nondeterministic communication complexity, in 24th STOC, 188–199.

P. Beame and P. McKenzie (2011): A note on Nechiporuk's method for nondeterministic branching programs, Manuscript, August 2011.

P. Beame and T. Pitassi (1996): Simplified and improved resolution lower bounds, in 37th FOCS, 274–282.

P. Beame, T. S. Jayram,[1] and M. Saks (2001): Time-space tradeoffs for branching programs, J. Comput. Syst. Sci. 63(4), 542–572.

P. Beame, M. Saks, X. Sun, and E. Vee (2003): Time-space trade-off lower bounds for randomized computation of decision problems, J. ACM 50(2), 154–195.

P. Beame, T. Pitassi, N. Segerlind, and A. Wigderson (2006): A direct sum theorem for corruption and a lower bound for the multiparty communication complexity of set disjointness, Comput. Complexity 15(4), 391–432.

P. Beame, T. Pitassi, and N. Segerlind (2007): Lower bounds for Lovász–Schrijver systems and beyond follow from multiparty communication complexity, SIAM J. Comput. 37(3), 845–869.

R. Beigel (1995): The polynomial method in circuit complexity, in 8th CCC, 82–95.

R. Beigel and J. Tarui (1994): On ACC, Comput. Complexity 4, 350–366.

A. Beimel and B. Chor (1994): Universally ideal secret sharing schemes, IEEE Trans. on Info. Theory 40(3), 786–784.

A. Beimel and A. Gál (1999): On arithmetic branching programs, J. Comput. Syst. Sci. 59, 195–220.

A. Beimel, A. Gál, and M. Paterson, (1996): Lower bounds for monotone span programs, Comput. Complexity 6, 29–45.

---

[1] Formely Jayram S. Thathachar.

Ben-Or and N. Linial (1990), Collective coin flipping, in Randomness and Computation, Academic Press, 91–115.

E. Ben-Sasson and S. Kopparty (2009): Affine dispersers from subspace polynomials, 41st STOC, 65–74.

E. Ben-Sasson and A. Wigderson (2001): Short proofs are narrow - resolution made simple, J. ACM 48(2), 149–169.

E. Ben-Sasson and N. Zewi (2010): From affine to two-source extractors via approximate duality, ECCC Report Nr. 144.

C. Berg and S. Ulfberg (1999): Symmetric approximation arguments for monotone lower bounds without sunflowers, Comput. Complexity 8(1), 1–20.

S. J. Berkowitz (1982): On some relationships between monotone and non-monotone circuit complexity. Technical Report, University of Toronto.

S. L. Bezrukov (1994): Isoperimetric problems in discrete spaces, in Extremal Problems for Finite Sets, Bolyai Soc. Math. Stud., vol. 3, János Bolyai Math. Soc., 59–91.

O. Beyersdorff, N. Galesi, and M. Lauria (2010): A lower bound for the pigeonhole principle in tree-like Resolution by asymmetric Prover-Delayer games, Inf. Proc. Lett. 110(23), 1074–1077.

A. Blake, A. (1937): Canonical expressions in boolean algebra, PhD thesis, University of Chicago.

N. Blum (1984): A boolean function requiring $3n$ network size, Theor. Comput. Sci. 28, 337–345.

M. Blum and R. Impagliazzo (1987): Generic oracles and oracle classes, in 28th FOCS, 118–126.

M. Blum and S. Micali (1984): How to generate cryptographically strong sequences of pseudorandom bits, SIAM J. Comput. 13(4), 850–864.

B. Bollobás (1965): On generalized graphs, Acta Math. Acad. Sci. Hungar. 16, 447–452.

B. Bollobás (1978): *Extremal Graph Theory*. Academic Press, New York.

B. Bollobás and A. Thomason (1981): Graphs which contain all small graphs, European J. Combin. 2, 13–15.

J. A. Bondy (1972): Induced subsets, J. Combin. Theory Ser. B 12, 201–202.

M. L. Bonet and S. R. Buss (1994): Size-depth tradeoffs for Boolean formulae, Inf. Proc. Lett. 48, 151–155.

M. L. Bonet and N. Galesi (1999): A study of proof search algorithms for resolution and polynomial calculus, in 40th FOCS, 422–432.

M. Bonet, T. Pitassi, and R. Raz (1997): Lower bounds for cutting planes proofs with small coefficients, J. Symbolic Logic 62(3), 708–728.

M. L. Bonet, J. L. Esteban, N. Galesi, and J. Johannsen (2000): On the relative complexity of resolution refinements and cutting planes proof systems, SIAM J. Comput. 30(5), 1462–1484.

R. B. Boppana (1986): Threshold functions and bounded depth monotone circuits, J. Comput. Syst. Sci. 32(2) (1986), 222–229.

R. B. Boppana (1985): Amplification of probabilistic Boolean formulas, in 26th FOCS, 20–29.

R. B. Boppana (1994): The decision-tree complexity of element distinctness, Inf. Proc. Lett. 52(6), 329–331.

R. B. Boppana (1997): The average sensitivity of bounded-depth circuits, Inf. Proc. Lett. 63(5), 257–261.

R. B. Boppana and M. Sipser (1990): The complexity of finite functions, in Handbook of Theoretical Computer Science, Volume A: Algorithms and Complexity (A), 757–804.

J. L. Bordewijk (1956): Inter-reciprocity applied to electrical networks, Appl. Sci. Res. B: Electrophysics, Acoustics, Optics, Mathematical Methods, 1–74.

A. Borodin, J. von zu Gathen, and J. Hopcroft (1982): Fast parallel matrix and GCD computations, in 23rd FOCS, 65–71.

A. Borodin, A. Razborov, and R. Smolensky (1993): On lower bounds for read-k times branching programs, Comput. Complexity 3, 1–18.

R. P. Brent, D. J. Kuck, and K. Maruyama (1973): The parallel evaluation of arithmetic expressions without divisions, IEEE Trans. Computers, vol. C-22, 532–534.

G. S. Brodal and T. Husfeld (1996): A communication complexity proof that symmetric functions have logarithmic depth, BRICS Tech. Rep. RS-96-1, 3 pp.

W. G. Brown (1966): On graphs that do not contain a Thompson graph, Can. Math. Bull. 9, 281–285.

N. H. Bshouty, R. Cleve, and W. Eberly (1991): Size-depth tradeoffs for algebraic formulas, 32nd FOCS, 334–341. Jornal version in SIAM J. Comput. 24(4), 682–705.

S. Bublitz (1986): Decomposition of graphs and monotone size of homogeneous functions, Acta Inform. 23, 689–696.

H. Buhrman and R. de Wolf (2002): Complexity measures and decision tree complexity: a survey, Theor. Comput. Sci. 288(1), 21–43.

J. Buresh-Oppenheim, N. Galesi, S. Hoory, A. Magen, and T. Pitassi (2006): Rank bounds and integrality gaps for cutting planes procedures, Theory of Computing 2(1), 65–90.

S. Buss (1987): Polynomial size proofs of the propositional pigeonhole principle, J. Symbolic Logic 52, 916–927.

S. R. Buss and P. Clote (1996): Cutting planes, connectivity and threshold logic, Archive for Mathematical Logic 35, 33–63.

S. Buss and T. Pitassi (1998): Resolutions and the weak pigeonhole principle, in 11th Int. Workshop on Computer Science Logic, CSL'97, Springer LNCS, vol. 1414, 149–156.

R. Canetti and O. Goldreich (1993): Bounds on tradeoffs between randomness and communication complexity, Comput. Complexity 3, 141–167.

C. Cardot (1952): Quelques résultats sur l'application de l'algébre de Boole á la synthése á relais, Ann. Telecomm. 7(2), 75–84.

A. Chakrabarti, Y. Shi, A. Wirth, and A. C. Yao (2001): Informational complexity and the direct sum problem for simultaneous message complexity, in 42nd FOCS, 270–278.

A. K. Chandra, M. Furst, and R. J. Lipton (1983): Multi-party protocols, in 15th STOC, 94–99.

A. V. Chashkin (1994): On the complexity of Boolean matrices, graphs and their corresponding Boolean functions, Discrete Math. and Appl. 4(3), 229–257.

A. V. Chashkin (1997): On the average time for the computation of the values of Boolean functions, Diskretn. Anal. Issled. Oper. 4(1) 60–78 (in Russian).

A. V. Chashkin (2000): On the mean time for computing the values of elementary Boolean functions, Discrete Math. and Appl. 11(1), 71–81.

A. V. Chashkin (2004): On the average monotone complexity of Boolean functions, Diskretn. Anal. Issled. Oper. 11(4) 68–80 (in Russian).

A. V. Chashkin (2006): On the realization of partial boolean functions, in Proc. of 7-th Int. Conf. on Discrete Models in the Theory of Control Systems, 390–404 (in Russian).

A. Chakrabarti, S. Khot, X. Sun (2003): Near-optimal lower bounds on the multi-party communication complexity of set disjointness, in 18-th CCC, 107–117.

A. Chattopadhyay and A. Ada (2008): Multiparty communication complexity of disjointness, ECCC Report Nr. 2.

A. Chattopadhyay, A. Krebs, M. Koucký, M. Szegedy, P. Tesson, D.Thérien (2007): Languages with bounded multiparty communication complexity, in Springer LNCS, vol. 4393, 500–511.

S. Chaudhuri and J. Radhakrishnan (1996): Deterministic restrictions in circuit complexity, in 28th STOC, 30–36.

D. Y. Cherukhin (2005): The lower estimate of complexity in the class of schemes of depth 2 without restrictions on a basis, Moscow Univ. Math. Bull. 60(4), 42–44.

D. Y. Cherukhin (2008): Lower bounds for depth-2 and depth-3 boolean circuits with arbitrary gates, in Springer LNCS, vol. 5010, 122–133.

D. Y. Cherukhin (2008b): Lower bounds for boolean circuits with finite depth and arbitrary gates, ECCC Report Nr. 32.

T. Y. Chow (2011): Almost-natural proofs, J. Comput. Syst. Sci. 77(4), 728–737.

F. R. K. Chung (1990): Quasi-random classes of hypergraphs, Rand. Struct. Algorithms 1, 363–382.

F. R. K. Chung and P. Tetali (1993): Communication complexity and quasi randomness, SIAM J. Discrete Math. 6, 110–123.

F. R. K. Chung, Z. Füredi, R. L. Graham, and P. D. Seymour (1988): On induced subgraphs of the cube, J. Combin. Theory Ser. A 49(1), 180–187.

V. Chvátal (1973): Edmonds polytopes and a hierarchy of combinatorial problems, Discrete Math. 4, 305–337.

V. Chvátal, W. Cook, and M. Hartmann (1989): On cutting-plane proofs in combinatorial optimization, Lin. Algebra and Appl. 114/115, 455–499.

S. Cook and R. Reckhow (1979): The relative efficiency of propositional proof systems, J. Symolic Logic 44, 36–50.

W. Cook and S. Dash (2001): On the matrix-cut rank of polyhedra, Math. of Oper. Res. 18, 25–38.

S. Cook and C. Dwork (1982): Bounds on the time for parallel RAM's to compute simple functions, in 14th STOC, 231–233.

W. Cook, C. R. Coullard, and Gy. Túran (1987): On the complexity of cutting plane proofs, Discrete Appl. Math. 18, 25–38.

W. Craig (1957): Three uses of the Herbrand–Gentzen theorem in relating model theory and proof theory, J. Symbolic Logic 44, 36–50.

S. Dantchev and S. Riis (2001): Tree resolution proofs of the weak pigeon-hole principle, in 16th CCC, 69–75.

S. Dash (2005): Exponential lower bounds on the length of some classes of branch-and-cut proofs, Math. of Operations Research 30(3), 678–700.

S. Dash (2010): On the complexity of cutting plane proofs using split cuts, Oper. Res. Lett. 38(2), 109–114.

M. Davis and H. Putnam (1960): A computing procedure for quantification theory, J. ACM 7(3), 210–215.

J. A. Dias de Silva and Y. O. Hamidoune (1994): Cyclic spaces fro Grassmann derivatives and additive theory, Bull. London Math. Soc. 26, 140–146.

R. De Wolf (2006): Lower bounds on matrix rigidity via a quantum argument, in 33rd ICALP, Springer LNCS, vol. 4051, 62–71.

E. Demenkov and A. S. Kulikov (2011): An elementary proof of 3n-o(n) lower bound on the circuit complexity of affine dispersers, ECCC Report Nr. 26.

E. Demenkov, A. Kojevnikov, A. S. Kulikov, and G. Yaroslavtsev (2010): New upper bounds on the Boolean circuit complexity of symmetric functions, Inf. Proc. Lett. 110(7), 264–267.

D. Dolev, C. Dwork, N. Pippenger, and A. Wigderson (1983): Superconcentrators, generalizer and generalized connectors with limited depth, in 15th STOC, 42–51.

A. Drucker (2011): Limitations of lower-bound methods for the wire complexity of boolean operators, ECCC Report Nr. 125.

M. Dubiner and U. Zwick (1992): Amplification and percolation, in 33rd FOCS, 258–267.

P. E. Dunne (1984): *Techniques for the analysis of monotone Boolean networks*. Ph.D. Thesis, Theory of Computation Report 69, Dept. of Computer Science, Univ. of Warwick.

P. E. Dunne (1988): *The complexity of Boolean networks*. Academic Press Professional, Inc., San Diego, CA.

P. Dymond and M. Tompa (1985): Speedups of deterministic machines by synchronous parallel machines, J. Comput. Syst. Sci. 30(2), 149–161.

A. Ehrenfeucht and D. Haussler (1989): Learning decision trees from random examples, Information and Computation 82, 231–246.

H. Ehlich and K. Zeller (1964): Schwankung von Polynomen zwischen Gitterpunkten, Mathematische Zeitschrift 86, 41–44.

F. Eisenbrand and A. S. Schulz (2003): Bounds on the Chvátal rank of polytopes in the 0/1-cube, Combinatorica 23(2), 245–261.

P. Erdős (1947): Some remarks on the theory of graphs, Bull. Amer. Math. Soc. 53, 292–294.

P. Erdős (1961): Graph theory and probability II, Canad. J. Math 12, 346–352.

P. Erdős (1962): On circuits and subgraphs of chromatic graphs, Mathematika 9, 170–175.

P. Erdős (1967): Some remarks on chromatic graphs, Colloq. Mathematicum 16, 253–256.

P. Erdős and R. Rado (1960): Intersection theorems for systems of sets, J. London Math. Soc. 35, 85–90.

P. Erdős, A. Hajnal, and J. Moon (1964): A problem in graph theory, Amer. Math. Monthly 71, 1107–1110.

P. Erdős, R. L. Graham, and E. Szemerédi (1976): On sparse graphs with dense long paths, in Computers and Math. with Appl, 365–369, Pergamon, Oxford.

C. M. Fiduccia (1973): *On the algebraic complexity of matrix multiplication*. PhD Thesis, Brown University.

M. J. Fischer (1974): The complexity of negation-limited networks–a brief survey, in Springer LNCS, vol. 33, 71–82.

M. J. Fischer (1996): Lectures on network complexity. Tech. Rep. TR-1104, Department of Computer Science, Yale University.

M. J. Fischer, A. R. Meyer, and M. S. Paterson (1982): $\Omega(n \log n)$ lower bounds on length of Boolean formulas, SIAM J. Comput. 11, 416–427.

R. Fleischer, H. Jung, and K. Mehlhorn (1995): A communication-randomness tradeoff for two-processor systems, Information and Computation 116, 155–161.

J. Forster (2002): A linear lower bound on the unbounded eror probabilistic communication complexity, J. Comput. Syst. Sci. 65(4), 612–625.

J. Forster, M. Krause, S. V. Lokam, R. Mubarakzjanov, N. Schmitt, and H.-U. Simon (2001): Relations between communication complexity, linear arrangements, and computational complexity, in Springer LNCS, vol. 2245, 171–182.

P. Frankl and R. M. Wilson (1981): Intersection theorems with geometric consequences, Combinatorica 1(4):357–368.

E. Friedgut (1998): Boolean functions with low average sensitivity depend on few coordinates, Combinatorica 18(1), 27–35.

J. Friedman (1986): Constructing $O(n \log n)$ size monotone formulae for the k-th elementary symmetric polynomial of n Boolean variables, SIAM J. Comput. 15, 641–654.

J. Friedman (1993): A note on matrix rigidity, Combinatorica 13(2), 235–239.

Xudong Fu (1998): Lower bounds on sizes of cutting planes proofs for modular coloring principles, in DIMACS Series in Discrete Math. and Theor. Comput. Sci., vol. 39, 135–148.

M. Furst, J. Saxe, and M. Sipser (1984): Parity, circuits and the polynomial time hierarchy, Math. Syst. Theory 17, 13–27.

A. Gál (2001): A characterization of span program size and improved lower bounds for monotone span programs, Comput. Complexity 10(4), 277–296.

A. Gál and J. T. Jang (2011): The size and depth of layered Boolean circuits, Inf. Proc. Lett. 111(5), 213–217.

A. Gál and P. Pudlák (2003): A note on monotone complexity and the rank of matrices, Inf. Proc. Lett. 87(6), 321–326.

A. Gál, K. A. Hansen, M. Koucký, P. Pudlák, and E. Viola (2011): Computing error correcting codes in bounded depth. Manuskript.

S. B. Gashkov (1978): The depth of Boolean functions, Problemy Kibernetiki 34, 265–268 (in Russian).

S. B. Gashkov (1987): On one method of obtaining lower bounds on the monotone complexity of computing polynomials, Moscow Univ. Math. Bull. 5, 7–13.

S. B. Gashkov (2007): Remark on minimization of depth of Boolean circuits, Moscow Univ. Math. Bull. 62(3), 87–89.

S. B. Gashkov and I. S. Sergeev (2010): On the complexity of linear Boolean operators with thin matrixes, J. Appl. Industrial Math. 5(2) (2011), 202–211.

S. B. Gashkov and I. S. Sergeev (2011): On one method of obtaining bounds on the complexity of arithmetical circuits computing real polynomials, Matematicheskii Sbornik (in Russian, submitted).

M. Goldmann and J. Håstad (1992): A simple lower bound for monotone clique using a communication game, Inf. Proc. Lett. 41(4), 221–226.

M. Goldmann, J. Håstad, and A. A. Razborov (1992): Majority gates vs. general weighted threshold gates, in 7th CCC, 2–13.

O. Goldreich (2001): *The foundations of cryptography*. Volume 1. Cambridge University Press

O. Goldreich (2008): *Computational Complexity: A Conceptual Perspective*. Cambridge University Press.

O. Goldreich, S. Goldwasser, and S. Micali (1986): How to construct random functions, J. ACM 33(4), 792–807.

S. Goldwasser and S. Micali (1984): Probabilistic encryption, J. Comput. Syst. Sci. 28(2), 270–299.

R. E. Gomory (1963): An algorithm for integer solutions of linear programs, in Recent Advances in Mathematical Programming, McGraw-Hill, 269–302.

C. Gotsman and N. Linial (1992): The equivalence of two problems on the cube, J. Combin. Theory Ser. A 61(1), 142–146.

R. L. Graham and J. Spencer (1971): A constructive solution to a tournament problem, Canad. Math. Bull. 14, 45–48.

F. Green, J. Köbler, K. W. Regan, T. Schwentick, and J. Toran (1995): The power of the middle bit of a # P function, J. Comput. Syst. Sci. 50(3) 456–467.

M. Grigni and M. Sipser (1995): Monotone separation of logarithmic space from logarithmic depth, J. Comput. Syst. Sci. 50(3), 433–437.

M. I. Grinchuk (1987): On the complexity of elementary periodical functions realized by switching circuites, in Springer LNCS, vol. 278, 163–166.

M. I. Grinchuk (1988): On the complexity of computing Boolean cyclic matrices by rectifier networks, Izvestiya Vysshikh Uchebnykh Zavedenii, Matematika 7, 39–44 (in Russian).

M. I. Grinchuk (1989): On the switching network size of symmetric Boolean functions. PhD thesis, Moscow State University, Moscow, 1989 (in Russian).

M. I. Grinchuk (1996): On the complexity of realization of boolean functions in three classes of circuits in the basis consisting of all symmetric functions, Diskretn. Anal. Issled. Oper. 3(1), 3–8 (in Russian).

M. I. Grinchuk and I. S. Sergeev (2011):Thin circulant matrices and lower bounds on the complexity of some Boolean operators,J. Appl. Industrial Math. (to appear).

V. Grolmusz (1994): The BNS lower bound for multi-party protocols is nearly optimal, Information and Computation 112(1), 51–54.

V. Grolmusz (1999): Harmonic analysis, real approximation, and the communication complexity of boolean functions, Algorithmica 23, 341–353.

V. Grolmusz and G. Tardos (2003): A note on non-deterministic communication complexity with few witnesses, Theory of Comput. Syst. 36(4), 387–391.

A. Gronemeier (2009): Asymptotically optimal lower bounds on the NIH-multi-party information complexity of the AND-function and disjointness, in Proc. of 26-th Int. Symp. on Theoretical Aspects of Comput. Sci. (STACS), 505–516.

M. Grötschel, L. Lovász, and A. Schrijver (1981): The ellipsoid method and its consequences in combinatorial optimization,Combinatorica 1, 169–197.

H. D. Gröger and G. Turán (1991), On linear decision trees computing boolean functions, in 18th ICALP, Springer LNCS, vol. 510, 707–718.

H. D. Gröger and G. Turán (1993): A Linear lower bound for the size of threshold circuits, Bull. of the EATCS 50, 220–221.

A. Hajnal (1965): A theorem on k-saturated graphs, Canad. Math. J. 17, 720–724.

A. Hajnal, W. Maass, P. Pudlák, M. Szegedy, and G. Turán (1993): Threshold circuits of bounded depth, J. Comput. Syst. Sci. 46(2), 129–154.

A. Haken (1985): The intractability of resolution, Theor. Comput. Sci. 39, 297–308.

A. Haken (1995): Counting bottlenecks to show monotone P≠NP, in 36th FOCS, 36–40.

A. Haken, and A. Cook, S. (1999): An exponential lower bound for the size of monotone real circuits, J. Comput. Syst. Sci. 58(2), 326–225.

M. M. Halldórsson, J. Radkhakrishnan, and K. V. Subrahmanyam (1993): Directed vs. undirected monotone contact networks for threshold functions, in 34th FOCS, 604–613.

G. Hansel (1964): Nombre minimal de contacts de fermeture necessaires pour realiser une function booleenne symetrique de n variables, C. R. Acad. Sci. 258(25), 6037–6040 (in French).

G. Hansel (1966): Résultats concernant le nombre minimal de contacts nécessaire pour réaliser certaines fonctions booléennes symétriques, C. R. Acad. Sci. Paris. Sér. A-B 262, A679–A681 (in French).

J. Hartmanis and L. A. Hemachandra (1991): One-way functions, robustness and non-isomorphism of NP-complete classes, Theor. Comput. Sci. 81(1), 155–163.

G. H. Hardy, J. E. Littlewood, and G. Polya (1952): *Inequalities*. Cambridge University Press, 1952.

D. Harnik and R. Raz (2000): Higher lower bounds on monotone size, in 32nd STOC, 378–387.

J. Håstad (1986): *Computational Limitations for Small Depth Circuits*. MIT Press.

J. Håstad (1989): Almost optimal lower bounds for small depth circuits, in Advances in Computing Research, vol. 5, 143–170.

J. Håstad (1994): On the size of weights for threshold gates, SIAM J. Discrete Math. 7(3), 484–492.

J. Håstad (1998): The shrinkage exponent is 2, SIAM J. Comput. 27, 48–64.

J. Håstad and M. Goldmann (1991): On the power of small-depth threshold circuits, Comput. Complexity 1(2), 113–129.

J. Håstad and A. Wigderson (2007): The randomized communication complexity of set disjointness, Theory of Computing 3(1), 211–219.

J. Håstad, S. Jukna, and P. Pudlak (1995): Top-down lower bounds for depth-three circuits, Comput. Complexity 5, 99–112.

J. Håstad, R. Impagliazzo, L. Levin, and M. Luby (1999): A pseudorandom generator from any one-way function. SIAM J. Comput. 28(4), 1364–1396.

P. Hatami, R. Kulkarni, and D. Pankratov (2011): Variations on the sensitivity conjecture, Theory of Computing, ToC Library Graduate Surveys 4 (2011), 27 pages.

T. P. Hayes (2001): Separating the k-party communication hierarchy: an application of the Zarankiewicz problem. Manuscript, November 2001. Available at http://www.cs.unm.edu/ hayes/papers/.

E. A. Hirsch (2000): SAT local search algorithms: worst-case study, J. Autom. Reasoning 24(1/2), 127–143.

L. Hodes and E. Specker (1968): Lengths of formulas and elimination of quantifiers, in Contibutions to Math. Logic (Noth-Holland, Amsterdam), 175–188.

T. Holenstein (2006): Pseudorandom generators from one-way functions: a simple construction for any hardness, in Springer LNCS, vol. 3876, , 443–461.

H. J. Hoover, M. M. Klawe, and N. Pippenger (1984): Bounding fan-out in logical networks, J. ACM 31(1), 13–18.

J. E. Hopcroft and R. M. Karp (1973): An $n^{5/2}$ algorithm for maximum matching in bipartite graphs, SIAM J. Comput. 2, 225–231.

P. Hrubes, S. Jukna, A. Kulikov, and P. Pudlák (2010): On convex complexity mesures, Theor. Comput. Sci. 411, 1842–1854.

H. Huang and B. Sudakov (2011): A counterexample to the Alon-Saks-Seymour conjecture and related problems, Combinatorica (to appear).

N. Immerman (1988): Nondeterministic space is closed under complementation, in 3rd CCC, 112–115.

R. Impagliazzo and N. Nisan (1993): The effect of random restrictions on formula size, Rand. Struct. Algorithms 4(2), 121–134.

R. Impagliazzo, T. Pitassi, and A. Urquhart (1994): Upper and lower bounds for tree-like cutting planes proofs, in Proc. of 9th Ann. IEEE Symp. on Logic in Computer Science, 220–228.

R. Impagliazzo, R. Paturi, and M. E. Saks (1997): Size-depth tradeoffs for threshold circuits, SIAM J. Comput. 26(3), 693–707.

R. W. Irving (1978): A bipartite Ramsey problem and Zarankiewicz numbers, Glasgow Math. J. 19, 13–26.

K. Iwama and H. Morizumi (2002): An explicit lower bound of $5n - o(n)$ for boolean circuits, in Springer LNCS, vol. 2420, 353–364.

J. W. Jordan and R. Livné (1997): Ramanujan local systems on graphs, Topology 36(5), 1007–1024.

S. Jukna (1995): A note on read-k times branching programs, RAIRO Theoret. Informatics and Appl. 29(1), 75–83.

S. Jukna (1999): Combinatorics of monotone computations, Combinatorica, 9(1) (1999), 1–21.

S. Jukna (2004): On the minimum number of negations leading to super-polynomial savings, Inf. Proc. Lett. 89(2), 71–74.

S. Jukna (2005): On the P versus NP intersected with co-NP question in communication complexity, Inf. Proc. Lett. 96(6), 202–206.

S. Jukna (2006): On graph complexity, Combinatorics, Probability and Computing 15, 855–876.

S. Jukna (2008): Expanders and time-restricted branching programs, Theoret. Comput. Sci. 409 (3), 471–476.

S. Jukna (2008b): Entropy of operators or why matrix multiplication is hard for depth-two circuits, Theory of Comput. Syst. 46(2) (2010), 301–310.

S. Jukna (2009b): A nondeterministic space-time tradeoff for linear codes, Inf. Proc. Lett. 109(5), 286–289.

S. Jukna (2010): Representing (0,1)-matrices by depth-2 circuits with arbitrary gates, Discrete Math. 310, 184–187.

S. Jukna and A. Razborov (1998): Neither reading few bits twice nor reading illegally helps much, Discrete Appl. Math. 85:3, 223–238.

S. Jukna and A. S. Kulikov (2009): On covering graphs by complete bipartite subgraphs, Discrete Math. 309(10), 3399–3403.

S. Jukna and G. Schnitger (2011): Min-rank conjecture for log-depth circuits, J. Comput. Syst. Sci. 77(6), 1023–1038.

S. Jukna, A. Razborov, P. Savický, and I. Wegener (1999): On P versus NP∩co-NP for decision trees and read-once branching programs, Comput. Complexity 8(4), 357–370.

J. Kahn, G. Kalai, and N. Linial (1988): The influence of variables on boolean functions, in 28th FOCS, 68–80.

G. Kalai and S. Safra (2006): Threshold phenomena and influence: Perspectives from mathematics, computer science, and economics, in Computational Complexity and Statistical Physics, Oxford Univ. Press, 25–60.

B. Kalyanasundaram and G. Schnitger (1992): The probabilistic communication complexity of set intersection, SIAM J. Discrete Math. 5(4), 545–557.

R. Kannan (1981): A circuit size lower bound, in 22th FOCS, 304–309.

M. Karchmer (1993): On proving lower bounds for circuit size, in Proc. 8th Ann. IEEE Symp. on Structure in Complexity Theory, 112–118.

M. Karchmer and A. Wigderson (1990): Monotone circuits for connectivity require super-logarithmic depth, SIAM J. Discrete Math. 3,255–265.

M. Karchmer and A. Wigderson (1993a): On span programs, in Proc. of 8th Ann. IEEE Conf. on Structure in Complexity Theory, 102–111.

M. Karchmer and A. Wigderson (1993b): Characterizing non-deterministic circuit size, in 25th STOC, 532–540.

M. Karchmer, E. Kushilevitz, and N. Nisan (1995): Fractional covers and communication complexity, SIAM J. Discrete Math. 8(1), 76–92.

M. Karchmer, I. Newman, M. Saks, and A. Wigderson (1994): Non-deterministic communication complexity with few witnesses, J. Comput. Syst. Sci. 49(2), 247–257.

R. M. Karp and R. J. Lipton (1980): Some connections between nonuniform and uniform complexity classes, in 12th STOC, 302–309.

J. Katz and Y. Lindell (2007): Introduction to modern cryptography: principles and protocols. Chapman & Hall.

V. M. Khrapchenko (1971): A method of obtaining lower bounds for the complexity of π-schemes, Math. Notes Acad. of Sci. USSR 10 (1972) 474–479.

V. M. Khrapchenko (1972): The complexity of the realization of symmetric functions by formulae, Math. Notes Acad. of Sci. USSR 11, 70–76.

V. M. Khrapchenko (1976): On the complexity of realization of symmetric boolean functions by formulas over finite bases, in Problemy Kibernetiki 31, 231–234 (in Russian).

V. M. Khrapchenko (1978): On a relation between the complexity and the depth, in Metody Diskretnogo Analiza in Synthezis of Control Systems, vol. 32, 76–94.

L. S. Khasin (1969): Complexity bounds for the realization of monotonic symmetrical functions by means of formulas in the basis $\vee$, &, $\neg$, Soviet Phys. Dokl. 14 (1969), 1149–1151.

M. Klawe, W. J. Paul, N. Pippenger, and M. Yannakakis (1984): On monotone formulae with restricted depth,in 16th STOC, 480–487.

D. J. Kleitman (1966): Families of non-disjoint subsets, J. Combin. Theory Ser. A 1, 153–155.

D. J. Kleitman and G. Markowsky (1975): On Dedekind's problem:the number of isotone Boolean functions II, Trans. of Amer. Math. Soc. 213, 373–390.

B. M. Kloss and V. A. Malyshev (1965): Complexity bounds for some classes of functions, Vestnik Moscow Univ., Mathematics4 (1965), 44–51 (in Russian).

M. Kochol (1989): Efficient monotone circuits for threshold functions, Inf. Proc. Lett. 32(3), 121–122.

J. Kollár, L. Rónyai, and T. Szabó (1996): Norm-graphs and bipartite Turán numbers, Combinatorica 16(3), 399–406.

A. D. Korshunov (1977): Solution of Dedekind's problem on the number of monotonic Boolean functions, Soviet Math. Dokl. 18 (1977), 442–445.

A. D. Korshunov (1981): The number of monotone Boolean functions, Problemy Kibernetiki 38, 5–108 (in Russian).

A. D. Korshunov (2003): Monotone Boolean functions, Russian Mathematical Surveys 58(5), 929–1001.

E. Koutsoupias (1993): Improvements on Khrapchenko's theorem, Theor. Comput. Sci. 116(2), 399–403.

J. Körner (1973): Coding of an information source having ambiguous alphabet and the entropy of graphs, in Trans. 6-th Prague Conf. on Information Theory, Academia, 441-425.

L. G. Kraft (1949): *A device for quantizing, grouping, and coding amplitude modulated pulses.* Cambridge, MA: MS Thesis, Electr. Eng. Dept, MIT.

J. Krajíček(1997): Interpolation theorems, lower bounds for proof systems, and independence results for bounded arithmetic, J. Symbolic Logic 62(2), 457–486.

E. G. Krasulina (1987): On the realization of monotone symmetric Boolean functions by contact networks, Vestnik Moskov. Univ. Ser. XV Vychisl. Mat. Kibernet. 2, 69–71 (in Russian).

E. G. Krasulina (1988): Complexity of the realization of monotone symmetric Boolean functions by switching circuits, Mat. Voprosy Kibernetiki 1, 140–167 (in Russian).

M. Krause (1996): Geometric arguments yield better bounds for threshold circuits and distributed computing, Theor. Comput. Sci. 156(1-2), 99–117.

M. Krause and S. Waack (1995): Variation ranks of communication matrices and lower bounds for depth-two circuits having nearly symmetric gates with unbounded fan-in, Math. Syst. Theory 28(6), 553–564.

M. Krause and P. Pudlák (1997): On the computational power of depth-2 circuits with threshold and modulo gates, Theor. Comput. Sci. 174(1–2), 137–156.

R. E. Krichevskii (1964): Complexity of contact circuits realizing a function of logical algebra, Soviet Physiscs Doklady 8, 770–772.

R. E. Krichevskii (1965): Minimal network of closing contacts for a function of the algebra of logic, Diskret. Analiz (Akad. Nauk SSSR Sib. Otd. Inst. Mat., Novosibirsk) Nr. 5, 89–92 (in Russian).

M. P. Krieger (2007): On the incompressibility of monotone DNFs, Theory of Comput. Syst. 41(2), 211–231.

E. Kushilevitz and Y. Mansour (1991): Learning decision trees using the Fourier spectrum, in 23rd STOC, 455–464.

E. Kushilevitz and N. Nisan (1997): *Communication complexity.* Cambridge University Press.

E. Kushilevitz and E. Weinreb (2009a): On the complexity of communication complexity, in 41th STOC, 465-474.

E. Kushilevitz and E. Weinreb (2009b): The communication complexity of set-disjointness with small sets and 0-1 intersection, in 50th FOCS, 63–72.

E. Kushilevitz, N. Linial, and R. Ostrovsky (1999): The linear-array conjecture in communication complexity is false, Combinatorica 19, 241–254.

O. Lachish and R. Raz (2001): Explicit lower bound of $4.5n - o(n)$ for boolean circuits, in 33th STOC, 399–408.

S. Laplante, T. Lee, and M. Szegedy (2006): The quantum adversary method and formula size lower bounds, Comput. Complexity 15(2), 163–196.

T. Lee and A. Shraibman (2007): Lower bounds in communication complexity, Foundations and Trends in Theoret. Comput. Sci. 3(4), 263–399.

T. Lee and A. Shraibman (2009): Disjointness is hard in the multiparty number-on-the-forehead model, Comput. Complexity 18(2), 309–336.

T. Lengauer (1990): VLSI Theory, in Handbook of Theoretical Computer Science, Vol. A, 835–868.

M. Li and P. Vitányi (2008): *An Introduction to Kolmogorov Complexity and Its Applications*. 3rd Edition,Springer Verlag.

R. Lidl and H. Niederreiter (1986): *Introduction to Finite Fields and their Applications*. Cambridge University Press.

N. Linial, Y. Mansour, and N. Nisan (1993): Constant depth circuits, Fourier transforms and learnability, J. ACM 40, 607–620.

R. J. Lipton (2010): *The P=NP Question and Gödel's Lost Letter*. Springer-Verlag.

R. J. Lipton and R. Sedgewick (1981): Lower bounds for VLSI, in 13th STOC, 300–307.

S. V. Lokam (2003): Graph complexity and slice functions, Theory of Comput. Syst. 36(1), 71–88.

S. V. Lokam (2009): Complexity lower bounds using linear algebra, Foundations and Trends in Theoretical Computer Science 4(1–2), 1–155.

L. Lovász (1975): On the ratio of optimal integral and fractional covers, Discrete Math. 13(4), 383–390.

L. Lovász (1977): Flats in matroids and geometric graphs, in Combinatorial Surveys, P. J. Cameron (ed.), Academic Press, 45–86.

L. Lovász (1979a): On the Shannon capacity of a graph, IEEE Trans. on Information Theory 25, 1–7.

L. Lovász (1979b): Determmants, matchings and random algorithms, in 7th FCT, L. Budach (ed.), Akademie-Verlag, Berlin, 565–574.

L. Lovász and M. Saks (1988): Lattices, Möbius functions and communication complexity, in 29th FOCS, 81–90.

L. Lovász and M. E. Saks (1993): Communication complexity and combinatorial lattice theory, J. Comput. Syst. Sci. 47(2), 322–349.

L. Lovász, D. B. Shmoys, and E. Tardos (1995): Combinatorics in computer science, in Handbook of Combinatorics, R. Graham, M. Grötschel, and L. Lovász (eds.), Elsevier Science, vol. 2, 2003–2038.

L. Lovász, M. Naor, I. Newman, and A. Wigderson (1995): Search problems in the decision tree model, SIAM J. Discrete Math. 8(1), 119–132.

S. A. Lozhkin (1976): Asymptotic behavior of Shannon functions for the delays of schemes of functional elements, Math. Notes Acad. of Sci. USSR 19(6), 548–555.

S. A. Lozhkin (1981): On the relation between the depth and the size of equivalent formulas for monotone functions of logic algebra, Problemy Kibernetiki 38, 269–271 (in Russian).

S. A. Lozhkin (1983): On the depth of functions of propositional calculus in some bases, Annales Univ. Budapest, Sectio Computatorica, vol. IV, 113–125.

S. A. Lozhkin (1996): Tighter bounds on the complexity of control systems from some classes, Mat. Voprosy Kibernetiki 6, 189–214 (in Russian).

S. A. Lozhkin (2005): On minimal π-circuits of closing contacts for symmetric functions with threshold 2, Discrete Math. and Appl. 15(5), 475–477.

S. A. Lozhkin and A. A. Semenov (1988): On a method for compressing information and on the complexity of the realization of monotone symmetric functions, Soviet Math. 32(7) (1988), 73–85.

A. Lubotzky, R. Phillips, P. Sarnak (1988): Ramanujan graphs, Combinatorica 8(3), 261–277.

O. B. Lupanov (1956): On rectifier and switching-and-rectifier schemes, Dokl. Akad. Nauk SSSR 111, 1171–1174 (in Russian).

O. B. Lupanov (1958a): A method of circuit synthesis, Izvesitya VUZ, Radiofiz, vol. 1, 120–140 (in Russian).

O. B. Lupanov (1958b): On the synthesis of contact schemes, Dokl. Akad. Nauk SSSR 119(1), 23–26 (in Russian).

O. B. Lupanov (1960): On the complexity of realization of functions of propositional calculus by formulas, Problemy Kibernetiki 3, 61–80 (in Russian).

O. B. Lupanov (1961): On realization of functions of propositional calculus by formulas of bounded depth over the basis $\{\&, \vee, \neg\}$, Dokl. Akad. Nauk SSSR 136(5), 1041–1042 (in Russian). Extended version in Problemy Kibernetiki 6 (1961), 5–14 (in Russian).

O. B. Lupanov (1962): On one class of circuits (formulas with partial memory), Problemy Kibernetiki 7, 61–114.

O. B. Lupanov (1963): On the synthesis of some classes of control systems, Problemy Kibernetiki 10, 63–97 (in Russian).

O. B. Lupanov (1965): On a certain approach to the synthesis of control systems—the principle of local coding, Problemy Kibernetiki 14, 31–110 (in Russian).

O. B. Lupanov (1965b): On computing symmetric functions of propositional calculus by switching networks, Problemy Kibernetiki 15, 85–100 (in Russian).

O. B. Lupanov (1970): On the influence of the depth of formulas to their complexity, Kibernetika 2, 46–49 (in Russian).

O. B. Lupanov (1973): On the synthesis of threshold circuits, Problemy Kibernetiki 26, 109–140 (in Russian).

O. B. Lupanov (1977): On realization of functions of propositional calculus by circuits of "bounded depth" over the basis $\{\&, \vee, \neg\}$, in Collection of works in Mathematical Cybernetics, Vol. 2 (Computer Center of Akad. of Sci. of USSR), 3–8 (in Russian).

F. J. MacWilliams and N. J. A. Sloane (1997): *The theory of error-correcting codes*. Elsevier, North-Holl.

G. A. Margulis (1973): Explicit constructions of concentrators, Probl. Peredachi Inf. 9, 71–80 (in Rusian). English transl. in Problems of Inf. Transm. (1975), 323–332.

A. A. Markov (1957): On the inversion complexity of systems of Boolean functions, Dokl. Akad. Nauk SSSR 116, 917–919 (in Russian). English transl. in J. ACM 5(4) (1958), 331–334.

A. A. Markov (1962): Minimal relay-diode bipoles for monotonic symmetric functions, Problemy Kibernetiki 8, 117–121 (in Russian). English transl. in Problems of Cybernetics 8 (1964), 205–212.

F. Mayer auf der Heide (1984): A polynomial linear search algorithm for the n-dimensional knapsack problem, J. ACM 31(3), 668–676.

W. F. McColl and M. S. Paterson (1977): The depth of all Boolean functions, SIAM J. Comput. 6, 373–380.

B. McMillan (1956): Two inequalities implied by unique decipherability, IEEE Trans. Information Theory 2(4), 115–116.

K. Mehlhorn (1979): Some remarks on Boolean sums, Acta Inform. 12, 371–375.

K. Mehlhorn (1982): On the program size of perfect and universal hash functions, in 23rd FOCS, 170–175.

K. Mehlhorn and E. Schmidt (1982): Las Vegas is better than determinism in VLSI and distributed computing, in 14th STOC, 330–337.

G. Midrijanis (2005): Three lines proof of the lower bound for the matrix rigidity, CoRR http://arxiv.org/abs/cs/0506081.

M. Minsky and S. Papert (1988): *Perceptrons*, MIT press, Cambridge, 1988 (Expanded edition). First edition appeared in 1968.

B. S. Mitiagin and B. N. Sadovskii (1965): On linear boolean operators, Dokl. Akad. Nauk SSSR 165(4), 773–776.

M. Morgenstern (1994): Existence and explicit constructions of $q + 1$ regular Ramanujan graphs for every prime power q, J. Combin. Theory Ser. B 62(1), 44–62.

H. Morizumi (2009): Limiting negations in formulas, in 36th ICALP, Part I, Springer LNCS, vol. 5555, 701–712.

H. Morizumi (2009b): Limiting negations in non-deterministic circuits, Theor. Comput. Sci. 410(38-4), 3988–3994.

H. Morizumi and G. Suzuki (2010): Negation-limited inverters of linear size, IEICE Transactions on Information and Systems, vol. E93-D, No. 2, 257–262.

D. E. Muller (1956): Complexity in electronic switching circuits, IRE Trans. Comput. EC-5, 15–19.

K. Mulmuley, U. Vazirani, and V. Vazirani (1987): Matching is as easy as matrix inversion, Combinatorica 7, 105–114.

S. Muroga (1971): Threshold Logic and Its Applications, Wiley-Interscience.

E. I. Nechiporuk (1962): On the complexity of schemes in some bases containing nontrivial elements with zero weights, Problemy Kibernetiki 8, 123–160 (in Russian).

E. I. Nechiporuk (1963): On the synthesis of rectifier networks, Problemy Kibernetiki 9, 37–44 (in Russian).

E. I. Nechiporuk (1964): On the synthesis of threshold circuits, Problemy Kibernetiki 11, 49–62 (in Russian).

E. I. Nechiporuk (1965): On the complexity of rectifier networks realizing boolean matrices with undefined elements, Dokl. Akad. Nauk SSSR 163(1), 40–42 (in Russian).

E. I. Nechiporuk (1966): On a Boolean function, Soviet Math. Dokl. 7(4), 999–1000.

E. I. Nechiporuk (1969): On a Boolean matrix, Problemy Kibernetiki 21, 237–240 (in Russian). English transl. in Systems Theory Res. 21 (1970), 236–239.

E. I. Nechiporuk (1969a): On topological principles of self-correction, Problemy Kibernetiki 21, 5–102 (in Russian).

I. Newman (1991): Private vs. common random bits in communication complexity, Inf. Proc. Lett. 39(2), 67–71.

I. Newman and A. Wigderson (1995): Lower bounds on formula size of boolean functions using hypergraph-entropy, SIAM J. Discrete Math. 8(4), 536–542.

R. G. Nigmatullin (1983): *The Complexity of Boolean Functions*. Izd. Kazansk. Univ. (Kazan University Press, in Russian).

N. Nisan (1989): CREW PRAMs and decision trees, in 21st STOC, 327–335.

N. Nisan (1991): Pseudorandom bits for constant depth circuits, Combinatorica 11(1), 63–70.

N. Nisan (2002): The communication complexity of approximate set packing and covering, in 29th ICALP, Springer LNCS, vol. 2380, 868–875.

N. Nisan and M. Szegedy (1994): On the degree of boolean functions as real polynomials, Comput. Complexity 4, 301–313.

N. Nisan and A. Wigderson (1994): Hardness vs. randomness, J. Comput. Syst. Sci. 49(2), 149–167.

N. Nisan and A. Wigderson (1995): On rank vs. communication complexity, Combinatorica 15(4), 557–565.

C. van Nuffelen (1976): A bound for the chromatic number of a graph, Amer. Math. Monthly 83, 265–266.

R. O'Donnell and K. Wimmer (2007): Approximation by DNF: examples and counterexamples, in 34th ICALP, Springer LNCS, vol. 4596, 195–206.

E. A. Okolnishnikova (1982): On the influence of negation on the complexity of realization of monotone Boolean functions by formulas of bounded depth, in Metody Diskretnogo Analiza 38, 74–80 (in Russian).

E. A. Okolnishnikova (1991): Lower bounds on the complexity of realization of characteristic functions of binary codes by branching programs, in Diskretnii Analiz 51 (Novosibirsk, 1991), 61–83 (in Russian).

E. A. Okolnishnikova (2009): Lower bound for the computation complexity of BCH-codes for branching programs, Diskretn. Anal. Issled. Oper. 16(5), 69–77 (in Russian).

V. A. Orlov (1970): Realization of "narrow" matrices by rectifier networks, Problemy Kibernetiki 22, 45–52 (in Russian).

M. S. Paterson (1976): An introduction to Boolean function complexity, Asterique, 38/39, 183–201.

M. S. Paterson and L. G. Valiant (1976): Circuit size is nonlinear in depth, Theor. Comput. Sci. 2(3), 397–400.

M. Paterson and U. Zwick (1993): Shrinkage of de Morgan formulae under restriction, Rand. Struct. Algorithms 4(2), 135–150.

M. Paterson and U. Zwick (1993b): Shallow circuits and concise formulae for multiple addition and multiplication, Comput. Complexity 3, 262–291.

M. S. Paterson, N. Pippenger, and U. Zwick (1992): Optimal carry save networks, in Boolean Function Complexity, LMS Lecture Note Series, Vol. 169, 174–201, Cambridge Univ. Press.

R. Paturi and J. Simon (1986): Probabilistic communication complexity, J. Comput. Syst. Sci. 33(1), 106–123.

R. Paturi and M. E. Saks (1994): Approximating threshold circuits by rational functions, Information and Computation 112(2), 257–272.

R. Paturi, M. Saks, and F. Zane (2000): Exponential lower bounds for depth three boolean circuits, Comput. Complexity 9(1), 1–15.

W. Paul (1977): A 2.5n-lower bound on the combinational complexity of boolean functions, SIAM J. Comput. 6, 427-443.

N. Pippenger (1976): Information theory and the complexity of Boolean function, Math. Syst. Theory 10(2), 129–167.

N. Pippenger (1976b): The realization of monotone Boolean functions, preliminary version, in. 8th STOC, Lect. Notes in Comput. Sci. (ACM Press, New York), 204–210.

N. Pippenger (1977): Superconcentrators , SIAM J. Comput. 6, 298–304.

N. Pippenger (1980): On another boolean matrix, Theor. Comput. Sci. 11, 49–56.

N. Pippenger (1982): Superconcentrators of depth 2, J. Comput. Syst. Sci. 24, 82–90.

N. Pippenger (1990): Communication networks, in Handbook of Theoretical Computer Science, Volume A: Algorithms and Complexity (A), 805–834.

F. P. Preparata and D. E. Muller (1975): The time required to evaluate division-free arithmetic expressions, Inf. Proc. Lett. 3(5), 144–146.

F. P. Preparata and D. E. Muller (1976): Efficient parallel evaluation of boolean expressions, IEEE Trans. Computers, vol. C-25, N. 5, 548–549.

P. Pudlák (1984): A lower bound on complexity of branching programs, in Springer LNCS, vol. 176, 480–489.

P. Pudlák (1984b): Bounds for Hodes–Specker theorem, in Logic and Machines: Decision Problems and Complexity, Springer LNCS, vol. 171, 421–445.

P. Pudlák (1987): The hierarchy of Boolean circuits, Computers and Artificial Intelligence 6(5), 449–468.

P. Pudlák (1994): Communication in bounded depth circuits, Combinatorica 14(2), 203–216.

P. Pudlák (1997): Lower bounds for resolution and cutting plane proofs and monotone computations, J. Symbolic Logic 62(3), 981–998.

P. Pudlák and R. Impagliazzo (2000): A lower bound for DLL algorithms for k-SAT, in Proc. of 11th Ann. ACM-SIAM Symp. on Discrete Algorithms, 128–136.

P. Pudlák and V. Rodl (1992): A combinatorial approach to complexity, Combinatorica 12(2), 221–226.

P. Pudlák and V. Rödl (1994): Some combinatorial-algebraic problems from complexity theory, Discrete Math. 136(1–3), 253–279.

P. Pudlák and V. Rödl (2004): Pseudorandom sets and explicit constructions of Ramsey graphs, in Quad. Mat. 13, 327–346.

P. Pudlák and P. Savický (1993): On shifting networks, Theor. Comput. Sci. 116, 415–419.

P. Pudlák and J. Sgall (1997): An upper bound for a communication game related to time-space tradeoffs, in The Mathematics of Paul Erdős, R. Graham and J. Nesetril (eds.), vol. I, Springer, 393–399.

P. Pudlák and J. Sgall (1998): Algebraic models of computation and interpolation for algebraic proof systems, in DIMACS Series in Discrete Math. and Theor. Comput. Sci., vol. 39, 279–295.

P. Pudlák and Z. Vavrín (1991): Computation of rigidity of order n^2/r for one simple matrix, Comment. Math. Univ. Carolinae 32, 213–218.

P. Pudlák, V. Rödl, and P. Savický (1988): Graph Complexity, Acta Inf. 25(5), 515–535.

P. Pudlák, V. Rödl, and J. Sgall (1997): Boolean circuits, tensor ranks, and communication complexity, SIAM J. Comput. 26(3) 605–633.

J. Radhakrishnan (1997): Better lower bounds for monotone threshold formulas, J. Comput. Syst. Sci. 54(2), 221–226.

J. Radhakrishnan and A. Ta-Shma (2000): Bounds for dispersers, extractors, and depth-two superconcentrators, SIAM J. Discrete Math. 13(1), 2–24.

R. Raz (2000): The BNS–Chung criterion for multi-party communication complexity, Comput. Complexity 9, 113–122.

R. Raz (2001): Resolution lower bounds for the weak pigeonhole principle, J. ACM 51(2), 115–138

R. Raz and A. Shpilka (2003): Lower bounds for matrix product in bounded depth circuits with arbitrary gates, SIAM J. Comput. 32(2), 488–513.

R. Raz and B. Spieker (1995): On the Log-Rank conjecture in communication complexity, Combinatorica 15(4), 567–588.

R. Raz and A. Wigderson (1989): Probabilistic communication complexity of Boolean relations, in 30th FOCS, 562–567.

R. Raz and A. Wigderson (1992): Monotone circuits for matching require linear depth, J. ACM 39(3), 736–744.

R. Raz and A. Yehudayoff (2011): Multilinear formulas, maximal-partition discrepancy and mixed-sources extractors, J. Comput. Syst. Sci. 77(1), 167–190.

A. A. Razborov (1985a): Lower bounds for the monotone complexity of some boolean functions, Soviet Math. Dokl. 31, 354–357.

A. A. Razborov (1985b): A lower bound on the monotone network complexity of the logical permanent, Math. Notes Acad. of Sci. USSR 37(6) (1985), 485–493.

A. A. Razborov (1987): Lower bounds on the size of bounded-depth networks over a complete basis with logical addition, Math. Notes Acad. of Sci. USSR 41(4), 333–338.

A. A. Razborov (1988): Bounded-depth formulae over the basis {&, ⊕} and some combinatorial problem, in Problems of Cybernetics, Complexity Theory and Applied Mathematical Logic (VINITI, Moscow) 149–166 (in Russian).

A. A. Razborov (1989a): On the method of approximations, in 21st STOC, 167–176.

A. A. Razborov (1989b): On rigid matrices. Preprint, Steklov Mathematical Institute (in Russian).

A. A. Razborov (1990): Applications of matrix methods to the theory of lower bounds in computational complexity, Combinatorica 10(1), 81–93.

A. A. Razborov (1990b): Lower bounds on the size of switching-and-rectifier networks for symmetric Boolean functions, Math. Notes Acad. of Sci. USSR 48(6), 79–91.

A. A. Razborov (1991): Lower bounds for deterministic and nondeterministic branching programs, in Springer LNCS, vol. 529, 47–60.

A. A. Razborov (1992a): On the distributional complexity of disjointness, Theor. Comput. Sci. 106(2), 385–390.

A. A. Razborov (1992b): On submodular complexity measures, in Boolean Function Complexity, London Math. Soc. Lecture Note Series 169, 76–83.

A. A. Razborov (1992c): The gap between the chromatic number of a graph and the rank of its adjacency matrix is superlinear, Discrete Math. 108, 393–396.

A. A. Razborov (1992d): On small depth threshold circuits, in Springer LNCS, vol. 621, 42–52.

A. A. Razborov (1995): Bounded arithmetics and lower bounds in boolean complexity, in Feasible Mathematics II, Proc. of Workshop (Cornell University, Ithaca, NY, USA, May 28–30, 1992), Birkhäuser, Boston, MA.

A. A. Razborov (1996): On small size approximation models, in The Mathematics of Paul Erdos I, Algorithms and Combinatorics, 385–392.

A. A. Razborov (2003): Resolution lower bounds for the weak functional pigeonhole principle, Theor. Comput. Sci. 303(1), 233–243.

A. A. Razborov and S. Rudich (1997): Natural proofs, J. Comput. Syst. Sci. 55(1), 24–35.

A.A. Razborov and A.A. Sherstov (2010): The sign-rank of AC^0, SIAM J. Comput. 39(5), 1833–1855.

A. Razborov and A. Wigderson (1993): $n^{\Omega(\log n)}$ lower bounds on the size of depth-3 threshold circuits with AND gates at the bottom, Inf. Proc. Lett. 45, 303–307.

N.P. Redkin (1973): Proof of minimality of circuits consisting of functional elements, Syst. Th. Res. 23, 85–110.

N.P. Redkin (2004): On the complexity of Boolean functions with a small number of ones, Discrete Math. and Appl. 2004, 14:6, 619–630.

O. Reingold (2008): Undirected connectivity in log-space, J. ACM 55 (4), Article 17, 24 pages.

R. Reischuk (1982): A lower time-bound for parallel random access machines without simultaneous writes. Tech. Rep. RJ3431, IBM, New York, 1982.

M.N.C. Rhodes (2009): On the Chvátal rank of the Pigeonhole Principle, Theor. Comput. Sci. 410(27–29), 2774–2778.

J. Riordan and C.E. Shannon (1942): The number of two terminal series-parallel networks, J. Math. Phys. 21, 83–93.

R.L. Rivest and J. Vuillemin (1976): On recognizing graph properties from adjacency matrices, Theor. Comput. Sci. 3(3), 371–384.

T.J. Rivlin and E.W. Cheney (1966): A comparison of uniform approximations on an interval and a finite subset thereof, SIAM J. Numer. Anal. 3(2), 311–320.

J.A. Robinson (1965): A machine-oriented logic based on the resolution principle, J. ACM 12(1), 23–41.

L. Rónyai, L. Babai, and M.K. Ganapathy (2001): On the number of zero-patterns of a sequence of polynomials, J. of AMS 16(3), 717–735.

A. Rosenbloom (1997): Monotone circuits are more powerful than monotone boolean circuits, Inf. Proc. Lett. 61(3), 161–164.

B. Rossman (2008): On the constant-depth complexity of k-clique, in 40th STOC, 721–730.

B. Rossman (2010): The monotone complexity of k-clique on random graphs, in 51th FOCS, 193–201.

D. Rubinstein (1995): Sensitivity vs. block sensitivity of boolean functions, Combinatorica 15(2), 297–299.

K.L. Rychkov (1985): A modification of Khrapchenko's method and its application to lower bounds for π-schemes of code functions, in Metody Diskretnogo Analiza, 42 (Novosibirsk), 91–98 (in Russian).

K.L. Rychkov (1994): Lower bounds on the complexity of parallel-sequential switching circuits that realize linear Boolean functions, Sibirsk. Zh. Issled. Oper. 1(4), 33–52 (in Russian).

K.L. Rychkov (2009): On the complexity of generalized contact circuits, Diskretn. Anal. Issled. Oper. 16(5), 78–87 (in Russian).

K.L. Rychkov (2010): A lower bound for the complexity of generalized parallel-serial contact circuits for a characteristic function of divisibility by q, Diskretn. Anal. Issled. Oper. 17(6), 68–76 (in Russian).

M. Santha and Ch. Wilson (1993): Limiting negations in constant depth circuits, SIAM J. Comput. 22(2), 294–302.

M. Sauerhoff (1999): Complexity theoretical results for randomized branching programs. PhD thesis, Univ. of Dortmund, Shaker-Verlag.

J.E. Savage (1976): The Complexity of Computing. Wiley, New York.

J.E. Savage (1998): Models of Computation: Exploring the Power of Computing. Addison-Wesley. The book is freely available for download at: http://www.cs.brown.edu/~jes/book/home.html.

P. Savický and S. Zák (1996): A large lower bound for 1-branching programs, ECCC Report Nr. 36.

P. Savický and S. Zák (1997): A lower bound on branching programs reading some bits twice, Theor. Comput. Sci. 172, 293–301.

P. Savický and A.R. Woods (1998): The number of Boolean functions computed by formulas of a given size, Rand. Struct. Algorithms 13(3–4), 349–382.

C. Schnorr (1974): Zwei lineare untere Schranken für die Komplexität Boolescher Funktionen, Computing 13, 155–171.

A. Schrijver (1980): On cutting planes, Anals of Discrete Math. 9, 291–296.

C. E. Shannon (1949): The synthesis of two-terminal switching circuits, Bell Systems Technical Journal 28, 59–98.

A. A. Sherstov (2010): On quantum-classical equivalence for composed communication problems, Quantum Information & Computation 10(5–6), 435–455.

Y. Shi (2000): Lower bounds of quantum black-box complexity and degree of approximating polynomials by influence of Boolean variables, Inf. Proc. Lett. 75(1–2), 79–83.

Y. Shi (2002): Entropy lower bounds for quantum decision tree complexity, Inf. Process. Lett. 81(1), 23–27.

M. A. Shokrollahi, D. A. Spielman, and V. Stemann (1997): A remark on matrix rigidity. Inf. Proc. Lett. 64(6), 283–285.

L. A. Sholomov (1969): On the realization of parially defined boolean functions by circuits, Problemy Kibernetiki 21, 215–226 (in Russian).

A. Shpilka and A. Yehudayoff (2010): Arithmetic circuits: A survey of recent results and open questions, Foundations and Trends in Theoret. Comput. Sci. 5(3–4), 207–388.

S. E. Shumilina (1987): On the realization of one symmetric function by formulas of special type, in Kombinatorno-Algebraicheskie Metody i ich Primenenie (Combinatorial-Algebraic Methods and their Application), Gorkij, 164–169 (in Russian).

H.-U. Simon (1983): A tight $\Omega(\log\log n)$-bound on the time for parallel RAM's to compute nondegenerated boolean functions, in Springer LNCS, vol. 158, 439–444.

R. K. Sinha and J. S. Thathachar (1997): Efficient oblivious branching programs for threshold and Mod functions, J. Comput. Syst. Sci. 55(3), 373–384.

M. Sipser (1985): A topological view of some problems in complexity theory, in Colloquia Mathematica Societatis János Bolyai 44, 387–391.

M. Sipser (1992): The history and status of the P versus NP question, in 24th STOC, 603–618.

D. R. Smith (1966): Bounds on the number of threshold functions, IEEE Trans. on Electronic Computers EC-15(6), 368–369.

R. Smolensky (1987): Algebraic methods in the theory of lower bounds for boolean circuit complexity, in 19th STOC, 77–82.

R. Smolensky (1990): On interpolation by analytic functions with special properties and some weak lower bounds on the size of circuits with symmetric gates, in 31st FOCS, 628–631.

J. Spencer (1995): Probabilistic methods, in Handbook of Combinatorics, R.L. Graham, M. Grotschel and L. Lovász (eds.), Elsevier Science, Vol. 2, 1786–1817.

P. M. Spira (1971), On time–hardware complexity tradeoffs for Boolean functions, in Proceedings of 4th Hawaii Symposium on System Sciences, Western Periodicals Company, North Hollywood, 525–527.

G. Stålmark (1996): Short resolution proofs for a sequence of tricky formulas, Acta Informatica 33(3), 277–280.

S. K. Stein (1974): Two combinatorial covering problems, J. Combin. Theory Ser. A 16, 391–397.

L. Stockmeyer (1977): On the combinational complexity of certain symmetric Boolean functions, Math. System Theory 10, 323–336.

B. A. Subbotovskaya (1961), Realizations of linear functions by formulas using $+, .., -$, Soviet Math. Dokl. 2, 110–112.

R. Szelepcsényi (1987): The method of forcing for nondeterministic automata, Bull. of EATCS 33, 96–99.

K. Tanaka and T. Nishino (1994): On the complexity of negation-limited Boolean networks, in 26th STOC, 38–47.

K. Tanaka, T. Nishino, and R. Beals (1996): Negation-limited circuit complexity of symmetric functions, Inf. Proc. Lett. 59(5), 273–279.

É. Tardos (1987): The gap between monotone and non-monotone circuit complexity is exponential, Combinatorica, 7(4), 141–142.

G. Tardos (1989): Query complexity, or why is it difficult to separate $NP^A \cap co - NP^A$ from P^A by a random oracle A? Combinatorica 8(4), 385–392.

P. Tesson (2003): Computational complexity questions related to finite monoids and semigroups, PhD Thesis, McGill University, Montreal.

G. A. Tkachov (1980): On complexity of realization of a sequence of boolean functions by schemas of functional elements and π-schemes under additional restrictions on the structure of the schemas, in Kombinatorno-Algebraicheskie Metody v Prikladnoj Matematike (Combinatorial-Algebraic Methods in Applied Mathematics), Gorkij, 161–207.

Shi-Chun Tsai (2001): A depth 3 circuit lower bound for the parity function J. Inf. Sci. Eng. 17(5), 857–860.

J. Tiekenheinrich (1984): A $4n$-lower bound on the mononotone network complexity of a one-output Boolean function, Inf. Proc. Lett. 18, 201–202.

G. C. Tseitin (1970): On the complexity of derivations in propositional calculus, in Studies in Constructive Mathematics and Mathematical Logics, Part II, Slisenko, A. O. (ed.), 115–125.

P. Turán (1941): On an extremal problem in graph theory, Math. Fiz. Lapok 48, 436–452.

Zs. Tuza (1984): Covering of graphs by complete bipartite subgraphs; complexity of 0-1 matrices, Combinatorica 4(1), 111–116.

A. B. Ugol'nikov (1976): The realization of monotone functions by networks of functional elements, Problemy Kibernetiki 31, 167–185 (in Russian).

D. Uhlig (1974): On a family of classes of easy computable boolean functions, Problemy Kibernetiki 28, 25–42 (in Russian).

D. Uhlig (1991): Boolean functions with a large number of subfunctions and small complexity and depth, in 8th FCT, 395–404.

A. Urquhart (1987): Hard examples for resolution J. ACM 34(1), 209–219.

A. Urquhart (2011): A near-optimal separation of regular and general resolution, SIAM J. Comput. 40(1), 107–121.

L. G. Valiant (1976): Graph-theoretic properties in computational complexity, J. Comput. Syst. Sci. 13(3), 278–285.

L. G. Valiant (1977): Graph-theoretic methods in low-level complexity, in Springer LNCS, vol. 53, 162–176.

L. G. Valiant (1983): Exponential lower bounds for restricted monotone circuits, in 15th STOC, 110–117.

L. G. Valiant (1984): Short monotone formulae for the majority function, J. Algorithms 5, 363–366.

L. G. Valiant (1986): Negation is powerless for Boolean slice functions, SIAM J. Comput. 15, 531–535.

L. G. Valiant (2004): The Log-Rank Conjecture and low degree polynomials, Inf. Proc. Lett. 89, 99–103.

L. G. Valiant and V. V. Vazirani (1986): NP is as easy as detecting unique solutions, Theor. Comput. Sci. 47, 85–93.

E. Viola (2009): On the power of small-depth computation, Foundations and Trends in Theoret. Comput. Sci. 5(1), 1–72.

H. Vollmer (1999): *Introduction to Circuit Complexity – A Uniform Approach*. Texts in Theoretical Computer Science, EATCS Series, Springer-Verlag, 1999.

H. Wang and C. Xang (2001): Explicit constructions of perfect hash families from algebraic curves over finite fields, J. Combin. Theory Ser. A 93, 112–124.

I. Wegener (1980): A new lower bound on the monotone network complexity of Boolean sums, Acta Inform. 15 147–152.

I. Wegener (1985a): On the complexity of slice functions, Theor. Comput. Sci. 38, 55–68.

I. Wegener (1985b): The critical complexity of all (monotone) Boolean functions and monotone graph properties, Information and Control 67(1–3), 212–222.

I. Wegener (1987): *The complexity of Boolean functions*. Wiley-Teubner.

I. Wegener (2000): *Branching programs and binary decision diagrams*. SIAM.

A. Weil (1948): Sur les courbes algébriques et les variétés qui s'en déduisent, Actualités Sci. Ind. 1041 (Herman, Paris).

R. Williams (2011): Non-uniform ACC circuit lower bounds, in 26th CCC, 115–125.

A. Wigderson (1993): The fusion method for lower bounds in circuit complexity, in Combinatorics, Paul Erdős is Eighty, vol. 1, János Bolyai Math. Soc., 453–468.

A. Wigderson (1994): NL/poly $\subseteq \oplus$L/poly, in 9th CCC, 59–62.

R. de Wolf (2008): A brief introduction to Fourier analysis on the boolean cube, Theory of Computing, Graduate Surveys 1, 1–20, http://theoryofcomputing.org.

M. Yannakakis (1991): Expressing combinatorial optimization problems by linear programs, J. Comput. Syst. Sci. 43, 441–466.

S. V. Yablonskii (1954): Realisation of the linear function in the class of π-schemes, Dokl. Akad. Nauk SSSR 94(5), 805–806 (in Russian).

S. V. Yablonskii (1959): On algorithmic difficulties of synthesis of minimal contact schemes, Problemy Kibernetiki 2, 75–121 (in Russian).

S. V. Yablonskii and V. P. Kozyrev (1968): Mathematical problems of cybernetics, in Information Materials of Scientific Council of Akad. Nauk SSSR on Complex Problem "Kibernetika", vol. 19a, 3–15.

S. Yajima and T. Ibaraki (1965): A lower bound on the number of threshold functions, IEEE Trans. on Electronic Computers EC-14(6), 926–929.

A. C. Yao (1979): Some complexity questions related to distributed computing, in 11th STOC, 209–213.

A. C. Yao (1981): The entropic limitations of VLSI computations, in 13th STOC, 308–311.

A. C. Yao (1982): Theory and applications of trapdoor functions, in 23rd FOCS, 80–91.

A. C. Yao (1983): Lower bounds by probabilistic arguments, in 24th FOCS, 420–428.

A. C. Yao (1985): Separating the polynomial time hierarchy by oracles, in 26th FOCS, 1–10.

A. C. Yao (1990): On ACC and threshold circuits, in 31th FOCS, 619–627.

S. V. Zdobnov (1987): On the complexity of linear function in the class of π-schemes, in Kombinatorno-Algebraicheskie Metody i ich Primenenie (Combinatorial-Algebraic Methods and their Application), Gorkij, 27–34 (in Russian).

S. Zák (1995): A super-polynomial lower bound for $(1, +k(n))$-branching programs, in Springer LNCS, vol. 969, 319–325.

E. J. Zacharova (1963): On synthesis of threshold circuits, Problemy Kibernetiki 9, 317–319 (in Russian).

U. Zwick (1991), An extension of Khrapchenko's theorem, Inf. Proc. Lett. 37, 215–217.

U. Zwick (1991b): A $4n$ lower bound on the combinatorial complexity of certain symmetric Boolean functions over the basis of unate dyadic Boolean functions, SIAM J. Comput. 20, 499–505.

Index

ACC circuit, 333
AC circuit. *See* Alternating circuit
Address function, 75
Adjacency matrix, 100, 576
Adleman's theorem, 15
Affine combination, 577
Affine dimension, 198
Affine space, 577
   dimension of, 577
Algebraic tiling number, 238
Alon-Saks-Seymour conjecture, 106
Alternating circuit, 343
Amplification of density, 217
Antichain, 210, 276
Approximate degree, 57
Approximate disjointness problem, 136
Approximation complexity, 26
Approximation lemma
   for AND and OR, 64
   for matrices, 367
Arithmetic branching program, 242
Arithmetic-geometric mean inequality, 320
Average degree, 76
Average depth, 436
Average sensitivity, 69, 351, 436
Average time, 15

Barrington's theorem, 450
BCH-code, 489
Biclique, 41
Binary decision diagram (BDD). *See* Branching program
Binary formula, 165, 172
Binary hypercube, 5, 352

Binary n-cube, 5
Bipartite complexity, 196
Bipartite formula complexity, 41
Bipartite graph, 576
Block sensitivity, 410
Blocking number, 284, 398
Blocking set, 283, 477
Bollobás' theorem, 253
Boolean function, 3, 230
   approximate degree of, 57
   approximation complexity of, 26
   block sensitivity of, 410
   certificate complexity of, 347, 410
   communication matrix of, 93
   decrease of, 290
   degree of, 57, 414
   dense, 471
   d-rare, 468
   dual of, 7, 241, 252
   evasive, 419
   graph of, 47
   k-fold extension of, 298
   m-mixed, 458
   monotone, 8, 245
   m-robust, 468
   negative input of, 260
   nondegenerate, 70
   partial, 26
   positive input of, 260
   random, 188
   rectangle-free, 471
   rectangle number of, 471, 481
   rectangular, 471
   robust, 477
   self-dual, 8, 568
   sensitive, 471
   sensitivity of, 69